—PATHOGENIC MICROORGANISMS

FUNGI

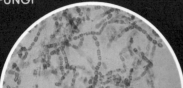

19. *Coccidioides immitis,* causative agent of coccidioidomycosis. Arthrospores.
20. *Coccidioides immitis* mature sporangium releasing spores in wet mount of pus.

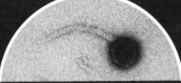

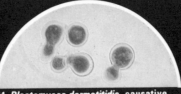

21. *Blastomyces dermatitidis,* causative agent of blastomycosis. Yeast phase.
22. *Blastomyces dermatitidis* in lung tissue mount. Direct fluorescent antibody technique.

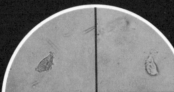

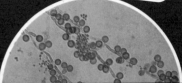

23. *Candida albicans,* causative agent of monilia vaginitis. Chlamydospores.
24. *Candida albicans* from a smear from a patient. Gram stain, high power.

PROTOZOA

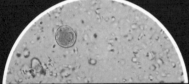

25. *Entamoeba histolytica,* causative agent of amebic dysentery. Immature cyst. Iodine stain.
26. *Entamoeba histolytica* immature cyst. Trichrome stain.

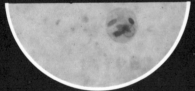

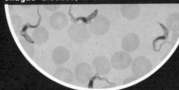

27. *Trichomonas hominis,* causative agent of trichomoniasis. Trophozoite. Unstained (left); iodine stain (right).
28. *Trypanosoma cruzi,* causative agent of Chagas' disease, in blood film.

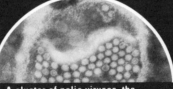

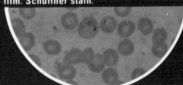

29. *Plasmodium falciparum,* causative agent of malaria. Immature gametocytes and rings in blood film.
30. *Plasmodium vivax,* causative agent of malaria. Growing trophozoite in blood film. Schuffner stain.

VIRUSES

31. Phage lamba, a virus of bacterium *Escherichia coli.*
32. Tobacco mosaic viruses, causative agents of tobacco mosaic disease in plants.

33. A cluster of polio viruses, the causative agents of polio.

34. Rotaviruses, causative agents of acute infectious diarrhea in infants.

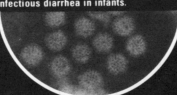

35. Enveloped herpes virus, causative agent of "cold sores" of mucous membranes.
36. Influenza viruses, causative agents of influenza ("flu").

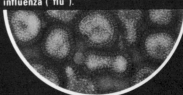

C.1

ELEMENTS OF MICROBIOLOGY

Michael J. Pelczar, Jr.
Professor Emeritus
University of Maryland

E. C. S. Chan
Associate Professor
McGill University

with the assistance of
Merna Foss Pelczar

McGRAW-HILL BOOK COMPANY
NEW YORK ST. LOUIS SAN FRANCISCO
AUCKLAND BOGOTÁ HAMBURG JOHANNESBURG LONDON MADRID MEXICO MONTREAL NEW DELHI
PANAMA PARIS SÃO PAULO SINGAPORE SYDNEY TOKYO TORONTO

ELEMENTS OF MICROBIOLOGY

234567890VHVH8987654321

This book was set in Primer by Black Dot, Inc. The editors were James E. Vastyan, Anne T. Vinnicombe, and James S. Amar; the designer was Hermann Strohbach; the production supervisor was Charles Hess. Von Hoffmann Press, Inc., was printer and binder.

Cover
Bacterial cell showing major structures external to the bacterial cell wall. Certain structures, for example, capsules, flagella, spores, and pili, are not common to all bacterial cells. (*Erwin F. Lessel, illustrator.*)

Illustration Credits
Dr. Erwin F. Lessel of Lederle Laboratories, Pearl River, New York, drew Figures 1-2, 1-12, 2-3, 2-5, 2-6, 2-7, 5-1, 5-3, 5-4, 5-5, 5-6, 5-8, 5-13, 5-14, 5-20, 5-22, 7-1, 7-5, 7-6, 7-8, 7-9, 7-10, 7-11, 7-12, 7-13, 7-14, 7-15, 7-16, 7-17, 7-18, 7-19, 7-20, 18-4, 20-5, 21-10, 22-3, 27-2, 27-6, 27-8, 27-16, 28-2, 28-6, 28-7, 28-10, 28-11, 28-12, 29-3, 29-7, 31-2, 32-1, 32-7, 32-10, 33-1, 33-4, 33-7, 33-8, 33-15, and the illustrations for Table 2-2 and the cover. The remaining new drawings were done by Fine Line Illustrations, Inc.

Endpapers
Figures 3, 4, 5, 10, 11, 14, 15, 17, 19, 20, 21, 22, 23, 24, 25, 26, 27, 28, 29, 30 courtesy of the Center for Disease Control, Atlanta, Georgia.
Figures 1, 2, 6, 7, 8, 9, 12, 13, 16, 18, 32, 36 courtesy of McGraw-Hill Book Company, from John H. Dustman and Terrance Lukas: DUSTMAN AND LUKAS SLIDES FOR MICROBIOLOGY, UNIT 5, 1st ed., 1976, New York.
Figures 33, 34, and 35 courtesy of Margaret Gomersall, McGill University.
Figure 31 courtesy of H. W. Ackermann, Laval University.

Library of Congress Cataloging in Publication Data

Pelczar, Michael Joseph, date
   Elements of microbiology

   Includes index.
   1. Microbiology. I. Chan, Eddie Chin Sun, date, joint author. II. Pelczar, Merna Foss, joint author. III. Title, DNLM: 1. Microbiology. QW4 P381e]
QR41.2.P39 616'.01   80-14735
ISBN 0-07-049240-9

CONTENTS

PREFACE

In the past several decades microbial cells have provided a useful model for the study of life processes because of their diversity, versatility, and ease of manipulation. Studies of microorganisms have contributed much of what we know today about genetics and metabolism. Microorganisms have also been the focus of increased interest because they may contribute solutions to some of the most pressing human problems, most of which can be traced to competition for finite amounts of resources and for limited space. Some of these problems, such as adequate energy or food supplies, environmental pollutants, and disease prevention and health maintenance, have already been tackled by microbiological technology. Certain microorganisms are being genetically engineered, that is, tailor-made, to produce in quantity alcohol for the alternative energy-source gasohol for our cars. Other microorganisms are being engineered to decompose specific pollutants. Some microorganisms are being looked at as sources of single-cell protein for feed, which farm animals then convert to eggs, milk, and meat, and when treated to resemble familiar foods, as protein substitutes for human consumption. Disease prevention was an early focus of microbiology—witness its use in the control of smallpox and rabies—and it continues as a major focus today.

The growing importance of microbiology as both a basic and applied science has contributed to the growth of interest in microbiology for a vast array of students—in home economics, forestry, agriculture, food and animal science, nutrition, allied-health sciences, nursing, liberal arts, and business. ELEMENTS OF MICROBIOLOGY is written for this diverse group of students. The themes of this book—what microorganisms are and what they do—reflect the two aspects of microbiology. Within these themes we have attempted to present the material clearly and concisely so that it will be understood by these students, most of whom are having their first academic experience with the world of microbes and most of whom have had limited exposure to science. We have exercised particular care in selecting illustrations and tables to convey important concepts and principles. Chapter 7 is an excellent example of this. The early pages of this chapter give students basic information on bacterial classification. The tabular lists and representative illustrations of each bacterial group which follow act as an aid and reference for later study. The clarity and precision of the illustrations of microbial cells here and throughout the book by Dr. Erwin F. Lessel (a microbiologist in his own right) will be a great aid to the student. We have arranged the material in a logical sequence so that it will be readily understood by students, but we are aware that the arrangement will not be acceptable to everyone. The instructor should feel no compulsion to follow the sequence we have used since the material can be presented with equal effectiveness when it is rearranged to fit individual preference. For example, Part Eight on "Microorganisms and Disease—Resistance to Infection" and Part Nine on "Microorganisms and Disease—Transmission of Pathogens" can easily be interchanged with or replaced by Part Ten on "Environmental and Applied Microbiology." Or Part Five on "Metabolism" and Part Six on "Genetics" can be taught in whole or in part depending on the interests of the instructor.

An Instructor's Manual is available for use with this text. It includes suggested lecture and laboratory schedules, part and chapter summaries, examination questions, and sources of audiovisual aids, cultures, culture media, reagents, and laboratory equipment. Our Laboratory Manual, LABORATORY EXERCISES IN MICROBIOLOGY, fourth edition, originally written to accompany our MICROBIOLOGY, fourth edition, is also a good accompaniment to ELEMENTS OF MICROBIOLOGY. New references for each experiment to the corresponding text material in ELEMENTS OF MICROBIOLOGY will be found in the Instructor's Manual.

We are extremely grateful to the many individuals and corporations who provided us with materials for this textbook. We also wish to thank our many reviewers: Phillip Achey, University of Florida; Frank L. Binder, Marshall University; John G. Chan, University of Hawaii at Hilo; L. R. DiLiello, SUNY College at Farmingdale; B. R. Funke, North Dakota State University; John L. Joy, CUNY/Kingsborough Community College; R. C. Kochersberger, Jamestown Community College; William L. Lester, Humboldt State University; Lloyd T. McAtee, Prince George's Community College; Charlotte D. Parker, University of Missouri, Columbia; Lavon Richardson, Oklahoma State University; Louis Shainberg, Mt. San Antonio College; Robert E. Sjogren, University of Vermont; and Fred D. Williams, Iowa State University. Their valuable suggestions, large and small, have contributed immeasurably to this effort. For their editorial expertise and patience, we wish to thank John Hendry and our McGraw-Hill editor, Anne Vinnicombe. Special thanks also to Carolyn Aks for research and editorial assistance and to Edna Khalil and Micheline Marano for such professional manuscript preparation.

Michael J. Pelczar, Jr.
E. C. S. Chan

PART ONE

INTRODUCTION TO MICROBIOLOGY

Microorganisms—organisms that can be seen only with the aid of the magnification of high-powered microscopes—were first seen and described about 300 years ago. Not until the 1870s, however, was their role as causative agents of disease understood and accepted. At approximately the same time it was established that microorganisms performed many vital functions in our environment.

A large and varied population of microorganisms exists, and they occur practically everywhere in nature. They are the most widely distributed and most abundantly occurring form of life on this planet. Indeed it has been calculated that the mass of microorganisms on earth exceeds the mass of all other organisms. Many millions of individual microorganisms are present in each gram of fertile soil. They are present in streams, lakes, rivers, and oceans. They are carried by air currents from the earth's surface to the upper atmosphere and from there borne hundreds of miles to new locations. They are on the surfaces of our bodies and in our mouths, noses, and other body orifices. A single sneeze can expel millions of microorganisms into the immediate environment. The bulk of material in feces is made up of microbial cells—billions of them per gram. Microorganisms occur most abundantly where nutrients, moisture, and temperature are suitable for their growth and multiplication.

Microorganisms are responsible for many diseases which have plagued civilizations for centuries. Prior to the understanding that infectious diseases are caused by microorganisms, populations were periodically devastated by outbreaks of diseases such as diphtheria, plague, and smallpox. Through applications of the discoveries made in microbiology, medicine has achieved its greatest successes in diagnosis, prevention, and cure of diseases. The dramatic decrease in deaths from infection, the doubling of the average life-span, and the survival of most children at birth are, to a great extent, the result of knowledge discovered from the study of microorganisms.

In Part I, the first four chapters, you will learn of some of the early work which led to the establishment of the science of microbiology. You will be introduced to the microbial world—the major groups of microorganisms. And you will learn in a general way how microorganisms are studied in the laboratory.

Louis Pasteur dictating a note to Marie Pasteur at Pont-Gisquet, where she established a home and laboratory for the entire research team investigating the diseases of silkworms. (Photograph by G. Serraz, reproduction by Institut Pasteur.)

HISTORY OF MICROBIOLOGY

Microbiology is the study of living organisms of microscopic size. The world of microorganisms consists of five groups of organisms: *bacteria, protozoa, viruses,* and the microscopic *algae* and *fungi.* In the science of microbiology one studies many aspects of these microorganisms (also called *microbes* or *protists*): their occurrence, their characteristics, their relationships with each other as well as with other groups of organisms, their control, and their importance to our health and welfare. Microorganisms, some helpful and others harmful, are intimately associated with our lives. Many are inhabitants of the human body. Some microorganisms cause disease and others are involved in such everyday human activities as the manufacture of wine, cheese, and yogurt, the production of penicillin, and the treatment processes associated with sewage disposal.

Microbiology is a comparatively young science. The world of microorganisms was discovered only about 300 years ago, and it was another 200 years before the real significance of microorganisms was understood and appreciated. During the last 40 years, microbiology has emerged as a very significant field of biology. Today microorganisms are used by researchers in the study of practically all major biological phenomena. (We shall note the reasons for this fact later in this chapter.)

3

Microscopes and the discovery of the microbial world

Although Antony van Leeuwenhoek (1632–1723), a Dutch student of natural history, was probably not the first to see the microbes we call bacteria and protozoa, he (Fig. 1-1) was the first to report his observations with accurate descriptions and drawings. Leeuwenhoek made these observations in the course of pursuing his hobby of grinding lenses and making microscopes. During his lifetime he made more than 250 microscopes, each consisting of a single home-ground lens mounted in brass and silver, the most powerful of which could magnify only 200 to 300 times (Fig. 1-2). These microscopes bear little resemblance to the compound light microscopes of today, which use two or more lenses in a system that is capable of magnifications of 1,000 to 2,000 times. But the lenses of Leeuwenhoek's microscopes were well made, and Leeuwenhoek had the openness of mind so very important in any investigator.

On June 9, 1675, Leeuwenhoek wrote in his diary, "Collected rainwater in a dish," and on June 10 he continued, "Observing this water I fancied that I discovered living creatures; but since they were so few and not clearly detectable, I could not accept this as truth." He returned to his observations on the next day and wrote, "I had no notion that I would see any living creatures, but on examining it I saw with wonder fully a thousand living creatures in one drop of water. The animalcules were of the smallest kind I had seen up to now." He made drawings of bacteria in rainwater, saliva, vinegar, and other substances and described them with fascinating word pictures. He related his exciting discoveries in a series of more than 300 letters to his friends in the Royal Society of London and the French Academy of Sciences.

A letter dated September 17, 1683, contains the first recorded drawings of bacteria. They were observed by Leeuwenhoek in a suspension of tartar scraped from his teeth. The thoroughness and accuracy of his observations

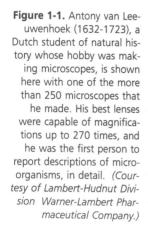

Figure 1-1. Antony van Leeuwenhoek (1632-1723), a Dutch student of natural history whose hobby was making microscopes, is shown here with one of the more than 250 microscopes that he made. His best lenses were capable of magnifications up to 270 times, and he was the first person to report descriptions of microorganisms, in detail. *(Courtesy of Lambert-Hudnut Division Warner-Lambert Pharmaceutical Company.)*

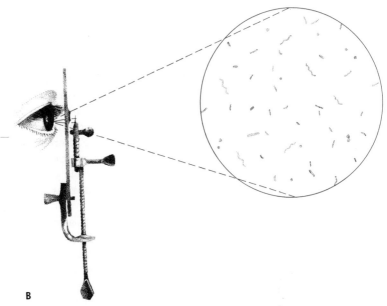

B

Figure 1-2. The Leeuwenhoek microscope. (A) Replica of a simple microscope made in 1673 by Leeuwenhoek. *(From the collection of the Armed Forces Institute of Pathology, Washington, D.C.)* (B) Side view of a Leeuwenhoek microscope, illustrating the manner in which observations of specimens were made. *(Erwin F. Lessel, illustrator.)*

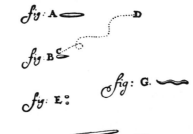

Figure 1-3. Leeuwenhoek's sketches of bacteria from the human mouth, from letter of September 17, 1683. The dotted line between B and D indicates motility (movement) of a bacterium. *(Courtesy of C. Dobell, Antony van Leeuwenhoek and His "Little Animals," Dover, New York, 1960.)*

are apparent in his drawings. He sketched bacterial cells (Fig. 1-3) as being spherical (these are now called *cocci*), cylindrical or rod-shaped (*bacilli*), or spiral (*spirilla*). Despite all the improvements in microscopy since Leeuwenhoek's time, we still recognize these same three general bacterial shapes.

Leeuwenhoek's observations, reported in enthusiastic letters, were read with interest, but the significance of his discoveries went unrecognized. Prior to the 1800s there was little awareness that microorganisms cause many diseases or that they cause chemical changes in countless materials in our environment.

Spontaneous generation versus biogenesis

The discovery of a world of organisms invisible to the naked eye spurred interest in a great debate of the time concerning the origin of life. Where did these microorganisms come from?

Spontaneous generation

Some people answered that microorganisms came into being as a result of the decomposition of dead plant or animal tissues. In other words, they thought that living organisms arose from nonliving materials upon decomposition. This concept, that life originates from nonliving material, is known as *spontaneous generation* or *abiogenesis* (*abio*, "nonliving"; *genesis*, "origin"). The idea of spontaneous generation dates back at least to the ancient Greeks who believed that decaying meat produced maggots and that flies and frogs sprang forth from mud under appropriate climatic conditions. Not only were there many who argued that microorganisms came into being through spontaneous generation, but there were numerous advocates for the spontaneous generation of worms, other insects, and even animals such as mice and frogs.

Disproof of spontaneous generation and acceptance of biogenesis

There were both champions and challengers of the theory that living things can originate spontaneously, each with a new and sometimes fantastic explanation or bit of experimental evidence. In 1749, John Needham (1713–1781), experimenting with cooked meat observed that there were microorganisms on the meat at the start of the experiment and concluded that they originated from the meat. About the same time, Lazaro Spallanzani (1729–1799), in an attempt to disprove spontaneous generation, boiled beef broth, a nutrient solution, in flasks for an hour and then sealed the flasks. No microbes appeared in the flasks. But his results, confirmed in repeated experiments, failed to convince Needham that microbes did not arise by spontaneous generation. Needham simply insisted that air was essential to the spontaneous generation of microbes and that because it had been excluded from the flasks in Spallanzani's experiments, no microbes appeared. This argument was answered independently some 80 or 90 years later by two other investigators, Franz Schulze (1815–1873) and Theodor Schwann (1810–1882). Schulze passed air through strong acid solutions into boiled meat broth in flasks and Schwann passed air through red-hot tubes into his flasks of boiled meat broth (Fig. 1-4A and B). Microbes did not

Figure 1-4. Sketches of apparatus used in a series of experiments to disprove spontaneous generation. The assumption made in each experiment was that microorganisms were carried on dust particles suspended in the air. If the air was treated so as to remove or kill the microorganisms before the air entered a nutrient solution, no growth of microorganisms would occur in the nutrient solution. (A) Microorganisms in air destroyed by strong acid. (B) Microorganisms in air destroyed by heat. (C) Microorganisms removed from air by filtration (passage through cotton). (D) Microorganisms removed by "settling" of dust particles in bottom of convoluted tubes.

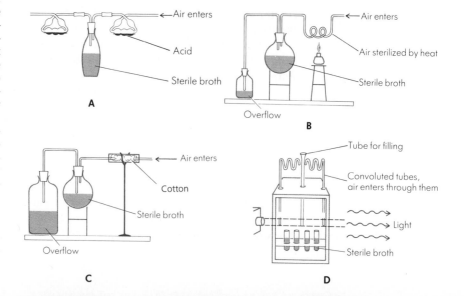

appear in either case. The acid in one experiment and the extreme heat in the other had killed them. But the die-hard advocates of spontaneous generation were still not convinced. Acid and heat altered the air so that it would not support growth, they said. About 1850 Schroder and von Dusch performed a more convincing experiment by passing air through cotton-filled tubes into flasks containing previously heated broth (Fig. 1-4C). Microbes were filtered out of the air by the cotton fibers and thus they were prevented from entering the flask and no microorganisms grew in the broth.

Some of the most important experimental evidence against spontaneous generation was provided by the work of John Tyndall in the early 1870s. He constructed a dustfree box (see Fig. 1-4D) and placed tubes of sterile broth within the box. The broth in the tubes remained sterile as long as the air in the box remained dustfree. The dust particles settled and were trapped in the gooseneck tubes leading into the box. This was proof that the microbes were carried on dust particles.

During the same period that these experiments were being performed a new figure was emerging in science, Louis Pasteur (1822–1895) (Fig. 1-5). Pasteur was educated as a chemist and came to national fame early in his career when he discovered the chemical structure of tartaric acid. He found that this compound existed in two forms, one the mirror image of the other—as the left hand is to the right hand. The chemical composition of the two compounds was the same in the number and types of atoms each contained, but their configuration was different. Pasteur later became interested in the wine industry and the changes produced during the fermentation process. This interest in fermentation forced him into the debate on spontaneous generation.

Fermentation is brought about by enzymes, substances produced by living cells which promote certain types of chemical reactions. For example, grape juice, when allowed to stand, ferments; alcohols and acids are produced. The questions arising from this were: Were these fermentation products—the alcohols and acids—produced by microorganisms present in the grape juice? Or did the microorganisms in the grape juice arise from the fermentation process, as the supporters of spontaneous generation claimed? Pasteur vigorously opposed the concept of spontaneous generation and realized that he could make little progress in his studies on fermentation until this concept was disproved. Accordingly he set out to carefully review the earlier work on the subject and then proceeded to design and conduct numerous experiments to document the fact that microorganisms could arise only from other microorganisms (biogenesis). He became embroiled in many debates with his opponents on the subject as he presented the results of his experiments.

One of the dedicated supporters of spontaneous generation during Pasteur's time was a French naturalist, Felix-Archimede Pouchet. In 1859 Pouchet published an extensive report "proving" its occurrence. But he reckoned without the ingenious, tireless, and stubborn Louis Pasteur. Irritated by Pouchet's logic and data, Pasteur performed experiments to end the argument for all time. He prepared nutrient solution in flasks with long, narrow openings—"goosenecks" (Fig. 1-6). He then heated the nutrient solutions and air—untreated and unfiltered—was allowed to pass in or out. No microbes appeared in the solution. The reason for this was the dust particles carrying microbes did not reach the nutrient solution—they settled out in the U-shaped part of the gooseneck tube and air currents were so reduced that the dust particles were not carried into the flasks.

Pasteur reported his results with a great flourish at the Sorbonne in Paris on April 7, 1864. His flasks would yield no sign of life he said:

For I have kept from them, and am still keeping from them, that one thing which is above the power of man to make; I have kept from them the germs that float in the air, I have kept from them life.

In his exuberance, Pasteur flung a few darts at those he disagreed with:

Figure 1-6. Pasteur's gooseneck flask which he used in experiments to disprove spontaneous generation. This has been preserved in the Pasteur Museum. *(From Institut Pasteur, Paris.)*

There is no condition known today in which you can affirm that microscopic beings come into the world without germs, without parents like themselves. They who allege it have been the sport of illusions, of ill-made experiments, vitiated by errors which they have not been able to perceive and have not known how to avoid.

With the acceptance of the concept of biogenesis, the path was cleared for Pasteur's future work. Pasteur could now move ahead with his studies on fermentation and later with his studies of microorganisms as a cause of disease.

The germ theory of fermentation

Many ancient cultures developed beverages and foods that were products of microbial fermentations. In Greece the production of wine by fermentation of fruit had been practiced for so long that the Greeks believed wine had been invented by Dionysus. A Mesopotamian clay tablet, written in Sumerian and Akkadian about 500 years before Christ, reveals that brewing had been a well-established profession in that culture for several thousand years. *Kiu*, a Chinese rice beverage and a kind of beer, has been traced back to 2300 B.C. The soy sauces of China and Japan, derived from fermented beans, have been made for centuries. For many hundreds of years, peoples of the Balkan countries have consumed fermented milk products. Central Asiatic tribesmen have long enjoyed koumiss, an alcoholic beverage made from fermented mare's or camel's milk. Indeed, anthropologists and historians know of no society in which fermentation has not been employed to make food, drink, or both.

During ancient times, people improved upon the quality of their fermented products through trial and error, unaware that the quality depended on providing or improving conditions for growth of the microorganisms to bring about fermentation. It was not until Pasteur's studies of the role of microorganisms in the fermentation process in wine-making that it was understood that microorganisms were responsible for fermentation.

In the 1850s Pasteur turned his attention to wine-making, a major industry of France. Having disproved spontaneous generation, thereby establishing that microorganisms are responsible for fermentation, he was ready to help French wine makers and brewers, who frequently encountered problems in making a good-quality product. By examining many batches of wine, he found microbes of different kinds. In good batches certain kinds predominated, in poor batches other kinds were present. Pasteur determined that by proper selection of the microbe the manufacturer might be assured of a consistently good and uniform product. To do this, microbes which were already in the juices had to be removed and the new fermentation started with a *culture*, a growth of microorganisms, from a vat of wine that had been satisfactory. Pasteur suggested that the undesirable types of microbes might be removed by heating—not enough to damage the flavor of the fruit juice but enough to kill the microbes. He found that holding the juices at a temperature of 62.8°C (145°F) for half an hour did the job. Today this process, *pasteurization*, is widely used in the fermentation industries. But we are most familiar with it in the dairy industry, where it is used to destroy disease-producing microorganisms in milk and milk products.

The germ theory of disease

Even before Pasteur had proved by experiment that bacteria are the cause of some diseases, many careful observers had expressed strong arguments for the *germ theory of disease*. In 1546 Fracastoro of Verona (1483–1553) suggested that diseases might be due to organisms too small to be seen, transmitted from one person to another. In 1762 von Plenciz of Vienna not only stated that living agents were the cause of disease but also suggested that different microorganisms were responsible for different diseases. The concept of *parasitism*, i.e., of an organism living in or on another organism from which it derives nutrients, was becoming quite widespread in the 1700s. This is reflected in the following bit of doggerel written by the great English satirist Jonathan Swift (1667–1745) early in the eighteenth century:

> So naturalists observe, a flea
> Hath smaller fleas that on him prey;
> And these have smaller fleas to bite 'em;
> And so proceed ad infinitum.

Oliver Wendell Holmes (1809–1894), a successful physician as well as a man of letters, in 1843 insisted that childbed fever (puerperal fever), a serious and often fatal infection of the mother after delivery, was contagious and that it was probably caused by microorganisms carried from one mother to another by midwives and physicians. At approximately the same time (the 1840s) the Hungarian physician Ignaz Phillip Semmelweis (1818–1865) was pioneering in the use of obstetrical procedures that would reduce chances of infections caused by microorganisms.

Pasteur's success in solving the problem of fermentation led to a request by the French government for him to investigate *pebrine*, a silkworm disease that was ruining the important French silk industry. The problem proved a difficult one, and for several years he struggled for a solution. Eventually, however, he isolated the microorganism (a protozoan) that caused the disease. Pasteur went even further and showed that silkworm farmers could eliminate the disease by selecting only healthy, disease-free worms to breed new crops of worms.

Pasteur next (in 1877) tackled the problem of anthrax, a disease of cattle, sheep, and sometimes human beings. After observing the anthrax-causing microbes from the blood of animals that had died of the disease, he grew them in laboratory flasks.

During the 1870s Robert Koch (1843–1910) was also busy with the anthrax problem in Germany. Koch, a quiet, meticulous physician, sometimes neglected his medical practice to pursue the fascinating new science of bacteriology. It was he who isolated the typical rod-shaped bacteria with squarish ends (bacilli) from the blood of sheep that had died of anthrax. He grew these bacteria in his laboratory, examined them microscopically to be sure he had but one kind present, and then injected them into mice to see if they became infected and developed symptoms of anthrax. From these mice he isolated bacteria like those he had originally found in sheep that died of anthrax. This was the first time a bacterium had been proven to cause an animal disease. Later Koch discovered the bacteria that caused tuberculosis and cholera.

The development of microbiological laboratory techniques and procedures

In their searches for microorganisms that might be responsible for different diseases, Koch and his colleagues developed several laboratory procedures which had tremendous impact on the development of microbiology. These included procedures for staining bacteria to make them more easily visible and techniques for *culturing* (growing) microbes in the laboratory. One important culturing technique which they developed was the use of a *medium* (plural, *media*), a substrate for growing bacteria, which solidifies and remains transparent at incubation temperatures (temperatures suitable for growth). Gelatin, which had been tried for this purpose, was unsatisfactory because it became liquid at body temperature. Other solid surfaces such as slices of potato or carrot had numerous disadvantages, including lack of nutrients for many microorganisms, particularly those associated with the human body. The problem was resolved by using an extract from certain marine algae. This extract, called agar-agar (frequently referred to simply as *agar*), could be dissolved in a nutrient solution and when it gelled would remain solid over a wide temperature range. We shall describe microbiological media and their use in Chap. 6.

The pure culture concept

The agar media provided an excellent means of separating a mixture of microorganisms into each of the different kinds present. The technique used for growing microorganisms on agar media permits them to grow at a distance from each other and permits each cell (microorganism) to accumulate a *colony*, a compact mass of cells which is visible to the naked eye (Fig. 1-7). All the cells in a colony are the same; it is assumed that they are

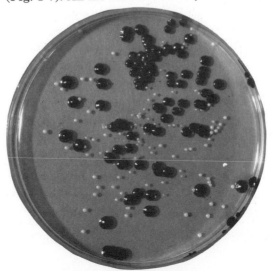

Figure 1-7. Colonies of microorganisms on nutrient agar medium contained in a petri dish. Note the two different kinds of colonies; they indicate two different kinds of bacteria.

the progeny of a single microorganism and hence represent what the microbiologist calls a *pure culture*. The use of agar in microbiological media, initially proposed by Koch's laboratory in the early 1880s, remains in wide use today.

Koch's postulates

The experiments of Koch and others in his laboratory proved that a particular microorganism caused a specific disease and led to the establishment of the criteria on which such a conclusion can be based. These criteria, known as *Koch's postulates*, have become and remain the guidelines for seeking evidence that a disease is caused by a specific microorganism. Koch's postulates are:

1 A specific microorganism can always be found in association with a given disease.
2 The microorganism can be isolated and grown in pure culture in the laboratory.
3 The pure culture of the microorganism will produce the disease when injected into a susceptible animal.
4 It is possible by laboratory procedures to recover the injected microorganism from the experimentally infected animal.

Prevention and treatment of microbial diseases

Once it was established that microorganisms are the cause of certain diseases, much attention was given to developing methods for the prevention and treatment of these diseases. The *etiological* (causative) *agents* for most of the bacterial diseases we know today—diseases that had plagued the world for centuries—were discovered in rapid succession between the years 1876 and 1898 (see Table 1-1).

The magnitude of human misery and devastation caused by microbial diseases prior to the latter part of the twentieth century is difficult to

Table 1-1. Discovery of etiological agents of microbial diseases, 1876–1898

DISEASE*	DATE CAUSATIVE AGENT DISCOVERED
Anthrax	1876
Gonorrhea	1879
Typhoid fever and malaria	1880
Wound infections	1881
Tuberculosis and glanders	1882
Diphtheria and cholera	1883
Typhoid fever	1884
Tetanus and swine erysipelas	1885
Bacterial pneumonia	1886
Meningitis and Malta fever	1887
Equine strangles	1888
Gas gangrene	1892
Plague	1894
Fowl typhoid	1895
Botulism (food poisoning)	1896
Bang's disease (bovine abortion)	1897
Dysentery and pleuropneumonia of cattle	1898

*All the diseases listed are caused by bacteria except for malaria, which is caused by a protozoan.

appreciate. Diseases such as plague, diphtheria, smallpox, and cholera literally devastated vast regions of the world. Great armies were brought to their knees by epidemics caused by microorganisms. Historical writings suggest that the outcome of most wars was determined by epidemics of disease rather than by brilliant generals.

Methods of prevention and treatment that have been introduced to control microbial diseases include *immunization* (e.g., vaccination), *antisepsis* (procedures to eliminate or reduce the possibility of infection), *chemotherapy* (treatment of the patient with chemicals), and public health measures (e.g., water purification, sewage disposal, and food preservation).

Immunization
Pasteur continued to make discoveries concerning the cause and prevention of infectious diseases. About 1880 he isolated the bacterium responsible for chicken cholera and grew it in pure culture. To prove that he really had isolated the bacterium responsible for this disease Pasteur made use of the fundamental techniques devised by Koch. He arranged for a public demonstration in which he repeated an experiment (Fig. 1-8) that had been successful in many previous trials in his laboratory: he inoculated (injected) healthy chickens with his pure cultures and waited for them to develop chicken cholera and die. But to his dismay, the chickens failed to get sick and die! Reviewing each step of the experiment, Pasteur found that he had accidentally used cultures several weeks old instead of the fresh ones grown especially for the demonstration. Some weeks later he repeated the experiment using two groups of chickens. The first group had been inoculated at the earlier demonstration with the old cultures that had proved ineffective; the second had not been previously exposed to these

Figure 1-8. The principle of immunization as demonstrated by Pasteur. Pasteur first inoculated chickens with cultures of chicken cholera bacteria which were several weeks old and these chickens remained healthy. Several weeks later he inoculated them with a fresh culture of chicken cholera bacteria. This fresh virulent culture did not make them sick, but it killed chickens that had not been inoculated with the earlier attenuated culture. This experiment demonstrated that the "old" culture of chicken cholera bacteria, even though unable to produce disease, was capable of causing the chickens to produce protective substances called antibodies in their blood.

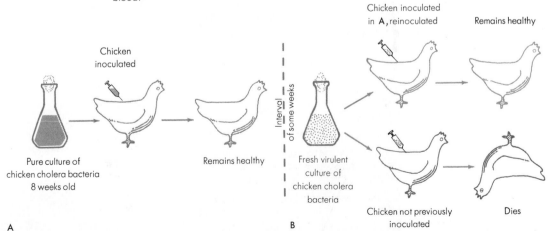

cultures. Both groups were inoculated with bacteria from fresh young cultures. This time the chickens in the second group got sick and died, but those in the first group remained healthy. This puzzled Pasteur, but he soon found the explanation. He discovered that in some way bacteria could lose their *virulence*, or ability to produce disease, after standing and growing old. But these *attenuated,* or less virulent, bacteria could still stimulate the host (in this case the chicken) to produce *antibodies*, substances that protect the host against infection due to subsequent exposure to the virulent organism.

Pasteur next applied this principle of inoculation with attenuated cultures to the prevention of anthrax, and again it worked. He called the attenuated cultures of bacteria *vaccines* (a term derived from the Latin *vacca,* "cow") and immunization with attenuated cultures of bacteria, *vaccination.* Pasteur thus honored Edward Jenner (1749–1823), who had successfully vaccinated a boy against smallpox in 1796 (Fig. 1-9). Jenner had learned that milkmaids who contracted cowpox from the cows they milked never subsequently contracted the much more virulent smallpox. From this information, Jenner hypothesized that a bout of cowpox somehow immunized one against smallpox. Accordingly, he tested this hypothesis by inoculating young James Phipps first with cowpox-causing material and later with smallpox-causing material. The boy did not get smallpox.

Pasteur's fame was by now well established throughout France, and many believed he could work miracles with bacteria and the control of infections. It was not surprising, then, that he was asked to make a vaccine for hydrophobia, or rabies, a disease transmitted to people by bites from rabid dogs, cats, and other animals. Being a chemist and not a physician, he paused before accepting this challenge. Once he agreed, he set out to make a vaccine. Despite the fact that the etiologic agent of rabies was not known, Pasteur proceeded on his strong intuition that it was a disease-

Figure 1-9. Edward Jenner vaccinating (inoculating) James Phipps with cowpox material which resulted in development of resistance to smallpox infection. *(From Culver Pictures, New York.)*

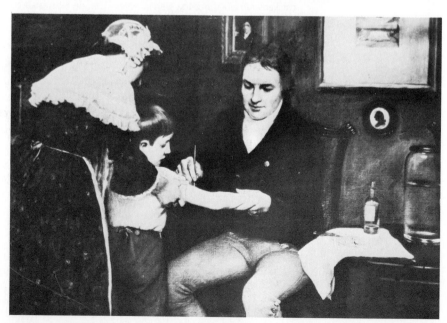

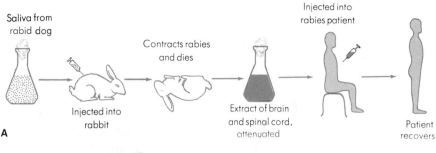

Saliva from rabid dog

Injected into rabbit

Contracts rabies and dies

Extract of brain and spinal cord, attenuated

Injected into rabies patient

Patient recovers

A

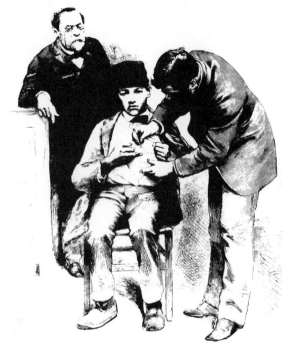

Figure 1-10. (A) Rabies vaccine was made by inoculating a rabbit with saliva from a rabid dog. Virus in the extract of the rabbit's spinal cord was attenuated before injection into a patient. (B) Pasteur (to the left) supervises inoculation of rabies vaccine by an assistant into patients bitten by mad animals. *(Courtesy of National Library of Medicine.)*

B

producing microorganism. Pasteur could produce the disease in rabbits by inoculating them with saliva from mad dogs. After an inoculated rabbit died, Pasteur removed the brain and spinal cord, dried them for several days, and then pulverized them and suspended the powder in a liquid (Fig. 1-10A). Inoculating dogs with this mixture protected them against rabies. But vaccinating dogs was quite different from treating a sick human (Fig. 1-10B). Since rabies was almost invariably fatal, it was only after a boy named Joseph Meister had been bitten by a rabid wolf that his family gladly permitted Pasteur to inoculate the boy with vaccine in hopes of saving his life. The worried Pasteur was as surprised as anyone else when after the crucial trial, which took several weeks, Joseph Meister did not die.

Phagocytosis

At about this time, Elie Metchnikoff (1845–1916) (Fig. 1-11), working in Pasteur's laboratory, observed that *leucocytes*, a type of cell in human blood, could ingest (eat) disease-producing bacteria found in the body. He called these defenders against infection *phagocytes* or, "eating cells," and

Figure 1-11. Elie Metchnikoff was the first person to recognize the role of phagocytes in combating bacterial infections. *(Courtesy of René-Dubos.)*

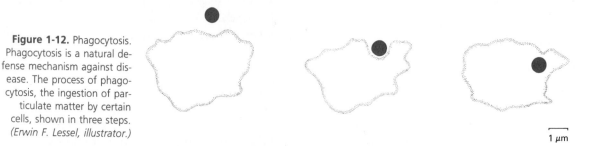

Figure 1-12. Phagocytosis. Phagocytosis is a natural defense mechanism against disease. The process of phagocytosis, the ingestion of particulate matter by certain cells, shown in three steps. *(Erwin F. Lessel, illustrator.)*

1 µm

the process of ingestion, *phagocytosis* (Fig. 1-12). From these observations Metchnikoff hypothesized that the phagocytes were the patient's first important line of defense against microorganisms which invade the body.

Antisepsis   The word *sepsis*, in a general sense, implies infection; *antisepsis* refers to measures which combat or prevent infection. We have already mentioned Semmelweis' introduction of aseptic methods during obstetrical procedures to reduce the incidence of childbed fever caused by microbes. In the 1860s an English surgeon, Joseph Lister (1827–1912), was searching for a way to keep microbes out of wounds and the incisions made by surgeons because deaths from these causes were appallingly high. In 1864, for example, Lister's records show that 45 percent of his own patients died after surgery. Disinfectants as such were unknown, but carbolic acid (phenol) was known to kill bacteria, so Lister used a dilute solution of this acid to soak surgical dressings and spray the operating room (Fig. 1-13). Wounds and incisions protected in this way rarely became infected and healed rapidly. So remarkable was his success that the technique was quickly accepted by other surgeons, and this antiseptic practice established the principle of present-day *aseptic techniques* used to exclude microbes from wounds or incisions. Today a large variety of chemical substances, such as alcohol and iodine solutions, and physical techniques, such as air filters and germicidal (ultraviolet) lamps, are used to reduce the number of microbes in such areas as operating rooms and nurseries for premature babies.

Chemotherapy

Chemicals have been used to treat disease for hundreds of years: mercury was used to treat syphilis as early as 1495 and cinchona bark (which contains quinine) was used in South America in the seventeenth century for the treatment of malaria. But it was not until the brilliant research and writing of the German physician Paul Ehrlich (1845–1915) that modern chemotherapy was launched. Ehrlich's objective was to develop chemicals that would kill specific microbes without harming the patient. To this end he systematically synthesized hundreds of compounds in his search for potential chemotherapeutic agents. In 1909 it was discovered that the arsenic compound *Salvarsan*, the 606th compound of a series synthesized by Ehrlich, kills the bacterium that causes syphilis.

Finally in the 1930s the dreams which Ehrlich had entertained about the

Figure 1-13. Operation performed with antiseptic precautions, using Lister carbolic spray (1882). The apparatus shown on the stool produces a spray of carbolic acid for the purpose of reducing infections. *(Courtesy of National Library of Medicine.)*

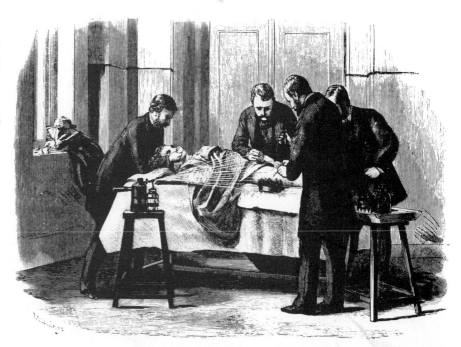

development of chemotherapeutic agents were fulfilled. Gerhard Domagk, a German scientist, discovered that a group of chemical compounds known as *sulfonamides* (popularly called *sulfa drugs*) were very effective for treatment of several bacterial infections. As we will explain more fully later, when we discuss chemotherapeutic agents in Chap. 21, these compounds inflict their damage on bacterial cells in a very specific manner and without harming the patient, just as Ehrlich had hoped.

Another dramatically effective chemotherapeutic agent, penicillin, was first noted at about this time. In 1929, years before sulfonamides became popular, Alexander Fleming (1881–1955), a Scottish bacteriologist working at St. Mary's Hospital in London, reported his finding of a substance produced by the mold *Penicillium notatum* which prevented the growth of bacteria on laboratory culture media. Fleming's report went virtually unnoticed until 10 years later, when scientists began searching for effective drugs to treat wounds and infections in military personnel. This substance, which Fleming called *penicillin* because it was produced by *Penicillium notatum*, aroused great interest when it became clear that it might cure some infections unaffected by sulfa drugs without the harmful side effects that some sulfonamides produce. But it was not until the 1940s that penicillin became available in sufficient amounts for clinical testing. Soon after it quickly acquired the title of "miracle drug" because it proved to be the most effective cure for a number of common diseases and the only cure for several diseases; responses to penicillin treatment were dramatic.

Penicillin is an *antibiotic*, a substance produced by one microorganism that in very small quantities is inhibitory to growth of other microorganisms. The discovery of penicillin opened the way for the discovery and commercial production of several other antibiotics. Now, of course, antibiotics are the major chemical compounds for the treatment of infectious diseases.

Widening horizons— applications of microbiology to nonmedical fields

As these dramatic discoveries in medical microbiology were being made, some scientists directed their attention to the role of microorganisms in fields other than medicine.

The field of soil microbiology started with the investigations of a Russian, Serge Winogradsky (1856–1953), who showed the importance of certain bacteria in maintaining soil fertility. He demonstrated that certain bacteria were capable of taking nitrogen (N_2) from the atmosphere, where it is abundant, and converting it into nitrogen compounds that plants can use as nutrients. In 1901, Martinus Beijerinck (1851–1931), a famous Dutch microbiologist, discovered other bacteria in soil that were important for soil fertility, again by synthesizing nitrogen compounds but by a process different from that discovered by Winogradsky. These discoveries triggered many other investigations on the relationships of microorganisms to soil fertility.

As we have already mentioned, Pasteur established the role of microorganisms in the fermentation industry through his studies on wines. Others were quick to extend his findings to other fermentation processes. A Dane, Emil Christian Hansen (1842–1909), started a business that provided desirable types of microorganisms for the manufacture of vinegar and dairy products (butter and cheeses).

Late in the nineteenth century, Thomas J. Burrill (1839–1916), at the University of Illinois, found that a disease of pear trees known as *fire blight* was caused by a bacterium. This discovery established a new area, *plant pathology* (the study of plant diseases), which attracted additional researchers. At about the same time, Erwin F. Smith (1854–1927), of the U. S. Department of Agriculture, was able to transmit *peach yellows*, another plant disease, from diseased to healthy plants. Transmission of virus diseases of plants by insects had also been observed by a Japanese farmer in 1894 and independently in 1907 by several American scientists. In 1892 Dmitri Ivanowski discovered that the causative agent of tobacco mosaic, a disease of tobacco plants, could be transmitted by the filtered extract (strained juice) from a diseased plant. The kind of filter used was known to be capable of preventing bacteria from passing through. The filtered juice contained no microscopically visible microorganisms. Further observations established that the filtered material contained a virus which was the infectious agent.

It is readily apparent from these examples and the early medical discoveries that microbiology developed initially as an applied science.

Table 1-2. Some basic (fundamental) and applied areas of microbiology	THE BIOLOGY OF MICROORGANISMS (BASIC BIOLOGY)	APPLICATIONS OF MICROBIOLOGY (APPLIED FIELDS)
	Cytology: The study of cell structures and their functions	*Medicine:* Etiology and diagnosis of infectious diseases
	Cultural characteristics: The appearance of microbial growth on various media	*Public health:* Measures to control the occurrence and spread of diseases
	Metabolism: All the chemical processes carried out by the organism	*Industry:* Fermentation products such as alcohols, acids, and pharmaceutical products such as antibiotics and vitamins
	Taxonomy: The naming and classification of organisms	*Milk and foods:* Quality control of milk and food products and production of cheese, yogurt, pickles, and other foods
	Genetics: The process by which the characteristics of the parent cell are transmitted to the daughter cell	*Agriculture:* Soil fertility, animal and crop diseases
	Microbial associations: How microorganisms, in their natural environment, affect each other	*Aerobiology:* Occurrence of microorganisms in air and spread of microorganisms through air
	Pathogenicity and virulence: The disease-producing ability of microorganisms	*Microbiology of domestic water and waste water:* Quality control of domestic water, treatment processes for waste water
	Immunology: Processes associated with host resistance to infection	*Marine microbiology:* The role of microorganisms in marine environments
		Exobiology: The search for life (microorganisms) in outer space

Figure 1-14. Nobel prize winners in physiology or medicine. In recent years many of the scientists who have been awarded the Nobel prize for their outstanding contributions to basic biology have used microorganisms in their research. A few of the Nobel Prize winners are shown here. (A) (1) Joshua and Marguerite Lederberg. (2) Edward Tatum. (3) George Beadle, who shared the Nobel prize for physiology or medicine with Joshua Lederberg and Tatum in 1958 for research in microbial genetics. (B) (1) Max Delbruck, (2) Alfred D. Hershey, and (3) Salvador E. Luria won the Nobel prize in 1969. Working individually but with frequent communication, they conducted research on the life processes of viruses that infect bacteria, helping to lay a foundation for much of modern molecular biology. Their investigations have also contributed to our knowledge of viral diseases in general.

Microbiology and biology

At the turn of the century the study of microbiology branched out in two different but complementary directions: the first was concerned with further studies to discover uses of microorganisms and the second was concerned with detailed studies of the biological characteristics of microorganisms (see Table 1-2). The results of research on the biological characteristics of microorganisms have proved invaluable to all the biological sciences. The reason for this is that early studies on the physiology and chemical makeup of microorganisms showed them to be fundamentally similar to all other living organisms. Because they possess many characteristics which make them ideal for use in research—among these are their rapid rate of growth (some bacteria reproduce every 10 to 15 min), their ease of growth and manipulation under laboratory conditions, and their great diversity which facilitates research into basic biological processes—they have frequently been studied for information about other organisms that could not be easily obtained by direct experiments on those organisms. Research with microorganisms has resulted in many outstanding contributions to biology, biochemistry, and medicine (Fig. 1-14).

Summary and outlook

The history of the development of microbiology as a science can be divided into three periods. The first period started with the revelation of a world of microorganisms by the microscopic observations of Leeuwenhoek in 1675. This aroused scientific curiosity about the origin of life. But not until after about the mid-1860s, when the theory of spontaneous generation was disproved and the principle of biogenesis accepted, was the knowledge of microorganisms more than speculative. During the next period, between 1860 and 1900, many important fundamental discoveries were made. The development of the germ theory of fermentation was followed by the development of the germ theory of disease in 1876. This spurred interest in laboratory techniques and procedures for the isolation and characterization of microorganisms. Within this period many of the etiological agents of microbial diseases were discovered and methods found to prevent and diagnose and treat these diseases. The discoveries in the field of medical microbiology revolutionized the practice of medicine.

Since 1900, the close alliance of microbiology, medicine, and other areas of applied microbiology has also grown. In addition, microorganisms have become model systems for the study of numerous biological processes basic to many or all living organisms.

New discoveries continue to be made. New species of microorganisms are continually being discovered and many biological processess are being studied via microorganisms.

Key terms

abiogenesis	antisepsis	biogenesis	culture
agar-agar	aseptic technique	chemotherapy	etiology
antibiotic	attenuated	cocci	fermentation
antibodies	bacilli	colony	immunization

Koch's postulates   parasitism   sepsis
leucocytes   pasteurization spirilla
medium (plural, media) phagocytes vaccination
microbiology   pure culture   virulence
microorganisms   putrefaction

Questions

1 Describe Leeuwenhoek's contributions to microbiology.
2 Compare the contributions of Pasteur with those of Koch. Comment on their differing conceptual approaches and the practical applications of their discoveries.
3 What is the relationship between the *germ theory of fermentation* and the *germ theory of disease*?
4 Why was it necessary to the development of microbiology to refute the theory of *spontaneous generation*?
5 Name an important contribution to microbiology made by each of the following persons:

Schulze   Ehrlich   Koch   Lister   Metchnikoff
Schwann   Holmes   Burrill Winogradsky

6 What occurs in the process of phagocytosis?
7 What criteria must be met to prove that a particular organism causes a specific disease? What are these criteria called?
8 Describe the importance of microorganisms in five applied areas of microbiology.
9 What are five of the most important historical events (scientific contributions) which led to the establishment of microbiology as a science?
10 Why are microorganisms frequently used for studies of biological phenomena?

A SURVEY OF THE MICROBIAL WORLD

Several distinct groups of microorganisms make up the microbial world. Most microorganisms are *unicellular*—that is, composed of a single cell. Some have characteristics of plant cells, others of animal cells, and still others of both plant and animal cells. Collectively, microorganisms are referred to as *protists*.

A major characteristic that distinguishes certain microbial groups from others is the organization of their cellular material. This difference, which is of fundamental importance, separates all protists into two major categories, the *procaryotes* and the *eucaryotes*. (All nonprotist organisms, including human beings, are also constructed of eucaryotic cells; more on this later.)

In this chapter we discuss the fundamental structural and functional unit of all living systems including protists, the *cell*. Next we consider the place of protists in the living world by briefly considering *biological classification*. This done, we introduce the procaryotic and eucaryotic cell types and the distinguishing features of each. We also introduce the major groups of procaryotic and eucaryotic organisms, again employing the tool of classification. Finally we take a first look at *viruses*—noncellular entities which nevertheless share several properties of unicellular organisms and which, for that reason, fall within the domain of microbiology.

The basic structural unit of life, the cell

The fundamental structural and functional unit of life is the *cell*. In unicellular microorganisms, the cell is not only the structural unit, it is the organism. Multicellular organisms, by contrast, are cells arranged into units which are integrated into a system or systems that together make up the living organism.

The word *cell* was first used more than two centuries ago, by an Englishman, Robert Hooke, in his descriptions (1665) of the fine structure of cork and other plant materials. The honeycomblike structure he observed in a thin slice of cork (see Fig. 2-1) was due to the intact cell walls of cells which were formerly living. But the concept of the cell as the structural unit of life, known as the *cell theory*, is credited to two Germans, Matthias Schleiden and Theodore Schwann, who in 1838–1839 described cells as the basic structural and functional units of all organisms. Schleiden and Schwann recognized that all cells, no matter what the organism, are very similar in structure. As the concept of the cell as the basic unit of life gained acceptance, investigators speculated on the nature of the substance contained within the cell. The term *protoplasm* (Greek *proto*, "first"; *plasm*, "formed substance") was introduced to characterize the living material of a cell. The protoplasm was thought to be responsible for all living processes. Later research with powerful microscopes and new methods of observation greatly advanced our knowledge of the structural organization within a cell.

As Schleiden and Schwann recognized more than a century ago, the cells of all living organisms share several basic similarities. Figure 2-2 compares

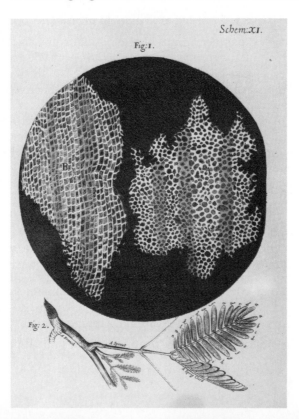

Figure 2-1. Robert Hooke's drawing of a thin slice of cork as he observed it under the microscope. This drawing was included in a report made to the Royal Society (London) in 1665. He is credited as being the first person to use the word *cell. (Courtesy of National Library of Medicine.)*

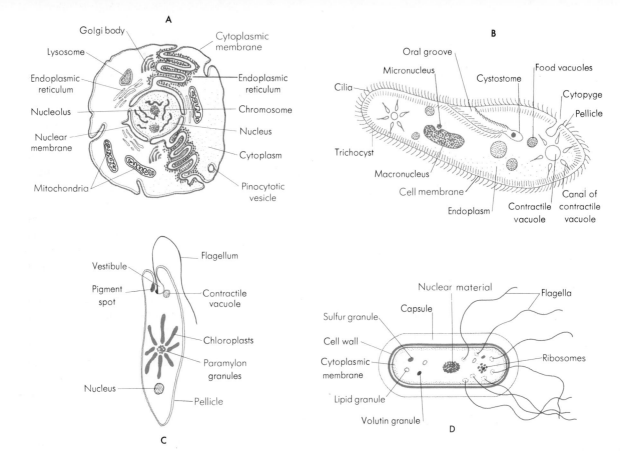

Figure 2-2. Composite drawings representing "typical" cells. (A) Parts found in a typical cell, (B) *Paramecium*, a single-celled animallike microorganism (a protozoan), (C) *Euglena*, a unicellular plantlike microbe (an alga), and (D) a bacterial cell.

a typical cell with several types of microbial cells. Each cell is enclosed by a *membrane*, an extremely thin layer of material which surrounds the substance of a cell; each contains a *nucleus* (plural, *nuclei*), the differentiated, nucleoprotein-rich, central protoplasm of a cell, or an equivalent nuclear body; and each contains *cytoplasm* (Greek *cyto*, "cell"; *plasm*, "formed substance"), within which there are various structural or particulate entities. The cytoplasm is generally regarded as all the nonstructural material contained within the membrane.

All living organisms have the following characteristics in common: (1) the ability to reproduce, (2) the ability to ingest or assimilate nutrients and metabolize them for energy and growth, (3) the ability to excrete waste products, (4) the ability to react to changes in the environment, sometimes called *irritability*, and (5) susceptibility to mutation (change in the characteristics of an organism). In the study of microbiology, we even encounter "organisms" which may represent the borderline of life, the viruses. Viruses provide an exciting opportunity to gain a better understanding of the nature of highly organized, complex organic substances that may bridge the gap between the animate and the inanimate.

Viruses, as we shall discuss later, are obligate parasites. That is, they are obligated to grow within an appropriate host cell—plant, animal, or microbe. They cannot multiply outside of an appropriate host cell. However, when a virus particle enters an appropriate living cell, it is able to create hundreds of identical virus particles, using the host cell's energy and biochemical machinery. A virus is an entity which is structurally simpler than a single cell, but it is made up of substances unique to life: *nucleic acids* (chemicals that make up genetic material) and *proteins* (complex nitrogenous substances found in various forms in animals and plants).

The place of microorganisms in the living world

In biology, as in any other field, *classification* means the orderly arrangement of the units under study into groups of larger units. Present-day classification in biology was established by the work of Carolus Linnaeus (1707–1778), a Swedish botanist. His books on the classification of plants and animals are considered the beginning of modern botanical and zoological *nomenclature*, a system of naming plants and animals. Nomenclature in microbiology, which came much later, was based upon the principles established for the plant and animal kingdoms.

Until the eighteenth century, the classification of living organisms placed all organisms into one of two kingdoms, plant or animal. But as investigators learned more and more about microorganisms, it became increasingly apparent that some of them did not fit properly into either of the two kingdoms. Some were plantlike in certain important respects but not in others. Some were animallike in certain characteristics but not in others. And some had structural and/or functional characteristics very different from those of either plants or animals. So it became clear that a two-kingdom system could not properly include all living organisms. Accordingly, new classification systems were proposed which introduced new kingdoms to accommodate microorganisms.

Haeckel's kingdom Protista

One of the early proposals to improve upon the two-kingdom classification was made in 1866 by the German zoologist E. H. Haeckel. He suggested that a third kingdom be added to include microorganisms. Haeckel called these organisms *protists* ("first life") and proposed that they form a new kingdom, the *Protista*, which would contain only unicellular organisms. Thus when we speak generally of protists we include bacteria, algae, fungi, and protozoa—but not the viruses because they are not cellular organisms. Bacteria are referred to as *lower protists;* the others—algae, fungi, and protozoa—are called *higher protists*.

There are, however, very fundamental differences between the cellular structures of lower protists and higher protists. These differences were discovered following the development of laboratory techniques which enabled researchers to observe the fine internal structures of these cells and identify the chemical makeup of subcellular structures. The lower protists were characterized as being *procaryotic* (prenuclear, i.e., nuclear material not enclosed within a membrane) cells and the higher protists as *eucaryotic* (containing a true or typical nucleus) cells. Later in the chapter we present a detailed comparison of procaryotic and eucaryotic cells.

Whittaker's five-kingdom system

A more recent classification scheme was suggested by R. H. Whittaker in 1969. It proposes five kingdoms as shown in Fig. 2-3. This scheme has several attractive features. It (1) suggests likely (or at least plausible) evolutionary relationships between the various groups; (2) distinguishes three main lines of nutrition; and (3) reflects the differences in cellular organization among three kingdoms, namely *Plantae, Fungi,* and *Animalia* (see Fig. 2-3). Microorganisms are found in three of the five kingdoms: kingdom *Monera,* comprising bacteria, including the cyanobacteria (formerly known as the blue-green algae); kingdom *Protista,* comprising microscopic algae and protozoa; and kingdom *Fungi,* comprising yeasts and molds. The viruses, being noncellular, are excluded from this scheme of living things.

Kingdom Procaryotae after *Bergey's Manual of Determinative Bacteriology*

Bergey's Manual of Determinative Bacteriology, 8th edition (1974), the standard reference for bacterial *taxonomy* (classification), has recognized the kingdom Monera of Whittaker but has called it kingdom *Procaryotae* because all procaryotic cells are also moneran cells. The kingdom Procaryotae has two divisions, one for the cyanobacteria and the other for the bacteria. The system of classification for bacteria proposed in *Bergey's Manual* is generally accepted and used internationally. This subject will be discussed in more detail in the next chapter. However, it is worth noting at this point that we will use the *Bergey's Manual* scheme for classification of bacteria throughout this book. Classification systems for microorganisms other than bacteria will be described in Parts III and IV.

It is important to understand that there is no manual of the classification of fungi, algae, or protozoa comparable with *Bergey's Manual* for bacteria.

Figure 2-3. A simplified schematic representation of Whittaker's five-kingdom system. *(Erwin F. Lessel, illustrator.)*

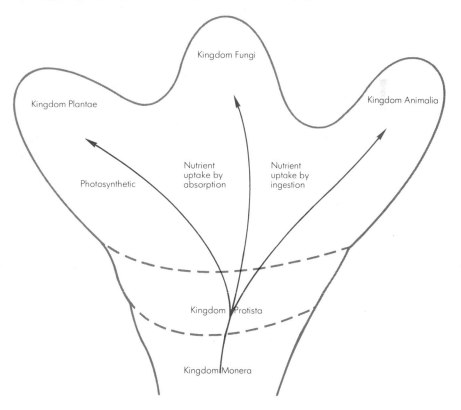

For each group of these eucaryotic protists there are publications by individuals who are authorities in classification and these publications serve as the references for classification.

Procaryotic and eucaryotic protists

We have said that there are fundamental differences in the internal structures of procaryotic and eucaryotic cells. These differences enable us to distinguish bacteria and cyanobacteria from other members of the microbial world. We wish to emphasize that this division of cells into procaryotes and eucaryotes is one of great significance. All organisms, including multicellular organisms, are made up of cells that are either procaryotic or eucaryotic.

Procaryotic cells

The principal features of the procaryotic cell can be summarized as follows:

1 No internal membranes separate the nucleus from the cytoplasm. Neither are there any internal membranes surrounding other structures or bodies within the cell.
2 Nuclear division is by *fission* (a simple, asexual process of division) and not by *mitosis* (a complex process of nuclear division which is found in eucaryotes).
3 The cell wall contains a kind of complex molecule called a *mucopeptide*. The mucopeptide provides the structural rigidity of the cell.

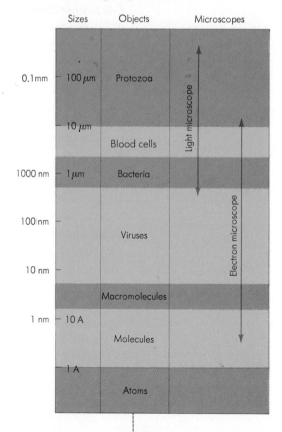

Figure 2-4. Units of measurement in microbiology. The basic unit of length is the meter (m). The relative size of microbes, molecules, and atoms is depicted here, together with an indication of the useful range of different types of microscope. *(Courtesy of A. J. Rhodes and C. E. van Rooyen, Textbook of Virology, Williams & Wilkins, Baltimore, 1968.)*

1 angstrom (A) $= 10^{-10}$ m = 0.0000000001 m
1 nanometer (nm) $= 10^{-9}$ m = 0.000000001 m
1 micrometer (μm) $= 10^{-6}$ m = 0.000001 m
1 millimeter (mm) $= 10^{-3}$ m = 0.001 m
1 m = 39.4 in

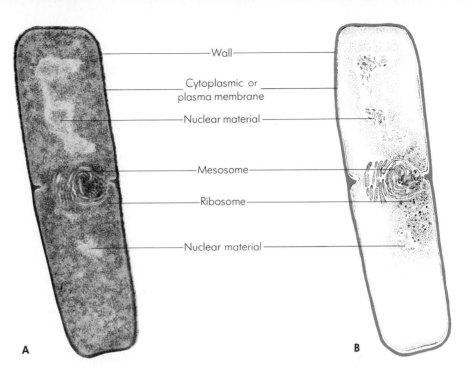

Wall

Cytoplasmic or plasma membrane

Nuclear material

Mesosome

Ribosome

Nuclear material

A

B

Figure 2.5. A typical procaryotic cell. (A) Electron micrograph of a bacterium, a procaryotic cell, (B) schematic representation of A. *(Erwin F. Lessel, illustrator.)*

Procaryotic cells are the simplest living cells in terms of structure and are believed to be the form of life which first appeared on earth. They are very small (see Fig. 2-4). One group of bacteria, the mycoplasmas, measure only 0.001 μm. However, the typical bacterial cell has a diameter of 0.5 to 1.5 μm and a length of 1.0 to 3.0 μm. The cyanobacteria are a few times larger than the bacteria. Figure 2-5 shows a typical procaryotic cell.

The material of the procaryotic cell (the cytoplasm and its contents) is surrounded by a *cytoplasmic membrane* (also called a *plasma membrane*), which controls passage of materials into and out of the cell. External to and covering the cytoplasmic membrane is a rigid cell wall (a protective coating) made up of chemical substances unique to procaryotes. Some procaryotes possess threadlike appendages which originate in the cytoplasm and extend beyond the cell wall. These are called *flagella* (singular, *flagellum*) and are responsible for the microorganism's *motility* (capability of movement). Some procaryotic cells have a mucilaginous, or slimy, covering around the rigid cell wall. This is called a *capsule* or *slime layer*.

Within the cytoplasmic area of the procaryotic cell the following substances can be observed:

Ribosomes: small particles consisting of proteins and *ribonucleic acids (RNAs)*, which are involved in the *synthesis* (manufacture) of new protein

Granules: deposits of various chemical substances which may serve as reserves of stored food

Nuclear material: strands of *deoxyribonucleic acid (DNA)*, the carrier of genetic information

Mesosomes: folds (or *invaginations*) of the cytoplasmic membrane into the cytoplasm

Eucaryotic cells

The eucaryotic cell is structurally more complex than the procaryotic cell. Remember, though, that both types of cells perform many of the same biological functions in order to grow, reproduce, and remain alive. Figure 2-6 shows a typical eucaryotic cell.

A major characteristic of the eucaryotic cell's internal structure—and one that distinguishes it from the procaryotic cell—is its extensive system of internal membranes. These membranes, called the *endoplasmic reticulum*, extend throughout the cytoplasm and partition parts of the cell by enclosing certain structures or sites of biochemical activity. These *membrane-bound structures*, as they are called, are also referred to as *organelles* ("little organs") because they perform specialized functions within the cell, in much the same way that organs (complex muticellular structures) perform specialized functions in multicellular living systems.

Let us now describe the main elements of the eucaryotic cell in terms of their appearance and function. These descriptions, together with the illustrations in Figure 2-6, will acquaint you with the major parts of the eucaryotic cell and help you see how it differs from the procaryotic cell.

Figure 2-6. (A) Electron micrograph of the alga *Chlamydomonas reinhardii* (X 15,000), a eucaryotic cell. *(Courtesy of George E. Palade, The Rockefeller University, by permission of Holt, New York, publishers of Ariel G. Leowy and Philip Seikovitz, Cell Structure and Function, 1969.)* (B) Schematic representation of (A). *(Erwin F. Lessel, illustrator.)*

Endoplasmic reticulum. This complex membrane system extends throughout the cytoplasm and divides it into compartments and channels. Part of the endoplasmic reticulum surrounds the nucleus and forms the nuclear membrane. Some parts are coated with ribosomes.

Many functions are performed by the endoplasmic reticulum. It serves as a barrier between the various organelles and keeps them in constant relative position. It provides channels that direct the flow of materials in the cell. It is a source of additional internal membranes. And it provides a firm surface for the alignment of ribosomes which function in the formation of new proteins (*protein biosynthesis*).

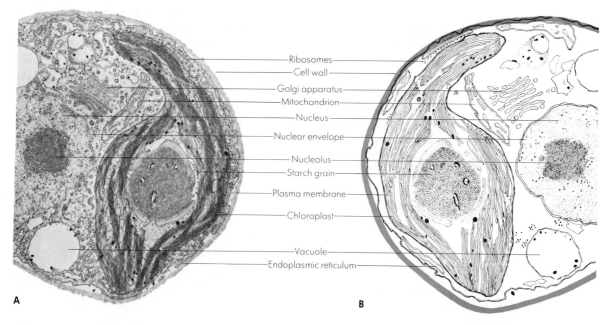

Ribosomes
Cell wall
Golgi apparatus
Mitochondrion
Nucleus
Nuclear envelope
Nucleolus
Starch grain
Plasma membrane
Chloroplast
Vacuole
Endoplasmic reticulum

A

B

Nucleus. This is a prominent, usually circular body surrounded by a double membrane, the *nuclear envelope* (Fig. 2-6). The nuclear envelope is continuous with the endoplasmic reticulum, which in turn has connections to the plasma membrane. The nuclear substance consists of DNA in the form of *chromosomes*, RNA, and proteins. Within the nucleus there are one or more dense bodies called *nucleoli* (singular, *nucleolus*). These are packed with RNA and are believed to be the sites of ribosomal RNA synthesis.

The nucleus is the principal location of genetic material, and, as such, functions as the control center of the cell.

Golgi apparatus. Also called the *Golgi complex*, this membranous organelle is made up of a group of flattened, disklike sacs, arranged in stacks like pancakes and surrounded by tubules and small vesicles. This structure, located in the endoplasmic reticulum region, packages and transports proteins and polysaccharides out of the cell. It also is the site for synthesis of new cell-wall material.

Mitochondria (singular, *mitochondrion*). These organelles enclosed within a double membrane function as the principal sites of energy production in cellular processes.

Chloroplasts. These are the organelles in plant cells which contain the green pigment *chlorophyll* and in which *photosynthesis* occurs (see Fig. 2-6). Photosynthesis is the process by which chlorophyll-containing organisms convert light energy into chemical energy.

Vacuole. This is a membrane-bound space within the cytoplasm, containing dilute solutions of various substances.

Microtubules and microfilaments. These very thin rods (*microtubules*, 250 nm; *microfilaments*, 40 to 80 nm) occur free or in bundles in the cytoplasm or within cytoplasmic structures. They maintain the cell's shape and promote orderly movement of components within organelles.

Flagella and cilia (singular, *cilium*). Flagella and *cilia* are appendages which extend outside the cell wall of certain bacteria, algae, fungi, and protozoa. These organelles generally function to move the organism. For this reason they are often called *locomotor organelles*. The flagella of eucaryotes are structurally more complex than those of procaryotes.

Cell walls. Some eucaryotic cells have cell walls, an external covering of the cytoplasmic membrane. Their structure consists of two main types of components: a network of *microfibrils* that gives rigidity to the cell wall, and a substance within which the microfibrils are embedded. The composition of these materials varies according to the type of organism. Protozoa do not have cell walls but do have a covering material called a *pellicle*.

The principal features of procaryotic and eucaryotic cells are summarized in Table 2-1.

Table 2-1. Some distinguishing characteristics of procaryotic and eucaryotic cells

CHARACTERISTIC	PROCARYOTIC CELL (BACTERIA, CYANOBACTERIA)	EUCARYOTIC CELL (ALGAE, FUNGI, PROTOZOA, PLANTS, ANIMALS)
Cell wall (when present):		
Peptidoglycan (murein, or mucopeptide) chemical component	+	−
Cytoplasmic region:		
Mesosomes	+	−
Mitochondria	−	+
Chloroplasts	−	±
Golgi structure	−	+
Endoplasmic reticulum	−	+
Membrane-bound vacuoles	−	+
Nuclear material:		
Membrane-bound	−	+
Sexual reproduction:	Rare	+

Major groups of microorganisms

As stated in Chap. 1, the major groups of microorganisms are bacteria (including the cyanobacteria), fungi, protozoa, the microscopic algae, and viruses. What follows is an overview of some major characteristics of each group. Each is discussed in greater detail in subsequent chapters.

Procaryotic protists

The bacteria and cyanobacteria are procaryotic protists.

Bacteria. Bacteria are typical procaryotic cells. They are unicellular and do not contain membrane-bound structures within the cytoplasm. Their single cells are characteristically spherical, rodlike, or spiral (Fig. 2-7). A typical bacterium measures approximately 0.5 to 1.0 μm in diameter and 1.5 to 2.5 μm in length. Reproduction is predominantly by simple binary fission, an asexual process. Some can grow at temperatures as low as 0°C (32°F); others thrive in hot springs, at temperatures of 90°C (194°F) or more. Most

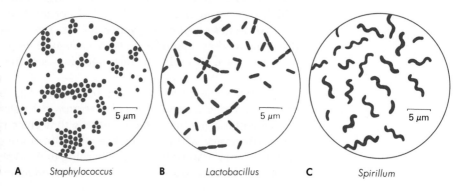

Figure 2-7. The cells of bacteria are generally either (A) spherical (cocci), (B) rodlike (rods or bacilli) or (C) spiral (spirilla). As you will see, however, there are many modifications of these three basic forms. *(Erwin F. Lessel, illustrator.)*

A *Staphylococcus*   B *Lactobacillus*   C *Spirillum*

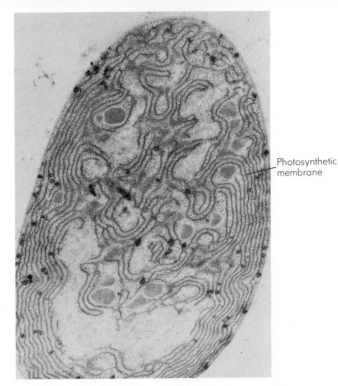

Photosynthetic membrane

Figure 2-8. Photomicrograph of the cyanobacterium *Anabaena azollae*. Most of the photosynthetic membranes are located near the periphery of the cell, but some extend into the midportions of the cell. *(Courtesy of Norma J. Lang and J Phycol,* **1**:127–134, 1965.)

grow at temperatures between these extremes. Bacteria cause a variety of chemical changes in the substances on which they grow; they can decompose many substances. They are essential to the maintenance of our environment in that they decompose material deposited on or in land and sea. Some types cause disease in animals (including humans), plants, and other protists. They are very widely distributed in and on the earth's surface, in the atmosphere, and in our everyday environment.

Cyanobacteria. The cyanobacteria are photosynthetic procaryotic organisms. That is, they contain chlorophyll and other pigments which enable them to carry out photosynthesis. Cyanobacteria are slightly larger than bacteria. They are unicellular and occur singly or in chains of cells, which are sometimes branched (Fig. 2-8). Reproduction is by simple binary fission, multiple fission, or a process of releasing specialized cells called *spores*.

Eucaryotic protists

The fungi, protozoa, and algae, like bacteria and cyanobacteria, are protists. However, they differ from bacteria and cyanobacteria in that their cells are eucaryotic; cells of bacteria and cyanobacteria are procaryotic. This distinction is of fundamental importance.

Fungi. Fungi are organisms which lack chlorophyll and have rigid cell walls. Some are unicellular; others are multicellular and exhibit some differentiation in their structural parts. They range in size and shape from single-celled microscopic yeasts to multicellular microscopic filamentous fungi (molds) to giant multicellular mushrooms and puffballs. Fungi reproduce by a variety of processes, both sexual and asexual. These methods of reproduction are described in Chap. 8.

Protozoa. Protozoa are single-celled eucaryotic protists which lack chlorophyll and cell walls. Protozoa vary greatly in size. Some are as small as 1 μm; others measure hundreds of micrometers and are visible to the naked eye. There is likewise considerable variation in their shapes. The protozoa are further examined in Chap. 9.

Algae. Algae are chlorophyll-containing eucaryotic protists. They range in size from single-celled organisms 5 to 10 μm long to the giant kelps, which may grow to a length of 100 ft. or more. The microscopic algae, which are of primary interest to the microbiologist, are mostly unicellular. Reproduction is mainly by simple asexual fission, but other modes of reproduction occur among some species. Algae grow widely in freshwater and marine water and in soil. These organisms are the subject of Chap. 10.

Viruses

Viruses are not cellular; their structure and composition are simpler than those of a procaryotic cell. They are not "free-living," but are *obligate parasites*; that is, they require another living cell for their propagation. They consist of a strand of nucleic acid, either DNA or RNA, wrapped in a layer of protein. In comparison with most microorganisms, viruses are very small. They range in size from about 20-25 nm to 200-300 nm. Viruses can not be seen by using light microscopy. Virus particles can be seen by electron microscopy, and they exhibit several shapes. Viruses are host specific. That is, they multiply only within a particular kind of living cell—plant, animal, or microbial. Viruses are discussed in Chaps. 11 and 12.

The main distinguishing characteristics of the major groups of microorganisms are presented in Table 2-2, and some typical species are shown in Fig. 2-9.

Where microorganisms are found

Terrestrial microorganisms

As stated earlier, microorganisms are ubiquitous in their occurrence; they are everywhere in the human environment. They are found in the soil; in aquatic environments, ranging from streams to oceans; and in the atmosphere. Local environmental conditions determine the characteristics of the microbial population. They may be present in extremely large numbers and in a great variety of types.

Contrary to popular belief, many more of these microorganisms are beneficial than are harmful. In the science of microbiology we attempt to learn more about the distribution of microorganisms in various environmental locations, the number and kinds that are present, and the roles they perform in the environment.

Table 2-2. Some characteristics of major groups of microorganisms *(Erwin F. Lessel, illustrator.)*

GROUP	MORPHOLOGY	SIZE	IMPORTANT CHARACTERISTICS	PRACTICAL SIGNIFICANCE
Bacteria		Typical: 0.5 to 1.5 μm by 1.0 to 3.0 μm Range: 0.2 by 100 μm	Procaryotic. Unicellular, simple internal structure. Grow on artificial laboratory media. Reproduction asexual, characteristically by simple cell division.	Some cause disease. Perform important role in natural cycling of elements which contributes to soil fertility. Useful in industry for manufacture of valuable compounds. Some spoil foods; some make foods.
Cyanobacteria		Range: 5.0 to 15 μm	Procaryotic. Unicellular. Cell structure like that of bacteria. Grow on artificial laboratory media. Reproduction asexual by simple cell division or production of spores. Contain chlorophyll and are photosynthetic.	Source of food for aquatic animals. Contribute to soil formation and enrichment.
Viruses		Range: 0.015 to 0.2 μm	Do not grow on artificial laboratory media; require living cells within which they are reproduced. All are obligate parasites. Electron microscopy required to see viruses.	Cause diseases in humans, other animals and plants. Also infect microorganisms.
Fungi: Yeasts		Range: 5.0 to 10.0 μm	Eucaryotic. Unicellular. Laboratory cultivation much like bacteria. Reproduction by asexual cell division, budding, or sexual processes.	Production of alcoholic beverages. Also used as food supplement. Some cause disease.
Fungi: Molds		Range: 2.0 to 10.0 μm by several mm	Eucaryotic. Multicellular with many distinctive structural features. Cultivated in laboratory much like bacteria. Reproduction by asexual and sexual processes.	Responsible for decomposition (deterioration) of many materials. Useful for industrial production of many chemicals including penicillin. Cause diseases of humans, other animals, and plants.
Protozoa		Range: 2.0 to 200 μm	Eucaryotic. Unicellular. Some cultivated in laboratory much like bacteria. Some are intracellular parasites. Reproduction by asexual and sexual processes.	Food for aquatic animals. Some cause disease.
Algae		Range: 1.0 μm to many feet	Eucaryotic. Unicellular and multicellular. Most occur in aquatic environments. Contain chlorophyll and are photosynthetic. Reproduction by asexual and sexual processes.	Important to the production of food in aquatic environments. Used as food supplement and in pharmaceutical preparations. Source of agar for microbiological media. Some produce toxic substances.

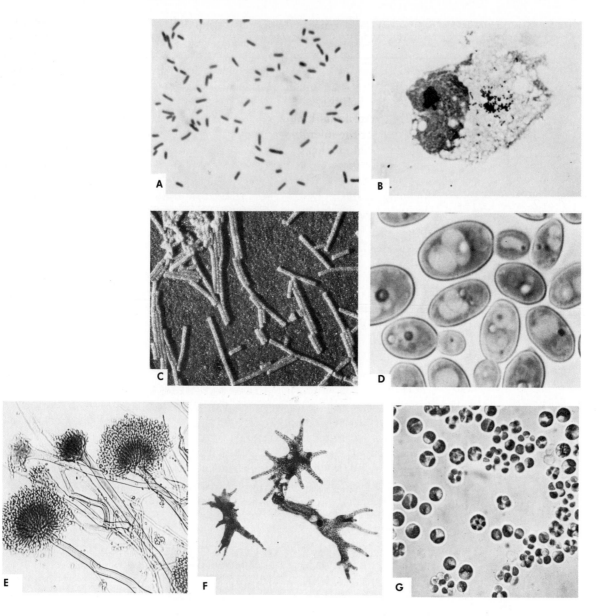

Figure 2-9. Morphological or structural features of various groups of microorganisms. (Note that this illustration is only intended to convey the impression of morphological diversity, not to show the size relationships between the different groups of microorganisms. The wide range in microbial sizes does not permit one to maintain a constant magnification and show meaningful morphological details at the same time. (A) Bacterium *Escherichia coli* (X 1,000). (B) Bacterium *Rickettsia tsutsugamushi* in mouse lymphoblast cell. Arrow points to rickettsias. *(Courtesy of R. Marilyn Bozeman.)* (C) Tobacco mosaic virus (X 100,000). *(Courtesy of Hitachi, Ltd., Tokyo.)* (D) Fungus-yeast *Candida utilis* (X 2,000 approximately). *(Courtesy of G. Svihla, J. L. Dainko, and F. Schlenk, J Bacteriol* **85**:399, 1963.) (E) Fungus-mold *Aspergillus* sp. *(Courtesy of Douglas F. Lawson.)* (F) Protozoan ameba. *(From Carolina Biological Supply Co.)* (G) Alga *Chlorella infusionum* (X 1,000). *(Courtesy of Robert W. Krauss.)*

The search for extraterrestrial microorganisms

Scientists, philosophers, poets, and of course science-fiction writers have speculated about the possiblity of life elsewhere in the universe. Interest in this question has been enhanced by the tremendous advances in spacecraft technology during the past two decades. These advances have made possible a new dimension of biological science—*exobiology*, the search for extraterrestrial life.

When astronauts landed on the moon in 1969, elaborate facilities were prepared for the microbiological examination of the lunar soil samples they brought back to earth. However, extensive examinations revealed no microorganisms. In 1977 a special life-seeking experimental unit was landed on Mars. It performed several experiments to search for microbial life. The results of these experiments, too, were negative. But there can be no doubt that the search for life on other planets—and elsewhere in the universe—will continue.

Summary and outlook

Early systems of classification placed all forms of life in either the plant or animal kingdoms. As more knowledge was obtained about microorganisms, it became increasingly apparent that although some showed characteristics only of plants or only of animals, others possessed characteristics of both plants and animals. This suggested the need for establishing a third kingdom (Protista) to include all unicellular microorganisms.

Further research with microorganisms, particularly investigations of their internal cell structure by electron microscopy, revealed a fundamental difference among them. Some, the bacteria and cyanobacteria, did not show, within the cytoplasm, structural units enclosed by membranes. These were designated procaryotic cells. Other microorganisms, the algae, fungi, and protozoa, as well as plant and animal cells, contained within their cytoplasm several membrane-bound structures. These cells were designated eucaryotic cells. This important distinction represents one of the great advances in modern-day microbiology.

Among the unresolved issues is the placement of viruses in any scheme of classification of life. Many question whether viruses are living agents and if they represent a connecting link between the complex organic nonliving molecules and the simplest microscopic cellular forms.

The search for new forms of microscopic life in our environment periodically results in the discovery of new microbial species. Such findings broaden our understanding of the world of microorganisms.

Key terms

capsule	cytoplasm	Golgi apparatus
cell	cytoplasmic membrane	granules
cell theory	deoxyribonucleic acid (DNA)	membrane
chloroplasts	endoplasmic reticulum	mesosomes
chromosomes	eucaryotes	microfilaments
cilia	exobiology	microtubules
classification	flagella	mitochondria

nomenclature	procaryotes	ribosomes
nucleus	protists	vacuole
organelles	protoplasm	
photosynthesis	ribonucleic acid (RNA)	

Questions

1 How does the classification scheme proposed by Whittaker differ from that proposed by Haeckel?

2 What contribution was made to the science of biology by Schleiden and Schwann? By Linnaeus?

3 Why are classification schemes important in microbiology and biology in general?

4 What is the fundamental difference between a procaryotic cell and a eucaryotic cell?

5 List several characteristics by which you can distinguish a procaryotic cell from a eucaryotic cell.

6 Name two groups of microorganisms which are procaryotic protists and two groups that are eucaryotic protists.

7 How do viruses differ from protists?

8 Name the major groups of microbes. List three distinctive characteristics of each group.

9 Where do microorganisms occur in nature?

CLASSIFICATION AND NAMING OF MICROORGANISMS

One goal of any scientific field is the organization and interpretation of factual information that is discovered in that field. So it is with microbiology. How can the many kinds of microorganisms be grouped into a pattern, or orderly system, that identifies similarities within a group as well as differences between groups? The study of organisms to establish a system of classification which best reflects all their similarities as well as all their differences is called *taxonomy*. Once an organism has been assigned to a taxonomic group, it is convenient to give it a name. Naming of microorganisms (*nomenclature*) provides a label or handle for convenient reference and communication.

If we are to develop a satisfactory scheme of classification, we must thoroughly understand the properties or characteristics of the subjects—in this case microorganisms—we wish to classify. In the preceding chapter we identified the major characteristics of microorganisms. In this chapter we acquaint you with the classification of microorganisms. We will use examples from classification schemes for bacteria; keep in mind that the classification systems for all microorganisms are somewhat similar.

The classification of microorganisms

Classification is a term related to and sometimes used interchangeably with taxonomy. Taxonomy is the science of classification or systematic arrangement of organisms into groups or categories called *taxa* (singular, *taxon*). However, the systematic arrangement of microorganisms requires that they be properly identified and named. The overall activity—classifying, naming, and identifying—is referred to as *microbial systematics*. These three processes, described below, are very much interdependent.

1 *Taxonomy (classification):* The orderly arrangement of units into groups of larger units. An analogy may be drawn to playing cards. Individual cards can be sorted first by color and then by suits; then within each suit cards can be arranged in numerical sequence, with face cards placed in order.

2 *Nomenclature:* The naming of the units characterized and delineated by classification. The same analogy may be used. Face cards are given names and perhaps even more than one name. For example, "jack" or "knave" refers to the same card. Fortunately, scientific nomenclature is the same in all languages.

3 *Identification:* Use of the criteria established for classification and nomenclature above to identify microorganisms by comparing the characteristics of the unknown with the known units. Identification of a newly isolated microorganism requires adequate characterization, description, and comparison with published descriptions of other similar microorganisms.

As we said earlier, the objective of a classification system is to group organisms in a way that reflects all their similarities as well as all their differences. From classification, the criteria necessary for identification of microorganisms are established. Classification also provides a means for determining evolutionary relationships among groups of microorganisms and for selecting microorganisms which may possess characteristics or abilities of special interest, such as antibiotic production.

Before 1700, organisms visible to the unaided eye were classified simply as either plant or animal. This practice was accepted by biologists as a basis for separating the living world into two kingdoms, *Animalia* and *Plantae*. In the 1750s these two kingdoms were subdivided into identifiable and related groupings by Carolus Linnaeus, a Swedish naturalist. An essential feature of the Linnean scheme is still used—the *binary* (two-part) *system of nomenclature*. More about naming organisms later.

Systems of classification are generally developed through international cooperation of scientists. By use of the Linnean schemes, classification systems were developed for the plant kingdom by botanists and for the animal kingdom by zoologists. Algae and fungi were included in the plant kingdom and protozoa in the animal kingdom. Many bacterial classification systems modeled after these older schemes were developed. In the last decade a classification scheme for viruses was proposed because by then sufficient data had been accumulated on which to base such a classification scheme.

The species concept
The basic unit or group in all classification systems of organisms, including microorganisms, is the *species*. We use this term frequently—but all too often with a feeling of unjustifiable authority. The truth is that the species concept is somewhat arbitrary and is not precisely defined; furthermore, it is usually subjective (based on individual judgment) in the field of microbiology. In general, a species is defined as a group of closely related individuals which are (1) distinguishable from the individuals of other similar groups and (2) all capable of interbreeding with other members of the group. The criterion of interbreeding is not easily and routinely applicable to microorganisms, particularly bacteria. So the latter part of the above-mentioned definition does not apply, and we have to fall back on an educated or experienced assessment by a researcher about how much alike a group of microorganisms must be in order to be designated a species. This is, then, a subjective decision made by the microbiologist. It follows that the more complete the characterization of a microorganism, the better the judgment about what constitutes a species.

Taxonomic categories (taxa)
A biologic classification system is based on a taxonomic hierarchy or ordering of groups or categories which places the species at one end and the kingdom at the other in the following sequence:

Species (plural, *species*): A group of closely related organisms (for our purposes microorganisms) in which the individuals within the group are alike in the greatest number of characteristics
Genus (plural, *genera*): A group of similar species
Family: A group of similar genera
Order: A group of similar families
Class: A group of similar orders
Phylum (plural, *phyla*) or division: A group of related classes
Kingdom: All the organisms within this hierarchy

The arrangement of species into a system of classification—for example, species → genus → family → order → class → phylum or division—might appear relatively easy and unequivocal. It is not. How similar must species be if they are to be included in the same genus? What are the boundaries of any particular genus, family, or order? These questions cannot be answered absolutely. Furthermore, the different taxa are not always equally useful. For example, the editorial board of the 8th edition of *Bergey's Manual* has concluded that "for most groups of bacteria, genera and species are the only categories that can now be recognized and defined with reasonable precision." The manner in which characteristics of bacteria overlap among species precludes establishing sharp lines of demarcation among taxonomic groups.

The species category is the most important group in this classification scheme. It provides the basis for the entire hierarchical structure.

Naming microorganisms—the binary system of nomenclature

Microorganisms, like all other forms of life, are named according to the *binary system of nomenclature* (Table 3-1). The primary purpose of a name is to provide a means of referring to a microorganism, not to describe it. Each organism is designated by a genus name and an ordinary or descriptive term referred to as the specific epithet, both of which are Latin or Latinized. The name of the genus is always capitalized; the specific epithet is always lower case. The two components together are called the scientific name (genus and specific epithet) and are always printed in italics—for example, *Neisseria gonorrhoeae*, the bacterium that causes gonorrhea.

Table 3-1. Examples of taxonomic names as applied to species in the animal, plant, and microbial kingdoms

TAXA	EXAMPLES OF TAXA			
	Lion*	Dandelion*	Ameba*	Tubercle bacillus*
Kingdom	Animalia	Plantae	Protista	Procaryotae
Phylum (or Division)	Chordata	Tracheophyta	Sarcodina	Bacteria
Class	Mammalia	Angiospermae	Rhizopoda	—
Order	Carnivora	Campanulales	Amoebida	Actinomycetales
Family	Felidae	Compositae	Amoebidae	Mycobacteriaceae
Genus	*Felis*	*Taraxacum*	*Amoeba*	*Mycobacterium*
Species	*F. leo*	*T. officinale*	*A. proteus*	*M. tuberculosis*

*Common name.

Codes of nomenclature

To achieve consistent and uniform naming of organisms, internationally accepted rules for naming organisms have been established that are followed by biologists in all countries. Such rules for plants and animals were established in the early 1900s by botanists and zoologists. The International Code of Zoological Nomenclature was first published in 1901; the International Code of Botanical Nomenclature was first published in 1906. In 1947 the International Association of Microbiological Societies adopted an International Code of Nomenclature of Bacteria and Viruses. The code, now known as the International Code of Nomenclature of Bacteria, is modified continually in an effort to improve and clarify its rules and regulations. The most recent edition was published in 1975.

Principles of nomenclature

The zoological, botanical, and bacteriological codes are based on certain common principles. Some of the most important are:

1 Each distinct kind of organism is designated as a species.
2 The species is designated by a Latin binary combination to provide it with a uniform, internationally understood label.
3 The nomenclature of organisms is regulated by the appropriate international overseeing organization—in the case of bacteria, the International Association of Microbiological Societies.
4 A law of priority ensures the use of the oldest available legitimate name for an organism. This means that the earliest name given to a microorganism, provided that the proper procedure was followed, is the correct name.

5 Designation of categories is required for classification of organisms.

6 Criteria are established for the formation and publication of new names.

Scientific names and common names

Scientific names of organisms are formed in accordance with the rules of the binary system of nomenclature, as we noted earlier. Organisms with which we have become familiar and to which we frequently refer usually acquire common names. Some examples of organisms frequently referred to by their common names are listed below, together with their scientific names. (In many instances the common name comes into use before an organism is given a scientific name.)

COMMON NAME	SCIENTIFIC NAME
Dog	*Canis familiaris*
Housefly	*Musca domestica*
White oak	*Quercus alba*
Bread mold	*Neurospora crassa*
Gonococcus	*Neisseria gonorrhoeae*
Tubercle bacillus	*Mycobacterium tuberculosis*

The advantages of using common names are convenience and more effective communication by doctors with patients. For example, in conversation in the laboratory or with a lay person it is more convenient to refer to the causative agent of tuberculosis as the "tubercle bacillus" rather than to say *Mycobacterium tuberculosis*. Common names are sometimes derived from generic names, for example, pseudomonads from *Pseudomonas*.

Recent developments in microbial taxonomy

Microbial taxonomy is not a static subject. Classification schemes are continuously evolving as more information is obtained and as different methods are developed for interpreting the data. Two relatively new developments have emerged for use in microbial taxonomy which in different ways will make for more objective decisions. One of these is numerical taxonomy, the other is genetic taxonomy.

Numerical taxonomy

Numerical taxonomy, also referred to as *computer taxonomy*, is based on principles published many years ago and only recently applied to microbial taxonomy. Numerical taxonomy requires that a large amount of information be available about the microorganism—information on as many unrelated characteristics as one can obtain. Every characteristic is given equal weight in constructing taxa. Overall similarity is based on the proportion of characteristics which organisms have in common. In practice the microbiologist accumulates data for each culture. By the use of a computer the data from each culture are compared with those from every other culture. (The aid of a high-speed computer is required because otherwise the many thousands of comparisons of various characteristics would take an extremely long time.) The end result is that the microbiologist can calculate and numerically express the degree of similarity of every culture to every other culture. Taxa are established on the basis of agreed-upon degrees of similarity. Numerical taxonomy has two great advantages. One is that it can be made objective: the taxonomist's biases do not enter the procedure, and so the results (if the procedure is correctly applied) are not open to dispute. The other great advantage of numerical taxonomy is that its findings are reproducible, or repeatable: another taxonomist following the same procedure with the same data would obtain the same result.

Genetic taxonomy

As you will learn in Chaps. 16 and 17, much is now known about the genetic material of bacteria, namely DNA. There are laboratory procedures by which one can determine the base composition (guanine plus cytosine content, or GC) of a given microorganism's DNA and then compare it with the DNA base composition of other microorganisms. The degree of the relatedness or similarity of the DNAs of different microorganisms can also be determined by hybridization experiments. In this technique single strands of a microorganism's DNA are exposed to single strands of DNA from another microorganism. The degree to which these single strands reunite reflects their degree of similarity. (The significance of this will be clearer when you read Chaps. 16 and 17.)

Genetic information is very valuable in validating existing taxonomic groups and establishing new groups.

Changing taxonomic concepts

Once microorganisms have been assigned a place in a taxonomic system, is the decision final? No. Classification schemes in microbiology are modified periodically; earlier taxonomic arrangements yield to better ones based on newer knowledge. The following examples illustrate the nature of some of the changes which have occurred.

Bergey's Manual of Determinative Bacteriology, 8th edition (1974), is the generally accepted authority on bacterial taxonomy. Each of the eight editions, published in the years since 1923, have listed different numbers of species for various genera. Some examples, presented in Table 3-2, show that considerable changes have occurred over the years in the number of species placed in each of these genera. Why? There are a variety of reasons. Some microbiologists working in the field of taxonomy are referred to as "splitters"; they establish new species on the basis of slight differences between related groups. Other microbiologists working in taxonomy are referred to as "lumpers"; they do not consider small differences sufficient to warrant new species.

Another reason for these changes is associated with the accumulation of new information about the microorganisms. The new information may provide better evidence for establishing new species, eliminating some species, or doing both. Still another reason is increased interest in a group of microorganisms. Refer again to Table 3-2 and see what has happened in the genus *Streptomyces*. This "explosion" of new species came about because of the discovery in the 1940s that *Streptomyces* species produce

Table 3-2

EDITION OF BERGEY'S MANUAL	NUMBER OF SPECIES IN SELECTED GENERA				
	Bacillus	Actinomyces	Pseudomonas	Escherichia	Streptomyces
1st (1923)	75	64	20	22	0
2d (1925)	75	64	20	22	0
3d (1930)	93	70	31	29	0
4th (1934)	93	70	31	22	0
5th (1939)	34	62	31	2	0
6th (1948)	33	2	148	3	73
7th (1957)	25	3	149	4	149
8th (1974)	22	5	29	1	415

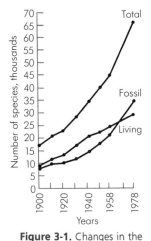

Figure 3-1. Changes in the number of species of protozoa identified since 1900. *(Courtesy N. D. Levine.)*

antibiotics. This discovery initiated a major worldwide search for these microorganisms in hopes of finding producers of new and better antibiotics.

A reduction in the number of species within a genus is illustrated with *Escherichia* (Table 3-2). The first four editions of *Bergey's Manual* listed more than 20 species; the eighth has only one. This reflects a change in the assessment of characteristics which justify a group being split into several species.

An example of changes associated with the taxonomy of protozoa is shown in Fig. 3-1. With protozoa, unlike bacteria, data can be obtained from fossil specimens as well as from the living organisms. This is possible because of their larger size and the nature of their skeletons. The fossilized specimens provide a record of specimens which occurred in earlier times. The data shown in Fig. 3-1 reveal that a large number of new protozoa have been discovered since the beginning of this century. Large numbers of new species have been identified in both fossil and living forms. Protozoologists feel that there are probably hundreds of thousands of yet-to-be-described species. Consequently, as new microorganisms are discovered, schemes of classification need to be modified to accommodate them.

Summary and outlook

The subject of microbial classification has a long history, and many taxonomic systems have been proposed over the years. A good system of microbial taxonomy is essential for organizing the science of microbiology. Science is not simply a collection of miscellaneous facts; it is the organization and interpretation of these facts into a system that reveals relationships among the various categories. And so it is with microbial taxonomy. A good system of taxonomy should reduce confusion and establish order. From a very practical standpoint, classification schemes provide a means for learning, simultaneously, the major characteristics of many species by learning the characteristics of the genus to which they belong. For example, *Bergey's Manual* (1974) describes 48 species of the genus *Bacillus* and 61 species of the genus *Clostridium*. Both genera belong to the family Bacillaceae. If you learn the criteria upon which this taxon is based, you will learn, at the same time, some of the major characteristics of 109 species of bacteria.

Microbial taxonomy is a dynamic rather than a static field. New microorganisms continue to be discovered and new knowledge is becoming available about microorganisms already classified. The most promising new information being made available comes from an analysis of the DNA of the microbial cell. This information is extremely important and valuable for determining the validity of taxonomic groups. In addition, the increased use of numerical, or computer, taxonomy will promote more objectivity in the establishment of taxonomic groups.

Key terms

binary system of nomenclature	genus	species
classification	nomenclature	taxon
genetic taxonomy	numerical taxonomy	taxonomy

Questions

1 What is the significance of an international code of nomenclature?

2 Why is it essential to classify organisms? What special problems arise in the classification of microorganisms?

3 Name several taxonomic categories of microorganisms and define each taxon.

4 *Bergey's Manual of Determinative Bacteriology* is now in its eighth edition. What is a major difference between this latest edition and all the earlier editions?

5 Considerable knowledge of bacterial genetics is being accumulated. How may this affect existing schemes of classification? Why?

6 How does the species concept among procaryotic organisms differ from that among eucaryotic organisms? Explain.

7 Numerical taxonomy is described as being an objective rather than a subjective system of classification. What is the basis for this statement?

8 Give examples of some common names used for microorganisms. What is the reason for having common names?

9 What is the difference between the following two words as written: bacillus, *Bacillus*?

METHODS IN MiCROBIOLOGY

It has been said that microbiology is a science defined more by the techniques it uses than by the subjects it studies. These techniques are many, and a great variety of laboratory instruments are used in applying them.

Microorganisms are studied in the laboratory for many purposes. The degree of detail in which the microorganisms are studied depends on the purpose of the laboratory examination. For example, the microbiologist may only want to determine the major type or types of microorganisms present in a specimen—say fungi, bacteria, or both. Or the microbiologist may have to identify each species in a specimen taken from a patient so as to provide information to assist in the diagnosis of a disease—in which case the microbiologist must examine the microorganism's characteristics in great detail. Or to take yet another example, the microbiologist may want to determine exactly how a certain fundamental biological process, for example the synthesis of a protein, is carried out by a microbial cell.

Techniques are available to determine the size, shape, and structure of individual cells, as well as how the cells are grouped. There are procedures for growing (culturing) microorganisms in the laboratory. Some of these procedures require very special conditions, such as the complete absence of free oxygen. Great advances have been made in equipment for the microbiological laboratory since the early 1900s. Today's instruments can, for example, identify in great detail the chemical composition of a microbial cell, as well as the chemical compounds which a cell produces.

In this chapter we will discuss some of the laboratory procedures microbiologists use to study microorganisms. We shall emphasize microscopy and microscopic methods.

●

Microscopes and microscopy

The microscope, the most often used and the most useful instrument of the microbiological laboratory, provides the magnification, or apparent enlargement, that enables one to see organisms and structures invisible to the unaided eye. Microscopes permit a wide range of magnifications—from one hundred times to hundreds of thousands of times.

The two categories of microscopes available are *light* (or optical) and *electron* microscopes. These categories differ in the principle on which magnification is based. Light microscopes, all of which employ a system of optical lenses, include (1) bright-field, (2) dark-field, (3) fluorescence, and (4) phase-contrast microscopes. Electron microscopes, as the name suggests, employ a beam of electrons in place of light waves to produce the magnified image.

As a beginning student, you will perform most, if not all, of your examinations with the bright-field microscope, the most widely employed instrument for routine microscopic work. The other types of microscopy are used for research, for procedures in the diagnostic microbiology laboratory, and for other special purposes. Each type is especially useful for the examination of some specific morphological characteristic, as described later in this chapter.

Bright-field microscopy

In bright-field microscopy, the microscopic field, or area observed, is brightly lighted and the objects being studied appear darker than their background. Figure 4-1 shows the main parts of a typical bright-field microscope and the path the light rays follow to produce magnification of the object. Generally, microscopes of this type produce a maximum useful magnification of about 1,000 diameters. With some modifications, including higher-powered eyepieces, this magnification can be increased. However, a magnification of 1,000 to 2,000 diameters is the limit of useful magnification obtainable with such an instrument.

Microscope lenses. Microscopes like the one shown in Fig. 4-1 are compound microscopes; magnification is obtained by use of a system of lenses, in contrast to a simple microscope like Leeuwenhoek's, which used only a single lens. As you can see from Fig. 4-1B, lenses are present in the condenser, the objective(s), and the eyepiece (ocular). The condenser lens focuses a cone of light on the specimen field. Some of the light rays in this cone of light pass directly into the objective lens to form the background light or bright field. The light rays which strike the objects (microorganisms) in the specimen and become "bent" are brought to focus by the

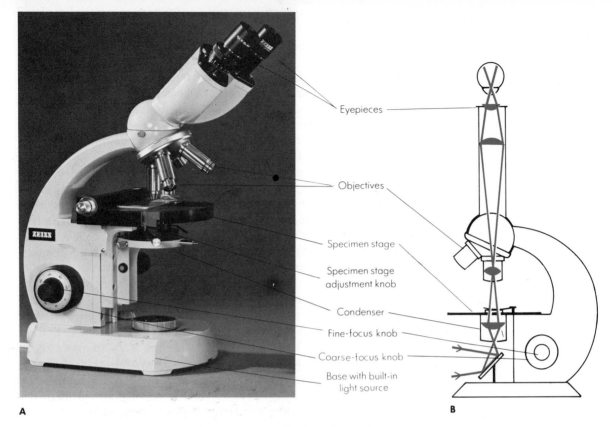

Eyepieces

Objectives

Specimen stage

Specimen stage
adjustment knob

Condenser

Fine-focus knob

Coarse-focus knob

Base with built-in
light source

A   B

Figure 4-1. A modern compound light microscope. (A) Identification of parts. (B) Cutaway sketch of student microscope showing optical parts and path of light. *(Courtesy of Carl Zeiss, New York.)*

objective lens to form an image of the object. The image is enlarged by the ocular lens. Thus it is the objective lens system that gives the initial magnification, which becomes further magnified by the ocular lens system.

Microscope objectives. Microscopes commonly used in microbiology are usually equipped with three objectives, each providing a different degree of magnification, which are mounted on a platform called a *turret* that can be rotated to move any one of them into alignment with the condenser. The total magnification obtainable with any one of these objectives is determined by multiplying the magnifying power of the objective by the magnifying power of the eyepiece, which is generally 10 times (×10).

OBJECTIVE DESIGNATION	OBJECTIVE MAGNIFICATION	EYEPIECE MAGNIFICATION	TOTAL MAGNIFICATION
Low power	10	10	100
High power (high dry)	40	10	400
Oil-immersion	100	10	1000

The oil-immersion objective is so designated because a drop of immersion oil is placed on the specimen slide and the objective is immersed in the oil. The specimen is then brought into focus while there is contact between the specimen slide, the oil, and the front lens of the objective. The oil allows more light to enter the objective. Since the oil-immersion objective provides the highest magnification, it is commonly used for the microscopic examination of procaryotic cells. You must be extremely careful when focusing with the oil-immersion objective, because the specimen comes into focus when the objective's front lens is very near the surface of the slide. You can damage the objective lens or break the specimen slide if you are not careful.

Resolving power. A microscope's useful magnification is limited by its *resolving power*, that is, its ability to produce distinct images of two adjacent points (*points* here meaning objects or details of objects). The resolving power of a light microscope is determined by the wavelength of light and a property of the objective and condenser lens known as the *numerical aperture* (NA). The resolving power of a light microscope is calculated by the following formula:

$$\text{Resolving power} = \frac{\text{Wavelength of light}}{\text{NA of objective} + \text{NA of condenser}}$$

From this you can observe that resolving power can be increased by either decreasing the wavelength of light or increasing the NA. The wavelength of light used in light microscopy is limited; the visible light range is between 400 and 700 nm. Likewise the degree to which microscope objectives can be altered is limited; the maximum NA for the high dry objective is approximately 0.85 and that for the oil-immersion objective is slightly higher (1.2 to 1.4).

To illustrate this further, assume that the wavelength of light being used is 0.55 μm. This wavelength of light can be selected by using a green filter. The oil-immersion objective used has an NA of 1.25, and the condenser has an NA of 0.9. Substituting these values in the equation above, we have

$$\text{Resolving power} = \frac{0.55}{1.25 + 0.9} = 0.255 \ \mu\text{m}$$

A resolving power of 0.255 μm means that if two dots are less than 0.255 μm apart, they will appear as a single dot. They need to be spaced more than 0.255 μm apart to appear as separate distinct objects. Magnification in excess of the resolving power serves no useful function.

Dark-field microscopy

Dark-field microscopy is performed with the same kind of microscope used for bright-field microscopy except that it is equipped with a dark-field condenser and an objective with a low numerical aperture. This kind of condenser directs the rays of light into the specimen field at such an angle that only the rays striking an object in the specimen field are refracted (bent) and enter the objective, as shown in Fig. 4-2. Thus the object is brightly illuminated and stands out against its dark field (dark background), as shown in Fig. 4-3. Dark-field microscopy is particularly valuable, as we shall explain later, for the examination of living microorganisms. It is a useful technique for the identification of the bacterium which causes syphilis.

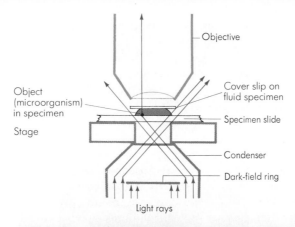

- Objective
- Object (microorganism) in specimen
- Stage
- Cover slip on fluid specimen
- Specimen slide
- Condenser
- Dark-field ring
- Light rays

Figure 4-2. Schematic representation of dark-field microscopy. By blocking a portion of the light rays that enter the condenser, the dark-field ring allows only those light rays which strike an object on the specimen slide to enter the objective. Hence the objects (microorganisms) become illuminated in a microscopic field which is otherwise dark.

Figure 4-3. Dark-field photomicrographs of *Treponema pallidum,* the causative agent of syphilis. Note the nature of the spirals and the length of each organism. *(Courtesy of American Society of Microbiology, LS 327.)*

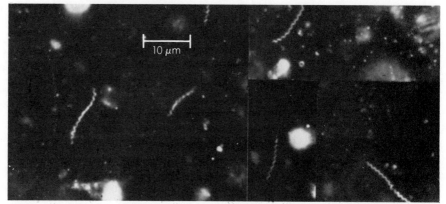

10 μm

Fluorescence microscopy

Fluorescence microscopy has become a very widely used and important procedure for the hospital or clinical laboratory. It is used to examine specimens that have been stained with fluorochrome dyes and makes possible rapid identification of microorganisms. These dyes absorb the energy of short, invisible light waves while emitting waves of a greater, visible wavelength. Such materials are called *fluorescent* and the phenomenon is termed *fluorescence.* This principle has been incorporated into

Figure 4-4. Fluorescence staining technique and microscopy. (A) The direct fluorescent antibody staining technique. When a bacterial cell is incubated with specific antibody conjugated (combined) with a fluorescent dye, the dye-antibody conjugate will cover the surface of the cell. The technique is performed on a glass slide, the excess fluorescent dye-antibody conjugate is washed off, and the preparation is examined by ultraviolet light microscopy. The bacterial cell will glow brilliantly as a result of fluorescence caused by the ultraviolet illumination of the dye-coated bacterial cell. Any bacterial cells not covered by the dye do not fluoresce and hence are not visible by this technique. (B) Photomicrograph of a fluorescent-stained *Proteus mirabilis* preparation as described above. *(Courtesy of Judith Hoeniger, F. M. Clinits, and E. A. Clinits, J Bacteriol,* **98***:226, 1969.)*

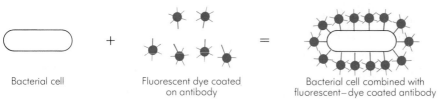

Bacterial cell + Fluorescent dye coated on antibody = Bacterial cell combined with fluorescent–dye coated antibody

A

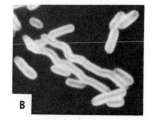

B

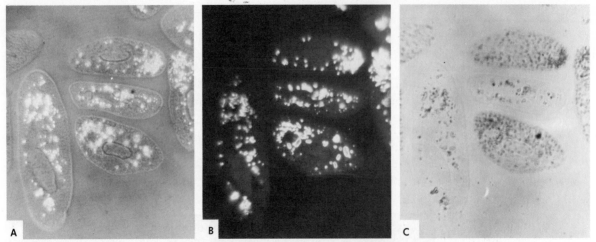

Figure 4-5. Phase-contrast microscopy compared with bright-field and dark-field microscopy. The same specimen of protozoa as seen by each method: (A) phase-contrast; (B) dark-field; (C) bright-field. (Courtesy of O. W. Richards, Research Department, American Optical Company.)

techniques which make it possible to identify microorganisms specifically by direct microscopic examination (see Fig. 4-4). The laboratory procedures involved can be carried out very quickly. For example, if a patient has a lesion and it is suspected that the lesion has been caused by the syphilis bacterium, one can examine fluid from the lesion microscopically by what is called the *fluorescein-antibody* technique. The syphilis bacterium, if present, can be identified positively by this procedure.

Figure 4-6. Transmission electron micrograph (TEM) of *Escherichia coli* (X 108,000). Inset is a photomicrograph of the same organism. *(Courtesy of I. D. J. Burdette and R. G. E. Murray, J Bacteriol* **119***:1039, 1974.)*

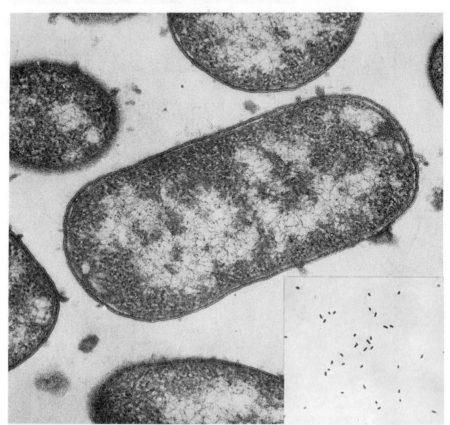

Phase-contrast
microscopy

Phase-contrast microscopy is a type of light microscopy which permits greater contrast between substances of different thickness or of different refractive index. This is accomplished through the use of a special condenser and objective which control the illumination of the object in a manner that accentuates slight differences in the thicknesses or refractive indexes of cellular structures. The differences are revealed in terms of differing degrees of brightness or darkness (better contrast), as shown in Fig. 4-5. With this technique you can locate structures within unstained cells not visible in bright-field microscopy. As you will read later in this chapter, microorganisms are frequently stained to make them more readily visible.

Table 4-1 compares the various types of microscopy.

Electron microscopy

Electron microscopy provides tremendously greater useful magnifications than can possibly be obtained by light microscopy. It can do so because of the greater resolving power obtainable from the very short wavelengths of the electron beams used instead of light for magnification. The electron beams used in electron microscopes have wavelengths in the range of 0.005 to 0.0003 nm, very short as compared with wavelengths of visible light used in light microscopy. The extremely short wavelength of the electron beam makes it possible to achieve resolving powers several hundred times greater than that obtainable with light microscopy. It is

Table 4-1. A comparison of different types of microscopy

TYPE OF MICROSCOPY	MAXIMUM USEFUL MAGNIFICATION	APPEARANCE OF SPECIMEN	USEFUL APPLICATIONS
Bright-field	1,000–2,000	Stained or un-stained; bacteria generally stained and appear color of stain	For gross morphological features of bacteria, yeasts, molds, algae, and protozoa
Dark-field		Generally unstained; appears bright or "lighted" in an otherwise dark field	For microorganisms (for example, the bacteria called spirochetes) that exhibit some characteristic morphological feature in the living state and in fluid suspension
Fluorescent		Bright and colored; color of the fluo-rescent dye	Diagnostic techniques whereby fluorescent dye aids in the identification of the microorganism
Phase-contrast		Varying degrees of "darkness"	For examination of cellular structures in living cells of the larger microorganisms, e.g., yeasts, algae, protozoa, and some bacteria
Electron	200,000–400,000	Bright on fluorescent screen	Examination of very small objects—viruses and the ultrastructure of microbial cells

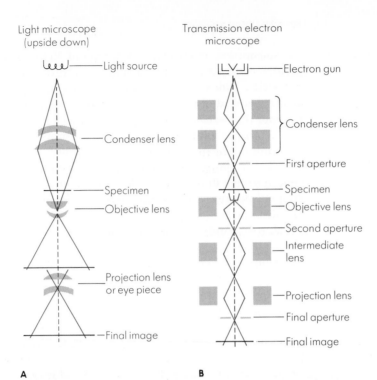

Light microscope (upside down)

Transmission electron microscope

Light source — Electron gun

Condenser lens — Condenser lens

First aperture

Specimen — Specimen

Objective lens — Objective lens

Second aperture

Intermediate lens

Projection lens or eye piece — Projection lens

Final aperture

Final image — Final image

A   B

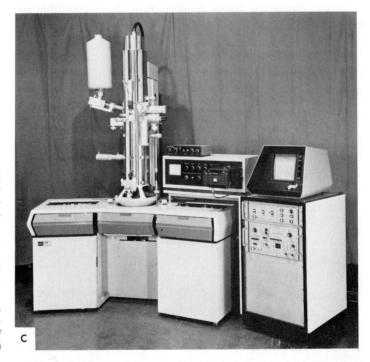

C

Figure 4-7. Diagrammatic comparison of imaging systems in (A) optical microscope, (B) transmission electron microscope. *(From L. A. Bulla, Jr., G. St. Julian, C. W. Hesseltine, and F. L. Baker, Scanning Electron Microscopy, in Methods in Microbiology, vol. 8, Academic Press, New York, 1973.)* (C) A high-resolution electron microscope. *(Courtesy of Hitachi-Perkin-Elmer.)*

possible by using electron microscopy to resolve objects in the range of 0.003 μm. Final magnifications approaching 1,000,000 times can be made when the photographed image is enlarged. A comparison of a light microscope (bright-field) photomicrograph and an electron microscope micrograph is shown in Fig. 4.6.

For electron microscopy, the specimen to be examined is prepared as an extremely thin dry film on small screens and introduced into the instrument at a point between the magnetic condenser and the magnetic objective (glass optical systems are not used in electron microscopy), which are comparable to the condenser and objective of the light microscope. The magnified image is viewed on a fluorescent screen or recorded on a photographic film by a camera attached to the instrument (Fig. 4-7).

Many techniques for the examination of microorganisms by electron microscopy have been developed. Among these are new staining methods, methods for slicing microbial cells into microscopically thin sections for examination, and radioactive techniques. These procedures apply to *transmission electron microscopy* (TEM). In this kind of microscopy, the electron beam passes through the specimen and the scattering of these electrons gives rise to the image. By 1960 a modification known as *scanning electron microscopy* (SEM) had been developed. This procedure exposes the specimen to a beam of electrons in a manner that makes it possible to obtain three-dimensional surface views of cells (see Fig. 4-8).

Preparations for light-microscopic examinations

Two general techniques are employed to provide material or specimens for microscopic examination. One is the suspension of organisms in a liquid. The other uses dried, fixed, and stained films or smears of the specimen.

Figure 4-8. Scanning electron micrographs (SEM). (A) SEM of *Treponema pallidum.* The treponemes were incubated with cells from rabbit testes and then examined by SEM. It was observed that large numbers of *T. pallidum* became attached to the rabbit testes cells. *(Courtesy of T. J. Fitzgerald, P. Cleveland, R. C. Johnson, J. N. Miller, and J. A. Stokes, J. Bacteriol, **130:** 1333, 1977.)* (B) SEM of the intestinal flora of the American cockroach. *(Courtesy of J. W. Bracke, D.L. Cruden, A. J. Markovetz, Antimicrobial Agents and Chemotherapy, **13**:115, 1978.)*

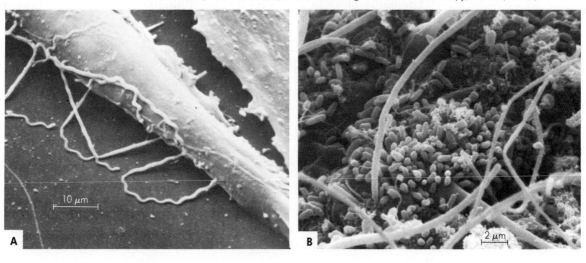

The wet-mount and hanging-drop techniques

Hanging-drop or wet preparations permit examination of living organisms suspended in fluid. Wet preparations are made by placing a drop of the fluid containing the organisms on a glass slide and covering the drop with a very thin piece of glass called a *cover slip*. To reduce the rate of evaporation and exclude air currents, the drop is usually ringed with petroleum jelly or a similar material to provide a seal between the slide and the cover slip. Special slides with a concave depressed area are available for hanging-drop preparations (see Fig. 4-9). Hanging-drop or wet preparations are especially useful when the morphology of the microorganism being examined may be distorted by treatment with heat or chemicals or when the microorganisms are difficult to stain. It is also the method of choice when certain life processes, such as motility or reproduction, are being observed.

When wet preparations are examined by bright-field microscopy, it is extremely important to adjust the light source appropriately. The intensity of the light can be diminished by use of special filters. Dark-field and phase-contrast microscopy offer a distinct advantage for examination of unstained microbial cells suspended in liquids; they provide a better contrast between the microorganism (or some part of it) and the background material. This technique is used in the clinical diagnostic laboratory, particularly for examination of specimens in which protozoa are suspected. Protozoan cells are generally larger than most bacteria and have a more elaborate internal structure. These structures, or intracellular parts, are important in characterizing the species and can be observed by using dark-field or phase-contrast microscopy.

Staining techniques

Many colored organic compounds (dyes) are used to stain microorganisms for microscopic examination. Staining procedures have been developed to:

1 Better observe the gross morphological appearance of microorganisms
2 Identify structural parts of the cells of microorganisms
3 Assist in identifying and/or differentiating similar organisms

The major steps in preparing a stained microbial specimen for microscopic examination are:

Figure 4-9. The hanging-drop technique. (A) A ring of petroleum jelly is made with a match stick around the depression on the slide. (B) A drop of the microbial suspension is placed at the center of a cover slip. (C) The cover slip is inverted and placed over the depression on the slide.

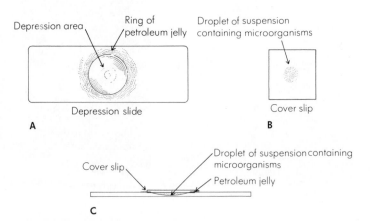

1 Placement of a *smear*, or thin film of specimen, on a glass slide
2 Fixation of the smear to the slide, usually by heating, which makes the microorganisms stick to the glass slide
3 Application of a single stain (*simple staining*) or a series of staining solutions or reagents (*differential staining*)

Simple staining. The coloration of bacteria or other microorganisms by applying a single solution of a stain to a fixed film, or smear, is termed *simple staining*. The fixed film is flooded with the dye solution for a specified period of time, after which the solution is washed off with water and the slide blotted dry. The cells usually stain uniformly. However, with some organisms, particularly when the dye is methylene blue, some granules in the interior of the cell appear more deeply stained than the rest of the cell.

Differential staining. Staining procedures that bring out differences between microbial cells or parts of a microbial cell are termed *differential staining* techniques. Differential staining techniques usually involve exposing the cells to more than one dye solution or staining reagent.

Gram staining. One of the most important and widely used differential staining techniques for bacteria is *gram staining*. In this process the fixed bacterial smear is subjected to the following solutions in the order listed: crystal violet, iodine solution, alcohol (a decolorizing agent), and safranin or some other suitable counterstain. Bacteria stained by the gram method fall into two groups. One of these, the *gram-positive bacteria*, retains the crystal violet dye and hence appears deep violet in color. The other group, the *gram-negative bacteria*, loses the crystal violet when washed with the alcohol, and when counterstained by the red dye safranin appears red in color. The steps in the procedure and results at each stage are summarized in Table 4-2.

Table 4-2.
The gram stain

| SOLUTIONS IN ORDER APPLIED | REACTION AND APPEARANCE OF BACTERIA | |
	Gram-positive	Gram-negative
1 Crystal violet (CV)	Cells stain violet	Cells stain violet
2 Iodine solution (I)	CV-I complex formed within cells; cells remain violet	CV-I complex formed within cells; cells remain violet
3 Alcohol	Cell walls dehydrated, pores shrink; permeability of cell wall and membrane decreases, CV-I complex cannot pass out of cells; cells remain violet	Lipid extracted from cell walls, pores enlarge, CV-I complex is removed from cell; cells become colorless
4 Safranin	Cells not affected, remain violet	Cells take up this stain, become red

Table 4-3. Summary of preparations for examination by light microscopy

TECHNIQUE	PREPARATION	APPLICATION*
A. Wet mount and hanging drop	Drop of fluid containing the organisms on glass slide or slip	Study morphology, internal cell structures, motility, or cell changes
B. Staining procedure	Suspension of cells is fixed to slide in film, usually by heat	Various staining procedures
1. Simple stain	Film stained with a single dye solution	Shows size, shape, and arrangement of cells
2. Differential stains:	Two or more reagents used in staining process	Can observe difference between cells or parts of cells
a. Gram	Primary stain (crystal violet) applied to film then treated with reagents and counterstained with safranin	Characterizes bacteria in one of two groups: 1. Gram-positive—deep violet 2. Gram-negative—red
b. Acid–fast	Film stained with carbol-fuchsin, decolorized and counterstained with methylene blue	Separate acid-fast bacteria, those not decolorized when acid solution is applied (eg., mycobacteria), from non-acid-fast bacteria, which are decolorized by acid.
c. Giemsa	Stain applied to blood smear or film of other specimens	Can observe: protozoa in blood smear; rickettsia (small parasitic bacteria) in certain cells of the host; nuclear material in bacteria
d. Spore	Primary stain (malachite green) applied with heat to penetrate spores; vegetative cells counterstained with safranin	Endospores can be seen in *Bacillus* and *Clostridium* species
e. Capsule	Smear stained following treatment with copper sulfate	Capsule can be observed as a clear zone surrounding cells of capsulated microorganisms
f. Flagella	Mordant acts to thicken flagella before staining	Observe flagella on bacteria
C. Negative staining	Specimen mixed with India ink and spread into thin film	Study morphology; staining procedure and reagents are very mild in their effect on the microorganism; called a negative stain because the microorganism is unstained and is made visible because the background is dark

* The bacterial structures referred to are described in Chap. 5.

Why does the gram-staining procedure stain some bacteria purple-violet and others red? The answer appears to be differences in their surface chemical structure. We shall discuss the mechanism of this staining reaction in Chap. 5 after you have had an opportunity to learn more about the chemical makeup of bacterial cells.

The gram-stain technique was first described in an 1884 publication by a Danish bacteriologist, Christian Gram. Gram developed this staining procedure while searching for a method to demonstrate the pneumococcus bacterium in the lung tissue of patients who had died of pneumonia. In addition he observed that some bacterial species could be differentiated by this staining procedure.

The gram stain is still one of the most widely used procedures for characterizing many bacteria. It is particularly valuable in the hospital diagnostic laboratory where information obtained from observation of a gram-stained specimen may give a quick clue to the organism causing an infection.

Many other differential stains are widely used in microbiology. These and other microscopic preparations are summarized in Table 4-3.

Pure culture techniques

The microbial population in our environment is both large and complex. Hundreds of different microbial species normally inhabit various parts of our body, including the mouth, intestinal tract, and skin. They are present in extremely large numbers. For example, a simple sneeze may disperse thousands of microorganisms. One gram of feces may contain millions of bacteria. Our environment—air, soil, water—likewise harbors a menagerie of microorganisms. A proper study of the microorganisms in these habitats requires techniques for sorting out the complex mixed population, or *mixed culture,* into separate, distinct species as *pure cultures.* A pure culture consists of a population of cells all derived from a single parent cell.

Cultivation and isolation of pure cultures

Microorganisms are cultivated in the laboratory on nutrient materials called *media* (singular, *medium*). A great many media are available; the kind used depends on many factors, one of which is the kind of microorganism to be grown.

Suppose we want to isolate pure cultures of bacteria from our mouth. We proceed by collecting some saliva in a sterile container. We then inoculate, or plant, a sample of the saliva upon an appropriate medium in such a way that individual microbial cells are located on or in the medium apart from the other microbial cells. The material which is inoculated onto the medium is called *inoculum*. By inoculating a nutrient agar-type medium by the *streak-plate method* or the *pour-plate method* (Table 4-4), cells are individually separated. Individual microbial cells, on incubation, reproduce so rapidly that within 18 to 24 h they produce a visible mass of cells, called a *colony*. The colony is visible to the naked eye. Each different colony may represent a different kind of microorganism; each colony is presumably a pure culture of a single kind of microorganism. If two microbial cells

Table 4-4. Methods for the isolation of pure cultures of microorganisms

TECHNIQUE	SCHEMATIC REPRESENTATION

Streak plate:

The inoculum is streaked over the surface of a nutrient agar-type medium, in a petri dish, with a transfer needle in the manner shown in the diagram in **A**. Somewhere along the streaked lines cells will be sufficiently separated that they will grow into isolated colonies as shown on the plate in **B**.

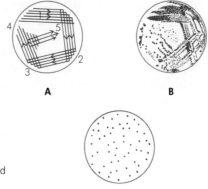

A   B

Spread plate:

A drop of inoculum is placed on the center of a nutrient agar-type medium, in a petri dish, and by use of a sterile bent glass rod, the inoculum is spread over the surface of the medium. The same glass rod can be used to inoculate a second plate to insure adequate "spreading-out" of cells. Isolated colonies will appear on some plates.

Pour plate:

Step 1: One loopful—using a transfer needle with a loop on the end—of original suspension is transferred to tube **A** (sterile liquid, cooled agar medium). Tube **A** is rolled between the hands to effect thorough mixing of inoculum. Similar transfers are made from **A** to **B** and **B** to **C**. Step 2: Contents of each tube are poured into separate petri dishes. Step 3: After incubation, plates are examined for isolated colonies. From a plate containing them, pure cultures of microorganisms can be isolated by transferring a portion of a colony to a tube of sterile medium.

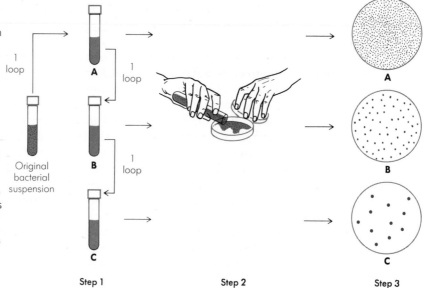

Step 1   Step 2   Step 3

Enrichment culture:

To improve the chances of isolating an organism which has some unique physiological or biochemical characteristic, one may precede the plating—streak, spread, or pour—by passing the inoculum through a series of transfers into a medium of a composition (and conditions of incubation) that

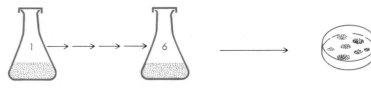

Successive transfer in medium with desired "nutrient"

Isolation of colonies on agar-medium containing same "nutrient"

will favor growth of the desired microorganism. The enrichment procedure in effect increases the proportion of the desired organism over that present in the initial inoculum.

Table 4-4. (Continued)

TECHNIQUE	SCHEMATIC REPRESENTATION
Single cell isolation: By means of a micromanipulator, one can use a microprobe to pick a single microorganism from a fluid suspension of cells while examining this preparation microscopically. The single cell is then transferred to a sterile medium.	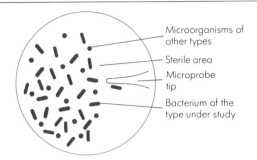

from the original inoculum fall too closely together on the agar medium, the colony resulting from each may intermix with, or at least adjoin, the other, and so the observable mass of cells will not be a pure culture.

A more direct method for isolating a single microorganism is by using a piece of micromanipulator equipment called a *microscopic probe* to remove a single cell from a fluid suspension of cells. This micromanipulator is used, of course, in conjunction with a microscope.

Several techniques for isolation of pure cultures are outlined in Table 4-4.

Maintenance and preservation of pure cultures

Once a microorganism has been isolated in pure culture, it may be necessary to maintain the culture in a living condition for some indefinite period of time. The fact is that most microbiology laboratories maintain a large collection of such cultures, often referred to as the *stock culture collection*. The American Type Culture Collection, located in Washington D.C., maintains thousands of species of microorganisms, including viruses. A catalog is available which gives in detail the history of each culture. The Center for Disease Control in Atlanta, Georgia, which is part of the U.S. Public Health Service, is another source of known cultures of microorganisms, particularly those which cause disease. As you will learn, these cultures are used for a variety of purposes in microbiological laboratories all over the world. For example, known cultures of microorganisms (reference cultures) from the Center for Disease Control are used by microbiologists in clinical laboratories to evaluate their laboratory procedures.

Many procedures are used to preserve and maintain cultures of organisms. The method of choice depends on the circumstances associated with the cultures. Is the culture to be kept for only a short time (months), or is it desirable to maintain it indefinitely (years)? For short-term maintenance, cultures can be stored at refrigeration temperatures [0 to 10°C (32 to 50°F)]; for long-term maintenance they are stored in liquid nitrogen at −196°C (−320°F). Or they can be dehydrated in a tube, while frozen and sealed under vacuum. This process is called *lyophilization* and is outlined in Fig. 4-10.

61

Techniques for characterizing microorganisms

Once you have obtained a pure culture of a microorganism, you are ready to perform a variety of laboratory examinations, each of which will contribute information about the microorganism. The extensiveness of the laboratory testing is determined by the questions you want to answer. You may wish to establish whether your culture is a new kind of microorganism or one which is already known. In either case you would need to characterize the culture extensively. You may be interested in finding a microorganism with some unique capability, such as rapid breakdown of cellulose. Accordingly, your testing would concentrate on finding out whether the culture degrades cellulose and, if so, at what rate. Whatever the objectives are, the characteristics sought and the tests performed are in the areas summarized

Table 4-5. Methods for characterizing microorganisms

MAJOR CHARACTERISTICS	METHODS
Morphological	Light and electron microscopic observations of specimens, either stained or unstained. Electron microscope techniques permit examination of ultrathin slices (sections) of microbial cells.
Nutritional	Determination of specific chemical substances and physical conditions (temperature, light, gases) that are needed to support the growth of the microorganism.
Cultural	Determination of appearance of microbial growth on various types of laboratory media, both liquid and solid.
Metabolic	Identification and measurement of chemical changes brought about by microorganisms. Some tests are simple to perform and merely provide a yes or no answer about whether the microorganism causes chemical changes in a particular substance—for example, whether it changes carbohydrates to acids. At the other extreme, there are methods which provide for identification of most chemical compounds that are involved in any metabolic process; this makes it possible to reconstruct step by step how the microorganism causes these changes. You will see examples of this in Chaps. 14 and 15.
Chemical composition	Determination of chemical makeup of cell components. Techniques are available to break cells apart and to recover (isolate) from the resulting mixture specific cellular components such as cell-wall fragments, nuclear material, and membranes. Laboratory procedures are available by which the chemical makeup of each component can be determined.
Antigenic composition	Characterization of microorganisms, particularly bacteria and viruses, by study of *antigens*, chemical substances present on their surface. Antigens are chemical substances of a microorganism which when injected into an animal will initiate the production of chemical substances (antibodies) that can be identified by laboratory procedures. Antigens and antibodies are part of the complex immunological system. The application of antigen-antibody procedures to microbiology is explained in Chap. 26.
Pathogenic	Determination of the disease-producing potential of a microbial culture is performed by inoculation of animals or plants with pure cultures of the microorganism.

Figure 4-10. Lyophilization process for preservation of cultures. (A) Small cotton-plugged vials containing frozen suspensions of the organisms are placed in the glass flask, which is attached to a condenser. The condenser is connected to a high-vacuum pump, and this system dehydrates the cultures. (B) After the cultures have been dehydrated, the vials are removed, placed individually in a larger tube, covered with asbestos packing, and sealed under vacuum. *(Courtesy of American Type Culture Collection.)*

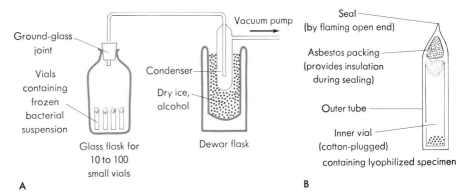

in Table 4-5. Some of these tests are relatively easy to perform. They might be simple visual observations and descriptions. Others are much more complex; for example, it may be necessary to determine not only what chemical changes the microorganism produces but also how these changes are brought about. This calls for the use of biochemical techniques which have been developed specifically to identify products of metabolism. The fact is that some aspects of microbiology are highly dependent on biochemical methodology.

The problems we experience presently with certain aspects of microbial taxonomy can be attributed to our incomplete knowledge about certain groups of microorganisms. With more detailed and complete knowledge about microbial species, derived from testing, it is possible to develop more satisfactory classification systems.

Summary and outlook

Laboratory studies of microorganisms are carried out for a variety of purposes. For example, you may be interested in how many microbes are present in a specimen. You may wish to know the exact identity of each different microorganism present. Or you may be concerned with some fundamental biological process as carried out by a microorganism. In general, the methods available to the microbiologist permit the characterization of microorganisms as follows: morphological characteristics, nutritional requirements, cultural characteristics, metabolic characteristics, chemical composition, and pathogenicity. New techniques continue to be developed and new laboratory equipment continues to become available by which one can obtain more detailed and specific information about each characteristic of a microorganism. Just as electron microscopy opened a new dimension in microbial morphology, advances in biochemical techniques have resulted in more precise and detailed understanding of the chemical processes performed by microorganisms.

The equipment of the microbiology laboratory continues to be improved. This makes it possible to identify additional, more detailed specific characteristics of microorganisms. Many procedures have become automated and their results are routinely analyzed by computer programs.

Key terms

differential staining
fluorescence
gram-negative bacteria
gram-positive bacteria
gram stain
inoculum
light microscopes

lyophilization
media
metabolism
mixed cultures
numerical aperture
pathogens
pour-plate technique

pure cultures
resolving power
simple stain
stock-culture collection
streaked-plate technique

Questions

1 What are the usual magnifications obtainable with light microscopes using different objectives? What determines the maximum limit of useful magnification?

2 Compare the basic principle of magnification in light microscopy and in electron microscopy. Compare the resolving powers obtainable with the electron microscope and the light microscope. What accounts for the difference?

3 What is the function of oil when used with the oil-immersion objective?

4 Assume that a yeast cell is examined by (*a*) bright-field, (*b*) phase-contrast, and (*c*) dark-field microscopy. Describe the likely differences in appearance of the cell when viewed by these methods.

5 Give an example of a situation in which it would be preferable to examine a specimen by the wet or hanging-drop technique rather than by staining.

6 Why is the gram stain so widely used in microbiology?

7 Distinguish, with examples, the difference between a simple stain and a differential stain.

8 What is the difference between a *pure culture* and a *mixed culture*? How are pure cultures isolated?

9 Assume that you have a culture which is supposed to be a pure culture and someone questions whether it really is pure. What kinds of tests or observations could you perform to get evidence to confirm that it is a pure culture?

10 Describe what kinds of laboratory data or information must be obtained in order to characterize microorganisms with reference to each of the following properties:

morphological   cultural   chemical composition
nutritional   biochemical   pathogenic

References for part one

Buchanan, R.E., and N. E. Gibbons (eds.): *Bergey's Manual of Determinative Bacteriology*, 8th ed., Williams & Wilkins, Baltimore, 1974. *This volume is the standard international reference book on the classification and taxonomy of bacteria. Each major group of bacteria is described, and recognized species are characterized in detail.*

Bulloch, W.: *The History of Bacteriology*, Oxford University Press, London, 1938. *The most complete and authoritative history of the development of bacteriology; it includes an extensive bibliography and a long list of biographical notices of some of the early workers in bacteriology.*

Clark, Paul F.: *Pioneer Microbiologists of America,* University of Wisconsin Press, Madison, 1961. *The people cited made microbiology a science in the United States. The author's entertaining style makes this an enjoyable as well as informative book, and his acquaintance with many of those he writes about adds a personal touch.*

Collins, C. H., and P. M. Lyne: *Microbiological Methods,* 4th ed., Butterworth's, Boston, 1976. *As the title states, this volume contains procedures for the performance of a wide variety of microbiological techniques.*

Dowling, Harry F.: *Fighting Infections, Conquests of the Twentieth Century,* Harvard University Press, Cambridge, Mass., 1977. *This is the story of the progress made during this century in the control of infectious diseases. It is an interestingly written history of the diagnosis, prevention, and treatment of infectious diseases and the social and economic forces that made this all possible.*

Gerhardt, P. (ed.): *Manual of Methods for General Bacteriology,* American Society for Microbiology, Washington, D. C., 1980 *A very comprehensive coverage of the general procedures used in the microbiology laboratory.*

Jennings, R. K., and R. F. Acker: *The Protistan Kingdom,* Van Nostrand Reinhold, New York, 1970. *This small book dealing with the protists and viruses is both informative and entertaining. It may provide a stimulus for further study by students and nonstudents alike.*

Postgate, J.: *Microbes and Man,* Penguin, Baltimore, 1975. *A very good introduction to the world of microorganisms. The role of microorganisms in our environment is covered, particularly the changes which they bring about and which are important to our well-being. Other topics include microbes in evolution and future roles for microorganisms.*

Wilson, M. B.: *The Science and Art of Basic Microscopy,* American Society for Medical Technology, Bellaire, Texas, 1976. *This short (61-page) manual provides a "simple, concise, practical explanation of the microscope and how it works".*

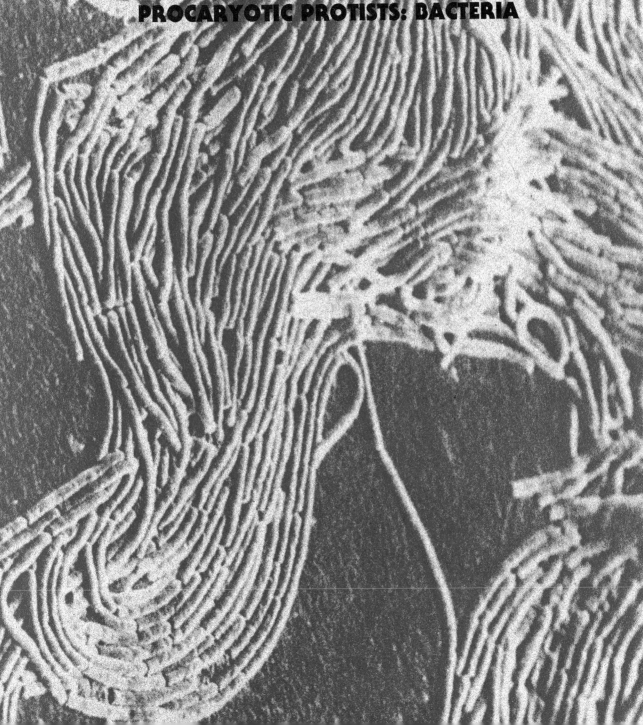

PART TWO

PROCARYOTIC PROTISTS: BACTERIA

The science of microbiology as we know it today was, to a large extent, initially built on knowledge and techniques gained from the study of bacteria. This began during the Pasteur era when the importance of bacteria in disease and in our environment started to be understood and appreciated.

One of the major problems which confronted microbiologists over many years was the development of criteria that would clearly distinguish a bacterium from other protists. This situation was greatly clarified as new information became available about the internal structure of microbial cells.

Part II provides a characterization of some major properties of the bacteria. It will serve to acquaint you with the great diversity of morphological and physiological types of bacteria—all of which are procaryotes.

Scanning electron micrograph of *Bacillus cereus* var. *mycoides*. (Courtesy of K. N. Crane and I. L. Roth, University of Georgia.)

5

THE MORPHOLOGY AND FINE STRUCTURE OF BACTERIA

Knowledge of the morphology and fine structure of bacteria was acquired in two different spans of time. The observations made by Leeuwenhoek with his simple microscope revealed the gross appearance of microorganisms, including bacteria. His carefully recorded drawings of what we now recognize as bacteria revealed their cell shapes to be either round, rodlike, or spiral. Further improvements in light microscopy, including staining techniques, made it possible to observe with more accuracy the characteristic shapes of these cells, their size, some of their external structures, and their patterns of arrangement. The size, shape, and arrangement constitute the *gross morphologic* characteristics of the cells of a bacterial species.

In the early 1940s the electron microscope became available as a laboratory instrument for microbiologists. The tremendously higher resolution obtainable with this new instrument, together with new methods of preparing the specimens, made it possible to identify structural parts of an individual cell. For example, techniques were developed whereby a single bacterial cell could be cut into thin slices and these slices (or sections) then examined by electron microscopy. These studies revealed the structural parts of the bacterial cells and what we now refer to as their *fine structure* in contrast to their gross morphology. Along with advances in electron microscopy, techniques were developed for disintegrating bacterial cells, isolating the cellular components, and analyzing their chemical composition. Thus we have witnessed an era in which the structural parts of the bacterial cell have been revealed, the chemical composition of these parts has been ascertained, and knowledge of the functions carried out by the fine structures of the cell has been acquired. Specialized fields of study have emerged, such as biochemical cytology which relates form to function.

In this chapter we will begin with a description of the gross morphology of bacteria, followed by a characterization of their fine structure. Knowledge of these properties of bacteria not only provides an aid to the identification of species, but also provides invaluable information for many other aspects of the study, use, and control of microorganisms.

Gross morphology of bacterial cells

Bacterial cells vary considerably in their length; cells of some species may be 100 times longer than cells of other species. There is more uniformity in their width than there is in their length.

Size

The unit of bacterial measurement is the micrometer (μm), which is equivalent to $1/1000$ mm (10^{-3} mm), or $1/25400$ in. The bacteria most frequently studied in an introductory microbiology laboratory course measure approximately 0.5 to 1.0 by 2.0 to 5.0 μm. For example, the spherical bacteria staphylococci and streptococci have diameters ranging from 0.75 to 1.25 μm. Average-sized rod forms such as typhoid and dysentery bacteria have a width of 0.5 to 1 μm and a length of 2 to 3 μm. Cells of some bacterial species are very long; they may exceed 100 μm in length and range from 1.0 to 2.0 μm in diameter. A group of bacteria known as the *mycoplasmas* are characteristically very small—so small that they are barely visible by light microscopy. They are also *pleomorphic*; that is, their morphology varies considerably. Their size ranges from 0.1 to 0.3 μm.

Although bacteria are extremely minute, it is possible to measure their dimensions with relative ease and accuracy. For this purpose, the microscope is equipped with an *ocular micrometer*, a disk with etched equidistant lines. The distance between the lines is determined beforehand by reference to a *stage micrometer*, an instrument which functions as a ruler for microscopic work (Fig. 5-1A). Examination of bacteria through a microscope equipped with an ocular micrometer shows the ruled lines imposed on the microorganisms (Fig. 5-1B) in such a way that the length and width of cells can be conveniently determined (Fig. 5-1C).

It is difficult to grasp the extremely small size of a bacterium in the quantitative terms given above. The following examples may help. A volume of 1 in^3 contains approximately 9 trillion average-sized rod-shaped bacteria. Calculations show that approximately 1 trillion (1,000,000,000,000, or 10^{12}) bacteria weigh a mere 1 g (1/454 lb). For the most part you will examine bacteria at a magnification of 1,000 times; the common housefly magnified to the same extent would appear to be more than 30 ft long.

Figure 5-1. Measuring bacterial cells. (A) The stage micrometer is placed on the microscope stage and observed under a low-power objective. (B) The ocular micrometer is now inserted in the eyepiece of the microscope. When viewed under the oil-immersion objective (high-power), the scale of the ocular micrometer is superimposed (top) on the scale of the stage micrometer (below). The divisions of the ocular micrometer are calibrated by comparison to the stage micrometer. (C) The calibrated ocular micrometer is used to measure bacterial cells. *(Erwin F. Lessel, illustrator.)*

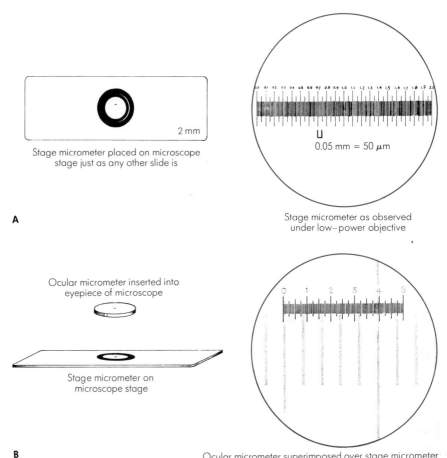

A

Stage micrometer placed on microscope stage just as any other slide is

0.05 mm $= 50$ μm

Stage micrometer as observed under low-power objective

Ocular micrometer inserted into eyepiece of microscope

Stage micrometer on microscope stage

B

Ocular micrometer superimposed over stage micrometer as viewed under the oil-immersion objective. In this case, 78 ocular micrometer divisions equal 50 μm. Therefore each of the smallest ocular micrometer divisions equals 0.64 μm. (Illustration slightly enlarged for clarification.)

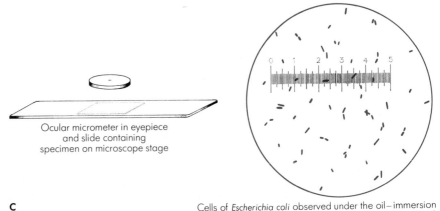

Ocular micrometer in eyepiece and slide containing specimen on microscope stage

C

Cells of *Escherichia coli* observed under the oil-immersion objective with the ocular micrometer in place. The cells of *E. coli* ordinarily measure 0.6 by 2.0 to 3.0 μm.

A distinctive feature of bacterial cells is revealed when you calculate the ratio of their surface area to their volume. For bacteria this value is very high compared with those of larger organisms. In practical terms this means that the content of a bacterial cell has a tremendous exposure to the interface between the cell wall and the nutrients in the environment. This characteristic, in part, accounts for the high rate of metabolism and growth of bacteria.

Shape   Individual bacterial cells are either ellipsoidal, spherical, rodlike (cylindrical), or spiral (helical). Each of these features is important in characterizing the morphology of a species.

Spherical or ellipsoidal bacterial cells are designated as *cocci* (singular, *coccus*). As noted below, cocci appear in several characteristic arrangements depending on the species.

Cylindrical, or rodlike, bacterial cells are designated *bacilli* (singular, *bacillus*). There are considerable differences in the length and width of the various species of bacilli (see Fig. 5-2). The ends of some bacilli are squarish, others are rounded, and still others are tapered or pointed like the end of a cigar. Sometimes the bacilli remain attached to each other, end to end, to give the appearance of a chain.

Spiral-shaped bacteria, or *spirilla* (singular, *spirillum*), occur predominately as unattached individual cells. Included in this morphologic group are the *spirochetes*, some of which cause serious diseases in humans. The

Figure 5-2. Typical rod-shaped bacteria (bacilli). Note differences in length and width. (A) *Clostridium sporogenes*, 0.6 to 0.3 μm by 3.0 to 7.0 μm; (B) *Pseudomonas* sp., 0.5 to 1.0 μm by 2.0 to 3.0 μm; (C) *Bacillus megaterium*, 0.2 to 1.5 μm by 2.0 to 4.0 μm; (D) *Salmonella typhi*, 0.6 to 0.7 μm by 2.0 to 3.0 um. *(Courtesy of J. Nowak, Documenta Microbiologica, Erster Teil, Bakterien, Gustav Fischer Verlag KG, Stuttgart, 1927.)*

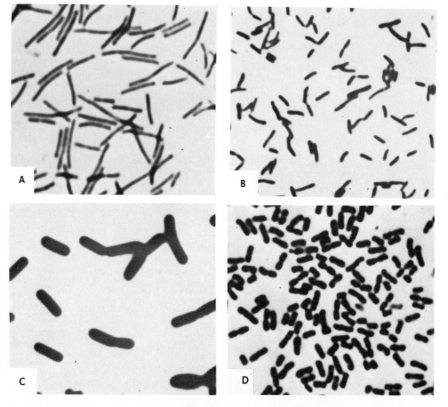

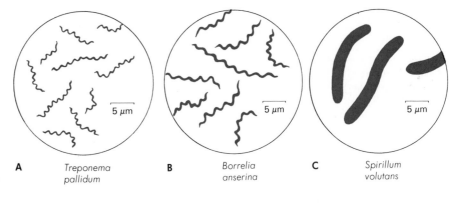

Figure 5-3. Spiral-shaped bacteria. Note differences in frequency of turns or waves, height of spirals, and length of cells. (A) *Treponema pallidum;* (B) *Borrelia anserina;* (C) *Spirillum volutans. (Erwin F. Lessel, illustrator.)*

A   *Treponema pallidum*   B   *Borrelia anserina*   C   *Spirillum volutans*

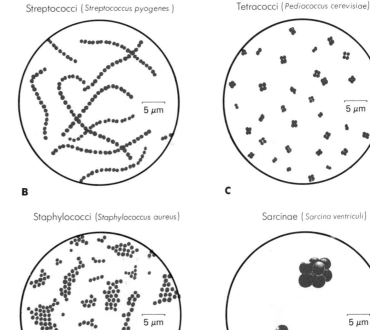

Diplococci (*Streptococcus pneumoniae*)

Streptococci (*Streptococcus pyogenes*)

Tetracocci (*Pediococcus cerevisiae*)

Staphylococci (*Staphylococcus aureus*)

Sarcinae (*Sarcina ventriculi*)

Figure 5-4. Patterns of arrangement of cocci. (A) Diplococci: cells divide in one plane and remain attached predominantly in pairs. (B) Streptococci: cells divide in one plane and remain attached to form chains. (C) Tetracocci: cells divide in two planes and characteristically form groups of four cells. (D) Staphylococci: cells divide in three planes in an irregular pattern, producing "bunches" of cocci. (E) Sarcinae: cells divide in three planes in a regular pattern, producing a cuboidal arrangement of cells. *(Erwin F. Lessel, illustrator.)*

individual cells of different species exhibit striking differences in length, number, and amplitude of spirals and in rigidity of cell walls. For example, some spirilla are short, tightly coiled spirals; others are very long and exhibit a series of twists and curves. Short, incomplete spirals are known as *comma bacteria,* or *vibrios.* Figure 5-3 shows several types of spiral bacteria. Note particularly the wavy cell forms, the frequency of turns or waves, the height of the spirals, and the overall length of the cell.

Arrangement   Certain species of bacteria exhibit patterns of arrangement of their cells, such as pairs, clusters, chains, or filaments.

Cocci exhibit several different arrangements, as shown in Fig. 5-4. Each of these arrangements is characteristic of particular species of cocci. Thus,

Figure 5-5. Patterns of arrangement of bacilli. (A) Palisade layer arrangement. *(Corynebacterium diphtheriae);* (B) Rosette arrangement *(Caulobacter vibrioides);* (C) Chains *(Bacillus cereus). (Erwin F. Lessel, illustrator.)*

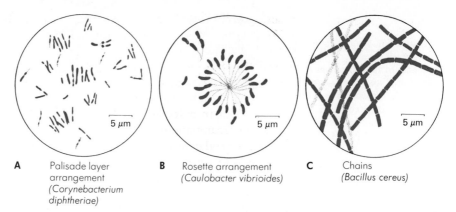

A Palisade layer arrangement (*Corynebacterium diphtheriae*)

B Rosette arrangement (*Caulobacter vibrioides*)

C Chains (*Bacillus cereus*)

in describing the morphology of cocci, special attention should be given to their pattern of arrangement. It should be borne in mind that rarely are all the cells of a given species arranged in exactly the same pattern. The predominant arrangement is the important characteristic.

Bacilli do not arrange themselves in a variety of patterns like those characteristic of cocci. Some species, however, do exhibit a tendency to arrangement of their cells. The bacillus which causes diphtheria tends to

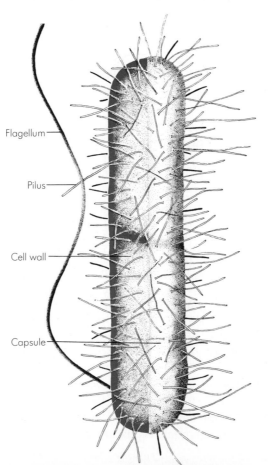

Flagellum

Pilus

Cell wall

Capsule

Figure 5-6. Major structures external to the bacterial cell wall. Certain structures, for example, capsules, flagella, spores, and pili, are not common to all bacterial cells. *(Erwin F. Lessel, illustrator.)*

produce groupings of cells lined side by side like matchsticks (palisade arrangement). The tubercule bacillus may occur in an arrangement of three bacilli that gives the impression of a Y structure. Cells of some aquatic species (see Chap. 35) arrange themselves into a rosette pattern. This is possible because the individual cells produce a very fine stalk called a *holdfast*, which can attach to the surface of an object. Cells of other species occur in pairs (*diplobacilli*) or in chains (*streptobacilli*). The pair and chain arrangements may be attributable more to a stage of growth or to certain cultural conditions than to a morphological characteristic. Some of these cellular arrangements are shown in Fig. 5.5.

Spiral-shaped bacteria occur predominantly as single cells.

Most bacteria studied in the laboratory of an introductory course in microbiology are of the coccus, bacillus, or spirillum morphology. As you will see in Chap. 7, where we describe all the major groups of bacteria, there are species, particularly those from aquatic and soil environments, which have a very different gross morphology.

Fine structure of bacterial cells

Examination of a bacterial cell by modern microscopic techniques reveals that there are structures outside the cell wall. Other structures or bodies are found enclosed within the cell wall. Some structural parts are common to all cells, such as the cell wall and the cytoplasmic membrane. Other structures are present only on or in cells of certain species.

Structures external to the cell wall

Let us assume that with the aid of a high-powered microscope you bring into focus a single bacterial cell. The major structures external to the cell wall that you are likely to see are flagella, pili, and capsules. These are shown schematically in Fig. 5-6.

Flagella. The extremely thin, hairlike appendages that protrude through the cell wall and originate from the basal body, a granular structure just beneath the cell membrane in the cytoplasm, are called *flagella*. Flagella are responsible for the motility of bacterial cells. The flagellum has three parts: the basal body, a hooklike structure, and a long filament outside the cell wall (see Fig. 5-7). The length of a flagellum is usually several times that of the cell, but its diameter is only a small fraction of the diameter of the cell, for example, 10 to 20 nm. The flagellum is built from protein subunits; this protein is called *flagellin*.

Not all bacteria possess flagella; many species of bacilli and spirilla do have them, but flagella are rarely found on cocci. For bacteria with flagella, the pattern of attachment of these appendages, as well as the number attached (see Fig. 5-8), is used in classifying these bacteria into certain taxonomic groups. For example, among the gram-negative rod-shaped bacteria, the genus *Pseudomonas* is characterized by flagella located at the end of the cell (*monotrichous* or *lophotrichous* flagella), and the genus *Escherichia* by flagella surrounding the cell (*peritrichous* flagella). These arrangements are shown in Fig. 5-9.

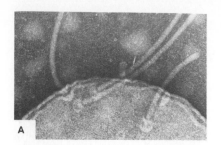

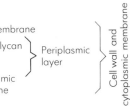

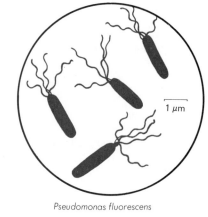

Filament

Hook

Basal body

Outer membrane
Peptidoglycan layer
Cytoplasmic membrane

Periplasmic layer

Cell wall and cytoplasmic membrane

C

10 nm

Figure 5-7. The mechanism of attachment of flagella within a bacterial cell is shown in this series. The bacterium is *Pseudomonas aeruginosa*. (A) Prior to electron microscopic examination, the bacterial cells were partially lysed and then stained to make the point of flagellar attachment (basal structure) more visible (magnification approximately X 80,000). (B) Isolated flagella showing basal structure attached at one end. (C) Model of basal structure illustrating its structure and attachment within a gram-negative bacterium. *(Courtesy of T. Lino, University of Tokyo.)*

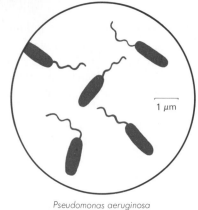

Pseudomonas aeruginosa

1 μm

A

Pseudomonas fluorescens

1 μm

B

Figure 5-8. Bacterial arrangements of flagella. (A) Monotrichous, a single flagellum; (B) Lophotrichous, a cluster of flagella; (C) Amphitrichous, flagella, either single or clusters, at both ends; (D) Peritrichous, surrounded by flagella. *(Erwin F. Lessel, illustrator.)*

Spirillum serpens

1 μm

C

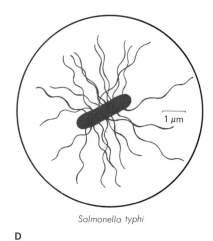

Salmonella typhi

1 μm

D

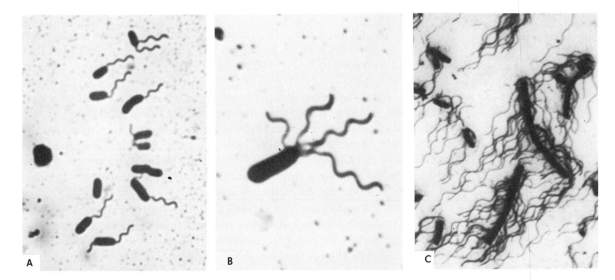

Figure 5-9. Bacterial flagella as seen in stained films. (A) (× 2,000), and (B) (× 4,000) are *Pseudomonas* species exhibiting characteristic monotrichous or lophotrichous flagellation, a single flagellum or a cluster of flagella at one end. (C) *Salmonella typhi* showing peritrichous flagella, stained preparation. *(From General Biological Supply House, Inc.)*

In their natural state flagella are too small to be seen with the light microscope. However, by special staining procedures which employ a mordant (a substance that fixes a dye to a surface), their diameters can be increased sufficiently to make them visible by light microscopy, as shown in Fig. 5-9. Indirect evidence for the presence of flagella can be obtained by examining bacteria in a wet preparation for evidence of motility.

Pili (fimbriae). Many gram-negative bacteria have filamentous appendages that are not flagella. These appendages, called *pili* (singular, *pilus*) or *fimbriae* (singular, *fimbria*), are smaller, shorter, and more numerous than flagella (Fig. 5-10). Pili can be seen only by electron microscopy. They have no function for motility; they are found on nonmotile as well as motile species. There are, however, several functions associated with different types of pili. One kind, known as the *F pilus* (or *sex pilus*), serves as the port of entry of genetic material during bacterial mating (see Chap. 17).

Figure 5-10. Fimbriated bacteria. (A) *Shigella flexneri:* dividing bacilli with numerous fimbriae surrounding the cells (X 20,000). (B) *Salmonella typhi:* dividing bacilli with numerous fimbriae and a few flagella (the very long appendages) (X 12,500). *(Courtesy J. P. Duguid and J. F. Wilkinson and The Society for General Microbiology; Symposia XI, 1961.)*

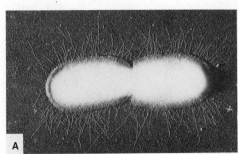

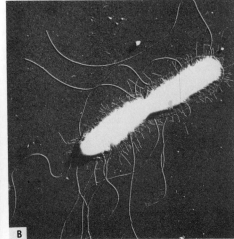

Some pili function as a means of attachment to various surfaces. This capability of pili to adhere to surfaces is important. This helps some bacteria to attach themselves to animal and plant tissues from which they can obtain nutrients.

Capsules. Some bacterial cells, for example, pneumococci which cause pneumonia, are surrounded by a layer of viscous material referred to as a *capsule* or *slime layer*. The size of these capsules is markedly influenced by the medium in which the bacterium is grown. In some instances the thickness of the capsule is only a fraction of the diameter of the cell; in other cases the capsule is much larger than the cell. Examples of capsulated and slime-producing bacteria are shown in Fig. 5-11.

The exact relationship of the capsule to the rest of the cell is not known. There is reason to believe that the material making up the capsule is secreted from the cell and that because of its viscosity it will not readily diffuse away from the cell and hence coats the cell wall. More soluble slime material excreted by the cell dissolves in the medium in which the organism is growing. Production of certain types of capsular material may greatly increase the viscosity of the medium in which the organism is cultured. For example, capsulated bacteria growing in milk will produce stringiness in the milk.

Bacterial capsules are important both to the bacterium and to other organisms. For bacteria, they provide a protective covering and also serve as a reservoir of stored food. The capsules of certain disease-producing bacteria increase the infective capacity of the bacteria. If the bacterium loses its capsule completely, it may lose its virulence and hence its ability to produce infection. Capsulated bacteria are also responsible for the development of nuisances such as slime in some industrial processes. The accumulation of slime in manufacturing equipment can clog filters, produce an undesirable coating on pipes or other equipment, and/or affect the quality of the final product. The organism shown in Fig. 5-11B, for

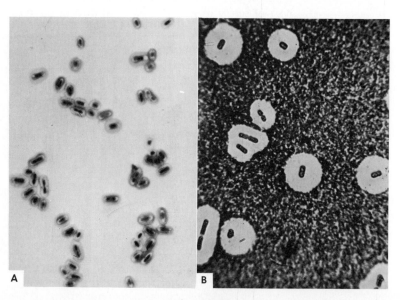

Figure 5-11. Capsuled bacteria: (A) *Klebsiella pnuemoniae. (From General Biological Supply House, Inc.)* (B) Capsulated slime-forming bacterium isolated from a papermill operation. Note the extremely large capsules (white areas) around each of the cells. *(Courtesy of P. M. Borick, Wallace and Tierman, Inc.)*

A   B

example, was found to be responsible for the development of slime in a paper-mill plant.

Sheaths. Some species of bacteria, particularly those from freshwater and marine environments, are enclosed with a *sheath*, or tubule. The sheaths are composed of insoluble metal compounds, such as ferric and manganic oxides, precipitated around the cells as products of their metabolic activity. These insoluble metal compounds are formed by the cell from soluble compounds of iron or manganese present in the environment. The sheath can become extended around many cells aligned end to end, giving the impression of filamentous growth (see Fig. 7-5). Actually, the cells enclosed by the sheath occur individually. Periodically they emerge from an open end of the sheath and initiate the sheath-forming process anew.

The sheath is not a vital part of the cell. The sheathed bacteria, as you will see in Chap. 7, constitute a major group of microorganisms. They are abundant in freshwater habitats that are rich in organic matter, as well as in polluted streams and in sewage disposal plants.

Stalks. Certain bacterial species are characterized by the formation of a semirigid appendage called a *stalk* which extends from the cell. The diameter of the appendage is smaller than that of the cell from which it is produced (see Fig. 7-6). The stalk has an adhesive substance at its distal end, that is, the end distant from the cell, which enables the cell to attach to solid surfaces. Stalked bacteria are found in freshwater and marine environments in which the ability to attach to solid surfaces is important for bacterial growth and survival.

The cell wall   Beneath such extracellular substances as capsules or slimes and external to the cytoplasmic membrane, which is a component of every living cell, is the *cell wall*, a very rigid structure which gives shape to the cell (see Fig. 5-12). The rigidity of the cell wall can be readily demonstrated by subjecting bacteria to very high pressures, very low temperatures, or other severe physical conditions. Most bacterial cells retain their original shapes during and after such treatment.

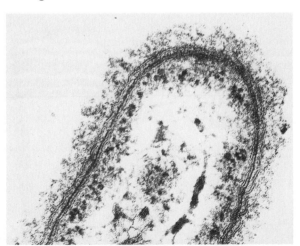

Figure 5-12. Thin section of a bacterial cell showing multiple surface-layer structures (cell wall and cytoplasmic membrane). *(Courtesy of Helen Crane. From F. L. Crane and J. D. Hall, Ann NY Acad Sci, **195**:24, 1972.)*

The thickness of the cell wall of most bacteria thus far studied ranges from 10 to 35 nm; however, some cell walls are considerably thicker. The cell wall constitutes a significant portion of the total dry weight of the cell. Depending on the species and the culture conditions, it may account for as much as 10 to 40 percent of the dry weight of the organism. Bacterial cell walls are essential for growth and division. With the exception of the mycoplasmas, all bacteria have rigid cell walls.

Isolation of cell-wall fragments. Several laboratory techniques have been developed by which researchers can isolate fragments of bacterial cell walls from other cellular material. Scientists have analyzed the cell-wall material for its chemical makeup and from this information they have been able to determine the exact structure of the cell wall.

The composition of the bacterial cell wall is extremely important in distinguishing bacteria from other protists, as well as in distinguishing among the groups of bacteria.

Figure 5-13. Schematic interpretation of cell walls from electron microscope observation: (A) gram-positive bacteria. Outer layer not always wavy and intermediate layer not always demonstrable. *(Erwin F. Lessel, illustrator.)*

Chemical composition of cell walls. *Peptidoglycans* provide the cell wall's rigid structure. These very large polymers (large molecules made up of repeating units) are composed of three kinds of building blocks: (1) *N-acetylglucosamine* (AGA), (2) *N-acetylmuramic acid* (AMA), and (3) a peptide consisting of four or five amino acids, namely L-alanine, D-

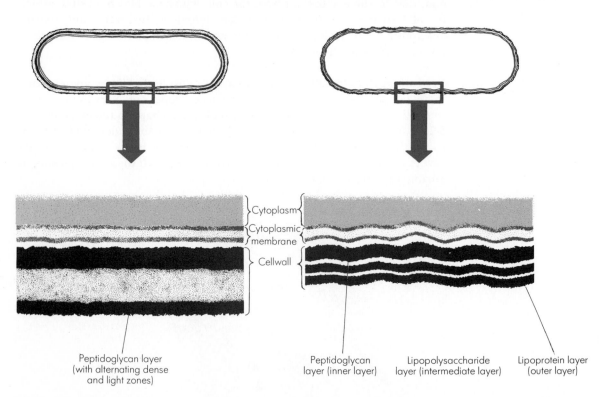

Cytoplasm

Cytoplasmic membrane

Cellwall

Peptidoglycan layer (with alternating dense and light zones)

Peptidoglycan layer (inner layer)

Lipopolysaccharide layer (intermediate layer)

Lipoprotein layer (outer layer)

A Gram-positive bacteria

B Gram-negative bacteria

alanine, D-glutamic acid, and either lysine or diaminopimelic acid. The manner in which these subunits are linked to form this polymer is discussed in Chap. 15. The intact cell wall also contains other chemical constituents, such as teichoic acids, protein polysaccharides, lipoproteins, and lipopolysaccharides, which are linked to the peptidoglycan.

Peptidoglycans along with two other cell-wall constituents, *diaminopimelic acid* and *teichoic acid,* are unique to procaryotes. Peptidoglycans, however, do vary in their chemical composition and structure from bacterial species to bacterial species. *N*-Acetylglucosamine and *N*-acetylmuramic acid are constant constituents of peptidoglycans, but variations do occur in the amino acids present and the nature of the linkages between these amino acids.

Another important discovery made during the course of identifying the chemical composition of bacterial cell walls was that several of the amino acids in the peptide in peptidoglycans exist in the D configuration. This is contrary to their appearance in proteins, where they exist only in the L configuration.

Differences in composition and structure of cell walls. Figure 5-13 and Table 5-1 highlight the significant differences in cell-wall composition and structure between gram-positive and gram-negative bacteria. These differences are important to understand because, it is now believed, they are responsible for the manner in which these two groups of bacteria respond to various treatments and agents, such as the gram stain and certain antibiotics.

Table 5-1. Some characteristics of gram-positive and gram-negative bacteria

CHARACTERISTIC	RELATIVE DIFFERENCES	
	Gram-positive	Gram-negative
Cell-wall structure	Thick (15–80 nm) Single layer (monolayer)	Thin (10–15 nm) Triple layer (multilayer)
Cell-wall composition	Low in lipids (1–4%) Peptidoglycan present in monolayer; major component, accounts for over 50% of dry weight in some bacterial cells Teichoic acids	High in lipids (11–22%) Peptidoglycan present in inner rigid layer; small amount, accounts for about 10% of dry weight No teichoic acids
Susceptibility to penicillin	More susceptible	Less susceptible
Growth is retarded by basic dyes, e.g., crystal violet	Growth is markedly retarded	Growth is retarded less
Nutritional requirements	Relatively complex in many species	Relatively simple
Resistance to physical disruption	More resistant	Less resistant

As we mentioned in Chap. 4, the gram stain is one of the most important staining techniques in microbiology. However, a precise explanation of why cells respond differently to this process is still lacking. The most plausible explanations for the mechanism of the gram stain are based on the structure and composition of the bacterial cell wall. Gram-negative bacteria contain a higher percentage of lipid, a fat or fatlike substance, than do gram-positive bacteria. Gram-negative bacterial cell walls are also thinner than gram-positive bacterial cell walls. Experimental evidence suggests that during the staining procedure the treatment of gram-negative bacteria with ethanol (alcohol) extracts the lipid, which results in increased porosity or permeability of the gram-negative cell wall. Thus the crystal violet–iodine (CV-I) complex, which had entered the cell wall during an earlier step in the staining process, can be extracted. The gram-negative organism is thereby decolorized. The cell walls of gram-positive bacteria, because of their lower lipid content, become dehydrated during treatment with ethanol. Their pore size decreases, their permeability is reduced, and the CV-I complex cannot be extracted.

Another similar explanation is also based on permeability differences between the two groups of bacterial cell walls. Gram-negative bacterial cell walls have a much smaller amount of peptidoglycan, and this peptidoglycan has less extensive cross-linking than that in the walls of gram-positive bacteria. Therefore, the pores in the peptidoglycan of gram-negative bacteria remain sufficiently large even after ethanol treatment to allow the CV-I complex to be extracted. In gram-positive bacteria, the CV-I complex is trapped in the wall following ethanol treatment, which presumably decreases the diameter of the pores in the cell-wall peptidoglycan.

These two explanations are not mutually exclusive, and it is likely that both may contribute to the mechanism of the gram stain. Furthermore, if gram-positive cells are treated with the enzyme lysozyme to remove the cell wall following staining with CV-I complex, the resulting structures will be stained by the CV-I complex. They are, however, easily decolorized by ethanol. All this is evidence that the cell wall in gram-positive bacteria is the site of retention of the crystal violet stain.

Gram-negative bacteria differ from gram-positive bacteria in other ways also. Gram-negative bacteria are more susceptible to antibiotics such as streptomycin. Gram-positive bacteria are generally more susceptible to the antibiotic penicillin and less susceptible to disintegration by mechanical treatment (such as exposure to very high pressures or ultrasonic vibrations) or by exposure to certain enzymes.

Functions of surface structures of the bacterial cell are summarized in Table 5-2.

Structures internal to the cell wall

The most common structures that may occur within the bacterial cell wall are shown in Fig. 5-14. A description of each of these structures follows.

Cytoplasmic membrane. Immediately beneath the cell wall is a thin membrane called the *cytoplasmic membrane*, also referred to as the

Table 5-2. Functions of surface structures of bacterial cells

STRUCTURE	FUNCTION	CHEMICAL COMPOSITION
Flagella	Locomotion	Protein
Pili	Conjugation tube Cell adhesion	Protein
Capsules and extra cellular material	Protective covering Cell adhesion Reserve food	Polysaccharides, polypeptides
Cell wall	Protective covering Permeability	Peptidoglycan, teichoic acids, polysaccharides, lipid, and protein
Cytoplasmic membrane and mesosome	Semipermeable covering Transport mechanism Cell division Synthesis of biological macromolecules	Lipid, protein

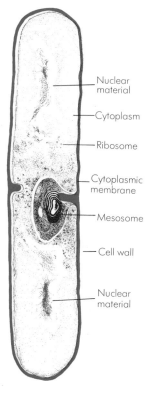

Nuclear material

Cytoplasm

Ribosome

Cytoplasmic membrane

Mesosome

Cell wall

Nuclear material

Figure 5-14. Major structures which occur within the bacterial cell wall. Certain structures, for example, granules or inclusion bodies, are not common to all bacterial cells. *(Erwin F. Lessel, illustrator.)*

protoplasmic membrane or simply the *plasma membrane*. Estimates of its thickness, based on electron micrographs of thin sections, are of the order of 7.5 nm.

The cytoplasmic membrane is extremely important, because it controls the passage of chemical substances in solution into and out of the cell. It is a remarkable accomplishment that the microscopic cell, floating as it does in an extremely complex and changing chemical environment, is able to take up and retain appropriate amounts of nutrients and to discharge excess nutrients and/or waste products.

The cytoplasmic membrane also provides the biochemical machinery to move mineral ions, sugars, amino acids, electrons, and other metabolites through the membrane. These substances in solution, or *solutes*, pass through the membrane by either *passive diffusion* or *active transport*. An explanation of these two important processes follows.

Passive diffusion (osmosis). Passive diffusion is nonspecific—a distinction is not made among the solutes that pass through the membrane. It is a process by which chemical substances move across the membrane from an area of higher to an area of lower concentration. Passive diffusion acts to equalize the concentration of solutes on both sides of the membrane.

Active transport. Active transport, in contrast to passive diffusion, is highly specific; that is, solutes are acted upon selectively. In addition it allows a higher concentration of a solute to be built up within the cell than exists outside the cell. Active transport is extremely important for bacterial cells in environments such as the ocean, where nutrients are present in small amounts. These cells must be able to concentrate nutrients in order to grow. The process of active transport involves an intricate mechanism within the cytoplasmic membrane—compounds called *membrane carriers* together with biochemical reactions that provide energy accomplish this kind of transport.

Figure 5-15. Bacterial membranes. Mesosome structure in a bacillus as demonstrated in an electron micrograph of an ultrathin section. Note that it is continuous with the cytoplasmic membrane. *(Courtesy of A. Ryer and C. Frebel.)*

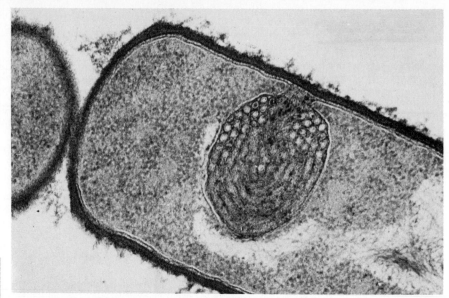

Figure 5-15. Bacterial membranes. Mesosome structure in a bacillus as demonstrated in an electron micrograph of an ultrathin section. Note that it is continuous with the cytoplasmic membrane. *(Courtesy of A. Ryer and C. Frebel.)*

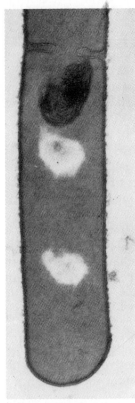

Figure 5-16. Electron micrograph of thin section of *Bacillus subtillis* showing nuclear material (light-appearing areas) in addition to cell wall, cytoplasmic membrane, mesosome, and initial stage of cross-wall formation (× 35,000). *(Courtesy of S. F. Zane and G. B. Chapman, Georgetown University.)*

Mesosomes. The cytoplasmic membrane, by folding inward or *invaginating* into the cytoplasm, produces a structure called a *mesosome* (see Fig. 5-15). Mesosomes are always continuous with the cytoplasmic membrane. They frequently are found to originate at the point where the membrane starts invagination prior to cell division, and they become attached to the nuclear region. Mesosomes are believed to function in cell-wall synthesis and the division of nuclear material.

Cytoplasm and structures within cytoplasm. The cell material contained within the cytoplasmic membrane is divided into: (1) the *cytoplasmic area*, granular in appearance, which is rich in RNA; (2) the *chromatinic* or *nuclear area*, which is rich in DNA; and (3) the *fluid portion*, which contains dissolved nutrients and particulate matter called *inclusion bodies*.

Cytoplasmic area. Densely packed throughout the cytoplasmic area are RNA-protein particles called *ribosomes,* which are the sites of protein biosynthesis. Ribosomes are found in all cells, eucaryotic and procaryotic.

Nuclear area. The bacterial cell, unlike the cells of eucaryotic organisms, lacks discrete chromosomes, the mitotic apparatus used for cell division, a nucleolus, and a nuclear membrane. The nuclear material or DNA in a bacterial cell occupies a position near the center of the cell and is attached to the mesosome–cytoplasmic membrane system (Fig. 5-16). This material is the entire genetic apparatus, or *genome*, of the bacterium and consists of a single, circular, chromosome in which all genes are linked. The nuclear material of the bacterium is referred to as the *chromatin body, nucleoid,* or *bacterial chromosome.*

Cytoplasmic inclusions. Different kinds of chemical substances can accumulate and form granules and globules in the cytoplasm called *inclusion bodies.* For example, some species of sulfur bacteria accumulate large amounts of sulfur, which appear as globules within the cytoplasm (Fig.

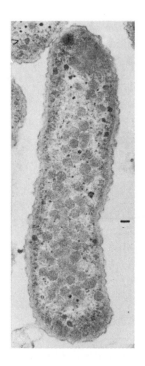

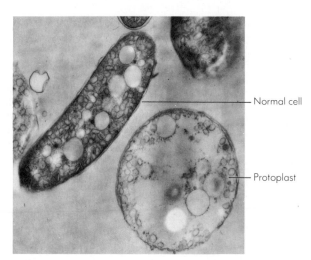

Figure 5-18. Electron micrograph of normal cell (top) and protoplast (bottom) of *Rhodospirillum rubrum* (× 65,000). *(Courtesy of E. S. Boatman and H. C. Douglas, Electron Microscopy, vol. 2, Fifth International Congress of Electron Microscopy (Philadelphia), Academic Press, New York, 1962.)*

Figure 5-17. Electron micrograph of a thin section of *Thiobacillus thioparus* showing granules or inclusion bodies which consist of sulfur. *(Courtesy of J. M. Shivley, G. L. Decker, and J. W. Greenawalt, J Bacteriol, **101**:620, 1970.)*

5-17). Other inclusion bodies found in bacteria are composed of polyphosphate, lipid, glycogen, or starch.

Protoplasts and spheroplasts

If the cell wall is removed from a bacterium, the remaining body enclosed by the fragile cytoplasmic membrane normally will burst because of the great difference between the concentration of dissolved substances inside and outside the cell. This is called *osmotic shock*. Experimental procedures have been devised to remove the cell wall from a bacterium while maintaining the viability of the bacterium. This viable structure, which amounts to the cytoplasmic contents surrounded by the cytoplasmic membrane, is called a *protoplast*. Protoplasts assume a spherical shape, since they possess no rigid outer cell wall (see Fig. 5-18). They may be characterized as follows: completely lacking a cell wall, nonmotile, spherical, unable to divide, unable to form new cell walls, and not susceptible to infection by *bacteriophages*, viruses that infect bacteria. In the case of gram-negative bacteria, which have multilayered cell walls, the removal of the peptidoglycan layer may leave some outer-layer material still attached to the cytoplasmic membrane. In these cases the cells are referred to as *spheroplasts*, because they are not completely cell-wall-free.

Spores

Certain species of bacteria produce *spores*, either external to the vegetative cell *(exospores)* or within the vegetative cell *(endospores)*. These are metabolically dormant bodies produced at a late stage of cell growth that, under appropriate conditions, germinate and produce the original kind of growing, or *vegetative*, cell. Spores are resistant to many physical and chemical agents.

Exosperes

Several species of bacteria produce external spores. *Streptomyces*, for example, produce a chain of spores (called *conidia*) which is borne at the end of a *hypha*, a vegetative filament (see Fig. 5-19). This process is similar to the spore-forming process in some fungi, as described in Chap. 8.

Endospores

The endospore is unique to bacteria. It is a thick-walled, highly refractive, and highly resistant body produced by all species of *Bacillus, Clostridium* and *Sporosarcina*.

Bacteria capable of producing endospores may grow and reproduce for many generations as vegetative cells. However, at some stage in the growth of a spore-forming bacterium there occurs, within the vegetative cytoplasm, synthesis of new protoplasm destined to become the spore.

The major steps in the process can be outlined as follows:

1 Realignment of DNA material into filaments and invagination of the cell membrane near one end of the cell, forming a structure called the *forespore*
2 Development of a series of layers which cover the forespore, namely, a *spore cortex* followed by a multilayered *spore coat*
3 Release of the free spore as the mother cell undergoes lysis

Figure 5-19. Spore production by *Streptomyces virido-chromogens*. This bacterium produces a chain of spores (called conidia) in the form of coils which develop at the ends of vegetative filaments called hyphae. These structures, conidia and hyphae, are characteristic of fungus structures (see Chap. 8). *(Courtesy of Mary P. Lechevalier.)*

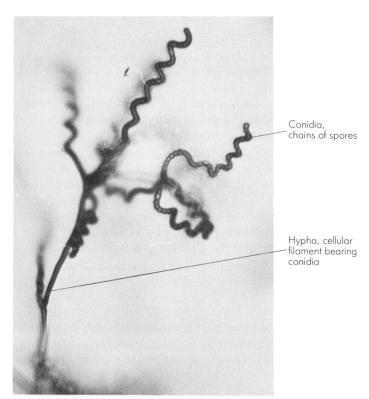

Conidia, chains of spores

Hypha, cellular filament bearing conidia

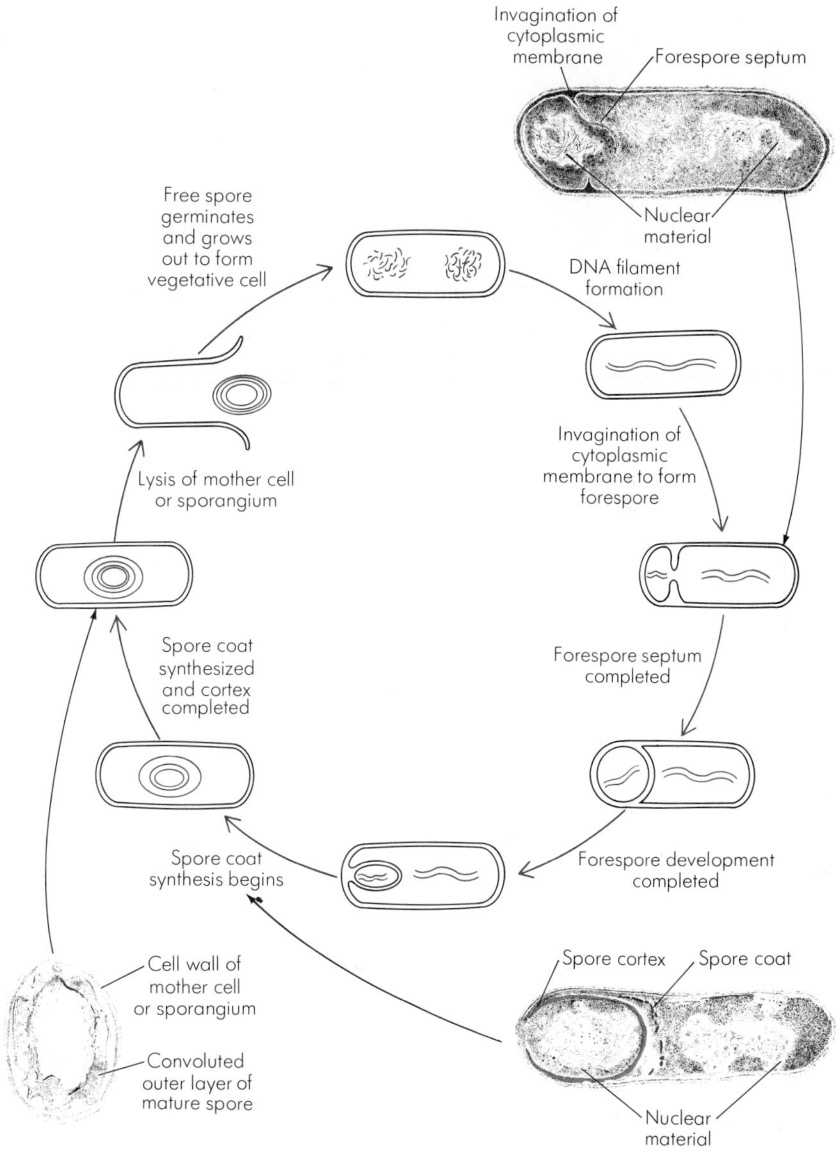

Figure 5-20. Structural changes in the bacterial cell during sporulation. *(Erwin F. Lessel, illustrator; redrawn, with modifications, from L.E. Hawker and A. H. Linton, Microorganisms—Function, Form and Environment, 2d ed., Univ. Park Press, Baltimore, 1979.)*

Some of the cytological changes associated with the process of endospore formation are shown in Fig. 5-20.

One of the unique features of bacterial endospores is their chemical makeup. All bacterial endospores contain large amounts of *dipicolinic acid*, a substance which is undetectable in vegetative cells. In fact, it accounts for 5 to 10 percent of the endospore's dry weight. Large amounts of calcium are also present in endospores, and it is believed that a Ca^{2+}–dipicolinic acid–peptidoglycan complex constitutes the cortical layer.

The mature spore has the capability of reproducing a vegetative cell. The steps involved in this process are (1) activation of the spore by heat or

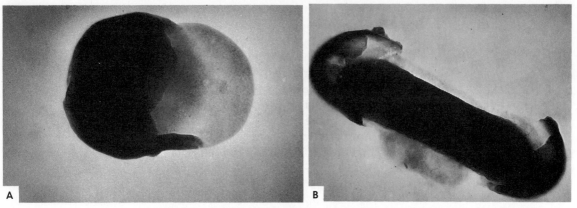

Figure 5-21. Germinating spore from culture of *Bacillus mycoides:* (A) grown 2 h at 35°C (× 44,000), (B) grown 1 3/4 h at 35°C (× 46,000). The two halves of the severed spore coat appear at the ends of the vegetative cell. *(SAB photos LS 203 and 204 courtesy of G. Knaysi, R. F. Baker, and J. Hillier, J. Bacteriol,* **53***: 525, 1947.)*

aging, (2) germination, and (3) growth into a vegetative cell. Figure 5-21 shows a germinating spore.

The location within cells and the size of endospores during their formation is not the same for all species. For example, some spores are *central*, i.e., they are formed in the middle of the cell; others are *terminal*, i.e., formed at the end; and still others are *subterminal*, i.e., formed a short distance from the end. The diameter of the spore may be larger or smaller than that of the vegetative cell. Hence endospore presence, location, and size are useful in characterizing and identifying bacteria. Some examples of differing locations and sizes of endospores within bacterial cells are shown in Fig. 5-22.

Compared with vegetative cells, spores are extremely resistant to adverse physical conditions such as high temperatures and desiccation and to chemical agents such as disinfectants. Some microbiologists have attributed this resistance to the impermeable spore coat, which in turn is associated with the dipicolinic acid–calcium–peptidoglycan complex. This unique survival ability will be discussed in Part VII, which deals with the inhibition and destruction of microorganisms.

Figure 5-22. Location, size, and shape of endospores in cells of various species of *Bacillus* and *Clostridium. (Erwin F. Lessel, illustrator.)*

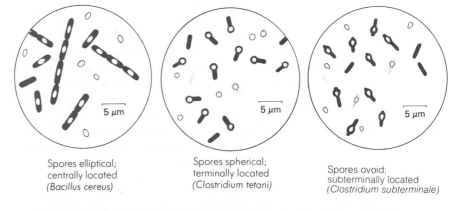

Spores elliptical;
centrally located
(*Bacillus cereus*)

Spores spherical;
terminally located
(*Clostridium tetani*)

Spores ovoid;
subterminally located
(*Clostridium subterminale*)

Summary and outlook

The bacterial cell has been examined in meticulous detail during recent years. It is not an exaggeration to say that it has been taken apart, structure by structure and molecule by molecule. One of the major objectives of this kind of research is to identify which life processes are associated with the various structural components of the bacterial cell. The structures which make up the bacterial cell have been characterized physically and chemically and a great deal has been learned about the function or role of each of these structures in the life processes of the cell. Developments in microscopy, particularly electron microscopy, have enabled researchers to identify increasingly smaller aggregates of materials within the bacterial cell. At the same time, new means of breaking the cell apart and of separating and isolating each of its components have been devised. Once the components are separated, they can be analyzed to determine their chemical makeup. This kind of precise information permits the microbiologist to identify specifically where life processes occur within the cell. It is an integration of knowledge of biochemical processes and of the structure and chemical composition of the morphological sites.

The knowledge gained from the study of the structure of bacterial cells and of the functions of these structures has considerable application to life processes of all other cells. There is considerable unity, or similarity, in the major biological phenomena among all organisms—from microbes to humans. Because of the ease of experimenting with and manipulating bacteria, they have become a favorite experimental model for investigating the mechanism of biological processes that occur in humans. Results gained from research with bacteria can halp to establish a blueprint of the details of processes in higher forms of life.

A more practical application of knowledge about the composition and structure of bacteria is the preparation of material for use in immunizing humans to certain diseases. Research has demonstrated that a certain part of the bacterial cell, for example the capsule or the cell wall, is the important material for injection into human subjects to build up their resistance.

Key terms

bacteriophage	gram-negative	peritrichous flagella
capsule	gram-positive	pili
conidia	macromolecule	protoplast
cytoplasmic membrane	mesosome	ribosomes
endospore	micrometer	sheath (bacterial)
exospore	osmosis	spheroplasts
fimbriae	osmotic shock	spore
flagella	peptidoglycan	

Questions

1 Describe the characteristic arrangements of cocci.
2 Describe some characteristic arrangements of bacilli that are typical for certain species.

3 What features would you note in describing a spiral-shaped bacterium?

4 What generalizations are appropriate in comparing the cell sizes of cocci, bacilli, and spiral-shaped bacteria?

5 Identify all the structures of a bacterium external to the cell wall.

6 In terms of cell-wall structure and composition, explain the difference between gram-negative and gram-positive bacteria.

7 Identify all the structures found within the cell wall of a bacterium.

8 What functions are associated with the following cell structures: capsule, cell wall, cytoplasmic membrane, nucleus, mesosome, spore, flagellum, and pili?

9 What is the significance of the peptidoglycan material in the structure of the bacterial cell?

10 Distinguish between a free spore, an exospore, and an endospore.

11 What bacterial structure contains large amounts of dipicolinic acid and calcium?

CULTIVATION, REPRODUCTION, AND GROWTH OF BACTERIA

To study bacteria in the laboratory one must be able to grow them in pure culture. To do this, one must understand the kinds of nutrients that the bacteria require, as well as the kinds of physical environment which provide optimum conditions for growth.

There is no one set of conditions satisfactory for the laboratory cultivation of all bacteria. Bacteria have great diversity in both nutritional and physical requirements. Some bacteria have simple nutrient requirements while other have very elaborate requirements. Some species grow at temperatures as low as 0°C (32°F), while others grow at temperatures reaching 75°C (167°F). Some require free oxygen, while others are inhibited by it. For this reason, conditions need to be adjusted so that they are favorable for the particular group of bacteria being studied.

Once satisfactory conditions for cultivation have been provided, it is possible to observe and measure the reproduction and growth of the bacteria, to determine the effect of a variety of conditions on both the reproduction and the growth of the bacteria, and to determine what changes the bacteria produce in the environment in which they grow.

In this chapter we will identify the conditions that are necessary for the cultivation of various kinds of bacteria, discuss the means by which they reproduce, describe their pattern of growth, and describe how their growth can be measured.

Nutritional requirements

All forms of life, from microorganisms to human beings, share certain nutritional requirements in terms of the chemicals necessary for their growth and normal functioning. The following observations illustrate this point and also reveal the great diversity of nutritional types found among bacteria.

1 All living organisms require a source of energy. Some forms of life, such as green plants, can utilize radiant energy or light and are designated *phototrophs*. Others, such as animals, rely on *oxidation* (the loss of electrons from an atom) of chemical compounds for their energy. These are called *chemotrophs*. All living organisms are either phototrophs or chemotrophs, and both these nutritional types exist among bacteria.

2 All living organisms require carbon in some form; all require at least small amounts of carbon dioxide, but most of them also require some organic carbon compounds, such as sugars and other carbohydrates. Plants use carbon dioxide and convert it to carbohydrates by photosynthesis. Many bacteria also require only carbon dioxide for their carbon source. Nutritionally speaking, all such organisms are *autotrophs*. If they obtain their energy from light, they are *photoautotrophs*, and if they obtain their energy by oxidizing chemical compounds, they are *chemoautotrophs*. Other bacteria are nutritionally similar to animals in that they are unable to use carbon dioxide as their sole carbon source and are dependent on autotrophs for the production of carbohydrates and other organic compounds, which they use as food. Microorganisms that require organic compounds as their carbon source are *heterotrophs*.

3 All living organisms require nitrogen in some form. Plants utilize nitrogen in the form of inorganic nitrogen salts such as potassium nitrate (KNO_3), whereas animals require organic nitrogen compounds such as proteins and their degradation products, namely peptides and certain amino acids. Bacteria are extremely versatile in this respect; some types use atmospheric nitrogen, some thrive on inorganic nitrogen compounds, and others require nitrogen from organic nitrogen compounds.

4 All living organisms require sulfur and phosphorus. The typical animal requirement for sulfur is satisfied by organic sulfur compounds; the typical plant requirement for sulfur is satisfied by inorganic sulfur compounds. Some bacteria require organic sulfur compounds, some are capable of utilizing inorganic sulfur compounds, and some can even use elemental sulfur. Phosphorus is usually supplied by phosphates, i.e., salts of phosphoric acid.

5 All living organisms require several metallic elements, namely, sodium, potassium, calcium, magnesium, manganese, iron, zinc, copper, and cobalt, for normal growth. Bacteria are no exception. The amounts required are usually mere traces and measured in parts per million.

6 All living organisms contain *vitamins* (specific organic compounds essential for growth) and vitaminlike compounds. Most vitamins function in forming substances which activate enzymes—substances that promote a chemical change. Animals, including human beings, must be furnished these substances in their diets. Bacteria present a variable pattern in this aspect of nutrition. Although all bacteria require vitamins in their normal metabolic processes, some are capable of manufacturing (synthesizing) their entire requirement of vitamins from other compounds in the medium. Others will not grow unless one or more vitamins are furnished preformed in the medium.

7 All living organisms require water for metabolic functions and growth. For bacteria, all nutrients must be in solution before they can enter the bacterium.

Nutritional types of bacteria

The major nutritional types of bacteria are shown in Table 6-1. Examples of media which will satisfy the nutritional requirements of a typical autotroph and a typical heterotroph are shown in Table 6-2.

Bacteriological media

Types of media

The great diversity of nutritional types among bacteria is matched by the large numbers of different media available for their cultivation. The kinds of media available may be grouped in several ways, as is shown in Table 6-3.

Media of the type shown in Table 6-2, chemically defined media, are not used for routine cultivation of bacteria. Instead, certain complex substances such as peptones, meat extract, and sometimes yeast extract are dissolved in water in various amounts, resulting in media which support the growth of a wide variety of bacteria and other microorganisms. Agar is included as a solidifying agent when a solid medium is desired. A description of these raw materials is given in Table 6-4. Examples of a relatively simple liquid and a solid medium that support the growth of

Table 6-1. Major nutritional types of bacteria

TYPE*	SOURCE OF ENERGY FOR GROWTH	SOURCE OF CARBON FOR GROWTH	EXAMPLE OF GENUS
Phototroph			
Autotroph	Light	CO_2	*Chromatium*
Heterotroph	Light	Organic compound	*Rhodopseudo-monas*
Chemotroph			
Autotroph	Oxidation of inorganic compound	CO_2	*Thiobacillus*
Heterotroph	Oxidation of organic compound	Organic compound	*Escherichia*

*These categories represent the major types. Many different nutritional types exist in each category.

Table 6-2. Composition of media for the growth of an autotroph and a heterotroph

A. AUTOTROPH (*Thiobacillus thiooxidans*)—a chemoautotroph	
Powdered sulfur	10 g
$(NH_4)_2SO_4$	0.4 g
KH_2PO_4	4.0 g
$CaCl_2$	0.25 g
$MgSO_4 \cdot 7H_2O$	0.5 g
$FeSO_4$	0.01 g
Water	1,000 ml
CO_2	

B. HETEROTROPH (*Escherichia coli*)—a chemoheterotroph	
$NH_4H_2PO_4$	1 g
Glucose	5 g
NaCl	5 g
$MgSO_4 \cdot 7H_2O$	0.2 g
K_2HPO_4	1 g
H_2O	1,000 ml

Table 6-3. Some ways of characterizing (classifying) bacteriological media

BASIS FOR CLASSIFICATION	DESIGNATION	EXAMPLES
Source of nutrients	Natural	Milk
	Artificial	Mixture of chemicals
Physical state	Solid-irreversible	Coagulated blood serum
	Solid-reversible	Nutrient agar
	Semisolid	"Soft agar"
	Liquid	Nutrient broth
Identification of chemical constituents	Complex (chemical composition unknown)	Nutrient agar (a mixture of complex organic materials solidified with agar)
	Chemically defined or synthetic (chemical composition known)	Ammonium sulfate, glucose, salts medium
Support of growth of *fastidious* bacteria, i.e., species with elaborate nutritional requirements	Enriched media	Heart infusion broth
Difference of growth; several types grow but one type will have a distinctive appearance	Differential media	Eosin-methylene blue agar
Selective growth, i.e., inhibition of one type without inhibition of another	Selective media	Deoxycholate agar
Quantitative measurement of vitamins or antibiotics	Assay media	Vitamin B_{12} assay medium

many common heterotrophs are nutrient broth and nutrient agar (see Table 6-5). The addition of yeast extract (about 5 g/liter) to each of these formulas improves the nutritional quality of the medium, since yeast extract contains several of the B vitamins and other growth-promoting substances.

Table 6-4. Characteristics of several complex materials used as ingredients of media

RAW MATERIAL	CHARACTERISTIC	NUTRITIONAL VALUE
Beef extract	An aqueous extract of lean beef tissue concentrated to a paste	Contains the water-soluble substances of animal tissue, which include carbohydrates, organic nitrogen compounds, water-soluble vitamins, and salts
Peptone	The product resulting from the digestion of proteinaceous materials, e.g., meat, casein, and gelatin; digestion of the protein material is accomplished with acids or enzymes; many different peptones (depending upon the protein used and the method of digestion) are available for use in bacteriological media; peptones differ in their ability to support growth of bacteria	Principal source of organic nitrogen; may also contain some vitamins and sometimes carbohydrates, depending upon the kind of proteinaceous material digested
Agar	A complex carbohydrate obtained from certain marine algae; processed to remove extraneous substances	Used as a solidification agent for media; agar dissolved in aqueous solutions gels when the temperature is reduced below 45°C; agar not a source of nutrient to the bacteria
Yeast extract	An aqueous extract of yeast cells, commercially available as a powder	A very rich source of the B vitamins; also contains organic nitrogen and carbon compounds

Preparation of media

Natural media, e.g., skimmed milk, pose no problem of preparation for use as media; they are merely dispensed into suitable containers such as test tubes or flasks and sterilized before use. Media of the nutrient-broth or agar type are prepared by dissolving the required individual ingredients or, more conveniently, by adding water to a commercial powdered-medium product which contains all the necessary nutrients. Practically all media are available commercially in powdered form, as well as ready-made to use in petri dishes, tubes, or bottles.

The preparation of bacteriological media other than natural media involves the following steps:

1 Each ingredient, or the complete dehydrated medium, is dissolved in the appropriate volume of distilled water.
2 The pH (degree of acidity or alkalinity) of the fluid medium is determined and adjusted (by the addition of either an alkaline or acid solution) to the value which is optimum for the growth of the kind of bacteria to be cultivated. The pH is determined by using pH indicators (see Table 6-6) or a pH meter.
3 The medium is dispensed into suitable containers such as tubes, flasks, or bottles and closed with cotton plugs or plastic or metal caps prior to sterilization.
4 The medium is sterilized, generally by autoclaving, a process using heat under steam pressure.

Table 6-5. Composition of nutrient broth and nutrient agar

Nutrient broth	
Beef extract	3 g
Peptone	5 g
Water	1,000 ml
Nutrient agar	
Beef extract	3 g
Peptone	5 g
Agar	15 g
Water	1,000 ml

Table 6-6. Characteristics of some pH indicators used in microbiology

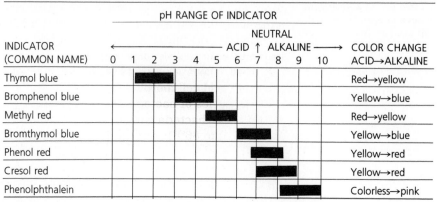

INDICATOR (COMMON NAME)	pH RANGE OF INDICATOR	COLOR CHANGE ACID→ALKALINE
Thymol blue	1–3	Red→yellow
Bromphenol blue	3–4.5	Yellow→blue
Methyl red	4.5–6	Red→yellow
Bromthymol blue	6–7.5	Yellow→blue
Phenol red	7–8.5	Yellow→red
Cresol red	7.5–9	Yellow→red
Phenolphthalein	8.5–10	Colorless→pink

*The term pH is a symbol for the degree of acidity or alkalinity of a solution; $pH = \log(1/[H^+])$ where $[H^+]$ represents the hydrogen-ion concentration. The pH of distilled water is 7.0 (neutral); vinegar, 2.25; tomato juice, 4.2; milk, 6.6; sodium bicarbonate (0.1 *N*), 8.4; milk of magnesia, 10.5.

Table 6-7. Physical conditions affecting bacterial growth

PHYSICAL CONDITIONS	TYPES OF BACTERIA (PHYSIOLOGICAL GROUPS)	CULTURAL (INCUBATION) CONDITIONS
Temperature (range of growth): minimum and maximum; optimum somewhere within range, depending on species	Psychrophiles	0–30°C (19.4–86°F)
	Mesophiles	25–40°C (77–104°F)
	Thermophiles:	
	Facultative (optional) thermophiles	25–55°C (77–131°F)
	Obligate thermophiles	45–75°C (113–167°F)
Gaseous requirements	Aerobes	Grow only in presence of free oxygen
	Anaerobes	Grow only in absence of free oxygen
	Facultative anaerobes	Grow in presence or absence of free oxygen
	Microaerophiles	Grow in presence of minute quantities of free oxygen
Acidity or alkalinity (pH)	Most bacteria associated with animal and plant life	Optimum pH 6.5–7.5
	Some exotic species	Minimum pH 0.5; Maximum pH 9.5
Light	Photosynthetic (autotrophs and heterotrophs)	Source of light
Salinity	Halophiles (obligate halophiles)	High concentration of salt, 10–15% NaCl

Physical conditions required for growth

In addition to supplying the proper nutrients for the cultivation of bacteria, it is also necessary to provide the physical conditions which permit optimum growth. Just as bacteria vary greatly in their nutritional requirements, so do they exhibit diverse responses to the physical conditions in their environment. The successful cultivation of various types of bacteria requires a combination of the proper nutrients and the proper physical environment.

Temperature   Since all the processes of growth are dependent on chemical reactions and since the rates of these reactions are influenced by temperature, the pattern of bacterial growth can be profoundly influenced by temperature. Temperature also affects the rate of growth and the total amount of growth of the organism. Variations in temperature may also alter certain metabolic processes and cell morphology.

Each species of bacteria grows at temperatures within a certain range. On this basis bacteria can be classified as: *psychrophiles*, which grow at 0 to 30°C (32 to 86°F); *mesophiles*, which grow at 25 to 40°C (77 to 104°F); and *thermophiles*, which grow at 50°C (122°F) or higher.

The growth responses of these groups to various temperatures is shown in Fig. 6-1 and Table 6-7. The temperature of incubation which allows for most rapid growth during a short period of time (12 to 24 h) is known as the *optimum-growth temperature*.

Gaseous atmosphere   The principal gases that affect bacterial growth are oxygen and carbon dioxide. Bacteria display a wide variety of responses to free oxygen and it is convenient to divide them into four groups on this basis: *aerobic* (oxygen-requiring organisms), *anaerobic* (grow in absence of molecular oxygen), *facultatively anaerobic* (grow under either aerobic or anaerobic conditions), and *microaerophilic* (grow best in small amounts of atmospheric oxygen). The gaseous requirements of these groups are shown in Table 6-7. Figure 6-2 shows diagrammatically how these four groups can be distin-

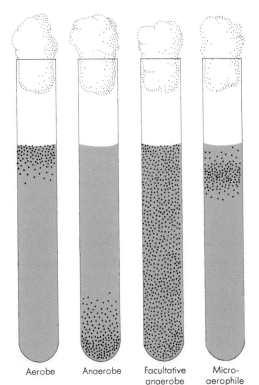

Aerobe Anaerobe Facultative anaerobe Micro-aerophile

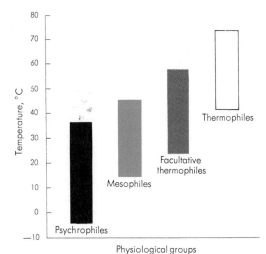

Figure 6-1. Approximate temperature range for growth of various bacteria. Some thermophiles in hot springs grow at 80°C (176°F).

Figure 6-2. Schematic illustration of the growth of bacteria in deep agar tubes, showing differences in response to atmospheric oxygen.

BACTERIA	pH RANGE FOR GROWTH		
	Lower Limit	Optimum	Upper Limit
Thiobacillus thiooxidans	0.5	2.0–3.5	6.0
Acetobacter aceti	4.0–4.5	5.4–6.3	7.0–8.0
Staphylococcus aureus	4.2	7.0–7.5	9.3
Azotobacter spp.	5.5	7.0–7.5	8.5
Chlorobium limicola	6.0	6.8	7.0
Thermus aquaticus	6.0	7.5–7.8	9.5

Source: Data from *Bergey's Manual of Determinative Bacteriology*, 8th ed., Williams & Wilkins, Baltimore, 1974.

guished by their patterns of growth in an agar medium contained in test tubes that are approximately two-thirds filled with the medium. When the medium is dispensed in this manner, (the samples thus prepared are called *agar-deeps*), diffusion of oxygen from the atmosphere through the medium is restricted, so that the bottom region of the medium is devoid of oxygen, that is, anaerobic.

Some bacteria are not only anaerobic but are also very sensitive to oxygen; exposure to oxygen will kill them. An anaerobic environment can be provided by the methods shown in Fig. 6-3.

Acidity or alkalinity (pH)

For most bacteria the optimum pH for growth lies between 6.5 and 7.5. However, some species can grow in very acid or very alkaline conditions, as is shown in Table 6-8. For most species, the minimum and maximum pH values are between 4 and 9.

When bacteria are cultivated in a medium originally adjusted to a given pH, for example, 7, it is very likely that this pH will change as a result of acidic or basic compounds produced during growth. This shift in pH may be so great that continued growth of the organism is inhibited. Shifts in pH can be prevented by incorporating a *buffer* into the medium. Buffers are compounds or pairs of compounds that can resist changes in pH. A combination of phosphate salts, such as KH_2PO_4 and K_2HPO_4, is widely employed in bacteriological media for this purpose. Some of the nutritional ingredients of the medium, such as peptones, also possess buffering capacity. The extent to which a medium should or may be buffered depends on its intended purpose and is limited by the buffering capacity of the compounds used.

Miscellaneous conditions

Temperature, the gaseous environment, and pH are the major physical factors to be taken into consideration in establishing optimum conditions for the growth of most species of bacteria (Table 6-7). But some groups of bacteria have additional requirements. For example, *photoautotrophic* (photosynthetic) organisms must be provided with a source of illumination, since light is their source of energy. Bacterial growth may also be influenced by conditions of either *osmotic pressure* (the force or tension built up when water diffuses through a membrane) or *hydrostatic pressure* (fluid tension). Certain bacteria, called *halophilic* bacteria and found in brines, containers of salt, salted foods, ocean water, and the great salt lakes, grow only when the medium contains a very high concentration of salt. Seawater contains 3.5 percent sodium chloride; in salt lakes, the concentration of NaCl may reach 25 percent. Microorganisms that require NaCl for growth are called *obligate halophiles*—they will not grow unless the salt concentration is high; those that will grow in a solution of NaCl but do not require it are called *facultative halophiles*—they grow in environments of high or low salt concentration. This represents a response to osmotic pressure. Bacteria have been isolated from the deepest ocean trenches, where the hydrostatic pressure is measured in tons per square inch.

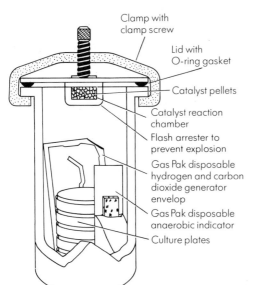

Clamp with clamp screw

Lid with O-ring gasket

Catalyst pellets

Catalyst reaction chamber

Flash arrester to prevent explosion

Gas Pak disposable hydrogen and carbon dioxide generator envelop

Gas Pak disposable anaerobic indicator

Culture plates

A

Figure 6-3. (A) The GasPak Anaerobic System. The two basic components of the GasPak Anaerobic System are the GasPak hydrogen + carbon dioxide generator envelope and a room-temperature palladium catalyst. Water is added to the GasPak envelope and hydrogen is produced. The hydrogen reacts with oxygen on the surface of the catalyst to form water and, subsequently, to produce an anaerobic condition. Carbon dioxide is also generated by the GasPak envelope in sufficient volume to support growth of fastidious anaerobic microorganisms which sometimes fail to grow or grow poorly in the absence of this gas. An anaerobic indicator strip (a pad saturated with methylene blue solution) changes from blue to colorless in the absence of oxygen. *(Courtesy of Bioquest, Division of Becton, Dickinson and Company.)* (B) The GasPak Anaerobic System with innoculated petri dishes, the GasPak generator envelope, and the anaerobic indicator strip. *(Courtesy of Bioquest, Division of Becton, Dickinson and Company.)* (C) An anaerobic workbench which provides a complete anaerobic environment in which one can perform microbiological operations as well as incubate inoculated media. *(Courtesy of The Germfree Laboratories, Inc.)*

B

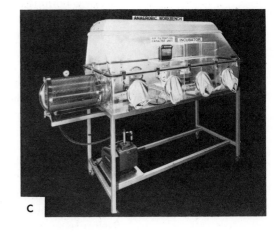

C

Choice of media and conditions of incubation

From this brief description of the nutritional characteristics of bacteria, you should recognize two essential steps for their cultivation in the laboratory: (1) inoculation (or planting) of bacteria on a medium of suitable nutritional content, and (2) incubation of the inoculated medium under appropriate physical conditions.

In order to make the proper selection of media and physical conditions, you must be able to answer such questions as these:

1 Are the bacteria to be isolated aerobes or anaerobes?
2 Does the specimen contain autotrophic and heterotrophic bacteria, and if so, are both types to be cultivated?
3 Does the specimen contain thermophilic, mesophilic, or psychrophilic organisms?

You may ask the question: How can these things be known beforehand? Some knowledge about groups of bacteria, experience, and some rational judgments make answers possible. For example, autotrophic bacteria would hardly be expected to be of significance in specimens taken from an animal body. The fact that none of the disease-producing bacteria are obligate thermophiles would automatically exclude this type of incubation in a laboratory procedure for the isolation of pathogens. When examining a specimen for a particular bacterial species, such as the typhoid bacterium, prior knowledge of its characteristics will make possible selection of the proper medium and incubation conditions for its culture. For example, the typhoid bacterium is heterotrophic, aerobic, and mesophilic. Its nutritional requirements are relatively simple. On the other hand, if the problem were to grow all the bacteria present in a specimen of milk, provisions would need to be made to satisfy a wide range of requirements. Cultural conditions would have to provide for aerobes, anaerobes, psychrophiles, mesophiles, and thermophiles, as well as heterotrophs, both *fastidious* (having complex nutritional requirements) and *nonfastidious*. Such an examination would require the inoculation of several different media as well as incubation under varied conditions.

Reproduction of bacteria

When bacteria are inoculated into a suitable medium and incubated under optimum conditions for growth, a tremendous increase in the number of cells occurs within a relatively short time. With some species the maximum population (the largest crop of cells obtainable) develops within 24 h; populations may reach 10 to 15 billion bacterial cells per milliliter. An asexual cell-division process causes this multiplication in numbers.

Cell division (transverse binary fission)

The most common process of reproduction in the usual growth cycle of bacterial populations is *transverse binary fission*. Transverse binary fission is an asexual reproductive process; a single cell divides into two cells, called *daughter cells*, after the development of a transverse cell wall.

Division of cells by fission is common to all actively growing cells in plants and animals. However, in multicellular plants and animals the asexual division of cells results in the growth of the individual plant or animal. With bacteria, the process results in two new organisms; each of the two new bacteria can repeat the process, as shown in Fig. 6-4.

Other processes

Most of the bacteria that are associated with our bodies and those that you will study in the general microbiology laboratory reproduce almost exclusively by transverse binary fission. However, some species can reproduce by additional processes, including production of reproductive spores; fragmentation of filamentous growth, with each fragment giving rise to growth; and budding. Descriptions of such species are given in Chap. 7.

New-cell formation

The microscopic dimensions of a bacterial cell make it difficult to observe directly all the intricate intracellular changes that take place in transverse binary fission. Research microbiologists are interested in determining precisely what happens in the parent cell as it evolves to the stage at which it divides into two new cells. Much has been learned about the process.

Highly specialized electron microscope techniques such as the examination of ultrathin stained sections have revealed some of the cytological changes (see Fig. 6-5). It has become apparent that the synthesis of substances by a single cell at a certain stage of the life cycle cannot be

Figure 6-5. Septum formation in *Escherichia coli* as seen by electron microscopy of thin sections at different stages (A, B, and C) of division. Note the nuclear material (white area) partitioned to each half of the cell. *(Courtesy of I. D. J. Burdett and R. G. E. Murray, J Bacteriol* **119***:1039, 1974.)*

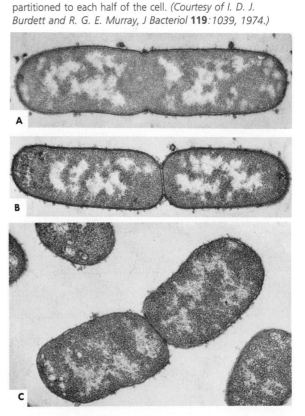

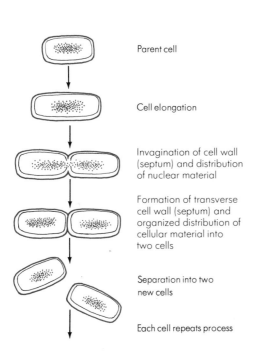

Parent cell

Cell elongation

Invagination of cell wall (septum) and distribution of nuclear material

Formation of transverse cell wall (septum) and organized distribution of cellular material into two cells

Separation into two new cells

Each cell repeats process

Figure 6-4. Bacterial multiplication by transverse binary fission (schematic illustration).

determined by currently known methods. The amounts produced are too small. To overcome this difficulty, techniques have been developed by which the researcher can, for a short time, maintain all the cells in a population at the same stage of their life cycle. Such a population is designated a *synchronous culture*. The fact that all cells are at the same stage of development allows one to measure the amounts of new-cell substances produced by a mass of these cells and then to determine by chemical analysis and calculation how much of each substance was made by a single cell. The experimental condition of synchrony is of short duration; after several divisions the cells revert to random reproduction behavior.

Results of studies on the process of cell division have revealed the following:

1 There is an increase in the amount of nuclear material, which separates into two units—one for each new daughter cell.
2 The cell wall and the cell membrane grow out and the cell membrane grows (extends) into the cytoplasm at a point halfway to the long axis of the cell. Two layers of cell-wall material are synthesized at this interface.
3 Mesosome formation becomes more pronounced. The mesosomes are associated with the formation of the septum (dividing cell wall) and also establish a linkage with the nuclear region.

Growth

Concept of bacterial growth

The term *growth* as commonly applied to bacteria and other microorganisms usually refers to changes in the crop of cells (increase in the total mass of cells) rather than to changes in an individual organism. More frequently than not, the inoculum contains thousands of organisms; growth denotes the increase in number or mass beyond that present in the original inoculum. During the phase of *balanced growth*, which we will describe later, the increase in bacterial mass is directly related (proportional) to increases in other cellular constituents, such as DNA, RNA, and protein. Therefore it is possible to develop a measurement for growth in a variety of ways.

Growth rate and generation time

As we have already indicated, the characteristic mode of bacterial reproduction is transverse binary fission; one cell divides, producing two cells. Thus, if we start with a single bacterium, the increase in population is by geometric progression:

$$1 \longrightarrow 2 \longrightarrow 2^2 \longrightarrow 2^3 \longrightarrow 2^4 \longrightarrow 2^5 \dots 2^n$$

or, in everyday arithmetic,

$$1 \longrightarrow 2 \longrightarrow 4 \longrightarrow 8 \longrightarrow 16 \longrightarrow 32 \dots$$

(The 2^n is merely an algebraic shorthand standing for the ultimate number of the progression—in this case, the eventual maximum number of cells in the population.) The time interval required for the cell to divide—or, what is the same thing, for the population to double—is known as the *generation time*. Not all species of bacteria have the same generation time. For some, such as *Escherichia coli* (a bacterium which is commonly present in the intestinal tract of humans and other animals), it may be as short as 15 to 20 min; for others it may be many hours. Nor is the generation time the same for a particular species of bacterium under all conditions. The generation time is strongly dependent on the adequacy of nutrients in the medium and on the appropriateness of the physical conditions.

It is possible to determine the generation time of bacteria by direct microscopic examination. The more practical and common method, however, is to inoculate a medium with a known number of bacteria, allow them to grow under optimum conditions, and determine the population at periodic intervals. The experimental data required to calculate the generation time are: (1) the number of bacteria present at the beginning, that is, in the inoculum; (2) the number of bacteria present at the end of a given time interval; and (3) the time interval.

Normal growth cycle
(growth curve)

Assume that a single bacterium is inoculated into a medium and multiplies at a constant rate. If we used the theoretical number of bacteria which should be present at various intervals of time and then plotted the data in two ways (logarithm of number of bacteria versus time and arithmetic number of bacteria versus time), we would obtain the curves shown in Fig. 6-6. However, this does not represent the normal pattern of growth but rather one selected portion of the normal-growth curve, namely, the *exponential*, or *logarithmic, phase* of growth (commonly referred to as the *log phase*). Here, the population increases regularly, doubling at regular time intervals (the generation time) during incubation; this growth phase is also referred to as *balanced growth*. In reality, when we inoculate a fresh medium with a given number of cells, determine the bacterial population intermittently during an incubation period of 24 h (more or less), and plot the logarithms of the number of cells versus time, we obtain a curve of the type illustrated in Fig. 6-7. From this it can be seen that there is an initial

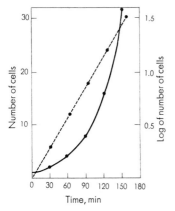

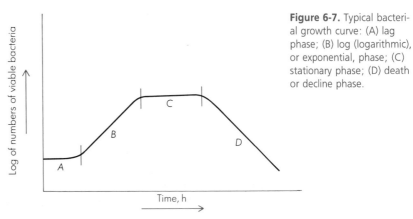

Figure 6-6. Hypothetical bacterial growth curve obtained by assuming that one bacterial cell is inoculated into a medium and divisions occur regularly at 30-min intervals (generation time). The dashed line shows the logarithm of the number of bacteria versus time; the solid line shows the arithmetic number of bacteria versus time.

Figure 6-7. Typical bacterial growth curve: (A) lag phase; (B) log (logarithmic), or exponential, phase; (C) stationary phase; (D) death or decline phase.

period of what appears to be no growth (*lag phase*) followed by a period of rapid growth (*log phase*), then by a leveling off (*stationary phase*), and finally by a decline in the viable population (*death* or *decline phase*). Between each of these phases there is a transitional period (curved portion). This represents the time that elapses before all the cells enter the new phase. The generation time of a bacterium can be calculated by the following formula:

$$G = \frac{t}{3.3 \log (b/B)}$$

where G = generation time

     t = time interval between measurement of number of cells in the population at one moment in the log phase (B) and then again at a later point in time (b)

     B = initial population

     b = population after time t

     $\log = \log_{10}$

     $3.3 = \log_2$-to-$\log_{10}$ conversion factor

Additional characteristics associated with each of the four phases of growth are summarized in Table 6-9.

As we stated previously, it is possible to manipulate the growth of a culture in several ways. For example, it is experimentally possible to keep all the cells in exactly the same stage of growth (*synchronous growth*) for a prolonged period of time. It is also possible to prolong the log phase of growth by providing a continuous supply of nutrients with simultaneous removal of the old (used) medium. This is called *continuous culture*.

Quantitative measurement of growth

We have seen that the term *growth* as used with bacteria refers to the changes in total population and not to the changes in an individual organism. Furthermore, under conditions of *balanced growth* there is an orderly increase in all cellular constituents. Consequently, growth can be determined not only by measuring the number of cells but also by measuring the quantities of various cellular constituents (RNA, DNA, protein) as well as certain metabolic products.

Many laboratory techniques are available for measuring bacterial growth; the apparatus ranges from simple equipment such as a slide with a

Table 6-9. Some characteristics of bacterial growth in each phase of the growth cycle

GROWTH PHASE	CHARACTERISTICS
Lag	No increase in population
	Cells undergo change in chemical composition and increase in size; increase of intracellular substances
Logarithmic or exponential	Cells double (divide) at constant rate
	Mass double at same rate
	Metabolic activities constant
	Condition of *balanced growth*
Stationary	Accumulation of toxic products and/or exhaustion of nutrients
	Some cells die while others grow and divide
	Number of viable cells forms a plateau or levels off
Decline or death	Cells die more rapidly than new cells are produced
	Death rate accelerates, becoming exponential
	Depending on species, all cells die within days to months

Table 6-10. Summary of methods for measuring bacterial growth

METHOD	SOME APPLICATIONS
Microscopic count	Enumeration of bacteria in milk and vaccines
Plate count	Enumeration of bacteria in milk, water, foods, soil, cultures, etc.
Membrane or molecular filter	Same as plate count
Turbidimetric measurement	Microbiological assays, estimation of cell crop in broth, cultures, or aqueous suspensions
Nitrogen determination	Measurement of cell crop from heavy culture suspensions to be used for research in metabolism
Weight determination	Same as for nitrogen determination
Measurement of biochemical activity	Microbiological assays

stained smear from a measured volume (microscopic count) to sophisticated electronic devices that measure photoluminescence of compounds produced by some species of bacteria. Some of these techniques and their applications are summarized in Table 6-10.

Significance of growth measurements

Before we can evaluate or interpret growth responses of bacteria in various media or under different conditions, we must be able to express growth in quantitative terms. In microbiology the term *growth* is interpreted in several ways. For example, we may judge a certain set of cultural conditions equally good because the bacteria grow rapidly in early stages, but the final total cell crop may not be as large as it would be under another set of conditions in which growth proceeds initially at a slower rate. Such a situation is shown schematically in Fig. 6-8, where the growth of the same bacterial species is compared in two different media. Growth occurs in both media. However, if we measured growth at time A, we would conclude that growth is best in medium II; measured at time B, growth would be equally good in both media; and at time C, growth would be better in medium I. If we were interested in rapid early growth of the organism, we would select medium II; if we were primarily interested in a large cell crop, we would

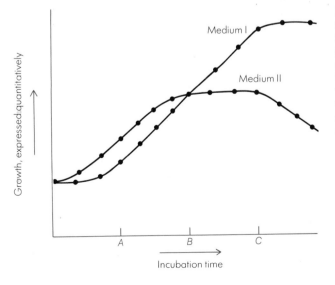

Figure 6-8. Quantitative measurement of growth is significant for interpretation of various growth responses. Hypothetical growth response of the same bacterium in media of two different compositions, I and II. Compare the cell crops, or amount of growth, at times A, B, and C in the two different media.

select medium I. Thus, the term *growth*, particularly in assessing the adequacy of a particular medium to support growth, has several interpretations.

Summary and outlook

Bacteria, as a group, live and grow under a wide range of conditions. Some species thrive in deposits in the deepest trenches of the ocean, others in arctic soil, still others in hot springs. To isolate and cultivate these bacteria under laboratory conditions it is necessary to provide the nutrients and the physical conditions that will satisfy the requirements of the particular type of bacteria being studied. Accordingly, a large variety of media is used in microbiology, in combination with various physical conditions for incubation.

Knowledge that has been gained through studies of the requirements for bacterial growth is oftentimes applicable to growth processes in higher forms of life. For example, several of the B vitamins that are now known to be required by humans were first discovered in studies on bacterial nutrition.

Bacterial growth can be manipulated and controlled in various ways that are convenient to the researcher who is investigating fundamental phenomena associated with growth processes. Cells can be studied at all stages of growth, microscopically and chemically, in an attempt to correlate the substances synthesized during growth with the structures that appear within the cell. The results obtained with bacterial cells provide the clues or guidelines for similar studies of cells in other forms of life, including humans.

Key terms

aerobes	halophiles	psychrophiles
anaerobes	heterotrophs	stationary phase
autotrophs	lag phase	synchronous culture
chemotrophs	log phase	thermophiles
death and decline phase	mesophiles	transverse binary fission
facultative anaerobe	osmotic pressure	
generation time	phototrophs	

Questions

1 List the nutritional requirements, in terms of chemical substances, that are needed for all forms of life for growth.
2 What are the ingredients of nutrient broth? What nutrients does each ingredient provide?
3 Compare the spectrum of nutritional requirements for bacteria as a group with (*a*) the nutritional requirements of plants and (*b*) the nutritional requirements of animals.
4 Outline a suitable laboratory procedure, including the kind of medium and the conditions of incubation, for the isolation of a heterotrophic, thermophilic, anaerobic bacterium.

5 Name and describe the predominant method of bacterial reproduction.

6 How does the term *growth* as used in microbiology differ from the same term when applied to plants and animals?

7 Is the generation time of all bacteria the same? Can the generation time for a particular species vary? Explain.

8 Draw a bacterial-growth curve, label all parts, and describe the characteristics of the culture in each phase.

9 Bacterial growth can be measured in several ways. Name four of these methods and compare them in terms of (*a*) practical applications, (*b*) advantages, and (*c*) limitations of use.

7

THE MAJOR GROUPS OF BACTERIA

In this chapter we present a brief description of the major groups of bacteria. Our purpose is to provide you with a concept of the range of characteristics of the bacterial world. More detailed descriptions of many species in each group, particularly those that are important as causative agents of disease, are presented in later chapters. In addition you will find descriptions of the more commonly encountered genera in Appendix B.

Bergey's Manual of Determinative Bacteriology

The standard reference for the classification and identification of bacteria is *Bergey's Manual of Determinative Bacteriology*, now in its 8th edition. The 8th edition represents a radical departure in the scheme of classification from all earlier editions. For example, in the first seven editions all species were allocated to orders, with intermediate taxa such as families, tribes, and genera. The impression was given that the bacterial world could be classified by a scheme which attempted to reveal natural or evolutionary relationships. However, microbiologists and particularly bacteriologists have recognized for a long time that these neat hierarchical schemes are hypothetical at best. Adequate knowledge of all bacteria is not available to permit the development of a natural classification scheme that reflects the actual evolutionary relationships. Accordingly, the 8th edition of *Bergey's Manual* makes no attempt to provide a complete hierarchy. Instead, all bacteria are divided into 19 groups (*Bergey's Manual* identifies them as parts). Each group is based upon several readily determined criteria. As more knowledge is obtained about each group, it will be possible to reconsider the relationships among them. However, this requires research and time to assimilate the new data. To quote from *Bergey's Manual:* *"Haste is unwise; all previous classifications seem to have suffered infinite rearrangement due to insufficient information."*

It is important to point out that there is no counterpart of *Bergey's Manual* for fungi, protozoa, or algae. For each of these other categories of microorganisms there is more than one authoritative publication or classification; there is no *one* agreed-upon international system of classification, together with a description of all the recognized species.

Kingdom Procaryotae

Bergey's Manual, 8th edition, recognizes *Procaryotae* as the kingdom to which bacteria belong. This designation stems from the results of research performed during the last several decades on cell structure. The recognition of procaryotic cell organization was instrumental in establishing the distinction between bacteria and other microorganisms. The overriding characteristic common to the world of bacteria is their procaryotic structure. (The properties of a procaryotic cell were described in Chap. 2.) Procaryotae is a kingdom defined by cellular characteristics and not organismal properties. The kingdom is divided into two divisions, the *Cyanobacteria* and the *Bacteria.*

The division Cyanobacteria

The Cyanobacteria, prior to the 8th edition of *Bergey's Manual*, were designated *blue-green algae.* The fact that they are procaryotes identifies them as members of the kingdom Procaryotae. However, they are phototrophic procaryotes that perform photosynthesis in a manner similar to that used by green plants and unlike that of other photosynthetic bacteria: the process is aerobic and the electron donor is water. Their photopigments include chlorophyll and phycobiliproteins. Some occur singly and others consist of straight or branched chains of cells or filaments. Reproduction may be by binary fission, by multiple fission, or by serial release of exospores. Filamentous forms may reproduce by fragmentation of the filament or by terminal release of short, motile chains of cells.

The division Bacteria

Microorganisms in this division are unicellular and sometimes show simple arrangements. Multiplication is characteristically by binary fission. In some, motility is due to the presence of flagella; and in others, it is due to other mechanisms. Endospores are produced by some species. With few exceptions the individual cells are surrounded by a rigid cell wall which is made up of peptidoglycans.

Those species capable of photosynthesis perform the process differently than do the cyanobacteria. In the division Bacteria the process is anaerobic and the electron donor is a substance other than water. There is also a difference in the kinds of photosynthetic pigments contained in the cells of these two divisions, that is, Cyanobacteria and Bacteria. For bacteria the common photosynthetic pigments are bacteriochlorophylls.

Criteria for establishing groups of bacteria

As we stated earlier, the bacteria recognized in the current edition of *Bergey's Manual* are arranged in 19 groups based on a few readily determined criteria.

In characterizing the various groups it is important to understand that all characteristics are not equally important for all groups. For example, the gram-stain reaction is important for rods and cocci but it is not a distinguishing characteristic for spirochetes. The presence and arrangement of flagella are important for some groups but not others. For some groups, biochemical characteristics are more significant than morphological characteristics. Accordingly, in the characterization of the various groups of bacteria, you should not expect to see the same characteristics described and used uniformly for each group. Instead, you will see that each group is characterized by those features which are most distinctive for that group, i.e., those that readily set the group apart from others. For a detailed characterization of species which belong to each group it is necessary to consult *Bergey's Manual*. In addition, *Bergey's Manual* can be used to help identify newly discovered bacteria.

Major groups of bacteria

In the following pages we will describe the major distinctive characteristics of each of the groups of bacteria as they are designated in *Bergey's Manual*. Each of these groups contains various taxonomic categories; some begin with orders, others with families; however, all conclude with genera and species. The characterization of all currently recognized (or accepted) species of bacteria is contained in *Bergey's Manual*.

Group 1. The phototrophic bacteria

The phototrophic bacteria are morphologically different organisms all of which contain a chlorophyll-like pigment, *bacteriochlorophyll*. They are photosynthetic. In these bacteria photosynthesis occurs under anaerobic conditions, and the electron donor is not water. These bacteria occur predominantly in aquatic environments.

Selected characteristics

Cell shapes: spherical, rod, vibrio, or spiral (see Fig. 7-1)
Gram-negative
Multiplication by binary fission

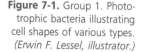

Figure 7-1. Group 1. Phototrophic bacteria illustrating cell shapes of various types. *(Erwin F. Lessel, illustrator.)*

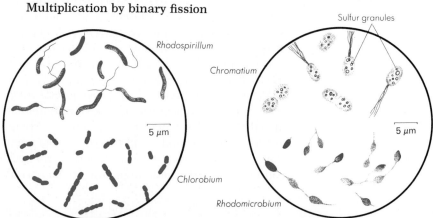

Motile by flagella or nonmotile
Photosynthetic, process occurs under anaerobic conditions, and oxygen
 is not produced
Bacteriochlorophyll, a photosynthetic pigment, present in all cells
Pigmented: purple-violet, purple, red, orange-brown, green
Habitat: aquatic environments

Group 2.
The gliding bacteria

This group is represented by some unusual morphological types. One kind, the *Myxobacterales*, also referred to as myxobacters, produce what are

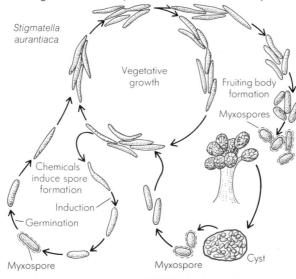

*Stigmatella
aurantiaca*

Vegetative
growth

Fruiting body
formation

Myxospores

Chemicals
induce spore
formation

Induction

Germination

Myxospore

Myxospore

Cyst

Figure 7-2. Group 2. Gliding bacteria. Life cycle of *Stigmatella aurantiaca* showing vegetative cells, spores, and fruiting body. *(After H. Reichenbach, from Martin Dworkin, ''The Myxobacterales,'' in A. I. Laskin and H. A. Lechevalier (eds.), Handbook of Microbiology, Vol. 1, CRC Press, Inc., 1973.)*

Figure 7-3. Group 2. Gliding bacteria. Stages in the fruiting body formation of the myxobacter *Chondromyces crocatus*. Early stages: (A) Initial stages of vegetative cell aggregation; (B) Fried-egg stage showing orientation of peripheral cells; (C) bulb formation and development of stalk. Late stages: (D) Initial stages of sporangia (spore) formation; (E) sporangia formation after elongation of stalk to maximum length. Structures range in size from approximately 10 to 60 μm. *(From P. L. Grilione and J. Pangborn, J Bacteriol **124**: 1558, 1975.)*

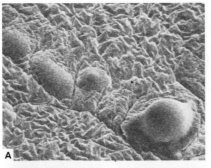

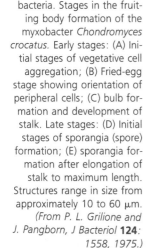

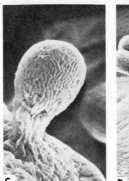

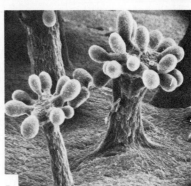

Figure 7-4. Group 2. Gliding bacteria. Species of the order *Cytophagales:* (A) *Flexibacter polymorphus.* Cells collected on the surface of a Nucleopore membrane filter (× 730). *(Courtesy of H. F. Ridgeway, Jr., Scripps Institution of Oceanography.)* (B) Filaments of the gliding bacterium *Herpetosiphon giganteus* on agar showing "bulbs" (bright spherical enlarged regions) (× 500). *(Courtesy of Hans Reichenbach.)* (C, D) *Simonsiella* sp. showing cells arranged in apposition to form filaments with free faces of terminal cells rounded. (C) Scanning electron micrograph (× 2,200); (D) transmission electron micrograph of thin section (× 20,000). *(Courtesy of J. Pangborn and Daisy Kuhn.)* (E) Trichomes attached to a common object are illustrated in this photomicrograph of *Thiothrix* sp. (× 420). *(Courtesy of F. E. Palmer and E. J. Ordal.)* (F) *Vitreoscilla* cell morphology and pattern of cell arrangement of *Vitreoscilla.* *(Courtesy of V. B. D. Skerman.)*

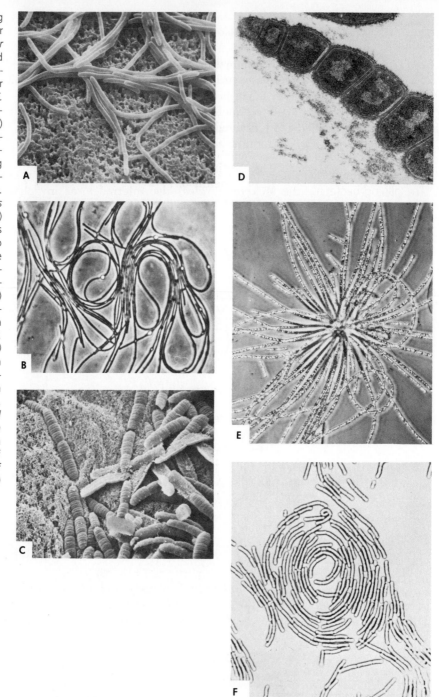

called fruiting bodies (spore-producing structures) composed of slime and cells. These fruiting bodies are often brightly colored and can grow to macroscopic dimensions (see Figs. 7-2 and 7-3). Individual cells can glide on solid surfaces but they do not possess flagella. The mechanism producing this motion is not understood. Another kind, the *Cytophagales*, also exhibit a gliding form of motion. The cells are rod-shaped and may occur singly, in filaments, or in arrangements like those shown in Fig. 7-4. These bacteria have been studied sporadically and to a limited degree. There is much that we do not know about them.

Selected characteristics

> Cell shapes: rod, spherical, or filamentous
> Gram-negative
> Motile by means of slow, gliding movement on surfaces; no locomotor organelles
> Cells may be embedded in slime
> Some produce fruiting bodies (Figs. 7-2 and 7-3)
> Habitat: soil, decomposing plant material, aquatic environments

Group 3.
The sheathed bacteria

This group is characterized by rod-shaped cells that are surrounded by sheaths, so that the individual cells have the appearance of being contained in a tube (see Fig. 7-5). The material composing the sheath differs among species. These bacteria occur in water, sewage, and industrial waste waters. As with the gliding bacteria, there is a great deal that we do not know about them.

Selected characteristics

> Cells enclosed in sheaths made up of deposits of insoluble compounds of iron and manganese
> Cell shapes: rod or filamentous
> Motile by flagella or nonmotile
> Some develop *holdfasts* (suckerlike bases) by which they adhere to surfaces
> Gram-negative
> Habitat: aquatic environments, sludge

Group 4.
Budding and/or
appendaged
bacteria

Bacteria in this group possess several distinctive structural characteristics. Some produce a filamentous outgrowth called a *prostheca* (plural, *prosthecae*) from the body of the cell. Multiplication by budding occurs at the tips of some prosthecae. Other bacteria in this group produce holdfasts. These structures appear at one end of the cell and are composed of cell-wall

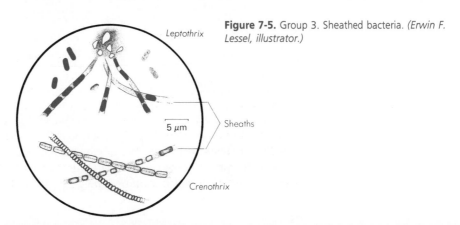

Figure 7-5. Group 3. Sheathed bacteria. *(Erwin F. Lessel, illustrator.)*

Leptothrix

5 μm

Sheaths

Crenothrix

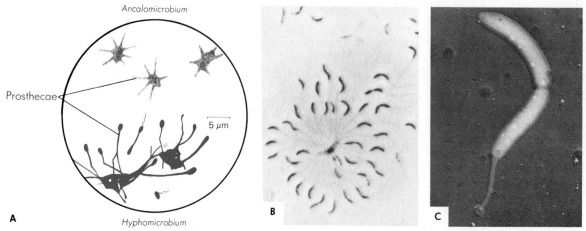

A

B

C

Ancalomicrobium

Prosthecae

5 µm

Hyphomicrobium

Figure 7-6. Group 4. Budding and/or appendaged bacteria. (A) Budding bacteria showing prosthecae. *(Erwin F. Lessel, illustrator.)* (B) Stalked (appendaged) bacteria. *Caulobacter* cells attached to a common holdfast and exhibiting a rosette pattern. *(Courtesy of V. B. D. Skerman.)* (C) Stalked (appendaged) bacteria. Stalk structure of *Caulobacter*. Note flagellum at opposite end (× 13,000). *(Courtesy of A. L. Houwink and W. van Iterson, Biochem Biophys Acta, **5**:10, 1950.)*

Figure 7-7. An electron micrograph of *Treponema pallidum* growing in rabbit tissue culture. The treponema cells adhere to the rabbit cells. *(Courtesy of Thomas Fitzgerald and J Bacteriol **130**: 1333, 1977.)*

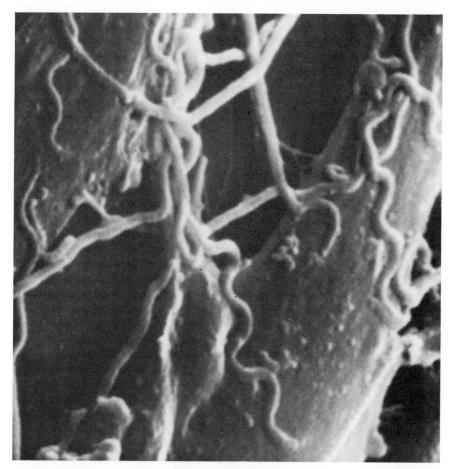

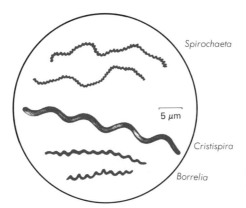

Figure 7-8. Group 5. The spirochetes. Note the differences in the nature of the spirals in the different cells. *(Erwin F. Lessel, illustrator.)*

and membrane material with some adhesive material at the end. They enable the bacteria to attach themselves to surfaces.

Selected characteristics

Cells with prosthecae or holdfasts (see Fig. 7-6)
Multiplication by budding and fission
Some species motile by polar flagella, other species nonmotile
Cell shapes: spherical, oval, bean, rod with pointed ends; some exhibit hyphal growth (filaments)
Habitat: soil, aquatic environments

Group 5.
The spirochetes

These bacteria are characterized by slender, flexuous, helically coiled cells. Various species range in length from 3 to 500 μm. They are motile by various means. Some are saprophytes and others are parasites. The causative agent of syphilis, *Treponema pallidum*, belongs to this group. *T. pallidum* is an obligate parasite which grows only in living tissue (see Fig. 7-7).

Selected characteristics

Cell walls: flexuous (nonrigid)
Cell morphology: slender, helically coiled (spiral); size, shape of ends, and degree of coiling are distinctive (see Fig. 7-8)
Multiplication by transverse fission
Motile either by rapid rotation about the long axis of the spiral or by flexion of cells; corkscrew motion
Many species are gram-negative
Habitat: soil and aquatic environments; any tissue or vascular organ of the body, including the genital regions and central nervous systems of humans and other animals
Pathogenicity: some species are pathogenic to humans and other animals

Group 6.
Spiral and curved
bacteria

These bacteria, like the spirochetes, are helically coiled, but they are rigid rather than flexuous. Some species, the *vibrios*, have less than one complete turn and so have the shape of a comma. Some are free-living in aquatic environments. Others are saprophytic or parasitic. *Campylobacter*

115

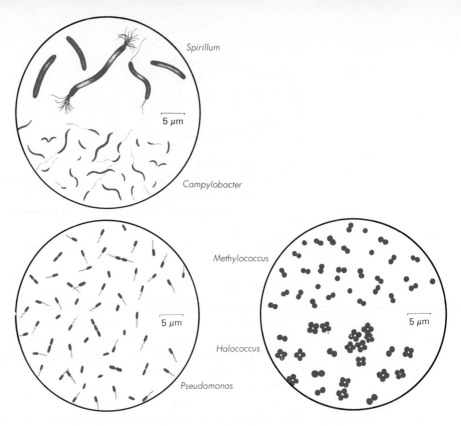

Figure 7-9. Group 6. Spiral and curved bacteria. *(Erwin F. Lessel, illustrator.)*

Figure 7-10. Group 7. Gram-negative aerobic rods and cocci. *(Erwin F. Lessel, illustrator.)*

fetus, one species in this group, is the causative agent of abortion in cattle and other animals and may also infect humans.

Selected characteristics

Cells walls: rigid

Cell shapes: helically coiled rods, some with one or more complete turns (see Fig. 7-9)

Motile by flagella

Gram-negative

Habitat: aquatic environments; reproductive organs, intestinal tract, and oral cavity of animals (including humans)

Pathogenicity: Some species pathogenic for animals (including humans)

Group 7. Gram-negative aerobic rods and cocci

This group contains a very large number of bacteria that are similar in morphology and gram-stain reaction and are aerobic but are very diverse in their metabolic characteristics. In morphological terms they can be considered representative of a typical bacterial cell; that is, the cells appear singly and have dimensions of approximately 0.5–1.0 by 1.5–4.0 μm. Because many species in this group are so similar morphologically, it is necessary to use biochemical characteristics for their differentiation.

Some species are pathogenic. *Brucella* species cause abortion in animals and may also infect humans. *Francisella tularensis*, the causative agent of tularemia (rabbit fever) belongs to this group. Tularemia, primarily a disease of rodents, may be transmitted to humans from rabbits, e.g., if the blood of a rabbit enters a cut or scratch.

Selected characteristics

> Cell morphology: rod, oval, spherical; typical dimensions for bacteria, that is, 0.5–1.0 μm by 1.5–3.0 μm (see Fig. 7-10)
> Motile by flagella or nonmotile
> Aerobic
> Gram-negative
> Distinctive metabolic characteristics of various species: some can fix atmospheric nitrogen; some can oxidize one-carbon compounds such as methane or methanol; some can break down large varieties of compounds
> Habitat: soil and aquatic environments, salt brines
> Pathogenicity: some species pathogenic for humans and other animals

Group 8. Gram-negative facultatively anaerobic rods

Many of the very common bacteria belong to this physiologically diverse group. Because so many of the species within this group are morphologically similar, it is necessary to use a large number of additional tests (biochemical, physiological, and/or serological) to identify a species. Many elaborate schemes of laboratory testing have been developed specifically to characterize and identify bacteria in this group.

One of the most widely studied species of bacteria, *Escherichia coli*, is a member of this group. This bacterium, because it is a normal inhabitant of the intestinal tract of humans and other animals, is used extensively as an indicator of pollution.

Bacteria in this group are also responsible for many intestinal tract (enteric) infections of humans and other animals, as well as for some plant diseases. Some examples are *Shigella* spp., which cause dysentery; *Salmonella* spp., which cause typhoid fever and other enteric infections; *Yersinia pestis*, which causes plague; and *Vibrio cholerae*, which causes cholera. Species of the genus *Erwinia* cause many plant diseases.

Selected characteristics

> Cell morphology: typically short rod (0.5–1.0 by 1.0–3.0 μm); considerable similarity in morphology of cells among taxa (see Fig. 7-11)
> Motile, cells are peritrichous (i.e., flagella uniformly distributed over cell surface) or nonmotile

Figure 7-11. Group 8. Gram-negative facultatively anaerobic rods. *(Erwin F. Lessel, illustrator.)*

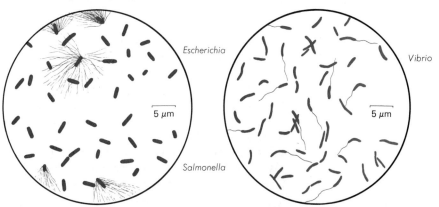

Biochemical characteristics; numerous changes produced in substrates, and this information provides principal means of differentiation and identification of species

Facultatively anaerobic

Gram-negative

Habitat: aquatic environments, soil, food, urine, feces

Pathogenicity: many species pathogenic for humans and other animals; some pathogenic for plants

Group 9.
Gram-negative
anaerobic rods

The cells of bacteria in this group are likely to appear in many different shapes (*pleomorphic*). They are also obligate anaerobes. Various species have been isolated from a wide variety of sources such as the oral cavity of humans, the intestinal tract (where they are the predominant organisms) and feces of humans and other animals, the rumen (compartment of the stomach) of cattle and sheep, and infected tissue. These bacteria, as a group, have not been characterized in the same detail as have those in other groups. Extensive biochemical data on a culture are required to identify a species as belonging to this group.

Selected characteristics

Cell morphology: rod, straight or curved, exhibiting considerable pleomorphism (the existence of different forms in the same species) (see Fig. 7-12)

Motility: cells are peritrichous or monotrichous (one flagellum); some species are nonmotile

Biochemical characteristics: numerous products from fermentation of glucose

Strict (obligate) anaerobes: some species very sensitive to free oxygen

Habitat: natural cavities of humans and other animals; also intestinal tract of insects

Pathogenicity: some species pathogenic for humans and other animals

Group 10.
Gram-negative cocci
and coccobacilli

This is a relatively small group of bacteria with two common pathogenic species, namely, *Neisseria gonorrhoeae*, the causative agent of gonorrhea, and *Neisseria meningitidis*, which causes meningitis. The gram reaction and cell morphology are important characteristics for the identification of

Figure 7-12. Group 9. Gram-negative anaerobic bacteria. *(Erwin F. Lessel, illustrator.)*

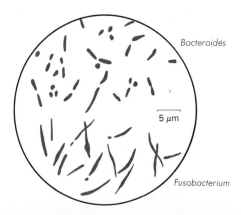

Bacteroidés

5 μm

Fusobacterium

these bacteria. For example, if a gram-stained preparation made from a urethral discharge reveals gram-negative diplococci, this is good evidence of infection with the gonococcus. *Moraxella*, another genus in this group, is associated with eye infections. Some species are saprophytes and others are parasites.

Selected characteristics

Cell morphology: cocci, in pairs (diplococci) and in masses: some (short rods) *coccobacilli*, occur singly and in pairs (see Fig. 7-13)
Nonmotile
Gram-negative
Aerobic
Biochemical characteristic: limited ability to break down various compounds (carbohydrates, proteins, etc.)
Habitat: on mucous membranes of humans and other animals
Pathogenicity: some species pathogenic for humans, notably *Neisseria gonorrhoeae* and *Neisseria meningitidis*

**Group 11.
Gram-negative
anaerobic cocci**

These bacteria are cocci and show considerable variation in size. The cell diameter for various species ranges from 0.3 to 2.5 μm. They are regarded as parasites of humans and other animals, in which they are present in large numbers in the respiratory and intestinal tracts. The establishment of species within this group is based largely on the biochemical characteristics of a culture. These bacteria are not regarded as pathogens.

Selected characteristics

Cell morphology: very small (0.3 to 0.5 μm) to larger (2.5 μm) spherical cells in pairs, masses, or chains (see Fig. 7-14)
Nonmotile
Anaerobic
Biochemical characteristics: metabolize (break down) carbohydrates and fatty acids
Habitat: the respiratory and intestinal tracts of humans and other animals
Parasitic

Figure 7-13. (Left). Group 10. Gram-negative cocci and coccobacilli. *(Erwin F. Lessel, illustrator.)*

Figure 7-14. (Right). Group 11. Gram-negative anaerobic cocci. *(Erwin F. Lessel, illustrator.)*

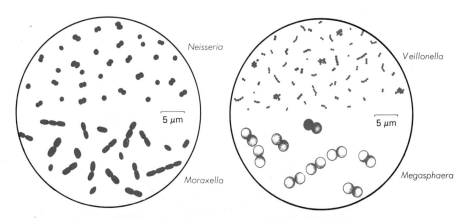

Group 12.
Gram-negative
chemolithotrophic
bacteria

The distinctive characteristic which is common to this group of bacteria is their ability to derive energy from the oxidation of inorganic chemical substances. They are chemolithotrophic. Identification of members within this group is based on the kind of compound utilized for energy, such as ammonia or nitrite nitrogen, sulfur or sulfur compounds, and iron or manganese.

These bacteria are alike in that they are all gram-negative, but they vary in morphology. The shape of cells of different species may be rodlike, spherical, or spiral. Some are motile and others are nonmotile. They occur widely in soil and aquatic environments and are very important in carrying out biochemical changes in these environments. They are not pathogenic.

Selected characteristics

Autotrophic (energy derived from oxidation of inorganic compounds, such as ammonia and nitrites, sulfur and reduced sulfur compounds, or iron and manganese)

Cell morphology; spherical, rod, spiral; multilayered membranes in some species; sulfur-oxidizing bacteria may store sulfur granules (see Fig. 7-15)

Motile by flagella or nonmotile

Gram-negative

Habitat: soil, sewage, aquatic environments; natural environments with large amounts of sulfur, iron, or manganese, such as acid mine waters and sulfur springs

Group 13.
Methane-producing
(methanogenic)
bacteria

The unifying feature of this group of bacteria is their ability to produce the gas methane. Methane is produced under anaerobic conditions.

Morphologically, the methanogenic bacteria are very diverse (see Fig. 7-16).

They are widely distributed in nature, where they are found in aquatic environments, particularly muddy sediments of natural waters and sewage digesters. They are also present in large numbers in the rumen of cattle.

Figure 7-15. Group 12. Gram-negative chemolitho-trophic bacteria. *(Erwin F. Lessel, illustrator.)*

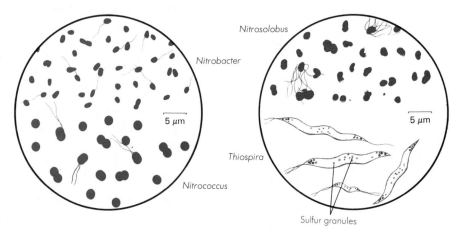

Nitrosolobus

Nitrobacter

5 μm

5 μm

Thiospira

Nitrococcus

Sulfur granules

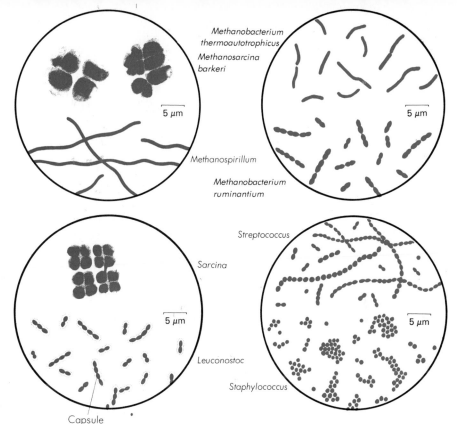

Figure 7-16. Group 13. Methane-producing bacteria. *(Erwin F. Lessel, illustrator.)*

Methanobacterium thermoautotrophicus

Methanosarcina barkeri

Methanospirillum

Methanobacterium ruminantium

Sarcina

Streptococcus

Leuconostoc

Staphylococcus

Capsule

Figure 7-17. Group 14. Gram-positive cocci. *(Erwin F. Lessel, illustrator.)*

Selected characteristics

Autotrophic or heterotrophic: energy derived from oxidation of hydrogen or formate or acetate with formation of methane and CO_2

Cell morphology: spherical, rod, spiral

Motile by polar flagella or nonmotile

Gram-positive or gram-negative

Anaerobic

Some thermophilic species

Habitat: gastrointestinal tract of animals, sediments of aquatic environments, and sewage

Group 14. Gram-positive cocci

Many important pathogenic species for humans and other animals are included in this group. They are all gram-positive cocci, but they exhibit differences in the arrangement of their cells, as shown in Fig. 7-17. Microscopic examination of a gram-stained preparation from a clinical specimen can be useful in the identification of the bacterium responsible for an infection. The major pathogens in this group are staphylococci and streptococci. Many species are saprophytic. Some species are important in the food and dairy industries.

Selected characteristics

Cell morphology: cocci occurring singly or in pairs, chains, packets, or clusters

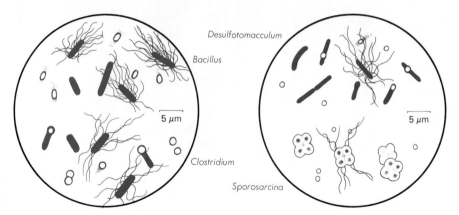

Figure 7-18. Group 15. Endospore-forming rods and cocci. *(Erwin F. Lessel, illustrator.)*

Nonmotile

Gram-positive

Facultatively anaerobic or microaerophilic

Heterotrophic: wide range of nutrient requirements

Habitat: soil; freshwater; skin and mucous membranes of warm-blooded animals, including humans

Pathogenicity: several species are important pathogens in animals (including humans); many are saprophytic

Group 15. Endospore-forming rods and cocci

The predominant and distinguishing characteristic of these bacteria is their ability to produce endospores. Most species are rod-shaped. Some are aerobic (genus *Bacillus*) and others are anaerobic (genus *Clostridium*).

These bacteria and their endospores are widely distributed in soil and are carried by dust particles in air. The endospores, because of their high resistance to heat, can survive for long periods of time.

Some species cause infections of humans and other animals, including insects.

Selected characteristics

Cell morphology: rods, except for one species in which cells are spherical and in packets (see Fig. 7-18)

Motile by flagella or nonmotile

Gram reaction: most species are gram-positive

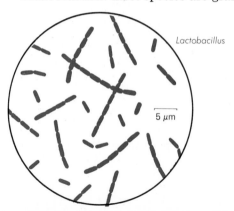

Figure 7-19. Group 16. Gram-positive asporogenous rod-shaped bacteria. *(Erwin F. Lessel, illustrator.)*

Aerobic, facultatively anaerobic, anaerobic, or microaerophilic
Endospores: produced by all species
Habitat: soil, air, aquatic environments, intestinal tract of animals (including humans)
Pathogenicity: a few species are pathogenic for animals (including humans) and some cause food poisoning

Group 16.
Gram-positive,
asporogenous, rod-
shaped bacteria

This group is comprised almost entirely of the lactobacilli. These are non-sporeforming rod-shaped bacteria that are closely associated with milk and milk products. They are capable of fermenting milk sugar (lactose) to lactic and other acids. They are present in fermenting animal and plant products and in the oral cavity, vagina, and intestinal tract of humans and other animals. They are not regarded as pathogenic.

Selected characteristics

Cell morphology: bacilli occurring singly or in chains (see Fig. 7-19)
Nonmotile
Gram-positive
Anaerobic or facultatively anaerobic
Metabolic characteristic: lactic acid is a characteristic end product of fermentation
Habitat: dairy products, grain and meat products, water, sewage, and fermented products; oral cavity, vagina, and intestinal tract of animals (including humans)

Group 17.
Actinomycetes and
related organisms

This is a large and very diverse group of bacteria. A unifying feature is the pleomorphism of their cells and a tendency to form branching filaments (hyphae). In some taxonomic groups (families) the hyphae mass together to form a mycelium. Many pathogenic species are members of this group, including *Corynebacterium diphtheriae*, *Mycobacterium tuberculosis*, and *Actinomyces israelii*. The last is the causative agent of *actinomycosis*, a disease of humans and other animals (also called *lumpy jaw* or *madura foot*).

Species of the genus *Streptomyces*, which belongs to this group, are well known for their antibiotic-producing ability. Most antibiotics available today for therapeutic use are produced by species of *Streptomyces*.

Selected characteristics

Cell morphology: very diverse and pleomorphic; irregular rod shapes, filaments, and branched filaments; mycelial structures (see Fig. 7-20)
Nonmotile
Gram-positive
Aerobic, facultatively anaerobic, or anaerobic
Habitat: soil, aquatic environments, air; and animals (including humans)
Pathogenicity: Many important pathogens for animals (including humans) and plants

Figure 7-20. Group 17. Actinomycetes and related organisms. *(Erwin F. Lessel, illustrator.)*

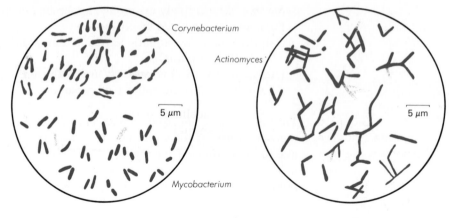

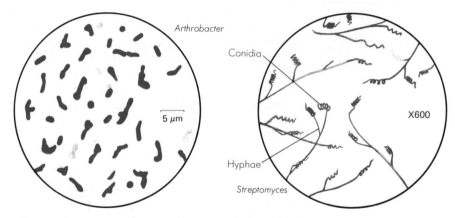

Group 18.
The rickettsias

The *rickettsias* are among the smallest bacteria, ranging in size from 0.3 to 0.7 μm wide by 1.0 to 2.0 μm long. They are gram-negative and nonmotile. Distinctive features of the rickettsias are their obligate parasitic nature and their relationship with blood-sucking arthropods such as fleas, lice, and ticks. Rickettsias grow only in the presence of another living cell (animal or insect); they are obligate intracellular parasites. They are the causative agents of several diseases, namely, typhus fever, Rocky Mountain spotted fever, scrub typhus, and Q fever. They are generally transmitted by blood-sucking insects. Insects become infected when they ingest blood from an infected individual. An infected insect, feeding upon a healthy individual, infects that individual directly with its mouth parts as it breaks through the skin or later via its feces, which enter through the broken skin layers.

Included in this group are the *chlamydias*. The chlamydias are also obligate intracellular parasites. However, their multiplication is characterized by a change of the small, rigid-walled infectious form (elementary body) into a larger, thin-walled, noninfectious form that divides by fission. Chlamydia species are the causative agents of trachoma, lymphogranuloma venereum, urethritis, psittacosis, ornithosis, and other diseases.

Selected characteristics

> Cell morphology: short rods or ovals, often pleomorphic; some form small coccoid bodies (elementary bodies) which develop into larger bodies in a characteristic life cycle (see Fig. 7-21)
> Gram-negative
> Nonmotile
> Obligate intracellular parasites (laboratory cultivation in tissue culture system or animals)
> Habitat: insect carriers, birds, and mammals (including humans)
> Pathogenicity: many important pathogens of humans and other animals

Group 19.
The mycoplasmas

The distinctive characteristic of the mycoplasmas is that they lack a true cell wall. The cells are enclosed by a membrane, but this membrane does not contain the structural units, the muramic and diaminopimelic acids which are present in the cell walls of other bacteria and which impart rigidity to the cell wall. The cells range in size from very small (125 to 250 nm) spherical or slightly ovoid bodies to slender filaments ranging in length from a few micrometers to 100 μm. They are very pleomorphic. They are parasitic and pathogenic, infecting a wide range of animals, including humans.

Selected characteristics

> Cell morphology: lack true cell wall; cellular content enclosed by a triple-layered, nonrigid, membrane. Some cells very small (0.2 μm); highly pleomorphic (Fig. 7-22)

Figure 7-21. Group 18. The Rickettsias. (A) *Rickettsia akari* in smear of peritoneal scraping of infected laboratory mouse (× 940). (B) *R. tsutsugamushi* in cytoplasm of infected cell (× 940). (C) *R. prowazekii* in yolk-sac culture (× 1,500). (D) *R. typhi* in yolk-sac culture (× 1,000). *(From N. J. Kramis and the Rocky Mountain Laboratory, U.S. Public Health Service.)*

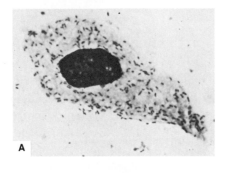
A

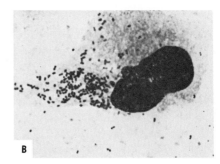
B

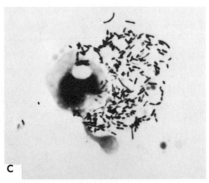
C

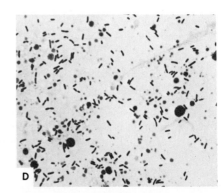
D

Usually nonmotile
Gram-negative
Facultatively anaerobic
Reproduction: binary fission, fragmentation, budding
Habitat: mucous membranes of respiratory tract and lower genital tract
Pathogenicity: parasites and pathogens of wide range of mammals and
  birds; some are possible pathogens of plants

Summary and outlook

The classification of bacteria in *Bergey's Manual of Determinative Bacteriology* is generally accepted internationally. This manual is revised periodically to take advantage of new knowledge which becomes available through research with microorganisms and through new techniques of analyzing the resultant data. The present 8th edition of *Bergey's Manual* divides all bacteria into 19 parts (groups), each of which is characterized by distinctive morphological or metabolic features. Emphasis is placed on grouping bacteria with common, readily identifiable characteristics. No attempt is made to organize them to reflect a scheme of evolutionary

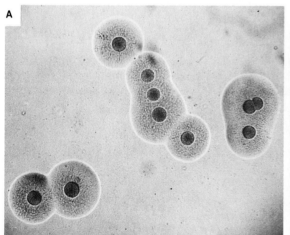

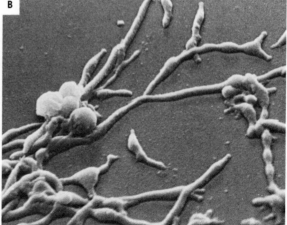

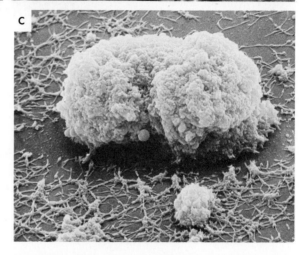

Figure 7-22. Group 19. The Mycoplasmas. (A) The mycoplasmas. Colonies of *Mycoplasma molare* showing typical "Fried egg" appearance. *(Courtesy of S. Rosendal and Int J Syst Bacteriol* **24***:125, 1974.)* (B) Scanning electron micrograph of *M. pneumoniae* from a 6-day culture showing irregular forms, crossing filaments, and piling up of spherical organisms probably representing an early stage of colony formation (× 27,600). (C) Scanning electron micrograph of a 6-day culture of *M. pneumoniae* showing dense network of filamentous forms growing on the surface and the rounded appearance of the organisms in the colonies (X 2,650). [*(B) and (C) Courtesy of G. Biberfeld and P. Biberfeld, J Bacteriol,* **102***:855, 1970.*]

development, as was done in all previous editions. The reason is that, in too many instances, our knowledge about microorganisms is incomplete.

Bacteria, as revealed by the brief descriptions presented here of the 19 groups, exhibit great diversity. No other group of organisms has a similar tremendous range in morphologic, physiologic, and metabolic characteristics. For example, the rickettsias are intracellular parasites, totally dependent on a host cell for some vital process or product. On the other hand, bacteria of the genus *Thiobacillus* derive energy from the oxidation of sulfur and derive their carbon from carbon dioxide. The mycoplasmas form elementary bodies, the smallest of which are beyond resolution with the light microscope. On the other hand streptomycetes grow into filaments longer than 100 μm. They are all bacteria.

It is reasonable to expect that the 9th edition of *Bergey's Manual*, when published, will contain many new taxa as well as changes in taxonomy. This is not an unusual expectation, since our knowledge of bacteria keeps increasing and we have to reassess our original ideas. In the future, taxonomic decisions will become more objective and less subjective as our base of knowledge grows and as the resulting increased amount of data for each species is subjected to computer analysis.

Key terms

bacteriochlorophylls
chemolithotrophic
facultatively anaerobic
fission
fruiting bodies

holdfasts
hypha
methanogenic
parasitic
photosynthesis

pleomorphic
Procaryotae
prostheca
saprophytic
sporangia

Questions

1 The major divisions of bacteria are presented as parts in *Bergey's Manual of Determinative Bacteriology*, 8th edition. How does this differ from the earlier editions?

2 Which groups are made up of bacteria with similar morphology and which are made up of bacteria with very different morphology?

3 Name 10 different genera of bacteria which have very distinctive morphologic characteristics. Describe the morphologic characteristic for each genus.

4 Name 10 different genera of bacteria having the same general morphologic characteristics. Describe these morphologic characteristics.

5 Name a taxonomic group of bacteria which has a characteristic related to (*a*) algae, (*b*) filamentous fungi, (*c*) protozoa. Describe the relationship in each case.

6 How do mycoplasmas differ from other bacteria?

7 Describe the distinctive characteristics of the myxobacters.

8 Compare the dimensions of bacteria in group 2 (gliding bacteria) with those in group 8 (gram-negative facultatively anaerobic rods).

9 Why is it reasonable to expect significant differences in the content of the next edition of *Bergey's Manual*?

References for part two

Buchanan, R. E., and N. E. Gibbons (eds.): *Bergey's Manual of Determinative Bacteriology*, 8th ed., Williams & Wilkins, Baltimore, 1974. *This volume is the standard international reference book on the classification and taxonomy of bacteria. Each major group of bacteria is described, and recognized species are characterized in detail.*

Karp, G., *Cell Biology*, McGraw-Hill, New York, 1979. *A good source of additional and more detailed discussion of cell structure and function. Well illustrated.*

Klainer, A. S., and Irving Geis: *Agents of Bacterial Disease*, Harper & Row, New York, 1973. *Profusely illustrated with scanning electron micrographs and other photomicrographs, together with detailed diagrammatic illustrations of the more common bacteria that cause disease. In addition, there are schematic diagrams of several bacteriological phenomena, e.g., cell structure, cell-wall formation, mode of action of antimicrobial agents. There is a minimum of text material.*

Lamanna, C., M. F. Mallette, and L. N. Zimmerman: *Basic Bacteriology, Its Biological and Chemical Background*, 4th ed., Williams & Wilkins, Baltimore, 1973. *Provides extensive information on bacterial nutrition and on physical conditions influencing growth.*

Lennette, E. H., E. H. Spaulding, and J. P. Truant (eds.): *Manual of Clinical Microbiology*, 3d ed., American Society for Microbiology, Washington, D.C., 1980. *Sections 97 and 98 contain a general description of culture media, a very extensive listing of media with their ingredients, and their use.*

Olds, R. J.: *Color Atlas of Microbiology*, Year Book Medical Publishers, Chicago, 1975. *A collection of photographs of the commonly encountered bacteria and fungi to illustrate their morphology and the appearance of their growth on various laboratory media.*

PART THREE

EUCARYOTIC PROTISTS: FUNGI, PROTOZOA, ALGAE

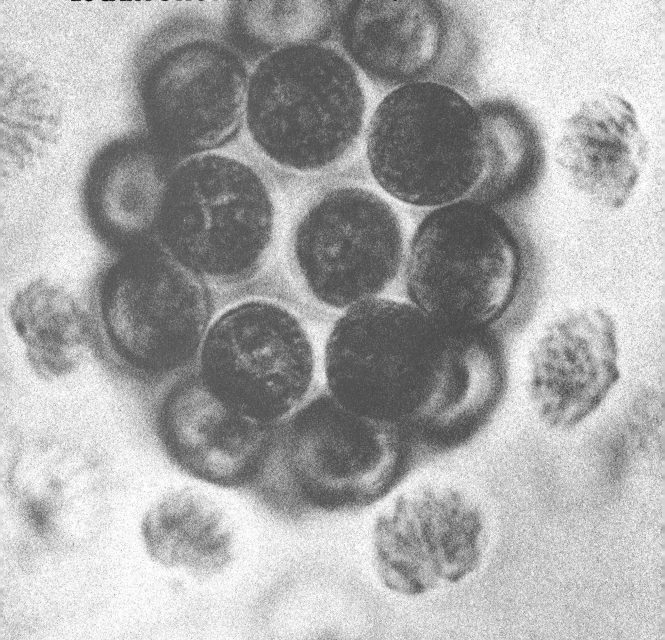

The procaryotic protists (discussed in Part II) are the bacteria and the cyanobacteria. (The viruses, even though they are not considered cells, share many of the characteristics of procaryotes.) These protists lack a nuclear membrane, their chromosome consists of deoxyribonucleic acid with little or no protein, and division is amitotic. Such protists are procaryotes, or "prenuclear organisms."

The eucaryotic protists comprise the fungi, the protozoa, and the algae. Their cells contain a nucleus surrounded by a nuclear membrane, their chromosomes consist of deoxyribonucleic acid complexed with large amounts of protein and are more than one in number. These protists are called eucaryotes, or organisms with a "true nucleus."

Part III covers the biology of the eucaryotic protists, including their importance in medicine and other applied areas.

The fungi comprise the molds and yeasts. The molds are filamentous in morphology while the yeasts are generally unicellular. They are heterotrophic organisms. The protozoa occur as single cells and are also heterotrophic, but they are distinguished by their ability to move and their lack of cell walls. For these reasons they are said to have animal characteristics. The algae occur in both macroscopic and microscopic forms. Their distinctive characteristic, which they share with green plants, is the ability to carry out photosynthesis.

Sexual reproduction in the green alga *Eudorina elegans*. Sperm packets of a male strain (strain 62 m) surrounding a female colony (strain 62 f). Magnification ×1300. *(Courtesy of M. Goldstein, McGill University.)*

THE FUNGI

The appearance of fungi is familiar to each and every one of us. We have seen the blue and green growth on oranges, lemons, and cheeses; the white, furry outgrowths on stale bread and jam; the mushrooms and toadstools in the fields and woods. These are bodies of various fungi. Thus fungi have a variety of appearances, depending on the species. The study of fungi is called *mycology*. Fungi comprise the molds and yeasts. While molds are filamentous, yeasts are usually unicellular.

The importance of fungi

The fungi are heterotrophic organisms—they require organic compounds for nutrition. When they feed on dissolved dead organic matter, they are known as *saprophytes*. Saprophytes decompose complex plant and animal remains, breaking them down into simpler chemical substances that are returned to the soil, thereby increasing its fertility. Thus they can be quite beneficial to humans. But they can also be harmful to us when they rot timber, textiles, food, and other materials.

Saprophytic fungi are also important in industrial fermentations, for example, the brewing of beer, the making of wine, and the production of antibiotics such as penicillin. The leavening of dough and the ripening of some cheeses also depend on fungal activity.

131

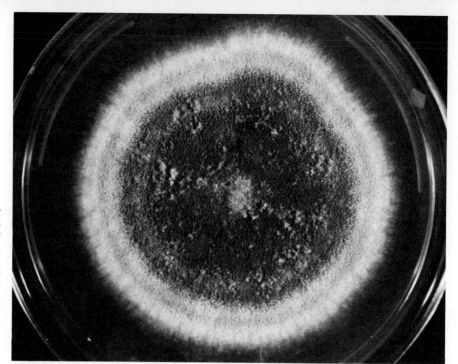

Figure 8-1. A mold colony growing in a petri dish. Note the filamentous growth of the organism. The powdery appearance is due to the presence of thousands of asexual spores or conidia. The species shown belongs to the genus *Penicillium*, the same genus of mold that produces the antibiotic penicillin.

Some fungi, even though they are saprophytic, can also invade living hosts and thrive as *parasites*. As parasites, fungi cause diseases in plants and animals, including humans. However, of the approximately 500,000 species of fungi, only about 100 are pathogenic to humans. Mortality from fungal infections other than skin diseases is very high. This may be due to wrong or belated diagnosis in the course of the disease or to the nonavailability of medically effective nontoxic antibiotics. Many pathogenic fungi, for example, *Histoplasma capsulatum*, which causes histoplasmosis (mycotic infection of the reticuloendothelial system involving many organs), can also live as saprophytes. Such fungi exhibit *dimorphism*; that is, they can exist either in a unicellular form like that of a yeast or in a filamentous form like that of a mold. The yeast phase is exhibited when the organism is a parasite or pathogen in tissue and the mold form when the organism is a saprophyte in soil or in a laboratory medium. The laboratory identification of such fungal pathogens is often dependent on the demonstration of dimorphism.

Morphology

In general, yeast cells are larger than most bacteria, but the smallest yeasts are not as large as the largest bacteria. Yeasts vary considerably in size, ranging from 1 to 5 μm in width and from 5 to 30 μm or more in length. They are commonly egg-shaped, but some are elongated and some spherical. Each species has a characteristic shape, but even in pure culture there is considerable variation in the size and shape of individual cells, depending on age and environment. Yeasts have no flagella or other organs of locomotion.

The body, or *thallus*, of a mold (Fig. 8-1) consists essentially of two parts: the *mycelium* and the *spores* (resistant, resting, or dormant cells). The mycelium is a complex of several filaments called *hyphae* (singular, *hypha*). Each hypha is about 5 to 10 μm wide, as compared with a bacterial cell, which is usually 1 μm in diameter.

A common cytoplasm exists along the length of each hypha. Hyphae occur in three morphological types (Fig. 8-2):

1 Nonseptate, or *coenocytic*. These hyphae have no cross walls or *septa* (singular, *septum*).
2 Septate with uninucleate cells. Septa divide the hypha into compartments or cells with a single nucleus in each compartment. There is a central pore in each septum which allows for the migration of nuclei and cytoplasm from one compartment to another. Although each compartment of a septate hypha is not bounded by a membrane as is a typical cell, each compartment is customarily referred to as a cell.
3 Septate with multinucleate cells. Septa divide the hypha into cells with more than one nucleus in each compartment.

Mycelia can be either vegetative or reproductive. Some hyphae of the vegetative mycelium penetrate into the medium in order to obtain nourishment for the organism. Reproductive mycelia are responsible for spore production and usually extend into the air from the medium. The mycelium of a mold may be a loosely woven network or it may be an organized, compact structure, as in mushrooms.

Reproduction   Most parts of a mold are potentially capable of growth and multiplication. Inoculation of a minute fragment on a medium is sufficient to start a new individual. This is done by planting the inoculum on fresh medium with the aid of a transfer needle, a method similar to that used for bacteria. One difference in the method is that the needle used for molds is stiffer and has a flattened tip for cutting the mycelium.

Fungi reproduce naturally by a variety of means, either asexually by fission, budding, or spore formation or sexually by fusion of the nuclei of

Figure 8-2. Three types of hyphae: (A) nonseptate (coenocytic); (B) septate with uninucleate cells; (C) septate with multinucleate cells.

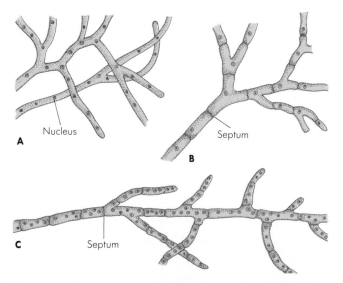

two parent cells. In *fission*, a cell divides to form two similar daughter cells. In *budding*, a daughter cell develops from a small outgrowth of the parent cell.

Asexual spores, whose function is to disseminate the species, are produced in large numbers. There are many kinds of asexual spores (Fig. 8-3):

1 *Conidiospores* or *conidia* (singular, *conidium*). Small, single-celled conidia are called *microconidia*. Large, multicelled conidia are called *macroconidia*. Conidia are formed at the tip or side of a hypha.
2 *Sporangiospores*. These single-celled spores are formed within sacs called *sporangia* (singular, *sporangium*) at the end of special hyphae (*sporangiophores*). *Aplanospores* are nonmotile sporangiospores. *Zoospores* are motile sporangiospores, their motility being due to the presence of flagella.
3 *Oidia* (singular, *oidium*) or *arthrospores*. These single-celled spores are formed by disjointing of hyphal cells.
4 *Chlamydospores*. These thick-walled single-celled spores are highly resistant to adverse conditions. They are formed from cells of the vegetative hypha.
5 *Blastospores*. The buds on yeast cells are blastospores.

Figure 8-3. Asexual spore types in fungi. *(Redrawn from the McGraw-Hill Encyclopedia of Science and Technology, 1977, vol. 5, p. 117.)*

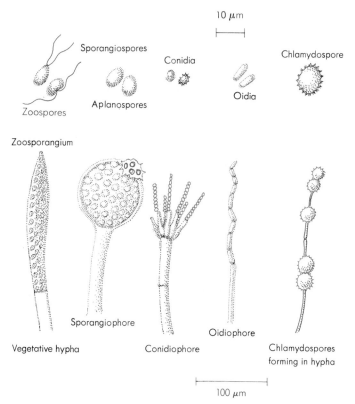

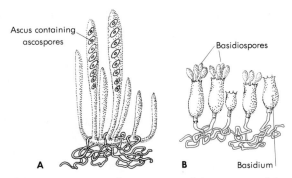

Ascus containing ascospores

Basidiospores

A   B   Basidium

Figure 8-4. Some sexual spore types and the structure of the corresponding reproductive mycelia in fungi. (A) Ascospores and (B) basidiospores.

Sexual spores, which are produced by the fusion of two nuclei, form less frequently, later, and in smaller numbers than do asexual spores. Also, they form only under certain conditions. There are several types of sexual spores:

Figure 8-5. The sexual process of basidiospore formation. A basidium with one nucleus from each parent. The basidium assumes the shape characteristic of that species and generally produces four tapering processes, the *sterigmata* (singular, *sterigma*). The four nuclei, produced after nuclear fission from meiosis, now move toward the sterigmata and form the basidiospores.

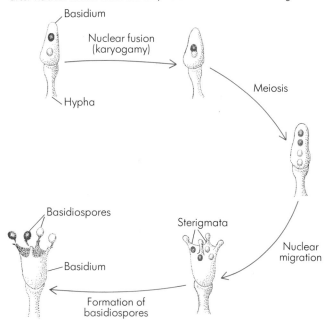

Basidium

Nuclear fusion (karyogamy)

Hypha

Meiosis

Basidiospores

Sterigmata

Basidium

Nuclear migration

Formation of basidiospores

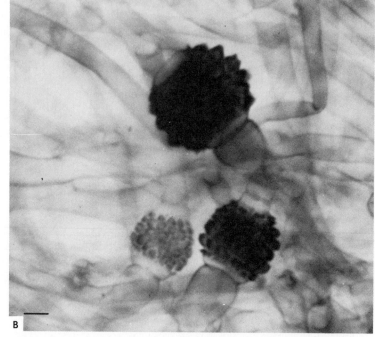

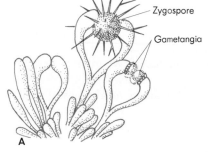

Figure 8-6. (A) The formation of zygospores. *(Redrawn from The McGraw-Hill Encyclopedia of Science and Technology, 1977, vol. 5, p. 118.)* (B) Zygospores in *Mucor hiemalis*. Sexual reproduction in *Mucor hiemalis* occurs when two sexually compatible mating types, + and −, come into contact with each other and produce zygospores. Zygospores of different ages are shown, the oldest one being darkest, largest, and roughest. Bar equals 0.01 mm. *(Courtesy of L. Kapica and E. C. S. Chan.)*

1 *Ascospores.* These single-celled spores are produced in a sac called an *ascus* (plural, *asci*). There are usually eight ascospores in each ascus (Fig. 8-4A).
2 *Basidiospores.* These single-celled spores are borne on a club-shaped structure called a *basidium* (plural, *basidia*) (Fig. 8-4B). Their formation is illustrated in Fig. 8-5.

Figure 8-7. The formation of oospores. *(Redrawn from The McGraw-Hill Encyclopedia of Science and Technology, 1977, vol. 5, p. 118.)*

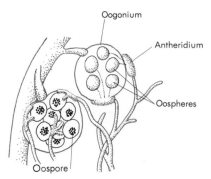

3 *Zygospores*. Zygospores are large, thick-walled spores formed when the tips of two sexually compatible hyphae, or gametangia, of certain fungi fuse together (Fig. 8-6A). These spores are illustrated in Fig. 8-6B.

4 *Oospores*. These are formed within a special female structure called the *oogonium*. Fertilization of the eggs, or *oospheres,* by male gametes formed in an *antheridium* gives rise to oospores. There are one or more oospheres in each oogonium. This process is illustrated in Fig. 8-7.

Asexual and sexual spores may be surrounded by highly organized protective structures called *fruiting bodies*. Asexual fruiting bodies have names like *acervulus* and *pycnidium*. Sexual fruiting bodies have names like *perithecium* and *apothecium*. Some of these structures are shown in Fig. 8-8.

Although a single fungus may produce asexual and sexual spores by several methods at different times and under different conditions, the spores are sufficiently constant in their structures and in the method by which they are produced to be used in identification and classification.

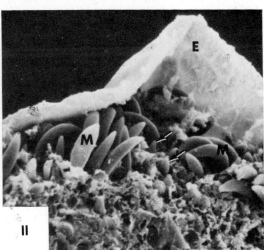

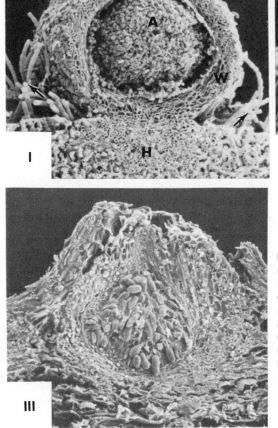

Figure 8-8. Fungal fruiting bodies as seen by scanning electron microscopy. (I) Longitudinal section of a perithecium (ascocarp) of *Ceratocystis fimbriata*. The perithecial wall (W) and hyphae (H) are clearly seen. Other structures seen are ascospores (A) within the perithecial cavity and conidiophores (arrows), which are specialized aerial conidia-bearing hyphae. (X 250). (II) Cross section of a subepidermal acervulus of *Marsonina juglandis* in black walnut leaf. Mature (M) and immature (arrows) conidia are exhibited. The host epidermis (E) is clearly seen (X 1,500). (III) Section of a pycnidium of *Dothiorella ribis* in apple bark tissue showing conidia compacted in the mucilaginous matrix (X 200). *(Courtesy of M. F. Brown and H. G. Brotzman, University of Missouri.)*

Physiology   Fungi are better able to withstand unfavorable environmental conditions than most other microorganisms. For example, yeasts and molds can grow in a substrate or medium containing concentrations of sugars that inhibit most bacteria; this is why jams and jellies may be spoiled by molds but not by bacteria. Also, yeasts and molds generally can tolerate more acidic conditions than most other microbes.

Yeasts are facultative; that is, they can grow under both aerobic and anaerobic conditions. Molds are strictly aerobic microorganisms. Fungi grow over a wide range of temperature, with the optimum for most saprophytic species from 22 to 30°C (71.6 to 86°F); pathogenic species have a higher temperature optimum, generally 30 to 37°C (86 to 98.6°F). Some fungi will grow at or near 0°C (32°F) and thus can cause spoilage of meat and vegetables in cold storage.

Fungi are capable of using a wide variety of materials for nutrition. However, they are heterotrophic. Unlike some bacteria, they cannot use inorganic carbon compounds, such as carbon dioxide. Carbon must come from an organic source, such as glucose. Some species can use inorganic compounds of nitrogen, such as ammonium salts. But all fungi can use organic nitrogen; this is why culture media for fungi usually contain peptone, a hydrolyzed protein product. A summary of the physiological characteristics of fungi in comparison with those of bacteria is found in Table 8-1.

Classification   The classification of fungi is based primarily on the characteristics of the sexual spores and fruiting bodies present during the sexual stages of their life cycles. Fungi with known sexual stages are termed *perfect fungi.* However, many fungi produce sexual spores and fruiting bodies only under certain exacting environmental conditions, if they produce them at all.

Table 8-1. Comparative physiology of fungi and bacteria

CHARACTERISTIC	FUNGAL	BACTERIAL
Optimum pH	3.8–5.6	6.5–7.5
Optimum temperature	22–30°C (71.6–86°F) (saprophytes) 30–37°C (86–98.6°F) (parasites)	20–37°C (68–98.6°F) (mesophiles)
Gases	Strictly aerobic (molds) Facultative (yeasts)	Aerobic→anaerobic
Light (for growth)	None	Some photosynthetic groups
Sugar concentration in laboratory medium	4–5%	0.5–1%
Carbon	Organic	Inorganic and/or organic
Cell-wall structural components	Chitin, cellulose, or glucan	Peptidoglycan
Antibiotic susceptibility	Resistant to penicillins, tetracyclines, chloramphenicol; sensitive to griseofulvin	Resistant to griseofulvin; sensitive to penicillins, tetracyclines, chloramphenicol

Table 8-2. Selected major characteristics of the classes of fungi

FEATURE	CLASSES			
	Phycomycetes	Ascomycetes	Basidiomycetes	Deuteromycetes (Fungi Imperfecti)
Mycelium	Nonseptate or coenocytic	Septate	Septate	Septate
Asexual spores	Sporangiospores, occasionally conidia	Conidia	Conidia	Conidia
Sexual spores	Zygospores, oospores	Ascospores	Basidiospores	Unknown
Natural habitat	Water, soil, animals	Soil, plants, animals	Soil, plants	Soil, plants, animals

Thus the complete life cycles, with sexual stages, for many fungi are as yet unknown. For those fungi without known sexual stages, designated *imperfect fungi*, characteristics other than their sexual stages must be used for classification. These include the morphology of their asexual spores and mycelia. The imperfect fungi are provisionally placed in a special class, Deuteromycetes or Fungi Imperfecti, until their sexual stages are found. Then they can be reclassified and placed into one of the other classes. Therefore, based on the manner and characteristics of reproduction, there are four classes of true or filamentous fungi in the kingdom Fungi: Phycomycetes, Ascomycetes, Basidiomycetes, and Deuteromycetes. The main characteristics of these classes of fungi are outlined in Table 8-2.

Class Phycomycetes

Members of this class are often referred to as the *lower fungi* because they are generally considered to be "primitive" on the evolutionary scale. They are such a large and heterogeneous group of microorganisms that some taxonomists divide class Phycomycetes into six separate classes. A common feature among them is lack of septa in the hyphae; this characteristic distinguishes them from members of the other three classes, Ascomycetes, Basidiomycetes, and Deuteromycetes, in the taxonomic scheme that we use in this text.

The Phycomycetes of medical importance are actually common fungi of air and soil, including the common bread molds *Mucor* and *Rhizopus*. Most of these fungi belong to the more highly developed genera of class Phycomycetes and reproduce both asexually and sexually. They are

Figure 8-9. *Rhizopus stolonifer,* the common bread mold. This fungus forms rootlike hyphae, or rhizoids; vegetative, or nonsexual, hyphae which penetrate the substrate; and fertile hyphae which produce sporangia at the tips of spore-producing hyphae, or sporangiophores. Stolons are rootlike filaments which connect individual organisms.

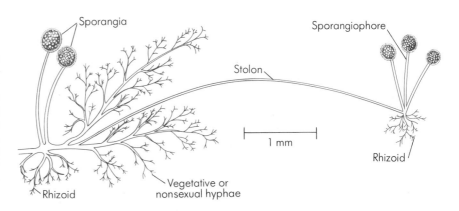

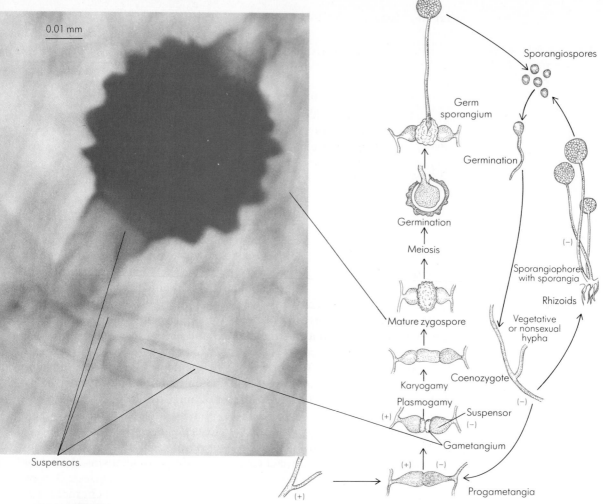

0.01 mm

Sporangiospores

Germ sporangium

Germination

Germination

Meiosis

Sporangiophores with sporangia

Mature zygospore

Rhizoids

Vegetative or nonsexual hypha

Karyogamy

Coenozygote

Plasmogamy

(−)

(+)

Suspensor
(−)

Gametangium

(+)

(−)

(+)

(−)

Progametangia

(+)

Suspensors

Figure 8-10. Life cycle of *Rhizopus stolonifer.* When the wall of the sporangium disintegrates, sporangiospores are released. A sporangiospore germinates to develop into an organism with many vegetative hyphae. Rhizoids which penetrate into the medium are also formed. Directly above the rhizoids, one or more sporangiophores are produced. The top of each sporangiophore develops into a sporangium containing sporangiospores. This completes the asexual portion of the life cycle. Sexual reproduction requires two sexually compatible types (+ and −). When these types come into contact with one another, copulating branches called progametangia are formed. A septum then forms near the tip of each progametangium, separating it into two cells, a terminal gametangium and a suspensor cell. The walls of the two contacting gametangia dissolve at the point of contact, the two protoplasts mix (plasmogamy), and the + and − nuclei fuse (karyogamy) to form many zygote nuclei. The structure which contains them is called the coenozygote. The wall around the coenozygote thickens and its surface becomes black and warty, forming the mature zygospore, which lies dormant for 1 to 3 or more months. The zygospore germinates to form a new organism; meiosis takes place during the germination process. Bar equals 0.01 mm. *(Photomicrograph courtesy of L. Kapica and E. C. S. Chan, McGill University.)*

Figure 8-11. Black piedra is a fungus infection of hair characterized by dark brown or black nodules on the hair shaft. It is caused by *Piedraia hortai. (Courtesy of Everett S. Beneke, Michigan State University.)*

opportunistic pathogens; this means that they do not cause disease in healthy people but do cause *mycoses* (fungal infections) in *compromised hosts*, i.e., people already weakened with debilitating diseases. Such infections can be *systemic* (generalized throughout the body), *lymphatic* (involving the lymphatic system), or *subcutaneous* (under the skin).

The Phycomycetes have a well-developed mycelial thallus. Fertile hyphae produce sporangia at the tip of sporangiophores. In the *Rhizopus* thallus, besides the vegetative hyphae and the sporangia, there are multiple-branched, short, rootlike hyphae, called *rhizoids* (Fig. 8-9).

Sexual reproduction in some genera takes place by simple fusion of the tips of multinucleate hyphae (Fig. 8-10). These tips consist of terminal swellings of hyphal branches. This pattern of sexual reproduction is common to the genera of medical importance: *Mucor*, *Absidia*, and *Rhizopus*. These three can cause *fulminating* (sudden, severe, and rapid) infections of tissues and rapidly invade the central nervous system.

Class Ascomycetes

Members of this class are characterized by the formation of asci in which ascospores are produced (Fig. 8-4). Some Ascomycetes form fruiting bodies or *ascocarps* which surround the asci and their ascospores (Fig. 8-8). Most of the approximately 15,000 species of Ascomycetes are saprophytes. Of the parasitic species, some are the causative agents of plant disease. Potato blight and wheat rust are just two examples. Another species, *Piedraia hortai*, causes a hair infection in humans called *black piedra*; the organism is infectious in its perfect or ascospore-forming stage (Fig. 8-11).

Many yeasts are Ascomycetes since they produce ascospores. A simple pattern of ascospore formation is seen in the life cycle of the common yeast *Schizosaccharomyces* (Fig. 8-12). Asexually, this genus of yeast repro-

Figure 8-12. Life cycle of the common yeast *Schizosaccharomyces*. It reproduces asexually by binary fission. Sexual reproduction is by conjugation of compatible cells with the subsequent formation of ascospores. (Haploid: half the number of chromosomes characteristic of a species; diploid: the number of chromosomes characteristic of a species; zygote: a diploid cell resulting from the fusion of two haploid cells.)

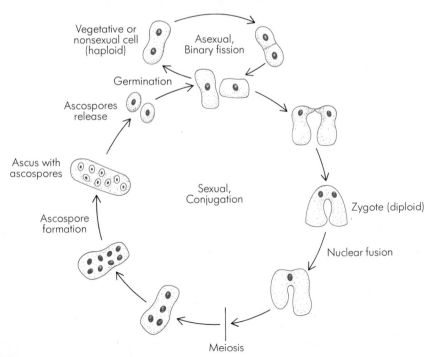

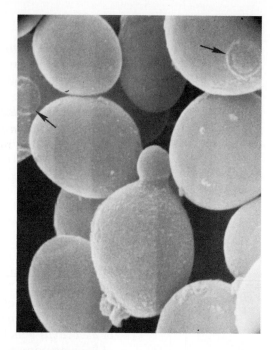

Figure 8-13. *Saccharomyces cerevisiae* showing bud and bud scars (arrows), (X 1,600) *(Courtesy of R. G. Kessel and C. Y. Shih, Scanning Electron Microscopy in Biology, Springer-Verlag, Berlin, 1974.)*

Figure 8-14. The morphology of yeasts varies widely. (A) *Saccharomyces cerevisiae* with cells appearing as vegetative forms, budding cells, and spores (× 73). *(Courtesy of George Svihla and with permission of The Microscope and Crystal Front.)* (B) *Saccharomyces ludwigii* (about X 95). *(From George Svihla, Argonne Natl Lab Annu Rep, 1965.)* (C) *Geotrichum candidum* (about X 110). (D) *Pichia membranaefaciens* (× 88). [(C) and (D), courtesy of George Svihla.]

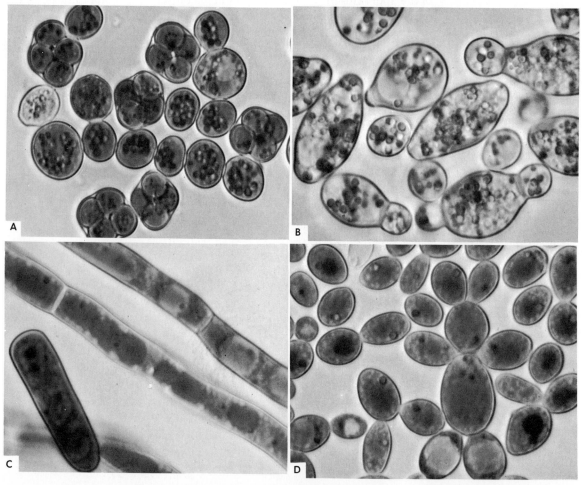

duces by transverse binary fission. Other yeasts in this class, such as the yeast *Saccharomyces cerevisiae* (used for making bread, wine, and beer), reproduce asexually by budding (Fig. 8-13). The morphology of the yeasts varies widely, as shown in Fig. 8-14. Asexual reproduction in the filamentous Ascomycetes is by the production of large numbers of conidia.

Candida albicans (Fig. 8-15) causes the condition known as *candidiasis*, a disease of the mucous membranes of the mouth (thrush), vagina, and

Figure 8-15. *Candida albicans,* a yeast pathogenic for humans. (A) Note pseudomycelia and blastospores (▶) in a urine sample from an infected patient. (B) The yeast also forms chlamydospores (▷) besides pseudomycelia and blastospores (▶), when grown on a special medium in the laboratory. *(Courtesy of L. Kapica and E. C. S. Chan, McGill University.)*

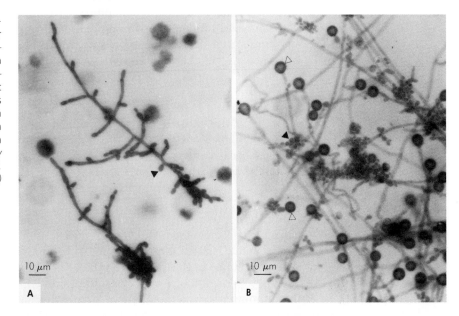

Figure 8-16. *Filobasidiella neoformans* cell. A budding cell is seen emerging from the parental cell. B = bud, C = capsule, CW = cell wall, M = mitochondria, N = nucleus, V = vacuole. The bar indicates 1 μm. *(Courtesy of Phyllis C. Braun, Georgetown University.)*

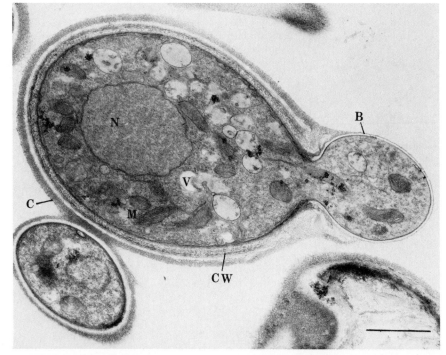

alimentary tract. More serious infections can involve the heart (endocarditis), the blood (septicemia), and the brain (meningitis). This organism may be present as a saprophyte on the above-mentioned mucous membranes in most people without causing disease. However, if the host is debilitated by a disease such as pneumonia or if competing bacteria are suppressed, as in prolonged antibiotic therapy, *Candida albicans* can cause infection.

Class Basidiomycetes

The Basidiomycetes are characterized by the production of basidiospores borne externally at the tip or side of a basidium (Fig. 8-4). Well-known Basidiomycetes include mushrooms, bracket fungi of trees, and the rusts and smuts that destroy cereal grains. Mushrooms are fruiting bodies, or *basidiocarps*, that bear the basidia and their basidiospores. Of the approximately 12,000 species of Basidiomycetes, none was implicated in human disease until recently. The sexual or perfect stage of *Cryptococcus neoformans* was found in 1975; it is now called *Filobasidiella neoformans* (Fig. 8-16). It is the most important basidiomycetous pathogen of humans, causing cryptococcosis, a systemic or generalized mycotic infection involving the bloodstream as well as the lungs, central nervous system, and other organs. Many mushrooms are extremely toxic; their *mycotoxins*, or fungal poisons, may cause death upon ingestion.

Class Deuteromycetes

This class includes fungi for which a perfect or sexual stage of reproduction has not been discovered. However, for reasons of convenience and because their conidial stages are so distinct and familiar, many species are still considered to be in this class even though their sexual stages are now well known.* Molds of the genera *Penicillium* and *Aspergillus* are classified as Deuteromycetes even though ascospore-forming stages have been found in some species. They have typical and distinctive conidial heads, as shown in

*Actually, many pathogenic fungi have two names because for many years most of them were known by their asexual stage only. They have an imperfect name, assigned when they were first described, and a later perfect name. For example, the common dermatophyte or ringworm fungus has the imperfect name *Trichophyton mentagrophytes*; its perfect-stage name is *Arthroderma benhamiae*.

Figure 8-17. Conidial heads of a few genera of fungi. Note the different arrangements of conidia, which are useful in identification.

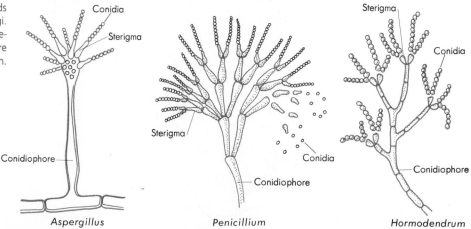

Conidia
Sterigma
Conidiophore
Aspergillus

Sterigma
Conidia
Conidiophore
Penicillium

Sterigma
Conidia
Conidiophore
Hormodendrum

Fig. 8-17. Fig. 8-18 shows the fine structure of the conidial head of *Aspergillus nidulans* by scanning electron microscopy.

Most of the fungi that are pathogenic for humans are Deuteromycetes. They form asexual spores, often several varieties within the same species, that aid in their laboratory identification. Many of them have a yeastlike parasitic phase as well as a mycelial saprophytic phase. Some of these pathogens are *Histoplasma capsulatum*, which causes histoplasmosis; *Blastomyces* spp., which cause blastomycosis; and *Coccidioides immitis*, which causes coccidioidomycosis; all these diseases are systemic infections. Histoplasmosis is an intracellular mycosis of the reticuloendothelial system, involving lymphatic tissues, lungs, spleen, central nervous system, and other organs of the body. Blastomycosis is a chronic respiratory infection, which can spread to the lungs, bones, and skin. Coccidioidomycosis can be mild or severe (see Fig. 8-19). Mild cases have lesions restricted to the upper respiratory tract and lungs. In severe cases the fungus spreads to visceral organs, bones and joints, skin, and subcutaneous tissue.

Figure 8-18. The conidial head of *Aspergillus nidulans* (X 1,250). *(Courtesy of R. G. Kessel and C. Y. Shih, Scanning Electron Microscopy in Biology, Springer-Verlag, Berlin, 1974.)*

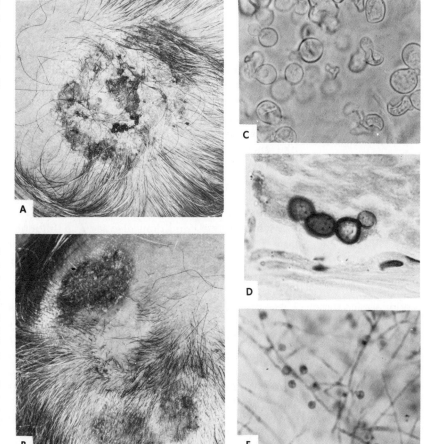

Figure 8-19. A case of North American blastomycosis caused by *Blastomyces dermatitidis*. (A) Lesion of a patient. (B) Spreading lesion 7 months later. (C) Budding cells in a biopsy section. (D) Budding cells in a smear of pus from microabscesses. (E) On Sabouraud agar at 25°C (77°F), the organism exhibits mycelium with conidia instead of the yeastlike morphology in (C) and (D). *(Courtesy of L. Kapica, McGill University.)*

The slime molds

The slime molds are a heterogeneous assemblage of microorganisms. They have both animal and plant characteristics. The acellular, creeping, somatic, or vegetative phase is definitely animallike in its structure and physiology; the reproductive structures are plantlike, producing spores covered by definite walls. This combination of animallike and plantlike phases in a life cycle is a distinguishing characteristic of slime molds.

There are four distinct types of slime molds, which differ in structure and physiology and have characteristic life histories. They are the *true* or *plasmodial slime molds* (Myxomycetes), *the endoparasitic slime molds* (Plasmodiophoromycetes), the *net slime molds* (Labyrinthulales), and the *cellular slime molds* (Acrasiales).

The slime molds have always defied classification. In one classification scheme they are treated as an independent phylum in the kingdom Protista. In another scheme they are included as a major taxon within the kingdom Fungi. In yet another classification scheme, the group of Myxomycetes has a taxonomic standing equal to that of the true fungi; the Acrasiales and the Labyrinthulales are treated separately as a group because they are "organisms of uncertain affinity," while the Plasmodiophoromycetes are considered as a class of the true fungi.

The true slime molds

The distinguishing feature of this group of slime molds is that their vegetative phase is a naked, multinucleate mass of protoplasm called a *plasmodium*. It varies greatly in size and color and changes shape as it creeps over the substrate on which it grows. It feeds by ingesting bacteria, spores of other fungi, and other small particulate organic matter present in the soil, dead leaves, or rotting logs on which it grows.

Under suitable environmental conditions the plasmodium moves about like an ameba, taking in food and increasing in size. When conditions become unfavorable for growth, the organism may become inactive. It is then converted into a dormant stage called a *sclerotium*, which is thick and hard rather than viscous or slimy. Upon return of favorable growth conditions, it becomes plasmodial again.

Environmental conditions also seem to influence the initiation of sporulation. This process is manifested by the formation of fruiting bodies. The fruiting bodies release flagellated spores, which can lose their flagella and develop into new plasmodial individuals.

The endoparasitic slime molds

These organisms have a naked multinucleate plasmodium which develops in the living tissues of a host plant. Infection occurs when a zoospore penetrates the rootlet of the host plant and immediately becomes a *myxameba* (naked ameboid organism), which develops into a plasmodium. Increase in size of the plasmodium causes swelling of the host plant roots, thus giving the name *clubroot* to the disease. As the infection proceeds, spores are formed which inhabit the host plant cells until their death and decay. Then the spores are liberated and go on to infect new seedlings. The roots of the host plants, such as potatoes and cabbages, may be damaged, leading to general stunting and premature death of the host plants.

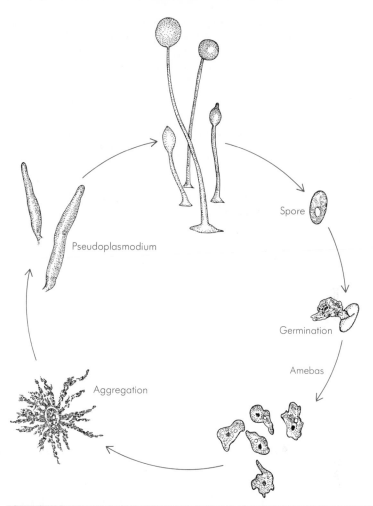

Figure 8-20. The life cycle of the cellular slime mold *Dictyostelium discoideum.* Amebas migrate into aggregation centers, becoming associated end to end in chains (see Fig. 8-21). They go through several multicellular stages (see Fig. 8-22) to form a fruiting body. The spores disperse to find a suitable environment before they germinate to form amebas which begin the life cycle anew.

Spore

Germination

Amebas

Aggregation

Pseudoplasmodium

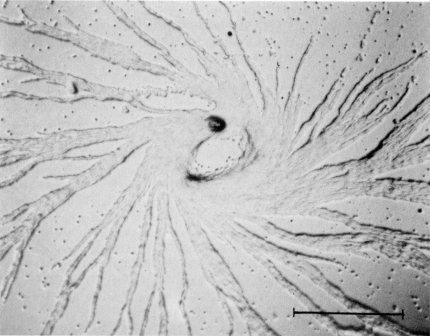

Figure 8-21. A spiral aggregation center of the cellular slime mold *Dictyostelium discoideum.* The amebas have become organized into chains and move toward a center. Bar equals 1,000 μm. *(Courtesy of T. L. Steck, University of Chicago, Science,* **204** *(4398):1163, 1979. Copyright 1979 by the American Association for the Advancement of Science.)*

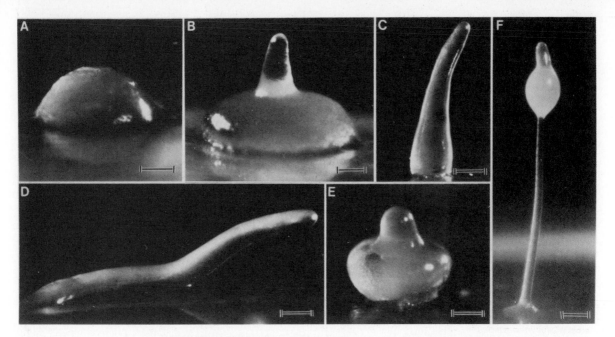

Figure 8-22. Major stages in the morphogenesis of *Dictyostelium discoideum*. (A) The first multicellular structure to form from aggregation is a broad hemispherical mound. (B) Emergence of the apical papilla (nipple). (C) Elongation into a cylindrical column. (D) The column reclines to form a slug. (E) The slug changes to form a sombrero shape. (F) A stalk is formed, resulting in a fruiting body. Bar equals 100 μm. *(Courtesy of T. L. Steck, University of Chicago, from Science,* **204** *(4398):1163, 1979. Copyright 1979 by the American Association for the Advancement of Science.)*

The net slime molds

The net slime molds are so named because of a slime they deposit in a fine network on their growing surface. Their cells are mostly oval or spindle-shaped. Net slime molds are most common in marine environments, where they grow as parasites or saprophytes on marine algae.

The cellular slime molds

These organisms are free-living and ameboid; their plasmodia are not multinucleated. They have an interesting life cycle (see Figs. 8-20 to 8-22). These organisms are ubiquitous in soil, where they feed on bacteria.

Summary and outlook

Mycology is the study of the nonphotosynthetic eucaryotic protists called fungi. The fungi comprise the yeasts and molds. Yeasts are usually unicellular; the molds are filamentous. All of them are heterotrophic microorganisms; some are saprophytes while others are parasites. Fungi reproduce by many means, including budding, fission, or sporulation. Spores can be produced asexually or sexually and they can be surrounded by fruiting bodies. Fungi can grow in environmental conditions unfavorable for most other microorganisms; these include the presence of acids and high sugar concentrations. The classification of fungi is based largely on morphological characteristics, especially on those structures related to reproduction, namely, asexual and sexual spores and their fruiting bodies. However, the identification of the unicellular yeasts, like bacteria, requires evaluation of many physiological characteristics and biochemical reactions, especially on sugars.

There are four classes of fungi: Phycomycetes, Ascomycetes, Basidiomycetes, and Deuteromycetes. Most fungal pathogens for humans are found in class Deuteromycetes. Although not a single taxonomic group, slime molds constitute an assembly of unique microorganisms that have amebalike characteristics as well as a *morphogenetic* (form-changing) life cycle.

The fungi will continue to be subjects for fundamental scientific studies, especially those concerned with *morphogenesis* (the process by which cells are organized into tissue structures). They will become increasingly important in commercial processes to supply useful products, including antibiotics such as penicillin. But there is a need for more antifungal agents, corresponding to antibacterial agents such as antibiotics, that are less toxic and more effective in the therapy of mycotic diseases. The increasing awareness of mycotoxins and their toxicity will demand greater control of mold infestations of food products. Perhaps fungal viruses (*mycophages*) can be used as a form of biological control, especially of those fungi that devastate crops and trees.

Medical mycology was previously concerned with a few pathogenic fungal species. More and more opportunistic saprophytic fungi are now implicated in infections of compromised patients. The major problems that remain in medical mycology are better clinical recognition, improvement in laboratory techniques, and more effective therapy of fungal infections.

Key terms

antheridium	mycelium (plural, mycelia)
aplanospore	mycology
arthrospore	mycophage
ascocarp	mycoses
ascospore	mycotoxin
ascus (plural, asci)	myxameba
basidiocarp	oidium (plural, oidia)
basidiospore	oogonium
basidium (plural, basidia)	oosphere
blastospore	oospore
budding	parasite
chlamydospore	perfect stage
coenocytic	plasmodium
compromised host	rhizoid
conidiospore	saprophyte
conidium (plural, conidia)	sclerotium
dimorphism	septum (plural, septa)
fission	sporangiophore
fruiting body	sporangiospore
hypha (plural, hyphae)	sporangium (plural, sporangia)
imperfect stage	subcutaneous
lymphatic	systemic
macroconidia	thallus
microconidia	zoospore
morphogenesis	zygospore

Questions

1 In what ways are fungi important to humans?

2 Yeasts, like molds, are fungi. How are they different from molds? How do they resemble bacteria?

3 Explain the difference between sexual spores and asexual spores with special reference to their formation.

4 Compare the physiology of fungi with that of bacteria.

5 What is unique about the class Deuteromycetes that makes it different from other classes of fungi?

6 What criteria are used for the classification of fungi?

7 What is customarily referred to as a cell in a filamentous mold?

8 What properties should a medium have in order to favor the isolation of fungi over bacteria?

9 Discuss the classification of the slime molds.

10 Describe the life cycle of the cellular slime mold *Dictyostelium discoideum.*

THE PROTOZOA

Protozoa (singular, *protozoan*), from the Greek *protos* and *zoon*, meaning "first animal," are eucaryotic protists that occur as single cells and may be distinguished from other eucaryotic protists by their ability to move at some stage of their life cycle and by their lack of cell walls. The study of protozoa is called *protozoology*. Protozoa are predominantly microscopic in size. *Colonies* of protozoa sometimes form; these are aggregates of independent cells.

There are more than 64,000 known species of protozoa. Approximately 32,000 are fossils, 22,000 are free-living forms, and 10,000 are parasites. Of the latter, only a few species cause disease in humans, but those species represent serious health hazards to millions of people.

Locomotion is a very important criterion in class differentiation in protozoa. Amebas move by·extending fingerlike protrusions, or *pseudopodia*, from their bodies. Ciliates move by means of the motion of tiny hairs, or *cilia*, surrounding the cells. The flagellates propel themselves by means of flagella, usually located at the end of the cell. The sporozoa move by gliding (flexing their bodies), since they do not have external organelles of locomotion.

The importance of protozoa

Protozoa serve as an important link in the *food chain* of communities in aquatic environments. For example, in marine waters, *zooplankton* (animallike organisms) are protozoa that feed on the photosynthetic *phyto-*

151

plankton (plantlike organisms). They in turn become food for larger marine organisms. This can be represented as follows:

Light energy → phytoplankton → zooplankton → carnivores
     (Primary producers)   (Primary consumers)   (Secondary consumers)

Also of particular importance in the ecological balance of many communities, in wetlands as well as aquatic environments, are the saprophytic and bacteria-feeding protozoa. They make use of the substances produced and organisms involved in the final decomposition stage of organic matter. This can be represented by the following sequence:

Dead bodies of producers → decomposition by fungi → ingestion
and consumers and their   and bacteria   by protozoa
excretion products
including feces

There are some protozoa that cause disease in animals, including humans. They multiply within the host much as bacteria do. Some live only as obligate parasites and may produce chronic or acute diseases in humans. Some well-known protozoan diseases in humans are intestinal amebiasis, African sleeping sickness, and malaria.

Morphology

The size and shape of protozoa show considerable variation. Some are oval or spherical, others are elongated, while still others are *polymorphic* (having morphologically different forms at different stages of the life cycle). Some protozoa are as small as 1 μm in diameter; others, like *Amoeba proteus* (Fig. 9-1), measure 600 μm or more. Certain common ciliates reach 2,000 μm or 2 mm and are thus easily visible without magnification.

A typical protozoan cell is enclosed by a cytoplasmic membrane. Many have an outer layer of cytoplasm, the *ectoplasm*, which can be distinguished from the inner cytoplasm, or *endoplasm*. Most cellular structures are found in the endoplasm.

Every protozoan cell has at least one nucleus. Many protozoa, however, have multiple nuclei throughout the greater part of their life cycle. In the ciliates, one large macronucleus and one small micronucleus are present. The macronucleus controls metabolic activities and growth and regeneration processes, and the micronucleus controls reproductive activity.

Over the cytoplasmic membrane of the cell is the *pellicle*. In some species of ameba the pellicle is a thin, diffuse layer. The pellicle of a ciliate is thick and often variously ridged and structured. Many protozoa form skeletal structures that give rigidity to their cells. These loose-fitting coverings external to the pellicle are called *shells* or *tests*. They have an organic matrix reinforced by inorganic substances such as calcium carbonate or

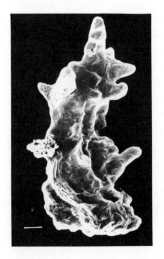

Figure 9-1. *Amoeba proteus* photographed with a scanning electron microscope. By this technique the pseudopods and other structures of the cell are made remarkably clear. The bar represents 10 μm. *(Courtesy of Eugene B. Small and Donald S. Marszalek, Science, **163**: 1064–1065, 1969. Copyright 1969 by the American Association for the Advancement of Science.)*

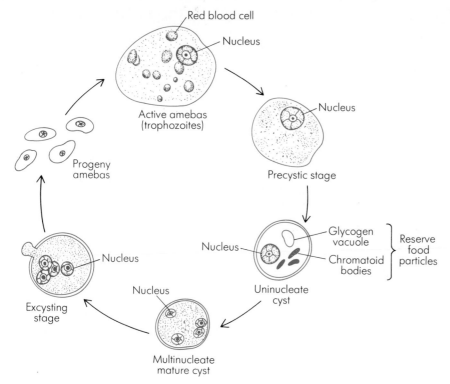

Figure 9-2. Life cycle of *Entamoeba histolytica*, the parasitic ameba of humans that causes amebic dysentery. Food or water contaminated with infective mature cysts is ingested. Excystation occurs in the host, releasing progeny amebas, which become active amebas, called trophozoites, in the intestine. Penetration of the intestinal mucosa by the trophozoites with subsequent invasion of the portal circulation may result in infection of the liver and other organs. Continued multiplication by binary fission and tissue destruction result in abscesses. Thus the accompanying diarrhea is often tinged with blood. Cysts are passed in the feces, which can infect other humans.

Figure 9-3. Binary fission in *Paramecium multimicronucleatum*. Note the transverse constriction furrow which extends inward from the equator so that one organism is eventually divided into two daughter organisms, each of which will eventually achieve the volume of the parent (X 550). *(Courtesy of R. G. Kessel and C. Y. Shih, Scanning Electron Microscopy in Biology, Springer-Verlag, Berlin, 1974.)*

silica. The presence of a pellicle, rather than a cell wall, as a covering is one of the major distinguishing characteristics of this group of protists.

Many protozoa can form *cysts*, which are temporary sheaths. In this way the vegetative forms, or *trophozoites*, protect themselves from environmental hazards, such as desiccation and depletion of food or gastric acidity within the host. The developmental stages of parasitic species which are transmitted to another host are always ensheathed by a resistant cyst (see Fig. 9-2).

Reproduction Protozoa reproduce by a variety of asexual and sexual processes. Asexual reproduction occurs by simple cell division or fission. The daughter cells may be equal or unequal in size. If there are two daughter cells, then the division process is *binary fission* (Fig. 9-3); if many daughter cells are formed, it is *multiple fission*. Fission can take place transversely (crosswise) or longitudinally (lengthwise) along the cell. Budding, a form of asexual reproduction, is also common.

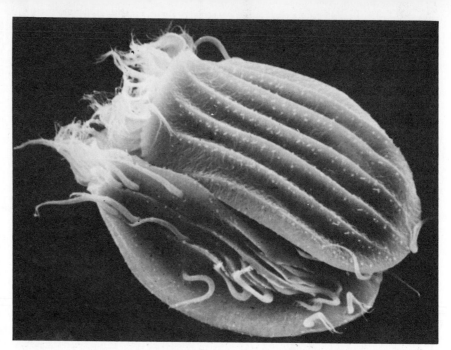

Figure 9-4. Pair of *Euplotes aediculatus* cells during conjugation. They are united by their apposed ventral surfaces (× 710). *(Courtesy of John A. Kloetzel, University of Maryland-Baltimore County.)*

Sexual reproduction occurs in various groups of protozoa. *Conjugation*, which is a temporary physical union of two individuals during which nuclear material is exchanged, is found exclusively in ciliates (Fig. 9-4).

Some protozoa have complex reproductive cycles in which part of the life cycle must occur in vertebrate hosts while another part of the cycle must occur in other hosts. For example, many species of trypanosomes spend part of their life cycle in the circulatory systems of vertebrate hosts and another part in blood-sucking invertebrates, such as insects.

Physiology

The vegetative, or *trophic*, stages of free-living protozoa occur in all aquatic environments, sand, soil, and decaying organic matter. They have been found in the polar regions, at very high altitudes, and even in the warm waters (30 to 56°C or 86 to 132.8°F) of hot springs. However, most protozoa have an optimum temperature for growth of 16 to 25°C (60.8 to 77.0°F), with a maximum at 36 to 40°C (96.8 to 104.0°F). The encysted stage can withstand a greater temperature variation than the trophic stage.

Some protozoa can tolerate a wide range of pH, for example, from pH 3.0 to 9.0. However, for the majority a pH range of 6.0 to 8.0 is optimal for maximum metabolic activity.

For those protozoa which have photosynthetic pigments (such protozoa are considered algae by some biologists) light is essential. But protozoa are generally nonphotosynthetic. Some protozoa obtain dissolved organic nutrients through their cytoplasmic membranes, as do bacteria. Other protozoa are *holozoic*; that is, they ingest food as solid particles through a mouth opening (Fig. 9-5). The ingested food is usually bacteria, algae, or other protozoa. After ingestion, the food becomes enclosed in a vacuole and enzymes degrade the complex substances into a soluble form which can be assimilated. Ingested material which is not broken down to a soluble form in the vacuole may be expelled from the cell through an anal pore or may remain in the vacuole, which then moves to the cell surface, where it breaks open to expel the waste material from the cell.

If the protozoan is a parasite, it may feed on the host's cells and tissue fluids. It may even enter the host's cells, living on the cytoplasm and the nuclei. As a result of these activities, the host may develop pathological conditions.

Sometimes, the interaction may be of mutual benefit to the two associated organisms. Such an association is called *mutualism*. For example, certain flagellates live in the gut of the termite and digest the cellulose in wood to a usable form for the termite. If deprived of these flagellates, the termite dies; if the flagellates are removed from the termite gut, they too perish. Thus the flagellates are provided with a protected environment and a supply of food.

Most protozoa are obligate aerobes or facultative anaerobes. Only a few obligate anaerobic species have been reported.

Classification

The protozoa (phylum Protozoa) may be classified into four major classes (or subphyla, depending on the authority) based on their form of locomotion. They are the flagellates, the amebas, the ciliates, and the sporozoa (Table 9-1). Protozoa of medical importance are found in all four classes.

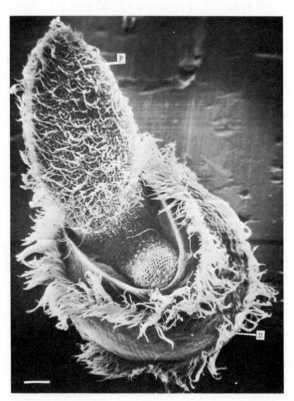

Figure 9-5. A scanning electron micrograph of a paramecium (P) being ingested by a *Didinium* sp. (D) Note that the cilia on the paramecium are being lysed as the paramecium enters the pharynx of the predator. The bar represents 10 μm. *(Courtesy of Eugene B. Small and Donald S. Marszalek, Science, **163**:1064–1065, 1969. Copyright 1969 by the American Association for the Advancement of Science.)*

Table 9-1. The major classes of protozoa

MAJOR GROUPS (COMMON NAME)	MODE OF LOCOMOTION	MODE OF REPRODUCTION	OTHER CHARACTERISTICS
Mastigophora (flagellates)	Flagella (one or more)	Longitudinal binary fission; sexual reproduction in some groups	Nutrition is phototrophic, heterotrophic, or both
Sarcodina (amebas)	Pseudopods mainly	Binary fission; no sexual reproduction	Most species free-living; heterotrophic
Ciliata (ciliates)	Cilia (numerous)	Transverse binary fission; sexual reproduction by conjugation	Most species free-living; heterotrophic
Sporozoa (sporozoa)	Movement by gliding or nonmotile; no external locomotor appendages	Multiple fission; may have flagellated microgametes in sexual reproduction	All species parasitic

The flagellates

The flagellates are divided into two groups: the plantlike forms, or *phytoflagellates*, and the animallike forms, or *zooflagellates*. Phytoflagellates contain chlorophyll and are photosynthetic. Zooflagellates are heterotrophic. All divide by longitudinal fission, and some have a sexual stage of reproduction.

Unlike the amebas, the flagellate cytoplasm is surrounded by a well-defined pellicle that helps to give shape to the organism (Fig. 9-6). Besides flagella, undulating membranes protrude from the organism. Both the flagella and the protruding membranes are used for locomotion and/or for gathering food (Fig. 9-7).

A number of flagellates infect humans, causing genital, intestinal, and systemic diseases. Most intestinal flagellates have a trophic and an encysted stage. The encysted stage, when present, is the stage which is transmissible to humans. Otherwise the trophic stage is infective. Intestinal flagellates are found in the small intestine as well as in the cecum and colon. Some, like *Giardia lamblia*, the only intestinal protozoan that causes

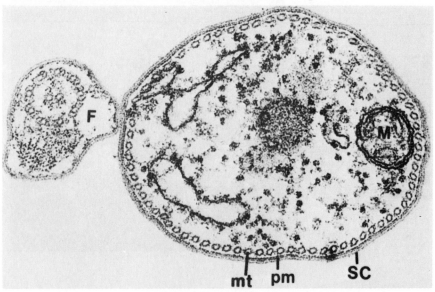

Figure 9-6. Electron micrograph of *Trypanosoma rhodesiense* (cause of African sleeping sickness). Note the presence of the pellicle or surface coat (SC), plasma membrane (pm), mitochondrion (M), microtubules (mt), and flagellum (F). (× 97,300). *(Courtesy of George C. Hill, Colorado State University, Science 202: 763–765, 1978. Copyright 1978 by the American Association for the Advancement of Science.)*

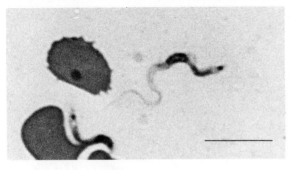

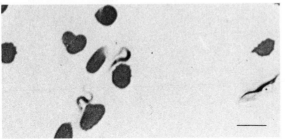

Figure 9-8. Flagellated protozoa. (A) *Euglena gracilis* are solitary, free-living flagellates with chlorophyll. (B) *Giardia lamblia,* a parasite found in the human intestine where it may cause dysentery. (C) *Trichomonas hominis,* a parasite found in the human intestine. Its role as a cause of disease has not been established. (D) *Trypanosoma rhodesiense,* the cause of African sleeping sickness. (E) *Trypanosoma cruzi,* the cause of Chagas' disease, or South American trypanosomiasis. (F) *Codosiga,* a colonial flagellate with a transparent protoplasmic collar into which food particles are whipped by the action of the flagella. (G) *Trichonympha,* a protozoan that inhabits the intestines of termites, where it converts wood cellulose into soluble substances that can be utilized by the termite. *(Redrawn after Ralph Buchsbaum, Animals without Backbones, rev. ed., The University of Chicago Press, Chicago, 1948.)*

Figure 9-7. Trypanosomes exhibiting flagella and undulating membranes in a blood smear. Bar equals 10 μm. *(Courtesy of Z. Ali-Khan and E. C. S. Chan, McGill University.)*

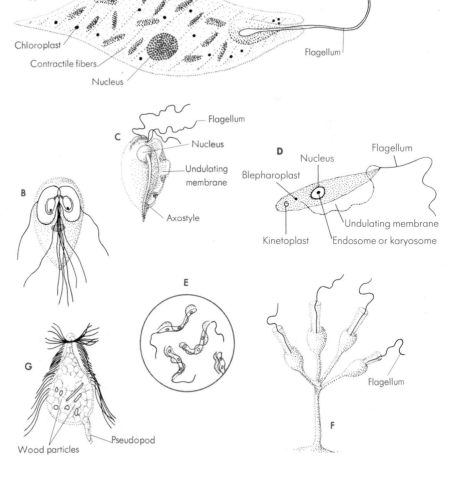

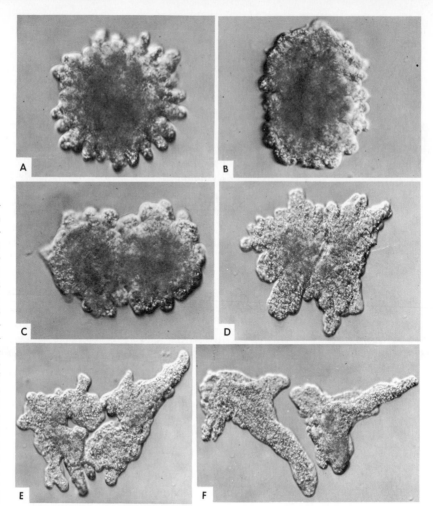

Figure 9-9. Amebas reproduce by fission, as shown in this series of photomicrographs of the dividing ameba. Total elapsed time from (A) through (F) is 21 min. Intermediate stages photographed at (B) 6 min, (C) 8 min, (D) 15 min, (E) 18 min, and (F) 21 min. Reproductive material is concentrated in a band at the equator of the nucleus; the chromosomes divide, half of each going to one of the two new nuclei. As the cell divides, one of the newly formed nuclei goes to each daughter cell; reproduction is complete when the cells are completely separated. *(Courtesy of Carolina Biological Supply Company.)*

dysentery or diarrhea, are found mainly in the duodenum. Their transmission occurs primarily through contaminated food or drink and by hand-to-mouth contact. Other so-called intestinal flagellates are found in genital tracts. *Trichomonas vaginalis* causes one type of vaginitis, that is, inflammation of the vagina with discharge and accompanying burning and itching. The organism has no encysted stage and is spread as a venereal disease.

In addition to the intestinal flagellates, a second group, the hemoflagellates (or blood and tissue forms), are transmitted by blood-sucking insects to humans, where they produce severe and sometimes fatal infections. The genera of these organisms are *Trypanosoma* and *Leishmania*. Trypanosomiasis includes African sleeping sickness, while leishmaniasis involves lesions of the skin or visceral organs depending on the species.

Figure 9-8 shows the morphology of some flagellates.

The amebas   Amebas derive their name from the Greek word *amoibe*, meaning "change," because their shapes are constantly changing. Amebas use pseudopodia, or "false feet," which are really protoplasmic extensions, to move on a surface and to engulf food particles. Such food particles are enclosed in vacuoles, where they are digested. Figure 9-9 shows the typical morphology of amebas as well as their mode of reproduction, binary fission. Some amebas have the capacity to develop cysts.

Species of the genus *Entamoeba* inhabit the intestinal tract of vertebrates. Many of them, such as *Ent. gingivalis*, which lives in the human mouth, and *Ent. coli*, which inhabits the human intestine, are harmless. However, one species, *Ent. histolytica*, is the cause of amebic dysentery or amebiasis in humans. Transmission is by ingestion of the mature cyst.

Amebic infections of humans are usually confined to the intestine. But the blood sometimes carries the amebas to other organs of the body, causing abscesses in the liver, lung, spleen, pericardium, and brain.

The ciliates　　The ciliates may be divided into two groups: those that have cilia over only part of the cell (Fig. 9-10) and those that have cilia distributed evenly over the entire cell (Fig. 9-11). (The common *Paramecium* is an example of the latter group.) The cilia are responsible for rapid movement through aqueous environments. Also, when located around oral grooves or cavities, they create a whirlpool effect which helps in food collection.

Most ciliates divide by transverse binary fission. Sexual reproduction is by conjugation of two cells (Fig. 9-4). Also, as mentioned earlier, they possess at least one macronucleus and one or more micronuclei per cell.

Figure 9-10. Ventral surface of a *Euplotes aediculatus* cell. (A) Cirri. Each cirrus is composed of 80 to 100 individual cilia that are not fused but beat as a functional unit in locomotion. (B) Cilia. Two to three long rows of cilia that function in locomotion as well as in food collection. (C) Buccal cavity. (D) Cell "mouth." (× 512). *(Courtesy of John A. Kloetzel, University of Maryland-Baltimore County.)*

Figure 9-11. Ciliary beat of *Paramecium bursaria*. The ciliary rows on the dorsal surface are illustrated. Note that by special techniques the pattern of ciliary coordination and the form of ciliary beat have been preserved for scanning electron microscopy (× 457). *(Courtesy of R. G. Kessel and C. Y. Shih, Scanning Electron Microscopy in Biology, Springer-Verlag, Berlin, 1974.)*

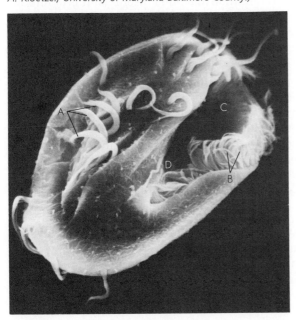

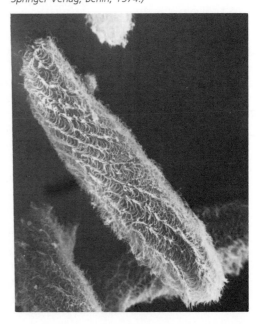

Most ciliates are free-living. *Balantidium coli*, a parasite, is the only species causing disease (bloody diarrhea) in humans. It lives in the gastrointestinal tract of some vertebrates and both trophozoites and cysts are passed in the feces. The encysted stage is transmissible to humans because it can survive in soil and water for some time. Swine are usually regarded as having an important role in human infection. Humans may ingest water or food contaminated with cysts present in swine feces.

The sporozoa

All sporozoa are parasitic for one or more animal species. Adult forms have no organs of motility but all are probably motile by gliding at one stage of their life cycle. They cannot engulf solid particles but feed on the host's cells or body fluids.

Many have complicated life cycles, certain stages of which may occur in one host and other stages in a different host. They all produce spores at some time in their life history. Their life cycles exhibit an *alternation of generations* of sexual and asexual forms, such that the intermediate host usually harbors the asexual forms and the final host the sexual forms. Sometimes humans serve as hosts to both forms.

Toxoplasmosis and malaria are the major human diseases caused by sporozoa. *Toxoplasma gondii* is the etiologic agent of toxoplasmosis. The symptoms of this disease vary greatly depending on the location of the parasites in the body. They can mimic, for example, the symptoms of meningitis and hepatitis. *Toxoplasma gondii* is the most widespread of the parasites that infect vertebrates. More than 50 percent of adults in the United States have been infected at some time, but the disease in humans is usually mild and asymptomatic. Spontaneous recovery usually follows. However, transplacental infection, that is, infection of the human embryo, may occur with serious consequences. The result may be birth of a stillborn child or a child with mental retardation and other disorders. It is of interest that the parasite will undergo its sexual reproductive cycle only in the intestinal cells of members of the cat family, including the domestic cat.

The most important sporozoa are those that cause malaria (see also Chap. 31). Malaria is a mosquito-borne disease of humans caused by sporozoa belonging to the genus *Plasmodium* which infect the liver and red blood cells. The final host for the parasite is the female anopheline mosquito; sexual reproduction of the parasite occurs in this host. Malaria has been one of the greatest killers of humans through the ages. At the present time, it has been conservatively estimated that 150 million people in the world have the disease and that about 1.5 million of them will die of it.

Four species of *Plasmodium* cause the following forms of malaria in humans:

P. vivax: benign tertian malaria (intermittent chills and fever at 48-h intervals, or every other day)

P. ovale: benign tertian malaria (symptoms same as for *P. vivax*)

P. malariae: benign quartan malaria (chills and fever at 72-h intervals, or every third day)

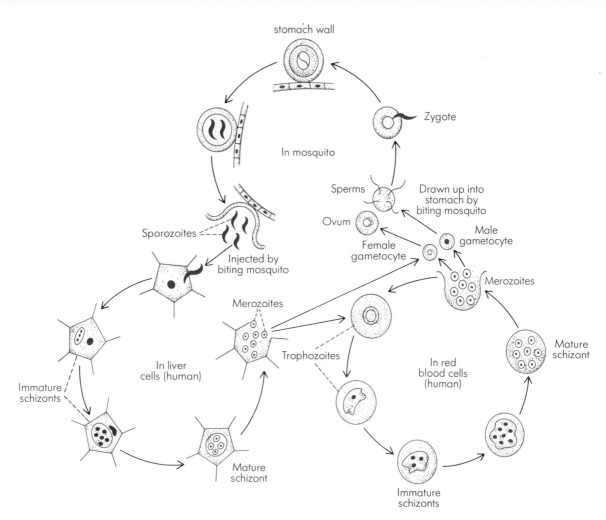

stomach wall

In mosquito

Zygote

Sperms

Drawn up into stomach by biting mosquito

Ovum

Male gametocyte

Female gametocyte

Merozoites

Sporozoites

Injected by biting mosquito

Merozoites

Trophozoites

In red blood cells (human)

Mature schizont

In liver cells (human)

Immature schizonts

Mature schizont

Immature schizonts

Figure 9-12. Life cycle of *Plasmodium* species that cause malaria. Sexual reproduction occurs in the mosquito. Asexual reproduction takes place in humans in liver cells as well as in red blood cells. Sporozoites injected by the mosquito's bite enter liver cells via the blood stream and multiply asexually *(schizogony)*. The resulting merozoites enter the red blood cells. Gametocytes formed from the merozoites are drawn up from the blood by the biting mosquito and transformed into gametes in the insect's stomach. The zygotes encyst externally on the stomach wall and form oocysts where, by asexual multiplication *(sporogony)*, many sporozoites are formed which then invade the mosquito's salivary glands. From there they are injected by the mosquito's bite into another human victim.

    P. falciparum: malignant tertian malaria (irregular chills and fever; if untreated, frequently fatal)

    The complex life cycle of the malaria-causing *Plasmodium* is illustrated in Fig. 9-12.

Summary and outlook

Protozoology is the study of protozoa, eucaryotic protists that exhibit movement at some stage of their life cycle, and that do not have cell walls. Some do, however, have a thickened pellicle or a rigid shell over the cytoplasmic membrane. There are free-living protozoa as well as parasitic ones. The free-living forms are very important in maintaining ecological balance in many natural environments, particularly aquatic environments. Some of the parasites are pathogenic for humans.

Protozoa have great diversity in size and shape and some species are polymorphic. Many of them can encyst, and cysts are important in the transmission of protozoan diseases. They are structurally more complex and generally larger than procaryotic protists.

Reproduction in protozoa is by a variety of asexual and sexual processes, depending on the species and environmental conditions. Some protozoa have very complex life cycles.

The various protozoa obtain nourishment by many means. Some are photosynthetic; others absorb dissolved nutrients; and still others ingest solid food particles.

There are four major groups of protozoa, characterized mainly by their mode of locomotion. These groups are the amebas, the ciliates, the flagellates, and the sporozoa. Protozoa of medical importance are found in all four groups.

Protozoa are of greatest medical concern in the tropics and subtropics because their transmission to animals and humans is enhanced by poor sanitation and high human-population densities, as well as by *insect vectors* or carriers of disease. However, threats of increased incidence of protozoan infections in temperate zones are very real. This situation is brought about by the greater mobility of people and thus of parasitic protozoa because of jet travel and the widespread prohibition of proven insecticides.

Key terms

conjugation	insect vector	polymorphic	test
cyst	multiple fission	protozoan (plural,	trophic stage
ectoplasm	mutualism	protozoa)	trophozoite
endoplasm	pellicle	protozoology	zooflagellate
food chain	phytoflagellate	pseudopodia	zooplankton
holozoic	phytoplankton	shell	

Questions

1 Why is this group of protists called protozoa?
2 Explain how locomotion is used as a criterion in class differentiation of protozoa.
3 Describe the roles played by protozoa in a natural environment.
4 Are all parasitic protozoa pathogenic for humans? Explain with appropriate examples.
5 Morphologically, how is a protozoan different from a fungus?
6 Compare the modes of asexual reproduction of protozoa with those of bacteria.
7 Describe several ways in which protozoa obtain nourishment.
8 Compare and contrast the general modes of reproduction of the four major groups of protozoa.

THE ALGAE

Algae (singular, *alga*) range in size from a few micrometers to many meters in length. They contain chlorophyll as well as other pigments for carrying out photosynthesis and are widespread in nature. They occur in almost all environments where there is sunlight. They vary widely in morphology and other characteristics.

Most algae are microscopic in size. This means that the techniques employed by the microbiologist to study bacteria and fungi may be used also to study these microalgae. The study of algae is called *phycology*.

The importance of algae

Most algae are aquatic organisms. Since 70 percent of the earth's surface is covered with water, it is probable that as much carbon is *fixed* (captured as carbon dioxide and changed to organic carbon compounds such as sugars) through photosynthesis by algae as is fixed by all land flora. Tiny floating algae constitute part of the *phytoplankton* (suspended flora of the sea) and serve as an important food source for other organisms, including large ones such as whales. These algae form the base or beginning of most aquatic food chains because of their photosynthetic activities and are therefore called *primary producers* of organic matter.

In addition, in all environments, algae produce oxygen during photosyn-

163

thesis. This gas is used by animals as well as other organisms for aerobic respiration. The oxygen so produced is also important in problems of pollution control and waste disposal. For example, in some areas organic matter from sewage is introduced into shallow ponds for oxidation. The oxidation is carried out by bacteria whose activities are enhanced by the oxygen provided by associated algae.

Algae are also present in soil even if their presence is not so obvious. They are probably important in stabilizing and improving the physical properties of soil by aggregating particles and adding organic matter.

Algae are used by humans in many ways. In countries where the red and brown seaweeds (algae) are plentiful, they are used as fertilizer. Diatomaceous earth, which is essentially the remains of dead algae called diatoms (accumulated over geologic periods in huge deposits in various parts of the world), is used as a polishing agent in many polishes. It has been used also for making heat-insulating material and in some kinds of filters.

Many algae synthesize vitamins A and D. As the algae are eaten by fish, the vitamins are stored in the organs (for example, liver) of fish, from which they can be extracted or used directly as a rich source of vitamins for human consumption (such as cod-liver oil). Green algae contain appreciable amounts of vitamins B_1, C, and K.

Algae are used as food, especially in the Orient. The Japanese cultivate and harvest *Porphyra*, a red alga, as a food crop. Red algae yield two important polysaccharide products, carrageenan and agar. They are used as emulsifying, gelling, and thickening agents in many of our foods. (Agar, as you have learned, is used as a solidifying agent in microbiological media.) From the giant kelps (brown algae) found along the southern California coasts, alginates are extracted and widely used in food and pharmaceutical products in much the same way as carrageenan and agar.

Although few algae are pathogenic, one, *Prototheca*, has been reported to be a probable pathogen of humans. It has been found in systemic and subcutaneous infections, as well as in *bursitis*, an inflammation of the joints (see Figs. 10-1 and 10-2). Some airborne algae have been implicated in inhalant allergies. Several species are parasitic on higher plants. For example, the green alga *Cephaleuros* attacks leaves of tea, coffee, pepper, and other tropical plants, causing considerable damage.

Some aquatic algae produce toxins which are lethal to fish and other animals. These toxins are either extracellular or are liberated from the alga

Figure 10-1. *Prototheca wickerhamii* in histological section from an infected joint of a patient. *(Courtesy of L. Kapica, McGill University.)*

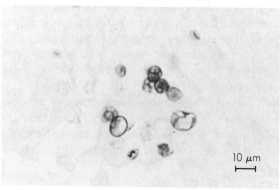

10 μm

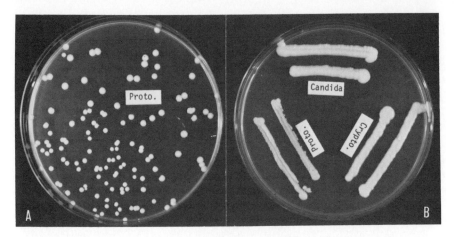

Figure 10-2. *Prototheca* cells are colorless and do not carry out photosynthesis. However, the presence of starch-containing inclusions constitutes significant evidence for their algal nature. They resemble yeast cells in their gross growth characteristics on media, which can lead to errors in laboratory diagnosis. (A) Colonies of *P. wickerhamii* on Sabouraud agar. (B) Streaked cultures of *P. wickerhamii* (Proto.), *Candida albicans* (Candida), and *Cryptococcus (Filobasidiella) neoformans* (Crypto.). The latter two species are yeasts. *(Courtesy of L. Kapica, McGill University.)*

Figure 10-3. The green alga *Chlamydomonas* in the vegetative and the palmelloid state (see text p. 171). Usually the cells in the palmelloid state are nonflagellated and are embedded in a gelatinous matrix. Flagella reappear and the cells swim away when favorable conditions return. *(Courtesy of R. G. Kessel and C. Y. Shih, Scanning Electron Microscopy in Biology, Springer-Verlag, Berlin, 1974.)*

Figure 10-4. Spherical colonies of *Volvox aureus.* Each spherical colony may contain up to 20,000 single-celled flagellates embedded in a gelatinous sphere. The individuals are usually joined by cytoplasmic threads. One parental colony has developing daughter colonies. *(Courtesy of Carolina Biological Supply Company.)*

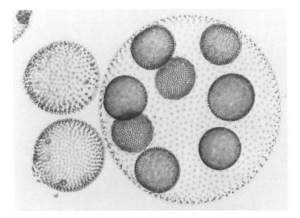

by bacterial decomposition of algal *blooms*, which are very dense populations of colored algae covering several square kilometers of the sea. Certain marine algae (dinoflagellates belonging to the genera *Gymnodinium* and *Gonyaulax*) cause death of aquatic animals by producing a *neurotoxin*, or nerve poison. These toxins are among the most potent toxins known. Certain toxins of some blooms are concentrated within the digestive glands or siphons of filter-feeding bivalve mollusks (clams, mussels, scallops, oysters, etc.) and cause paralytic shellfish poisoning when ingested by humans.

Morphology Many algal species occur as single cells that may be spherical, rod-shaped, club-shaped, or spindle-shaped. They may or may not be motile. A typical unicellular green alga is shown in Fig. 10-3. Other algal species form

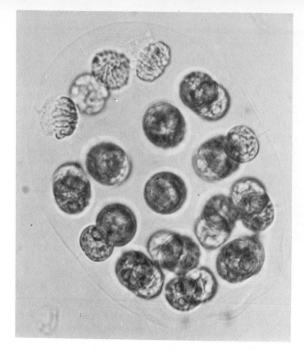

Figure 10-5. A colony (circumscribed by thin outer line) of the green alga *Eudorina elegans* var. *carteri*. This photomicrograph shows an early stage in sexual reproduction just after sperm-packet formation (top four less dense cells). Other cells are eggs. (× 650). *(Courtesy of M. Goldstein, McGill University.)*

Figure 10-6. The fine structure of an algal cell is revealed by this near-median longitudinal section of a cell of *Ochromonas danica*. The single Golgi body (g) lies anterior to the nucleus (n). The endoplasmic reticulum (ER) is visible on the right side of the section. A single chloroplast (C) is shown, but it appears to be two chloroplasts because of the site of sectioning. A large starch-containing vacuole (v) occupies almost half of the cell. Numerous mitochondria (m) are present in the peripheral cytoplasm of the cell (× 12,120). *(Courtesy of Sarah P. Gibbs, McGill University.)*

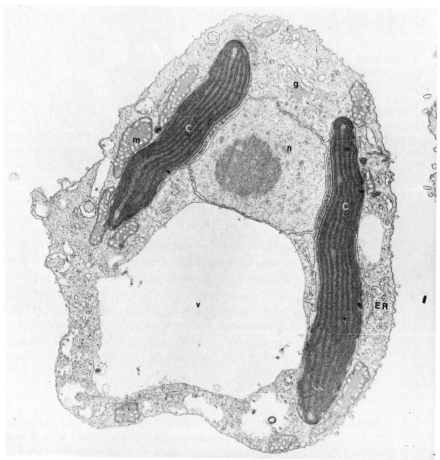

multicellular colonies (Figs. 10-4 and 10-5). Some colonies are simply aggregations of single, identical cells that cling together after division; others are composed of different kinds of cells specializing in particular functions. Other algae are multicellular and are quite large and complex in morphology.

Algae, like all other eucaryotic protists, contain a membrane-bound nucleus. Other inclusions are starch and starchlike grains, oil droplets, and vacuoles. One or more chloroplasts occur per cell. They may be ribbonlike or in the form of discrete disks as in green plants. Within the chloroplast matrix are found flattened membranous vesicles called *thylakoids*. The thylakoid membrane contains chlorophylls and accessory pigments and is the site of the light reactions in photosynthesis. The fine structure of an algal cell is shown in Fig. 10-6.

The motile algae have flagella occurring singly, in pairs, or in clusters at the anterior or posterior end of the cell. Other structures occurring in some algae include exterior spines or knobs and stalks by which they are anchored to some object.

Reproduction

Algae reproduce either sexually or asexually. Some species are limited to one or the other of these processes, but many have complicated life cycles involving both types of reproduction.

Asexual reproduction includes the simple binary fission by which bacteria also reproduce. A new algal organism may even start from a fragment of an old multicellular organism from which it has broken off. However, most asexual reproduction is more complex than this and involves the production of unicellular spores. Among these are *akinetes*, which are essentially vegetative cells that have thickened walls and thus are better able to withstand desiccation and other conditions not conducive to vegetative cell growth. Many asexual spores of aquatic algae have flagella and are motile. These are called *zoospores*. Nonmotile spores, or *aplanospores*, are more likely to be formed by the terrestrial types of algae.

All forms of sexual reproduction are found among the algae. In these processes there is a conjugation of *gametes* (sex cells) forming a *zygote*. If the gametes are morphologically similar, the process is said to be *isogamous*. If the gametes differ in size, the process is *heterogamous*. In the higher forms of algae, the sexual cells become more characteristically male and female. The *ovum* (female egg cell) is large and nonmotile, and the male gamete (sperm cell) is small and actively motile. This type of sexual process is called *oogamy*. If the male and female gametes occur in the same individual of the species, the individual and species are said to be *bisexual*. If the male and female gametes are produced by different individuals, these individuals are *unisexual*.

Physiology

The algae are aerobic photosynthetic organisms. They are found wherever sufficient light, moisture, and simple nutrients are available to sustain them.

Table 10-1. Comparative
physiology of algae and
bacteria

CHARACTERISTIC	ALGAL	BACTERIAL
Optimum pH	4–11	6.5–7.5
Optimum temperature	20–30°C	20–37°C
Gases	Aerobic	Aerobic → anaerobic
Light (for growth)	All (with few exceptions)	Some photosynthetic groups
Carbon	Inorganic for most	Organic and inorganic
Cell-wall structural components	Mostly cellulose; replaced by xylans and mannans in some	Peptidoglycan

Some species of algae grow on the snow and ice of polar regions and mountain peaks, sometimes in such abundance that the landscape becomes colored by the pigments of their cells. Some algae grow in hot springs at temperatures as high as 70°C (158°F), although the optimal growth temperatures of these thermal algae are between 50 and 54°C (122 to 130°F). The rocks in certain hot springs in Yellowstone National Park are colored bluish green by them. Some freshwater algae have adapted their metabolism to the high salt concentrations found in the brine lakes of the arid southwestern United States.

Marine algae adapt to variations in salt concentration in various parts of the sea. These algae are not normally found in northern ocean waters at depths greater than 150 to 180 ft (45.7 to 54.9 m) because insufficient sunlight reaches deeper waters in these latitudes. In the clearer, warmer tropical waters, where the sunlight is more direct and has a longer average daily period, they may be found at depths as great as 600 ft (183 m). These and other factors are responsible for the *zonation* phenomenon, or stratification of certain kinds of algae at certain depths and locations in the ocean.

Some algae are adapted to moist soil, the bark of trees, and even the surface of rocks, which the algae degrade, making the decomposition products available for soil building and enrichment.

Algae have three kinds of photosynthetic pigments: chlorophylls, carotenoids, and phycobilins. All the photosynthetic pigments are found in chloroplasts in algae. All algae have chlorophyll a, which is present in all photosynthetic organisms other than photosynthetic bacteria. Other chlorophylls are b, c, d, and e. They are distinguished from each other by slight differences in molecular structure, and these in turn determine the wavelengths of light that each chlorophyll type can absorb as energy. There are two kinds of carotenoids, carotenes and xanthophylls. There are two kinds of phycobilins, phycocyanin and phycoerythrin. The presence of the other pigments may mask the green color of chlorophyll; for example, some algae are brown because they have a relative abundance of xanthophylls and carotenes that mask the green color reflected by chlorophyll. Other algae appear purplish or reddish because of their phycobilin content. (It should be noted that some algae are colorless, do not photosynthesize, and are considered to be protozoa by some scientists.)

As a result of their photosynthetic activity, algae store various reserve food products as granules or globules in their cells. For example, green algae store a plantlike starch. Other algae may store other kinds of starchlike carbohydrates. Some algae store oils or fats.

Table 10-1 summarizes some physiological characteristics of algae and compares them with those of bacteria.

Classification

Although scientists do not agree on the details of algal classification, algae are generally classified on the basis of the following characteristics:

1 Pigments: their chemical composition
2 Reserve food products: their chemistry
3 Flagella (if present): their number and morphology
4 Cell walls: their chemistry and physical features
5 Cellular organization
6 Life history (the complete series of changes in an organism) and reproduction

A summary of the more significant characteristics in the algal divisions (some call these phyla) is shown in Table 10-2. The major divisions of algae of microbial dimensions will be discussed briefly. Only a few outstanding characteristics of each group will be mentioned.

Chlorophycophyta

This large and diverse group of organisms, called green algae, consists mainly of freshwater species. However, some species are found in seawa-

Table 10-2. Significant characteristics* of major algal taxonomic groups (divisions)

DIVISION (COMMON NAME)	STORAGE PRODUCTS	FLAGELLA AND DETAILS OF CELL STRUCTURE
Chlorophycophyta (green algae)	Starch, oil	Mostly nonmotile (except one order), but some reproductive cells may be flagellated
Rhodophycophyta (red algae)	Floridean starch (like glycogen)	Nonmotile; agar and carrageenan in cell walls
Chrysophycophyta (golden algae)	Starchlike carbohydrates; oils	Flagella: 1 or 2 equal or unequal; in some, cell surface covered by characteristic scales
Phaeophycophyta (brown algae)	Starchlike carbohydrates; mannitol	Flagella: 2 lateral, unequal; alginic acid in cell walls
Bacillariophycophyta (diatoms)	Starchlike carbohydrates; oils	Flagella: 1 in male gametes, apical; cell in 2 halves; walls silicified with elaborate markings
Euglenophycophyta (euglenoids)	Starchlike carbohydrates; oils	Flagella: 1, 2, or 3 equal, slightly apical; gullet present; no cell walls but have elastic pellicle
Cryptophycophyta (cryptomonads)	Starch	Flagella: 2 unequal, lateral; gullet in some species; no cell walls
Pyrrophycophyta (dinoflagellates, phytodinads)	Starch; oils	Flagella: 2 lateral: 1 trailing, 1 girdling
Xanthophycophyta (yellow-green algae)	Starchlike carbohydrates; oils	Flagella: 2 unequal, apical

*Principal pigments of differentiation have been omitted from this table for simplification.

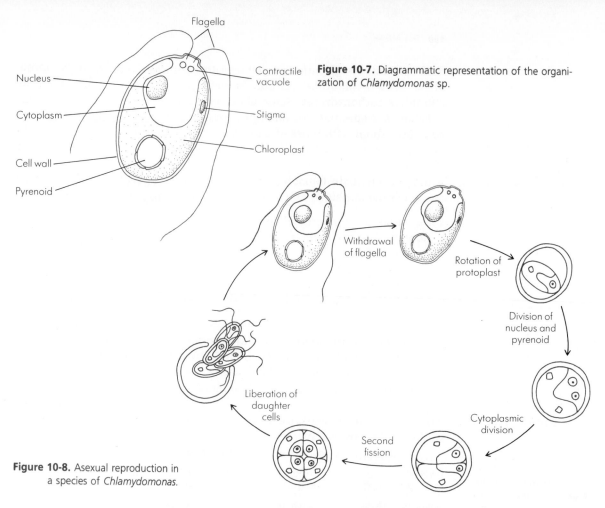

Figure 10-7. Diagrammatic representation of the organization of *Chlamydomonas* sp.

Flagella

Nucleus

Cytoplasm

Cell wall

Pyrenoid

Contractile vacuole

Stigma

Chloroplast

Withdrawal of flagella

Rotation of protoplast

Division of nucleus and pyrenoid

Cytoplasmic division

Second fission

Liberation of daughter cells

Figure 10-8. Asexual reproduction in a species of *Chlamydomonas*.

ter, and some are terrestrial. The majority of green algae contain one chloroplast per cell. The chloroplasts often contain centers of starch formation called *pyrenoids*.

There are many single-celled, colonial, and other morphological types of green algae. Many unicellular green algae exhibit motility due to the presence of flagella. Some species have holdfasts which anchor them to submerged objects or aquatic plants.

Green algae reproduce by fission, by formation of asexual flagellated zoospores, or by isogamous and heterogamous sexual modes.

Chlamydomonas is considered the typical green alga (Fig. 10-3). It is a unicellular, motile green alga and is widely distributed in soils and freshwater. Its cellular organization is shown in Fig. 10-7. It ranges in size from 3 to 30 μm in common forms, and it is motile except during cell division. Motility derives from the presence of two flagella. Each cell has one nucleus and a single large chloroplast that in most species is cup-shaped. The cell wall contains cellulose. The *red eyespot* (*stigma*) is the site of light perception and controls the phototactic response (movement toward light) of the organism. It also has a pyrenoid. The *contractile vacuole* is for forcing excess water from the cell.

In asexual reproduction the free-swimming individual becomes nonmotile by withdrawing its flagella and undergoes longitudinal fission of the protoplast (organism within the cell wall) to form two, four, or eight daughter protoplasts. The daughter cells develop two flagella each and construct new cell walls. They are then liberated from the parent cell wall. This cycle may be repeated indefinitely in laboratory culture or in nature (Fig. 10-8).

In some cases, the daughter cells do not develop flagella and are unable to escape. Instead, they keep on multiplying within a more or less gelatinized matrix. The masses of cells so formed are called *palmelloid* stages (Fig. 10-3). The formation of such stages is determined by environmental conditions which are generally favorable to growth but not to motility. Any individual cell, however, can develop flagella and escape from the mass. Palmelloid stages occur in many algae either as predominant or as occasional phases of development.

Under certain conditions, sexual reproduction occurs in many species of *Chlamydomonas*. One of the conditions for sexual reproduction is the compatibility of mating types within the same species. When this is present, an aggregation of individuals follows. From these aggregates, pairs of gametes emerge for gametic union to form zygotes.

In addition to motile unicellular algae such as *Chlamydomonas*, nonmotile unicellular green algae are widely distributed. One of the most important of these is *Chlorella*, which has served as a useful biological

Figure 10-9. Scanning electron micrograph of a desmid, *Cosmarium* sp. The cells are uninucleate and composed of two halves with a constriction in the middle of the cell. A single nucleus lies embedded in the middle of the cytoplasm that separates the two chloroplasts into semicells (SC) (× 800). *(Courtesy of R. G. Kessel and C. Y. Shih, Scanning Electron Microscopy in Biology, Springer-Verlag, Berlin, 1974.)*

system in the laboratory for many investigations on photosynthesis and supplemental food supplies.

Prototheca, usually considered a colorless *Chlorella*, has been found to be pathogenic for humans (see Figs. 10-1 and 10-2).

Desmids (Fig. 10-9) are green algae found in a variety of attractive shapes and designs. Each cell is made up of two symmetrical halves (semicells) containing one or more chloroplasts.

Chrysophycophyta

These are golden-brown algae. The species of Chrysophycophyta are predominantly flagellates, but some are ameboid by virtue of pseudopodial extensions of the protoplasm. The naked ameboid forms can ingest particulate food by means of pseudopodia. Nonmotile coccoid and filamentous forms are also included in this division.

The Chrysophycophyta differ from the green algae in their frequent incorporation of silica either in their walls or as scales. Most species are unicellular, but some form colonies. The characteristic color of chrysophytes is due to the masking of their chlorophyll by brown pigments. Reproduction is commonly asexual by binary fission but occasionally sexual by isogamy.

Ochromonas is a unicellular genus with two flagella, one long and one short. One species is remarkably versatile in its nutrition. It may grow autotrophically by using light energy or heterotrophically by absorption of solutes or engulfment of particulate matter. The fine structure of a chrysophyte alga is shown in Fig. 10-10. Although many chrysophytes are to some degree ameboid, *Chrysamoeba* is a strongly ameboid genus with flagella (see Fig. 10-11).

Figure 10-10. The fine structure of a chrysophyte alga *Olisthodiscus luteus*. (Some workers consider it a xanthophyte.) The cell has a number of peripherally located discoidal chloroplasts surrounding a large central nucleus and scattered mitochondria with tubular cristae. (C = chloroplast, N = nucleus, Nu = nucleolus, M = mitochondrion, G = Golgi apparatus.) (× 6,200). *(Courtesy of Sarah P. Gibbs, McGill University.)*

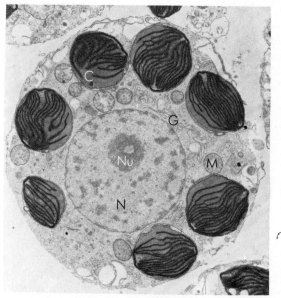

Figure 10-11. Algae of Chrysophycophyta: (A) *Ochromonas* sp.; (B) *Chrysamoeba* sp.

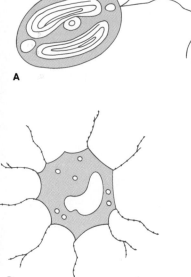

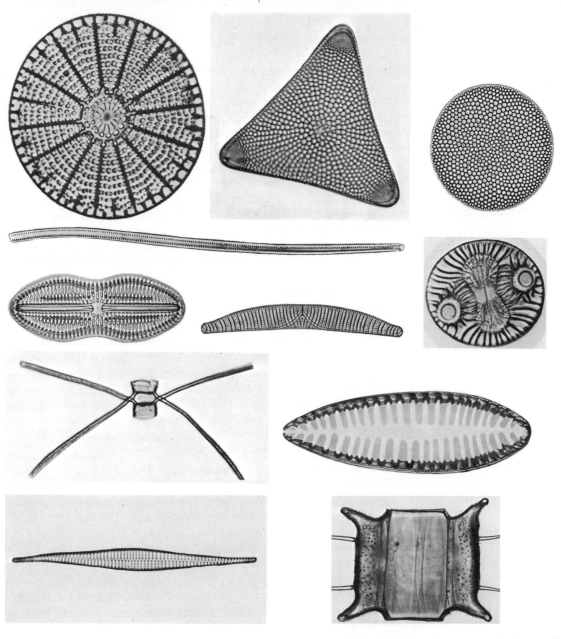

Figure 10-12. Diatoms are unicellular algae found abundantly in freshwater and saltwater. Their hard, silica-containing walls consist of two valves which fit together like a petri dish and its cover. They occur in myriads of shapes, many with beautiful surface designs. Magnifications range from × 400 to × 800. *(From Johns-Manville Research Center.)*

Bacillariophycophyta This group consists of the diatoms found in both freshwater and saltwater and in moist soil. Abundant in cold waters, diatoms are the most plentiful form of plankton in the Arctic. The thousands of species of diatoms provide an ever-present and abundant food supply for aquatic animals. Diatoms are either unicellular, colonial, or filamentous and occur in a wide variety of shapes (Fig. 10-12). Each cell has a prominent, single nucleus and massive ribbonlike, or smaller lenslike, plastids. They produce shells (cell walls) containing silica, some of which are very beautiful. Deposits of these shells resulting from centuries of growth are called *diatomite* or *diatomaceous earth*. Although diatomite from prehistoric times is found in Oregon, Nevada, Washington, Florida, and New Jersey, the world's largest and most productive commercial source is at Lompoc, California. It is used in insulating materials; as a filter for clarifying fruit juices, cane sugar, and beverages; in cosmetic bases; and in abrasives.

Euglenophycophyta

These unicellular algae are actively motile by means of flagella. They reproduce by longitudinal binary fission. Dormant cysts are formed from all types. *Euglena* is widely distributed and occurs in soil as well as in water, where it forms a velvety film. Its characteristics are both animallike and plantlike.

   The *Euglena* cell is not rigid. It has no cell wall containing cellulose. The outer membrane is pliable and stretchable. An anterior "gullet" (Fig. 10-13) is present even though no particulate food is ingested through it; the contractile vacuole discharges into it. Certain species develop a prominent red eyespot. Contractile vacuoles and fibrils (microtubules) are also present in the cell. All these are animal attributes. On the other hand, the organism carries out photosynthesis in chloroplasts and is facultatively autotrophic, and these are plant attributes. The majority of *Euglena* can assimilate organic substances during photosynthesis. A few types can even ingest particulate food through transient openings adjacent to the gullet.

Cryptophycophyta

These algae are called the *cryptomonads*. They possess two unequal flagella. They are generally flattened, slipper-shaped cells occurring singly, some being walled and others naked. Food reserves are stored as starch. Cells divide longitudinally. Sexual reproduction has not been observed.

Pyrrophycophyta

This division includes the dinoflagellates, which are motile, and the phytodinads, which are nonmotile but which produce zoospores with flagella. Both dinoflagellates and phytodinad zoospores have flagella which emerge from a common point on the cell. Some dinoflagellates have prominent cell walls made up of plates, which may contain cellulose. Others, for example the *Gymnodinium*, have no cell walls. An important constituent of plankton, dinoflagellates live in both freshwater and salt-water. Several genera (for example, *Gonyaulax*) can occur as massive growths called "blooms" or "red tides" in marine environments, commonly in the Gulf of Mexico, off the Pacific coast of North America, and, recently, off New England. Blooms of certain species are frequently toxic to fish and humans but not to shellfish which feed on them. Ingestion of such shellfish can lead to food poisoning and death. Other marine dinoflagellates are luminescent. Reproduction is largely by asexual cell division.

Xanthophycophyta

These yellow-green algae may represent a condition midway between the Chrysophycophyta and the Chlorophycophyta. The motile cell typically has two flagella of unequal length, and the cell wall often incorporates silica. Nonmotile members also exist. Genera of this division may be unicellular, colonial, filamentous, or tubular. The storage products are oils.

Vaucheria is a well-known member of this division and is widely distributed on moist soil and in water. Both freshwater and marine species are known. The tubular body is slightly branched. Sexual reproduction is oogamous (involving gametes of unequal size).

Lichens

A lichen is a composite organism consisting of an alga or a cyanobacterium and a fungus growing together in *symbiosis* (the living together of two or more organisms). It is generally agreed that this association is mutualistic; i.e., each partner benefits. Apparently the alga or cyanobacterium provides the fungus with food, particularly carbohydrates produced by photosynthesis, and possibly vitamins. The fungus probably absorbs, stores, and supplies water and minerals required by the photosynthetic partner, as well as providing a supporting and protective framework. The fungus is able to obtain nourishment from its partner by means of tiny rootlike projections called *haustoria* which penetrate the photosynthetic cell.

With very few exceptions, each lichen consists of a single species of

Figure 10-14. Examples of two types of lichens: (A) foliose lichen; (B) fruticose lichen (*Courtesy of Carolina Biological Supply Company.*)

Figure 10-13. Left: Schematic representation of a euglenoid; Right: *Euglena acus.* (*Courtesy of Carolina Biological Supply Company.*)

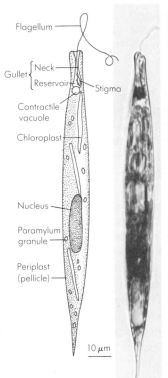

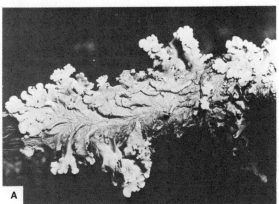

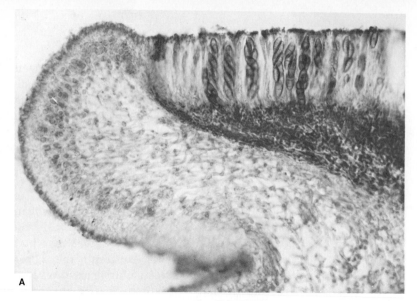

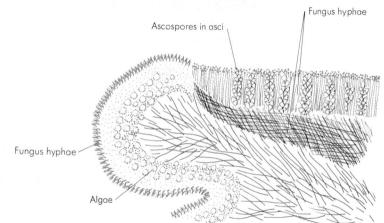

Ascospores in asci

Fungus hyphae

Fungus hyphae

Algae

Figure 10-15. The structure of lichens is suggested by the two parts of this illustration. (A) A portion of a vertical section through a foliose lichen thallus, *Physica* sp. (× 150). *(From General Biological Supply House.)* (B) Detail of algal and fungal relationships in a foliose lichen.

fungus and a single species of alga or cyanobacterium. Over 27 different genera of algae and cyanobacteria have been found in lichens. The commonest is the green unicellular alga *Trebouxia*. The other two most common genera are *Trentepohlia* (a filamentous alga) and *Nostoc* (a cyanobacterium). These three genera account for the photosynthetic partner in over 90 percent of all lichen species. Most lichen fungi are Ascomycetes although a few are Basidiomycetes.

Lichens are widely distributed in nature. They grow on rocks (Fig. 10-14), tree bark, and other substrates generally unsuitable for the growth of plants. Many lichens are resistant to extremes of temperature and are thus able to grow at the low temperatures found at high altitudes and in polar environments. Consequently, lichens, such as the reindeer moss, are an important source of food in the Arctic regions. They also withstand long periods of desiccation.

Morphologically, a lichen is made up of a layer of tightly woven fungal mycelia; below this is a layer of algal or cyanobacterial cells, and below that another layer of fungus (Fig. 10-15). The bottom layer may attach to the substratum directly or by means of short twisted strands of hyphae called *rhizines*, which serve as anchors.

There are two kinds of lichens, *foliose* (leaflike) and *fruticose*. Foliose lichens grow closely appressed to the substratum or even within its surface. Fruticose lichens have an erect, shrublike morphology and can be about 10 cm high.

Lichens reproduce by vegetative processes. Fragments of the thallus can give rise to new individuals. Lichens may produce "reproductive bodies" called *soredia*, which are knots of fungal hyphae containing some cells of the partner. Lichens grow slowly in nature. Many species grow less than a centimeter per year.

Summary and outlook

Algae are aerobic photosynthetic eucaryotic organisms. They contain chlorophyll a, other chlorophylls, and other photosynthetic pigments. The pigments are located in chloroplasts. Algae are ubiquitous in habitat, as long as sunlight is available together with moisture and simple nutrients. Algae may be unicellular or multicellular and may be arranged in colonies, filaments, or other multicellular forms. Some are microscopic; others are macroscopic.

Algae reproduce by asexual or sexual means. They use many modes of each type of reproduction. Some algae have complex life cycles which include both asexual and sexual modes.

The classification of algae is based on several criteria. Among the most important ones are type of pigments, nature of storage products, and nature of flagellation. Algae are generally assigned to nine major divisions, although this number varies with different authorities.

As a group algae are not important infectious agents for humans. But some colorless algae (for example, *Prototheca*) have been implicated as pathogens. Of public health concern is the production of toxins by some algae in "red tides" or blooms. Besides killing fish, these toxins can cause paralytic shellfish poisoning when oysters and other shellfish are eaten. Studies on these toxins may one day lead to their use for medicinal purposes.

Some algae live in symbiosis with other living organisms. For example, the lichen is a composite organism consisting of a fungus and a photosynthetic symbiont, an alga or a cyanobacterium.

Our concern with the environment may lead to the increased use of algae as a source of oxygen for the oxidative decomposition of dissolved organic matter in bodies of water. They may even be used as biological indicators of water pollution. The capacity of algae to produce oxygen and to carry out photosynthetic activity, as well as the ease of cultivation of unicellular forms, makes them extremely attractive for use in space travel. Research related to this has been carried out. Algae can supply oxygen for the space traveler in the closed confines of the spaceship. At the same time, they can remove carbon dioxide, thus purifying the air, and they can also serve as a renewable source of fresh food.

Only the Japanese have used algae (seaweeds) widely as a food. Perhaps algae will be a future source of food for the West—for either human or other animal consumption.

Key terms

akinete	isogamous	soredia
algal bloom	neurotoxin	stigma (red eyespot)
bisexual organism	oogamy	symbiosis
contractile vacuole	palmelloid	thylakoid
diatomite	phycology	unisexual organism
foliose lichen	primary producer	zonation phenomenon
fruticose lichen	pyrenoid	
heterogamous	rhizine	

Questions

1 How would you distinguish unicellular algae from photosynthetic bacteria?

2 Why are algae important in the food chain of aquatic environments?

3 Discuss the various uses of algae that make them commercially important.

4 Distinguish between a typical algal colony and a bacterial colony.

5 Compare the asexual reproduction of algae with that of bacteria.

6 How do the physiological characteristics of algae determine their occurrence?

7 Explain why all algae are not green in color.

8 Discuss the significant characteristics used as a basis for algal classification.

9 Which attributes does *Euglena* share with plants and which does it share with animals?

References for part three

Alexopoulos, C. J., and H. C. Bold: *Algae and Fungi*, Macmillan, New York, 1967. *Contains basic information for understanding these two groups of eucaryotic microorganisms.*

Beck, J. W., and J. E. Davies: *Medical Parasitology*, 2d ed., Mosby, St. Louis, 1976. *A manual for the teaching of medical parasitology to medical students, technologists, and practicing physicians. Part I is on medical protozoology. Protozoa of medical importance are covered in some detail.*

Bold, H. D., and M. J. Wynne: *Introduction to the Algae: Structure and Reproduction*, Prentice-Hall, Englewood Cliffs, N.J., 1978. *A comprehensive, well-written text for the teaching of general phycology. Very well illustrated with line drawings and both photomicrographs and electron micrographs.*

Emmons, C. W., C. H. Binford, J. P. Utz, and K. J. Kwon-Chung: *Medical Mycology*, 3d ed., Lea & Febiger, Philadelphia, 1977. *A classic text on medical mycology. The biology of the organisms is also well covered.*

Grell, K. G.: *Protozoology*, Springer-Verlag, Heidelberg, 1973. *A reference text on the biology of protozoa. Well illustrated.*

Ross, I. K.: *Biology of the Fungi*, McGraw-Hill, New York, 1979. *An undergraduate text on the biology of fungi with emphasis on their developmental and regulatory aspects.*

PART FOUR

VIRUSES

Viruses are microorganisms so small that they can only be seen at magnifications provided by the electron microscope. They pass through the pores of filters which do not permit the passage of bacteria. Viruses also reproduce only in the cells of animals, plants, and other microorganisms. For this reason, they are referred to as *obligate intracellular parasites*.

Why are viruses dependent on host cells for reproduction? Because viruses largely lack metabolic machinery of their own, viruses are unable to generate energy or synthesize proteins. They depend on the host cells to carry out these vital functions.

However, like the host cells, viruses do have genetic information for reproduction and for taking over the host cell's energy-generating and protein-synthesizing systems. This information is in the viral genes.

Actually, viruses in transit from one host cell to another are small packets of genes. Unlike host cells, the viral genetic material is either DNA or RNA not both DNA and RNA. The genetic material is enclosed in a highly specialized protein coat of varying design. The coat protects the genetic material when the virus is outside of any host cell and serves as an entry vehicle when the virus enters a particular host cell. The structurally complete mature and infectious virus is called the *virion*.

During reproduction in the host cells viruses may cause disease. In fact, viruses cause the most common acute infectious illnesses of humans, and there is growing evidence to suggest that they may cause many different chronic diseases as well. The average person experiences a few viral infections a year and well over 200 infections in a lifetime. Obviously, the economic cost of virus infections, due to medical care and loss of productivity, is billions of dollars each year.

Bacterial viruses, or *bacteriophages,* have provided the microbiologist with a model for the study of viruses *(virology)* and molecular biology. It has been said that the study of bacteriophages stands in relation to biology as the study of gravitation stands to classical physics! We shall discuss the bacteriophages in Chap. 11 and follow with a look at the viruses tha infect animals and plants in Chap. 12.

Bacteriophages—Bacterial viruses—of *Bacillus thuringiensis.* Magnification × 396,000. *(Courtesy of H. W. Ackermann, Laval University.)*

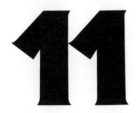

BACTERIAL VIRUSES

Bacteriophages (or simply *phages*), viruses that infect bacteria, were discovered independently by Frederick W. Twort in England in 1915 and by Felix d'Herelle at the Pasteur Institute in Paris in 1917. Twort observed that bacterial colonies sometimes underwent *lysis* (dissolved and disappeared) and that this lytic effect could be transmitted from colony to colony. Even high dilutions of material from a lysed colony that had been passed through a bacterial filter could transmit the lytic effect. However, heating the filtrate destroyed its lytic property. From these observations Twort cautiously suggested that the lytic agent might be a virus. D'Herelle rediscovered this phenomenon in 1917 (hence the term Twort-d'Herelle phenomenon) and coined the word *bacteriophage*, which means "bacteria eater." He considered the filtrable agent to be an invisible microbe, for example, a virus, that was parasitic for bacteria.

Since the bacterial hosts of phages are easily cultivated under controlled conditions, demanding relatively little in terms of time, labor, and space compared with the maintenance of plant and animal hosts, the bacteriophage has received considerable attention in viral research. Furthermore, since bacteriophages are the smallest and simplest biological entities known which are capable of *self-replication* (making copies of themselves), they have been used widely in genetic research. Of impor-

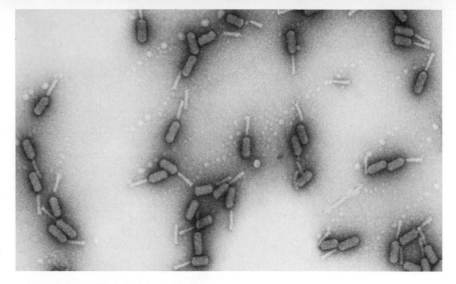

Figure 11-1. The unique morphology of many bacteriophages is exemplified by this phage of *Salmonella newport*. Each phage has a tail through which it inoculates the host cell with viral nucleic acid. Note that some tails have broken off from the phage heads in this preparation (× 59,400). *(Courtesy of H.-W. Ackermann, Laval University.)*

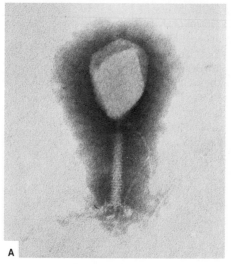

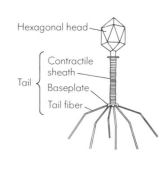

Hexagonal head

Tail {
Contractile sheath
Baseplate
Tail fiber

A   B

Figure 11-2. (A) Electron micrograph and (B) diagrammatic representation of the coliphage T2 virion. Magnification in micrograph (× 189,000). *(Courtesy of H.-W. Ackermann, Laval University.)*

tance too have been studies on the bacterium-bacteriophage interaction. Much has been learned about host-parasite relationships from these studies, which have provided a better understanding of plant and animal infections with viral pathogens. Thus the bacterium-bacteriophage interaction has become the model system for the study of viral pathogenicity. This chapter describes the biology of these important viruses.

General characteristics

Bacterial viruses are widely distributed in nature. Phages exist for most, if not all, bacteria. With the proper techniques, these phages can be isolated quite easily in the laboratory.

Bacteriophages, like all viruses, are composed of a nucleic acid core surrounded by a protein coat. Bacterial viruses occur in different shapes although many have a tail through which they inoculate the host cell with viral nucleic acid (Fig. 11-1).

There are two main types of bacterial viruses: *lytic*, or *virulent*, and *temperate (lysogenic)*, or *avirulent*. When lytic phages infect cells, the cells respond by producing large numbers of new viruses. That is, at the end of the incubation period the host cell bursts or lyses, releasing new phages to infect other host cells. This is called a *lytic cycle*. In the temperate type of infection, the result is not so readily apparent. The viral nucleic acid is carried and replicated in the host bacterial cells from one generation to another without any cell lysis. However, temperate phages may spontaneously become virulent at some subsequent generation and lyse the host cells. In addition, there are some filamentous phages which simply leak out of cells without killing them.

Morphology and structure

Morphological groups of phages

The electron microscope has made it possible to determine the structural characteristics of bacterial viruses. All phages have a nucleic acid core covered by a protein coat, or *capsid*. The capsid is made up of morphological subunits (as seen under the electron microscope) called *capsomeres*. The capsomeres consist of a number of protein subunits or molecules called *protomers*. Figure 11-2 shows the fine structure and anatomy of a common morphological form of the bacteriophage, one with a head and a tail.

Bacterial viruses may be grouped into six morphological types (Fig. 11-3):

A This most complex type has a hexagonal head, a rigid tail with a contractile sheath, and tail fibers.

B Similar to A, this type has a hexagonal head. However, it lacks a contractile sheath, its tail is flexible, and it may or may not have tail fibers.

Figure 11-3. Schematic representation of morphological types of bacteriophages. See text for further details. [*As described by D. E. Bradley, "Ultrastructures of Bacteriophages and Bacteriocins," Bacteriological Review **31**: 230–314 (1967).*]

Type	A	B	C	D	E	F
Morphology	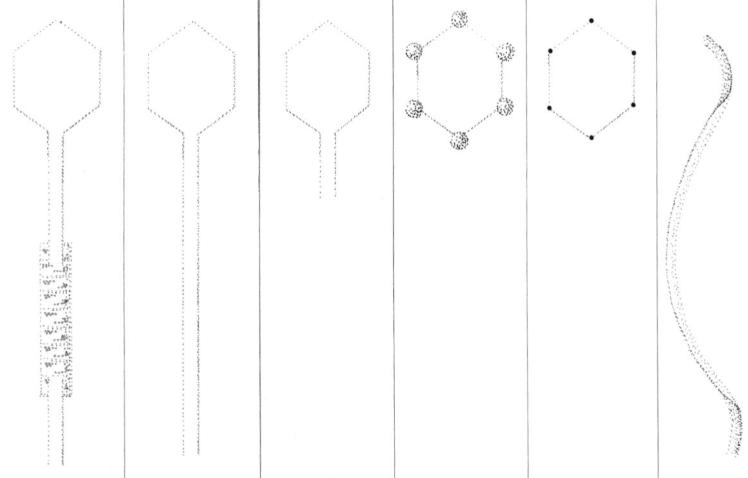					
Nucleic acid type and number of strands	DNA, 2	DNA, 2	DNA, 2	DNA, 1	RNA, 1	DNA, 1

C This type is characterized by a hexagonal head and a tail shorter than the head. The tail has no contractile sheath and may or may not have tail fibers.

D This type has a head with no tail, and the head is made up of large capsomeres.

E This type has a head with no tail, and the head is made up of small capsomeres.

F This type is filamentous.

Types A, B, and C show a morphology unique to bacteriophages. The morphological types in groups D and E are found in plant and animal (including insect) viruses as well. The filamentous form of group F is found in some plant viruses.

Phage structure

Phages, like all viruses, occur in two structural forms having *cubic* or *helical* symmetry. In overall appearance, cubic phages are regular solids or, more specifically, *polyhedra* (singular, *polyhedron*); helical phages are rod-shaped. In many bacteriophages the heads are polyhedral but the tails are rod-shaped (Fig. 11-2).

Polyhedral phages are *icosahedra*; that is, the capsid has 20 facets, each of which is an equilateral triangle. These facets come together to form 12 vertices. In the simplest capsid, there is a capsomere at each of the 12 vertices; this capsomere is surrounded by five other capsomeres (Fig. 11-4A). Other capsids may be composed of hundreds of capsomeres, but they are based on this simple model. The elongated heads of phages are derivatives of the icosahedron.

Rod-shaped viruses have their capsomeres arranged helically and not in stacked rings (Fig. 11-4B).

Phage nucleic acids

Different morphological types of phages are also characterized by having different nucleic acid types (Fig. 11-3). All tailed phages contain double-stranded DNA. The phages with large capsomeres (group D) and the filamentous ones (group F) have single-stranded DNA. Group E phages have single-stranded RNA. The DNAs of some phages are circular under certain conditions. (Circular simply means a *closed loop* because the molecule is in the form of a folded loose coil packed inside the capsid.) The DNA of phage φX174 is circular both in the virion and in the host cell. The DNA of phage *lambda* (λ) is linear in the virion, but on entering the host cell, the ends join to form a circle (Fig. 11-5). It may be noted at this point that bacteriophages are named as above by code symbols assigned by investigators. Although serving the practical needs of the laboratories, this is a haphazard method of nomenclature. Figure 11-6 identifies some of these phages.

Some bacteriophages of *Escherichia coli*

The most extensively studied group of bacteriophages are the *coliphages*, so-called since they infect the nonmotile strain B of *Escherichia coli*. They are designated T1 to T7. All these phages are composed almost exclusively

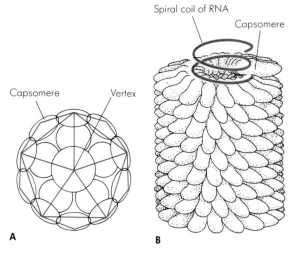

Capsomere   Vertex

A

Spiral coil of RNA

Capsomere

B

Figure 11-4. (A) Diagram of the simplest icosahedral capsid. The colored triangular outlines delineate the icosahedral symmetry. The circles represent capsomeres. (B) Diagram of a rod-shaped virus with helical symmetry. The capsomeres are arranged helically around a hollow core containing a spiral coil of RNA.

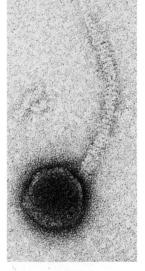

Figure 11-5. The morphology of phage *lambda* as seen by electron microscopy (× 297,000). *(Courtesy of H.-W. Ackermann, Laval University.)*

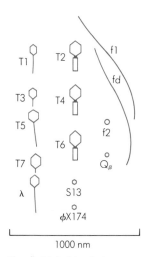

T1   T2   f1

T3   T4   fd

T5   f2

T6

T7   Q$_\beta$

λ   S13

φX174

1000 nm

Figure 11-6. Morphology and relative sizes of some bacteriophages.

of DNA and protein in approximately equal amounts. Except for T3 and T7, all have a tadpole shape with polyhedral heads and long tails. The tails of T3 and T7 are very short (see Fig. 11-6). The T phages range from about 65 to 200 nm in length and 50 to 80 nm in width. The continuous, or circular, molecule of double-stranded DNA (about 50 μm long or about 1,000 times as long as the phage itself) is tightly packed in the protein head.

There are other bacteriophages for *E. coli* whose morphology and chemical composition are very different from those of the T phages. The f2 phage, for example, is much smaller than the T phages and has a single-stranded linear molecule of RNA, rather than DNA. It has no visible tail appendage.

There are also coliphages which possess single-stranded DNA. Morphologically they can be either icosahedral (cubic symmetry) or filamentous (helical symmetry). An icosahedral phage with circular single-stranded DNA is φX174.

Filamentous coliphages (Fig. 11-7) were discovered long after the

Figure 11-7. Electron micrograph of If1 phage that infects *Escherichia coli*. The phage adsorbs specifically to I-pili, a type of surface appendage of cells (× 24,750). *(Courtesy of R. L. Wiseman, The Public Health Research Institute of the City of New York, Inc.)*

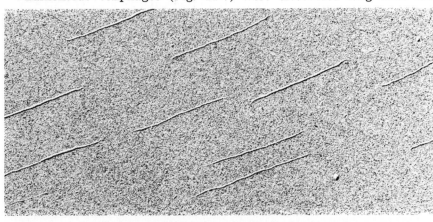

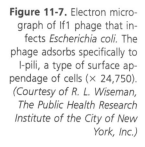

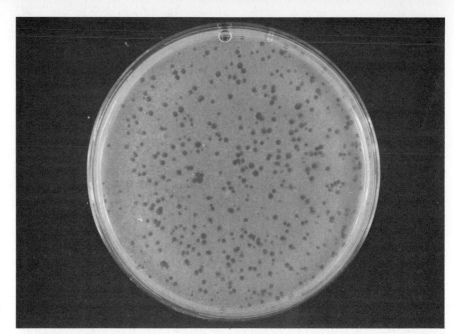

Figure 11-8. Plaques (clear zones) are formed when a bacterial growth in a petri dish is lysed by a bacteriophage. *(Courtesy of C. Alfieri and E. C. S. Chan, McGill University.)*

tadpole-shaped phages were known. Filamentous phages are continuously produced by viable and reproducing bacteria (and without the host-cell lysis characteristic of the virulent icosahedral phages). Such filamentous phages for *E. coli* include the fd and f1 bacteriophages (Fig. 11-6). They all have circular single-stranded DNA.

Isolation and cultivation of bacterial viruses

Bacterial viruses are easily isolated and cultivated on young, actively growing cultures of bacteria in broth or on agar plates. In liquid cultures, lysing of the bacteria may cause a cloudy culture to become clear, whereas in agar-plate cultures, clear zones, or *plaques*, become visible to the unaided eye (Fig. 11-8).

The principal requirement for isolation and cultivation of phages is that optimal conditions for growth of the host organisms be provided. The best and most usual source of bacteriophages is the host habitat. For example, coliphages or other phages pathogenic for other bacteria found in the intestinal tract can best be isolated from sewage or manure. This is done by centrifugation or filtration of the source material and addition of chloroform to kill the bacterial cells. A small amount (such as 0.1 ml) of this preparation is mixed with the host organism and spread on an agar medium. Growth of phage is indicated by the appearance of plaques in the otherwise opaque growth of the host bacterium as shown in Fig. 11-8.

Medical use of virulent phages

Virulent phages have been used in the detection and identification of pathogenic bacteria. Strains of bacteria are characterized by their resistance or susceptibility to infection by specific virulent bacteriophages. The resulting pattern of lysis of a bacterial strain by different phage types gives an indication of the identity of the bacterium. This process is termed *phage-typing* and is routinely used for the identification of bacterial pathogens such as staphylococci and typhoid bacilli. In this way phages serve as a tool for medical diagnosis and for following the spread of a disease in a community.

Reproduction of bacterial viruses

Much of what is known about bacteriophage reproduction has come from studies of the virulent even-numbered T phages (T2, T4, T6) of *E. coli*. We shall use these as a model for discussing phage reproduction.

Adsorption and penetration

The lysis of bacteria by bacteriophages is summarized in Fig. 11-9. The first step in reproduction of a bacteriophage is adsorption. Here the tip of the

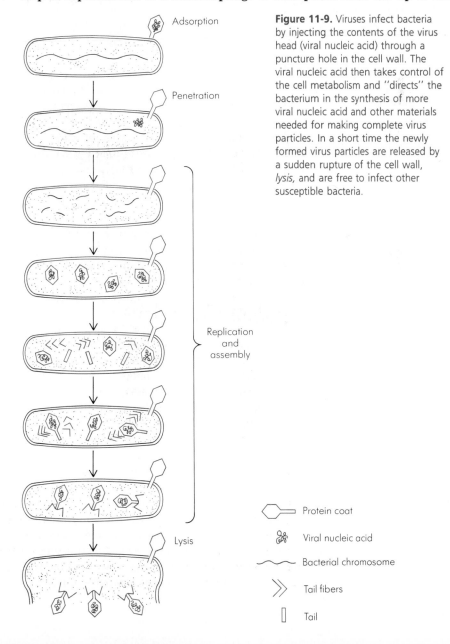

Figure 11-9. Viruses infect bacteria by injecting the contents of the virus head (viral nucleic acid) through a puncture hole in the cell wall. The viral nucleic acid then takes control of the cell metabolism and "directs" the bacterium in the synthesis of more viral nucleic acid and other materials needed for making complete virus particles. In a short time the newly formed virus particles are released by a sudden rupture of the cell wall, *lysis,* and are free to infect other susceptible bacteria.

Adsorption

Penetration

Replication and assembly

Lysis

Protein coat

Viral nucleic acid

Bacterial chromosome

Tail fibers

Tail

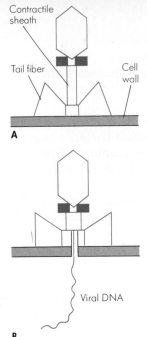

Contractile sheath

Tail fiber

Cell wall

A

Viral DNA

B

Figure 11-10. Penetration of the host cell by a bacteriophage. (A) The fibers anchor the virus in place against the cell wall. (B) The tail sheath contracts, the core penetrates into the cell, and the viral DNA is injected into the cell.

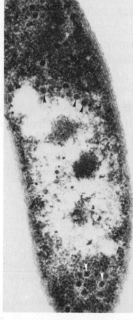

Figure 11-11. The intracellular location of a phage is shown. This cell of *Pseudomonas cepacia* 383 was sampled 5 min prior to culture lysis. Electron-dense particles (arrows) occupy the ribosome-rich regions of the cell. Note the hexagonal shape of some particles. *(Courtesy of T. G. Lessie, University of Massachusetts.)*

virus tail becomes attached to the cell wall. Attachment is specific in that certain viruses and susceptible bacteria have complementary molecular configurations at their opposing receptor sites.

If too many phages are attached to the bacterium and penetrate it, there may be premature lysis (lysis from without), which is not accompanied by the production of new virus.

The actual penetration of phage into the host cell is mechanical, but it may be facilitated by an enzyme, *lysozyme*, carried on the tail of the phage which digests the cell wall. Penetration is achieved when: (1) the tail fibers of the virus attach to the cell and hold the tail firmly against the cell wall; (2) the tail sheath contracts, driving the tail core into the cell through the cell wall and cell membrane; (3) the virus injects its DNA as a syringe injects a vaccine (Fig. 11-10). The protein coat, which forms the phage head, and the tail structure of the virus remain outside the cell.

In the case of bacteriophages which do not have contractile sheaths, it is still not clear how the nucleic acid enters the cell. The RNA and filamentous coliphages both are adsorbed on pili, which are thin appendages extending from the cell surface. The nucleic acid probably passes down the inside of the pilus for a short distance and then the pilus retracts into the cell, thus initiating injection of nucleic acid.

Replication, assembly, and lysis

Studies on coliphages have revealed some aspects of the reproduction and structure of phages and the mechanism of genetic control over their propagation. As already indicated, the viral material entering the cell is nucleic acid (either RNA or DNA), which carries the information necessary for the synthesis of new virus particles. Immediately after injection of the viral nucleic acid into the host cell, the virus takes charge of the host's metabolic machinery, causing it to manufacture viral nucleic acids rather than bacterial nucleic acids. About 25 min after initial infection, some 200 new bacteriophages have been assembled (see Fig. 11-11), and the bacterial cell bursts, releasing the new phages to infect other bacteria and begin the cycle over again. As shown in Fig. 11-12, the first stage is the latent period (time during which no infectious virus is demonstrated) followed by the rise period (due to cell lysis and release of a large number of phages). Similar life cycles have been demonstrated for a number of other viruses.

Lysogeny

Not all infections of bacterial cells by phages proceed as described above to produce more viral particles and terminate in lysis. An entirely different relationship, known as *lysogeny*, may develop between the virus and its bacterial host (Fig. 11-13). In lysogeny the viral DNA of the temperate phage, instead of taking over the functions of the cell's genes, is incorporated into the host DNA and becomes a prophage in the bacterial chromo-

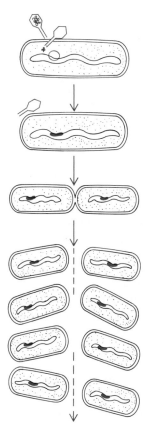

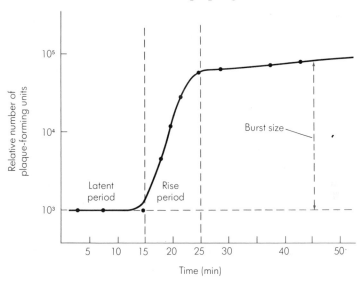

Figure 11-12. One-step-growth curve of plaque-forming units. In a one-step growth experiment, after adsorption of the virus on the host, the suspension is diluted to such an extent that virus particles released after the first round of replication cannot attach to uninfected cells; thus only one round of replication can occur. Each plaque-forming unit is equal to one phage particle in the original suspension.

Infective virus
(bacteriophage)

Viral nucleic
acid

Bacterial
chromosome

Figure 11-13. Lysogeny is a process in which the viral nucleic acid does not usurp the functions of the host bacterium's synthetic processes but becomes an integral part of the bacterial chromosome. As the bacterium reproduces, viral nucleic acid is transmitted to the daughter cells at each cell division. In the lysogenic state the virus is simply one of the bacterial genes. Under certain natural conditions or artificial stimuli (such as exposure to ultraviolet light), the synthesis of virus may take over, and lysis occurs.

some, acting as a gene. In this situation the bacterium metabolizes and reproduces normally, the viral DNA being transmitted to each daughter cell through all successive generations. Sometimes, however, for reasons unknown, the viral DNA is removed from the host's chromosome and the lytic cycle occurs. This process is called *spontaneous induction*. Infection of a bacterium with a lysogenic phage can be detected by the facts that the bacterium is resistant to infection by the same or related phages and that it can be induced to produce phage particles. A change from lysogeny to lysis can sometimes be induced by irradiation with ultraviolet light or by exposure to some chemical. A good part of our knowledge on lysogeny comes from studies on the coliphage *lambda* (λ).

In the prophage state, all the phage genes except one are repressed or prevented by a regulatory mechanism from expressing themselves. The gene that is expressed is an important one because it codes for the synthesis of a repressor molecule which makes the cell resistant to lysis initiated either by the prophage or by lytic infection by other viruses. Radiation or chemicals may induce release of the prophage from the host genome. Now, the phage can behave like a lytic phage.

No RNA phages have yet been shown to be temperate. It is possible that temperate RNA phages exist because the phage can form a DNA copy of the RNA genome, which can then be integrated into the bacterial chromosome.

Medical aspects of lysogeny

Diphtheria is caused by the bacterial pathogen *Corynebacterium diphtheriae*. Its capacity to cause disease is directly related to its ability to produce a toxin. It can only produce toxin when it carries a temperate phage. In the same way, only those streptococci which carry a temperate phage can produce the erythrogenic (rash-producing) toxin of scarlet fever. In another known instance, some types of toxins of botulism are produced by *Clostrodium botulinum* as a result of lysogeny. This phenomenon in which a prophage is able to make changes in the properties of a host bacterium in lysogeny is termed *lysogenic conversion*.

Bacterial lysogeny is a good conceptual model for the study of oncogenic or cancer-producing viruses since these viruses also have the capacity of perpetuating their genomes in infected cells.

Summary and outlook

Bacteriophages (also called phages for short) are viruses that infect bacteria and can only reproduce in them. The relative convenience of handling and the simplicity of the phage-bacterium interaction have made it a model system for the study of viral pathogenicity as well as many fundamental problems in biology, including cellular and molecular biology and immunology.

Phages, like all viruses, essentially consist of a nucleic acid core enclosed in a protective protein coat. They have no enzymes capable of energy production or organelles such as ribosomes for independent replication. They exist in several morphological types and in two structural forms (cubic and helical). In the typical phages, the hexagonal-shaped heads are usually complex derivatives of the icosahedral three-dimensional structure; the tails are helical. Different morphological types have different kinds of nucleic acids.

The reproduction of virulent bacterial viruses involves the following general sequence: adsorption of phage particle; penetration of nucleic acid; replication of virus nucleic acid; assembly of new phage particles; and release of these phage particles in a burst, with concomitant lysis of the host cell. Virulent phages have been used in the detection and identification of pathogenic bacteria.

Some phages do not cause cell lysis but their nucleic acids are integrated into the host DNA as a prophage. These are temperate phages. Such phages are the basis for the production of exotoxins by some bacteria, such as those causing diphtheria, scarlet fever, and botulism.

The bacteriophage-bacterial system is and will continue to be a useful model for the study not only of host-parasite relationships but also of fundamental problems in biology. This is especially so in the field of genetics because the bacteriophage is an organism with a minimum of genetic characters, is *haploid* (unpaired chromosome), and is easily grown for the rapid detection of rare events (like a change in genome) in a large population. Further, studies on control of differentation and morphogenesis will focus on the bacteriophage since it is the smallest and simplest of self-replicating entities.

Apart from their research applications, phages are used routinely and widely for the detection and identification of pathogenic bacteria. Their adverse effects are also important: in certain industrial processes that make use of bacteria (for example, cheese making and antibiotic production), phage infection of the bacteria can lead to great economic losses. Proper control of such infections maintains efficient processing.

Another application in which phages may see wider use is the measurement of radiation dosage. Since the radiation susceptibilities of certain phages are known very accurately, it is possible to incorporate them into a material that is to be irradiated. The radiation dosage may then be calculated from the degree of destruction of the bacteriophage. This gives a direct measure of the biological effectiveness of a radiation treatment that may be impossible to calculate from physical and chemical data in complex systems.

Key terms

bacteriophage	lysogeny	self-replication
capsid	lysozyme	spontaneous induction
capsomere	lytic cycle	temperate (lysogenic,
coliphage	lytic (virulent) phage	avirulent) phage
helical	phage-typing	virion
icosahedral	plaque	virology
lysis	prophage	
lysogenic conversion	protomer	

Questions
1 Why are bacteriophages suitable subjects for research in genetics?
2 What features distinguish a bacteriophage from a bacterial cell?
3 Compare the processes of infection by a temperate phage and a virulent phage.
4 Describe the morphological and structural features that are unique to bacteriophages.
5 Discuss the coliphages with respect to morphology and type of nucleic acid.
6 Describe the technique by which one isolates phages.
7 Explain the process and application of phage-typing.
8 What is the relationship of temperate phages to the disease-producing ability of some bacteria?

ANIMAL AND PLANT VIRUSES

Animal and plant viruses vary greatly in size (Fig. 12-1) and shape (Figs. 12-2 and 12-3), but they do not have the tadpole morphology characteristic of some bacteriophages. Size and shape are characteristic properties of each type of virus. Virion sizes range from 10 to 300 nm.

Much of the basic biology of bacteriophages discussed in Chap. 11 is also applicable to animal and plant viruses. The special properties of animal and plant viruses which differ from those of phages will be discussed in this chapter.

Structure and composition   Like bacteriophages, animal and plant virions are composed of a central core of nucleic acid surrounded by a capsid, which is made up of capsomeres. All virions possess a true symmetry of structure, as explained in Chap. 11. But in some animal viruses the *nucleocapsid* (nucleic acid and capsid) is covered by an outer membrane called the *envelope*, which is made of lipoproteins and conceals this symmetry. Virions that have envelopes are sensitive to lipid solvents such as ether and chloroform. Their

193

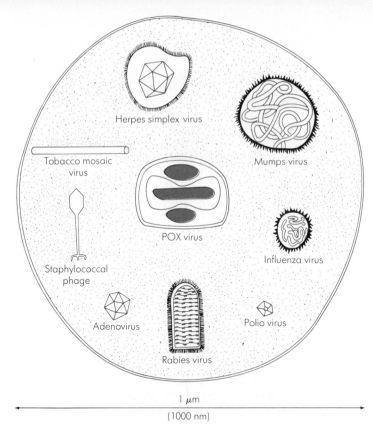

Figure 12-1. Comparative sizes and morphology of some viruses. The large circle gives an indication of the relative size of a coccoidal bacterial cell. *(Redrawn from a drawing in Pharmaceutical Microbiology, W. D. Hugo and A. D. Russell (eds.), Blackwell Scientific, Oxford, 1977.)*

Herpes simplex virus

Mumps virus

Tobacco mosaic virus

POX virus

Influenza virus

Staphylococcal phage

Adenovirus

Rabies virus

Polio virus

1 μm
(1000 nm)

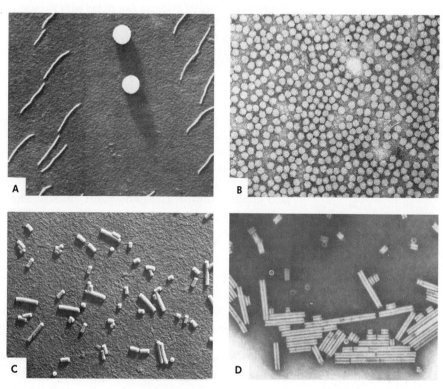

Figure 12-2. Electron micrographs of plant viruses. (A) Potato virus X particles appear as flexuous rods 513 nm long, (× 17,000). Also shown are two latex spheres used in electron microscopy to show relative sizes. (B) Tomato ringspot virus is icosahedral in structure. (× 150,000). (C) and (D) Tobacco rattle virus particles appear as both long and short rods. Both lengths are necessary to establish infection. (× 23,600 and × 25,400). *(Courtesy of M. K. Corbett, University of Maryland.)*

A

B

C

D

Figure 12-3. Electron micrographs of some animal viruses. (A) A cluster of polioviruses, the cause of poliomyelitis. (B) Rotavirus particles, the cause of acute infectious diarrhea a major cause of death in very young infants. (C) An enveloped herpesvirus. It is a persistent virus in humans and occasionally manifests its presence in "fever blisters" or "cold sores" of mucous membranes. (All micrographs are at the same magnification, × 200,000). *(Courtesy of Margaret Gomersall, McGill University.)*

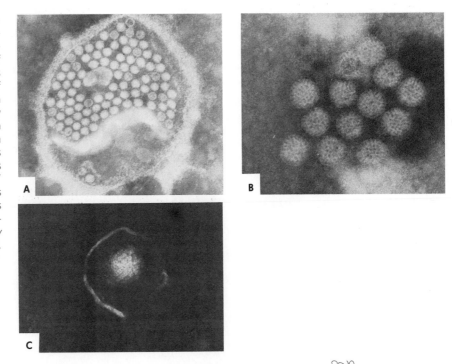

Figure 12-4. Schematic diagram of virion shapes and components. The capsomeres make up the capsid. The capsid and nucleic acid core form the nucleocapsid.

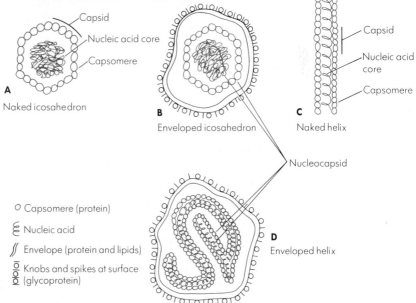

infectivity is inactivated by these solvents. Nonenveloped viruses are referred to as *naked* virions. These viruses are not affected by lipid solvents.

Morphology   Animal and plant viruses may be classified into four groups on the basis of overall morphology as follows:

1 *Icosahedral*: Examples of such viruses (see Fig. 12-4) are polioviruses and adenoviruses (Fig. 12-5), which cause poliomyelitis and respiratory infections, respectively.

2 *Helical*: Rabies virus is an example of such a virus (see Fig. 12-4). Many plant viruses are helically shaped; an example is shown in Fig. 12-6.

3 *Enveloped*: The internal nucleocapsid of these viruses (see Fig. 12-4), which may be either icosahedral or helical, is surrounded by a membranous envelope. Some envelopes have surface projections called *spikes* (Fig. 12-7). These are made of glycoproteins (proteins with carbohydrate

Figure 12-5. An adenovirus with icosahedral morphology. (× 200,000). *(Courtesy of Margaret Gomersall, McGill University.)*

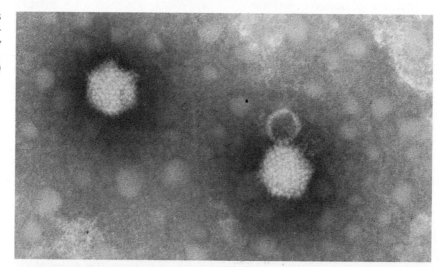

Figure 12-6. Electron micrograph of a chromium-shadowed preparation of cymbidium mosaic virus. Virus particles are flexuous rods 480 nm long. (× 33,000). *(Courtesy of M. Kenneth Corbett, University of Maryland.)*

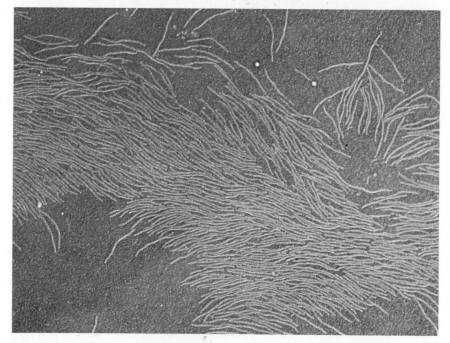

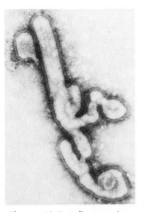

Figure 12-7. Influenza virus. Note the fringes of spikes on the surfaces of the virions. (× 55,300). *(Courtesy of Margaret Gomersall, McGill University.)*

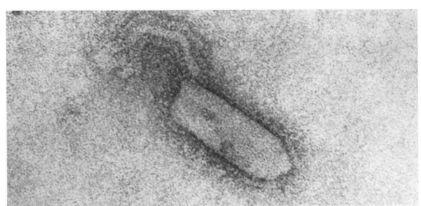

Figure 12-8. Vesicular stomatitis virus is bullet-shaped. Shown also is an irregular handle, which may represent remains of a disrupted end. (× 200,000). *(Courtesy of Margaret Gomersall, McGill University.)*

groups). Their presence generally is correlated with the ability of the virion to *agglutinate* (clump) with erythrocytes or red blood cells. Enveloped virions are pleomorphic (have varying shapes) since the envelope is not rigid. In an enveloped helical virus, such as the influenza virus, the nucleocapsid is coiled within the envelope (Fig. 12-7).

4 *Complex*: Some viruses have a complex structure. For example, the vesicular stomatitis virus (a pathogen of cattle) is bullet-shaped (Fig. 12-8), and the outside of the virion has spikes resembling those on envelopes. Poxviruses (such as vaccinia virus, the avirulent or noninfectious virus used to vaccinate against smallpox) do not possess clearly identifiable capsids but have several coats surrounding the nucleic acid (Fig. 12-9).

Nucleic acids

Like bacteriophages, animal and plant viruses contain either DNA or RNA. Both are not found in the same virion. This, of course, is in contrast to all cellular forms of life which, without exception, contain both types of nucleic acid in each cell. There are four possible kinds of nucleic acid: single-stranded DNA, single-stranded RNA, double-stranded DNA, and double-stranded RNA (Table 12-1). All four types have been found in animal viruses. In plant viruses, single- and double-stranded RNA, as well as single-stranded DNA, have been found.

In addition, the structure of the nucleic acid in the virion may be either linear or circular. For example, simian vacuolating virus 40 (SV40), found in monkey kidney cells, has circular double-stranded DNA while herpesvirus has linear double-stranded DNA.

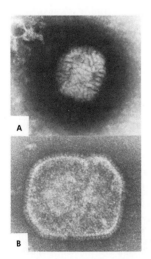

Figure 12-9. Vaccinia virus, a poxvirus with complex morphology. (A) Whole (coated) virus showing surface tubules. (× 200,000). *(Courtesy of Margaret Gommersall, McGill University.)* (B) Immature virion obtained from an infected cell showing a bounding membrane with subunit projections. (× 160,000). *(Courtesy of K. B. Easterbrook, Dalhousie University.)*

Table 12-1. Occurrence of types of nucleic acids in viruses

VIRUS	NUCLEIC ACID			
	DNA		RNA	
	Single-stranded	Double-stranded	Single-stranded	Double-stranded
Animal	+	+	+	+
Plant	+		+	+
Bacterial	+	+	+	

Other chemical components

Protein. Protein is the other major chemical component of viruses. It makes up most of the capsid. Many viruses have now been found to contain an enzyme or enzymes that function in the replication of their nucleic acid components. Some virions may contain a specific enzyme that uses the viral RNA as a model for synthesizing a second RNA strand that can direct host cells to make virus. The RNA tumor viruses contain an enzyme that synthesizes a DNA strand using the viral RNA genome as a template.

Lipid. A wide variety of lipid (fatty) compounds have been found in viruses. These include phospholipids, glycolipids, neutral fats, fatty acids, fatty aldehydes, and cholesterol. Phospholipid is the predominant lipid substance and is found in the viral envelope.

Carbohydrate. All viruses contain carbohydrate since the nucleic acid itself contains ribose or deoxyribose. Some enveloped animal viruses, such as the influenza virus and other myxoviruses, have spikes made of glycoprotein on the envelope.

Reproduction of viruses (viral replication)

Virus particles outside a host cell have no independent metabolic activity and are incapable of reproduction by processes characteristic of other microorganisms (Table 12-2). Multiplication takes place by *replication*, in which the viral protein and nucleic acid components reproduce within susceptible host cells.

The whole process of infection can be generalized as follows: The virion attaches to a susceptible host cell at more or less specific sites. Either whole virus or viral nucleic acid penetrates to the inside of the cell. If whole virus has penetrated the cell, uncoating of the virus must take place to release the nucleic acid. Reproduction of the virus takes place in the cytoplasm, the nucleus, or both. The viral protein and nucleic acid components are assembled into virus particles and released from the host cell. The steps of virus infection are therefore: (1) attachment or adsorption, (2) penetration and uncoating, (3) component replication and biosynthesis, (4) assembly and maturation, and (5) release.

Attachment (adsorption)

The attachment process occurs in two steps. The first one involves preliminary attachment by ionic bonds or charges and is easily reversed by a shift in pH or salt concentration. The second step appears to involve firmer attachment and to be irreversible. In contrast to the marked specificities of attachment with animal and bacterial viruses, plant viruses do not seem to require specific receptor sites.

Penetration and uncoating

The penetration of animal viruses into attached cells occurs by one of two mechanisms. One mechanism consists of engulfment of whole virions by the cells in a phagocytic process called *viropexis*, followed by uncoating or removal of the capsid. This takes place in the phagocytic vacuoles and is due to the action of enzymes called *lysosomal proteases*. The other mechanism occurs in the enveloped viruses; the viral lipoprotein envelope fuses with the host cell's surface membrane. This fusion results in the release of the viral nucleocapsid material into the cytoplasm of the host cell. Uncoating again occurs within the host cell.

Plant viruses penetrate host cells through transient pores (called *ectodesmata*) which protrude through the cell wall at intervals and communicate to the exterior of the cells. These pores function for the purpose of water and nutrient uptake as well as for secretion of substances such as waxes. Whole virus particles are apparently engulfed at these points. Also, insects can accidentally inoculate plant viruses into cells during feeding. Sometimes this is a purely mechanical process; at other times the virus is found in the insect tissue and may even multiply there. Insect feeding is probably the most important means of plant virus transfer in nature. Once the virus is inside the plant cell, uncoating occurs.

Viral component replication and biosynthesis

Shortly after penetration, there follows a period of time called the *latent period*. It is during the latent period that uncoating of the virion takes place, followed by replication of the viral nucleic acid and synthesis of viral

Table 12-2. Comparison of viruses with some bacteria

MICROORGANISM	MULTIPLICATION	DIAMETER, nm	CHEMICAL COMPOSITION	INHIBITION BY ANTIBIOTICS
Typical bacteria	In vitro* fluid and solid media, cell surfaces, or intracellularly, by binary fission	1000–3000	Complex proteins, carbohydrates, fats, etc.; DNA and RNA; peptidoglycan in cell wall	Yes
Mycoplasmas	Like typical bacteria but by budding rather than fission	150–1000	Like typical bacteria but without cell walls	Yes
Rickettsiae	In living cells only and by binary fission	250–400	Like typical bacteria	Yes
Viruses	In living cells only and by synthesis from pools of constituent chemicals	10–300	Either RNA or DNA and protein; some may have lipid and/or carbohydrate components	No

*In vitro means "in glass," i.e., in laboratory vessels. (This is in contrast to in vivo, which means "within a living organism.")

protein. Viral-nucleic-acid replication may take place in the cell nucleus or in the cytoplasm, depending on the specific virus. Viral proteins are synthesized in the cytoplasm, where the *ribosomes*, or protein-synthesizing organelles of the cell, are located. For these processes the cell provides energy, enzymes, precursor or "building-block" molecules, and other biosynthetic machinery.

Viral nucleic acids are synthesized from constituent nucleotides, the building blocks of nucleic acids, using replicase enzymes coded for by the nucleic acid of the infecting virion. The exact mechanism of synthesis varies with the virus and the type of nucleic acid involved. Other enzymes which are not available in the host, as well as viral structural proteins, are also coded for by the viral nucleic acids and then produced by the host cell's biosynthetic machinery.

Assembly and maturation
When a critical number of viral components have been synthesized, they are assembled into mature virus particles in the nucleus and/or cytoplasm of the infected cell (depending on the type of virus).

Release
The mechanism of release of virions from a host cell varies with the type of virus. In some animal virus infections, the host cells lyse, releasing the virions. This happens in poliovirus infection. In other animal and plant viruses, the host cells are not destroyed. The viruses leave the cells by special channels (tubules) over an extended period of time. Still other viruses leave cells by budding or extrusion. In budding out, enveloped animal viruses acquire a portion of the host-cell membrane in the process. Some components of the host-cell membrane are incorporated into the envelope. The release of enveloped influenza virus is an example of this budding type of release. Presumably the same kind of mechanism operates in the case of enveloped plant viruses.

The yield of virus particles per cell varies with the virus, the cell, and the growth conditions. The average yield of plant and animal virions ranges from several thousand to about a million per cell, compared with the two hundred or so of a bacterial T phage.

As an example of the viral replicative process in the eucaryotic cell, Fig. 12-10 shows the replication of the herpes simplex virus (Fig. 12-11), which is the cause of "fever blisters" or "cold sores." There is a considerable degree of organization associated with this replicative process. As can be seen in Fig. 12-10, the events related to biochemical replication occur in both the nucleus and the cytoplasm, with assembly of the virion initiated in the nucleus. The nucleocapsids of these viruses then migrate to the cytoplasmic membrane, where the mature viruses appear to bud off by a process that is the reverse of the penetration step.

Classification
Animal viruses have been classified in several ways. One early means of classification was based on virus tissue affinity, for example, neurotropic (nerve-tissue) viruses and dermatropic (skin-tissue) viruses. As methods of

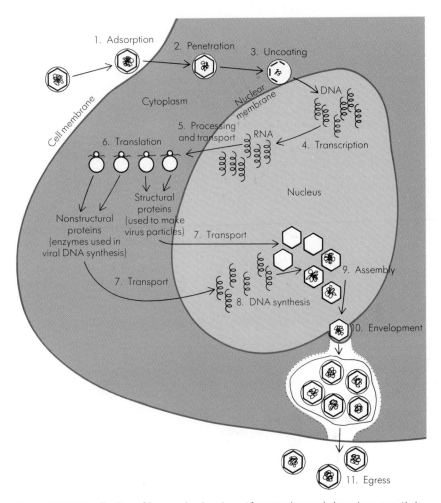

Figure 12-10. Replication of herpes simplex virus. After attachment (adsorption: step 1) the virus particles are taken into the cell (penetration: step 2) by engulfment or possibly by fusion of the viral envelope with the cell membrane. After the virus enters the cell, the envelope and protein coat are removed by cellular enzymes (uncoating: step 3) and the DNA of the virus is released into the nucleus. One strand of the viral DNA is transcribed into an RNA copy (transcription: step 4). Suitable transcripts (now considered to be viral messenger RNA molecules), are processed and transported (step 5) into the cytoplasm, where the RNA code is translated into an amino acid code (translation: step 6). During this process both proteins which will subsequently be used to construct new virus particles (structural proteins) and proteins which act as enzymes involved in DNA metabolism (nonstructural proteins) are synthesized and transported (step 7) from the cytoplasm into the nucleus. New viral DNA is now synthesized in the nucleus (DNA synthesis: step 8) by the nonstructural proteins and possibly some cellular enzymes. Assembly (step 9) of the structural protein subunits around the viral DNA results in the formation of a noninfectious virus particle lacking an outer coat or envelope. The final envelope, which enables the particle to be infectious, is acquired from the nuclear membrane by a budding-off process (envelopment: step 10). The virus particles are now transported from the vicinity of the nucleus to the periphery of the cell (egress: step 11). The exact mechanism by which this is accomplished is unclear. For a better understanding of nucleic acid and protein synthesis, see Chapter 16. [*After a drawing by C. Shipman, Jr., and S. Marty-Everhardus. From "An Age-old Problem: The Control of Viruses," University of Michigan Research News*, **26**:1–7, (1975).]

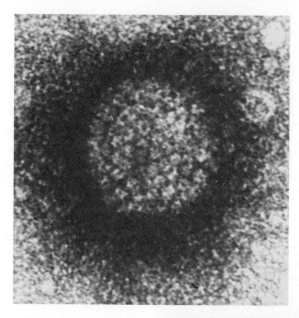

Figure 12-11. Herpes simplex virus (without envelope) showing individual capsomeres. (× 340,000). *(Courtesy of Dr. Peter Gill, McGill University.)*

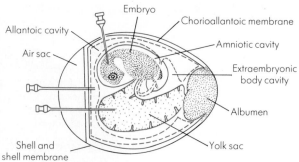

Figure 12-12. Diagrammatic representation of section of embryonated hen's egg 10 to 12 days old. The hypodermic needles show the routes of inoculation of the yolk sac, allantoic cavity, and embryo (head).

Figure 12-13. Embryonated hen's egg is used for the cultivation of many mammalian viruses.

measuring physical, chemical, and biological characteristics of viruses have developed, information has been accumulated to formulate a classification scheme based on these properties for all viruses. The properties are summarized in Table 12-3. Table 12-4 shows a classification scheme for animal viruses based on these criteria. As can be seen, animal viruses have been assigned to different genera, which are then grouped into families. No species names have been formulated as yet, probably because our knowledge of viruses is still insufficient for this level of taxonomy. The scheme provides great convenience and utility but does not attempt to show *phylogenetic* (evolutionary) relationships among the virus families. Plant

Table 12-3. Properties used for classification of viruses	PRIMARY CHARACTERISTICS	SECONDARY CHARACTERISTICS
	Nucleic acid: 　RNA: Single- or double-stranded 　DNA: Single- or double-stranded Structure of virus particle: 　Helical capsid: naked or enveloped 　Icosahedral capsid: naked or enveloped 　Complex structure (e.g., tail) Number of capsomeres Size of virion Susceptibility to inactivating agents Site of replication: 　Nucleus 　Cytoplasm	Host range: 　Species of host 　Specific host tissues or 　cell types Mode of transmission (e.g., feces) Specific surface structures (antigenic properties)

Table 12-4. Outline of classification of animal viruses

NUCLEIC ACID	SYMMETRY	ENVELOPE	VIRION SIZE, nm	FAMILY	EXAMPLE OF GENERA	HUMAN DISEASES CAUSED BY FAMILY
RNA	Icosahedral	No	18–30	Picornaviridae	*Enterovirus* *Rhinovirus*	Intestinal infections, polio, colds
		No	60–80	Reoviridae	*Reovirus*	Respiratory and intestinal infections
		Yes	35–80	Togaviridae	*Alphavirus*	Yellow fever, encephalitis
	Helical	Yes	80–120	Orthomyxoviridae	*Influenzavirus*	Influenza
		Yes	150–300	Paramyxoviridae	*Mobillivirus*	Mumps, measles, respiratory infections
		Yes	60–180	Rhabdoviridae	*Lyssavirus*	Rabies
	?	Yes	100	Retroviridae	None	Animal tumors
	?	Yes	100–300	Arenaviridae	*Arenavirus*	Meningitis
	?	Yes	90–100	Bunyaviridae	*Bunyavirus*	Encephalitis
		Yes	120	Coronaviridae	*Coronavirus*	Respiratory infections
DNA	Icosahedral	No	18–24	Parvoviridae	*Parvovirus*	Coinfection with adenovirus
		No	40–50	Papovaviridae	*Papillomavirus*	Warts
		No	70–80	Adenoviridae	*Mastadenovirus*	Respiratory infections
		Yes	110	Herpesviridae	*Herpesvirus*	Cold sores, genital herpes, shingles, chicken pox
	Complex	Complex coat	230×300	Poxviridae	*Orthopoxvirus*	Smallpox

viruses have descriptive group names but no families or genera; an example is the cucumber mosaic virus group.

Cultivation of animal viruses

Embryonated hen's eggs

Since viruses can grow only in living cells, one of the most economical and convenient methods for cultivating a wide variety of viruses is the chick-embryo technique (Fig. 12-12). Fertile or embryonated hen's eggs incubated for 5 to 12 days can be inoculated through the shell aseptically. The opening may be sealed with paraffin wax and the egg incubated at 36°C (96.8°F) for the time required for growth of the virus. The egg may be

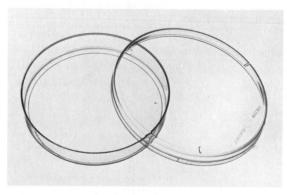

Figure 12-14. Containers used for the cultivation of cell-tissue cultures. Shown are a tissue-culture flask, a tissue-culture tube, and a tissue-culture dish. *(Courtesy of BioQuest, Division of Becton, Dickinson and Company, Cockeysville, Maryland.)*

inoculated on the chorioallantoic membrane, where some viruses, for example, vaccinia, will grow and produce lesions or pocks. The yolk sac and the embryo also can be used to grow viruses. The chick-embryo technique has been used in the production of viruses for vaccines against smallpox, yellow fever, influenza, and other diseases and in immunologic tests and other studies (Fig. 12-13).

Tissue cultures Tissue cultures of viruses began with the cultivation of minced chick embryos in serum or salt solutions. This led to the use of pure tissue cultures of animal cells in which the viruses would grow. Animal cells can now be grown in a manner similar to that used for bacterial cells. When animal cells are cultured in plastic or glass containers (Fig. 12-14), they attach themselves to the surface of the vessel and continue to divide until the whole surface area covered by the medium is occupied. A monolayer of cells is produced and is used for the propagation of viruses. Different tissue cells are more effective for cultivation of some viruses than others. This approach has made it possible to cultivate many viruses in pure culture in large quantities for study and for the commercial production of vaccines. It is also widely used for the isolation and propagation of viruses from clinical material.

Vaccines prepared from tissue cultures have an advantage over those prepared from embryonated hen's eggs in minimizing the possibility of a patient developing hypersensitivity or allergy to egg albumin. The Salk poliomyelitis vaccine, which is produced in tissue culture, was developed

Figure 12-15. Light microscope view of tissue cultures used for the cultivation of viruses. (1A) Normal human lung fibroblast cell culture. (1B) Human lung fibroblast cell culture infected with varicella (chicken pox) virus. Characteristic cytopathic effect (CPE) seen: rounded and enlarged cells with splitting of cell sheet (layer) and stranding. (2A) Normal monkey kidney cell, line GL V3A. (2B) Monkey kidney cell, line GL V3A, infected with poliovirus. CPE seen: cell shrinkage with retracted margins exhibiting angular shapes. (3A) Normal rabbit cell, line RK13. (3B) Rabbit cell, line RK13, infected with rubella (German measles) virus. CPE seen: discrete foci or centers of aggregated cells. Original magnification × 400. (Courtesy of A. F. Doss, McGill University.)

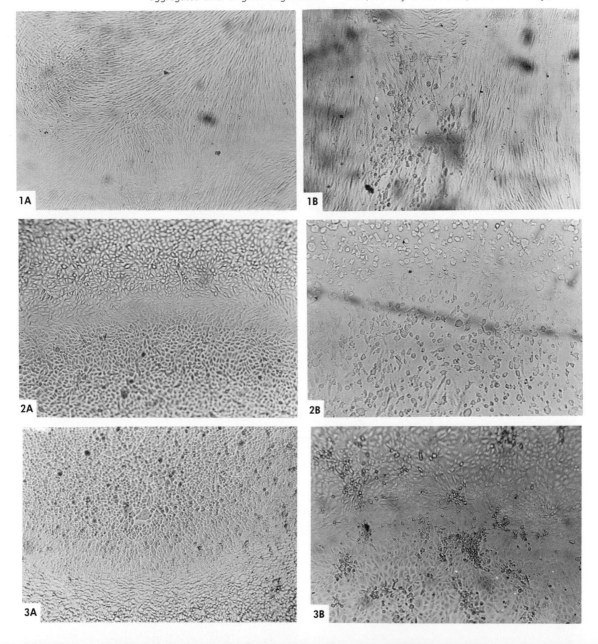

after basic research had shown that the poliovirus would grow satisfactorily on monkey kidney-cell cultures. HeLa cell cultures, originally isolated from a tumor in a human patient and cultivated since then as a *line* or serially propagated population of cells, have been useful in cultivating many viruses previously difficult or impossible to grow. Growth of viruses in tissue culture is shown in Fig. 12-15. It can be seen that the tissue structure deteriorates as the virus multiplies. This deterioration is called the *cytopathic effect* (CPE).

Animals   Some viruses cannot be cultivated in cell culture and must be grown in animals. Mice, guinea pigs, rabbits, and primates are used for this purpose. Animal inoculation is also a good diagnostic tool because the animal can show typical disease symptoms and *histological* (tissue) sections of infected tissue can be examined microscopically.

Figure 12-16. Virus particles in a suspension may be enumerated by means of a plaque assay. Shown are plaques of reovirus on a monolayer of L-929 mouse fibroblast cells in tissue-culture dishes. Note that increasing dilution of the virus suspension results in decreasing plaque numbers in the dishes. *(Courtesy of Collette Oblin, McGill University.)*

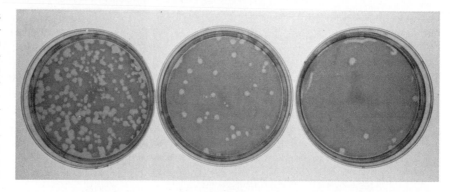

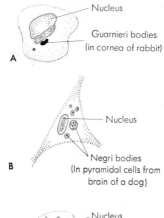

A

B — Negri bodies (In pyramidal cells from brain of a dog) / Nucleus

Figure 12-17. Inclusion bodies produced by viruses in certain host tissues. (A) Guarnieri bodies of variola (smallpox) virus in the cytoplasm of rabbit corneal cells; (B) Negri bodies in the cytoplasm of Purkinje cells (nerve cells of the brain) infected with rabies virus; (C) Bollinger bodies in the cytoplasm of cells infected with fowl pox virus; (D) intranuclear inclusions in epithelial cells of rabbit cornea inoculated with herpes virus.

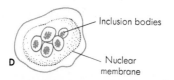

Effects of virus infection on cells

Cell death

Disease symptoms in the host due to virus infection vary from none to massive destruction of infected cells leading to cell death. In cell tissue culture, groups of killed cells (*plaques*) have been used in the enumeration of viruses because the number of plaques is proportional to the number of infectious virus particles present. Each virion gives rise to a single plaque, just as a bacterium gives rise to a single colony (Fig. 12-16).

Other effects of cell infection by viruses include formation of giant cells (*polykaryotes*), genetic changes such as chromosomal breakage, induction of *interferon* (a protein) production by the infected cell that prevents infection of healthy cells (see Chap. 25), and formation of *inclusion bodies*.

Inclusion bodies

Before it was possible to study the morphology of viruses at the high magnifications provided by the electron microscope, investigators had observed intracellular structures, or inclusion bodies, associated with virus diseases (Fig. 12-17). In 1887 J. B. Buist noted small particles in the cytoplasm of cells surrounding the lesions of smallpox. These he called *elementary bodies*. E. Paschen made the same observation independently in 1906. It is now known that these *Paschen bodies* are aggregates, or colonies, of virions growing in the cytoplasm of the host cell. In 1892, G. Guarnieri reported having seen small round particles in the cytoplasm of similar cells. These *Guarnieri bodies* are also thought to consist of aggregates of unassembled virus subunits and intact virions.

Characteristic inclusion bodies are found in the cytoplasm of certain nerve cells and in the Purkinje cells of the cerebellum in cases of rabies infection. Finding these typical inclusions (called *Negri bodies* for their discoverer) is diagnostic for the disease.

Inclusion bodies have been found in connection with many other virus diseases. They occur in the cytoplasm in most pox diseases (smallpox, sheep pox, fowl pox), rabies, molluscum contagiosum, and others. Intranuclear inclusions are found in chicken pox, herpes, and the polyhedral diseases of insects. Intranuclear and intracytoplasmic inclusions may be found in the same cell in instances of multiple infection. Some inclusions are useful in establishing diagnosis while the significance of others is not yet known.

Inclusion bodies are for the most part characteristic of the virus causing the infection and even suggest definite pathological changes in the cell. It is generally true, however, that inclusion bodies are aggregates of unassembled virus subunits and intact virions in infected cells. It is experimentally possible to remove them from the cell and use them as inoculum to infect other cells.

Progressive and fatal diseases associated with viruses

There are still progressive or gradually extending diseases which usually terminate in death that are poorly understood and require much research. Some of these are or may be caused by viruses, such as classic slow virus diseases and cancer.

Classic slow virus diseases

These diseases have a *slow* progressive course usually with a fatal outcome. (Incubation periods are measured in years!) They are caused by transmissible agents whose properties and behavior (for example, unusual resistance to ultraviolet radiation and heat) suggest an *unconventional* or *atypical* virus.

There are four classic slow virus diseases. Each may be described as a neurological disease. They are *Kuru* and *Creutzfeld-Jakob disease* of humans and *scrapie* and *transmissible mink encephalopathy* (TME) of animals. (The term *encephalopathy* describes these diseases well because each involves widespread destructive cerebral changes due to degeneration without inflammation.)

Scrapie is so named because the diseased animal tends to scrape against fixed objects. It is a chronic infection of the central nervous system of sheep. TME, a disease found in mink farms in the United States, may have arisen from mink fed on the meat of sheep contaminated with scrapie virus. It is also possible that skunks and raccoons harbor this virus naturally.

Kuru and Creutzfeld-Jakob disease are similar degenerative diseases of the human central nervous system. Fortunately, both are rare, and Kuru is restricted to the Foré people of New Guinea. Kuru was spread among these people by ingestion of infectious human brain tissue during ritual cannibalism of the dead as a mourning ceremony. Now that ritual cannibalism has been eliminated, the disease has been declining in prevalence and should soon disappear. (Dr. D. Carleton Gajdusek of the National Institute of Neurological Diseases and Strokes studied this disease extensively and was awarded the 1976 Nobel prize in medicine for his pioneering work.) Unlike Kuru, which is restricted geographically, Creutzfeld-Jakob disease has a worldwide distribution. The onset generally occurs between the ages of 35 and 65.

Cancer

More than 100 clinically distinct types of cancer are recognized, each having a unique set of symptoms and requiring a specific course of therapy. However, almost all of them can be grouped into four major categories:

1 *Leukemias*: Abnormal numbers of white blood cells (leukocytes) are produced by the bone marrow
2 *Lymphomas*: Abnormal numbers of lymphocytes (a type of leukocyte) are produced by the spleen and lymph nodes
3 *Sarcomas*: Solid tumors growing from derivatives of embryonal mesoderm, such as connective tissues, cartilage, bone, muscle, and fat
4 *Carcinomas*: Solid tumors derived from epithelial tissues, the most common form of cancer; epithelial tissues are the internal and external body surface coverings and their derivatives, and thus include skin, glands, nerves, breasts, and the linings of the respiratory, gastrointestinal, urinary, and genital systems

Cancer has three major characteristics: *hyperplasia*, *anaplasia*, and *metastasis*. Hyperplasia is the uncontrolled proliferation of cells. Anaplasia is structural abnormality of cells (these cells also have a loss of or reduction in their functions). Metastasis is the ability of a malignant cell to detach itself from a tumor and establish a new tumor at another site within the host.

For a long time microbiologists entertained the idea that cancer might be caused by viruses. However, since the disease did not appear to be infectious, and there were no confirmed isolations of a causative virus, the idea became more and more unattractive. However, in recent years good evidence has been accumulated to show that some viruses do in fact cause cancer in animals. These findings revitalized the idea that human cancers might be caused by viruses because it is reasonable to expect that if viruses can cause cancer in other animals, then surely they can do so in humans.

Both RNA and DNA viruses have been found capable of infecting animals; in these animals the affected cells are *transformed*, resulting in the formation of tumors. (A transformed cell acquires properties distinctly different from uninfected cells or infected cells in which tumors are not produced.) Such tumor-inducing viruses are called *oncogenic* viruses.

A common characteristic of all oncogenic viruses is that the viral genome in some way becomes either integrated or intimately associated with the host DNA. The host cell does not lyse—a situation similar to the model of lysogeny in bacteria infected with temperate phages. It should be noted that if the viral genome is RNA, it serves as a template for the synthesis of a DNA molecule complementary to it; the enzyme *reverse transcriptase* is responsible for this synthesis. The enzyme may also synthesize a second strand of DNA complementary to the first strand. This results in a double-stranded DNA molecule synthesized from the viral RNA and called a *provirus*, which now can be integrated into the host DNA. In this way transformation and tumors are induced in host cells.

Oncogenic DNA viruses. Polyoma virus is endemic in wild and laboratory mouse populations. SV40 virus cannot induce tumors in the monkey (natural host) but can do so in rodents in the laboratory. The Epstein-Barr virus (EBV), a herpesvirus, has been consistently associated with certain human *neoplasias*, or tumors. It was discovered by Epstein and Barr in 1964 in cultured Burkitt's lymphoma cells. Burkitt's lymphoma is a cancer of the lymphoid system. The EBV has also been implicated in infectious mononucleosis (regarded by some authorities as a self-limiting leukemia) as well as in nasopharyngeal carcinoma.

Other herpesviruses, such as herpes simplex viruses (HSV) types 1 and 2, have also been implicated in certain human cancers. Cancers of the lip or mouth have been associated with HSV 1. Cervical cancer has been associated with HSV 2.

Oncogenic RNA viruses. Oncogenic RNA viruses (also called *oncornaviruses*) are generally divided into three main classes, A, B, and C, on the basis of gross morphological characteristics. Type A RNA viruses are not infectious and have not been found outside the cell. They are actually a very small group of viruslike particles encapsulated by protein shells, rather than by lipid-containing membranes as are types B and C. Type A particles found in the cytoplasm of the cell are believed to be immature forms of type B viruses while those in the cisternae (cell spaces) are thought to be immature type C particles. Type B RNA viruses have been found capable of producing tumors in mouse mammary glands. Circumstantial evidence encourages many investigators to believe that a similar virus is involved in human breast cancer. Type C RNA viruses constitute the most important class, and most have been shown to infect a large number of animal species. They cause leukemias, lymphomas, and sarcomas.

So far, no RNA viruses have been isolated from human cancer cells, but some circumstantial evidence implicates them in this disease. For example, molecular components have been found in human leukemic cells which are related to similar components in RNA tumor viruses.

Research aimed at isolating a causative viral agent of human cancer or at least finding an association between viruses and human cancer continues. Although no results showing a confirmed viral etiology for human cancers have been obtained, much fundamental knowledge on tumor biology has accumulated and many conceptual hypotheses concerning carcinogenesis have been formulated.

Summary and outlook

Animal and plant viruses are minute obligate intracellular parasites. Each virion has a nucleic acid central core surrounded by a capsid. Some virions also have an envelope. Morphologically, animal and plant viruses may be icosahedral, helical, enveloped, or complex.

The process of viral replication begins with adsorption of the virion on a host cell. This is followed by penetration and uncoating, biosynthesis of viral components, and assembly and maturation of virions. The process is terminated by release of viruses from the host cell.

Based on physical, chemical, and biological characteristics, animal viruses are now classified into families and genera. Many methods are available to cultivate these viruses. These include the use of embryonated chicken eggs, tissue cultures, and animals.

Upon infection by viruses, cells may die, form inclusion bodies, produce interferon, or suffer some form of cellular damage.

In the area of medical virology, current developments point to a very challenging and promising future. We shall see increased attention given to improving public health procedures for those common viral diseases for which immunization is practical. Studies will continue on developing vaccines for the other viral diseases. In addition, new chemotherapeutic antiviral agents will be sought for the treatment of viral diseases. They will be used against viruses just as antibiotics are used against bacteria.

There are progressive and usually fatal diseases which are still poorly understood. One group of these comprises the classic slow virus diseases. Such virus infections persist for years, ultimately killing the host. The new methods and concepts used to study these slow virus diseases may now be used for research on many other chronic degenerative diseases. Such work should yield information on the possible role of slow viruses in other diseases, including diabetes and multiple sclerosis.

Cancers represent another group of diseases which are progressive and fatal. To date no viruses have been shown unequivocally to be the cause of human cancer. However, the large number of virus-induced tumors in mammals and birds have suggested that some types of cancers may have a viral etiology. Once a viral agent has been isolated and cultivated, a vaccine against cancer can be developed from the virus.

Key terms

agglutinate	nucleocapsid	replication
cell line	oncogenic virus	reverse transcriptase
cytopathic	oncornavirus	slow virus
envelope	phylogenetic	spike
inclusion bodies	polykaryote	transformed cell
interferon	provirus	viropexis
naked virion		

Questions

1 Draw a simple virion, identifying the following structures: capsid, capsomere, nucleocapsid, nucleic acid, envelope, spike.
2 Discuss three properties of viruses that distinguish them from other miroorganisms.
3 Compare the mode of multiplication of viruses with that of some representative bacteria.
4 Compare the overall morphology of a typical animal virus with that of a typical bacteriophage.
5 What structural components of a virion consist of protein?
6 Discuss the replication of herpesvirus.
7 In what ways do plant viruses differ from animal viruses in the process of reproduction?
8 Explain how viruses are classified at the present time.
9 Why do we think that viruses may be the cause of cancer in humans?
10 How do the methods used for cultivating viruses differ from those used for cultivating bacteria?
11 What are inclusion bodies? Where are they found? What is their significance?
12 What new concepts have we learned about virology from an understanding of slow virus diseases?
13 Explain how the behavior of a temperate phage is similar to that of an oncogenic virion.

References for part four

Horne, R. W.: *Virus Structure*, Academic, New York, 1974. *A short monograph of 52 pages that describes the physical characteristics of isolated viruses that represent typical structural groups.*

Hughes, Sally S.: *The Virus: A History of the Concept*, Science History Publications, New York, 1977. *Concise and highly readable history of virology in relatively nontechnical language.*

Knight, C. Arthur: *Molecular Virology*, McGraw-Hill, New York, 1974. *A condensed and integrated presentation of the basic principles of molecular virology for the serious student of microbiology. A highly practical chapter on the purification of viruses is included.*

Luria, S. E., J. E. Darnell, Jr., D. Baltimore, and A. Campbell: *General Virology*, 3d ed., Wiley, New York, 1978. *A well-organized text on the biology of viruses and their activities.*

Primrose, S. B.: *Introduction to Modern Virology*, Blackwell Scientific, Oxford, 1974. *An introductory paperback on the biochemical and genetic aspects of virology.*

Sanders, F. Kingsley: *The Growth of Viruses*, Oxford University Press, London, 1975. (Available from Carolina Biological Supply Company, Burlington, N.C.) *A short monograph of 16 pages that describes the structure of viruses as well as their behavior within cells. It is well illustrated with diagrams and electron micrographs.*

Volk, W. A.: *Essentials of Medical Microbiology*, Lippincott, Philadelphia, 1978. *A medically oriented microbiology text. The essentials of medical microbiology are covered, with 12 chapters on virology.*

Wagner, E. K.: "The Replication of Herpes Virus," *Amer. Sci.* **62**(5): 584–593, September–October, 1974. *A popular and informative review of the herpesvirus and postulates regarding the mechanism of viral tumor formation.*

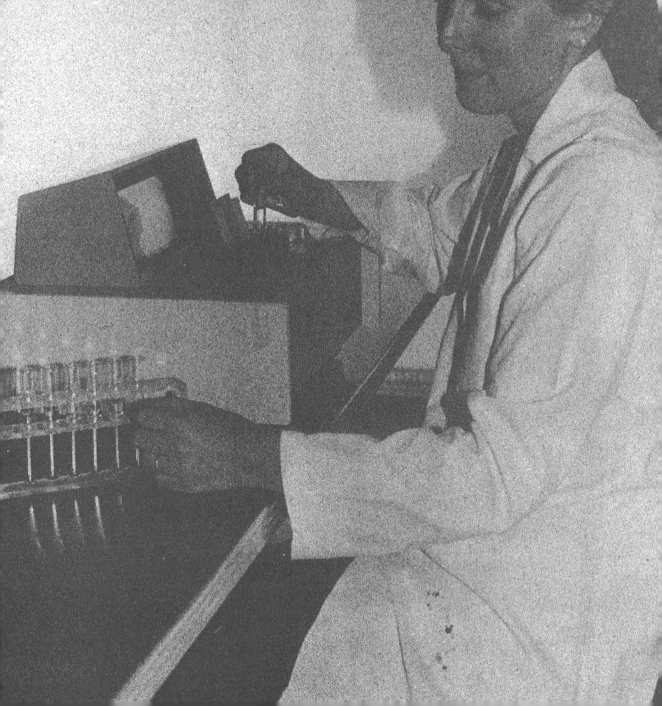

PART FIVE
METABOLISM OF MICROORGANISMS

In order to stay alive, grow, and reproduce, a cell must be able to perform cellular activities involving many chemical reactions and energy changes. The totality of these transformations is termed *metabolism*. The cell may have to alter nutrients in the environment before they can be absorbed. It must effect additional changes in these nutrients once they are inside. Some of the assimilated materials or nutrients are incorporated into compounds which form part of the cell structure. Other materials are broken down to provide energy for many cellular functions.

All these activities involve many processes with complex chemical reactions. It is the purpose of Part V to explain these cellular processes. We will discuss first (in Chap. 13) the biological catalysts, the *enzymes*, which are essential for all the chemical reactions in the cell, and how the enzymes are regulated. An understanding of this is important because the control of cellular metabolism ultimately involves the regulation of enzyme activity. Then we will consider how energy is produced by the cell (Chap. 14) and how this energy is utilized (Chap. 15).

A spectrophotometer being used to follow the progress of a microbial enzyme-catalyzed reaction. Experiments such as these yield information on the rates of enzyme-catalyzed reactions and how these rates depend on enzyme and substrate concentrations and other conditions such as temperature and pH.

ENZYMES AND THEIR REGULATION

Chemical activities carried out by the cell are very complex. This is understandable, given the great variety of materials used as nutrients by the cell on the one hand and the variety of substances synthesized into cell constituents on the other. How does the cell carry out all these activities? The answer lies in the action of *enzymes*, substances present in the cell in minute amounts and capable of effecting all the changes associated with cellular (and life) processes. There can be no life without enzymes.

There are thousands of different enzymes in an average cell. These enzymes and their activities in the living cell must be coordinated in such a way that they deliver appropriate products to the right place, in the right amount, at the right time, and with a minimum expenditure of energy. This coordination is brought about by enzyme regulation or control. This chapter describes some major characteristics of enzymes and the ways in which they are regulated.

Some characteristics of enzymes

Enzymes are *biological catalysts*. Catalysts, even in small amounts, have the unique ability to speed up chemical reactions without themselves being used up or altered after the reaction. For example, hydrogen and oxygen do not combine to any appreciable extent under normal atmospheric conditions. If, however, the two gases are allowed to touch colloidal platinum, they react instantly to produce water. The platinum is the catalyst in this example; it greatly increases the speed of this water-forming reaction without being used up in the reaction or altered after the reaction.

However, unlike the above example of the inorganic platinum, enzymes are organic compounds produced by living cells. This is why enzymes are called biological or organic catalysts or catalytic agents. Catalysts also exhibit *specificity*. This means that a certain catalyst will effect only a certain type of reaction.

Although all enzymes are initially produced within the cell, some are excreted through the cell wall and can function outside the cell. Thus two types of enzymes are recognized: *extracellular enzymes*, or *exoenzymes* (functioning outside the cell) and *intracellular enzymes*, or *endoenzymes* (functioning within the cell). The principal function of exoenzymes is to effect whatever changes in the nutrients in the environment are necessary to enable these nutrients to enter the cell. For example, amylases break down starch to smaller sugar units. Intracellular enzymes synthesize cellular material and also break down nutrients to provide the energy requirements of the cell. For example, hexokinases catalyze the phosphorylation of glucose and other hexoses (simple sugars) in the cell.

The general characteristics of enzymes are the same no matter whether they are produced by the cells of microorganisms, humans, or other forms of life. In fact, cells from organisms that are very different may contain some enzymes which are identical or at least have similar functions. For example, many of the enzyme reactions taking place in a yeast cell are identical to those in a human muscle cell.

Typically, one enzyme molecule can catalyze the conversion of 10 to 1,000 molecules of *substrate* (compound acted on by an enzyme) per second. Enzyme-catalyzed reactions are often from several thousand to more than a million times faster than the same reactions without enzymes. Calculations have shown that the breakdown of proteins in human digestive processes would take more than 50 years instead of a few hours without the aid of enzyme action!

Chemical and physical properties of enzymes

Enzymes are pure proteins or proteins combined with other chemical groups. Like all proteins, they are denatured by heat, are precipitated by ethanol or high concentrations of inorganic salts such as ammonium sulfate, and do not diffuse (*dialyze*) through semipermeable or selective membranes. Enzyme proteins are very large molecules: molecular weights from approximately 10,000 to 1 million have been determined for different enzymes.

Many enzymes consist of a protein combined with a low-molecular-weight organic molecule referred to as a *coenzyme*. The protein portion is the *apoenzyme*. When united, the two portions form the complete enzyme, called the *holoenzyme*.

Apoenzyme   + coenzyme   → holoenzyme

Inactive	Inactive	Active
Protein	Organic molecule	
High molecular weight	Low molecular weight	
Nondialyzable	Dialyzable	

The integral part of some coenzymes is a vitamin. Several of the B vitamins have been identified as the main components of coenzymes, (Table 13-1).

In some instances the nonprotein portion of an enzyme may be a metal, for example, iron, in the enzyme catalase. Thus many enzymes require the addition of metal ions (Mg^{2+}, Mn^{2+}, Fe^{2+}, Zn^{2+}, etc.) to activate them. These ions are regarded as inorganic coenzymes or *cofactors*. Sometimes both a cofactor and a coenzyme (organic) are required before an enzyme becomes active.

A large number of enzymes have been extracted from cells and, by a combination of physical and chemical techniques, have been obtained in chemically pure form. Urease was the first enzyme to be isolated in a purified crystalline form in 1926.

Enzyme molecules are exceedingly efficient in accelerating the transformation of substrate to end product. As we noted earlier, a single enzyme molecule can effect the change of as many as 1,000 molecules of substrate per second. This ability, together with the fact that the enzyme is not consumed or altered after the reaction, reveals why very small quantities of enzymes are sufficient for cellular processes.

However, enzymes are unstable. Their activity may be significantly decreased or destroyed by a variety of physical or chemical conditions. Great differences exist among enzymes in this respect. Some may be inactivated by very minor changes in the environment, such as short exposure to room temperature.

The two most striking characteristics of enzymes are (1) their high catalytic efficiency and (2) their high degree of specificity for substrates. A single enzyme may react with only a single substrate or, in some instances, with a particular chemical grouping on chemically related substrates. This means that cells usually produce a different enzyme for every compound

Table 13-1. Some vitamins and their coenzyme forms

VITAMIN	COENZYME
Thiamine (B_1)	Cocarboxylase
Riboflavin (B_2)	Riboflavin adenine dinucleotide
Niacin	Nicotinamide adenine dinucleotide
Pyridoxine (B_6)	Pyridoxal phosphate
Folic acid	Tetrahydrofolic acid

they metabolize. Furthermore, each enzyme causes a one-step change in the substrate. For example, yeasts change glucose to alcohol and carbon dioxide. The initial reactants and the final products of the reaction are shown in the following overall equation:

Glucose + yeast cells   →   alcohol + carbon dioxide

Substrate   Source of   End products
            enzymes

This transformation is accomplished not by a single enzyme but by a group of enzymes, an *enzyme system*. More than a dozen individual enzymes work in sequence, each causing a chemical reaction which produces a single specific change in the product formed by the preceding enzyme reaction. The last reaction of the many in the system yields the final products.

Over 1,000 different enzymes are known, and well over 150 have been crystallized (purified enough to form crystals). Many more remain to be discovered. Exciting new areas of enzyme involvement have been under investigation in recent years, including the self-regulating nature of many enzyme systems (to be discussed later), the genetic control of enzyme function and synthesis, and the role of enzymes in *development* and *differentiation* (processes by which cells are changed into different kinds of structures and tissues).

How enzymes are named and classified

Enzyme nomenclature has been formalized by international agreement under the auspices of the Commission on Enzymes of the International Union of Biochemistry. However, common or trivial names still remain in general use because of their familiarity and convenience. Let us examine briefly how enzymes are formally named and classified.

The suffix -*ase* identifies an enzyme and should be used only for single enzymes, for example, succinate dehydrogenase. When it is desired to name a complex of several enzymes on the basis of the overall reaction catalyzed by it, the word *system* should be used, for example, the succinate oxidase system, which catalyzes the oxidation of succinic acid by oxygen in several steps and by several individual enzymes. For classification only single enzymes, not enzyme systems, are considered.

The type of chemical reaction catalyzed is the basis for the classification and naming of enzymes. It is this specific property that distinguishes one enzyme from another. Two names are recommended for each enzyme: a working or trivial name and a systematic name. The trivial name is shorter and more convenient to use; in many cases, it is the name already in current use. The formal or systematic name is formed in accordance with definite rules, identifies the substrate(s), and specifies the type of reaction catalyzed. For example, hexokinase is the trivial name of ATP:hexose phosphotransferase, the enzyme that adds a phosphate group to glucose. (In addition, a method of coding enzymes by numbers is available.) According to international classification, enzymes are grouped into six

Table 13-2.
Major classes of enzymes

CLASS NO.	CLASS	CATALYTIC REACTION
1	Oxidoreductases	Electron-transfer reactions (transfer of electrons or hydrogen atoms)
2	Transferases	Transfer of functional groups (functional groups include phosphate, amino, methyl, etc.)
3	Hydrolases	Hydrolysis reactions (addition of a water molecule to break a chemical bond)
4	Lyases	Addition' to double bonds in a molecule as well as nonhydrolytic removal of chemical groups
5	Isomerases	Isomerization reactions (reactions in which one compound is changed into an isomer, i.e., a compound having the same atoms but differing in molecular structure)
6	Ligases	Formation of bonds with cleavage or breakage of ATP (adenosine triphosphate)

major classes (see Table 13-2). As we noted above, these groupings are based on the type of chemical reactions which the enzymes catalyze.

The nature and mechanism of enzyme action

Most enzyme reactions may be represented by the following overall reaction:

Enzyme E + substrate S ⇌ enzyme-substrate complex ES ⇌
product P + enzyme E

The enzyme E and substrate S combine to give an enzyme-substrate complex ES, which then breaks up to yield the product P. The enzyme is not used up in the reaction but is released for further reaction with another substrate molecule. This process is repeated many times until all the available substrate molecules are consumed.

Central to the theory of the mechanism of enzyme action is the concept of *activation* of the substrate following formation of the enzyme-substrate (ES) complex. Activation allows the substrate to be changed by enzyme action. The activation of the substrate molecule takes place because of high chemical affinity of the substrate for certain areas of the enzyme surface called the *active sites*. A strain or distortion is produced at some linkage in the substrate molecule, making it *labile* (unstable), and it undergoes a change determined by the particular enzyme. The altered molecules lack affinity for the active sites and hence are released. The enzyme is then free to combine with more substrate to repeat the action (see Fig. 13-1). Almost

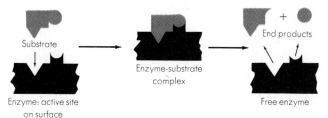

Figure 13-1. Enzyme-substrate reaction, depicted schematically. The substrate is attracted to and combines with the enzyme at the active site on the surface of the enzyme molecule. The chemical groupings of the substrate are strained by this attraction, and cleavage results. The cleavage products are released from the enzyme, and the enzyme is free to combine with more substrate and repeat the process.

all intracellular enzymes have more than one active site per molecule. For example, lactate dehydrogenase has four. α-Chymotrypsin, in common with other secreted extracellular enzymes, has only one active site.

The main function of an enzyme is to lower the *activation energy* barrier to a chemical reaction. Activation energy refers to the amount of energy required to bring a substance to the reactive state. The enzyme combines with the substance (substrate) to produce a transition state requiring less activation energy for the chemical reaction to proceed. Figure 13-2 illustrates this concept.

This discussion applies to substrates undergoing degradation and being utilized for energy. The same type of explanation could be used to describe *synthesis*, or building up of complex compounds from simpler ones. Thus, two different molecules might become attached to adjacent sites on the enzyme surface. A unique activation by the enzyme would lead to establishment of a bond between the two molecules, thereby creating a new compound of the original two substrates. The product has little affinity for the active sites and hence is released, thereby freeing the active sites to repeat the process with another two molecules of substrate. The active site on an enzyme surface is actually a very small area, which means that large regions of the protein (which has hundreds of amino acids) do not contribute to enzyme specificity or enzyme action. Thus only a relatively few amino acids must be directly involved in the catalytic process—perhaps even less than five! It should also be emphasized that the "fit" between a portion of the enzyme surface and the substrate is not a static one. Rather it is a dynamic interaction in which the substrate induces a structural change in the enzyme molecule, as a hand changes the shape of a glove.

Figure 13-2. The main function of an enzyme is to lower the activation-energy, the energy necessary to attain the transition state of a chemical reaction. As can be seen, an enzyme-catalyzed reaction has a lower activation energy and so requires less energy for the reaction to proceed.

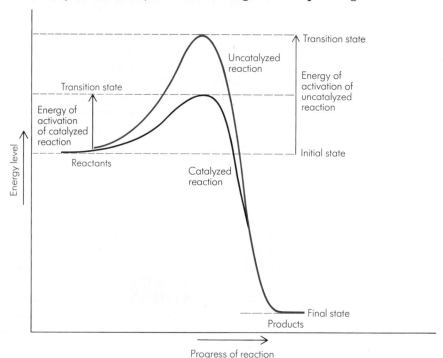

Conditions affecting enzyme activity

Among the conditions affecting the activity of an enzyme are the following:

1 Concentration of enzyme
2 Concentration of substrate
3 pH
4 Temperature

Generally speaking, there is an optimum relation between the concentrations of enzyme and substrate for maximum activity, as illustrated in Figs. 13-3 and 13-4. Likewise, each enzyme functions optimally at a particular pH and temperature (Figs. 13-5 and 13-6). From these illustrations, it is

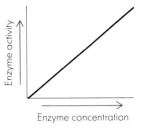

Figure 13-3. Effect of enzyme concentration on rate of enzyme activity. With highly purified enzymes there exists, within limits, a linear relationship between amount of enzyme present and level of activity. (Note that enzyme activity is a measure of the disappearance of reactants or appearance of products of the reaction catalyzed.)

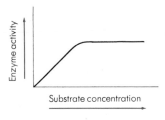

Figure 13-4. Effect of substrate concentration on enzyme activity. Note that the rate increases rapidly with initial increases in substrate concentration. Further increases in substrate concentration have no effect on rate; the rate becomes independent of substrate concentration.

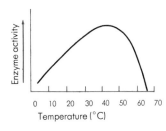

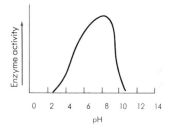

Figure 13-5. Effect of pH on enzyme activity. Maximum activity occurs at a particular pH, and deviations from it result in decreased activity. (Note that not all enzymes exhibit optimum activity at the same pH.)

Figure 13-6. Effect of temperature on enzyme activity. Starting at a low temperature, the enzyme activity increases with increasing temperature until the optimum activity is reached. Further increases in temperature result in decreased enzyme activity and eventual destruction of the enzyme. (Note that not all enzymes exhibit optimum activity at the same temperature.)

clear that deviations from the optimal conditions result in significant reduction of enzyme activity. This is characteristic of all enzymes. Extreme variations in pH can even destroy the enzyme, as can high temperatures; boiling for a few minutes will *denature* (destroy) most enzymes. Extremely low temperatures, for all practical purposes, stop enzyme activity but do not destroy the enzymes. Many enzymes may be preserved by holding them at temperatures around 0°C or lower.

We know that the growth of microorganisms is influenced by a variety of physical and chemical conditions. Since enzymes are responsible for catalyzing the reactions associated with life processes, it follows that the conditions mentioned above affect the enzymes and thereby the growth response. Just as there is an optimal pH and temperature for growth, there is also an optimal pH and temperature for the activity of each enzyme and for the amount of each enzyme the cell will produce. This does not mean, however, that the values are the same in each enzyme; they may not be. The reason for the discrepancies is that during growth, activity or response is measured in terms of the total activities required for growth when all enzymes and enzyme systems are functioning harmoniously in the cell. Optimum conditions must be estimated in terms of what is best for the entire cellular system. In assessing the activity of an isolated and purified specific enzyme, the situation is entirely different. The enzyme is no longer in its normal environment and thus is not influenced by, or integrated into, the multitude of reactions which occur within the cell. Hence, the optimum conditions for activity of one enzyme are not necessarily optimum for other enzymes or for the functioning of an entire cell.

Inhibition of enzyme action

The activity of an enzyme can be inhibited (slowed down or stopped) by chemical agents in several different ways. We are concerned here with chemical substances that inhibit in a more subtle fashion than denaturation (destruction) of the protein portion of the enzyme. Specific information about how these chemical agents (naturally occurring or otherwise) exert their detrimental action contributes to our knowledge of how enzymes work, suggests new types of chemicals which may be useful for inhibiting microorganisms, and in other ways contributes to our understanding of life processes.

Enzyme inhibition may be classified into *nonreversible* and *reversible* types. Nonreversible inhibition usually involves the modification or inactivation of one or more functional groups of the enzyme.

There are two major types of reversible inhibition, namely, *competitive* and *noncompetitive*. *Competitive inhibition* can be reversed by increasing the substrate concentration, whereas *noncompetitive inhibition* cannot be. We will present a few examples.

The enzyme succinic dehydrogenase accomplishes the transfer of hydrogen atoms from succinic acid to a suitable acceptor compound; in the reaction shown below, the acceptor, methylene blue, is a compound that does not occur naturally within a cell.

$$
\begin{array}{l}
\text{COOH} \\
| \\
\text{CH}_2 \\
| \quad + \text{MB} \quad \rightleftharpoons \\
\text{CH}_2 \\
| \\
\text{COOH}
\end{array}
\quad
\begin{array}{l}
\text{COOH} \\
| \\
\text{CH} \\
\| \quad + \text{MB·H}_2 \\
\text{CH} \\
| \\
\text{COOH}
\end{array}
$$

Succinic acid　Methylene blue　Fumaric acid　Reduced methylene blue (colorless)

COOH
|
CH₂
|
COOH

Malonic
acid

This reaction can be inhibited by chemical compounds which have a structure similar to that of succinic acid. One of these inhibitors is malonic acid, which has the structure shown in the margin.

Because of its structural similarity to succinic acid, malonic acid attaches itself to the enzyme site where succinic acid normally attaches. The succinic acid is "blocked out" by the malonic acid, but since the malonic acid is not activated by the enzyme, there is no reaction. In this example there is *competition* for the same enzyme site by two different molecules (see Fig. 13-7). This type of enzyme inhibition is referred to as competitive inhibition.

Figure 13-7. Competitive inhibition (schematic) between malonic acid and succinic acid. Note that each molecule has a structually similar fragment. Since this portion of either molecule can fit or combine with the active site on the enzyme surface, there is competition between the two substrates for this site. Because this enzyme is specific for succinic acid, if the malonic acid occupies the site, further activity is blocked, as malonic acid is not changed by this enzyme.

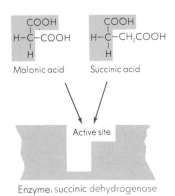

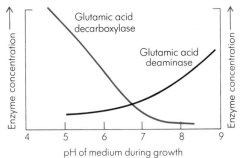

Figure 13-8. Variations in the concentrations of glutamic acid decarboxylase and glutamic acid deaminase present in *E. coli* with variations in the pH of the medium during growth. Note that at a low pH (acidic medium) glutamic acid decarboxylase predominates and at a high pH (basic medium), glutamic acid deaminase predominates.

Certain chemical agents have a high affinity for metal ions, forming complexes with the metal. As already mentioned, many enzymes require a metal ion for their activity. Cyanide is a strong inhibitor of iron-containing enzymes by virtue of the fact that it "ties up" the iron, depriving the enzyme of an essential component. Similarly, fluoride inhibits enzymes which require calcium or magnesium by binding these metals. (It may be noted that the relatively small amount of fluoride which we find in tooth enamel as a result of water fluoridation does not exert this inhibitory action in vivo.) This type of enzyme inhibition is referred to as noncompetitive inhibition since the inhibitor is not competing with the substrate for an active site on the enzyme surface.

Conditions affecting enzyme formation

The enzymatic content of animal-tissue cells is relatively constant, since they are in an environment which in terms of physical and chemical conditions is subjected to little change. But the bacterial cell is exposed to an ever-changing environment. For example, *Escherichia coli* may grow in an acid or alkaline medium (from pH 4.5 to pH 9.5), at room temperature or above body temperature, aerobically or anaerobically. Cells of *E. coli* grown at the extremes of these conditions do not contain the same kinds or amounts of enzymes (see Fig. 13-8). The fact that the environment influences the formation of enzymes should not be misconstrued to mean that there is no consistent enzyme pattern for a given organism. Organisms manifest changes in reaction to environment only within certain limits. It is important to recognize their capacity for undergoing these changes, however, when examining organisms. When organisms are studied physiologically, these studies must be performed under certain established conditions, which include the composition of the medium in which the cells are grown as well as the physical conditions during incubation.

With respect to the presence of substrate and enzyme formation, enzymes may be divided into two groups as follows:

Constitutive enzymes: These are always produced by the cell. Some of the enzymes of glycolysis, or sugar breakdown, are constitutive types. They are found in essentially similar amounts regardless of the concentrations of their substrates in the medium.

Induced enzymes: These are produced by the cell only in response to the presence of a particular substrate; they are produced, in a sense, only when needed. The process is referred to as *enzyme induction*, and the substrate (or a compound structurally similar to the substrate) responsible for evoking formation of the enzyme is an *inducer*. An example of an inducible enzyme is β-galactosidase; its inducer is the sugar lactose.

In reality, the distinction presented by these definitions is more operational than literal. Induced enzymes are believed to exist in noninduced cells but in relatively low quantities. Likewise, production of constitutive enzymes can often be enhanced when their specific substrates are present. The technique of evoking new enzyme formation through use of inducers has been extensively exploited in research designed to elucidate the mechanism of enzyme formation.

Determination of enzyme activity

The enzymatic activity of bacteria (or of any cells or tissues) can be determined by a variety of techniques. Some procedures require special, elaborate instruments; others require only a test tube and a few reagents. All are based on a few simple principles. In order to carry out a quantitative assay of enzyme activity, it is necessary to know the following:

1 The nature of the reaction catalyzed
2 What cofactors and coenzymes are required
3 The required concentrations of both substrate and cofactor or coenzyme
4 The optimum pH
5 The optimum temperature
6 A simple analytical method for determining the disappearance of substrate or the appearance of products of the reaction

The substrate concentration should be above the saturation level so that the initial reaction rate is proportional to enzyme concentration alone. Coenzymes and cofactors should also be added in excess. Doing this ensures that the true limiting factor is the enzyme concentration (at its optimum pH and temperature). Generally, measurement of reaction product formation is more accurate than measurement of the disappearance of substrate. Such determinations of enzymatic activity have provided us with a wealth of knowledge concerning cells and their chemical reactions.

The nature and mechanisms of enzyme regulation

We have said that a living cell has thousands of different enzymes. Each one of them is an effective catalyst for one or more chemical reactions. But they all act together in a coordinated manner so that all the chemical activities in the living cell are integrated with one another. One obvious consequence of this is that the living cell synthesizes and catabolizes materials as required for normal metabolism and growth. All this requires precise control mechanisms of cellular metabolism. Indeed, all major metabolic pathways have the capacity for self-regulation.

The control of cellular metabolism ultimately involves the regulation of enzyme activity. In a microbial cell, such as a bacterium, the existence of cellular regulatory mechanisms is all the more important because of the absence of supracellular controls, such as neural and hormonal controls, which are operational in the tissue cells of higher organisms.

Enzyme activity can be regulated in two ways: (1) *direct control* of catalysis and (2) *genetic control*. Direct control of enzyme activity can be the control of the catalytic mechanism itself or control exerted through a coupling of the catalytic mechanism with other processes. Genetic control includes the phenomena of enzyme induction and repression, which are discussed later in the chapter.

Direct control of the catalytic mechanism itself

Direct control of the catalytic mechanism is effected by altering the concentrations of substrate or reactants. For example, as the substrate concentration increases, the reaction rate increases until a limiting value is reached. And as the product accumulates, the reaction rate decreases. In addition, the concentrations of coenzymes and cofactors can exert controlling influences in the cell.

This type of control can also be effected by compartmentalization within the cell. That is, enzymes may be bound to various internal structures, especially membranes and macromolecules, so that enzymes and substrates are not in direct contact.

In some microbes, highly specific *proteolytic* (protein-degrading) enzymes break down other enzymes which are no longer required for metabolic reactions.

Direct control through coupling with other processes

The direct control of enzyme activity by coupling with other processes usually implies regulation by *ligands* (molecules capable of binding to an enzyme) which do not participate in the catalytic process itself. Such ligands are structurally unrelated to the substrate of the regulatory enzyme. There are many variations of this kind of control. Some of these are discussed below.

Feedback inhibition. In *feedback inhibition*, the regulatory ligand is the end product of a *metabolic pathway* which can shut off its own synthesis by inhibiting the activity of one of the early enzymes in its own biosynthetic pathway. (A metabolic pathway is a series of steps in the chemical transformation of organic molecules.)

It has been shown that after every branch point in a metabolic pathway (which may be compared with a fork in a road) the early enzymes in each branch are subjected to feedback inhibition control. In most cases, as in the biosynthesis of amino acids and *nucleotides* (building blocks of nucleic acids), it is the first enzyme in the branch that is controlled by the end products. Thus feedback inhibition is a process that microorganisms as well as other cells use to prevent overproduction of low-molecular-weight intermediates such as amino acids and purine and pyrimidine nucleotides. The enzymatic activity of glutamine synthetase, for example, is influenced by nine different compounds.

Precursor activation. In precursor activation, the precursor or the first metabolite of a pathway is the regulatory ligand. It activates, or stimulates the activity of, the last enzyme in the sequence of reactions.

Energy-link control. The regulatory ligands involved in energy-linked reactions are adenylates, such as adenosine triphosphate (ATP), or other purine or pyrimidine nucleotides. Some enzymes appear sensitive to the absolute concentrations of ATP, adenosine diphosphate (ADP), or adenosine monophosphate (AMP); others seem sensitive to the ratio of a given two of these nucleotides. In general, enzymes responsible for energy production are inhibited by high energy charge (e.g., high concentration of ATP), while certain key biosynthetic enzymes are stimulated. Such a control is very important in balancing energy production and utilization.

A scheme showing these three mechanisms for the regulation of enzyme activity by ligands is shown in Fig.13-9. The two techniques that have been especially useful in elucidating these mechanisms are the analysis of the properties of isolated enzymes (enzymes removed from cells) and observation of the behavior of mutant bacteria with specific enzyme defects. These mutant studies have been crucial in establishing that the regulatory mechanisms postulated from isolated enzyme studies do indeed function in vivo.

Figure 13-9. Some mechanisms for the regulation of enzyme activity by direct control through a coupling of the catalytic mechanism with other processes. Feedback inhibition, precursor activation, and energy-link control are shown. See text for fuller explanation.

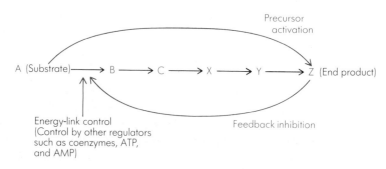

Binding properties of regulatory enzymes. Not all enzymes are *regulatory* enzymes that are subject to direct control of their activity. But those that are make possible the control and integration of the activity of metabolic pathways.

Regulatory enzymes are subject to the action of the regulatory metabolite, which may be either an *activator* or an *inhibitor*. The presence of an activator leads to increased affinity of the enzyme for substrates, resulting in enhanced enzyme activity. An inhibitor, of course, leads to diminished affinity of the enzyme for substrates, causing decreased enzyme activity.

Since there is a specificity of fit between the substrate molecule and the active site of an enzyme, an activator or inhibitor must bind to the enzyme at some site other than its catalytic site. Therefore regulatory enzymes contain two or more binding sites, and at least one of these is specific and catalytic for the substrate. Regulatory enzymes are also called *allosteric* enzymes because the site on the enzyme molecule where an *effector* (inhibitor or activator) acts is different from the catalytic site. Thus the *allosteric site* is one that regulates the activity of the enzyme.

While direct control over enzymatic activity is responsible for the instantaneous or moment-to-moment control of intermediary metabolism (intermediate steps in a metabolic process) the mechanisms of enzyme *induction* and *repression* come into play as the cell adapts to a changed environment. These mechanisms are effective at the level of the synthesis of the enzyme protein itself.

Genetic control: enzyme induction and repression

Besides direct control of catalysis, enzymes are also regulated by genetic control. This involves the *induction* and *repression* of enzyme synthesis at the genetic level.

When an inducer, some low-molecular-weight substance, either the substrate or a compound related to the substrate of the enzyme-catalyzed reaction, is required for enzyme synthesis to occur, the process is called *induction*. When other low-molecular-weight substances, either products or related compounds of the particular reaction, act as *corepressors* by preventing synthesis of the enzyme, this phenomenon is termed *repression*. Corepressors function by combining with a protein, the *repressor*, to form an active complex which combines with the operator gene (site at which repressor molecule binds) to prevent *m*RNA (messenger ribonucleic acid) synthesis by the structural genes. The repressor is also capable of combining with an inducer to form an inactive complex, in which case synthesis of *m*RNA can proceed (see Fig. 13-10). For more detailed coverage of *m*RNA and enzyme protein synthesis, see Chapter 16.

To synthesize a specific enzyme, an organism must have *structural genes* for it on the chromosome. Structural genes specify the structures of enzymes in terms of amino acid sequences. They do not control the rate at which enzymes are produced. Genetic control of the rate of enzyme synthesis is directed by *regulator genes*.

In many bacteria the structural genes governing the biosynthesis of enzymes of a particular metabolic pathway are positioned in the exact order of the sequence of reactions in that pathway. This means that the ordering

Figure 13-10. The Jacob-Monod model of gene control for the *lac* operon. (A) Repression of *m*RNA synthesis from *lac* operon. In the absence of inducer, the repressor (the product of the *i* gene) binds to the corepressor. The repressor-corepressor then binds with the *o* gene to prevent transcription of the *z, y,* and *a* genes. (B) Induction of *m*RNA synthesis from *lac* operon. In the presence of inducer, the repressor binds to the inducer and can no longer combine with the *o* gene. The lac operon is no longer repressed, and transcription of the *z, y,* and *a* genes takes place. (C) Positive control of enzyme synthesis. Presence of cylic AMP (cAMP) activates the catabolite gene activator protein (CAP) which in turn activates transcription of the *lac* operon.

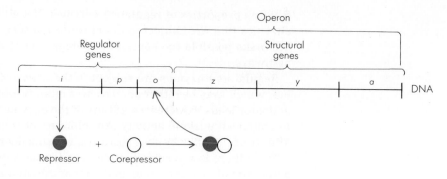

A Repression of *m*RNA synthesis

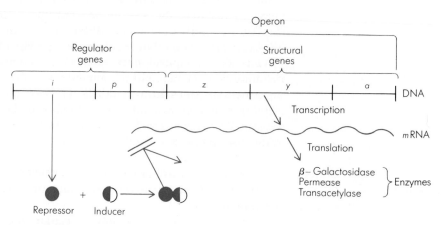

B Induction of *m*RNA synthesis

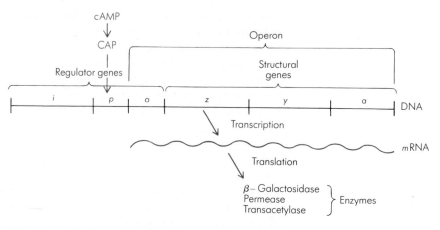

C Positive control of enzyme synthesis

of the sequential reactions in the metabolic pathway is dictated by the chromosome. A group of such consecutive genes forming an operational unit was named an *operon* by Francois Jacob and Jacques Monod; it includes the associated operator gene.

Keeping in mind the above definitions, we can better understand the genetic control of enzyme synthesis by discussing the lactose (*lac*) operon or β-galactosidase system of *Escherichia coli*. This operon is now by far the best understood part of any cellular genome. There are other bacterial operons, but they are less well understood and are different in structure from the *E. coli lac* operon.

The *lac* operon. When inducers such as lactose or other β-galactosides are added to a culture of *E. coli*, there is a 1,000-fold increase in the rate of synthesis of the enzymes β-galactosidase (which hydrolyzes lactose to glucose and galactose), β-galactoside permease (which transports lactose into the cell), and thiogalactoside transacetylase (the physiological function of which is unknown). The genes for these three proteins are linked together on the *E. coli* chromosome. These are shown in Fig. 13-10 as *z, y,* and *a,* coding for β-galactosidase, galactoside permease, and transacetylase, respectively. In the absence of control, the rate of enzyme production would be constant and depend only on the structural genes (such as *z, y,* and *a*), amino acid levels, activating enzymes, and other substances. However, the control of the rate of enzyme synthesis is directed by the regulator genes designated *i, p,* and *o,* shown in Fig. 13-10, where *i* is the repressor gene, *p* the promoter gene, and *o* the operator gene. The *i* gene codes for a repressor protein which binds to the DNA of the operator *o* gene in conjunction with the corepressor, thus preventing *transcription*, that is, the synthesis of *m*RNA (see Fig. 13-10A). The promoter gene *p* is considered to be the site on the DNA where the RNA polymerase enzyme, catalyzing the synthesis of *m*RNA, binds, and is thus the site where the specific *lac m*RNA (responsible for the biosynthesis of the specific enzymes of the operon) synthesis begins. Let us discuss the functioning of the Jacob-Monod model of gene control for the *lac* operon as it is now understood.

1 Genes function as templates or blueprints for the transcription of *m*RNA. Using the protein-synthesizing machinery of the cell (*ribosomes*), the *m*RNA directs the synthesis of *polypeptides* (long chains of amino acids) in a process called *translation* .
2 The genes *z, y,* and *a* operate as a single unit of transcription, which is initiated at *p*.
3 Transcription of the operon is both negatively and positively controlled.

Negative control is mediated by the *lac* repressor which with the corepressor forms the repressor-corepressor complex which binds to the *o* gene and blocks transcription. The corepressor in this case may be glucose or galactose, the products of galactosidase activity. Inducers, such as lactose, stimulate *lac m*RNA synthesis by binding to the repressor and reducing its affinity for the operator (see Fig. 13-10B). Both repression and induction of enzyme synthesis are negative control systems because, in either case, the synthesis of enzyme can proceed only when the repressor is removed from its blocking site on the *o* gene.

Positive control of enzyme synthesis is said to occur when an association between a protein and a part of the regulatory region of an operon (not necessarily the operator gene) is essential for expression of related structural genes in the operon. Expression of the *lac* operon is inhibited when a more efficient source of carbon, such as glucose, is present in the medium. The presence of glucose results in a decreased concentration of intracellular cyclic AMP (adenosine-3′,5′-monophosphate). Cyclic AMP is necessary for efficient expression of the *lac* operon since it activates the catabolite gene activator protein (CAP), which in turn activates transcription of *lac* mRNA by RNA polymerase at the promoter site (see Fig. 13-10C).

Thus both cyclic AMP and a specific inducer acting in concert are necessary for the synthesis of many inducible enzymes in *E. coli*. Little enzyme is made if either is absent.

Summary and outlook

There are thousands of different enzymes in every cell that specifically catalyze all the essential chemical activities of a cell. Enzymes are proteins and may need to be combined with coenzymes or cofactors to be active. They have formal names as well as identifying numbers although their trivial names are in common use.

The activity of enzymes on their substrates takes place at the active sites on the enzyme surface. This process lowers the activation-energy barrier to a chemical reaction and allows it to proceed at a rapid rate. Many conditions affect enzyme activity. Among these are pH, temperature, and concentrations of enzyme and substrate. Enzyme activity may also be inhibited by certain inhibitors.

There are two kinds of enzymes in a cell, induced enzymes and constitutive enzymes. Their presence is governed to a limited extent by environmental conditions. In order for enzymes to function in a coordinated manner, enzyme activity has to be regulated.

Enzyme regulation is exerted either by genetic control or by direct control of catalysis. Direct control may be imposed on the catalytic mechanism itself or may be effected through coupling with other processes. Enzymes subject to this kind of regulation are allosteric enzymes.

Genetic control involves the induction and repression of enzyme synthesis. This type of regulation is exerted at the level of regulator genes. The *E. coli lac* operon provides an example of genetic control.

The future will obviously see the isolation and purification of new enzymes, especially those that are particularly labile in vitro. Certain novel enzymes will be found in greater numbers. For example, many enzymes called *isozymes*, occur in different structural forms but possess identical catalytic activities. Also, a number of *multifunctional enzymes* have been described in which two or more activities are located on a single polypeptide chain. We will also find more multienzyme complexes in which groups of individual enzymes are physically associated and function together and often catalyze a series of directly connected reactions. Finally, the mechanisms by which these enzymes and enzyme complexes are regulated will be determined.

Key terms

activation energy	corepressor	isozyme
active site	effector	ligand
allosteric enzyme	energy-link control	metabolic pathway
allosteric site	enzyme	operon
apoenzyme	enzyme system	precursor activation
catalyst	feedback inhibition	regulator gene
coenzyme	holoenzyme	repression
cofactor	induced enzyme	repressor
competitive inhibition	inducer	structural gene
constitutive enzyme	induction	substrate

Questions

1 Explain two distinctive properties that characterize an enzyme.
2 Why are vitamins important for cellular activities?
3 Once a cell has synthesized a sufficient initial amount of enzymes for a particular reaction, does it need to synthesize more during its life-span? Give reasons for your answer.
4 When should one use the trivial name and when the systematic name of an enzyme?
5 The main function of an enzyme, from the point of view of its mechanism of action, is said to be to lower the *activation energy barrier* to a chemical reaction. Explain the meaning of this statement.
6 Are the optimum conditions determined in vitro for the activity of a particular enzyme the existing conditions within the cell? Are these same optimum conditions necessary in vivo?
7 In the regulation of enzyme activity, what advantages, if any, does direct control of catalysis have over genetic control?
8 What kind of control is *feedback inhibition*? Why does the cell employ this kind of regualtion?
9 What does *energy-link* control accomplish in the metabolism of a cell?
10 Explain why the allosteric property is necessary for the activity of a regulatory enzyme.
11 Distinguish between *negative* control and *positive* control of transcription, using the *E. coli lac* operon as a model.

DISSIMILATION REACTIONS
AND ENERGY RELEASE

Metabolism is all the chemical reactions performed by a cell which produce energy and which utilize energy for the synthesis of cell components and for cellular activities, such as movement. Those chemical reactions that release energy by the breakdown of nutrients are called *dissimilation* reactions. They constitute the *catabolic*, or degradative, activities of the cell. Those chemical reactions that use energy for synthesis and other cell functions are called *assimilation*, or *anabolic*, reactions. Thus, dissimilation reactions produce energy, and assimilation reactions utilize energy.

When a cell breaks certain chemical bonds during metabolism, the energy released is available to perform useful biological work. During the life of a cell, this work is extensive and varied (see Chap. 15). Nonphotosynthetic heterotrophic microorganisms obtain energy from the dissimilation of organic compounds. Nonphotosynthetic autotrophic microorganisms obtain energy from the *oxidation* (the removal of electrons or hydrogen atoms) of inorganic compounds. Photosynthetic microorganisms obtain energy from light. In this chapter we will examine some of the ways in which microbes obtain or produce energy.

The coupling of cellular reactions

During the course of any chemical reaction, energy is either released or utilized. The amount of energy liberated or used up during the course of a reaction is referred to as the *free-energy change* of the reaction, which is represented by the symbol ΔG, where Δ means "change in" and G is the *free energy*. This free-energy change can be defined as the amount of available or useful energy liberated or used up in a reaction.

ΔG is expressed in terms of *calories*. However, this is merely a convenience since the free energy does not always occur as heat but may be in the form of chemical energy. If the ΔG of a chemical reaction has a negative value, such as $-8,000$ calories (cal), the reaction releases energy. Such a reaction is called an *exergonic* reaction. If the ΔG of a reaction has a positive value, such as $+3,000$ cal, the reaction requires energy. This is called an *endergonic* reaction.

In a cell there are many endergonic chemical reactions that take place during metabolism. These reactions do not proceed spontaneously; the energy released in exergonic reactions is used to "drive" them. This is done by *coupling* reactions. Living organisms have developed a characteristic way of coupling exergonic reactions with endergonic reactions to provide energy to drive endergonic reactions—by use of a *common reactant*. This can be explained by the following example. Consider these two general reactions where compound A is converted to B and compound C is converted to D:

$$A \rightarrow B \quad \Delta G = -10,000 \text{ cal}$$
$$C \rightarrow D \quad \Delta G = +5,000 \text{ cal}$$

The first reaction is exergonic since the ΔG is negative. The energy released from this reaction can be used to drive the second reaction (C→D), which is endergonic. This is accomplished by coupling the two reactions as follows:

$$A + Y_1 \rightarrow B + Y_2 \quad \Delta G = -2,000 \text{ cal} \quad \text{(Compound } Y_2 \text{ traps energy.)}$$
$$C + Y_2 \rightarrow D + Y_1 \quad \Delta G = -3,000 \text{ cal} \quad \text{(Compound } Y_2 \text{ releases energy.)}$$

Y is a reactant common to both reactions. In the first reaction, the overall ΔG of $-2,000$ cal indicates that 8,000 of the original 10,000 cal were used for the conversion of Y_1 to Y_2; that is, 8,000 cal were trapped, or conserved, in Y_2. In the second reaction, Y_2 was converted back to Y_1, releasing the previously trapped 8,000 cal to drive the endergonic conversion of C→D. Thus the overall ΔG of the second reaction is $(+5,000) + (-8,000)$, or $-3,000$ cal (which is a negative value, and therefore the reaction could proceed); that is, more energy was released by the first reaction than is required to make the second reaction go. The common reactant Y is referred to as an *energy-rich* or a *high-energy-transfer* compound. Such a compound contains potential energy that is released when it donates some portion of itself to water (in the process called *hydrolysis*) or to some other acceptor molecule (when it changes from Y_2 to Y_1 in the example above).

A variety of energy-rich compounds exist in cells. The energy in these compounds is contained in them as a whole. Breakage of certain bonds in

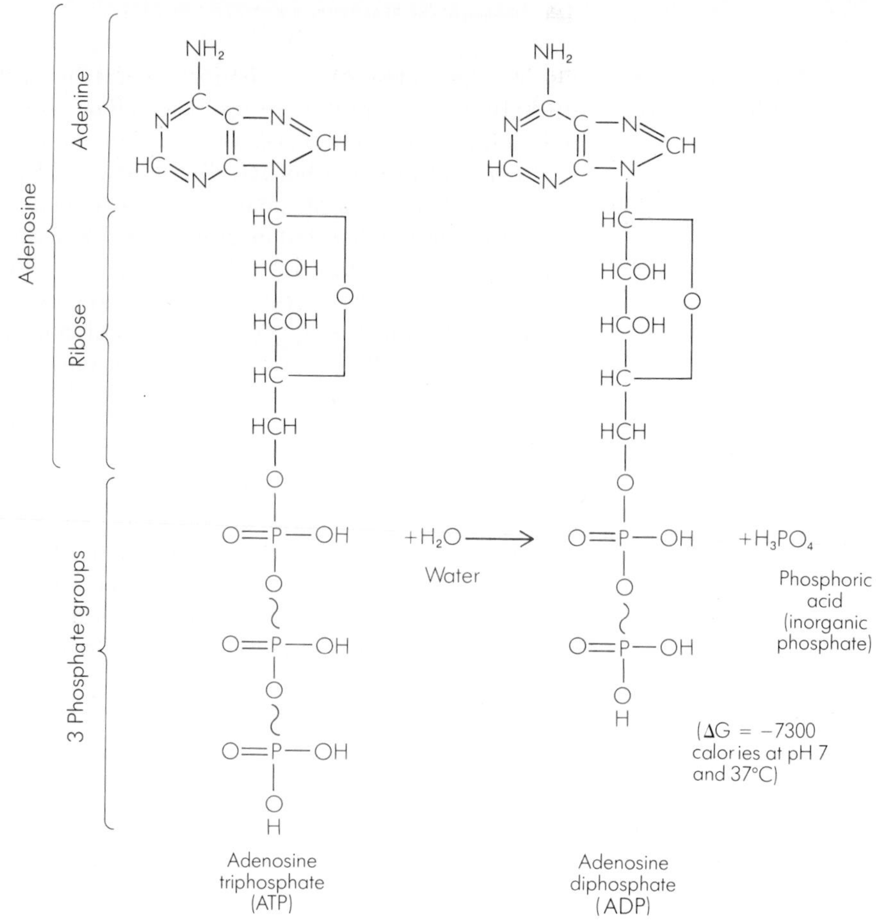

Overall reaction:
ATP + H₂0 → ADP + H₃PO₄

$\}$ = "High-energy bond"

Figure 14-1. Hydrolysis of adenosine triphosphate.

these compounds releases some of the conserved energy. An energy-rich compound is analogous to a mousetrap. When set, the trap has great energy, and the energy is contained in the taut spring. Tripping the catch releases the energy in the spring.

Table 14-1 lists some of the energy-rich compounds found in cells. Of these, adenosine triphosphate (ATP) is by far the most important. ATP is

Table 14-1. Some high-energy-transfer compounds found in cells, with their standard free-energy changes upon hydrolysis.

COMPOUND	$\Delta G^{\circ\prime}$, kcal
Adenosine triphosphate (ATP)	−7.3
Adenosine diphosphate (ADP)	−7.3
Guanosine triphosphate (GTP)	−7.3
Guanosine diphosphate (GDP)	−7.3
Uridine triphosphate (UTP)	−7.3
Cytidine triphosphate (CTP)	−7.3
Acetyl phosphate	−10.1
1,3-Diphosphoglyceric acid	−11.8
Phosphoenolpyruvic acid (PEP)	−14.8

234

the "energy currency" of the cell. It is the medium by which energy is exchanged between exergonic and endergonic reactions. It should be noted that all the compounds shown in Table 14-1 can transfer their energy directly or indirectly to ATP. An example of a direct transfer is

1,3-Diphosphoglyceric acid + ADP $\rightleftharpoons$ 3-phosphoglyceric acid + ATP

An example of indirect transfer is:

(i) Succinyl-coenzyme A + Pi + GDP $\rightleftharpoons$ succinic acid + GTP + coenzyme A
Inorganic
phosphate

(ii) GTP + ADP $\rightleftharpoons$ GDP + ATP

Energy is released from ATP by hydrolysis, as shown in Fig. 14-1. The compound ADP is also a high-energy-transfer compound, since its hydrolysis also liberates a large quantity of energy.

Oxidation and energy production

Cells derive energy from nutrients by a series of chemical reactions, some of which are oxidations. During oxidation, energy is released and energy-rich chemical bonds, like those of ATP, may be formed to store the released energy.

Oxidation, the loss of electrons by a molecule, is always accompanied by *reduction*, the gain of electrons by another molecule. When one speaks of an oxidation, one is referring to the transfer of an electron from one molecule to another; so whenever one molecule is oxidized, another molecule is reduced.

Frequently oxidation reactions are *dehydrogenations*, reactions involving the loss of hydrogen atoms (H). A hydrogen atom consists of a proton (H^+) plus an electron (e^-), so a compound which loses a hydrogen atom has lost an electron and thus has been oxidized.

An *oxidant* (an *oxidizing agent*) will accept electrons and will therefore be reduced, as illustrated by the following examples. Ferric ion is an oxidizing agent in the following reaction; it accepts electrons and is reduced to the ferrous ion:

$$Fe^{3+} + e^- \rightarrow Fe^{2+}$$
Ferric ion   Electron   Ferrous ion

Fumaric acid, an intermediate in metabolism, is another example of an oxidizing agent. In the following reaction, it accepts hydrogen atoms and becomes reduced to succinic acid:

```
COOH                    COOH
|                       |
CH                      CH₂
‖      + 2e⁻ + 2H⁺ →    |
CH                      CH₂
|                       |
COOH                    COOH
Fumaric acid            Succinic acid
```

A *reductant* (a *reducing agent*) donates electrons and is oxidized in the process. Ferrous ion is a reducing agent; it donates electrons and becomes oxidized to ferric ion:

$$Fe^{2+} \quad \rightarrow \quad Fe^{3+} \quad + \quad e^-$$

Ferrous ion   Ferric ion   Electron

From these examples, one can see that the reverse of an oxidation reaction is a reduction reaction, and vice versa. Moreover, in each reaction, a *pair* of substances is involved, for example, ferrous ion and ferric ion, succinic acid and fumaric acid. Each pair of such substances is referred to as an *oxidation-reduction (O/R) system.*

One O/R system may tend to accept electrons from another O/R system. This tendency to accept electrons can be expressed by the *electromotive potential* (E_0' of an O/R system), which is measured electrically under standardized conditions of comparison and expressed in volts. The higher

Figure 14-2. Glycolysis, the pathway for breakdown of glucose to pyruvic acid. Note that one molecule of fructose-1,6-diphosphate is split into 2 three-carbon compounds that are in equilibrium with each other. Thus as the glyceraldehyde-3-phosphate is oxidized to 1,3-diphosphoglyceric acid, the dihydroxyacetone phosphate is converted to more glyceraldehyde-3-phosphate. Regulation by means of allosteric enzymes is also shown by broken arrows. Plus denotes activation while minus denotes inhibition by compounds listed.

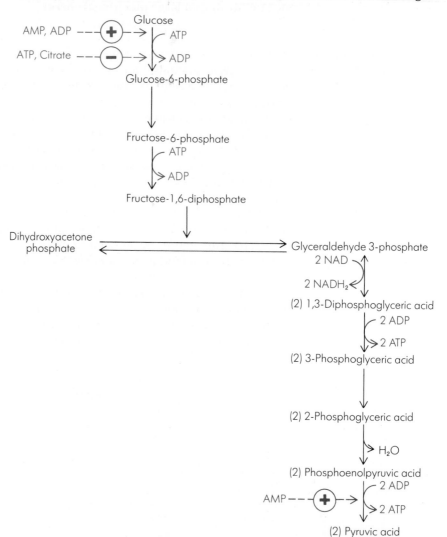

the E_0' value of the O/R system, the greater is the tendency of the system to accept electrons (or the greater is the oxidizing ability of the system). Consequently, any system listed in Table 14-2 can oxidize any other system listed above it, but not below it, under standard conditions. Such relationships are very important in understanding the orderly sequence in which biological oxidations occur.

When one O/R system oxidizes another, energy is released in direct proportion to the difference in E_0' values. If the voltage difference is large, an amount of free energy sufficient to drive the synthesis of a high-energy-transfer compound such as ATP may be liberated. If the free energy released is not sufficiently high, then whatever energy is released is dissipated as heat.

With the understanding that electron transfer is intimately associated with energy production, we can now examine the ways in which microorganisms produce energy. In general, energy production by microorganisms, particularly bacteria, may be divided into three categories: *anaerobic energy production*, *aerobic energy production*, and *photosynthetic energy production*.

Energy production by anaerobic processes

Glycolysis

Heterotrophic bacteria can use a variety of organic compounds as energy sources. These compounds include carbohydrates, organic and fatty acids, and amino acids. For many microorganisms the preferred compounds are carbohydrates, especially the six-carbon sugar glucose.

The breakdown of glucose can occur by *glycolysis*. Glycolysis is one of the most important pathways used by the cell to produce energy. Glycolysis does not require the presence of oxygen and may be present in both aerobic and anaerobic cells.

In glycolysis, as shown in Fig. 14-2, fructose-1,6-diphosphate formed from glucose is split into 2 three-carbon units (dihydroxyacetone phosphate and glyceraldehyde-3-phosphate), and they are subsequently oxidized to pyruvic acid. At the step where glyceraldehyde-3-phosphate is oxidized, a pair of electrons (two hydrogen atoms) is removed. In the absence of oxygen, this pair of electrons may be used to reduce pyruvic acid to lactic acid or ethanol. In the presence of oxygen, this pair of electrons may enter the electron transport chain (see pages 240 and 241).

For each molecule of glucose metabolized, two molecules of ATP are used

Table 14-2. Component O/R systems of the respiratory chain, with their corresponding E_0' values

O/R SYSTEM	E_0' AT pH 7 AND 25°C
$NAD^+/NADH + H^+$	−0.32
Flavoprotein/flavoprotein-H_2	−0.03
Cyt b−Fe^{3+}/cyt b−Fe^{2+}	+0.05
CoQ/CoQ-H_2	+0.10
Cyt c_1−Fe^{3+}/cyt c_1−Fe^{2+}	+0.21
Cyt c−Fe^{3+}/cyt c−Fe^{2+}	+0.23
Cyt a/a_3−Fe^{3+}/cyt a/a_3−Fe^{2+}	+0.54
Oxygen/water	+0.82

Note: The relative position of CoQ to Cyt b is uncertain.

up and four molecules of ATP are formed. Therefore for each molecule of glucose metabolized by glycolysis, there is a net yield of two ATP molecules. This is shown in Fig. 14-2.

The overall reaction of glycolysis can be summarized as follows:

$$C_6H_{12}O_6 + 2NAD + 2ADP + \quad 2P_i \quad \rightarrow 2CH_3COCOOH + 2NADH_2 + 2ATP$$

Glucose $\qquad\qquad\qquad$ Inorganic $\quad$ Pyruvic acid
$\qquad\qquad\qquad\qquad\quad$ phosphate

The pentose phosphate pathway

This pathway for carbohydrate breakdown involves the formation of six-carbon sugar phosphates (hexose monophosphates) and five-carbon sugar phosphates (pentose phosphates). Since this pathway involves some reactions of the glycolytic pathway, it has been viewed as a "shunt" of glycolysis. Glucose can be oxidized by the *pentose phosphate pathway* with the liberation of electron pairs, which may enter the electron transport chain. However, this cycle is not generally considered a major energy-yielding pathway in most microorganisms. The pentose phosphate pathway is used principally for biosynthesis since it provides pentose phosphates for use in nucleotide synthesis. Although it can provide energy to the cell as an alternate pathway for the oxidation of glucose, it is also a mechanism for obtaining energy from five-carbon sugars. Like the glycolytic pathway, the pentose phosphate pathway occurs in both procaryotes and eucaryotes.

There are other pathways for carbohydrate breakdown, for example the Entner-Doudoroff pathway, which is common to both aerobic and anaerobic procaryotes. Some microbes make use of only one of these pathways, and others use more than one, either at the same time or at different times. The pathways that microbes use and when they use them depend on the microbes' inherited capabilities and the environment in which they find themselves.

Fermentation

Anaerobes also produce energy by reactions called *fermentations* which employ organic compounds as electron donors and acceptors. Facultative anaerobic bacteria and obligately anaerobic bacteria employ many different kinds of fermentations to produce energy. The *lactic fermentation* is a typical example. *Streptococcus lactis*, the bacterium responsible for the normal souring of raw milk, dissimilates glucose to lactic acid, which accumulates in the medium as the sole fermentation product. How does this happen? By glycolysis (Fig. 14-2), one molecule of glucose is converted to two molecules of pyruvic acid with concomitant production of two $NADH + H^+$. The pyruvic acid is converted to lactic acid in the following reaction:

$$
\begin{array}{ccc}
\text{COOH} & & \text{COOH} \\
| & & | \\
2\ \text{C}{=}\text{O} + 2NADH + 2H^+ \rightleftharpoons 2\ \text{H}{-}\text{C}{-}\text{OH} + 2NAD^+ \\
| & & | \\
\text{CH}_3 & & \text{CH}_3 \\
\text{Pyruvic} & & \text{Lactic} \\
\text{acid} & & \text{acid}
\end{array}
$$

Table 14-3. Bacteria grouped according to major products of glucose dissimilation	GROUPS WITH EXAMPLES OF SOME GENERA	REPRESENTATIVE PRODUCTS

	Lactic acid bacteria
	  Streptococcus
	  Lactobacillus
	  Leuconostoc
		Lactic acid only or lactic acid plus acetic acid, formic acid, and ethyl alcohol; species producing only lactic acid are *homofermentative*, and those producing lactic acid plus other compounds are *heterofermentative*

	Propionic acid bacteria
	  Propionibacterium
	  Veillonella
		Propionic acid plus acetic acid and carbon dioxide

	Coli-aerogenes-typhoid bacteria
	  Escherichia
	  Enterobacter
	  Salmonella
		Formic acid, acetic acid, lactic acid, succinic acid, ethyl alcohol, carbon dioxide, hydrogen, 2,3-butylene glycol (produced in various combinations and amounts depending on genus and species)

	Acetone, butyl alcohol bacteria
	  Clostridium
	  Eubacterium
	  Bacillus
		Butyric acid, butyl alcohol, acetone, isopropyl alcohol, acetic acid, formic acid, ethyl alcohol, hydrogen, and carbon dioxide (produced in various combinations and amounts depending on species)

	Acetic acid bacteria *Acetobacter*	Acetic acid, gluconic acid, kojic acid

Insufficient energy for ATP synthesis results from this reaction.

In other carbohydrate fermentations, the initial stages of glucose dissimilation frequently, but not always, follow the scheme of glycolysis. Differences in carbohydrate fermentations usually occur in the ways the resulting pyruvic acid is used. Thus pyruvic acid is the "hub" of carbohydrate fermentations. Figure 14-3 illustrates the variety of products resulting from the metabolism of pyruvic acid.

Most heterotrophic bacteria produce several end products of the types indicated in Fig. 14-3 from glucose dissimilation, but no single species produces all these end products. The types listed represent a summary of what can be expected when one takes an inventory of the end products of glucose dissimilation by all heterotrophs. Actually, it is possible to group microorganisms on the basis of their products of fermentation (the lactic acid group or the propionic acid group of bacteria, for example, as shown in Table 14-3). Such designations are established on the basis of the major end products of carbohydrate fermentation. From this it is evident that not all microorganisms metabolize the same substrate in exactly the same manner. For example, *Streptococcus lactis* and *Escherichia coli* both ferment glucose but by quite different pathways of fermentation, as shown in Fig. 14-4.

However, some anaerobes do not have a functional glycolytic system. They may have carbohydrate fermentation pathways that use the pentose-phosphate pathway and the Entner-Doudoroff pathway. Fermentations of noncarbohydrate substrates, such as amino acids, involve highly specific pathways.

Figure 14-3. Pyruvic acid is regarded as the key compound in the dissimilation of glucose, as shown in this schematic illustration.

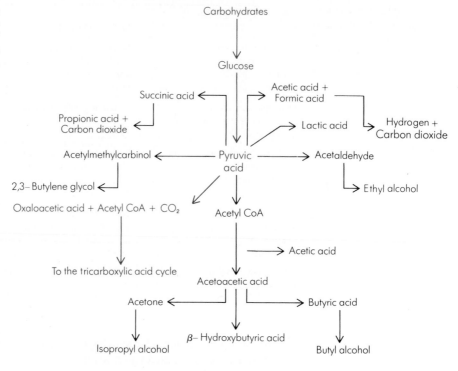

Figure 14-4. Glucose is fermented by many different bacteria and in many different ways. (A) *E. coli* fermentation of glucose results in a mixture of products, whereas (B) *S. lactis* fermentation of glucose produces lactic acid almost exclusively.

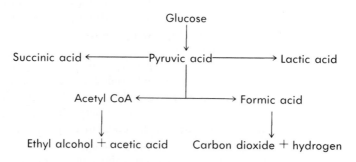

A   Glucose fermentation by *Escherichia coli*

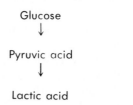

B   Glucose fermentation by *Streptococcus lactis*

Energy production by aerobic processes

The *electron-transport chain*, also known as the *cytochrome system* or *respiratory chain*, is a sequence of oxidation-reduction reactions for the

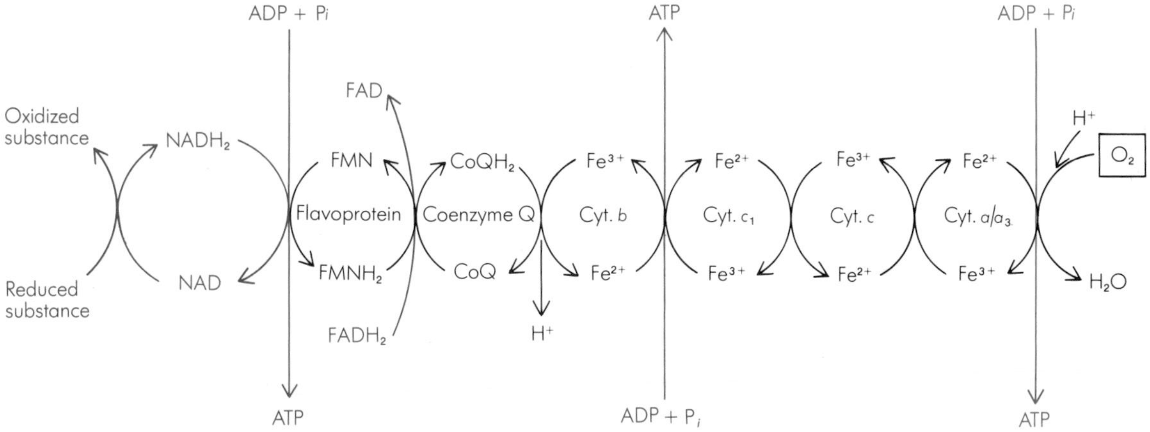

Figure 14-5. The electron-transport chain, showing sequential oxidation steps and points where sufficient energy is liberated to permit synthesis of ATP. Electron transfer is accompanied by a flow of protons (H^+) from $NADH_2$ through coenzyme Q but not in later steps involving cytochromes. Note that 3 ATPs are formed per molecule of $NADH_2$ reoxidized but only 2 ATPs per molecule of $FADH_2$ reoxidized.

The electron-transport chain

generation of ATP. The function of this sequence of reactions is to accept electrons from reduced compounds and transfer them to oxygen, with the resulting formation of water. At several steps in the chain, sufficient energy is liberated for ATP synthesis from ADP and inorganic phosphate. This ATP synthesis is called *oxidative phosphorylation* because high-energy phosphate bonds are formed. The electron-transport chain is shown in Fig. 14-5. As can be seen from the figure, when hydrogen atoms (electrons plus protons, H^+) are removed from organic substances by oxidation, they are transferred by dehydrogenases containing NAD (nicotinamide adenine dinucleotide) or NADP (nicotinamide adenine dinucleotide phosphate), flavoproteins containing FAD (flavin adenine dinucleotide) or FMN (flavin mononucleotide), and iron-containing cytochromes to molecular oxygen, with the resulting formation of water. Note also the synthesis of ATP at three specific points in the chain.

Electrons removed from inorganic substances (e.g., NO_2^-, H_2, and H_2S) by oxidation may also be fed into the electron-transport chain to obtain energy. This is how chemoautotrophic bacteria obtain ATP from the electron-transport chain.

The tricarboxylic acid cycle

The tricarboxylic acid (TCA) cycle is a sequence of reactions that generate energy in the form of ATP and reduced coenzyme molecules ($NADH_2$ and $FADH_2$). It also performs other functions. Many intermediates in the cycle are precursors in the biosynthesis of amino acids, purines, pyrimidines, etc. Thus the TCA cycle is an *amphibolic* cycle, which means that it functions not only in catabolic (breakdown) but also in anabolic (synthesis) reactions. The cycle is shown in Fig. 14-6.

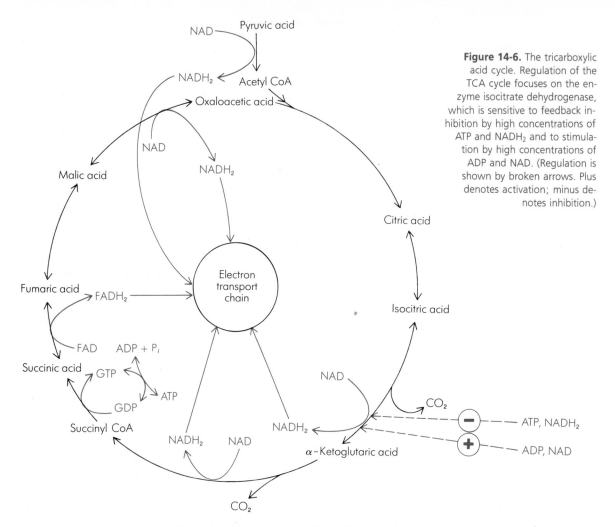

Figure 14-6. The tricarboxylic acid cycle. Regulation of the TCA cycle focuses on the enzyme isocitrate dehydrogenase, which is sensitive to feedback inhibition by high concentrations of ATP and $NADH_2$ and to stimulation by high concentrations of ADP and NAD. (Regulation is shown by broken arrows. Plus denotes activation; minus denotes inhibition.)

The overall reaction of the TCA cycle can be summarized as follows:

$$Acetyl\text{-}CoA + 3H_2O + 3NAD^+ + FAD + ADP + P_i \rightarrow$$
$$2CO_2 + CoA + 3NADH_2 + FADH_2 + ATP$$

Since the breakdown of glucose by glycolysis yields two acetyl-CoA molecules which can enter this cycle, the overall equation for the cycle, per glucose molecule broken down, is twice the above.

Energy yield in
aerobic respiration

We may now look at the energy yield from the aerobic breakdown of one molecule of glucose when the electrons stored in the reduced coenzyme molecules are fed into the electron-transport chain. As shown previously, the electrons are transferred stepwise from the coenzyme carriers to molecular oxygen, and this transfer is coupled to the generation of ATP by oxidative phosphorylation.

For each glucose molecule broken down, there are 12 reduced coenzymes to be oxidized: 2 $FADH_2$'s (1 from each turn of the TCA cycle) and 10 $NADH_2$'s (2 from glycolysis; 2 from the gateway step between glycolysis

Figure 14-7. ATP yield per glucose molecule broken down in aerobic respiration.

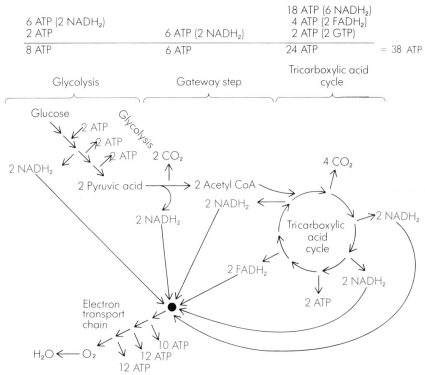

and the TCA cycle, i.e., pyruvic acid to acetyl-CoA; and 6 from two turns of the TCA cycle). Since 3 ATPs are produced from each $NADH_2$ and 2 ATPs from each $FADH_2$, there are 34 ATPs generated from the reduced coenzymes via oxidative phosphorylation through the respiratory chain. But the total yield of ATPs from the aerobic respiration of 1 glucose molecule is 38: 34 from the oxidation of reduced coenzymes, 2 from glycolysis, and 2 from the side reaction of the TCA cycle, that is, from 2 GTPs. The total ATP yield per glucose molecule from aerobic respiration is summarized in Fig. 14-7.

The complete oxidation of glucose via glycolysis, the TCA cycle, and the respiratory chain is summarized in this overall reaction:

$C_6H_{12}O_6 + 6O_2 \rightarrow 6CO_2 + 6H_2O$
Glucose

Catabolism of lipids

Glucose is the single most important source of energy for most cells. However, for many microorganisms, other substances such as lipids and proteins, may be used as alternate sources of energy. There is a general rule that governs their utilization: They are converted as quickly and efficiently as possible into intermediates of the glycolytic and TCA pathways so that a minimum number of additional enzymes is required to effect complete breakdown. This rule highlights the fact that the glycolytic pathway and the TCA cycle serve as a common center around which other catabolic pathways are built.

The breakdown of lipids or fats begins with the cleavage of triglycerides by the addition of water to form glycerol and fatty acids by means of enzymes called *lipases*:

$$
\begin{array}{c}
\underset{\text{Triglyceride}}{
\begin{matrix}
\text{H}_2\text{C}\!-\!\text{O}\!-\!\overset{\overset{\displaystyle\text{O}}{\|}}{\text{C}}\!-\!\text{R}_1 \\
\text{HC}\!-\!\text{O}\!-\!\overset{\overset{\displaystyle\text{O}}{\|}}{\text{C}}\!-\!\text{R}_2 \quad + \; 3\text{H}_2\text{O} \\
\text{H}_2\text{C}\!-\!\text{O}\!-\!\overset{\overset{\displaystyle\text{O}}{\|}}{\text{C}}\!-\!\text{R}_3
\end{matrix}}
\xrightarrow{\text{lipase}}
\begin{matrix}
\text{H}_2\text{C}\!-\!\text{OH} \\
\text{HC}\!-\!\text{OH} \\
\text{H}_2\text{C}\!-\!\text{OH}
\end{matrix}
\; + \;
\begin{matrix}
\text{HO}\!-\!\overset{\overset{\displaystyle\text{O}}{\|}}{\text{C}}\!-\!\text{R}_1 \\
\text{HO}\!-\!\overset{\overset{\displaystyle\text{O}}{\|}}{\text{C}}\!-\!\text{R}_2 \\
\text{HO}\!-\!\overset{\overset{\displaystyle\text{O}}{\|}}{\text{C}}\!-\!\text{R}_3
\end{matrix}
\end{array}
$$

Triglyceride   Glycerol Fatty acids
($-$R = hydrocarbon chain)

Glycerol as a component of fats can be converted into an intermediate of the glycolytic pathway (dihydroxyacetone phosphate) by the following reactions:

$$\text{Glycerol} + \text{ATP} \xrightarrow[\text{Mg}^{2+}]{\text{glycerol kinase}} \text{ADP} + \text{glycerol-3-phosphate}$$

$$\text{Glycerol-3-phosphate} + \text{NAD}^+ \xrightarrow{\substack{\text{glycerol phosphate} \\ \text{dehydrogenase}}} \text{dihydroxyacetone phosphate} + \text{NADH}_2$$

The dihydroxyacetone phosphate formed would be broken down by the mechanisms shown in Fig. 14-2. Fatty acids are oxidized by the successive removal of two-carbon fragments in the form of acetyl-CoA. The acetyl-CoA formed can then enter the TCA cycle, and the hydrogen atoms and their electrons enter the electron-transport chain, leading to oxidative phosphorylation.

There is more energy yield per gram of fat than per gram of carbohydrate. However, relatively few microbial species are effective in breaking down lipids of either simple or complex types, partly because of the limited solubility of lipids.

Catabolism of proteins Many heterotrophic microorganisms can degrade exogenous proteins, using the products as carbon and nitrogen energy sources. Since protein molecules are too large to pass through membranes, bacteria excrete exoenzymes called *proteases* that hydrolyze exogenous proteins to peptides.

Bacteria produce peptidases that break down peptides to the individual amino acids, which are then broken down according to the specific amino acid and the species or strain of bacteria breaking it down. This process may be shown as follows:

$$\text{Proteins} \xrightarrow{\text{proteases}} \text{peptides} \xrightarrow{\text{peptidases}} \text{amino acids}$$

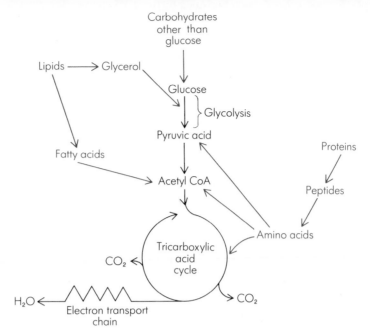

Figure 14-8. Metabolism of carbohydrates, lipids, and amino acids. As can be seen from the diagram, acetyl-CoA is a common intermediate of carbohydrate and lipid metabolism, and the TCA cycle is the common pathway for oxidation of carbohydrates, lipids, and amino acids.

Where amino acids are broken down, the carbon skeletons of the amino acids undergo oxidation to compounds that may enter the TCA cycle for further oxidation. Entry into the TCA cycle can be via acetyl-CoA, α-ketoglutaric acid, succinic acid, fumaric acid, or oxaloacetic acid.

An overall view of the dissimilation of carbohydrates, lipids, and proteins is shown in Fig. 14-8.

Anaerobic respiration in some bacteria

Some bacteria which are ordinarily aerobic can grow anaerobically if nitrate is present. For example, *Spirillum itersonii*, an aquatic bacterium, is dependent on oxygen unless potassium nitrate is added to the medium. In such cases nitrate essentially substitutes for oxygen as the final electron acceptor in the respiratory chain. This process is termed *anaerobic respiration*. The pathways for the dissimilation of the carbon and energy sources are identical with those in aerobic respiration, and electron transport occurs via a respiratory chain similar to that in aerobic cells. Oxygen is replaced as the terminal electron acceptor by nitrate. However, in some strict anaerobes, other inorganic compounds, such as carbon dioxide, or ions, such as sulfate ion, can be the terminal electron acceptors.

Energy production by photosynthesis

Plants, algae, and cyanobacteria are photoautotrophs. They use light as their source of energy and carbon dioxide as their sole source of carbon. In order for carbon dioxide to be useful for metabolism, it must first be reduced to carbohydrate. This process, by which light is used to convert carbon dioxide to carbohydrate, is called *photosynthesis*. The overall reaction can be written as

$$2H_2O + CO_2 \longrightarrow (CH_2O)_x + O_2 + H_2O$$

Carbohydrate

In the presence of light (radiant energy) and the green pigment chlorophyll

245

Here $(CH_2O)_x$ is a formula representing any carbohydrate.

This process has two important requirements: (1) a large amount of energy in the form of ATP, and (2) a large quantity of a chemical reductant, in this case water.

Several groups of bacteria—the photoautotrophic green and purple bacteria—are also characterized by their ability to perform photosynthesis. But unlike plants, algae, and cyanobacteria, they do not use water as their chemical reductant, nor do they produce oxygen as one of their end products of photosynthesis. The general equation for bacterial photosynthesis is:

$$2H_2A + CO_2 \longrightarrow (CH_2O)_x + 2A + H_2O$$

In the presence of   Carbohydrate
light (radiant energy)
and the green pigment
bacteriochlorophyll

Here H_2A represents the chemical reductant, such as the inorganic compounds H_2, H_2S, or $H_2S_2O_3$, or the organic compounds lactate or succinate. If H_2A in this equation stood for H_2S, then A would stand for S.

Both of the preceding equations represent the overall results of photosynthesis. A great deal has been learned about the specific chemical reactions involved in bacterial and plant photosynthesis. What follows is a look at the light-dependent energy-yielding processes involving bacteriochlorophyll in bacteria and chlorophyll in plants, algae, and cyanobacteria. What is presented is in accord with the latest results of many investigators but may require modification as further evidence is accumulated.

Cyclic and noncyclic photophosphorylation

Photosynthetic bacteria possess chlorophylls called bacteriochlorophylls that differ from those of plants in structure and in light-absorbing properties. Bacteriochlorophylls absorb light in the near-infrared region (660 to 870 nm). They are not contained in chloroplasts but are found in extensive membrane systems throughout the bacterial cell.

When a molecule of bacteriochlorophyll absorbs a quantum of light, the energy of the light raises the molecule to an excited state. In this excited state an electron is given off by bacteriochlorophyll. Bacteriochlorophyll thus becomes positively charged. It then serves as an electron trap or strong oxidizing agent.

The electron, carrying some of the energy absorbed from light, is transferred to an iron-containing heme protein known as *ferredoxin*. From there it is passed successively to ubiquinone, to cytochrome b, and to cytochrome f, and finally back to the positively charged bacteriochlorophyll. Essentially, the electron has gone around in a cycle, beginning with bacteriochlorophyll and returning to it. Therefore, the whole process is termed *cyclic phosphorylation*. This relatively simple process is illustrated in Fig. 14-9.

The energy released in the step between cytochrome b and cytochrome f is used for *photophosphorylation*—the generation of ATP from ADP and inorganic phosphate.

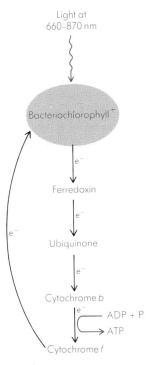

Light at
660–870 nm

Bacteriochlorophyll⁺

e^-

Ferredoxin

e^-

Ubiquinone

e^-

Cytochrome b

e^- — ADP + P
→ ATP

Cytochrome f

e^-

Figure 14-9. Cyclic photophosphorylation as it occurs in photosynthetic bacteria. The electron returns, at a lower energy state, to the bacteriochlorophyll, which had become positively charged after the initial ejection of the electron. No NADP is reduced and no external electron donor is necessary for this process.

Note that no NADP⁺ has been reduced in these reactions. The reduction of NADP⁺ in photosynthetic bacteria is accomplished not by photosynthesis but by using reducing power from constituents of the environment, such as H_2S and other inorganic and organic compounds. Such reduced compounds usually abound in the anaerobic environment of photosynthetic bacteria.

It may be added that light of higher energy than that absorbed by bacteriochlorophylls can contribute to bacterial photosynthesis since there are carotenoids and other accessory pigments in the bacterial cells which absorb light at shorter wavelengths and transfer the energy to the bacterio-chlorophylls.

In plants, algae, and cyanobacteria, noncyclic photophosphorylation occurs in photosynthesis. In this process, when a molecule in pigment system II (one of two systems of light reactions) absorbs light, this energy raises the molecule to an excited state and the molecule releases an electron. This electron is transferred to plastiquinone, to cytochrome b, to cytochrome f, and finally to pigment system I. Photophosphorylation occurs with generation of ATP from ADP and inorganic phosphate in the step between cytochrome b and cytochrome f. When pigment system I absorbs light, it releases an electron. This electron is transferred from ferredoxin, to flavoprotein, to NADP⁺. Photophosphorylation occurs again between the release of the electron from pigment system I to ferredoxin.

Figure 14-10. Noncyclic photophosphorylation as it occurs in green plants, algae, and cyanobacteria. In this process, electrons raised to a high energy state ultimately reduce NADP⁺ and are not recycled to the light-pigment systems. The protons necessary for reduction come from the dissociation of water, which results in evolution of oxygen. Electrons are restored to the pigments of system II from the OH⁻ ion of H_2O. The OH⁻ ion is split to e^-, H⁺, and ½O₂ by photolysis.

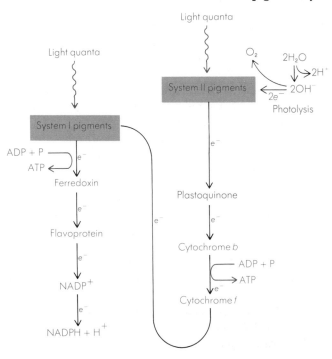

Also note that NADP$^+$ is reduced in this part of the process (see Fig. 14-10). This process differs from cyclic photophosphorylation because the electron lost by pigment system II is not cycled back to it. Instead, electrons are replaced in pigment system II by the light-generated breakdown of water called *photolysis*.

The mechanism of ATP synthesis

The chemical reactions that lead to the synthesis of ATP are now quite well understood. But how the transfer of electrons through the electron-transport chain is coupled to the synthesis of ATP is not very clear. Several alternate hypotheses have been proposed to explain how energy released during electron transport is conserved in the form of ATP. The prevailing theory is the *chemiosmotic hypothesis* advanced in 1961 by Peter Mitchell, a British biochemist. Mitchell was awarded the Nobel Prize for his work in this field in 1978. According to this theory, the flow of electrons through the system of carrier molecules releases energy which drives positively charged hydrogen ions (H$^+$), or protons, across the membranes of chloroplasts, mitochondria, and bacterial cells (Fig. 14-11). This movement of hydrogen ions results in the acidification of the medium that the organelle or cell is in and the generation of a *pH gradient* (a difference in pH) across the organelle or cell membrane. In addition, such hydrogen-ion movements lead to the formation of an *electric potential gradient* (a difference in charge) across the membrane (since an electric charge is carried by the proton). In this way, energy released during the transfer of electrons through the electron-transport chain is conserved as a "protonmotive force"; the electric potential gradients are produced by pumping hydrogen ions across the membrane.

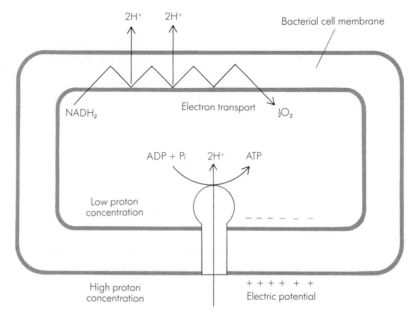

Figure 14-11. Mechanism of ATP synthesis. Flow of electrons through the electron-transport chain drives hydrogen ions across the membrane. This results in a high hydrogen ion concentration outside the cell and a low concentration inside the cell. This produces a pH or electrochemical gradient. ATP synthesis at the site of the ATPase complex (a knobbed structure on the membrane) is driven by the release of energy when hydrogen reenters the bacterial cell.

Following this first energy-conservation step, when the hydrogen ions reenter the organelle or cell, they are transported by the membrane-bound enzyme adenosine triphosphatase. The energy released on reentry drives the synthesis of ATP, the second energy-conservation step. This is shown in Fig. 14-11.

Summary and outlook

When microorganisms break down nutrients, the energy released is conserved in energy-rich compounds such as ATP. By coupling the breakdown of these compounds with endergonic reactions, the energy released on breakdown can be used to drive the endergonic reactions.

The mechanism that organisms use to produce energy is the oxidation-reduction reactions which transfer electrons from O/R systems of lower oxidation-reduction potential to those of higher oxidation-reduction potential, such as in the electron-transport chain.

Energy is produced by microorganisms from chemical compounds or from light. Many anaerobic and aerobic catabolic pathways exist that produce energy from chemical compounds. One of the most important of these pathways is glycolysis. Pyruvic acid is the end product of this pathway. Under aerobic conditions, pyruvic acid is oxidized to CO_2 and H_2O through the TCA cycle and electron-transport chain. Both glycolysis and the TCA cycle are also important in the catabolism of lipids and proteins. Under anaerobic conditions, pyruvic acid can be fermented to many organic end products. In addition, glycolysis and the TCA cycle are important in supplying carbon intermediates for biosynthesis of cellular material.

Photosynthetic organisms use light to produce energy in the form of ATP by means of cyclic and noncyclic photophosphorylation. The mechanism of ATP synthesis in the living cell is explained by Mitchell's chemiosmotic hypothesis.

In the future, catabolic pathways will be worked out for those organisms in which such pathways are now not known. There will be better understanding of the structural components of the cell and their relation to their special functions. Hypotheses pertaining to basic phenomena, such as Mitchell's chemiosmotic hypothesis, will keep evolving as new data are accumulated.

In the area of applications, one can foresee the use of microorganisms for the increased production of alcohol from raw materials such as molasses. This alcohol can be used for industrial purposes as well as being used as a gasoline extender. This practice of extending gasoline usage is already a reality in Brazil and in some parts of the United States. The capacity for photosynthesis by macroorganisms and microorganisms may also be harnessed for the production of fuel (cellulose in the form of wood or fiber). This type of fuel would be, moreover, a renewable resource. It may also be possible to manipulate such organisms to alter their storage products from photosynthesis to produce more hydrocarbons and less carbohydrates.

Key terms

aerobic respiration
amphibolic
anabolism
anaerobic respiration
assimilation reaction
bacteriochlorophyll
catabolism
chemiosmotic hypothesis
chlorophyll
common reactant
coupling mechanism
cyclic photophosphorylation
cytochrome system
dehydrogenation
dissimilation reaction
electric potential gradient
electromotive potential
electron-transport chain
  (respiratory chain)

endergonic
energy-rich compound
exergonic
fermentation
free-energy change
glycolysis
hydrolysis
noncyclic photophosphorylation
oxidant
oxidation
O/R system
oxidative phosphorylation
pentose phosphate cycle
pH gradient
photosynthesis
protonmotive force
reductant
reduction
tricarboxylic acid cycle

Questions

1 Explain how the energy released from an exergonic reaction can be used to drive an endergonic reaction.

2 Describe the role of ATP in energy exchanges within the cell.

3 Relate electron flow and ΔG to O/R systems of different E_0' values, using the electron-transport chain as a model.

4 Distinguish between anaerobic respiration, aerobic respiration, and fermentation.

5 (a) Is glycolysis the only catabolic pathway found in bacteria for the dissimilation of carbohydrates?
  (b) Is glycolysis found only in anaerobic cells? Explain.
  (c) How is the glycolytic sequence regulated?

6 Explain the ATP yield per glucose molecule from glycolysis.

7 Why is aerobic respiration considered to be a more efficient process for obtaining energy than fermentation? Explain.

8 What other substances besides carbohydrates may be used for the production of energy? Is there a general rule governing their utilization?

9 What is the source of reducing power in photosynthesis by green plants? In photosynthesis by bacteria?

10 Compare the photosynthetic process of the following groups of microorganisms: photosynthetic bacteria, algae, and cyanobacteria.

11 Why is pyruvic acid considered the hub of carbohydrate metabolism?

12 Explain the prevailing theory on the mechanism of ATP synthesis during electron transport.

ASSIMILATION REACTIONS
AND ENERGY UTILIZATION

The ATP formed by the dissimilation reactions of the microbial cell is expended in various ways. Much of it is used in the biosynthesis of cell structures such as the cell wall, cell membrane, and energy-storage granules. ATP is also required for synthesis of enzymes and other chemical substances, and maintenance of the physical and chemical integrity of the cell, including repair of cell damage. Other processes utilizing ATP include motility, heat production, transport of solutes across membranes, and even bioluminescence. A large portion of ATP from dissimilation reactions is utilized for metabolic processes that are not related to biosynthesis of cell material. We will examine some of these processes before discussing the utilization of energy in biosynthesis.

Energy utilization in nonbiosynthetic processes

Heat production

Microorganisms produce heat from their normal metabolic activities, which leads to a rise in the temperature of a culture. This may be observed most readily in very large scale cultures used in the fermentation industries, such as in the production of antibiotics. We have little information on what fraction of the free energy of microbial exergonic reactions is given off as heat.

Presumably, activity of the enzyme ATPase (the enzyme that breaks down ATP) is mainly responsible for the heat produced. The physiological role of this enzyme is somewhat vague, but it has been suggested that it serves to get rid of excess ATP and thus helps to regulate the energy metabolism of the cell. Loss of energy from the high-energy phosphate bonds in the form of heat also occurs in other ways. For example, when a cell forms an ester or amide bond in a molecule during synthesis, only about 3,000 cal are required. But a high-energy phosphate bond when broken down releases 12,000 cal. The energy not used in the ester or amide bond formation (9,000 cal) is released as heat.

Motility   Energy is needed to power the movement of cilia and flagella of motile microorganisms. It has been estimated that as much as 10 percent of the energy expended by some of these microbes is used for flagellar motion.

Evidence suggesting that ATP is required to power flagella comes from cytochemical studies on motile bacteria which reveal the presence of Mg-dependent ATPase activity at the membranous sites of flagella origin. Figure 15-1 shows negative stains of *Bacillus licheniformis*, with lead phosphate deposited at sites of ATPase activity. In carrying out this experiment, it was assumed that ATP was the sole substrate for the ATPase associated with the bacterial membrane. Breakdown of ATP by ATPase releases inorganic phosphate which reacts with soluble lead nitrate previously put in bacterial solution, leaving dark areas (deposits of insoluble lead phosphate) when viewed in the electron microscope. As Figure 15-1 shows, these dark areas are localized in the membrane in areas of flagellar attachment, indicating that ATPase activity is localized in these areas and therefore that ATP is the energy source for bacterial flagella movement.

Figure 15-1. Negative stains of *Bacillus licheniformis* with lead phosphate deposited at sites of ATPase activity. Arrows indicate some areas of activity which appear to be at the origin of flagella. (Courtesy of Z. Vaituzis and National Research Council of Canada. From Can. J. Microbiol. **19**:1265-1267, 1973.)

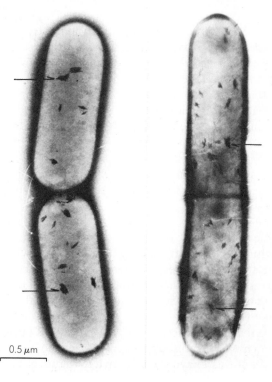

0.5 μm

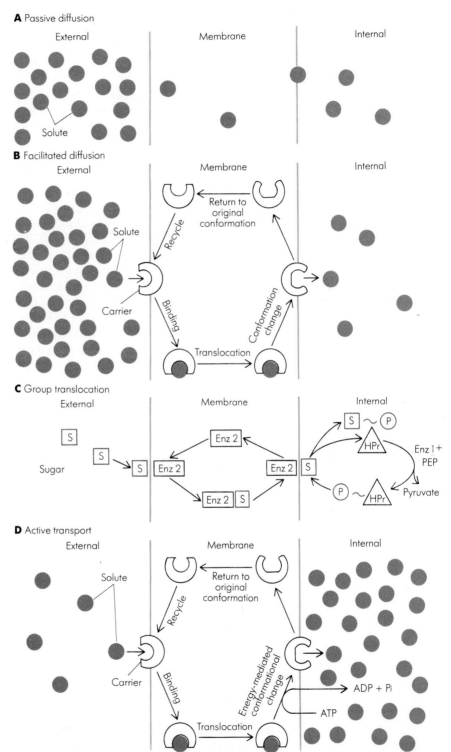

Figure 15-2. Mechanisms of nutrient transport into cells. (A) Passive diffusion; (B) facilitated diffusion; (C) group translocation; (D) active transport.

A Passive diffusion

External   Membrane   Internal

Solute

B Facilitated diffusion

External   Membrane   Internal

Solute

Return to original conformation

Recycle

Carrier

Binding

Translocation

Conformation change

C Group translocation

External   Membrane   Internal

Sugar

S   Enz 2

Enz 2   Enz 2

Enz 2   S

Enz 2   S

S ~ P

HPr

Enz 1 + PEP

P ~ HPr

Pyruvate

D Active transport

External   Membrane   Internal

Solute

Return to original conformation

Recycle

Carrier

Binding

Translocation

Energy-mediated conformational change

ADP + Pi

ATP

Transport of nutrients   Except for water and some lipid-soluble molecules, few compounds can enter the semipermeable cell membrane by *simple*, or *passive*, diffusion. In this process the solute crosses the membrane as the result of random molecular motion and does not interact specifically with any substance in the membrane (Fig. 15-2A).

Another mechanism by which substances cross the semipermeable cell membrane is *facilitated diffusion*. Here, a molecule of solute combines reversibly with a specific protein carrier molecule (called a *porter*) in the membrane, and the carrier-solute complex moves between the inner and outer surfaces of the membrane, releasing one solute molecule and returning to bind a new solute molecule. This type of diffusion always moves solute molecules from a region of higher solute concentration to one of lower solute concentration. See Fig. 15-2B.

Neither simple diffusion nor facilitated diffusion requires metabolic energy, nor does either lead to concentration or accumulation of solute against an electrochemical or osmotic gradient. Therefore, these two mechanisms are of little interest to us in the context of this chapter. However, two other mechanisms by which solutes cross membranes, *group translocation* and *active transport*, require metabolic energy.

In group translocation, the substance to be transported must be modified chemically before it is transported by the carrier into the cell. The process leading to chemical modification requires metabolic energy.

In active transport, the solute is transported by a carrier against an osmotic or electrochemical gradient at the expense of metabolic energy. Except for the need for metabolic energy to drive this transport, active transport is much like facilitated diffusion.

Group translocation. A good example of group translocation is the transport of certain sugars, such as glucose, fructose, and mannose, into bacteria. In this process, first a heat-stable carrier protein (HPr) is activated by transfer of a phosphate group from the high-energy compound phosphoenolpyruvate (PEP) inside the cell, as shown in Fig. 15-2C:

$$\text{PEP} + \text{HPr} \underset{\text{enzyme 1}}{\rightleftharpoons} \text{pyruvate} + \text{phospho-HPr}$$

At the same time, the sugar combines with enzyme 2 at the outer membrane surface and is transported to the inner membrane surface. Here it combines with the phosphate group carried by the activated HPr. The sugar-phosphate is released by enzyme 2 and enters the cell. This reaction can be summarized as follows.

$$\underset{\text{(Outside cell)}}{\text{Phospho-HPr} + \text{sugar}} \xrightarrow{\text{enzyme 2}} \underset{\text{(Inside cell)}}{\text{sugar-phosphate} + \text{HPr}}$$

The transport reaction only transports sugar into the cell, because the sugar phosphate within the cell has no affinity for the carrier protein.

HPr and enzyme 1 are soluble cytoplasmic proteins. HPr is of low molecular weight and has been highly purified. Enzyme 2 is membrane-bound and is specific for the particular sugars transported. It has been solubilized and partially purified.

Other known group-translocation processes include the uptake of adenine and butyrate at the exterior surface of the cell and conversion of them at the interior membrane surface to adenosine monophosphate (AMP) and butyryl-CoA, respectively.

Active transport. Almost all solutes, including sugars, amino acids, peptides, nucleosides, and ions, are taken up by cells through active transport. The three steps of active transport are:

1 Binding of a solute to a receptor site on a membrane-bound carrier protein.
2 Translocation of the solute-carrier complex across the membrane.
3 Coupling of translocation to an energy-yielding reaction to lower the affinity of the carrier protein for the solute at the inner membrane surface so that the carrier protein will release solute to the cell interior. This process is illustrated in Fig. 15-2D.

Several mechanisms have been proposed to explain the molecular basis of active transport of solutes in microorganisms. The accumulated evidence suggests that active transport may also be explained by Mitchell's chemiosmotic theory (see Chap. 14). In this case, energy released during the flow of electrons through the electron-transport chain or the splitting of a phosphate group from ATP drives protons out of the cell. This generates a difference in pH value and electric potential between the inside and the outside of the cell or across the membrane. This proton gradient gives rise to a protonmotive force which can be used to pump the solutes into the cell. When protons reenter the cell, the energy released on reentry drives the transport mechanism in the cell membrane, probably by inducing a conformational change in the carrier molecule so that its affinity for the solute is decreased and the solute is released into the cell interior. The link between active transport and metabolic energy generation is illustrated in Fig. 15-3.

Energy utilization in biosynthetic processes

We have seen how energy is expended by heat production, motility, and transport of nutrients into microbial cells. All these are nonbiosynthetic processes. Biosynthetic processes in the cell also require energy; energy

Figure 15-3. How release of metabolic energy is coupled to and drives active transport.

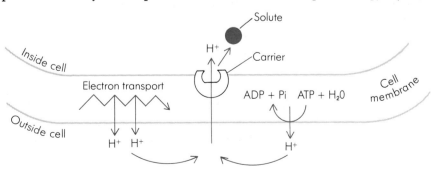

Table 15-1. Amino acid building blocks of protein, with standard abbreviations

Alanine (Ala)	Glycine (Gly)	Proline (Pro)
Arginine (Arg)	Histidine (His)	Serine (Ser)
Asparagine (Asp-NH$_2$, Asn)	Isoleucine (Ile)	Threonine (Thr)
Aspartic acid (Asp)	Leucine (Leu)	Tryptophan (Trp)
Cysteine (Cys)	Lysine (Lys)	Tyrosine (Tyr)
Glutamic acid (Glu)	Methionine (Met)	Valine (Val)
Glutamine (Glu-NH$_2$, Gln)	Phenylalanine (Phe)	

from ATP is used to convert one substance into another and to synthesize complex substances from simpler ones.

Conversion of substances

Amino acids, of which there are about 20 (Table 15-1), are the building blocks of proteins. The sequence in which they are linked determines the type of protein they form.

Consider the specific example of the synthesis of the amino acid proline by the bacterium *Escherichia coli*. Glutamic acid is the initial reactant. The steps involved are shown in Fig. 15-4. In the first step an acid group (—COOH) is reduced to an aldehyde group (—CHO). This requires two electrons from NADPH$_2$ and energy from ATP. The aldehyde group then spontaneously reacts with the amino group (—NH$_2$) on the same molecule, forming a ring. This step is followed by ring reduction to form proline.

Another example is the pathway for conversion of aspartic acid to lysine, methionine, and threonine. The conversion utilizes metabolic energy in the form of ATP, as shown in Fig. 15-5. These two examples illustrate the expenditure of energy in the interconversion of one substance to another.

Synthesis of macromolecules

Another level of biosynthesis is that of joining smaller molecules to form larger ones—the synthesis of macromolecules. We illustrate this process with the biosynthesis of bacterial cell-wall peptidoglycan.

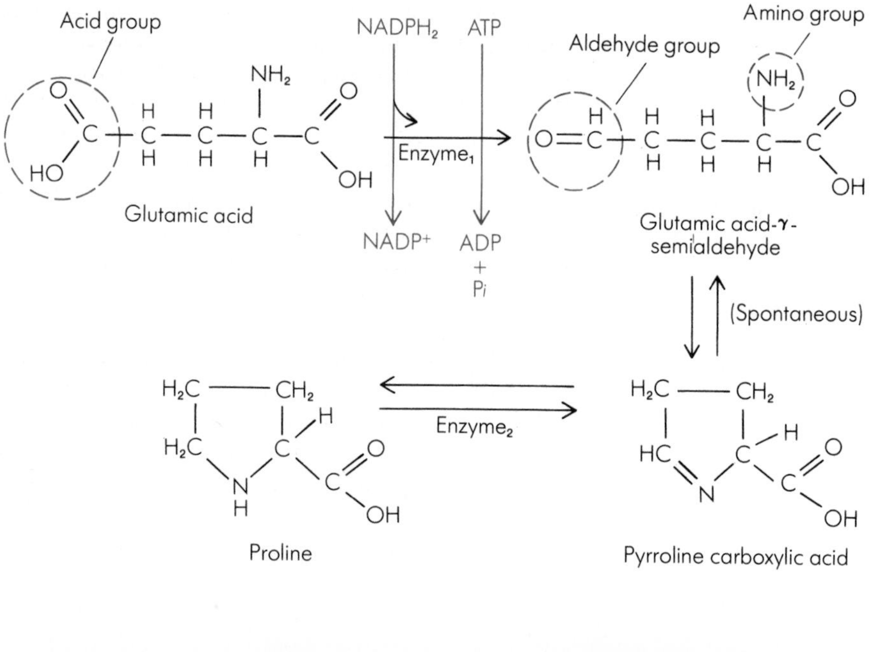

Figure 15-4. The biosynthesis of proline from glutamic acid in *E. coli*. Note the utilization of metabolic energy in the form of ATP and NADPH$_2$ in the initial step. (See text for details of synthetic steps.) Sections of the molecule reacting in each step are enclosed in color dashed lines.

Figure 15-5. The biosynthesis of methionine, threonine, and lysine: another example of metabolic energy utilization in the form of ATP and NADPH$_2$ in the interconversion of substances.

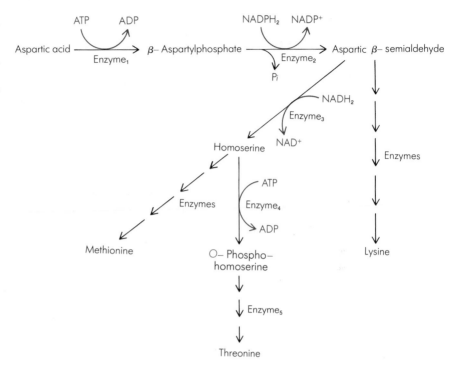

Structure of peptidoglycan. The rigid portion of a bacterial cell wall is made up of a polymer known as a *peptidoglycan, murein,* or *mucopeptide.* The walls of gram-positive bacteria contain a large proportion of peptidoglycan; those of gram-negative bacteria have a much smaller proportion. Peptidoglycans vary in their chemical composition and structure from species to species, but there are basic similarities. Peptidoglycans are very large polymers composed of three kinds of building blocks: (1) *acetylglucosamine* (AGA), (2) *acetylmuramic acid* (AMA), and (3) a *peptide* consisting of four or five amino acids of limited variety. Several of the amino acids exist in the D-isomer configuration, a configuration not usually found elsewhere in nature. A peptidoglycan can best be thought of as consisting of polysaccharide backbone chains composed of alternating units of AGA and AMA, with the short peptide chains projecting from the AMA units. Many of these peptide chains are cross-linked with each other, imparting great rigidity to the total structure. Figure 15-6 illustrates the basic structure of peptidoglycans. Some peptidoglycans differ in that the peptide chains may not be directly cross-linked to each other, being linked instead by another kind of peptide which forms a *bridge* between the terminal carboxyl (acid) group of one side chain and the free amino group of lysine or diaminopimelic acid (DPM) on the next side chain. For example, in *Staphylococcus aureus* a bridge composed of five glycine molecules can link two AMA peptides together. The type of cross-linking is usually characteristic of the species.

Activation of a peptidoglycan precursor. *Escherichia coli* can synthesize

Figure 15-6. General structure of peptidoglycans. Note the pentapeptide bridge for cross-linking. In some peptidoglycans, peptides extending from AMA are linked directly to each other without pentapeptide bridges. AMA = acetylmuramic acid; AGA = acetylglucosamine.

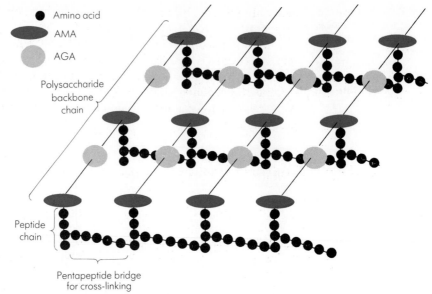

Amino acid

AMA

AGA

Polysaccharide backbone chain

Peptide chain

Pentapeptide bridge for cross-linking

Figure 15-7. Biosynthesis of acetylmuramic acid-UDP, a key precursor in the synthesis of peptidoglycans. All high-energy compounds are in color. ATP = adenosine triphosphate; UTP = uridine triphosphate.

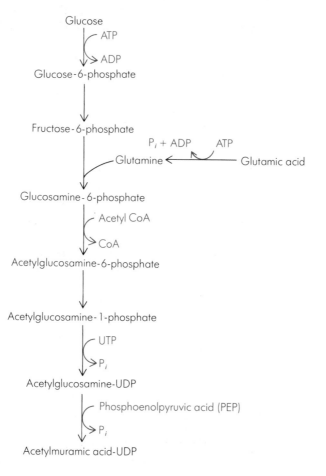

Glucose

ATP

ADP

Glucose-6-phosphate

Fructose-6-phosphate

P_i + ADP   ATP

Glutamine ← Glutamic acid

Glucosamine-6-phosphate

Acetyl CoA

CoA

Acetylglucosamine-6-phosphate

Acetylglucosamine-1-phosphate

UTP

P_i

Acetylglucosamine-UDP

Phosphoenolpyruvic acid (PEP)

P_i

Acetylmuramic acid-UDP

cell-wall peptidoglycan when grown in a simple medium of glucose, ammonium sulfate, and mineral salts. One of the early steps in this synthesis is the formation of an activated derivative of AMA. This process, which is shown in Fig. 15-7, requires energy at several points. The activation of sugars, such as acetyl glucosamine, by the attachment of a uridine diphosphate (UDP) to form a sugar-UDP precursor is not peculiar to AMA but is a general method involved in the biosynthesis of many kinds of polysaccharides.

Synthesis of peptidoglycan. After formation of the activated AMA, the synthesis of peptidoglycan proceeds as follows:

1 Amino acids are linked to the AMA portion of the activated precursor to form a short pentapeptide chain. Ribosomes are not involved, but each amino acid addition requires energy from the breakdown of ATP and the presence of Mg^{2+} or Mn^{2+} and a specific enzyme.
2 The AMA-UDP precursor is coupled to a membrane phospholipid called *bactoprenol.*
3 The AGA couples with AMA of the AMA-UDP precursor. This reaction requires the *activated* form of AGA, that is, the AGA-UDP derivative. In some organisms, the addition of bridging peptides takes place at this step.
4 The precursor, still linked to bactoprenol, is carried out of the cell through the cell membrane and is linked to a growing peptidoglycan chain in the cell wall. Peptide cross-linking may now occur, and the incorporation of the precursor into the growing peptidoglycan is thus completed. Figure 15-8 illustrates the steps in a typical peptidoglycan biosynthesis.

Figure 15-8. Biosynthesis of peptidoglycan in *Staphylococcus aureus*. Note that the second pentapeptide, added for cross-linking, has a different composition and is added by a different mechanism than the first. The item added in each step is indicated in color. AMA = acetylmuramic acid; AGA = acetylglucosamine.

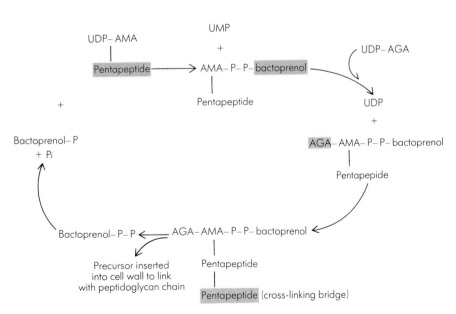

Organic synthesis in chemoautotrophic bacteria

Chemoautotrophic bacteria require no organic nutrients and utilize carbon dioxide as their sole source of carbon. These bacteria oxidize inorganic nutrients such as hydrogen, ammonia, nitrite, and thiosulfate to produce metabolic energy in the form of ATP and in some cases reducing power in the form of $NADPH_2$. Compared to other bacteria, such as heterotrophs, chemoautotrophic bacteria are at a considerable energetic disadvantage. Electrons entering the electron-transport chain from oxidations of inorganic nutrients by chemoautotrophic bacteria usually enter at a higher point in the electron-transport chain (the hydrogen bacteria *Hydrogenomonas* and the nonphotosynthetic sulfur bacteria are exceptions). Consider for example *Nitrobacter*. Because the E_0' of NO_3^-/NO_2^- is higher than the E_0' of $NADP/NADPH + H^+$, it is not possible for oxidation of nitrite (*Nitrobacter's* inorganic energy source) to be coupled with the production of reducing power in the form of $NADPH_2$ at the beginning of the electron-transport chain. By entering the chain at cytochrome *c*, whose E_0' is higher than NO_3^-/NO_2^-, *Nitrobacter* produces much less ATP than heterotrophs and no reducing power in the form of $NADPH_2$. How then do chemoautotrophs like *Nitrobacter* generate $NADPH_2$ for use with ATP in carbon dioxide fixation—the beginning of the biosynthesis of all organic compounds they require? These chemoautotrophs use a process called reversed electron flow or ATP-dependent $NADPH_2$ production. In this process, energy released on breakdown of ATP is used to drive electrons from the oxidation of the inorganic energy source to a E_0' at which they can subsequently reduce NAD^+ or $NADP^+$. Figure 15-9 shows how this occurs.

Figure 15-9. Production of energy from the oxidation of an inorganic source (NO_2^-) and the generation of reducing power (NADPH) by an ATP-dependent system. Diagram is representative of the activity of the chemoautotroph *Nitrobacter*.

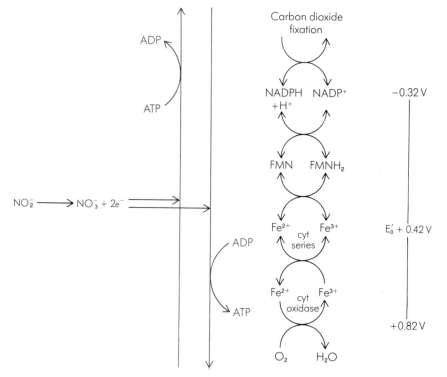

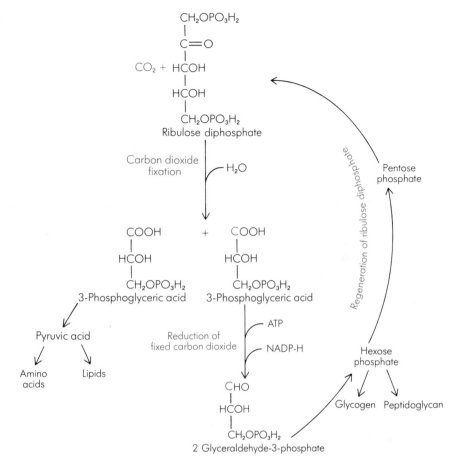

Figure 15-10. The Calvin cycle for carbon dioxide fixation in autotrophic organisms.

The principal method of carbon dioxide fixation in autotrophic bacteria is the Calvin cycle, which is illustrated in Fig. 15-10. In the Calvin cycle, carbon dioxide is fixed in a reaction with the acceptor molecule ribulose diphosphate. The primary product of carbon dioxide fixation is 3-phosphoglyceric acid, from which all other organic molecules of the cell are synthesized. However, carbon dioxide fixation is dependent on a supply of the acceptor molecule, ribulose diphosphate, and so most of the 3-phosphoglyceric acid produced must be used to regenerate ribulose diphosphate. Thus the process of carbon dioxide fixation is cyclic. Each turn of the cycle results in the fixation of one molecule of carbon dioxide. Various intermediates of the cycle are drawn off and enter different biosynthetic pathways.

This cycle of carbon dioxide fixation is complex. It shares certain reactions of the glycolytic and pentose phosphate pathways discussed in Chap. 14. Two reactions are specific to the cycle: the carbon dioxide fixation reaction and the reaction which regenerates the carbon dioxide acceptor ribulose diphosphate.

The overall reaction for the Calvin cycle is

$$6CO_2 + 12NADPH + 12H^+ + 18ATP + 12H_2O \rightarrow C_6H_{12}O_6 + 12NADP + 18ADP + 18P_i$$
$$\text{(glucose)}$$

Note the high utilization of reducing power and energy in this cycle.

Summary and outlook

Microorganisms use energy for nonbiosynthetic and biosynthetic processes. The nonbiosynthetic processes include heat production, motility by flagellar action, and transport of solutes into the cell. Transport that results in accumulation of solutes within the cell and requires energy is effected by group translocation and active transport mechanisms. Biosynthetic processes in the cell which utilize energy include the interconversion of chemical substances and the synthesis of macromolecules. Amino acid interconversion is illustrative of the first type of energy requirement in biosynthesis. Bacterial cell-wall peptidoglycan biosynthesis is an example of the second type.

Organic synthesis in chemoautotrophic bacteria requires more reducing power and metabolic energy for carbon dioxide fixation.

In the coming years, the greatest challenge to a better understanding of assimilation reactions and energy utilization is likely to be problems associated with molecular events at the cell membrane. At the applied level, microbes will be used more widely to convert domestic and industrial wastes into energy and into cell-protein which can be used as animal feed.

Key terms

acetylglucosamine
acetylmuramic acid
active transport
bactoprenol
Calvin cycle
carbon dioxide fixation
facilitated diffusion

group translocation
passive diffusion
peptide bridge
peptidoglycan (murein, mucopeptide)
porter
reversed electron flow

Questions

1 What evidence do we have that bacterial motility is energy-dependent?
2 How is group translocation similar to and different from active transport of solutes?
3 Using Mitchell's chemiosmotic theory, explain how the molecular mechanisms of active transport and ATP synthesis are coupled.
4 Explain why the interconversion of chemical substances utilizes metabolic energy.
5 Describe how AMA, AGA, and amino acids are linked in the structure of peptidoglycan.
6 How is energy utilized in the biosynthesis of peptidoglycan?

7 Explain why chemoautotrophs are at a considerable energetic disadvantage compared to chemoheterotrophs.

8 Why do chemoautotrotrophic bacteria require much reducing power? Explain how this reducing power is used.

References for part five

Cohen, P.: *Control of Enzyme Activity*, Chapman and Hall, London, 1976. (Distributed in the United States by Wiley, New York.) *A short paperback that deals with direct control of enzyme activity by changing the activity of preexisting enzyme molecules. Detailed examples of each type of control are given.*

Dawes, I. W., and I. W. Sutherland: *Microbial Physiology*, Blackwell Scientific, Oxford, 1976. *An introduction to microbial physiology which attempts to provide a broad perspective of the subject.*

Holloway, M. R.: *The Mechanism of Enzyme Action*, Blackwell Scientific, Oxford, 1976. *A short monograph summarizing the state of knowledge about the mechanism of enzyme action. Well illustrated with structural diagrams.*

Lehninger, A. L.: *Short Course in Biochemistry*, Worth, New York, 1973. *A shorter and more concise version of the more comprehensive textbook* Biochemistry. *Gives a clear description in a succinct manner of biochemical principles. Chapter 4 deals with enzymes.*

Mandelstam, J., and K. McQuillen (eds.): *Biochemistry of Bacterial Growth*, 2d ed., Blackwell Scientific, Oxford, 1973. *Contains contributions by many authors. Has a fairly comprehensive description of bacterial life in biochemical terms. The text is divided into three sections: the first section is a summary of the book; the second section is a somewhat more detailed description of the same material; the last section contains the detailed biochemistry.*

Rose, A. H.: *Chemical Microbiology*, 3d ed., Plenum, New York, 1976. *A general text on the subject of microbial physiology. Quite comprehensive and well referenced.*

Stanier, R. Y., E. A. Adelberg, and J. L. Ingraham: *The Microbial World*, 4th ed., Prentice-Hall, Englewood Cliffs, N.J., 1976. *A good general text on the physiology of microorganisms. More detailed and advanced explanations of various physiological and biochemical processes are provided.*

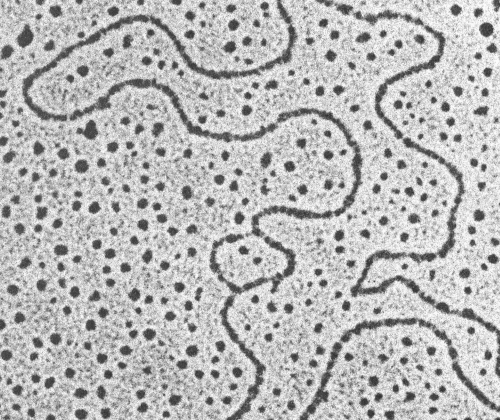

PART SIX
MICROBIAL GENETICS

Genetics is the study of the *inheritance* and *variability* of the characteristics of an organism, whether that organism is unicellular or multicellular. Research in genetics at the molecular level has identified deoxyribonucleic acid (DNA), the chemical substance making up chromosomes, as the hereditary substance. Much has been discovered about the molecular structure of DNA as well as its replication. The genetic code contained in it has been deciphered. We now have an appreciation of how genetic information is transmitted to control cellular growth and activity.

This knowledge has also led to the understanding that molecular defects in the information coded in DNA are the cause of many genetic diseases. These diseases include albinism (congential absence of pigment), sickle-cell anemia (hemolytic anemia), xeroderma pigmentosum (skin diseases resulting from exposure to sunlight), and alkaptonuria and phenylketonuria (metabolic diseases).

Like biochemical principles, genetic principles are universal. The study of microbial genetics has contributed much to what we now know about the genetics of all organisms. Procaryotic cells, especially bacteria, have been eminently useful in this regard; because procaryotes are organisms with single chromosomes, changes in their genetic material result in an immediate, observable expression of a change in characteristics. There is no masking effect of a change on a chromosome owing to the presence of the unaffected member of the paired chromosomes, a situation which may occur in organisms with paired chromosomes. Other distinct advantages in working with microbes for the study of genetics include the ease of imposing constant controlled conditions on microbes, the rapid growth rates of microbes, and the large populations of microbes which develop in cultures.

In Chap. 16, we will discuss the nature of genetic material. In Chap. 17, the genetics of bacteria will be considered.

A plasmid, a circular piece of DNA sometimes found in addition to chromosomal DNA in bacteria. The loop is a replication bubble or eye in the DNA. Bacterial plasmids play an important role in the developing field of genetic engineering. Magnification ×23,000. *(Courtesy of J. A. Hassell, McGill University.)*

THE NATURE OF GENETIC MATERIAL

Deoxyribonucleic acid (DNA) is the chemical substance responsible for transmission of hereditary information. But how? DNA makes up the chromosome of cells. In its structure is coded information for the synthesis of all the cell proteins. Discrete segments of DNA or chromosomes, called *genes*, code for each protein. This information is transmitted from cell to cell by *replication* of DNA.

Another type of nucleic acid, ribonucleic acid (RNA), is also present in each cell. It is similar to but not the same as DNA and it acts to process the information coded in DNA for protein synthesis by *transcription* and *translation*.

The structure of deoxyribonucleic acid

DNA is a long ropelike molecule (Fig. 16-1) usually composed of two strands, each wound around the other to form a double helix (Fig. 16-2). Each strand of the DNA helix is made up of *nucleotides* linked together to form a chain, a polynucleotide. Each nucleotide is constructed of three parts:

1 A nitrogen-containing ring compound called a nitrogenous base which is either a *purine* or a *pyrimidine*
2 A five-carbon sugar (pentose) called *deoxyribose*
3 A phosphate molecule

267

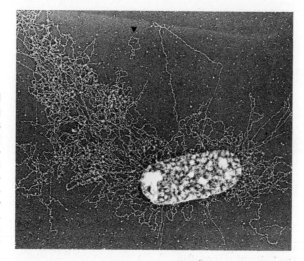

Figure 16-1. A disrupted *Escherichia coli* cell showing the ropelike DNA that has spilled out. Note the plasmid (top center pointer), a circular piece of DNA which is not part of the *E. coli* chromosome and which replicates separately from it. *(By permission from J. D. Griffith, The University of North Carolina.)*

These parts are linked together in the following sequence: nitrogenous base—deoxyribose—phosphate.

In DNA two kinds of purines, *adenine* and *guanine*, and two kinds of pyrimidines, *cytosine* and *thymine*, are found. The structures of these bases, as well as the structures of deoxyribose and phosphoric acid, are shown in Fig. 16-3. Since there are four kinds of bases, four kinds of nucleotides are found in DNA:

Deoxyadenosine-5′-monophosphate (adenine + deoxyribose + phosphate)
Deoxyguanosine-5′-monophosphate (guanine + deoxyribose + phosphate)
Deoxycytidine-5′-monophosphate (cytosine + deoxyribose + phosphate)
Thymidine-5′-monophosphate (thymine + deoxyribose + phosphate)

These four kinds of nucleotides are joined together in the polynucleotide strands of DNA by *phosphodiester linkages*; that is, each phosphate group links the number-3 carbon atom of one deoxyribose of a nucleotide to the number-5 carbon atom of the deoxyribose of the next nucleotide, with the phosphate group on the outside of the chain (Fig. 16-4). The result is a chain of alternating phosphate and sugar groups, with the nitrogenous bases projecting from the sugar groups (see Fig. 16-4). Weak bonds, known as *hydrogen bonds*, link the base on one chain and the base on the other chain. Two bases so linked are called a *complementary base pair*. Only two kinds of complementary base pairs are found in double-stranded DNA because of their hydrogen-bonding properties: adenine (A) and thymine (T), and guanine (G) and cytosine (C). As a consequence, the ratio of adenine to thymine or of guanine to cytosine in double-stranded DNA is always 1:1. Other ratios, such as adenine/cytosine, cytosine/thymine, or adenine + thymine/guanine + cytosine are not necessarily 1:1; their values are characteristic of the kind of organism from which the DNA has been extracted. Such values are used in the identification or taxonomic grouping of bacteria.

The failure to find a 1:1 ratio between adenine and thymine or guanine

Figure 16-2. Diagram of a short segment of DNA, showing how its two strands are wound to form a double helix. The two ribbons represent the sugar (S)-phosphate (P) backbones and the horizontal bars represent base (A, adenine; G, guanine; C, cytosine; T, thymine) pairs held together by hydrogen bonds.

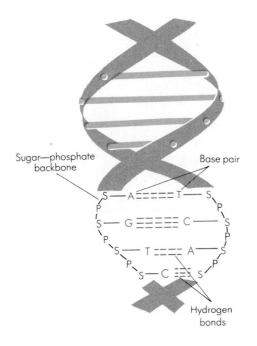

Sugar—phosphate backbone

Base pair

Hydrogen bonds

Figure 16-3. Building blocks of the nucleotides of DNA and RNA. Note that 2-deoxyribose, the sugar found in DNA, differs from ribose, the sugar found in RNA due to the absence of an —OH group on carbon 2. RNA uses uracil as a building block instead of thymine found in DNA.

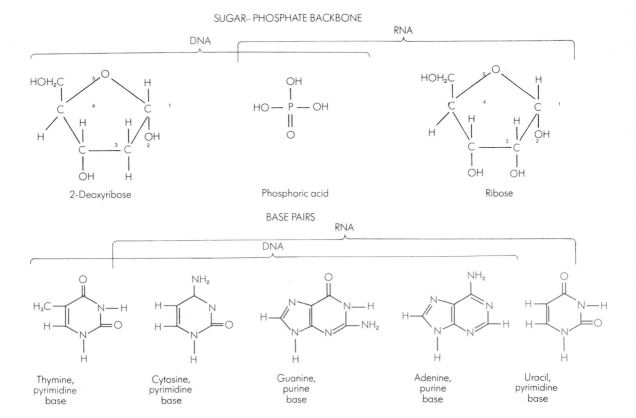

SUGAR– PHOSPHATE BACKBONE

DNA

RNA

2-Deoxyribose

Phosphoric acid

Ribose

BASE PAIRS

DNA

RNA

Thymine, pyrimidine base

Cytosine, pyrimidine base

Guanine, purine base

Adenine, purine base

Uracil, pyrimidine base

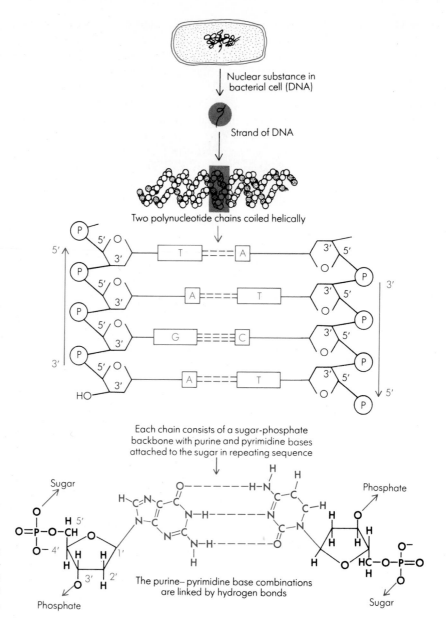

Nuclear substance in bacterial cell (DNA)

Strand of DNA

Two polynucleotide chains coiled helically

Each chain consists of a sugar-phosphate backbone with purine and pyrimidine bases attached to the sugar in repeating sequence

The purine–pyrimidine base combinations are linked by hydrogen bonds

Figure 16-4. DNA location in the cell, molecular configuration, and chemical structure as viewed in progressively greater detail.

and cytosine in certain viruses led to the discovery of single-stranded DNA in these organisms. These cases are rare exceptions to the general occurrence of double-stranded DNA.

The complementary base pairs hold the two strands of the DNA helix together by hydrogen bonding. There are two hydrogen bonds formed between each A-T pair, whereas there are three hydrogen bonds formed between each G-C pair. The complementarity of the purines and pyrimidines means that the sequence of bases on one strand dictates the sequence on the other strand. This is of critical importance in the replication (synthesis) of new strands of DNA during cell division. A consequence of the formation of the A-T and G-C pairs is that the two strands of the DNA helix are said to be *antiparallel*, which means that each strand runs in opposite directions so that one is terminated by a free 3′-hydroxyl group and the other by a 5′-phosphate group. We can also say that one strand runs from the 3′ end to the 5′ end, and the other from the 5′ end to the 3′ end. The numbers refer to the carbon atoms on the sugar residues (see Fig. 16-4).

The relationship of DNA to its low-molecular-weight components is shown in Fig. 16-5. The removal of the phosphate group from the nucleotide yields a *nucleoside* consisting of a pentose sugar linked to a nitrogenous base.

The structure of ribonucleic acid

The other naturally occurring nucleic acid is *ribonucleic acid* (RNA). It differs from DNA in that:

1 It is usually single-stranded.
2 The sugar component of the nucleotides which make up RNA is *ribose*, instead of deoxyribose. Ribose is identical to deoxyribose except for the presence of a hydroxyl group at the number-2 carbon atom.
3 The pyrimidine nitrogenous base uracil, instead of thymine, is found in the nucleotides which make up RNA. (See Fig. 16-3.)

The biosynthesis of DNA

The biosynthesis of nucleotides

Before the polynucleotide chains of DNA can be synthesized by bacteria (or any other organisms), an intracellular pool of nucleotides must be available. In certain bacteria, these nucleotides must be supplied preformed in the medium. But other bacteria can synthesize these nucleotides from very simple nutrients such as glucose, ammonium sulfate, and minerals. The conversion of simple nutrients into nucleotides for DNA synthesis involves a complex series of reactions, several of which require energy in the form of

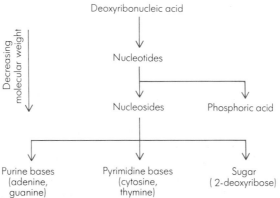

Figure 16-5. Breakdown of deoxyribonucleic acid into lower-molecular-weight components. DNA is made up of nucleotides which when you remove phosphoric acid yield nucleosides. Nucleosides can be broken down into bases and sugar (2-deoxyribose).

ATP. One of thse reactions is the formation of *activated forms* of nucleotides as direct precursors for synthesis of the polynucleotide chains of double-stranded DNA:

$$\text{Nucleotide} + \text{ATP} \xrightarrow{\text{kinase}} \text{nucleotide-phosphate} + \text{ADP} \tag{1}$$

$$\text{Nucleotide-phosphate} + \text{ATP} \xrightarrow{\text{kinase}} \text{nucleotide-diphosphate} + \text{ADP} \tag{2}$$

Energy in the form of ATP is provided. The activated nucleotides each have two phosphate groups attached to them from the breakdown of two ATPs.

The replication of DNA

The chromosome of a typical bacterium is a circular double-stranded DNA molecule, which has an approximate molecular weight of 2.5×10^9 daltons (a dalton is equal to the mass of one hydrogen atom). It has about 4×10^6 base pairs. If the chromosome were extended linearly in the double-helix form, it would measure approximately 1250 μm (1.25 mm), which is several hundred times longer than the bacterial cell that contains it.

Replication begins at a fixed specific site on the bacterial chromosome known as the *origin* (Fig. 16-6). The two strands of DNA separate at this site to form a Y-shaped structure, the juncture of which is called the *growing point*. Replication moves sequentially from the growing point either in one direction (*unidirectional replication*) or in both directions (*bidirectional replication*) (Fig. 16-6). The origin and the growing point are attached to the cell membrane and it is there that both strands are duplicated. Each of these strands has a base sequence complementary to one of the original strands of the DNA.

The *DNA polymerase* enzymes add nuleotides to the 3′-hydroxyl end of the replicating DNA strand or strands, thus synthesizing DNA strands in a 5′ to 3′ direction (Fig. 16-7). So far no polymerases have been found which replicate in the 3′ to 5′ direction. Since this is the case, how does DNA synthesis proceed sequentially from a single origin along both strands of DNA? DNA synthesis is *discontinuous*; that is, the strands are replicated in small fragments, called *Okazaki fragments*, in the 5′ to 3′ direction. These fragments are subsequently joined together by the enzyme *DNA ligase* as shown in Fig. 16-7.

In addition to the mechanism of DNA replication just described, the *initiation* of DNA replication requires a *primer*, a short sequence of RNA that is synthesized by *RNA polymerase* and is complementary to the DNA. In the presence of this primer, the DNA polymerase can start synthesizing deoxyribonucleotides. Once priming has occurred, the DNA polymerase then digests the RNA and replaces it with DNA. The participation of RNA

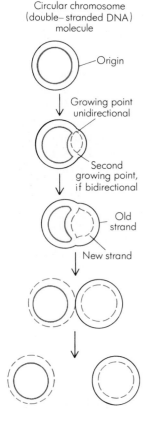

Circular chromosome
(double–stranded DNA)
molecule

Origin

Growing point
unidirectional

Second
growing point,
if bidirectional

Old
strand

New strand

Figure 16-6. The replication of the circular (double-stranded DNA) chromosome in a bacterium. Synthesis of DNA at the growing point is shown in greater detail in Fig. 16-7.

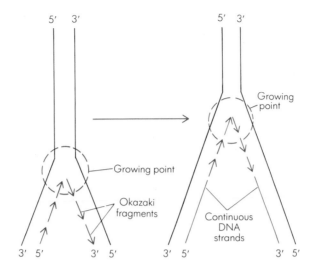

Figure 16-7. The discontinuous synthesis of DNA. Short segments, Okazaki fragments, of DNA are synthesized in the 5' to 3' direction by DNA polymerase. These fragments are represented by short color arrows. Subsequently the fragments are linked together by ligase to form continuous strands as the growing point moves farther along the DNA.

as primer appears to be extensive since each Okazaki fragment also contains a portion of RNA as primer.

When the growing points have moved along the entire length of the DNA molecule, two complete DNA molecules are formed (see Fig. 16-6). Each double-stranded molecule contains one of the original strands and one new strand; this mode of replication is termed *semiconservative*.

The biosynthesis of proteins

The building blocks of proteins

Just as nucleotides are building blocks of DNA, amino acids are building blocks of proteins. However, DNA consists of only four kinds of nucleotides, whereas proteins consist of about 20 kinds of amino acids (Table 15-1). Microorganisms differ widely in their ability to synthesize amino acids. For example, *Escherichia coli* can synthesize all of the amino acids required for protein synthesis, but lactic acid bacteria cannot and must be supplied with preformed amino acids.

There are thousands of different proteins in a bacterial cell. Each type of protein has its own specific sequence of amino acids and three-dimensional structure. The amino acids are joined together by *peptide bonds* to form a long chain. A peptide bond is formed as shown:

$$\underset{\text{Amino acid I}}{\underset{NH_2}{\overset{H \quad O}{R-C-C-OH}}} + \underset{\text{Amino acid II}}{\underset{\underset{OH}{C=O}}{\overset{H \quad H}{H-N-C-R'}}} \rightarrow \underset{\text{Dipeptide}}{\underset{\underset{OH}{\underset{C=O}{NH_2}}}{\overset{\overset{\text{Peptide bond}}{\overbrace{\qquad}}}{\overset{H \quad O \quad H \quad H}{R-C-C-N-C-R'}}}} + \underset{\text{Water}}{H_2O}$$

The chain of amino acids formed when a large number of amino acids are

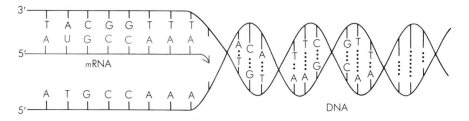

Figure 16-8. Transcription of DNA by DNA-dependent RNA polymerase. The chain of mRNA is being synthesized from the information in DNA. Note how complementarity is maintained. For example, where there is a G in DNA, a C is inserted in mRNA and where there is an A in DNA, a U is inserted in mRNA. The mRNA will serve as a pattern for protein synthesis (see Fig. 16-9). The growth of a mRNA chain proceeds in a 5'→3' direction in a manner similar to DNA replication. (A, adenine; T, thymine; C, cytosine; G, guanine; U, uracil.)

joined together by peptide bonds is called a *polypeptide chain*. Proteins consist of one or more polypeptide chains.

Transcription
Protein synthesis takes place on the ribosomes, which are large RNA-protein particles in the cytoplasm of the bacterial cell. (In eukaryotic cells the ribosomes are attached to the endoplasmic reticulum.) *Ribosomal RNA (rRNA)* consititutes about 90 percent of the total cellular RNA. Before protein synthesis can proceed, the coding of DNA must first be transferred to a substance that passes information from the DNA in the nuclear region to the ribosomes in the cytoplasm. This substance is known as *messenger ribonucleic acid (mRNA)*. The process or step in which a single-stranded mRNA is synthesized complementary to one DNA strand is called *transcription*. The synthesis of the polynucleotide chain of mRNA is catalyzed by the enzyme RNA polymerase. Just as they are required in DNA synthesis, activated ribonucleotides are also required as substrates for this enzyme.

Transcription is the first step in gene expression. This process, as shown in Fig. 16-8, involves separation of the two DNA strands, one of which serves as a template for the synthesis of a complementary strand of mRNA by DNA-dependent RNA polymerase. The strand of DNA selected for transcription is the one containing a specific *initiation* site and it is called the "sense" strand. Termination of mRNA synthesis is at those specific points along the DNA molecule which are recognized by RNA polymerase.

Translation
Translation, the next step in gene expression, is the process in which the genetic information now present in the mRNA molecule directs protein synthesis.

When the four different bases of the nucleotides of mRNA are arranged in sequences of three, each base *triplet*, called a *codon*, is capable of specifying a particular amino acid. Since there are four different bases, the number of sequences of three of them are 4^3, or 64, different sequences possible. These base triplets, each of which specifies a particular amino acid, constitute the genetic code (Table 16-1). The code is probably universal for all species of living organisms.

How is this code translated? Using Table 16-1, suppose the base sequence of *m*RNA is

CUUAGAAAAUUUAGUGGGACUUCU

The translation of this code into amino acids in a polypeptide chain, at a ribosome, would be

Leu-Arg-Lys-Phe-Ser-Gly-Thr-Ser

Five of the triplets, or codons, do not specify any amino acids. Of these, AUG and GUG are polypeptide-chain-initiating codons, and UAA, UAG, and UGA are polypeptide-chain-terminating codons. The latter three are called *nonsense codons*.

Another distinctive property of the genetic code is that the same amino acid may be coded for by more than one codon; that is, the code is *degenerate*. Furthermore, no "punctuation," or signal, is necessary to indicate the end of one codon and the beginning of the next. Therefore, the reading frame, or the sequence in which the genetic code is deciphered, must be correctly set at the beginning of the readout of a *m*RNA molecule. Reading then moves sequentially from one triplet to the next one without pause. If the reading frame is incorrectly set in the beginning, all codons will be out of step and lead to the formation of a *missense protein* with a deranged amino acid sequence (see Chap. 17).

The events occurring from DNA to RNA to protein is shown in Fig. 16-9.

Table 16-1. The genetic code for the base triplets of *m*RNA and the amino acids they code for*

FIRST BASE	SECOND BASE							THIRD BASE
	U		C		A		G	
U	UUU UUC	Phenylalanine	UCU UCC	Serine	UAU UAC	Tyrosine	UGU UGC	Cysteine	U C
	UUA UUG	Leucine	UCA UCG		UAA UAG	"Ochre" "Amber"	UGA UGG	"Umber" Tryptophan	A G
C	CUU CUC CUA CUG	Leucine	CCU CCC CCA CCG	Proline	CAU CAC	Histidine	CGU CGC CGA CGG	Arginine	U C
					CAA CAG	Glutamine			A G
A	AUU AUC AUA	Isoleucine	ACU ACC ACA ACG	Threonine	AAU AAC	Asparagine	AGU AGC	Serine	U C
	AUG	Methionine			AAA AAG	Lysine	AGA AGG	Arginine	A G
G	GUU GUC GUA GUG	Valine	GCU GCC GCA GCG	Alanine	GAU GAC	Aspartic acid	GGU GGC GGA GGG	Glycine	U C
					GAA GAG	Glutamic acid			A G

*The codons read in the 5' to 3' direction (left to right) on the *m*RNA. Codons UAA (ochre), UAG (amber), and UGA (umber) cause termination of synthesis of a protein chain. AUG and GUG are chain-initiating codons. Note that the same amino acid may be coded for by more than one codon (such a code is called *degenerate*). But no codon codes for more than one amino acid.

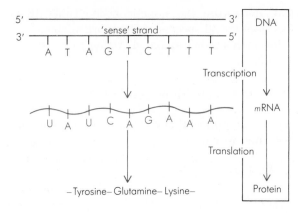

Figure 16-9. The events from DNA to *m*RNA to protein.

This will introduce us to a more detailed discussion of protein synthesis, which is a very elaborate biosynthetic process.

The process of protein synthesis

The first step in protein synthesis is the activation of amino acids. The amino acids are activated by amino acid–activating enzymes called *aminoacyl-tRNA synthetases*. This activation reaction requires energy in the form of ATP:

Amino acid + ATP   +   E   ⇌ amino acid-AMP-E +   P-P

Amino acid-activating enzyme   Aminoacyl-adenylate-E   Pyrophosphate

There is a specific activating enzyme for each kind of amino acid. The activated amino acids remain tightly bound to the enzyme after activation.

Next the activated amino acid binds to an RNA molecule called *transfer RNA* (*tRNA*), by the same enzyme now bound to the amino acid:

Amino acid-AMP-E + *t*RNA ⇌ amino acid-*t*RNA + AMP + E

The *t*RNA functions in protein synthesis to carry amino acids to, and recognize codons in, *m*RNA.

Transfer RNA is a single chain of about 80 nucleotides that is folded back upon itself and held in a cloverleaf arrangement by means of hydrogen bonding due to complementary base pairing. The general structure of *t*RNA is shown in Fig. 16-10. Three of the unpaired bases in *t*RNA form an *anticodon* triplet which specifically recognizes the complementary codon in *m*RNA for a specific amino acid. The terminal sequence of nucleotides is adenylic-cytidylic-cytidylic (ACC) and is found in all *t*RNA. The amino acid to be carried is linked to the terminal nucleotide containing adenine.

Like *m*RNA, *t*RNA is transcribed from a certain region of the DNA molecule by RNA polymerase. The *t*RNA molecules function as "adapters" into which specific amino acids are "plugged" so that they can be adapted to the nucleotide triplet language of the genetic code which is transcribed on the *m*RNA.

The *t*RNA now carries the amino acid to the *m*RNA attached to the

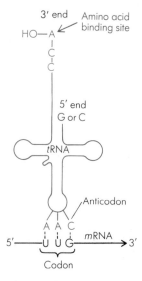

Figure 16-10. The cloverleaf structure of *t*RNA.

surface of the ribosome. Here the amino acid is added to a growing polypeptide chain. The surface of the ribosome may be viewed as the assembly point for protein synthesis.

Assembly of the protein chain on the ribosome. As mentioned before, the ribosome is the site of protein synthesis. Its *r*RNA is transcribed from certain portions of DNA by the same energy-requiring process used for synthesis of *m*RNA and *t*RNA. A ribosome is analogous to a tape-playing machine; just as the latter will produce any kind of sound, depending on the tape played, the ribosome will manufacture any kind of protein, depending on the kind of *m*RNA supplied.

The ribosomes of *Escherichia coli* have been studied extensively and have been found to consist of two subunits, each of which is identified by a sedimentation constant (S) determined by ultracentrifugation studies. The larger subunit is a 50-S particle, while the smaller subunit is a 30-S particle. Ribosomal subunits may associate or dissociate with each other. A 30-S and a 50-S subunit associate to form a 70-S ribosome. (S is the *Svedberg unit*, a measure of how fast a particle sediments during ultracentrifugation. The

Figure 16-11. Synthesis of a protein chain on a ribosome.

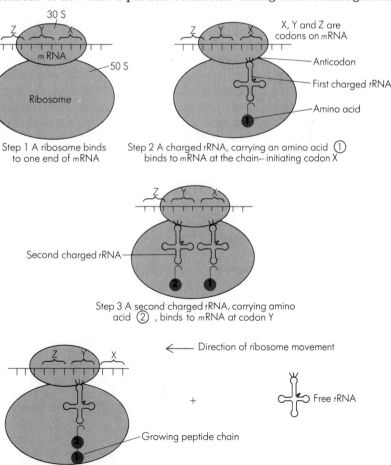

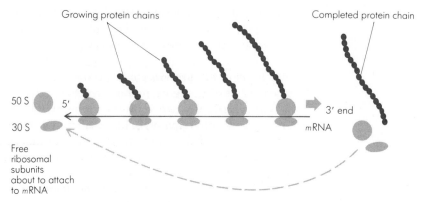

Growing protein chains

Completed protein chain

50 S

30 S

Free
ribosomal
subunits
about to attach
to mRNA

5'

3' end

mRNA

Figure 16-12. Schematic representation of a polysome during protein synthesis. The mRNA moves from right to left. The ribosomes move from the 5' end to the 3' end (left to right) of the mRNA, all reading simultaneously. In procaryotes, the life-span of mRNA is very short, only about 2 min.

association of a 30-S and a 50-S subunit to form a 70-S ribosome shows that the sedimentation behavior of the 70-S ribosome is not a simple addition of the units of the smaller particles.)

The synthesis of a protein chain on a ribosome is carried out as follows:

Step 1 A ribosome binds to one end of a mRNA molecule at a specific site. (The specificity here is important because it starts the translation of the mRNA in the correct reading sequence. See Fig. 16-11.)

Step 2 A charged tRNA carrying the first amino acid molecule then attaches to the chain-initiating codon (X) of the mRNA.

Step 3 Another tRNA carrying the second amino acid binds at the next codon (Y).

Step 4 The amino group of the amino acid on the second tRNA reacts with the active terminal carboxyl group on the amino acid of the first tRNA to form a dipeptide; the first tRNA is then released. The mRNA is moved along the ribosome to position the next codon (Z) in readiness for the tRNA carrying the third amino acid.

The process is continued until the peptide chain is complete. Termination takes place at one of the nonsense codons UAA, UAG, or UGA, and the chain dissociates from the last tRNA molecule.

A single molecule of mRNA is long enough for several ribosomes to read the molecule at the same time. When a number of 70-S ribosomes are actively engaged in protein synthesis on a strand of mRNA, this is called a *polysome* (see Fig. 16-12).

Summary and outlook

The genetic code is carried by DNA and it is passed on from generation to generation by DNA replication. The deciphering and use of the genetic code in the production of proteins, which are essential for cell growth and activity, involve two stages, transcription and translation. Transcription is the transfer of information from DNA to RNA, and translation is the translation of this information from RNA into protein.

Much remains to be discovered about the more intricate processes of DNA replication, transcription, and translation—the exact functions of some enzymes, for example. Each amino acid in a protein is coded for by a triplet of bases, or codon, and until recently we believed that each string of codons was read only one way. However, a gene was discovered in the chromosome of the bacteriophage φX174 that is read two different ways and thus produces two entirely different proteins. This discovery was prompted by an apparent discrepancy between the amount of DNA contained in the bacteriophage and the quantity of proteins it synthesizes. In other words, there is insufficient DNA to account for the 10 different proteins that the virus codes for, according to the traditional "one gene–one protein" rule. It was found that the gene for one protein was wholly contained within the gene that codes for another protein. Specifically, the first gene was the same as the last 60 percent of the second gene. The two genes are also read in different sequences. The smaller gene is read in a sequence shifted by one frame.

Discovery of these *overlapping genes* undermines the one gene–one protein concept. It remains to be seen how widely disseminated overlapping genes are among living cells and viruses. Discoveries of this nature will lead to a greater knowledge of how cellular activity is controlled.

Key terms

adenine	overlapping gene
anticodon	peptide bond
antiparallel	phosphodiester linkage
bidirectional replication	polynucleotide
codon	polypeptide
complementary base pair	polysome
cytosine	primer
deoxyribonucleic acid	purine
deoxyribose	pyrimidine
discontinuous synthesis	replication
DNA ligase	replication origin
DNA polymerase	ribonucleic acid
gene	ribose
growing point	ribosomal RNA
guanine	ribosome
hydrogen bond	RNA polymerase
inheritance	semiconservative replication
initiation site	Svedberg unit
messenger RNA	thymine
missense protein	transcription
nonsense codon	transfer RNA
nucleoside	translation
nucleotide	unidirectional replication
Okazaki fragment	variability

Questions   1 Give four reasons why bacteria are experimentally more useful than higher organisms for the study of genetics.
2 Why are the two strands in a DNA molecule said to be *antiparallel*?
3 Differentiate between a nucleoside and a nucleotide and explain how nucleotides are the building blocks of DNA.
4 Why do you think DNA base ratios are used in the characterization of microorganisms?
5 Describe the events that occur at the growing point during DNA replication.
6 Explain why the genetic code is said to be degenerate and how it determines the initiation and termination of protein synthesis.
7 Describe the different kinds of RNA in a cell and give their functions.
8 Summarize the four steps in the formation of a protein chain on a ribosome.

GENETICS OF BACTERIA

Genetics is a dynamic and rapidly advancing discipline. Its study is being pursued by thousands of scientists around the world. Genetic engineering, a new facet of genetic studies, offers society both the promise of beneficial developments and the possibility of disastrous consequences. Consider, for example, the hope of conquering all genetic diseases and the chances of changing a common, harmless microbe into a highly pathogenic form.

It was only a little more than a hundred years ago that the first serious study of genetics was undertaken by the Austrian botanist Gregor Mendel in his pea patch. In the 1860s he crossed strains of peas and studied the results of these crosses—changes in color, shape, size, and other properties of the peas. From these studies, he developed the basic laws of heredity.

At about the same time, a British naturalist named Charles Darwin introduced his theory of evolution. The theory is based on the principles of natural selection and survival of the fittest. It is obvious that in a changing environment over time, only those organisms that could adapt genetically to the changing environment would survive. Therefore, the ability to acquire new genetic traits confers a survival advantage to such organisms.

The laws of heredity are common to all life forms. They apply to humans and to the once most popular experimental organism in genetic studies, the fruit fly *Drosophila*. This tiny insect was easily and conveniently studied. Changes in the color of its eyes yielded much information on the location of genes in its chromosomes. However, about thirty years ago, *Drosophila* was replaced by the bacterium *Escherichia coli* as the favorite genetic research organism because it is even more easily and conveniently studied. *Escherichia coli* is now genetically the best understood of all organisms at the molecular level and is the organism of choice for many geneticists. This turn of events fostered the development of the field of *microbial genetics*. The microorganisms studied in this field include bacteria, yeasts, molds, and even viruses. This chapter will deal only with the genetics of bacteria.

The inheritance of characteristics and variability

A characteristic of all forms of life, from the standpoint of genetics, is the general stability or "likeness" in the characteristics of progeny and parent. We readily observe in our own species, for example, that some families regularly have black hair, brown eyes, and a certain shape of nose and chin, whereas other families have blond hair, blue eyes, and a different facial structure. In the same way, and in spite of their small dimensions, bacteria and other protists also transmit characteristics to their progeny. The very fact that we can identify species and even strains of bacteria implies that they are capable of transmitting genetic information from generation to generation with great accuracy.

However, in addition to the inheritance of characteristics, which accounts for the *constancy* exhibited by biological species, there is *variability* or change expressed in the progeny. These changes are associated with two fundamental properties of the cell or organism, namely, the *genotype* and the *phenotype*. The genotype refers to the genetic constitution of the cell. The phenotype is the expression of the genotype in observable properties characteristic of the cell or organism.

The genotype of a culture of cells remains relatively constant during growth. However, it can change by mutation. This change can result in an alteration in the observable properties, or phenotype, of the cells. So the genotype represents the heritable total potential characteristics of a cell, whereas the phenotype represents the characteristics expressed.

Phenotypic changes due to exvironmental changes

Bacteria, like the cells of higher organisms, carry more genetic information—their genotype—than is utilized or expressed at any one time. The extent to which this information is expressed depends on the environment. For instance, a facultative anaerobic bacterium will produce different end products of metabolism, depending on the presence or absence of oxygen during growth. The presence or absence of oxygen determines which enzymes function and which do not. Indeed in a given environment, knowledge of the factors which regulate genetic activity constitutes a very important aspect in the understanding of cell metabolism.

The outstanding feature of this type of phenotypic change is that it involves most of the cells in the culture. A phenotypic change of this type is not inherited; rather, it occurs when some condition of the environment changes. A return to the original phenotype occurs when the original environmental conditions are restored.

Figure 17-1 shows some phenotypic changes due to alterations in environmental conditions.

Genotypic changes

The genotype of a cell is determined by the genetic information contained in its chromosome. The chromosome is divided into *genes*. A gene is a functional unit of inheritance; it specifies the formation of a particular polypeptide as well as various types of RNA. Each gene consists of hundreds of nucleotide pairs. For instance, if a polypeptide chain contains 300 amino acids, then the gene coding for this polypeptide must contain 900 base pairs (three bases for each amino acid). This is the basis for the one gene–one polypeptide relationship. It has been estimated that the bacterial chromosome has the capacity to code for approximately 3,000 different proteins.

However, any gene is capable of changing or *mutating* to a different form so that it specifies formation of an altered or new protein which may in turn damage the characteristics of the cell (sometimes leading to its death). For example, the substitution of even one amino acid among several hundred in a polypeptide chain may cause the protein to be nonfunctional. A *mutation* is a change in the nucleotide sequence of a gene. This gives rise to a new genetic trait, or a changed genotype. A cell or an organism which shows the effects of a mutation is called a *mutant*. Thus we occasionally see sudden changes in familiar plants and animals. Now and then, an albino cat appears in a black litter, or a yellow pea appears among many green peas. The same sort of phenomenon occurs among microorganisms.

Mutations are rare events which occur at random and arise spontaneously with no regard to the requirements of the environment. Spontaneous

Figure 17-1. Morphological change in a strain of *Bacillus subtilis*. This phenotypic change resulted from a shift of growth temperature from 30 to 45°C (86 to 113°F). (A) 0 min at 45°C. (B) After 80 min at 45°C. (C) After 265 min at 45°C. (×3,280.) *(Courtesy of H. J. Rogers, The National Institute for Medical Research, Mill Hill, England.)*

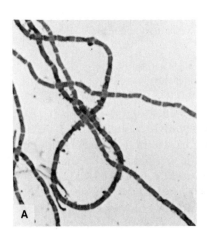

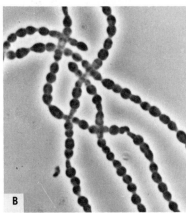

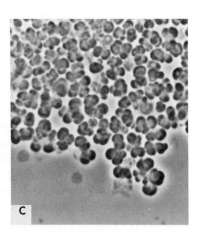

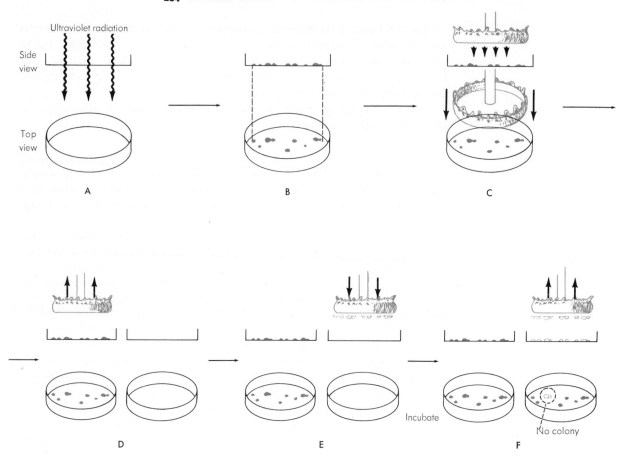

Figure 17-2. Replica plating is used for isolating nutritional mutants of *E. coli.* (A) Bacterial suspension placed in open half of petri dish and exposed to mutagenic agent, such as ultraviolet radiation. (B) Sample from (A) plated on surface of a "complete" medium such as nutrient agar. The plate is incubated; after incubation, the exact position of colonies on the plate is noted. (C) A sterile replica plating unit is gently pressed to the surface of plate (B), then raised (D), and then pressed to the surface of a plate of "minimal" agar medium (E). The positioning of the replica plating unit on the minimal agar must be precise, so that colony locations will be comparable on each of the two plates. The plates will be replicas of one another. The minimal agar in the plate in (E) consists of inorganic salts and glucose, nutrients which normally permit growth of *E. coli.* After incubation (F), colonies appear on the new plate at most, but not all, of the positions corresponding to locations of colonies on the original plate. It may be assumed that the organisms that failed to develop are nutritional mutants; that is, they are not able to grow on an inorganic salts-glucose medium, a characteristic which they originally possessed.

bacterial mutations may occur at a rate of only 1 mutation in 1 million bacterial cells to a rate of 1 mutation in 10 billion bacterial cells. Generally, the mutants in a cell population are masked by the greater numbers of unmutated cells. Isolating a mutant cell is like looking for the proverbial needle in a haystack. However, microbiologists have developed techniques which facilitate isolation of the few mutants from a large population of nonmutated (*wild-type*) cells. For example, an antibiotic can be incorporated in a medium to select for antibiotic-resistant mutants grown on the medium. Figure 17-2 shows a unique method, the *replica plating technique*, for the isolation of nutritional mutants.

Types of mutations

At the molecular level there are several ways in which changes in the purine-pyrimidine base sequence of a gene can occur, resulting in mutation. Two common types are *point mutations* and *frameshift mutations*.

Point mutations. Point mutations occur as a result of the substitution of one nucleotide for another in the specific nucleotide sequence of a gene. The substitution of one purine for another purine or one pyrimidine for another pyrimidine is termed a *transition* type of point mutation. A *transversion* is the replacement of a purine by a pyrimidine, or vice versa. This *base-pair substitution* may result in one of three kinds of mutations affecting the translation process:

1 The altered gene triplet produces a codon in the *m*RNA which specifies an amino acid different from the one present in the normal protein. This mutation is called a *missense mutation*. Such a protein may be functionally inactive or less active than the normal one. A good example of a missense mutation in humans is the disease sickle-cell anemia. A single base substitution in the codon for the sixth amino acid of normal hemoglobin A changes the sixth amino acid from glutamic acid to valine, thus forming the characteristic hemoglobin S of sickle-cell anemia. That is, GAG, which codes for glutamic acid, could have been changed to GUG for valine. Under low oxygen concentration the altered hemoglobin S molecules stack into crystals, giving the red blood cells a sickle shape.
2 The altered gene triplet produces a chain-terminating codon in *m*RNA, resulting in premature termination of protein formation during translation. This is called a *nonsense mutation*. The result is an incomplete polypeptide which is nonfunctional.
3 The altered gene triplet produces a *m*RNA codon which specifies the same amino acid because the codon resulting from mutation is a synonym for the original codon. This is a *neutral mutation*.

Frameshift mutations. These mutations result from an *addition* or *loss* of one or more nucleotides in a gene. This results in a shift of the reading frame. We saw in Chap. 16 that during protein synthesis the reading of the genetic code starts from one end of the protein template, *m*RNA, and is read in consecutive blocks of three bases. Frameshift mutations, therefore, generally lead to nonfunctional proteins, because an entirely new sequence of amino acids is synthesized from a frameshift reading of the nucleotide sequences of *m*RNA (which was transcribed from a mutation in the DNA of the cell). This type of mutation is illustrated in Fig. 17-3.

How mutations occur

Mutations most commonly occur during DNA replication. Some mutations occur as the result of damages inflicted by ultraviolet (UV) light or x-rays.

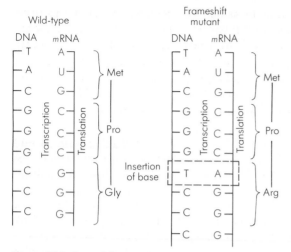

Figure 17-3. Frameshift mutation, as a result of insertion of a nucleotide in a gene. Insertion of a nucleotide in a gene results in the transcription of an additional nucleotide in mRNA. This results in a frameshift when codons are read during translation so all codons following the insertion are altered and all amino acids coded for are changed. A frameshift mutation as a result of deletion of a nucleotide would have essentially the same effect.

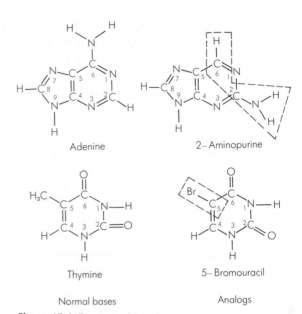

Figure 17-4. Two normal DNA bases and two base analog mutagens. 2-Aminopurine is an analog of adenine and can pair with thymine or cytosine. 5-Bromouracil is an analog of thymine and can pair with adenine or guanine. A dashed color box highlights the part of the analog which differs from the normal base.

Since these agents are an inescapable part of the environment (e.g., UV light is a component of sunlight), they probably account for many *spontaneous* mutations. However, mutation rates can be increased substantially by exposing a culture to such radiation. Any agent that increases the mutation rate is called a *mutagen*. Mutations obtained by use of a mutagen are said to be *induced*, rather than spontaneous, though they may differ only in frequency, not in kind. For example, UV light causes mutation under both natural and laboratory conditions. The number of mutants obtained by laboratory conditions is much higher, however, because of the high UV light dosage employed.

No one specific mechanism can be proposed for the mutagenic effect of x-rays. Since they can cause the rupture of many different kinds of chemical linkages, they probably damage DNA in a variety of ways. The major effect of UV light is to cause the formation of *dimers* by cross-linking between adjacent pyrimidine, especially thymine, residues. These cross-linked residues disrupt the normal process of replication.

The most revealing findings about mutation in recent years have come from studies on the mutagenic effects of various chemicals. There are two main types of mutagenic chemicals. The first consists of compounds that can react chemically with DNA. Since specificity of DNA replication depends upon purine-pyrimidine bonding, which results from hydrogen bonding between the amino and hydroxyl groups of the purines and pyrimidines, chemical modification of these amino and hydroxyl groups can cause mutation. Nitrous acid, which can remove amino groups from purines and pyrimidines, is such a mutagen. The second type of mutagenic chemicals consists of *base analogs*. These are chemicals sufficiently similar in structure to normal DNA bases to be substituted for them during DNA replication (Fig. 17-4). Although similar in structure, base analogs do not have the same hydrogen-bonding properties as the normal bases. They can therefore introduce errors in replication which result in mutations.

How mutations are repaired

We have said that DNA damage can occur by UV radiation, x-rays, and certain chemicals. Fortunately, cells contain specific enzymes which can repair damaged DNA. In this way, some affected cells can continue to function normally.

Many kinds of bacterial cells and yeasts have been shown to possess an efficient *photoreactivating* mechanism for repairing damage caused by UV radiation. This *photoreactivation* occurs when cells exposed to lethal doses of UV light are immediately exposed to visible light. A special enzyme, promoted by visible light, splits or unlinks the dimers formed because of exposure to UV light and restores normal activity to the damaged cells.

Some bacteria have enzymes, called *endonucleases* and *exonucleases*, that *excise* or cut out a damaged segment of DNA. Then the other enzymes, polymerases and ligases, repair the resulting break by filling in the gap and joining the fragments together. This mechanism is illustrated in Fig. 17-5.

Several repair-type diseases occur in humans. The most well known is *xeroderma pigmentosum*. In this genetic disease, the skin is extremely sensitive to UV light. In these individuals, minimal exposure to sunlight causes severe burns which eventually may lead to cancer of the skin. The defect involved is a faulty excision-repair mechanism, that is, an inability to excise pyrimidine dimers formed from exposure to UV light.

Figure 17-5. Excision repair of UV-light–damaged DNA containing a thymine-thymine dimer.

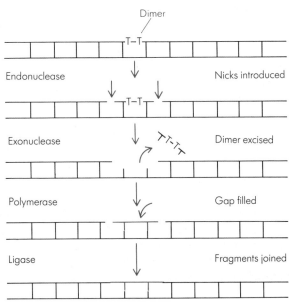

It may be noted that when DNA-containing pyrimidine dimers undergo replication in excision-repair defective bacteria, the synthesis of the new strand stops at the position of the dimer but continues at some distance after it. Thus a gap occurs in the newly synthesized DNA strand. These gaps can be filled by exchanging and combining fragments with accumulated DNA strands whose gaps do not overlap. Further, since replication is interrupted and therefore hindered at these gaps, strands with fewer gaps replicate faster and can evenutally dilute out damaged DNA.

Mutation rate

The rate of mutation is the probability that a gene will mutate at any particular cell division. Thus the mutation rate is generally defined as the average number of mutations per cell per division. It is expressed as a negative exponent per cell division. For example, if there is one chance in a million that a gene will mutate when the cell divides, the mutation rate for any single gene equals 10^{-6} per cell division. Generally, the mutation rate for any single gene ranges between 10^{-3} and 10^{-9} per cell division.

The mutation rate has some practical implications. Since genes mutate at random and independently of each other, the chance of two mutations in the same cell is the product of the single mutation rates of each. So, for example, if the mutation rate to penicillin resistance is 10^{-8} per cell division and that to streptomycin resistance is 10^{-6} per cell division, the probability that both mutations will occur in the same cell is $10^{-6} \times 10^{-8}$, or 10^{-14}. This mutation rate is very low. For this reason, it is a common practice to give two antibiotics simultaneously in the treatment of some diseases. For example, a combination of penicillin G and streptomycin has been of proven value in treating streptococcal infections. A cell which has become resistant to one antibiotic is still likely to be inhibited or killed by the other.

Types of bacterial mutants

Since all properties of living cells are ultimately gene-controlled, any cell characteristic may be changed by mutation. A large variety of bacterial mutants have been isolated and studied intensively. Some of the major types of mutants are as follows:

1 Mutants that exhibit an increased tolerance to inhibitory agents, particularly antibiotics (antibiotic- or drug-resistant mutants)
2 Mutants that demonstrate an altered fermentation ability or increased or decreased capacity to produce some end product
3 Mutants that are nutritionally deficient; that is, they require a more complex medium for growth than the original culture from which they were derived
4 Mutants that exhibit changes in colonial form or ability to produce pigments
5 Mutants that show a change in the surface structure and composition of the microbial cell (antigenic mutants)
6 Mutants that are resistant to the action of bacteriophages
7 Mutants that exhibit some change in morphological features, for example, the loss of ability to produce spores, capsules, or flagella (Fig. 17-6).

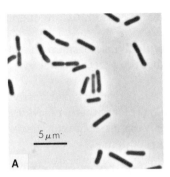 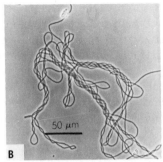 

Figure 17-6. Some mutants exhibit morphological changes. Mutants of *Bacillus subtilis* that are grossly deficient in the enzymes needed to separate daughter cells after division grow at normal rates as very long chains of unseparated cells. Under certain growth conditions these mutants also form helical structures. (A) Wild-type, phase-contrast photomicrograph; (B) mutants, phase-contrast photomicrograph; (C) mutants, scanning electron micrograph. Note helical structures in B and C. *(Courtesy of Jared E. Fein, McGill University.)*

It is evident from this list that all the characteristic features of bacteria are subject to alteration by the process of mutation. It is also apparent that some of the specific changes caused by mutation are similar or the same as those resulting from a change in environmental conditions. It is therefore necessary to ascertain experimentally that a change is really due to mutation and not to a response to the environment.

There are many practical implications associated with the occurrence of microbial mutants. The following examples illustrate this.

1 Some microorganisms are known to develop resistance to certain antibiotics because of mutation. This fact is of great importance in the treatment of disease, since antibiotics originally effective for the control of a bacterial infection become less effective or ineffective as antibiotic-resistant mutants appear (Fig. 17-7).
2 It is possible to isolate biochemical mutants capable of producing large yields of an end product. This is important in industry. For example, the

Figure 17-7. Antibiotics originally effective for control of a bacterial infection become less effective or ineffective as antibiotic-resistant mutants appear. Note how the use of antibiotics (color screen) actually selects for bacteria that have become antibiotic resistant due to chromosomal mutation. Bacteria sensitive to the antibiotic are killed or prevented from reproducing. *(Courtesy of SANDORAMA 1978/I, Sandoz Ltd., Basel, Switzerland, and G. Lebek, University of Berne.)*

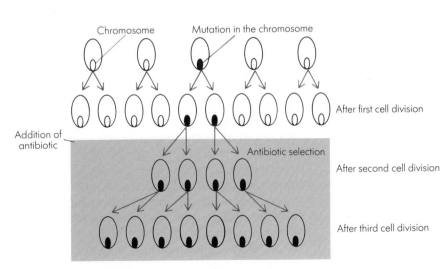

yield of penicillin in commercial production was dramatically increased through selection of mutant strains of *Penicillium*.

3 The maintenance of pure cultures of typical microorganism species requires that occurrence of mutation be prevented; otherwise, the culture will no longer be typical.

4 Microbial mutants have been extensively employed in the investigation of various biochemical processes, particularly biosynthetic reactions. For example, mutants with blocks or impairment at different enzymatic steps have been used to unravel metabolic sequences.

Many mutants, perhaps a majority, are able to revert to the wild-type condition by *reverse mutation*. This is a return to the original phenotype by the mutant cells. However, this may not *necessarily* be due to a precise reversal of the original mutation. Sometimes, the effect of the original mutation may be partially or entirely suppressed by a second mutation at a different site on the chromosome.

Bacterial recombination

Genetic recombination is the formation of a new genotype by reassortment of genes following an exchange of genetic material between two different chromosomes which have similar genes at corresponding sites. These are called *homologous chromosomes*. Progeny from recombination have combinations of genes different from those that are present in the parents. In bacteria, genetic recombination results from three types of gene transfer:

1 *Conjugation.* Transfer of genes between cells that are in physical contact with one another.

2 *Transduction.* Transfer of genes from one cell to another by a bacteriophage.

3 *Transformation.* Transfer of cell-free or "naked" DNA from one cell to another.

These three types of gene transfer are shown in Fig. 17-8.

In bacterial recombination the cells do *not* fuse, and usually only a portion of the chromosome from the *doner cell* (male) is transferred to the *recipient cell* (female). The recipient cell thus becomes a *merozygote*, a zygote that is a partial diploid. Once merozygote formation has occurred, recombination can take place.

The general mechanism for bacterial recombination is believed to take place as follows. Inside the recipient cell the donor DNA fragment is positioned alongside the recipient DNA in such a way that homologous genes are adjacent. Enzymes act on the recipient DNA, causing nicks and excision of a fragment. Then the donor DNA is integrated into the recipient chromosome in place of the excised DNA. The recipient cell then becomes the recombinant cell because its chromosome contains DNA of both the donor and the recipient cell. (The excised DNA pieces from the recipient chromosome are probably broken down by specific enzymes.) This general recombination mechanism is seen in Fig. 17-9.

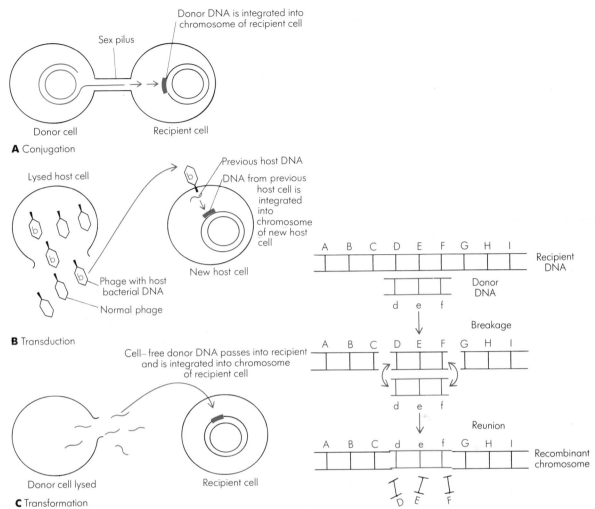

A Conjugation

B Transduction

C Transformation

Figure 17-8. Three types of gene transfer from which genetic recombination results. (A) Conjugation, the transfer of genes between cells in physical contact with each other, perhaps by a sex pilus; (B) transduction, the transfer of genes between cells by a bacteriophage; (C) transformation, the transfer of cell-free or ''naked DNA'' from one cell to another.

Figure 17-9. The breakage and reunion model of bacterial recombination. The donor DNA becomes integrated into the recipient DNA.

Conjugation

Conjugation, one of the three mechanisms for the transfer of genetic material from one bacterium to another, is dependent on cell-to-cell contact. While only very small fragments of the bacterial chromosome are transferred in transduction and transformation (to be discussed later), in conjugation it is possible for large segments of the chromosome, and in special cases the entire chromosome, to be transferred.

Bacterial conjugation was first demonstrated by Lederberg and Tatum in 1946. They combined two different mutant strains of *Escherichia coli* that lacked the ability to synthesize one or more essential growth factors and gave them an opportunity to mate. Then they plated the mixed culture on a minimal medium in which only wild-type strains could grow. When they

291

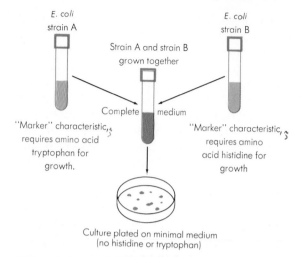

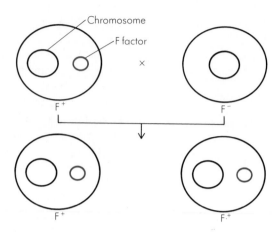

Figure 17-10. Experiments showing evidence of conjugation in bacteria as a method of genetic recombination. The A mutant strain of *E. coli* requires tryptophan for growth and the B mutant strain requires histidine for growth. Wild-type *E. coli* can produce both tryptophan and histidine. Growth of colonies of *E. coli* on the minimal medium after conjugation indicates that the conjugation of mutant strain A and B has resulted in recombinant wild-type strains.

Figure 17-11. During mating of an F+ and F− cell, the F+ cell replicates the sex or F factor and the copy is almost always transferred to the F− cell. Thus an F− cell usually becomes an F+ cell during mating.

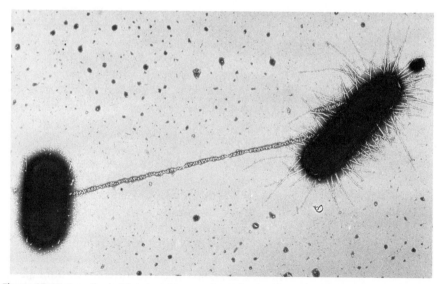

Figure 17-12. Sex pilus holding together a mating pair of *Escherichia coli*. The male cell (on the right) also has another type of pili besides the sex pilus. Small RNA bacteriophages adsorbed to the sex pilus may be seen as dots. (×25,000.) *(Courtesy of C. Brinton Jr., University of Pittsburgh.)*

found wild-type colonies, they knew that these must have been the result of genetic recombination by conjugation between the mutant strains. Figure 17-10 shows the principle of their experiments in a simplified form.

Sex factors. A clearer understanding of conjugation in bacteria came about with the discovery that there is sexual differentiation in *E. coli*; in other words, different mating types of the bacterium exist. Male cells contain a small circular piece of DNA, in the cytoplasm and not part of the chromosome, called the *sex* or *F factor* (fertility factor). These cells are referred to as F^+ and are donors in mating. Female cells are labeled F^- and lack this factor. They are recipient cells.

Crosses between two F^- strains do not yield recombinants. In $F^+ \times F^-$ crosses, the male replicates its sex factor, and one copy of it is almost always transferred to the female recipient. The F^- cell is converted to an F^+ cell and is itself capable of serving as a donor (see Fig. 17-11). Therefore, as long as the cells grow, the conjugation process can continue in an infectious way with repeated transfer of the sex factor. The transfer of the F factor in an $F^+ \times F^-$ cross occurs with a frequency which approaches 100 percent. But the formation of recombinants in an $F^+ \times F^-$ cross occurs at a low frequency—about one recombinant per 10^4 to 10^5 cells. Thus it is obvious that the transfer of the F factor is independent of the transfer of chromosomal genes.

Since the transfer of the F factor is independent, it follows that the F factor DNA replicates independently of the F^+ donor cell's normal chromosome. The F factor DNA is only sufficient to specify about 40 genes which control sex-factor replication and synthesis of sex pili. One or more sex pili are produced by each F^+ cell (Fig. 17-12). Sex pili seem to act to bind an F^- cell to an F^+ cell and then to retract into the F^+ cell, pulling the F^- cell into close contact. There is also some evidence that sex pili are tubules through which DNA passes from an F^+ to an F^- cell during conjugation.

High-frequency recombination strains. The study of conjugation in bacteria was made easier when new strains of cells were isolated from F^+ cultures

Figure 17-13. An Hfr cell arises from an F^+ cell in which the F factor becomes integrated into the bacterial chromosome. During mating of an Hfr and F^- cell, the F^- cell almost always remains F^-. This results because Hfr rarely transfers F factor to the F^- cell.

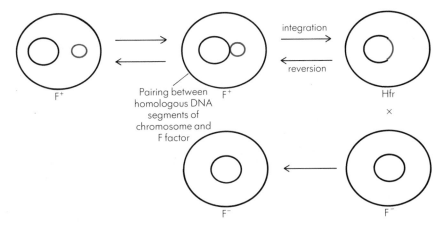

Figure 17-14. Simplified linkage map (color) of the circular chromosome of *E. coli* constructed from interrupted mating experiments using different Hfr strains. The arrows on the linkage map indicate the leading end and direction of entry of the chromosomes injected by each of the Hfr strains, the designations of which are shown inside the circle. This is determined by the position of the F factor in each of the Hfr strains. The numbers around the outside of the map show distances, in minutes, based on time of entry of each codon in experiments. (Note that the map distances in minutes are drawn relative to the Hfr strain H.)

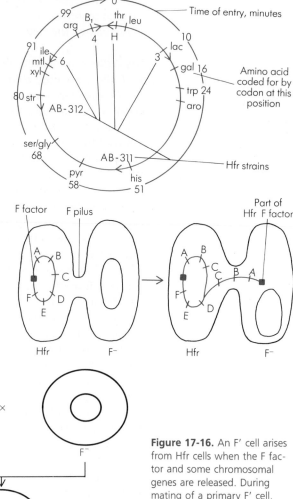

Figure 17-15. Mechanism for DNA transfer between Hfr and F⁻ cells. The Hfr chromosome begins replicating at the point of insertion of the F factor. Since the F factor can integrate in different positions of the Hfr bacterial chromosome, the first genes to enter an F⁻ cell will vary with different Hfr strains. As shown, the order of genes transferred is ABCDEF. In another Hfr strain, the F factor might be integrated between B and C. In this case the order of genes transferred would be CDEFAB.

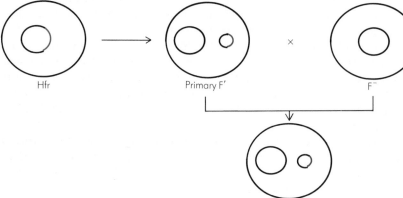

Figure 17-16. An F′ cell arises from Hfr cells when the F factor and some chromosomal genes are released. During mating of a primary F′ cell, with an f⁻ cell, the F⁻ cell becomes a secondary F′ cell because it now carries the F factor as well as the chromosomal genes from the primary F′ cell.

which underwent sexual recombination with F⁻ cells at a rate 10^8 times greater than F⁺ × F⁻ cells. These new donor strains were thus called *high-frequency recombination, or Hfr, strains.* Hfr cells arise from F⁺ cells in which the F factor becomes integrated into the bacterial chromosome. They differ from F⁺ cells in that the F factor of the Hfr is rarely transferred during recombination. Thus in an Hfr × F⁻ cross, the frequency of recombination is high and the transfer of F factor is low (Fig. 17-13); in an $F^+ \times F^-$ cross, the frequency of recombination is low and the transfer of F factor is high.

The order in which chromosomal material is transferred from an Hfr donor to an F^- recipient was determined by the *interrupted mating* experiments of Elie Wollman and François Jacob. An Hfr strain was mixed with an F^- strain, and at various times the conjugation was interrupted by breaking the cells apart in a high-speed blender. The cells were then plated on various types of selective agar media in order to select for recombinant cells which had received donor genes before mating was interrupted.

Interrupted mating experiments reveal the order of genes on a chromosome by the time of entry and the frequency of recombination of each *marker*, which is a detectable mutation serving to identify the gene at the *locus* or site where it occurs. Each gene enters the F^- cell at a characteristic time, and a *linkage map* of the Hfr chromosome can be constructed using time of entry as a measure. This is the principal method of learning where the genes are on a bacterial chromosome (Fig. 17-14). This is all possible because the Hfr chromosome is transferred to the F^- cell in a *linear* fashion (Fig. 17-15) even though it is a circular chromosome. During transfer the Hfr chromosome begins replicating at the point of insertion of the F factor. Since the F factor can integrate in different positions of the Hfr bacterial chromosome, the first genes to enter an F^- cell will vary with different Hfr strains (Fig. 17-15). This means that the integrated F factor serves as the point of chromosomal opening, and part of it serves as the origin of transfer. The 5′ end of the single DNA strand enters the F^- cell first. This origin of transfer is *not* the same as the chromosomal origin of DNA replication discussed in Chap. 16.

It takes about 100 min to inject a copy of the whole *E. coli* genome (i.e., the DNA strand comprising the F factor). Since conjugation is usually interrupted by accident before this can occur, the distal Hfr genes are rarely transferred. Since *all* the Hfr genes must be transferred in order to convert the recipient F^- cell to the male Hfr form, most recipients remain F^- after conjugation with Hfr cells.

Hfr cells can revert to the F^+ state. When this occurs, the sex factor is released from the chromosome and resumes its autonomous replication. Sometimes this detachment is not cleanly accomplished, so the F factor carries along with it some chromosomal genes. In this state it is termed an F′ factor, and the cell in which this has occurred is called an F′ cell (see Fig. 17-16). When such primary F′ cells are mated with F^- recipients, the sex factor is transferred very efficiently *together* with the added bacterial genes. The recipient cell then becomes a secondary F′ cell; it is a partial diploid for those genes it receives from the primary F′ cell. This process whereby bacterial genes are transmitted from donor to recipient as part of the sex factor has been termed *sexduction* by Jacob and Wollman (Fig. 17-16).

Extrachromosomal genetic elements

In addition to the normal DNA chromosome, *extrachromosomal genetic elements* are often found in bacteria. These elements are called *plasmids*

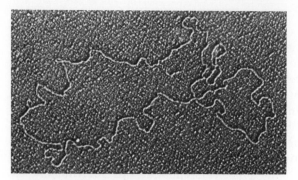

Figure 17-17. Bacterial plasmid shown as a molecule of looped DNA. The drug-resistant plasmid shown is called R28K, carries ampicillin resistance, and has a length of 21 μm. *(Courtesy of Michiko Egel-Mitani.)*

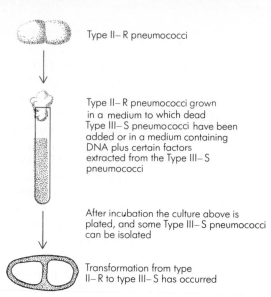

Type II– R pneumococci

Type II– R pneumococci grown in a medium to which dead Type III– S pneumococci have been added or in a medium containing DNA plus certain factors extracted from the Type III– S pneumococci

After incubation the culture above is plated, and some Type III– S pneumococci can be isolated

Transformation from type II– R to type III– S has occurred

Figure 17-18. Type transformation in pneumococci. Type II-R pneumococci (rough noncapsula-ted) are transformed into type III-S pneumococci (smooth capsulated) by incorporation of type III-S DNA into type II-R chromosome. The type III-S pneumococci, which possess capsules, are better able to resist being phagocytosed by white blood cells and so are more infectious.

and are capable of autonomous replication in the cytoplasm of the bacterial cell (Figs. 16-1 and 17-17). Plasmids are circular pieces of DNA that are extra genes. When an extrachromosomal element is capable of replicating autonomously and integrating into the bacterial DNA chromosome, it is called an *episome*. This behavior distinguishes an episome from a plasmid, because the latter does not integrate into the chromosome. Thus the F factor of *E. coli* is an episome because it can alternatively exist in the F⁺ or Hfr state.

Some bacteria have plasmids that are *bacteriocinogenic factors*. They determine the formation of *bacteriocins*, which are proteins that kill the same or other closely related species of bacteria. The bacteriocins of *E. coli* are called *colicins*; those of *Pseudomonas* spp. are called *pyocins*, and so on. Bacteriocins have proven useful for distinguishing between certain strains of the same species of bacteria in medical bacteriological diagnosis. Bacteria possess other kinds of plasmids called *resistance-transfer factors* or R factors which confer resistance to a number of antibiotics. Some of the R factors can be transferred to other cells by conjugation, hence the term *infectious resistance*.

Transformation Transformation is the process whereby *cell-free*, or "naked," DNA containing a limited amount of genetic information is transferred from one bacterial cell to another. The DNA is obtained from the donor cell by natural cell lysis or by chemical extraction. Once the DNA is taken up by the recipient cell, recombination occurs. Bacteria that have inherited markers from the donor cells are said to be transformed. Thus certain bacteria, when grown in the presence of dead cells, culture filtrates, or cell extracts of a closely related strain, will acquire and subsequently transmit a characteristic(s) of the related strain. This phenomenon is illustrated in Fig. 17-18, which shows the transformation of type II pneumococci into type III pneumococci. This "transforming principle" was identified as DNA by Avery, MacLeod, and McCarty in 1944. They defined DNA as the chemical substance responsible for heredity. The principal steps of transformation are shown in Fig. 17-19. Bacterial species that have been transformed include, besides *Streptococcus pneumoniae*, those in the genera *Bacillus*, *Haemophilus*, *Neisseria*, and *Rhizobium*.

After DNA entry into a cell, one strand is immediately degraded by deoxyribonucleases, while the other strand undergoes base pairing with a homologous portion of the recipient cell chromosome; it then becomes integrated into the recipient DNA (see Fig. 17-19). Since complementary base pairing takes place between one strand of the donor DNA fragment and a specific region of the recipient chromosome, only closely related strains of bacteria can be transformed.

Properties of recipient cells. Conditions suitable for uptake of donor DNA into recipient cells occur only during the late logarithmic phase of growth. During this period, the transformable bacteria are said to be *competent* to take up and incorporate donor DNA. Competent cultures probably produce an extracellular protein factor that apparently acts by binding or trapping donor DNA fragments at specific sites on the bacterial surface.

Figure 17-19. Principal steps in bacterial transformation.

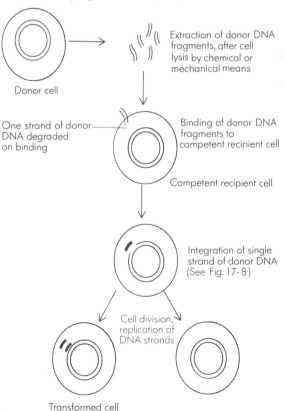

Donor cell

Extraction of donor DNA fragments, after cell lysis by chemical or mechanical means

One strand of donor DNA degraded on binding

Binding of donor DNA fragments to competent recipient cell

Competent recipient cell

Integration of single strand of donor DNA (See Fig. 17-8)

Cell division, replication of DNA strands

Transformed cell

Figure 17-20. Transduction. The phage P1 chromosome, after injection into the host cell, causes degradation of host chromosome into small fragments. During maturation of the virus particles, a few phage heads may envelop fragments of bacterial DNA instead of phage DNA. When this bacterial DNA is introduced into a new host cell, it can become integrated into the bacterial chromosome, thereby transferring several genes from one host cell to another.

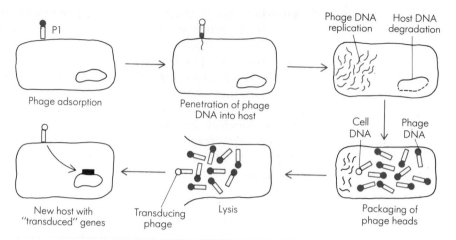

Phage adsorption

Penetration of phage DNA into host

Phage DNA replication   Host DNA degradation

Cell DNA   Phage DNA

Packaging of phage heads

New host with "transduced" genes

Transducing phage

Lysis

The significance of transformation as a natural mechanism of genetic change is questionable. It probably occurs following the lysis of a microbe and the release of its DNA into the environment. It is conceivable that transformation between bacterial strains of low virulence can give rise to transformed cells of high virulence. In any case, the phenomenon of transformation has proved to be extremely useful in genetic studies of bacteria in the laboratory, particularly in mapping the bacterial chromosome (because the frequency of transformation of two genes at the same time is an indication of the distance between these genes on the chromosome).

Transduction

Transduction is the process in which there is gene transfer from one bacterium to another by a bacteriophage. Some temperate bacteriophages (see Chap. 11), which do not ordinarily lyse the host cell, carry DNA that can behave as episomes in bacteria. When this viral DNA is integrated into the bacterial chromosome, it is called a *prophage*. When a lysogenic bacterium enters the lytic cycle, either spontaneously or because of induction by UV light, occasionally part of the bacterial chromosome can be accidentally assembled into the phage head. Upon subsequent infection of another bacterium by the virus, this bacterial chromosome fragment from the first cell is also injected as part of viral DNA. This chromosomal fragment undergoes base pairing with the host chromosome; it will integrate and become a permanent part of the infected bacterium's chromosome.

Generalized transduction. Generalized transduction occurs when a temperate phage transfers any of the genes of the bacterial chromosome or of a plasmid (Fig. 17-20). This is in contrast to *specialized transduction*, in which certain temperate phage strains can transfer only a few restricted genes of the bacterial chromosome. In generalized transduction, as the phage begins the lytic cycle, viral enzymes hydrolyze the bacterial chromosome into many small pieces of DNA. Any part of the bacterial chromosome may be incorporated into the phage head during phage assembly. For example, coliphage P1 can transduce a variety of genes in the bacterial chromosome. After infection a small proportion of the phages carry only bacterial DNA (see Fig. 17-20). Since this DNA matches the DNA of the new bacterium infected, this second bacterium will not become lysogenic for P1 phage. Instead, the injected DNA will be integrated into the chromosome of the recipient cell.

Transduction has been demonstrated in several bacterial species. This process provides a powerful tool for developing new bacterial strains, mapping bacterial chromosomes, and for many other genetic experiments.

Genetic engineering with microorganisms

Genetic engineering, or the biochemical manipulation of genes, became a reality in 1973 when techniques were developed for the isolation and joining of unlike pieces of DNA so that biologically active recombinant DNA molecules could be produced. These molecules, once produced, could be introduced into bacterial cells, such as *E. coli*, where they could replicate.

How is this done? Endonucleases can be used to cut short specific fragments of DNA, usually containing several genes, out of the chromosomes of any type of cell—plant, animal, or bacterial—or virus. These fragments have "sticky" or cohesive ends that will readily join together by base pairing of the single-stranded regions with other DNA strands. In this way, the fragments obtained from the chromosome of any cell or virion can be spliced into bacterial plasmids or phage genomes by means of other enzymes, such as polynucleotide ligases. The plasmids or phages can then replicate in their bacterial hosts. Sometimes new genetic material in plasmids or phage DNA also can integrate into the host-cell chromosome and act as a normal component of it. Essentially such bacterial cells have received foreign genes and are new organisms; their properties and capacities may differ profoundly from either host or donor. They are also self-perpetuating and can make multiple copies of themselves in a very short time. The main features of the biochemical manipulation of genes are shown in Fig. 17-21.

The importance of the new technology is the movement of genes across species lines, such as from animals and fruit flies to bacteria! This results in the creation of new, redesigned organisms.

There is concern that production of recombinant DNA molecules that are functional in vivo could prove biologically hazardous. If they are carried in a microbe like *E. coli*, which is a commensal bacterium in the human gut and can exchange genetic information with other types of bacteria, they might possibly become widely disseminated among human, bacterial, plant, or animal populations, with unpredictable results.

Of special concern is construction of new autonomously replicating bacterial plasmids that could, if not very carefully controlled, introduce genetic determinants for antibiotic resistance or bacterial toxin formation

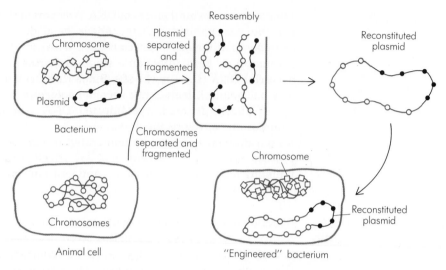

Figure 17-21. "Genetic engineering" with bacterial plasmids. The bacterial envelopes of cells are first dissolved in a detergent-like solution and the DNA is spilled out. The bacterial plasmids are then isolated from the chromosomal DNA in a centrifuge. These plasmids are cleaved with restriction enzymes and these opened plasmids are then mixed in a solution with genes (also removed by use of restriction enzymes) from the DNA of a plant, animal, virus, or bacterium which has also been extracted and opened. In the solution, DNA ligase cements the foreign genes into place in the opening of the plasmids resulting in recumbinant bacterial plasmids (loops). These reassembled recombinant bacterial plasmids are then placed in a cold calcium chloride solution containing bacterial cells. When the solution is suddenly heated, the envelopes of the bacterial cells become permeable allowing the recombinant plasmids to enter. Once inside the bacterial cells, they can remain as plasmids or can integrate with the host cell chromosome. [Adapted from BioScience **24**(12):692, 1974.]

into bacterial strains that do not presently carry such determinants. Experiments to link all or segments of the DNA from oncogenic or other animal viruses to autonomously replicating DNA elements, such as bacterial plasmids or other viral DNAs, also pose threats. Dissemination of such recombinant DNA molecules could possibly increase the incidence of cancer or other diseases.

But the new recombination techniques, if safely conducted, may also provide solutions to theoretical and practical biological problems. For example, the recombinant DNA method might be used to develop bacteria that can synthesize a wide variety of biological and chemical substances as yet unavailable on an industrial scale. Bacteria capable of producing the hormone insulin have been reported. Another bacterium, a *Pseudomonas* species, has been developed and patented because it can be very useful for cleaning up oil spills. (But what if it got loose in an oil well?) One can also think of the impact on agriculture. We appreciate the importance of nitrogen fixation by procaryotes in enhancing soil fertility. The genes for nitrogen fixation (*nif*) form a cluster on the *Klebsiella pneumoniae* chromosome and are transferable. They can integrate into or segregate from both chromosomal or plasmid DNA, and *nif*-bearing plasmids can acquire new properties by recombination. Therefore, exciting possibilities are now open for preparing new nitrogen-fixing bacteria, using plasmid-based *nif* genes associated with sex-factor activity. Further, it may even be possible to alter the genome of existing plants by the use of bacterial genes on a plasmid and make the plants themselves nitrogen-fixing.

Some scientists are even thinking seriously about treating human genetic diseases by replacing "bad" genes with normal ones. A wide range of genetic diseases could be treated by gene replacement as they made their appearance at various stages of life. Even if a genetic disease were diagnosed *in utero*, fetal cells might be reprogrammed in the uterus! Thus, we see that the future prospects of genetic engineering can be very bright, although hazardous if uncontrolled. Nonetheless these areas should be explored with the proper precautions. Many countries now have guidelines for biological and physical containment of recombinants; many microbiologists regard such guidelines as providing adequate levels of safety for recombinant DNA experiments.

Summary and outlook

Bacteria are capable of storing and transmitting genetic information to their progeny as are other living organisms. In addition to the inheritance of characteristics, bacteria also exhibit variability associated with their genotype and phenotype. Phenotypic changes may be due to the effect of the environment or the mutation in the genome. Mutations at the molecular level are exemplified by point mutations and frameshift mutations. Cells contain specific enzymes which can repair damaged DNA.

There are three main types of gene transfer in bacterial recombination. These are conjugation, transduction, and transformation. Conjugation is dependent on cell-to-cell contact and is governed by sex factors, which are episomes. Other extrachromosomal elements include plasmids such as resistance-transfer factors. Transduction is a type of gene transfer in which the donor DNA is carried from the donor to the recipient cell inside a phage. In transformation a naked DNA fragment is taken up by the recipient cell.

Techniques have been available since 1973 for the genetic engineering of bacterial cells. With these techniques, genes can be moved across species lines, resulting in the creation of new, redesigned microorganisms.

The outlook for molecular genetics remains as exciting as it has been for many years. The recent discovery of overlapping genes, for example, was mentioned in Chap. 16. Moreover, after 9 years of work, Har Gobind Khorana and his collaborators have synthesized in the laboratory the first complete and functional DNA gene for the incorporation of tyrosine in protein synthesis. Furthermore, researchers have now found evidence of "action at a distance" on DNA molecules. That is, a protein will bind to the DNA at one place, and this will affect gene expression at a position some distance away. This means that DNA is a far more "active" molecule than present concepts indicate and that mechanisms of control of gene expression may be more subtle than previously imagined. In genetic engineering, recombinant DNA technology places in human hands the capacity to redesign living organisms, which up to this time was possible only through the very slow process of evolution. Such new organisms created will be self-perpetuating and permanent. So far, only bacterial cells have been the recipients of new genes from other species. In the future, it is likely that simple animal virus DNA molecules will be used for the transfer of new DNA into animal cells. All these possibilities have vast implications and unforeseeable consequences.

Key terms

bacteriocin
bacteriocinogenic factor
base analog
base-pair substitution
cell-free DNA
colicin
competent cell
conjugation
endonuclease
episome
exonuclease
F factor
frameshift mutation
generalized transduction
genome
genotype
high-frequency recombination (Hfr) strain
induced mutation
infectious resistance
interrupted mating
linkage map
locus
marker
merozygote

missense mutation
mutagen
mutation
neutral mutation
nonsense mutation
phenotype
photoreactivation
plasmid
point mutation
prototroph
pyocin
replica plating technique
resistance-transfer factor
reverse mutation
sexduction
specialized transduction
spontaneous mutation
transduction
transformation
transition
transversion
variability
wild-type cell

Questions

1 Explain how variability can arise in a bacterial culture.
2 How do genotypic changes differ from phenotypic changes?
3 Describe some of the ways in which mutation is effected by mutagens.
4 With reference to point mutations and frameshift mutations, which type is likely to be more severe? Why?
5 Explain how cells can repair damage to DNA in the presence of light.
6 Why are two antibiotics given simultaneously in the treatment of some diseases?
7 What are the distinguishing characteristics of each of the three types of gene transfer?
8 Explain how the discovery of sexual differentiation contributed to an understanding of conjugation in *E. coli*.

9 What are extrachromosomal elements? What are some practical implications of their existence?

10 What are some arguments for and against genetic engineering?

References for part six

Fincham, J. R.: *Microbial and Molecular Genetics*, 2d ed., Hodder & Stoughton, London and Toronto, 1976. *A short and highly readable text on molecular genetics, mostly that of bacteria and phages. Essentially a book for undergraduate students.*

Lehninger, A. L.: *Short Course in Biochemistry*, Worth, New York, 1973. *A shorter and more concise version of the more comprehensive text* Biochemistry. *Gives a clear description in a succinct manner of biochemical principles.*

McElroy, W. D., and C. P. Swanson: *Modern Cell Biology*, 2d ed., Prentice-Hall, Englewood Cliffs, N.J., 1976. *A general text (paperback) that deals with the structure, function, and biochemistry of cells. Well written and very well illustrated.*

Sokatch, J. R., and J. J. Ferretti: *Basic Bacteriology and Genetics*, Year Book, Chicago, 1976. *The genetics of bacteria and viruses are well covered in this paperback text. Line drawings are very well done for easy understanding and learning.*

Woods, R. A.: *Biochemical Genetics*, Chapman & Hall, London, 1973. *A small paperback book that gives an outline of biochemical genetics. Microorganisms discussed include, besides bacteria, molds, yeasts and viruses.*

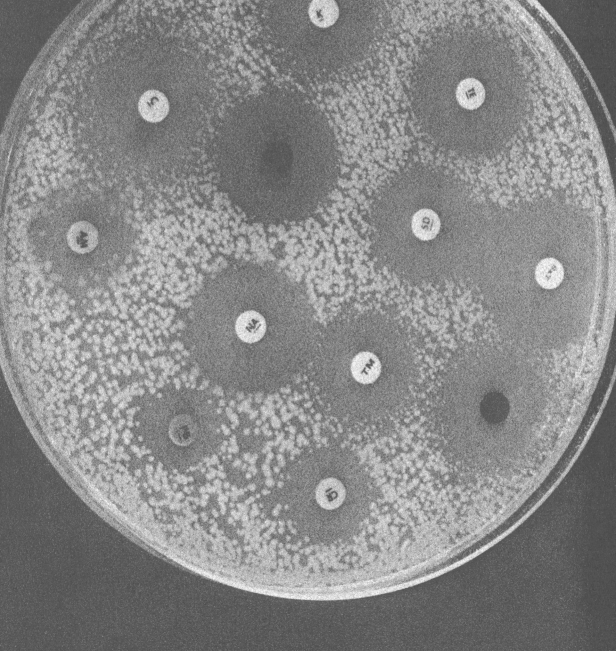

PART SEVEN

CONTROL OF MICROORGANISMS

In the preceding chapters we have described the morphological, cultural, and biochemical characteristics of microorganisms. To observe the characteristics of microorganisms, the appearance of their growth, or their capability for performing a variety of biochemical changes, it is necessary to provide for their growth under controlled conditions. The physical environment and chemical substances (nutrients) provided must be suitable to support growth.

In the following chapters our concern is in a different direction. We are interested in how we can control microorganisms. By control we mean the elimination, inhibition, or killing of microbial cells. Detrimental microorganisms must be controlled if their destructive action is to be prevented. In some instances this destructive action takes the form of spoilage or deterioration of foods or other materials; in other instances the destructive influence may be exhibited in infections and disease.

A great many techniques are available for the control of microorganisms. All of them, however, fall into the following three categories: removal of microorganisms; inhibition or killing using physical agents such as high or low temperatures or radiation; and inhibition or killing using one of many chemical substances.

To know the effectiveness of various methods used to control microorganisms is of great importance. This requires that suitable laboratory procedures be available to test and evaluate the various methods and substances.

In Part VII we discuss various aspects of microbiological control.

Prevention of growth around disks impregnated with antibiotics. Clear areas indicate absence of bacterial growth. *(Courtesy of Difco Laboratories.)*

FUNDAMENTALS OF CONTROL

Microorganisms can cause great harm and damage. They infect people,
other animals, and plants, producing diseases that range in seriousness
from mild infections to death. They contaminate food and, by producing
chemical changes in it, make it inedible or even poisonous. Microorganisms are also responsible for the spoilage of many materials, including
fabrics; leather; wooden structures, such as pilings, bridges, and houses;
electric insulation made of plastics; and other organic materials, even
jet fuel. The resulting economic losses can be very substantial. Thus it is
imperative that procedures be available to control microbial contamination and growth. The term *control*, as we use it in this book, refers to the
inhibition, killing, or removal of microorganisms.

In this chapter you will be introduced to the general principles and
terminology associated with the control of microorganisms.

Importance of   The principal reasons for controlling microorganisms can be **summarized**
microbial control   as follows:

1 To prevent transmission of disease and infection

307

2 To eradicate microorganisms from a host that is infected

3 To prevent deterioration and spoilage of materials by microorganisms

Microorganisms can be removed, inhibited, or killed by physical agents, physical processes, or chemical agents. A variety of techniques and agents are available; they act in many different ways, and each has its own limits of practical application. A *physical agent* is a physical condition or property which causes change. Temperature, pressure, radiation, and filters are examples of physical agents. A *physical process* is a procedure causing a change, for example, sterilization, incineration, and sanitation. A *chemical agent* is a substance (solid, liquid, or gas) which is characterized by a definite molecular composition and causes a reaction. Phenolic compounds, alcohols, chlorine, iodine, and ethylene oxide are examples of chemical agents.

Several specific terms are used to describe these agents and processes. The terms are important in labeling drugs and chemicals used against microorganisms. A judicial decision under the Pure Food and Drug Act states, "Language used in the label is to be given the meaning ordinarily conveyed by it to those to whom it was addressed." Therefore, both the manufacturer and the consumer must understand the precise meaning of these terms—the terms must be defined in lay language that might be found in an ordinary dictionary.

Definition of terms

Sterilization. The process of destroying all forms of life is called *sterilization*. A *sterile* object, in the microbiological sense, is free of living microorganisms. An object or substance is either sterile or nonsterile; it can never be semisterile or almost sterile.

Disinfectant. An agent, usually a chemical, that kills the vegetative cells but not necessarily the spore forms of disease-producing microorganisms is called a *disinfectant*. The term is commonly applied to substances used on inanimate objects. *Disinfection* is the process of destroying the vegetative cells but not necessarily the spores of infectious agents.

Antiseptic. A substance that opposes infection (*sepsis*) or prevents the growth or action of microorganisms either by destroying them or by inhibiting their growth and activity is called an *antiseptic*. The term is usually associated with substances applied to the body.

Sanitizer. An agent that reduces the microbial population to levels judged safe by public health requirements is called a *sanitizer*. Usually it is a chemical agent that kills 99.9 percent of the growing bacteria. Sanitizers are commonly applied to inanimate objects and are generally employed in the daily care of equipment and utensils in dairies and food plants, and for glasses, dishes, and utensils in restaurants. The process of disinfection would produce sanitization; however, in the strict sense, sanitization implies a sanitary or clean condition, which disinfection does not necessarily imply.

Germicide (Microbicide). An agent that kills the vegetative cells but not necessarily the resistant spore forms of germs is called a *germicide* (*microbicide*). In practice, a germicide is almost synonymous with a disinfectant, but a germicide is commonly used for all kinds of germs (microbes) for any application.

Bactericide. An agent that kills vegetative forms of bacteria (adjective, *bactericidal*). Similarly, the terms *fungicide*, *viricide*, and *sporicide* refer to agents that kill fungi, viruses, and spores, respectively.

Bacteriostasis. A condition in which the growth of bacteria is prevented (inhibited) is called *bacteriostasis* (adjective, *bacteriostatic*). Similarly, *fungistatic* describes the action of an agent that stops the growth of fungi. Agents that have in common the ability to inhibit growth of microorganisms are collectively designated *microbistatic* agents.

Antimicrobial Agent. A general term, *antimicrobial agent* refers to an agent that interferes with the growth and metabolism of microbes. In common usage the term denotes inhibition of growth, and when referring to specific groups of organisms, terms such as *antibacterial* or *antifungal* are frequently employed.

Some antimicrobial agents are used specifically for the treatment of infections. These are called *therapeutic* agents.

The pattern and rate of bacterial death

When one drops a suspension of bacteria into a bottle of hot acid or an incinerator, they may all be killed instantly or at least so fast that it is not possible to measure the death rate. Less drastic treatment, however, does not kill all the cells at the same instant; rather the cells are killed over a period of time at a constant *exponential* rate that is essentially the inverse of their exponential growth pattern.

Exponential death can be understood more easily in terms of a simple model. Imagine that each cell is a *target* and that a large number of bullets (chemical or physical agents) are being fired at them randomly; that is, no one is aiming directly at a target. Then common sense will dictate some rules about the way the bacteria will die. We assume for the sake of simplicity that a single hit kills a cell.

The probability of hitting a target is proportional to the number of targets, that is, the number of bacteria present. If we shoot randomly at many targets, we know intuitively that we have a good chance of hitting one. We also know that as time goes on, the number of targets not yet hit decreases steadily, and it becomes harder and harder to hit the remaining ones. (Hitting a target repeatedly does not count; a bacterium can be killed only once.) Let us take a simple numerical example. Assume that we have

Table 18-1. A theoretical case of disinfection

TIME	SURVIVORS	DEATHS PER UNIT TIME	TOTAL DEATHS	TOTAL DEATHS, %
0	1,000,000	0	0	0
1	100,000	900,000 = 90%	900,000	90
2	10,000	90,000 = 90%	990,000	99
3	1,000	9,000 = 90%	999,000	99.9
4	100	900 = 90%	999,900	99.99
5	10	90 = 90%	999,990	99.999
6	1	9 = 90%	999,999	99.9999

SOURCE: Modified from O. Rahn, *Physiology of Bacteria*, McGraw-Hill, New York, 1932.

an initial population of 1 million targets. We shower them with bullets for 1 min and manage to hit 90 percent, so that there are now 100,000 left. Then we shower them with bullets for 1 min more; but since we only have one-tenth as many targets as in the first round, we hit only one-tenth as many bacteria. In other words, this time we hit 90,000 of the targets and have 10,000 survivors. We shower these survivors with bullets for another minute; and since we have only one-tenth as many targets as in the last minute, we again hit only one-tenth as many bacteria, or 9,000. This pattern of striking targets (killing bacterial cells) repeats until there are no targets left, as shown in Table 18-1. But notice that it takes just as long to kill the last nine bacteria (or the last "nine-tenths" of a bacterium) as it does to kill the first 900,000. In fact, we can never be sure that we have killed the last one; all we can do is shoot at the targets long enough for there to be a good chance that the last one has been hit (killed).

The pattern of death among *Bacillus anthracis* spores exposed to a 5% phenol solution is shown in Fig. 18-1. The number of survivors is plotted against time; curve A is an arithmetic plot and curve B is a logarithmic plot. Both curves show that some portion of the population dies during any given unit of time. But the logarithmic plot also reveals that the death *rate*

Figure 18-1. The death curve of *B. anthracis* spores exposed to 5% phenol. Curve A: Number of survivors expressed arithmetically per unit volume plotted against time. Curve B: Logarithm of number of surviving bacteria per unit volume plotted against time.

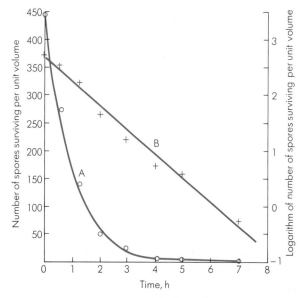

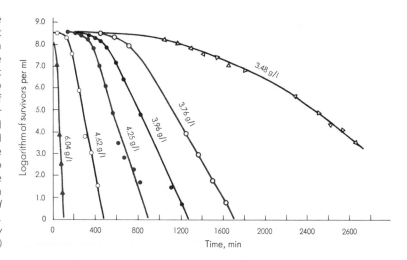

Figure 18-2. Increasing the concentration of disinfectant increases the speed with which bacterial cells are killed. In this experiment *E. coli* were exposed to various concentrations of phenol at 35°C. The number of survivors, expressed logarithmically, is plotted against time. The concentrations of phenol to which cells were exposed are shown above each graph line. *(From R. C. Jordan and S. E. Jacobs, J Hyg.* **44***:210, 1945, Cambridge University Press.)*

is constant. The points fall on a straight line, and the slope of the curve is a measure of the death rate.

Results such as those shown in Fig. 18-1 are obtained only when all conditions are kept strictly constant, including uniformity in the physiological condition of all the microorganisms in the population. Such standardized conditions do not occur in everyday situations such as those in a hospital ward or at a food processing plant. Cells of different ages or physiologic stages of growth exhibit differences in their susceptibility to an antimicrobial agent. As a consequence, the logarithmic plot of survivors would not follow the ideal pattern shown in Fig. 18-1.

Conditions influencing antimicrobial action

Many factors and conditions influence the rate at which microorganisms are inhibited or destroyed by an antimicrobial agent or process. All of these must be considered for the effective practical application of control methods.

Concentration or intensity of antimicrobial agent

Referring back to our bullet-target model, it is evident that the probability of hitting a target is proportional not only to the *number of targets present* but also to *the number of bullets shot*, that is, the concentration of the chemical agent or intensity of the physical agent. The more bullets we shoot in a given time, the faster the targets will be hit. If the targets are bacteria and the bullets are x-rays or ultraviolet light, it stands to reason that the cells will be killed faster as the *intensity* of the radiation increases. If the bullets are molecules of some chemical agent, the cells will be killed more rapidly as the *concentration* of the agent increases (up to a certain limit, of course). Figure 18-2 shows the relationship of increased concentrations of phenol to survival of bacteria. For example, a concentration of 4.25 g/liter requires approximately 10 h to achieve the same result obtained in approximately 1.5 h with a concentration of 6.04 g/liter.

Number of microorganisms

The longer we shoot, the more targets we hit; but the more targets we have, the longer it takes to hit them all, all other conditions remaining

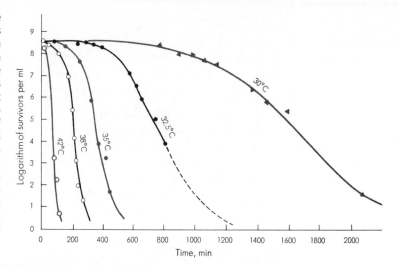

Figure 18-3. Increasing the temperature also increases the speed with which bacterial cells are killed when the concentration of the disinfectant remains constant. In this experiment *E. coli* were exposed to phenol at a concentration of 4.62 g/liter at temperatures between 30 and 42°C. The number of survivors, expressed logarithmically, is plotted against time. The temperatures to which cells were exposed are shown above each graph line. *(From R. C. Jordan and S. E. Jacobs, J Hyg.* **44***:210, 1945, Cambridge University Press.)*

constant. This is a rather obvious restatement of the exponential death pattern. It means that it takes time to kill the population; and if we have many cells, we must treat them for a longer time to be reasonably sure that they are all dead.

Temperature
Modest increases in temperature can greatly increase the effectiveness of a disinfectant or other chemical antimicrobial agent. For example, from the data shown in Fig. 18-3 you can see that an increase in temperature from 30 to 42°C (86 to 107.6°F) greatly increases the bactericidal action of phenol. This is explained by these facts: (1) chemical agents damage microorganisms through chemical reactions, and (2) the rate of a chemical reaction is accelerated by increased temperature.

Species of micro-organism
Species of microorganisms differ in their susceptibility to physical and chemical agents. We have already noted that in sporeforming species, the growing vegetative cells are much more easily killed than are the spore forms. In fact, bacterial spores are the most resistant of all living organisms in their ability to survive under adverse physical or chemical conditions. The relative resistance of bacterial spores in comparison with other microorganisms is shown in Table 18-2. Differences exist among species of microorganisms in terms of the susceptibility of their vegetative cells (as well as spores) to various chemical and physical agents.

Table 18-2. Resistances of bacterial and mold spores and of viruses, relative to the resistance of *Escherichia coli* as unity

STERILIZING AGENT	*Escherichia coli*	BACTERIAL SPORES	MOLD SPORES	VIRUSES AND BACTERIOPHAGES
Phenol	1	100,000,000	1–2	30
Formaldehyde	1	250		2
Dry heat	1	1,000	2–10	±1
Moist heat	1	3,000,000	2–10	1–5
Ultraviolet radiation	1	2–5	5–100	5–10

SOURCE: O. Rahn, *Bacteriol Rev*, 9:1, 1945.

Presence of organic material

The presence of extraneous organic matter can significantly reduce the effectiveness of an antimicrobial chemical agent by inactivating it or protecting the microorganism from it. For example, organic matter present in a disinfectant–microorganism mixture may result in:

1 Combining of the disinfectant with the organic material to form a product which is not microbicidal
2 Combining of the disinfectant with the organic material to form a precipitate, thus removing the disinfectant from possible combination with microorganisms
3 Accumulation of the organic matter on the microbial cell surface, providing, in effect, a coating which will impair the contact between disinfectant and cell

In practice, if serum or blood are present on an object being treated with an antimicrobial agent, these substances will inactivate some amount of the agent.

Acidity or alkalinity (pH)

Microorganisms present in material with an acid pH are destroyed at lower temperatures and in shorter time exposures than are the same microorganisms in an alkaline environment. In home canning, shorter times and lower temperatures are required for processing tomatoes and fruits (acid foods) than beans and corn (alkaline foods). For example, quart jars of tomatoes can be processed by boiling at 100°C (212°F) for approximately 45 min, whereas beans require 115°C (239°F) at 10-lb pressure for 25 min in a pressure cooker.

Mode of action of antimicrobial agents

The many processes and substances used as antimicrobial agents act in one of several ways. Knowledge of the precise manner (mode of action) by which the agent inhibits or kills microorganisms is extremely useful. For example, with this information one may be able to predict the best circumstances for its use and the kinds of microorganisms against which it is likely to be most effective. This knowledge may also be helpful in planning the design of new and more effective agents.

In a general way, one may assess the possible sites of action of an antimicrobial agent by reviewing the structure and composition of the microbial cell (Fig. 18-4). A normal living cell contains a multitude of enzymes responsible for metabolic processes as well as additional proteins, nucleic acids, and other compounds. A semipermeable membrane (cytoplasmic membrane) maintains the integrity of the cellular contents; the membrane selectively controls the passage of substances between the cell and its external environment. This membrane is also the site of some enzyme reactions. The cell wall provides a protective covering for the cell in addition to participating in certain physiological processes. Damage at any one of these sites may initiate changes leading to the death of the cell.

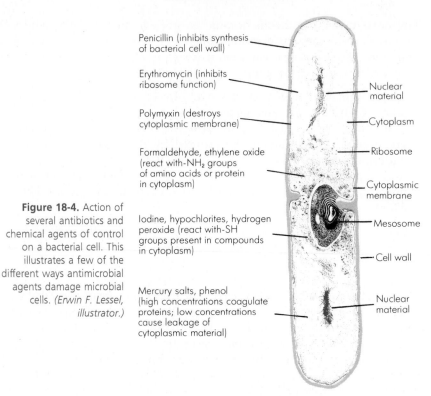

Penicillin (inhibits synthesis of bacterial cell wall)

Erythromycin (inhibits ribosome function)

Polymyxin (destroys cytoplasmic membrane)

Formaldehyde, ethylene oxide (react with-NH$_2$ groups of amino acids or protein in cytoplasm)

Iodine, hypochlorites, hydrogen peroxide (react with-SH groups present in compounds in cytoplasm)

Mercury salts, phenol (high concentrations coagulate proteins; low concentrations cause leakage of cytoplasmic material)

Nuclear material

Cytoplasm

Ribosome

Cytoplasmic membrane

Mesosome

Cell wall

Nuclear material

Figure 18-4. Action of several antibiotics and chemical agents of control on a bacterial cell. This illustrates a few of the different ways antimicrobial agents damage microbial cells. *(Erwin F. Lessel, illustrator.)*

These changes are listed below; they will be discussed in more detail in later chapters when we consider the mode of action of specific antimicrobial agents.

Damage to cell wall

The cell-wall structure may be damaged either by inhibiting its formation or by altering it after it has been formed.

Alteration of cell permeability

The cytoplasmic membrane keeps certain materials in the cell and regulates the inflow and outflow of other materials; it preserves the integrity of the cellular constituents. Damage to this membrane will result in the inhibition of growth or the death of the cell.

Alteration of protein and nucleic acid molecules

The viability of a cell is dependent upon the maintenance of protein and nucleic acid molecules in their natural state. A condition or substance which alters this state, i.e., denatures the proteins or nucleic acids, may irreparably damage the cell. High temperatures and high concentrations of some chemicals can bring about irreversible coagulation (denaturation) of these vital cellular constituents.

Inhibition of enzyme action

Each of the hundreds of different enzymes in the cell is a potential target for an inhibitor. Many chemical substances are known to interfere with biochemical reactions. This inhibition can result in a slowdown of metabolism or death of the cell.

Inhibition of nucleic acids and protein synthesis

The vital role of DNA, RNA, and proteins in the normal life processes of the cell means that any interference with their formation or function will seriously impair the cell.

The selection of antimicrobial agents or techniques

No single antimicrobial agent or technique is the best for all practical applications. For example, an agent or process suitable for the control of microorganisms in a food product is not likely to be satisfactory for use in biological material to be injected into a patient. Similarly, an agent suitable for application to an inanimate surface, such as a contaminated tabletop, is unlikely to be appropriate for application to the skin. Each situation must be assessed in terms of the result desired and the agent or method that will best achieve this result.

Summary and outlook

Our health and welfare depends in many instances upon our ability to control microbial populations. Control, in this context, means the inhibition, killing, or removal of the microbial population in or on some material or environment. Many different physical agents and methods or chemical agents are available for this purpose; the choice of procedure is dependent upon the circumstances. For example, does the process require that the microorganisms be killed, or only inhibited? Do all forms of microbial life need to be destroyed, or only the vegetative cells? How will the material being treated be affected by the antimicrobial agent? The answers to these and other questions will determine the technique best suited for dealing with a specific situation.

New agents and techniques for the control of microorganisms are continually being sought. In some instances this comes about because a new type of product which cannot be sterilized on a commercial scale by existing techniques demands a novel agent or technology. As an example, many plastic materials, such as syringes, containers for collection of patient specimens, petri dishes, and pipets, all of which came into widespread use during the last two decades in hospitals and laboratories, could not withstand high-temperature treatment. Thus, methods for large-scale applications of ethylene oxide as a sterilizing agent were devised and implemented. This has been supplemented by radiation-sterilization techniques.

A different need for the control of microorganisms surfaced during the United States space program explorations. One of the objectives of this program was to seek evidence for the existence of life on other planets. The possibility that extraterrestrial life exists required that any space vehicle which was to land on another planet had to be sterile, to prevent contamination with planet earth microorganisms. This required development of new technologies. Similarly, specimens returned from another planet needed to be held under rigid quarantine to protect the earth from possible contamination.

New regulations by health agencies require more stringent microbiological standards for a product, thus necessitating and promoting the development of new or modified microbial control procedures. As new agents and processes are proposed, they must be evaluated for their effectiveness.

Key terms

antimicrobial agent	disinfectant	sepsis
antiseptic	fungicide	sporicide
bactericide	germicide	sterilization
bacteriostasis	physical agent	
chemical agent	sanitizer	

Questions

1 What are the principal reasons for controlling microbial populations?
2 Distinguish between the following pairs of terms: *sterilization* and *disinfection*; *sanitization* and *disinfection*; *germicidal action* and *bactericidal action*; *bacteriostatic substances* and *bactericidal substances.*
3 Describe situations where it would be appropriate to (a) sterilize, (b) sanitize, (c) disinfect.
4 Assume that a disinfectant is added to a suspension of microorganisms. Do all cells die instantaneously? Explain your answer.
5 How does the term *death* as used in microbiology differ from the same term as used with higher organisms?
6 Enumerate the conditions that may influence the effectiveness of any physical or chemical agent when used to control microbial populations.
7 What effect may extraneous organic matter have on the effectiveness of a disinfectant?
8 Describe several possible modes of action of antimicrobial agents.
9 Of what practical significance is it to know how an agent kills microorganisms?
10 What factors should be taken into consideration when selecting a chemical compound (or a physical agent) for the control of microorganisms?

CONTROL OF MICROORGANISMS BY PHYSICAL AGENTS

Microorganisms can be controlled, that is, killed, inhibited, or removed from an environment, by various physical agents or processes. The physical agent or process used depends upon many factors and can only be decided upon after an evaluation of the specific circumstances. For example, if we wish to destroy the infectious microorganisms in a dead diseased animal, it might be appropriate to incinerate the animal. However, if we need to sterilize plastic bags that are to be used for the collection of blood, we would need to select a sterilizing process that would not damage the bags. Research and experience provide guidance in the selection of the most appropriate method.

In this chapter we discuss the physical agents and processes most widely used to control microorganisms. We describe how they inhibit or kill microorganisms, the factors which influence their effectiveness, and their practical applications.

317

High temperatures   High temperature combined with high humidity is one of the most effective methods of killing microorganisms. High temperature also can be applied as dry heat. There are important differences between the two techniques.

Moist heat kills microorganisms by coagulating their proteins. Moist heat is much more rapid and effective in killing microorganisms than is dry heat, which destroys microorganisms by oxidizing their chemical constituents. Two examples will illustrate the difference. Spores of *Clostridium botulinum*, some of which produce a deadly toxin that causes the often fatal food poisoning called *botulism*, are killed in 4 to 20 min by moist heat at 120°C (248°F), whereas about 2 h of exposure to dry heat at the same temperature is required. The spores of *Bacillus anthracis*, which can kill cattle, other grazing animals, and people, are destroyed in 2 to 15 min by moist heat at 100°C (212°F) (Table 19-1). But with dry heat, 1 to 2 h at 150°C (302°F) is required to achieve the same result (Table 19-2).

We have said that vegetative cells of bacteria are much more sensitive to heat than are their spores. Cells of most bacteria are killed in 5 to 10 min at 60 to 70°C (140 to 158°F) (moist heat). Most bacterial spores are killed only by temperatures maintained above 100°C (212°F) for extended periods (Table 19-2). Vegetative cells of yeasts and other fungi are usually killed in

Table 19-1. Some quoted destruction times of bacterial spores by moist heat

ORGANISM	DESTRUCTION TIME, min							
	AT 100°	AT 105°	AT 110°	AT 115°	AT 120°	AT 125°	AT 130°	AT 134°
Bacillus anthracis	2–15	5–10						
B. subtilis	Many hours			40				
A putrefactive anaerobe	780	170	41	15	5.6			
Clostridium tetani	5–90	5–25						
C. perfringens	5–45	5–27	10–15	4	1			
C. botulinum	300–530	40–120	32–90	10–40	4–20			
Soil bacteria	Many hours	420	120	15	6–30	4		1.5–10
Thermophilic bacteria		400	100–300	40–110	11–35	3.9–8.0	3.5	1
C. sporogenes	150	45	12					

SOURCE: G. Sykes, *Disinfection and Sterilization,* 2d ed., Lippincott, Philadelphia, 1965.

Table 19-2. Some quoted destruction times of bacterial spores by dry heat

ORGANISM	DESTRUCTION TIMES, min						
	AT 120°	AT 130°	AT 140°	AT 150°	AT 160°	AT 170°	AT 180°
Bacillus anthracis			Up to 180	60–120	9–90		3
Clostridium botulinum	120	60	15–60	25	20–25	10–15	5–10
C. perfringens	50	15–35	5				
C. tetani		20–40	5–15	30	12	5	1
Soil spores				180	30–90	15–60	15

SOURCE: G. Sykes, *Disinfection and Sterilization,* 2d. ed., Lippincott, Philadelphia, 1965.

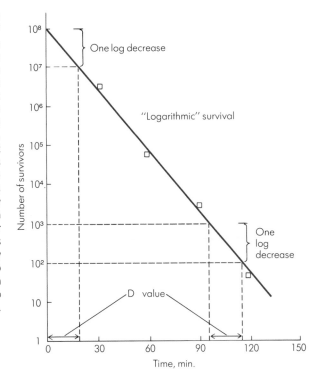

Figure 19-1. Graph illustrating the concept of decimal reduction time (D value), the time in minutes to reduce the microbial population by 90 percent. Stated differently, it is the time in minutes for the thermal-death-time curve to pass through one log cycle. The D value is independent of time when the response is logarithmic, that is, when the same length of time is required to acccomplish any given log decrease in number of survivors. For example, the D value in this illustration is approximately 20 min., the time required to reduce the survivors from 10^8 to 10^7, or from 10^7 to 10^6, and so on.

5 to 10 min by moist heat at 50 to 60°C (122 to 140°F); their spores are slightly more resistant. The susceptibility of viruses to heat is similar to that of mesophilic vegetative bacterial cells.

Thermal death time and decimal reduction time

Two terms are used to express the resistance of bacteria to heat, namely, *thermal death time* and *decimal reduction time*. Thermal death time refers to the shortest period of time required to kill a suspension of bacteria (or spores) at a given temperature and under specific conditions. Decimal reduction time refers to a specific reduction in the number of viable cells, that is, the time in minutes to reduce the population by 90 percent. Stated differently, it is the time in minutes for the thermal-death-time curve (Fig. 19-1) to pass through one logarithmic decrease (a 90 percent reduction in the microbial population).

From the definitions of these terms, it is clear that the time-temperature relationship is a critical one in terms of establishing the susceptibility of microorganisms to heat.

Application of high temperatures for killing microorganisms

Practical procedures by which heat is employed for killing microorganisms are conveniently divided into two categories: *moist heat* and *dry heat* .

Moist heat

Steam under pressure. Heat in the form of saturated steam under pressure is the most practical and dependable agent for sterilization. Steam under pressure provides temperatures considerably above those obtainable by boiling, as shown in Table 19-3. In addition, it has the advantages of rapid

Table 19-3. Temperature of steam under pressure

STEAM PRESSURE, lb/in²	TEMPERATURE, °C
0	100.0
5	109.0
10	115.0
15	121.5
20	126.5

SOURCE: J. J. Perkins, *Principles and Methods of Sterilization,* Charles C Thomas, Springfield, Ill., 1969.

heating, penetration, and abundant moisture, all of which facilitate the coagulation of proteins of microbial cells.

The sterilizing apparatus that employs steam under regulated pressure is called an *autoclave* (see Fig. 19-2). It is essentially a double-jacketed steam chamber filled with air-free saturated steam and maintained at a designated temperature and pressure for a selected period of time. In operating an autoclave it is absolutely essential that the air in the autoclave chamber be completely replaced by saturated steam. If any air is present, it will reduce the chamber temperature substantially below that reached by pure saturated steam under the same pressure. It is not the *pressure* that kills the organisms, but the high *temperature* of the steam. As we noted earlier, the pressure makes possible temperatures above that of boiling water—100°C or 212°F—just as it does in a pressure cooker used for home canning.

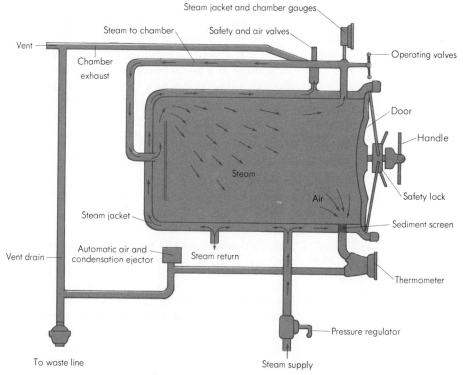

Figure 19-2. Autoclave, a pressure steam sterilizer. Cross-sectional view illustrating operational parts and path of steam flow. *(Courtesy of Wilmot Castle Company.)*

Table 19-4. Exposure periods required for aqueous solutions or liquids in various containers affording a reasonable factor of safety for sterilization by autoclaving

CONTAINER	MINUTES OF EXPOSURE AT 121–123°C (250–254°F)
Test tubes:	
18 × 150 mm	12–14
32 × 200 mm	13–17
38 × 200 mm	15–20
Erlenmeyer flasks:	
50 ml	12–14
500 ml	17–22
1,000 ml	20–25
2,000 ml	30–35
Fenwal flasks:	
500 ml	24–28
1,000 ml	25–30
2,000 ml	40–45
Milk-dilution bottle, 100 ml	13–17
Serum bottle, 9,000 ml	50–55

SOURCE: J. J. Perkins, *Principles and Methods of Sterilization*, Charles C Thomas, Springfield, Ill., 1969.

Figure 19-3. An autoclave being loaded for sterilization of hospital supplies. *(Courtesy of Amsco/American Sterilizer Company.)*

The autoclave is an essential unit of equipment in every microbiological laboratory, hospital sterilizing room, and other facility engaged in producing sterile products (Fig. 19-3). Generally, though not always, the autoclave is operated at a pressure of approximately 15 lb/in² at 121°C (249.8°F). The time of operation to achieve sterility depends on the nature of the material being sterilized, the type of container, and the volume of material. For example, 1,000 test tubes, each with 10 ml of liquid media, can be sterilized in 10 to 15 min at 121°C (249.8°F); the same quantity of media in 10 one-liter (1.06 quarts) containers would require 20 to 30 min at the same temperature to ensure sterilization (Table 19-4).

The characteristics of certain materials make it impractical for them to be sterilized by autoclaving. Substances immiscible in water, such as fats and oils, cannot be reached by the steam; hence, organisms contained in them will survive. Furthermore, some substances are altered or destroyed by exposure to high temperatures and consequently must be sterilized by other methods.

Fractional sterilization. Some bacteriological media and chemicals cannot be heated above 100°C (212°F) without being adversely affected. If, however, they can withstand the temperature of free-flowing steam (100°C), it is possible to sterilize them by *fractional sterilization*. In this process the material is heated to 100°C on three successive days with incubation periods in between. Resistant spores, if present, germinate during the incubation periods; on subsequent exposures to heat, the vegetative cells are destroyed.

Boiling water. Vegetative cells of microorganisms will be destroyed within 10 min in boiling water, but some bacterial spores can withstand this condition for many hours. The practice of exposing instruments for short periods of time to boiling water is more likely to disinfect than sterilize. Boiling water, then, is not reliable for sterilization.

Pasteurization. Milk, cream, and certain alcoholic beverages (beer and wine) are subjected to a controlled heat treatment which kills microorganisms of certain types but not others. *Pasteurized* milk is not sterile milk. The temperature selected for pasteurization is based on the thermal death time of the most resistant type of pathogen to be destroyed. Exposure of milk to excessively high temperature is undesirable because it produces off flavor. For many years *Mycobacterium tuberculosis* was thought to be the most heat-resistant pathogen likely to be carried by raw milk. Accordingly, pasteurization of milk employed a temperature of 61.7°C (143°F) for 30 min; *M. tuberculosis* is destroyed at 60°C (140°F) in 15 min. However, it was later discovered that a rickettsial organism, *Coxiella burnetii*, the cause of Q fever, may be present in milk, and it has a heat resistance greater than that of *M. tuberculosis*. As a result, the temperature for pasteurization of milk was raised to 62.8°C (145°F) for 30 min.

Dry heat   **Hot-air sterilization.** Dry-heat, or hot-air, sterilization is recommended where it is either undesirable or unlikely that steam under pressure will make direct and complete contact with the material to be sterilized. This is true of certain items of laboratory glassware, such as petri dishes and pipetes, as well as oils, powders, and some instruments. These articles are sterilized in an electric or gas oven. For laboratory glassware, a 2-h exposure to a temperature of 160°C (320°F) is sufficient for sterilization.

Incineration. The burning, or incineration, of material destroys microorganisms. Incineration is used for the destruction of carcasses, infected laboratory animals, and other infected material to be disposed of. Destruction of microorganisms by burning is also practiced routinely in the laboratory when the transfer needle is held in the flame of the bunsen burner. A note of caution should be added here. In sterilizing the transfer needle, care should be exercised to prevent *spattering* of droplets since some droplets are likely to carry viable organisms. Spattering can be greatly reduced or eliminated by drying the needle outside the flame before it is plunged into the flame or by placing a shield around the flame. Electrical incinerating devices designed to eliminate spattering are available which use heat from an electric coil.

Low temperatures   Temperatures below the optimum for growth depress the rate of metabolism; and if the temperature is sufficiently low, metabolism and growth cease. Low temperatures are useful for preservation of cultures because microorganisms have a unique capacity for surviving extreme cold.

Refrigeration

Cultures of some bacteria, yeasts, and molds grown on agar media in test tubes remain viable for months at refrigerator temperatures of about 4 to 7°C (39.2 to 44.6°F). This is a convenient method for preserving cultures of some, but not all, microorganisms.

Subzero temperatures

Bacteria and viruses can be maintained at −20°C (−4°F) (the temperature of mechanical freezers), −70°C (−94°F) (the temperature of dry ice, frozen CO_2), and even at −195°C (−319°F) (the temperature of liquid nitrogen). In fact, liquid nitrogen is frequently used for preserving cultures of many viruses and microorganisms, as well as stocks of mammalian tissue cells used in animal virology and many other types of research. In all these procedures, the initial chilling kills some of the cells, but the large number of survivors usually remain viable for long periods.

From these facts, it is apparent that low temperatures, however extreme, cannot be depended upon for disinfection or sterilization. Microorganisms maintained at freezing or subfreezing temperatures may be considered dormant; they have no detectable metabolic activity. This is the basis for the successful preservation of foods by the use of low temperatures.

Desiccation

Desiccation (drying) of the microbial cell and its environment greatly decreases or stops metabolic activity; death of some cells follows. In general, the time of survival of microorganisms after desiccation varies depending upon the following factors:

1 The kind of microorganism
2 The material in or on which the organisms are dried
3 The completeness of the drying process
4 The physical conditions to which the dried organisms are exposed, e.g., light, temperature, humidity

Species of gram-negative cocci such as *Neisseria gonorrhoeae* and *N. meningitidis* are very sensitive to desiccation; they die in a matter of hours. Streptococci are much more resistant; some survive weeks after being dried. The tubercle bacillus dried in sputum remains viable for even longer periods of time. Dried spores of microorganisms are known to remain viable indefinitely. These generalizations apply to air-drying.

In the process of *lyophilization*, described in Chap. 4, microorganisms are subjected to extreme dehydration in the frozen state and then sealed in a vacuum. Lyophilization is a process of preserving rather than destroying microorganisms. Lyophilized cultures of microorganisms remain viable for many years.

Osmotic pressure

Osmosis is diffusion through a semipermeable membrane separating two solutions of different solute concentrations. The process tends to equalize the concentrations of solute on either side of the membrane. As an illustration, assume that some bacterial cells are suspended in a solution containing a high concentration (20%) of sodium chloride. Water will pass from the region of lower concentration of dissolved substance (the interior of the cell has a low salt concentration) through the cytoplasmic membrane, which is semipermeable, and into the solution surrounding the cell. The cell thus becomes dehydrated; the effect produced is similar to that of drying the cells. This process is known as *plasmolysis*. In animal cells, which do not have rigid walls, an actual shrinkage of the cell can be observed as a result of plasmolysis. If the bacteria are placed in a solution containing considerably less than 1% sodium chloride, say 0.01%, the flow of water will be reversed; i.e., the water will flow from the solution into the cell. This process is called *plasmoptysis*. An *osmotic pressure* builds up inside the cell because of the large amount of water which accumulates there. If the cell membrane is elastic, as, for example, in red blood cells, this pressure will cause a swelling and may even burst the cell. Bacteria have very rigid cell walls which can withstand changes in osmotic pressure, so they usually do not exhibit striking changes in form or size when plasmolysis or plasmoptysis occurs.

A concentration of salt much above 1% is harmful to many bacteria but not, of course, to organisms indigenous to the ocean, where the salinity is about 3.5 to 4%, or to those found in the Dead Sea and the Great Salt Lake of Utah, where the salinity is about 29%. Some organisms are likewise capable of growing in high concentrations of sugar. Generally, microorganisms are inhibited by high concentrations of salt (10 to 15%) and sugar (50 to 70%). This inhibition is the basis for preservation of foods by "salting" or by using concentrated sugar solutions. The mechanism of microbial inhibition is plasmolysis: the cells are dehydrated and hence are unable to metabolize or grow. They may die or they may remain alive but in a dormant condition.

Radiation   Several kinds of radiation are lethal to microbial cells as well as to cells of other organisms. These kinds of radiation include portions of the electromagnetic spectrum (ultraviolet, gamma, and x-ray radiation, see Fig. 19-4) and cathode rays (high-speed electrons).

Ultraviolet light   The ultraviolet portion of the spectrum includes all radiation from 15 to 390 nm. Wavelengths around 265 nm have the highest bactericidal efficiency (see Fig. 19-5). Although the radiant energy of sunlight is partly composed of ultraviolet light, most of the shorter ultraviolet wavelengths are filtered out by the earth's atmosphere (ozone, clouds) and atmospheric pollutants (smoke). Consequently, the ultraviolet radiation at the surface of the earth is restricted to the span from about 280 to 390 nm. From this we may conclude that sunlight, under certain conditions, has microbicidal capacity, but to a limited degree.

Many lamps, called germicidal lamps, are available which emit a high

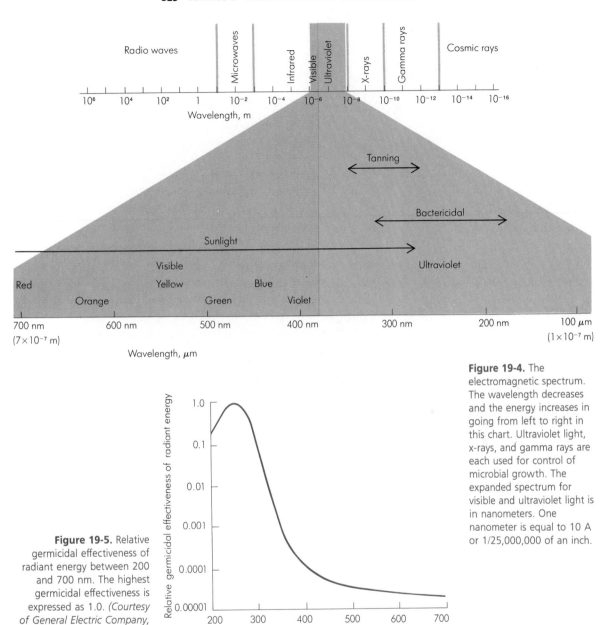

Figure 19-4. The electromagnetic spectrum. The wavelength decreases and the energy increases in going from left to right in this chart. Ultraviolet light, x-rays, and gamma rays are each used for control of microbial growth. The expanded spectrum for visible and ultraviolet light is in nanometers. One nanometer is equal to 10 A or 1/25,000,000 of an inch.

Figure 19-5. Relative germicidal effectiveness of radiant energy between 200 and 700 nm. The highest germicidal effectiveness is expressed as 1.0. *(Courtesy of General Electric Company, Lamp Division LD-11.)*

concentration of ultraviolet light in the most effective germicidal region of 260 to 270 nm. Germicidal lamps are widely used to reduce the microbial population in hospital operating rooms, aseptic filling rooms in the pharmaceutical industry, where sterile products are pipetted into vials or ampules, and in the food and dairy industries for treatment of contaminated surfaces. An example of the practical use of ultraviolet lamps is shown in Fig. 19-6.

Figure 19-6. Ultraviolet lamps (arrows) installed in refrigerated storage room for beef. Ultraviolet lamps installed in walk-in refrigerators irradiate the air killing airborne microorganisms and thus decreasing contamination of refrigerated materials by these microorganisms. *(Courtesy of Westinghouse Electric Corporation, Lamp Commercial Division, Bloomfield, NJ)*

Table 19-5. Median lethal dose of x-rays for various species of organisms

ORGANISM		MEDIAN LETHAL DOSE, rd*
Viruses:	Tobacco mosaic	200,000
	Rabbit papilloma	100,000
Bacteria:	*Escherichia coli*	5,000
	Bacillus mesentericus	130,000
Algae:	*Mesotenium*	8,500
	Pandorina	4,000
Protozoa:	*Colpidium*	330,000
	Paramecium	300,000
Vertebrates:	Goldfish	750
	Mouse	450
	Rabbit	800
	Rat	600
	Monkey	450
	Human (?)	400

*A *rad* (radiation absorbed dose), abbreviated rd, is the dose which delivers 100 ergs/g of irradiated material; it is equal to 6×10^{13} eV.

SOURCE: Modified from E. Paterson, in R. Paterson (ed.), *The Treatment of Malignant Disease by Radium and X-rays,* E. Arnold, London, 1948; *McGraw-Hill Encyclopedia of Science and Technology,* vol. 11, p. 244, McGraw-Hill, New York, 1977.

An important practical consideration in using ultraviolet light to destroy microorganisms is that it has very little ability to penetrate matter. Even a thin layer of glass filters out a large percentage of the light. Thus only the microorganisms on the surface of an object, where they are exposed directly to the ultraviolet light, are susceptible to destruction.

Ultraviolet light is absorbed by many cellular materials but most significantly by the nucleic acids, where it does the greatest damage. This phenomenon is discussed in Chap. 17.

X-rays

X-rays are lethal to microorganisms as well as to higher forms of life (see Table 19-5). Unlike ultraviolet radiations, they have considerable energy and penetration ability. However, they are impractical for routine methods of controlling microbial populations because their great penetrating ability makes protective shielding very difficult and they are hard to utilize efficiently. X-rays have been widely used experimentally to produce microbial mutants, as mentioned in Chap. 17.

Gamma rays

Gamma radiation, which is even more energetic than x-rays, is emitted from certain radioactive isotopes (or radioisotopes) such as ^{60}Co. As a result of the major research programs with atomic energy, large quantities of

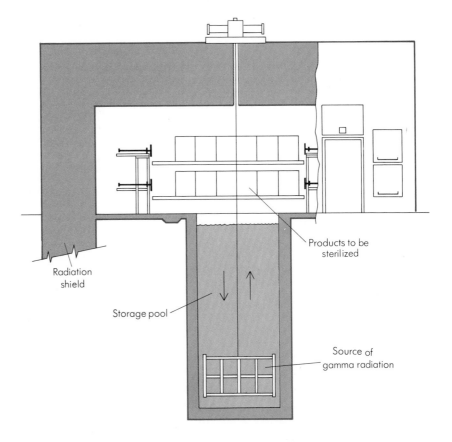

Figure 19-7. A commercial gamma-radiation sterilizer for heat-sensitive medical items. The product, packaged in its final shipping carton, is fed automatically into the shielded sterilizing chamber by a conveyor. The radioactive source, ^{60}Co (an isotope of cobalt), is raised from the storage pool and placed close to the packaged material. Timer mechanisms govern the length of exposure of the product to the gamma radiation emitted by the ^{60}Co. *(Courtesy of Atomic Energy of Canada, Ltd.)*

Radiation shield

Storage pool

Products to be sterilized

Source of gamma radiation

radioisotopes have become available as by-products of atomic fission. These isotopes are sources of gamma radiation. Gamma rays are similar to x-rays but have shorter wavelengths, and therefore higher energy. They have great penetrating power and are lethal to all life including microorganisms.

Because of their great penetrating power, high microbicidal effect, and greater efficiency as compared with that of x-rays, gamma rays are attractive for use in sterilization of materials of considerable thickness or volume, for example, packages of medical equipment or foods. A commercial gamma sterilizer is shown in Fig. 19-7.

Cathode rays (electron-beam radiation)

Special types of equipment have been designed which produce electrons called *cathode rays* or *electron beams*. These electrons of high intensities (millions of volts) are accelerated to extremely high velocities. These intense beams of accelerated electrons are microbicidal as well as having other effects on biological and nonbiological materials.

The electron accelerator, a device which produces the high-voltage electron beam, is used today for the sterilization of surgical supplies, drugs, and other materials. One of the advantages of this process, as with gamma-ray treatment, is that the material can be sterilized at room temperature after it has been packaged (see Fig. 19-8). Electron-beam radiation has limited power of penetration, but within its limits, sterilization is accomplished on very brief exposure.

Filtration

Some materials, particularly biological fluids such as animal serums, solutions of substances like enzymes, and some vitamins or antibiotics, are

Figure 19-8. A commercial high-voltage electron beam sterilizer for packaged products. The generator in this sterilizer has a capacity of several million volts. The packaged products move on a belt under the thin aluminum window. The high-voltage electron beam penetrates the packages and sterilizes the contents. *(Adapted from W. M. Urbain, Food Engineering, 25 February 1953.)*

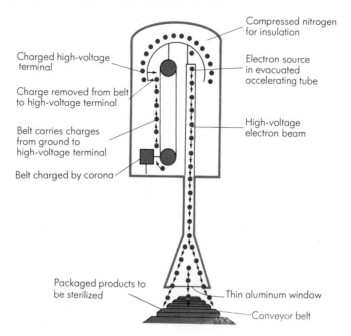

Compressed nitrogen for insulation

Charged high-voltage terminal

Electron source in evacuated accelerating tube

Charge removed from belt to high-voltage terminal

Belt carries charges from ground to high-voltage terminal

High-voltage electron beam

Belt charged by corona

Packaged products to be sterilized

Thin aluminum window

Conveyor belt

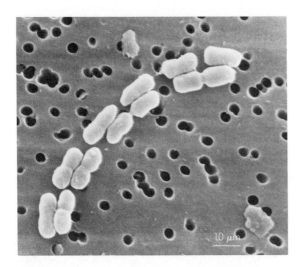

Figure 19-9. Bacterial cells from a marine sample on a membrane filter. *(Courtesy of Pall Corporation.)*

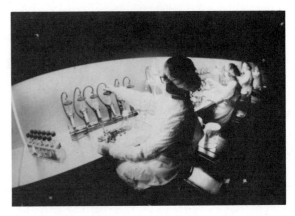

Figure 19-10. A series of membrane filters in use for sterilization of a biological liquid. *(Courtesy of Lilly Research Laboratories, Division of Eli Lilly and Co.)*

thermolabile, i.e., destroyed by heat. Likewise, other physical agents such as radiations are detrimental to these materials or in other ways impractical for sterilizing them. Accordingly, the option available is to sterilize by filtration.

Bacteriological filters

For many years a variety of bacteriological filters have been available to the microbiologist. The filtering material was a relatively thick pad composed of asbestos, diatomaceous earth, porcelain, or sintered glass. Removal of microorganisms by the filter pad was not solely a function of the pore size of the filters. Instead, it was a combination of pore size, the meshwork of fibrous or particulate material in the filter pad, and the electric charge of these materials.

In recent years a new type of filter for use in removing organisms has been developed and is now in widespread use. The filter material is prepared from a cellulose ester or other polymeric substances with pores of a precise and uniform dimension (see Fig. 19-9). Filters of this type can be produced with known porosities (pore diameters) ranging from approximately 0.01 to 10 μm. The filter material is very thin, approximately 150 μm in thickness, and hence it is referred to as a *membrane filter*. In use it is placed on a support base in a filtration unit. Membrane filters are used extensively in the laboratory and in industry to sterilize fluid materials (see Fig. 19-10). They have also been adapted to microbiological procedures for the identification and enumeration of microorganisms from water samples and other materials (see Chap. 35).

Air filters

The development of high-efficiency particulate air (HEPA) filters has made it possible to deliver clean (dust-free) air to an enclosure such as a cubicle or room. This type of air filtration together with a system of *laminar airflow* is now used extensively to provide dust- and bacteria-free air (Fig. 19-11). Air

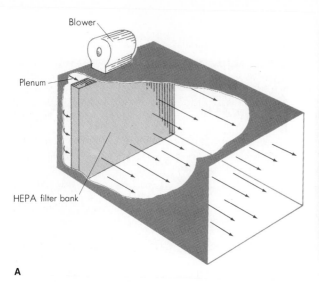

A

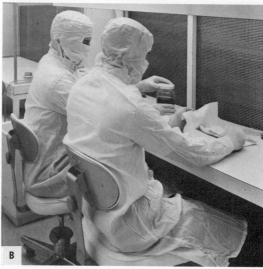

B

Figure 19-11. Laminar-airflow system. (A) Schematic drawing of horizontal laminar-flow tunnel. Arrows in tunnel denote parallel flow of air through a room. *(Redrawn from M. S. Favero, "Industrial Applications of Laminar Air-flow," in Developments in Industrial Microbiology, American Institute of Biological Sciences, Washington, 1970, vol. 11.)* (B) Laboratory personnel performing sterility test in laminar-airflow unit. *(Courtesy of B. Phillips, Becton, Dickinson & Company.)*

filters are used in microbiological transfer chambers to prevent contamination, in isolation areas to prevent spread of infection, and in rooms used to assemble miniaturized electronic equipment because contamination with particles as small as bacteria can ruin a component's performance.

Face masks
Gauze shields fitted with tapes or ties to enclose the mouth and nose are called face masks; they are used by the surgical team during operations as a filter to remove microorganisms during breathing so as not to contaminate the operating theater. Masks also provide protection for hospital personnel caring for patients with infectious diseases by filtering airborne microorganisms from the air during breathing.

Physical cleaning

Ultrasonics
High-frequency sound waves are used for disintegrating microbial cells and cleaning (removing) microorganisms from equipment, and in specialized diagnostic and surgical techniques. Of particular interest in relation to control of microorganisms is the action of high-frequency sound waves in ultrasonic cleaners, units filled with a liquid through which the sound waves pass. As the high-frequency sound waves travel through the liquid, they produce large numbers of small bubbles which, after they reach a certain size, collapse violently. This phenomenon is called *cavitation*; the force generated is responsible for removing any dirt or particles (including microorganisms) from surfaces of materials in the liquid. An ultrasonic cleaner is more efficient in removing organic material from instruments than is mechanical brushing.

Washing
Washing or scrubbing with soap is another physical means of removing microorganisms from surfaces. Scrubbing loosens dirt from objects and

330

Table 19-6. Application of physical agents for controlling microorganisms

METHOD	RECOMMENDED USES	LIMITATIONS
Moist heat:		
Autoclave	Sterilizing instruments, linens, utensils and treatment trays, media and other liquids	Ineffective against organisms in materials impervious to steam; cannot be used for heat-sensitive articles
Free-flowing steam or boiling water	Destruction of nonsporeforming pathogens; sanitizes bedding, clothing, and dishes	Cannot be guaranteed to produce sterilization on one exposure
Dry heat:		
Hot-air oven	Sterilizing materials impermeable to or damaged by moisture, e.g., oils, glass, sharp instruments, metals	Destructive to materials which cannot withstand high temperatures for long periods
Incineration	Disposal of contaminated objects that cannot be reused	Size of incinerator must be adequate to burn largest load promptly and completely; potential of air pollution
Radiation:		
Ultraviolet light	Control of airborne infection; disinfect surfaces	Must be absorbed to be effective (does not pass through transparent glass or opaque objects; irritating to eyes and skin; low penetration)
X-ray, gamma, and cathode radiation	Sterilization of heat-sensitive surgical materials and other medical devices	Expensive and requires special facilities for use
Filtration:		
Membrane filters	Sterilization of heat-sensitive biological fluids	Fluid must be relatively free of suspended particulate matter
Fiberglass filters (HEPA)	Air disinfection	Expensive
Physical cleaning:		
Ultrasonics	Effective in decontaminating delicate cleaning instruments	Not effective alone, but as adjunct procedure enhances effectiveness of other methods
Washing	Hands, skin, objects	Sanitizes; reduces microbial flora

from the skin and hands by friction. The soap removes the oily film which holds the bacteria to surfaces including the skin. Once this film is removed, microorganisms are flushed off by the flow of water.

A summary of the application of physical agents for the control of microorganisms is provided in Table 19-6.

Summary and outlook

A variety of physical agents or processes are available for controlling microbial populations. The control may be accomplished by killing the microorganisms, by inhibiting their growth and metabolism, or by physically removing them. The type of control to be employed is determined by the circumstances peculiar to each situation.

High temperature, particularly steam under pressure, is one of the most efficient and effective ways of sterilizing a product. However, certain materials used in the laboratory, home, and hospital are damaged by high temperatures. Alternative sterilization procedures, such as radiation, electron beams, or filtration, must be employed to sterilize materials that would be damaged by high temperatures.

As new materials are manufactured which require that they be sterile before use, studies need to be performed to determine the most satisfactory method for their sterilization. Such needs encourage research into new areas for microbiological control. This is particularly true when a product needs to be made available in a sterile condition on a commercial scale.

The national space program has had to cope with new problems of microbial control, such as the assembly of a unit to be sent into space that must be sterile and the assembly of miniaturized electronic equipment that can be fouled by contamination by particles no larger than microorganisms. Very specific methods for microbial control must be developed for these situations.

Key terms

autoclave	laminar airflow	radioisotopes
cathode rays	lyophilization	thermal death time
cavitation	membrane filter	thermolabile
decimal reduction time	osmosis	ultrasonics
desiccation	plasmolysis	ultraviolet rays
fractional sterilization	plasmoptysis	x-rays
gamma rays		

Questions

1 Compare the efficacy of dry heat and moist heat as means of sterilization.

2 Compare vegetative cells of bacteria with bacterial spores in terms of resistance to heat.

3 What is an autoclave? What kind of materials are usually sterilized by autoclaving?

4 What is the principle of fractional sterilization?

5 List some materials preferably sterilized in a hot-air oven rather than in the autoclave. Indicate, for each material, why dry-heat sterilization is the method of choice.

6 What is the effect of subzero temperatures on the survival of microorganisms?

7 Why is desiccation effective as a means of preservation? Is the antimicrobial action microbicidal or microbistatic? Give several examples of materials preserved by desiccation.

8 In what respect is microbial inhibition by desiccation similar to inhibition brought about by high concentrations of sugars or salts?

9 List several types of microbicidal radiations. What are the advantages, limitations, and practical applications of each?

10 What types of material are generally sterilized by means of bacteriological filtration devices? Why?

CONTROL OF MICROORGANISMS BY CHEMICAL AGENTS

Many chemical substances are capable of inhibiting or killing microorganisms. They range from heavy metallic elements such as silver and copper to complex organic molecules such as the quaternary ammonium compounds. The various substances exert their antimicrobial effect in different ways and on different groups of organisms. The effects they have upon the surfaces or materials to which they are applied also vary; some are compatible while others are destructive. Because of these and other variables, it is necessary to know the behavior of a chemical agent before using it for a particular practical application.

In this chapter we will characterize several classes of compounds which are used to control microbial populations, describe their modes of action, and indicate their practical applications. Chemicals used for the treatment of specific diseases (*chemotherapeutic agents*) will be discussed in Chap. 21.

As you read this chapter, it is important to keep in mind the conditions described in Chap. 18 which influence the effectiveness of an antimicrobial agent.

Characteristics of an
ideal disinfectant

No single chemical antimicrobial agent is best for all purposes. This is not surprising in view of the variety of conditions under which agents may be used, the differences in mode of action, and the many types of microbial cells to be destroyed. If there were an ideal disinfectant, it would have to possess a formidable array of characteristics. A single compound possessing these properties may never be found. Nevertheless, the specifications described below can be aimed for in the preparation of new antimicrobial compounds, and they should be considered in evaluating disinfectants for practical use.

1 *Antimicrobial activity.* The capability of the substance to kill microorganisms is the first requirement. The chemical, at a low concentration, should have a broad spectrum of antimicrobial activity; that is, it should kill many kinds of microbes.

2 *Solubility.* The substance must be soluble in water or other solvents to the extent necessary for effective use.

3 *Stability.* Changes over time in the substance as it stands on the shelf should be minimal and should not result in significant loss of antimicrobial action.

4 *Nontoxicity to humans and other animals.* Ideally, the compound should be lethal to microorganisms and noninjurious to humans and other animals.

5 *Homogeneity.* The preparation must be uniform in composition so that active ingredients are present in each application. Pure chemicals are uniform, but mixtures of materials may lack homogeneity.

6 *Noncombination with extraneous organic material.* Many disinfectants combine with proteins or other organic material. When such disinfectants are used in situations where there is considerable organic material, most of the disinfectant will be inactivated.

7 *Antimicrobial activity at room or body temperatures.* It should not be necessary to raise the temperature beyond that normally found in the environment where the compound is to be used.

8 *Ability to penetrate.* Unless the substance can penetrate the surface, its antimicrobial action is limited to the site of application. Sometimes, of course, surface action is all that is required.

9 *Noncorroding and nonstaining.* The compound should not rust or otherwise disfigure metals, nor should it stain or damage fabrics.

10 *Deodorizing ability.* Deodorizing while disinfecting is a desirable attribute. Ideally, the disinfectant itself should either be odorless or have a pleasant smell.

11 *Detergent capability.* A disinfectant which is also a detergent (cleaning agent) has the advantage that the cleaning action improves the effectiveness of the disinfectant.

12 *Availability and cost.* The compound must be available in large quantities at a reasonable price.

Selection of a chemical antimicrobial agent

Some of the factors to be considered in the selection of a chemical antimicrobial agent for practical application are:

1 *Nature of the material to be treated.* A chemical agent used to disinfect contaminated utensils might be quite unsatisfactory for application to the skin; i.e., it might seriously injure the tissue cells. Consequently, the substance selected must be compatible with the material to which it is applied.

2 *Type of microorganism.* Not all microorganisms are equally susceptible to the inhibitory or killing action of a specific chemical. Therefore, the agent selected must be known to be effective against the type of microorganism to be destroyed. For example, spores are more resistant than vegetative cells. Gram-positive and gram-negative bacteria differ in susceptibility; *Escherichia coli* (gram-negative) is much more resistant to cationic disinfectants than *Staphylococcus aureus* (gram-positive). Different strains of the same species also vary in their susceptibility to a given antimicrobial agent.

3 *Environmental conditions.* The factors discussed in Chap. 18—temperature, pH, time, concentration, and presence of extraneous organic material—may all have a bearing on the rate and efficiency of microbial destruction. The successful use of an antimicrobial agent requires an understanding of the influence of these conditions on the particular agent so that it can be employed under the most favorable circumstances.

Major groups of chemical antimicrobial agents

Some of the major groups of chemical antimicrobial agents are:

1. Phenol and phenolic compounds
2. Alcohols
3. Halogens
4. Heavy metals and their compounds
5. Detergents
6. Aldehydes
7. Gaseous chemosterilizers

Phenol and phenolic compounds

Phenol (carbolic acid), which was first used by Lister during the 1860s in his work to develop aseptic surgical techniques, has long been the standard against which other disinfectants are compared when evaluating their bactericidal activity. Many other disinfectants now available are much more effective and are active in considerably lower concentrations. Figure 20-1 shows the chemical structure of phenol and of a few of the numerous phenol derivatives that have been prepared for use as disinfectants; many of the latter are more effective than phenol itself. These compounds probably act primarily by denaturing cell proteins and damaging cell membranes. The cresols are several times more germicidal than phenol; o-phenylphenol and other highly substituted phenolic compounds are effective at high dilutions. Phenolics are either bactericidal or bacteriostatic, depending upon the concentration used. Bacterial spores and viruses are more resistant to them than are bacterial vegetative cells. Some phenolics are highly fungicidal. The antimicrobial activity of phenolics is reduced at

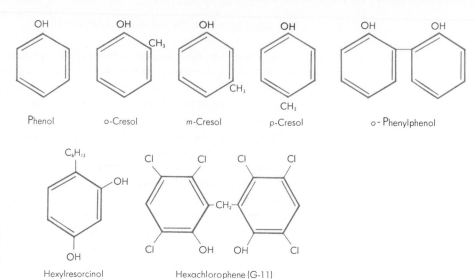

Figure 20-1. Phenol and some phenolic compounds which are used as disinfectants.

Phenol   o-Cresol   m-Cresol   p-Cresol   o-Phenylphenol

Hexylresorcinol   Hexachlorophene (G-11)

an alkaline pH and by organic material. Low temperatures and the presence of soap will also reduce their antimicrobial activity.

The phenolic compounds are among the best surface disinfectants for inanimate objects.

Alcohols

Ethyl alcohol in concentrations of 50 to 70% is effective against vegetative or nonsporeforming microorganisms.

Ethyl alcohol has low sporicidal activity. In his book *Disinfection and Sterilization*, G. Sykes notes that "there is one record of the survival of anthrax spores in alcohol for 20 years and another one of *Bacillus subtilis* spores for 9 years."

Methyl alcohol is less bactericidal than ethyl alcohol. It is also highly poisonous; even the fumes of this compound may produce permanent injury to the eyes. Thus, it is generally not employed as a disinfectant. The higher (longer carbon chain) alcohols—propyl, butyl, amyl, and others—are germicidal. However, alcohols of molecular weight higher than that of propyl alcohol are not miscible in all proportions with water and hence are not commonly used in disinfectants. Propyl and isopropyl alcohols in concentrations ranging from 40 to 80% are useful as skin disinfectants. Various other agents, such as iodine and glycerol, have been combined with alcohol; some have improved antibacterial efficiency.

Alcohol is effective in reducing the microbial flora on the skin and for the disinfection of oral thermometers. The comparative effectiveness of alcohol and other disinfectants applied to skin is shown in Fig. 20-2. Alcohol concentrations above 60% are effective against viruses; however, the effectiveness is influenced considerably by the amount of extraneous protein material in the mixture. The extraneous protein reacts with the alcohol, thus reducing the effectiveness of the alcohol against the virus.

Alcohols are protein denaturants, a property that may, to a large extent, account for their antimicrobial activity. They are also lipid solvents and hence may damage the cell membrane.

336

Halogens and their compounds

The halogen family consists of the elements fluorine, chlorine, bromine, and iodine. Chlorine and iodine are most widely used as antimicrobial agents.

Iodine. This is one of the oldest and most effective germicidal agents. It has been in use for more than a century, having been recognized in the *United States Pharmacopeia* in 1830. Pure iodine is only slightly soluble in water but readily soluble in alcohol and aqueous solutions of potassium or sodium iodide. Iodine is traditionally used as a germicidal agent in a form referred to as *tincture of iodine*—2% iodine plus 2% sodium iodide in approximately 50% alcohol. Iodine is also used in the form of substances known as *iodophors*, mixtures of iodine with surface-active agents which act as carriers and solubilizers for the iodine. One of these agents is polyvinylpyrrolidone (PVP); the complex with iodine is often called PVP-I. Iodophors slowly release iodine from the complex and thus possess the germicidal characteristics of iodine but have the additional advantages of nonstaining and lesser irritant properties.

Iodine is a highly effective agent and is unique in that it is effective against all kinds of bacteria, spores, fungi, and viruses. Iodine solutions are chiefly used for the disinfection of skin, particularly as a preoperative skin disinfectant.

Chlorine and chlorine compounds. Chlorine, either as a gas or in chemical combination (chlorine compounds), is one of the most widely used disinfectants. The compressed gas in liquid form is almost universally employed for the purification of municipal water supplies. Chlorine gas is difficult to handle and dangerous unless special equipment is available to dispense it.

There are available many compounds of chlorine which can be handled more conveniently than chlorine gas and which, properly used, are equally effective as disinfectants. Some of these are described below.

Hypochlorites. Calcium hypochlorite, $Ca(OCl)_2$ (also known as chlorinated

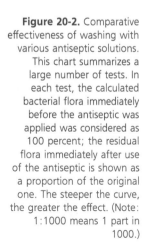

Figure 20-2. Comparative effectiveness of washing with various antiseptic solutions. This chart summarizes a large number of tests. In each test, the calculated bacterial flora immediately before the antiseptic was applied was considered as 100 percent; the residual flora immediately after use of the antiseptic is shown as a proportion of the original one. The steeper the curve, the greater the effect. (Note: 1:1000 means 1 part in 1000.)

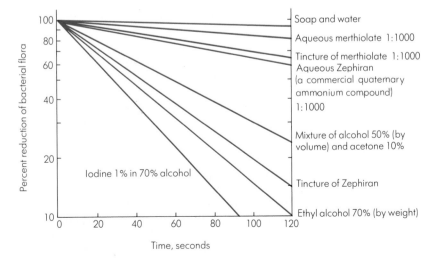

lime), and sodium hypochlorite, NaOCl, are compounds widely used domestically and industrially. They are available as powders or liquid solutions and in varying concentrations depending upon the recommended use. Products containing 5 to 70% calcium hypochlorite are used for sanitizing dairy equipment and eating utensils in restaurants. Solutions of sodium hypochlorite of a 1% concentration are used for personal hygiene and as a household disinfectant; higher concentrations of 5 to 12% are employed for household bleaches and disinfectants and for sanitizing agents used in dairy and food-processing establishments.

Chloramines. The chloramines are characterized by having one or more of the hydrogen atoms in an amino group of a compound replaced with chlorine. The simplest of these is monochloramine. Chloramine-T and azochloramide, two of several germicidal compounds in this category, have more complex chemical structures as shown here:

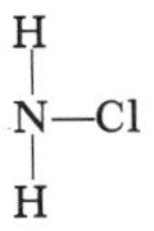

Monochloramine 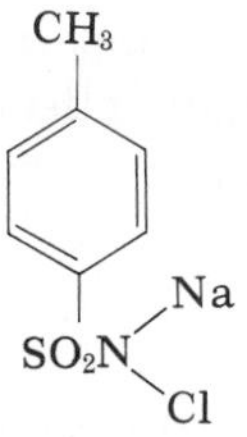　　　　Chloramine-T 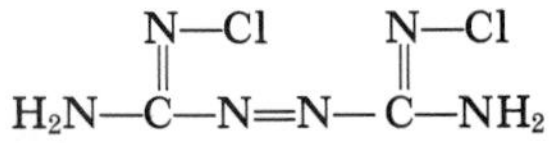　　　　Azochloramide

One of the advantages of the chloramines is their stability, which is much greater than the stability of hypochlorites. That is, they release chlorine over a long period of time.

The germicidal action of chlorine and its compounds comes through the hypochlorous acid formed when free chlorine is added to water:

$$Cl_2 + H_2O \longrightarrow HCl + HClO$$
　　　　　　　　　　Hypochlor-
　　　　　　　　　　ous acid

Similarly, hypochlorites and chloramines undergo hydrolysis, forming hypochlorous acid. The hypochlorous acid formed in each instance is further decomposed:

$$HClO \longrightarrow HCl + O$$
　　　　Formed from
　　　　chlorine,
　　　　hypochlorites,
　　　　chloramines

The oxygen released in this reaction (nascent oxygen) is a strong oxidizing agent, and it destroys microorganisms by its action on their cellular constituents. Chlorine and its compounds also kill microorganisms by direct combination of the chlorine with proteins of the cell.

Figure 20-3. Oligodynamic phenomenon. A zone of inhibition surrounds the silver coin placed on an inoculated petri dish.

Heavy metals and their compounds

Most of the heavy metals, either alone or in certain compounds, are detrimental to microorganisms. The most effective are mercury, silver, and copper.

Heavy metals (oligodynamic action). Extremely small amounts of certain metals, particularly silver, can exert a lethal effect upon bacteria; this is designated *oligodynamic action* from the two Greek words *oligos*, meaning "small," and *dynamis*, meaning "power." The phenomenon can be demonstrated in the laboratory by placing a clean piece of metal (silver or copper) on an inoculated plate. After incubation, a zone of inhibition (no growth) surrounds the metal (see Fig. 20-3). The amount of metal needed for this inhibitory effect is extremely small, only a few parts per million.

Heavy-metal compounds. Numerous compounds of heavy metals have germicidal or antiseptic activity. The most prominent antimicrobial compounds of heavy metals are those of mercury, silver, and copper. Mercuric chloride, once a popular disinfectant, is no longer used; however, several organic mercury compounds (Merthiolate, Mercurochrome, and Metaphen) are used as antiseptics. Silver nitrate has long been used for the prevention of gonococcus infections of newborn infants' eyes (neonatal gonococcal ophthalmitis). Copper compounds are widely used as fungicides in agriculture.

One mode of action of heavy metals and their compounds is through protein denaturation. In the case of mercuric chloride, the inhibition is directed at enzymes which contain the sulfhydryl grouping as illustrated below:

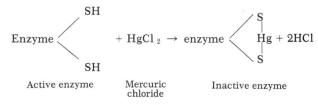

$$\text{Enzyme}\left\langle \begin{array}{c} \text{SH} \\ \\ \text{SH} \end{array} \right. + \text{HgCl}_2 \rightarrow \text{enzyme} \left\langle \begin{array}{c} \text{S} \\ | \\ \text{Hg} \\ | \\ \text{S} \end{array} \right. + 2\text{HCl}$$

Active enzyme   Mercuric chloride   Inactive enzyme

Detergents   Surface-tension depressants, or wetting agents, employed primarily for cleansing surfaces are called *detergents*. Soap is an example. It does not, however, work well in hard water. For this reason, many new more efficient cleaning agents have been developed; these are called *surfactants* or *synthetic detergents*. They do not form precipitates in alkaline or acid water, nor do they react with the minerals present in hard water to form deposits. Some soaps and other detergents are bactericidal.

Chemically, detergents are classified as follows:

1 Those which ionize with the detergent property resident in the *anion* are referred to as *anionic* detergents. Examples are:

$$[C_9H_{19}COO]^- Na^+ \qquad [C_{12}H_{25}OSO_3]^- Na^+$$

A soap   Sodium lauryl sulfate

The real value of soaps lies in the mechanical removal of microorganisms. Like other detergents, soaps reduce surface tension and thereby increase the wetting power of the water in which they are dissolved. Soapy water has the ability to emulsify and disperse oils and dirt. The microorganisms become enmeshed in the soap lather and are removed by the rinse water. Various chemicals are incorporated into some soaps to enhance their germicidal activity.

2 Those which ionize with the detergent property resident in the *cation* are referred to as *cationic* detergents. An example is:

$$\left[\bigcirc\!\!\!\!-N-C_{16}H_{33}\right]^- Cl^+$$

Cetylpyridinium chloride (Ceepryn)

Quaternary ammonium compounds. These cationic detergents are more germicidal than the anionic compounds. A large number of these compounds have been synthesized and evaluated for practical use as disinfectants, antiseptics, and sanitizers. Examples of three major types of quarternary ammonium compounds are benzalkonium chloride (Zephiran), benzethonium chloride (Phemerol), and cetylpyridinium chloride (Ceepryn).

The bactericidal power of the quaternaries is exceptionally high against gram-positive bacteria, and only slightly less effective against gram-negative organisms. Bactericidal concentrations range from dilutions of one part in a few thousand to one part in several hundred thousand. The quaternaries manifest bacteriostatic action in dilutions far beyond their effective bactericidal concentration. For example, the list of bactericidal action for a given compound may be at a dilution of 1:30,000; yet it may remain bacteriostatic in dilutions as high as 1:200,000. The antimicrobial characteristics of these compounds demonstrate the need to distinguish between static and lethal activity. This can be done by laboratory tests.

Quaternaries are also active against fungi and protozoa, but viruses appear to be more resistant. Although they are used extensively as skin antiseptics and as sanitizing agents in eating and drinking establishments, dairies, and food-processing plants, quaternaries have several limitations. They do not kill or inhibit bacterial and fungal spores, nor are they effective against the tubercle bacillus. Many *Pseudomonas*-type bacteria are resistant to them. The activity of quaternaries is reduced by soaps, anionic detergents, and organic material. Dilute solutions of them may become contaminated.

Various modes of antimicrobial action have been proposed for the quaternaries, including enzyme inhibition, protein denaturation, and disruption of the cell membrane causing a leakage of vital constituents.

Aldehydes

Glutaraldehyde and formaldehyde are two compounds in the aldehyde class which have various applications for controlling microorganisms.

Glutaraldehyde. This compound is a saturated dialdehyde with the formula

$$O=\overset{\overset{\displaystyle H}{|}}{C}-CH_2-CH_2-CH_2-\overset{\overset{\displaystyle H}{|}}{C}=O$$

A 2% solution of glutaraldehyde exhibits a wide spectrum of antimicrobial activity. It is effective against vegetative bacteria, fungi, bacterial and fungal spores, and viruses. It is used for sterilizing urological instruments, instruments with lenses, and other medical devices. However, a long exposure time is required for sterilization.

Formaldehyde. Formaldehyde is a gas that is stable only in high concentrations and at elevated temperatures. At room temperature it polymerizes, forming a solid substance. The important polymer is *paraformaldehyde*, a colorless solid which rapidly yields formaldehyde upon heating. Formaldehyde is also marketed in aqueous solution as *formalin*, which contains 37 to 40% formaldehyde. Formalin has very high antimicrobial activity (see Table 20-1); formaldehyde vapors sterilize when used in an enclosed area and under appropriate conditions. The undesirable characteristics of formaldehyde are its irritation of the skin and its noxious fumes.

Gaseous chemo-
sterilizers

A variety of products are now made from materials which cannot be sterilized by the use of high temperatures or liquid chemosterilizers. Chemical sterilization by gaseous agents is effective and practical for such items. In this process, the material is exposed to a gas in a confined area at room temperature. After treatment, the gas can be conveniently and easily removed. Plastic heat-sensitive items, such as syringes, test tubes, petri dishes, and pipetes, and enclosed areas, such as sealed-off rooms, can be sterilized by gases. The main agent currently used for gaseous sterilization is ethylene oxide.

Ethylene oxide. This is a relatively simple organic compound having the formula
$$H_2C-CH_2$$
$$\backslash\,/$$
$$O$$

Table 20-1. Evaluation of selected germicides

CLASS	USE CONCENTRATION	ACTIVITY LEVEL*
Ethylene oxide (gas) (in autoclave-type equipment at 55 to 60°C)	450–800 mg/liter	High
Glutaraldehyde, aqueous	2%	High
Formaldehyde + alcohol	8% + 60–70%	High
Formaldehyde, aqueous	3–8%	High to intermediate
Iodine + alcohol	0.5–70%	Intermediate
Alcohols	70–95%	Intermediate
Chlorine compounds	4–5%	Intermediate
Phenolic compounds	0.5–3%	Intermediate to low
Iodophors	75–150 ppm	Intermediate to low
Quaternary ammonium compounds	1:750	Low
Mercurial compounds	1:500–1,000	Low

INTERPRETATION OF ACTIVITY LEVEL*

	TEST ORGANISM					
	BACTERIA			FUNGI‡	VIRUSES	
	VEGETATIVE†	TUBERCLE BACILLUS	SPORES		LIPID AND MEDIUM-SIZED	NONLIPID AND SMALL-SIZED
High	+	+	+	+	+	+
Intermediate	+	+	−	+	+	+
Low	+	−	−	+	+	−

*+means that cidal effect can be expected.
†Common forms of bacterial cells, e.g., *Staphylococcus*.
‡Includes usual asexual spores but not necessarily dried chlamydospores and sexual spores.
Courtesy of E. H. Spaulding, Temple University.

It is a liquid at temperatures below 10.8°C (51.4°F). Above this temperature it vaporizes rapidly. Vapors of ethylene oxide in air are highly flammable, even in low concentrations. This objectionable feature has been overcome by preparing mixtures of ethylene oxide in carbon dioxide or dichlorodifluoromethane (Freon). These mixtures are available commercially.

Ethylene oxide is a powerful sterilizing agent. It is universally used for sterilizing heat- or moisture-sensitive materials in hospitals, industry, and laboratories. Bacterial spores, which are many times more resistant than vegetative cells as measured by other antimicrobial agents, show little resistance to destruction by ethylene oxide. Figure 20-4 illustrates the sporicidal action of this gas. An outstanding and desirable feature of ethylene oxide is its power to penetrate. It will pass through and sterilize large packages of materials, bundles of cloth, and even certain plastics. Ethylene oxide is explosive and toxic, however, and must be used with caution. Devices are available for its safe routine laboratory and hospital use. The commercially available apparatus for this purpose is essentially an autoclave modified to allow the chamber to be filled with the gas under controlled conditions. The concentration of ethylene oxide, the temperature, and the humidity are the critical factors which determine the time required to achieve sterilization. Modern autoclaves are equipped with controls to maintain the desired concentration of ethylene oxide and the proper temperature and humidity.

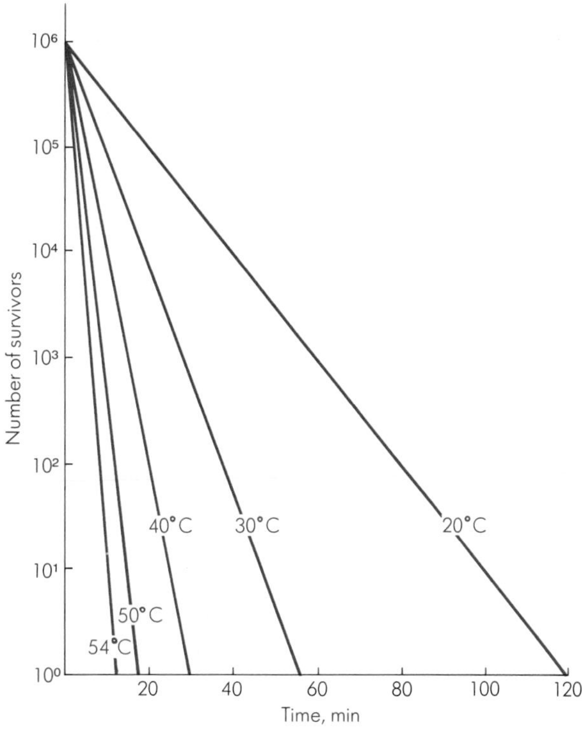

Figure 20-4. Decrease in number of *Bacillus subtilis* spores on paper strips surviving at various temperatures in gaseous ethylene oxide at 1,200 mg/liter and 40% relative humidity. *(From R. R. Ernst, "Ethylene Oxide Gaseous Sterilization for Industrial Applications," in G. B. Phillips and W. S. Miller (eds.), Industrial Sterilization, 1973, Duke University Press, Durham, NC.)*

An evaluation of the antimicrobial action of ethylene oxide and a comparison of its effectiveness with other agents is shown in Table 20-1.

The mode of action of ethylene oxide is believed to be alkylation reactions with organic compounds including enzymes and other proteins. *Alkylation* is the replacement of an active hydrogen atom in an organic compound with an alkyl group, such as replacing the hydrogen atom in a free carboxyl, amino, or sulfhydryl group. In this reaction, the ring in the ethylene oxide molecule splits, and the molecule attaches itself to the site where the hydrogen was originally. This reaction would inactivate an enzyme with a sulfhydryl group as shown:

$$H_2C\!\!-\!\!CH_2 + enzyme\!\!-\!\!SH \rightarrow enzyme\!\!-\!\!S\!\!-\!\!CH_2CH_2OH$$
$$\underset{O}{\diagdown\!\diagup} \qquad\qquad\qquad Inactive$$

Table 20-2 provides a summary of the various chemical antimicrobial agents (or classes of agents) with a general statement of their applicability for practical usage. The likely mode of action of the various agents is also described.

Evaluation of disinfectants and antiseptics

Laboratory evaluations of antimicrobial chemical agents are conducted by one of three general procedures. In each, the agent is tested against a selected microorganism referred to as the *test organism*. These procedures are:

1 A liquid, water-soluble antimicrobial agent is appropriately diluted and dispensed into sterile test tubes; to each is added a measured amount of

343

Table 20-2.
Major groups of chemical antimicrobial agents

MAJOR GROUP	MODE OF ACTION	ADDITIONAL CHARACTERISTICS
Phenol and phenolic compounds	Denature protein; damage cell membrane	Derivatives (hexylresorcinol) greatly reduce surface tension
Alcohols	Denature protein; damage cell membrane; dehydrating agents; detergent action	More carbons in the alcohol make it more germicidal
Halogens Iodine	Halogenation of tyrosine; inactivates enzymes and other proteins	Effective against bacteria and spores
Chlorine (and chlorine compounds)	Combines with proteins of cell membranes and enzymes	Chlorine used to disinfect water; chlorine compounds are easier to apply and have many applications
Aldehydes	Break hydrogen bonds; denature protein	Effective against all microorganisms except bacterial spores
Gaseous chemosterilizers	Ethylene oxide alkylates organic compounds; inactivates enzymes	Kill *all* forms of life
Quaternary ammonium compounds (cationic detergents)	Denature proteins; damage to cell membrane	More germicidal than other detergents; most bactericidal against gram-positive bacteria; fungicidal

the test organism. At specified intervals, a transfer is made from this tube into tubes of sterile media that are then incubated and observed for the appearance of growth. This procedure can also be used to determine the number of organisms killed per unit of time by performing a plate count on samples at selected intervals.

A highly standardized test procedure, namely the *phenol-coefficient method*, is performed in this general manner. The test microorganisms (a designated strain of *Staphylococcus aureus* or *Salmonella typhi*) are exposed to dilutions of pure phenol and to dilutions of the chemical agent being evaluated. The phenol coefficient is expressed as a number, and it is calculated by comparing the activity of a standard dilution of phenol with the activity of a dilution of the chemical being tested. More specifically, one determines the dilution of each substance (phenol and the test compound) that does not kill in 5 min but does kill all cells in 10 min. The calculation of the phenol coefficient is done in the following manner:

$$\text{Phenol coefficient} = \frac{\text{the greatest dilution of test compound that kills in 10 min but not in 5 min}}{\text{the greatest dilution of phenol that kills in 10 min but not in 5 min}}$$

(SPECIFIC) COMPOUNDS	RECOMMENDED USE	LIMITATIONS
Cresols (more germicidal than phenol); hexylresorcinol	General disinfectant	Microbial effectiveness limited; irritating and corrosive
Methyl (least bactericidal, most toxic); ethyl (least toxic, used in 50–70% concentration); propyl, butyl, amyl, etc.	Skin antiseptic; 60% concentration kills viruses if there is no extraneous organic material	Antiseptic
Tincture of iodine (dissolved in alcohol); iodophors (+ surface active agents)	Disinfects skin	Irritating to mucous membranes
Hypochlorites (sanitize utensils and equipment); chloramines (oxidizing agents)	Water disinfection	Inactivated by organic material; pH-dependent for effectiveness; objectionable taste and odor unless strictly controlled
Glutaraldehyde	Sterilizing instruments; fumigation	Stability limited; not sporicidal
Formaldehyde; aqueous formalin	Sterilizing instruments; fumigation; tissue preservation	Permeation poor; corrosive
Ethylene oxide	Sterilizing heat-sensitive materials, instruments, large equipment, and mattresses	Flammable; potentially explosive in pure form; comparatively slow action
Cetylpyridinium chloride; Zephiran; Phemerol	Skin disinfection; sanitizing agents	Not sporicidal

Hence, if the test compound at a dilution of 1:180 killed the test organism in 10 min but not in 5 min, and if a 1:90 dilution of phenol did the same thing, the phenol coefficient of the test compound would be $^{180}\!/_{90}$, or 2. It is necessary in this type of procedure to ascertain whether the inhibitory action is bactericidal or bacteriostatic.

2 The chemical agent is incorporated into an agar medium or broth, inoculated with the test organism, incubated, and then observed for (a) decrease in the amount of growth or (b) absence of growth, depending on which effect is important in the intended application.

3 An agar medium, in a petri dish, is inoculated with the test organism. The chemical agent is placed on the surface of the medium. After a period of incubation, the plate is observed for evidence of a zone of inhibition (no growth) surrounding the site where the chemical substance was applied. This method is particularly suitable for testing antimicrobial agents that are to be used in semisolid preparations such as jellies or salves. Liquid preparations are applied to absorbent paper discs, and the discs are then placed on the agar medium.

These procedures are illustrated schematically in Fig. 20-5.
It should be emphasized that no single microbiological test method is

Figure 20-5. Laboratory evaluation of chemical antimicrobial agents: (A) No growth (clear) or growth (turbid) in broth due to presence or absence of chemical agent; (B) increased growth on broth as concentration of chemical agent is decreased; (C) increased growth in nutrient agar plates as concentration of chemical agent is decreased; (D) inhibition of microbial growth by chemical agent applied to center of inoculated medium in petri dish; zone of inhibition develops if compound is active. (*Erwin F. Lessel, illustrator.*)

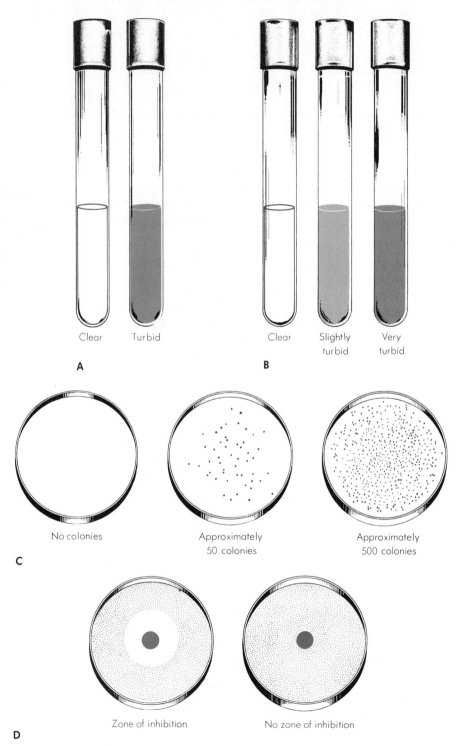

Clear   Turbid

A

Clear   Slightly turbid   Very turbid

B

No colonies   Approximately 50 colonies   Approximately 500 colonies

C

Zone of inhibition   No zone of inhibition

D

suitable for the evaluation of all germicidal chemicals for all applications recommended. Therefore, one must exercise care in selecting a test method for a specific chemical agent so that the results obtained will be meaningful and reproducible and lend themselves to some degree of practical interpretation. Since the conditions of a particular environment, such as temperature, pH, and presence of organic material, can greatly influence the antimicrobial action of a chemical substance, it is essential that the evaluation of its effectiveness be in terms of the likely practical conditions of use.

Summary and outlook

Thousands of chemical substances are available for use in controlling microorganisms. It is important to understand the distinctive characteristics of each of these substances in terms of the kinds of microorganisms it is active against as well as how it is influenced by the environment in which it is used. Each chemical substance has limitations in terms of its effectiveness; when used under practical conditions, these limitations need to be observed. Furthermore, the desired objective in terms of microbial control is not always the same; in some instances it may be necessary to kill all the microorganisms (sterilize), while in other cases it may be satisfactory to kill most, but not all (sanitize). Thus the selection of a chemical agent for practical use is also influenced by the antimicrobial result expected of it.

The mode of action by which chemicals inhibit or kill microorganisms varies; some alter the cell wall or the cell membrane structures; others inhibit the synthesis of vital cellular constituents or change the physical state of cellular material. Knowledge of the specific manner in which a chemical produces its antimicrobial effect is useful both in terms of assessment for practical usage and for suggesting possible improvements for the design of new substances.

The development of new products sometimes requires the development of new methods for their sterilization. For example, plastic medical devices cannot be sterilized by autoclaving without being destroyed, so commercial equipment using ethylene oxide was developed. New chemicals are continually being synthesized and evaluated for their antimicrobial ability in the hope of discovering more effective antimicrobial substances.

Key terms

aldehyde	detergent	quaternaries
alkylation	formaldehyde	tincture of iodine
anionic detergent	glutaraldehyde	
cationic detergent	iodophors	
chemical antimicrobial agent	oligodynamic action	

Questions

1 Why is no single chemical agent best for all situations requiring control of microbial populations?

2 What characteristics of a chemical agent should be considered when selecting it for disinfection of utensils in a hospital?

3 List the major conditions influencing the antimicrobial activity of chemical agents.

4 Compare the germicidal capability of phenol with that of other phenolic disinfectants.

5 What relationship exists between various concentrations of ethyl alcohol and germicidal activity? How do ethyl alcohol and the higher alcohols vary in effectiveness?

6 How effective are iodine preparations as antimicrobial agents? Chlorine preparations?

7 Define the term *oligodynamic action*. How can it be demonstrated in the laboratory?

8 Explain the term *cationic detergents*. What class of compounds of this category has come into prominence? What is their antimicrobial capacity?

9 Why has sterilization with ethylene oxide become a widely used practice? Why is this gas mixed with carbon dioxide or dichlorodifluoromethane (Freon) for use as a sterilizing agent?

10 A disinfectant is found to have a phenol coefficient of 2.0. What does this mean?

11 How would you demonstrate the antimicrobial activity of an antiseptic ointment?

12 What is the likely mode of action of the following chemicals in inhibiting or killing microorganisms: phenol, chlorine, alcohol, silver nitrate, ethylene oxide?

ANTIBIOTICS AND OTHER CHEMOTHERAPEUTIC AGENTS

Chemotherapeutic agents are chemical substances used for the treatment of infectious diseases (chemotherapy) or the prevention of disease (chemoprophylaxis). These substances are obtained from microorganisms or plants or are synthesized in the chemistry laboratory. In general, the naturally occurring chemical substances are distinguished from synthetic compounds by the name *antibiotics*.

For a chemical substance to be useful as a chemotherapeutic agent it must have *selective toxicity*. That is, it must inhibit or kill the parasite (or malignant cell) while causing little or no injury to the cells of the host. Other requirements for a practical chemotherapeutic agent are that it be capable of penetrating the cells and tissues of the host and that it not alter the host's natural defensive mechanisms.

In this chapter we will describe the characteristics of various chemotherapeutic agents and the clinical laboratory procedures performed to determine their antimicrobial activity.

History of chemotherapy

Quinine

Europeans used natural quinine from the bark of the cinchona tree of South America to treat malaria (a disease caused by protozoa of the genus *Plasmodium*) as early as 1630. It was used even earlier by South American Indians who relieved symptoms of malarial fever by chewing the bark of the cinchona tree. New synthetic compounds (quinacrine, chloroquine, paludrine, and primaquine) have replaced quinine for treatment of malaria.

Salvarsan

Syphilis, caused by the bacterium *Treponema pallidum*, is one of the first diseases for which a chemotherapeutic agent was used. Mercury was used to treat syphilis as early as 1495. But it was not until about 1910, when Paul Ehrlich synthesized an arsenical compound known as Salvarsan, that a specific drug capable of curing the disease without great danger to the patient was developed (Fig. 21-1). Ehrlich's contributions were especially important because his was the first systematic and deliberate search for a compound that had potent parasiticidal properties, low toxicity for humans and other animals, and good chemical stability. For this important discovery he and Elie Metchnikoff were awarded the Nobel Prize in physiology or medicine. Ehrlich's compound has been replaced in syphilis therapy byarsphenamine, neoarsphenamine, other arsenical compounds, and antibiotics.

Sulfonamides

Despite the effectiveness of Ehrlich's drug, no significant developments in the discovery of new chemotherapeutic agents occurred during the next two decades. In 1935, however, a breakthrough occurred when a group of researchers in Germany under the direction of Gerhard Domagk found that a particular dye (Prontosil) cured mice that were given lethal doses of hemolytic streptococci, bacteria that destroy red blood cells and cause "strep throat," scarlet fever, and other infections in humans.

Figure 21-1. Paul Ehrlich is generally regarded as having established chemotherapy as a science. His research in the early 1900s resulted in the synthesis of an arsenical compound (Salvarsan) for treatment of syphilis. His research represented a major contribution to the systematic search for new drugs.

After Domagk's reports in 1935 and confirmatory research in other countries—notably England, France, and later the United States—interest in chemotherapy reached an all-time high. French chemists at the Pasteur Institute studying the action of Prontosil on bacteria and attempting to improve it discovered that its antibacterial activity was due to the sulfanilamide component of the dye. Although sulfanilamide had been synthesized by the German chemist Paul Gelmo in 1908, it was the recognition of the antibacterial activity of sulfanilamide which sparked a search for related compounds having therapeutic value. By 1945, an estimated 5,488 derivatives of sulfanilamide had been made. Several of these have antimicrobial activity that makes them useful as chemotherapeutic agents. The important result of the search for new varieties of sulfonamides has been the development of drugs with increased antibacterial activities and fewer unfavorable reactions in the host animal.

Antibiotics
The word *antibiotic* has come to refer to a metabolic product of one organism that is detrimental or inhibitory to other microorganisms in very small amounts. Stated differently, an antibiotic is a chemical produced by one microorganism which is inhibitory to other microorganisms.

It has been known for many years that antagonisms exist between some microorganisms growing in close association in natural environments.

The first systematic search for, and study of, antibiotics made by A. Gratia and S. Dath about 1924 resulted in the discovery of actinomycetin in strains of actinomycetes, one of the major groups of bacteria found in soil. Actinomycetin was never used for the treatment of patients but was used to lyse cultures of bacteria for the production of vaccines. Since 1940, however, many valuable chemotherapeutic antibiotics have been isolated from the actinomycetes.

In 1929 Alexander Fleming (Fig. 21-2) noticed that an agar plate inoculated with *Staphylococcus aureus* had become contaminated with a mold and that the mold colony was surrounded by a clear zone, indicating inhibition of bacterial growth (Fig. 21-3). Because the mold was identified as a species of *Penicillium*, Fleming called the antibiotic *penicillin*. Although he isolated and identified the mold and studied its activities, the great importance of Fleming's observation was not realized until World War II when there was an urgent need for a better means of preventing fatal infection of war wounds. Investigators in England and the United States formed a research team to develop a method for large-scale production of penicillin. Both governments assigned top priority to accomplishing this objective. The inhibitory substance from Fleming's "contaminant mold" became a "miracle drug." The use of penicillin and other antibiotics has resulted in dramatic relief from the suffering and death of patients with infectious diseases. Earlier in this century the leading causes of death included bacterial diseases such as pneumonia, diphtheria, tuberculosis, and dysentery; therapy for the venereal diseases syphilis and gonorrhea was prolonged and uncertain. Today, these and many other dreaded diseases can be effectively treated with one of the antibiotic chemotherapeutic agents developed in this century.

Figure 21-2. Sir Alexander Fleming discovered the bacterial inhibitory properties of a metabolic product of *Penicillium notatum*. He called the substance penicillin. In 1929 this discovery opened the era of antibiotics. For his contributions, Fleming was knighted and shared the Nobel Prize in physiology or medicine in 1945 with Ernest B. Chain, a chemist, and Sir Howard W. Florey, a physician.

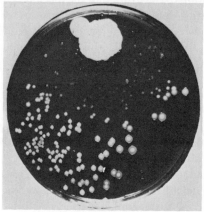

Figure 21-3. Fleming's original plate demonstrated the inhibition of *Staphylococcus aureus* (colonies at bottom) by a colony of *Penicillium notatum* (large white circle at top). This led to the discovery of penicillin. *(Courtesy of Prof. Robert Cruickshank.)*

Figure 21-4. In 1939 Rene Dubos isolated two antibiotics, gramicidin and tyrocidine, from a soil bacterium, *Bacillus brevis*.

In 1939 René Dubos (Fig. 21-4) isolated from New Jersey soil a culture of *Bacillus brevis* which produced a substance that killed many gram-positive bacteria. The cell-free extract produced from *B. brevis* was found to contain two active agents now known as *gramicidin* and *tyrocidine*. These successes were followed closely by the discovery of *streptomycin* by Selman Waksman (Fig. 21-5) and his associates.

Several thousand antibiotic substances have been isolated and identified since 1940. Of these, only a small number have proved useful for the treatment of disease. But the few effective substances have radically changed the entire medical practice of treating infectious diseases.

**Antibiotic chemo-
therapeutic agents**

An ideal antibiotic chemotherapeutic agent would have the following qualities:

1 It should have the ability to destroy or inhibit specific pathogenic microorganisms. The larger the number of different species affected the better. A *broad-spectrum antibiotic* is effective against many species.
2 It should not cause the development of resistant forms of the parasites.
3 It should not produce undesirable side effects in the host, such as allergic reactions, nerve damage, or irritation of the kidneys and gastrointestinal tract.
4 It should not eliminate the normal microbial flora of the host. To disturb the normal flora may upset the "balance of nature," permitting normally nonpathogenic microbes or pathogenic forms restrained by the normal flora to establish a new infection. Prolonged use of the broad-spectrum antibiotics, for example, may eliminate the normal bacterial flora but not *Monilia* (fungi) from the intestinal tract. Under these conditions, the *Monilia* may establish an infection.
5 It should be able to be given orally without inactivation by stomach acids, or by injection (parenterally) without binding to blood proteins.
6 It should have a high level of solubility in body fluids.
7 It should be possible to achieve concentrations of the antibiotic in the tissues or blood which are sufficiently high to inhibit or kill the infectious agent.

Penicillins

The first of the modern antibiotics, and still among the most useful and widely used, are the *penicillins*. Penicillins are a class of compounds with related structure and slightly different properties and activities. All penicillins have a common core, a fused β-lactam-thiazolidine ring; different side

Figure 21-5. Selman A. Waksman isolated the organism *Streptomyces griseus*, which produces the antibiotic streptomycin.

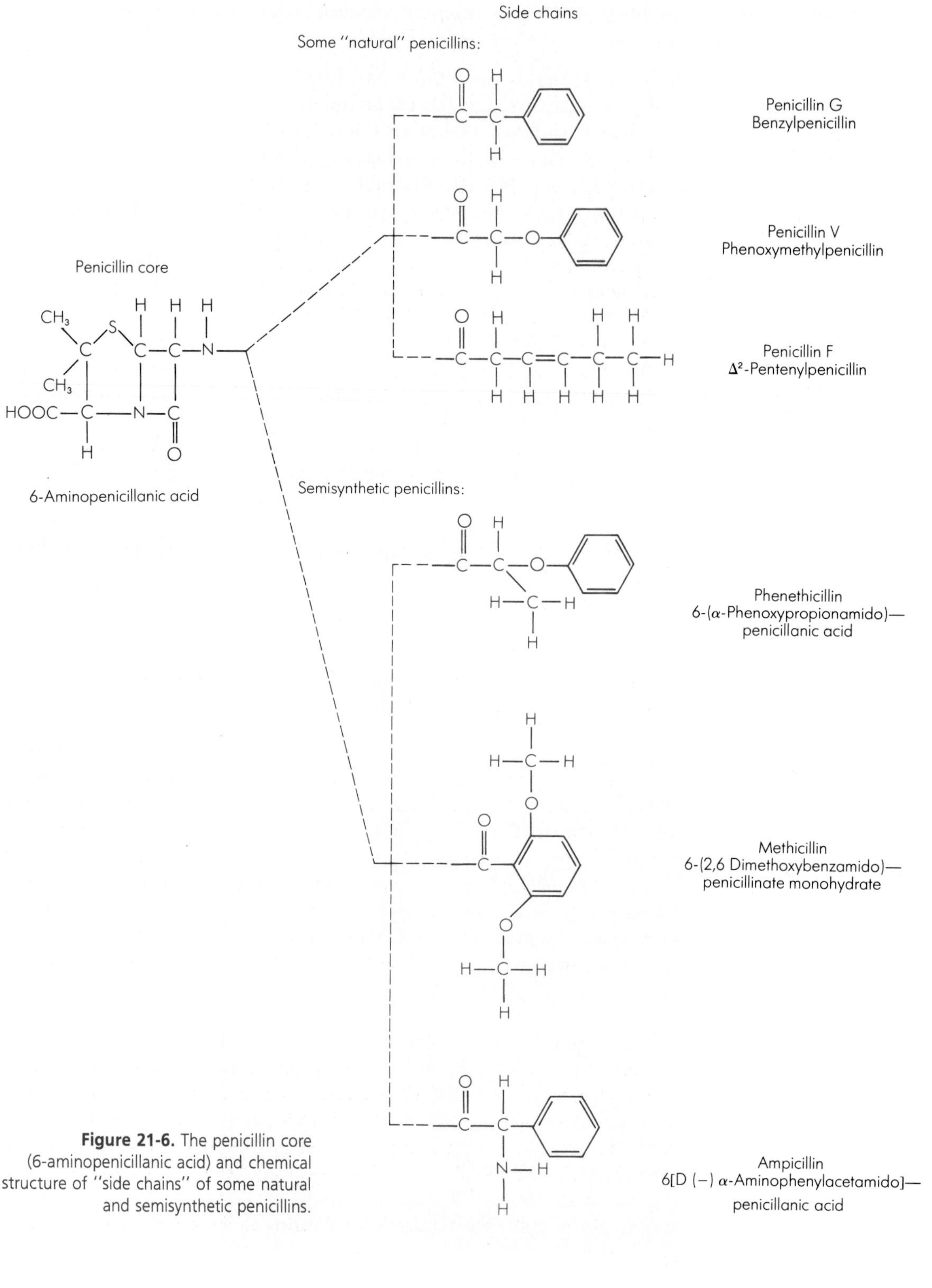

Side chains

Some "natural" penicillins:

Penicillin G
Benzylpenicillin

Penicillin V
Phenoxymethylpenicillin

Penicillin F
Δ^2-Pentenylpenicillin

Penicillin core

CH$_3$

CH$_3$

HOOC

6-Aminopenicillanic acid

Semisynthetic penicillins:

Phenethicillin
6-(α-Phenoxypropionamido)—
penicillanic acid

Methicillin
6-(2,6 Dimethoxybenzamido)—
penicillinate monohydrate

Ampicillin
6[D (−) α-Aminophenylacetamido]—
penicillanic acid

Figure 21-6. The penicillin core
(6-aminopenicillanic acid) and chemical
structure of "side chains" of some natural
and semisynthetic penicillins.

Figure 21-7. *Penicillium notatum.* A colony on agar-plate culture (X 3). (*Courtesy of Chas. Pfizer and Co.*)

chains give each penicillin its unique properties (Fig. 21-6). Chemically they are classified as β-lactam antibiotics.

Natural penicillins. Penicillins are produced during the growth and metabolism of certain fungi, namely *Penicillium notatum* (Fig. 21-7) and *P. chrysogenum*. Several different kinds of penicillin molecules are produced by the same culture. The two most important are penicillin G and penicillin V (see Fig. 21-6).

Natural penicillins can be prepared as salts of sodium, potassium, procaine, and other bases. The crystalline sodium or potassium salts are freely soluble in water. Natural penicillins are inactivated by heat, cysteine (a sulfur-containing amino acid constituent of many proteins), sodium hydroxide, penicillinase (any enzyme found in many bacteria that destroys penicillins), and hydrochloric acid, such as that in the stomach. They are effective against gram-positive bacteria, particularly pneumococci, beta hemolytic streptococci, and some staphylococci; some gram-negative bacteria (meningococci and gonococci); and the spirochete that causes syphilis.

Semisynthetic penicillins. During extensive research on the chemistry of the natural penicillins, it was discovered that they possessed a core or common structure identified as β-aminopenicillanic acid (Fig. 21-6). This led to the development of culturing techniques by which the core compound could be produced in quantity and then, by chemical reactions, different side chains could be added on. The resulting products are called *semisynthetic penicillins*.

One of the first semisynthetic penicillins to be produced for clinical use

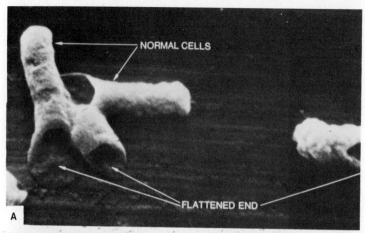

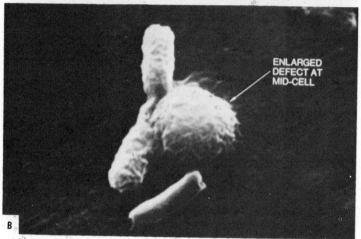

Figure 21-8. The morphologic changes, which occur in *Escherichia coli* as a result of exposure to penicillin, shown in this illustration, provide visual evidence that this antibiotic acts upon the cell wall. (A) Normal cells of *E. coli*. (B) *E. coli* cells after exposure to penicillin. Enlarged (bulged) defect indicates alteration of cell wall. *(Courtesy of Lilly Research Laboratories, Division of Eli Lilly and Company.)*

was *phenethicillin*. It is more readily absorbed than penicillin V and just as effective as penicillin G. Another of the semisynthetic penicillins, *methicillin*, is more resistant to penicillinase and therefore is less likely to be inactivated. A third semisynthetic penicillin, *ampicillin*, acts against many gram-negative bacteria as well as gram-positive species. It is strongly bactericidal and lacks toxicity, but it is not resistant to penicillinase, nor is it stable at an acid pH. The chemical structures of these three penicillins are shown in Fig. 21-6.

Mode of Action. Penicillin inhibits bacterial cell-wall formation by preventing the incorporation of N-acetylmuramic acid, produced within the cell, into the mucopeptide structure that normally constitutes the rigid bacterial cell wall. This type of action is consistent with the fact that penicillin acts only upon actively growing bacteria. Penicillin-sensitive bacterial cells grown in the presence of this antibiotic are abnormally large and have unusual shapes (see Fig. 21-8). Bacilli exposed to penicillin develop extrusions in their cell walls into which the cytoplasm flows. The cell loses its cytoplasm by lysis, and empty cytoplasmic membranes are left as "ghosts."

Cephalosporins

The cephalosporins are a group of antibiotics produced by a species of marine fungus, *Cephalosporium acremonium*. They are of the same chemical class as the penicillins, namely β-lactams. A large number of semisynthetic cephalosporins have been prepared and several are valuable for chemotherapy. They are active against many gram-positive and gram-negative bacteria. They are not destroyed by penicillinase, and some are stable at acid pH.

Mode of action. The cephalosporins, like the penicillins, exert their antibacterial effect by inhibiting the synthesis of the bacterial cell wall (see Fig. 21-9). They are bactericidal.

Streptomycin

Streptomycin is produced by *Streptomyces griseus* (Fig. 21-10), a soil bacterium isolated by Waksman and his associates, who reported on its antibiotic activity in 1944. Of particular significance was the discovery of its activity against tubercle bacilli; it became the primary drug for tuberculosis chemotherapy. Unfortunately, rapid development of resist-

Figure 21-9. The morphologic changes in *Staphylococcus aureus* produced by the antibiotic cephalothin. (A) Normal cells of *S. aureus*. (B) *S. aureus* cells after exposure to cephalothin. Note enlarged spherophasts—cells without walls. *(Courtesy of Lilly Research Laboratories, Division of Eli Lilly and Company.)*

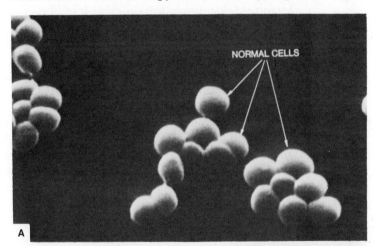
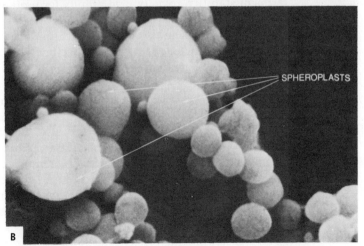

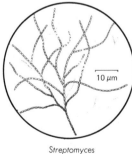

Streptomyces griseus

Figure 21-10. *Streptomyces griseus*, the organism that produces streptomycin. This antibiotic inhibits the growth of certain gram-negative pathogens and *Mycobacterium tuberculosis*. *(Erwin F. Lessel, illustrator.)*

ance to the drug and toxicity during prolonged use reduce its utility in treating tuberculosis. However, it is still considered to be one of several major drugs for tuberculosis therapy. Streptomycin is effective against many gram-positive and gram-negative bacteria.

Streptomycin belongs to the chemical class of antibiotics known as aminoglycosides. As the term suggests, these compounds contain amino sugars, and components of the molecule are linked by glycosidic bonds. Other antibiotics in this group are *dihydrostreptomycin, spectinomycin, neomycin, kanamycin, gentamicin, and tobramycin*. Dihydrostreptomycin has an antibacterial spectrum similar to streptomycin, but it is more toxic. Spectinomycin is recommended for treatment of gonorrhea in persons who are allergic to penicillin or who are infected with a penicillin-resistant strain of *Neisseria gonorrhoeae*. Kanamycin and gentamicin are active against a variety of gram-positive and gram-negative bacteria. Gentamicin is especially active against some strains of *Pseudomonas*. Neomycin is poorly absorbed from the intestinal tract; if given orally, its major effect is upon the intestinal flora. It is also used in the form of lotions and ointments for topical (local) application against skin and eye infections.

Mode of action. Streptomycin exerts its antimicrobial action through combination with and distortion of ribosome subunits, thus interfering with protein synthesis. Other antibiotics in this class—that is, other aminoglycosides—act in a similar manner.

Tetracyclines

Tetracycline, chlortetracycline, and oxytetracycline are generic names for three antibiotics having similar biological and chemical properties. As a group, they are commonly called the *tetracyclines*. They are produced by bacteria in the genus *Streptomyces*. They are regarded as broad-spectrum antibiotics, and their antimicrobial spectra are similar: organisms that are resistant to one are likewise resistant to the other two. They are used for the treatment of infections caused by many gram-negative and some gram-positive bacteria. There are some differences among these compounds in terms of stability, toxicity, and affinity for proteins.

Mode of action. The tetracyclines act by blocking the binding of RNA (aminoacyl transfer RNA) to a specific site on the ribosome during peptide-chain elongation. This action inhibits the synthesis of proteins.

Erythromycin

Erythromycin was discovered by Selman Waksman in 1952 in metabolic products from a strain of *Streptomyces erythreus* isolated from soil collected in the Philippines. It belongs to the chemical class known as *macrolide* antibiotics; other members are *oleandomycin* and *spiramycin*. Chemically, the macrolides are characterized by a molecular structure containing lactone rings linked through glycosidic bonds with amino sugars. Erythromycin is active against most gram-positive bacteria, some gram-negative bacteria (*Neisseria* spp. and *Bordetella pertussis*), and pathogenic spirochetes. Erythromycin resembles penicillin in its antibacterial spectrum and clinical usefulness, but it is also active against organisms that become resistant to penicillin and streptomycin. It is often prescribed to patients who have an allergy to penicillin.

Mode of action. Erythromycin interacts with ribosome subunits to prevent the normal sequence of reactions for protein synthesis.

Chloramphenicol (chloromycetin)

Chloramphenicol is a broad-spectrum antibiotic active against many gram-positive and gram-negative bacteria. Although it is relatively non-toxic to mammals when used therapeutically, it may cause serious abnormalities in the blood of some patients. Therefore, it is properly prescribed only in cases where other antibiotics are not effective.

Mode of action. Chloramphenicol combines with ribosome subunits, interfering with protein synthesis.

Polymyxin and bacitracin

Species of the genus *Bacillus* produce a group of antibiotics which have many common biological and chemical properties. The polymyxins are made by *Bacillus polymyxa*. Another antibiotic of this group is bacitracin, which is produced by *B. subtilis*. These antibiotics are classified chemically as polypeptides.

The polymyxins are active against many gram-negative bacteria including *Pseudomonas aeruginosa* frequently involved in infections of the urinary tract or persons with extensive skin burns. Bacitracin is active against gram-positive but not gram-negative bacteria; it is highly toxic, and hence its use is restricted to external application. The polymyxins are also toxic when administered internally; some preparations, however, have been formulated that are suitable for parenteral use, that is, by subcutaneous, intravenous, or intramuscular injection.

Mode of action. Bacitracin inhibits the synthesis of the bacterial cell-wall structure and may affect the integrity of the cytoplasmic membrane. Polymyxins, on the other hand, damage the cell-wall structure. The antibiotic combines with the cell membrane, causing disorientation of the lipoprotein constituents and preventing the membrane from functioning as an osmotic barrier.

Antifungal antibiotics (nystatin, griseofulvin, amphotericin B)

Nystatin is an antifungal agent useful in the therapy of nonsystemic fungal infections. It is produced by a strain of *Streptomyces noursei*. This antifungal antibiotic was discovered in 1950 by Elizabeth Hazen and Rachel Brown (see Fig. 21-11).

The antimicrobial activity of nystatin is restricted to yeasts and other fungi. It is particularly effective in treatment of infections—typically of the skin, nails, and vagina—caused by the yeast *Candida albicans*. *Thrush*, a disease affecting the oral mucous membranes, particularly in children, is also caused by this organism.

Mode of action. Nystatin damages yeast cells, as well as the cells of other fungi, by combining with sterols in the cell membrane. This causes a

Figure 21-11. Elizabeth Hazen (left) and Rachel Brown examine early samples of the first antifungal antibiotic, nystatin. *(Courtesy of Research Corporation.)*

Table 21-1. Bacterial and fungal metabolic products useful as antibiotics

ANTIBIOTIC	PRODUCED BY	ACTIVE AGAINST	MODE OF ACTION
Penicillins			Inhibit cell-wall synthesis
Penicillin G.	*Penicillium chrysogenum*	Gram-positive bacteria	
Ampicillin	*P. chrysogenum*	Gram-negative bacteria causing respiratory, intestinal, and urinary infections	
Methicillin	*Penicillium* sp.	Penicillinase-producing bacteria	
Cephalosporins	*Cephalosporium* spp.	Gram-negative and gram-positive bacteria	Act on cell wall—inhibit synthesis
Cephalothin			
Cephaloridine			
Cephaloglycin			
Cephalexin			
Aminoglycosides			Induce abnormal protein synthesis
Streptomycin	*Streptomyces griseus*	Tuberculosis infection	
Spectinomycin	*Streptomyces* sp.	Penicillin-resistant *Neisseria gonorrhoeae*	
Neomycin	*S. fradiae*	Inhibits intestinal bacteria	
Kanamycin	*S. kanomyceticus*	Most gram-negative bacteria except *Pseudomonas*	
Gentamicin	*Micromonospora purpurea*	Active against a variety of gram-positive and gram-negative bacteria including *Pseudomonas*	
Tetracyclines		Broad spectrum—many gram-positive and gram-negative bacteria, also organisms such as *Mycoplasma*, *Rickettsia*, and *Chlamydia*	Interfere with protein synthesis
Chlortetracycline	*Streptomyces aureofaciens*		
Tetracycline	*S. aureofaciens*		
Oxytetracycline	*S. rimosus*		
Erythromycin	*Streptomyces erythreus*	Commonly encountered gram-positive bacteria	Interferes with protein synthesis

disorganization in the molecular structure of the membrane followed by an impairment in its function. Sterols are not a component of bacterial membranes, so nystatin is not effective against bacteria.

Griseofulvin is obtained from *Penicillium griseofulvin*. It is used in the treatment of many superficial fungus infections of the skin and body surfaces, such as ringworm. It is also effective in the treatment of some systemic (involving the body as a whole) mycoses, but it is not active against the yeast *Candida albicans* or bacteria.

Amphotericin B is one of the few chemotherapeutic agents available for the treatment of systemic fungal infections. Its mode of action is very much like that described for nystatin.

In addition to the major characteristics of the chemotherapeutic antibiotics which have already been described, those of a few others are also summarized in Table 21-1.

Synthetic chemo-therapeutic agents In contrast to the antibiotics, which are synthesized in whole or part by living cells, there are other chemicals, synthesized completely in the chemistry laboratory, which are useful for treating certain diseases. The sulfonamides are one group of synthetic chemotherapeutic agents, the nitrofurans are a second. Some other specific compounds include isonicotinic acid hydrazide (isoniazid) and nalidixic acid.

ANTIBIOTIC	PRODUCED BY	ACTIVE AGAINST	MODE OF ACTION
Chloramphenicol (Chloromycetin)	*Streptomyces venezuelae*	Broad spectrum; severe gram-negative infections	Interferes with protein synthesis
Polypeptides Colistin (Polymyxin E)	*Bacillus colistinus*	Most gram-negative bacteria, including *Pseudomonas aeruginosa*	Deterioration of cell membrane
Polymyxin B	*B. polymyxa*	Gram-negative bacteria; less effective than colistin	Deterioration of cell membrane
Bacitracin	*B. subtilis*	Gram-positive, but not gram-negative, bacteria	Inhibits cell-wall formation
Lincomycin	*Streptomyces lincolnensis*	Commonly encountered gram-positive bacteria	Interferes with protein synthesis
Vancomycin	*Streptomyces orientalis*	Gram-positive bacteria, including penicillinase-producing staphylococci and enterococci	Interferes with protein synthesis
Viomycin	*Streptomyces griseus* var. *purpureus*	Tuberculosis infection	Interferes with protein synthesis
Rifamycin	*Streptomyces mediterranei*	Tuberculosis infection	Interferes with protein synthesis
Antifungal Antibiotics Nystatin	*Streptomyces noursei*	Fungal infections, particularly oral, skin, intestinal, and vaginal lesions due to *Candida*	Damages cell membrane
Griseofulvin	*Penicillium griseofulvin*	Fungal infections	Damages cell membrane
Amphotericin B	*Streptomyces nodosus*	Deep-seated mycotic infections	Interferes with membrane function

Sulfonamides   As we noted earlier in this chapter, the discovery that sulfonamide was effective for chemotherapy stimulated the synthesis of many new sulfonamide-like compounds. The structural formulas of some of these are shown in Fig. 21-12.

The search for new varieties of sulfonamides has resulted in the

Figure 21-12. Basic structure of the sulfonamides and structures of some specific compounds in this group.

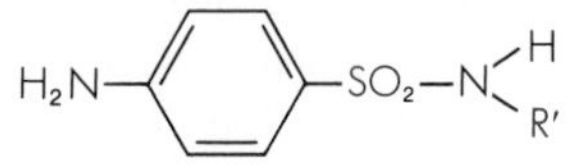

General structure of sulfonamides

The sulfonamides differ primarily by virtue of the different substituents in the R' position, as indicated below:

Sulfanilamide   R' = H

Sulfapyridine (N'-2-Pyridylsulfanilamide)   R' =

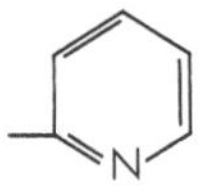

Pyridine

Sulfathiazole (N'-2-Thiazolysulfanilamide)   R' =

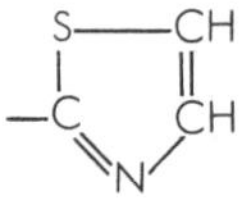

Thiazole

Sulfadiazine (N'-2-Sulfanilamidopyrimidine)   R' =

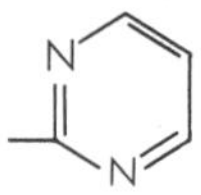

Pyrimidine

Sulfamerazine [N'-(4-Methyl-2-pyrimidyl)-sulfanilamide]   R' =

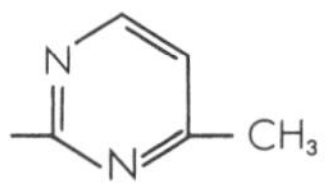

4-Methylpyrimidine

Sulfamethazine [N'-(4,6-Dimethyl-2-pyrimidyl)-sulfanilamide]   R' =

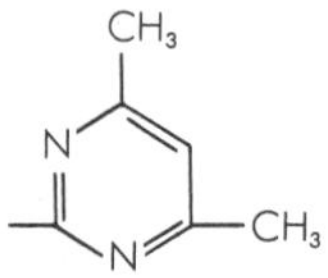

4,6-Dimethylpyrimidine

Sulfaguanidine (N'-Guanylsulfanilamide)   R' =

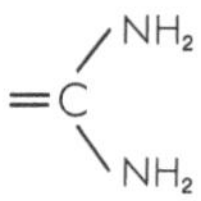

Guanidyl

development of drugs with increased antibacterial activity; the unfavorable reactions of the first compounds have been reduced. Some sulfonamides have been especially useful against certain types of infections. Sulfadiazine and sulfamerazine are extensively used because of their effectiveness in a wide range of bacterial infections and because they have lower toxicity than other sulfonamides.

The sulfonamides are especially useful in treating respiratory infections caused by streptococci and staphylococci, and urinary infections caused by gram-negative organisms. They are also valuable against rheumatic fever, wound infections, and endocarditis (an inflammation of the endocardium, the membrane lining the heart cavities and valves).

Mode of action. Many bacteria require p-aminobenzoic acid (PABA) to synthesize the essential coenzyme folic acid. PABA is a structural part of the folic acid molecule. Since the PABA molecule and a sulfonamide molecule are very similar, the sulfonamide acts by competing in the reaction with PABA and may block the synthesis of the essential cellular constituent folic acid (see Fig. 21-13). The cellular functions of the folic acid coenzyme include purine and pyrimidine synthesis. Lack of this coenzyme will obviously disrupt normal cellular activity. Sulfonamides will inhibit growth of cells which synthesize their own folic acid but will not interfere with the growth of cells (including mammalian host cells) which require preformed folic acid. This accounts for the selective antibacterial action of sulfonamides and makes them useful in the treatment of many infectious diseases.

This mode of action is an example of competitive inhibition (the general nature of which was discussed in Chap. 13) between an essential metabolite (PABA) and a metabolic analog (a sulfonamide).

Nitrofurans

The prototype of the nitrofuran derivatives is *furfural* (see Fig. 21-14), which can be prepared from corncobs, cornstalks, oat hulls, beet pulp, peanut hulls, and other vegetable by-products.

High antibacterial effect is conferred upon furfural by the addition of a nitro group in the 5-position of the furan ring (see Fig. 21-14).

As a class, the nitrofurans generally are effective against a broad spectrum of both gram-positive and gram-negative bacteria, several pathogenic protozoa, and some fungi which cause superficial infections in both humans and other animals.

Isonicotinic acid hydrazide (isoniazid)

Isoniazid has an important, though restricted, application in the therapy of disease. It functions by competitive inhibition and affects one group of microorganisms, the mycobacteria. It has proved to be very useful in the control of tuberculosis in humans and is most effective when given alternately with streptomycin. Because it is a structural analog of pyridoxine (vitamin B_6), isoniazid (see Fig. 21-15) can block pyridoxine-catalyzed reactions in some microorganisms. This accounts for its antimicrobial activity.

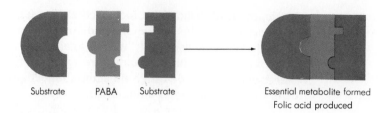

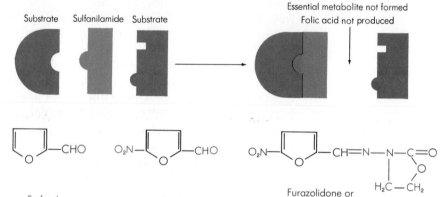

Figure 21-13. Sulfonamides inhibit growth of susceptible bacteria by blocking out PABA, which is essential for the synthesis of folic acid, a requirement for growth.

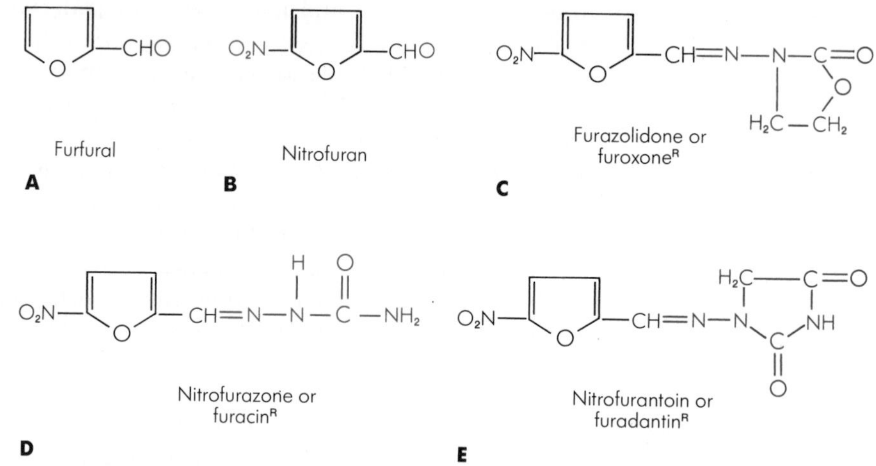

Figure 21-14. Furfural (A) is the prototype of nitrofuran compounds; (B), (C), (D), and (E) are chemotherapeutic derivatives of furfural.

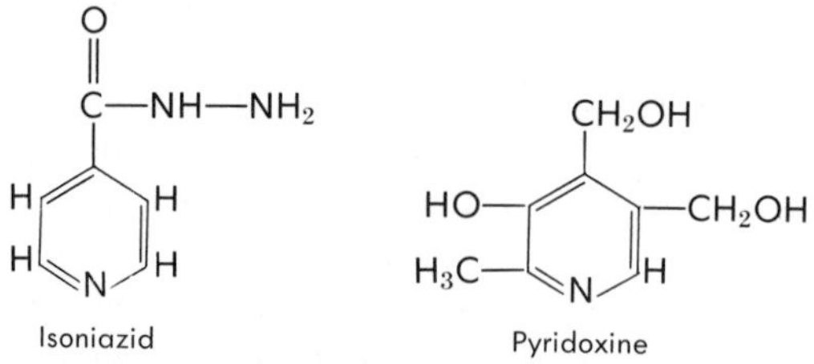

Figure 21-15. Isoniazid, a structural analog of pyridoxine (vitamin B_6), may prevent the growth of microorganisms by blocking pyridoxine-catalyzed reactions in the microbial cell.

Nalidixic acid

Nalidixic acid is a useful chemotherapeutic agent for urinary-tract infections caused by gram-negative bacteria. Its antimicrobial activity is attributed, at least in part, to inhibition of DNA synthesis.

Some characteristics of synthetic chemotherapeutic agents are summarized in Table 21-2.

Table 21-2. Synthetic chemotherapeutic agents

CHEMOTHERAPEUTIC AGENTS	MODE OF ACTION	SPECTRUM OF ACTIVITY
Sulfonamides Sulfanilamide Sulfapyridine Sulfathiozole and others	Interfere with folic acid metabolism; PABA analogs	Used in urinary-tract infections, therapy of nocardiosis, and upper respiratory-tract infections
Nalidixic acid	Inhibits DNA synthesis	Used principally in urinary-tract infections
Isonicotinic acid hydrazide	Blocks pyridoxine-catalyzed reactions (antimetabolite)	Treatment of tuberculosis
Nitrofurans		Most organisms commonly encountered in urinary-tract infections

Resistance to antibiotics

Development of drug resistance is just one example of nature's never-ending processes whereby organisms develop a tolerance for new environmental conditions. Drug resistance in a microorganism may be due to a preexisting factor in that microorganism, or the factor may be acquired. Penicillin resistance in an organism, for example, may result from the production of penicillinase, an enzyme which inactivates penicillin. On the other hand, some normally susceptible strains of bacteria may acquire resistance to penicillin. Acquired resistance is also due to penicillinase production by the genetically adapted strains of microorganism. In cultures of penicillin-sensitive bacteria, perhaps one organism in a hundred million may be a penicillin-resistant mutant. Normally the ratio of sensitive to resistant organisms is maintained, and no problem develops. When penicillin is present the sensitive strains do not reproduce. The resistant mutants do reproduce, however, and they eventually dominate the population. This has important clinical implications, and it is one of the practical reasons why extensive research was performed to develop synthetic penicillins which are not susceptible to the action of penicillinase.

Many organisms which do not produce penicillinase are also resistant to penicillin. This suggests that they possess alternative metabolic pathways or enzyme reactions not susceptible to inhibition by penicillin.

Transmission of drug resistance

When chemotherapeutic agents such as the sulfonamides and antibiotics were first used, development of bacterial resistance was extremely rare. Resistance became much more of a problem as the widespread use of antibiotics led to the elimination of susceptible organisms from the population while the numbers of resistant organisms could increase freely.

The initial appearance of resistant organisms was thought to be the result of a change in a single bacterial gene that conferred resistance to the bacterium. The evidence that this was taking place during sulfonamide therapy is not questioned. Another, more recent explanation for the development of resistance, at least in some gram-negative bacteria, is that resistant organisms have a gene whose function is to protect the bacterium from the bactericidal effect of a drug or antibiotic. For example, such a gene is responsible for penicillinase production by penicillin-resistant staphylococci. Some individuals of a bacterial species carry the resistant gene at the time of infection, and their propagation is encouraged, while the sensitive strains are inhibited or killed. The resistant gene may also be transmitted by conjugation, transformation, or transduction from other bacteria during antibiotic treatment.

Gene transfer between cells, as explained in Chap. 17, is accomplished by transformation, transduction, or conjugation. Most frequently the transfer of antibiotic resistance is by conjugation. This phenomenon was first reported independently by two Japanese scientists in 1958. They isolated both antibiotic-sensitive and antibiotic-resistant organisms of the same serotype from patients with enteric infections being treated with sulfonamides, tetracyclines, streptomycin, or chloramphenicol. They went on to demonstrate that the antibiotic resistance was caused by resistant genes in a reservoir of intestinal-tract *Escherichia coli* being transferred to *Shigella dysenteriae*, the cause of the infection. Since then, transfer of antibiotic resistance by bacterial conjugation has been observed in other organisms in other parts of the world.

We now know that these are resistance, or R, factors, in plasmids, which are small, extrachromosomal, self-replicating, extranuclear, DNA units (see Chap. 17).

The transmission of resistance factors in enteric infections is especially important in places where such infections are common. Organisms that are good recipients of the R factors from *E. coli* donors include species of *Enterobacter*, *Klebsiella*, *Salmonella*, and *Shigella*. Weak recipients are species of *Pasteurella*, *Proteus*, and *Serratia*.

Antibiotic resistance represents a serious problem for clinicians, and great effort is being made to understand the mechanisms involved and to prevent its occurrence. The development of resistance can be minimized by (1) avoiding the indiscriminate use of antibiotics where they are of no real clinical value, (2) refraining from the use of antibiotics commonly em-

Table 21-3. Basic sets of antimicrobial agents suggested for routine disk-susceptibility tests in clinical microbiology laboratories. In general, routine tests should include only one representative from each of the four groups

GROUP 1 STAPHYLOCOCCUS AUREUS	GROUP 2 ENTEROCOCCI	GROUP 3 ENTEROBAC- TERIACEAE	GROUP 4 PSEUDOMONADS
Penicillin G	Penicillin G	Ampicillin	Gentamicin
Oxacillin or methicillin	Ampicillin	Cephalothin	Carbenicillin
Cephalothin	Cephalothin	Kanamycin	Polymyxin B
Erythromycin	Erythromycin	Gentamicin	Kanamycin‡
Clindamycin	Chloramphenicol*	Polymyxin B	Chloramphenicol‡
Chloramphenicol*	Tetracycline*	Tetracycline	Tetracycline‡
Tetracycline*		Chloramphenicol	Sulfonamides†,‡
Gentamicin*		Nitrofurantoin†	
Kanamycin*		Nalidixic acid†	
		Sulfonamides†	

*Suggested only as secondary drugs.
†Only with isolates from urinary-tract infections.
‡Indicated for testing *Pseudomonas* species other than *P. aeruginosa* or for other nonfermentative gram-negative bacilli.
SOURCE: E. H. Lennette, E. H. Spaulding, and J. P. Truant (eds.): *Manual of Clinical Microbiology*, 2d ed., American Society for Microbiology, Washington, D.C., 1974.

ployed for generalized infections or topical applications, (3) using correct dosages of the right antibiotic to overcome an infection quickly, (4) using combinations of antibiotics of proven effectiveness, and (5) using a different antibiotic when an organism gives evidence of becoming resistant to the one used initially.

Determining the effectiveness of chemotherapeutic agents

Species and strains of species of microorganisms have varying degrees of susceptibility to different antibiotics. Furthermore, the susceptibility of an organism to a given antibiotic may change, especially during treatment. It is therefore important for the clinician to know the identity of the microbe and the specific antibiotic which may be expected to give the most satisfactory results in treatment (Table 21-3). The clinical microbiology laboratory must, therefore, make an accurate diagnosis and determine the susceptibility of the organism to various antibiotics. From time to time during the course of therapy, it may be necessary to make estimates of any change in the susceptibility of the pathogen to the drug, and possibly even to assay the antibiotic concentration in the body fluids.

Susceptibility tests

The susceptibility of a microorganism to antibiotics and other chemotherapeutic agents can be determined by either the *tube-dilution* or the *paper-disk-plate* technique. The tube-dilution technique determines the smallest amount of chemotherapeutic agent required to inhibit the growth of the organism in vitro. This amount is referred to as the MIC (minimal inhibitory concentration).

The paper-disk-plate method is the most commonly used technique for determining susceptibility of microorganisms to chemotherapeutic agents. Small paper disks impregnated with different drugs in specified amounts (commercially available) are placed on the surface of an inoculated plate. After incubation, the plates are observed for any zones of inhibition surrounding the disks (see Fig. 21-16). A zone of inhibition (a clear area) around the disk indicates that the organism was inhibited by the drug which diffused into the agar from the disk.

A single-disk method for susceptibility testing is currently recommended by the Food and Drug Administration (FDA). This is a highly standardized technique; the amount of antimicrobial agent contained in the disk is specified as well as the test medium, size of the inoculum, conditions of incubation, and other details. When the susceptibility test is performed in conformity with the FDA procedure, one can correlate the sizes of the zones of inhibition with the MIC of the drug for the microorganism in question.

Microbiological assays of antibiotics

The potency of antibiotic content (the amount of pure antibiotic) in samples can be determined by chemical, physical, and biological means. Biological tests offer the most convenient means of making such determinations.

Biological assay

Biological potency is expressed in terms of micrograms (or other units) as determined by comparing the amount of killing, or bacteriostasis, of a test

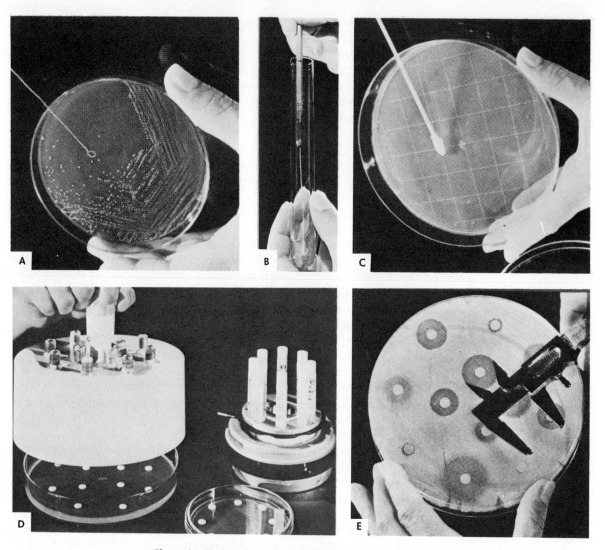

Figure 21-16. Steps in the laboratory procedure for the determination of the susceptibility of a bacterium to antibiotics. (A) A bacterial colony is "picked-off" a petri dish with a transfer needle, and (B) the colony is dispersed in a few milliliters of sterile broth. (C) A sterile cotton swab is immersed in the broth-suspension of bacteria and used to inoculate the surface of agar media in a petri dish. (D) A mechanical device places small paper disks, each impregnated with a different antibiotic, on the surface of the inoculated plate. (E) After incubation, the petri-dish culture is examined for evidence of zones of inhibition around the antibiotic discs. When the test is performed under carefully controlled conditions, there is a relationship between the size of the zone of inhibition (clearing) and the susceptibility of the bacterium to the antibiotic which produced the zone. *(Courtesy of Lilly Research Laboratories, Division of Eli Lilly and Company.)*

organism caused by the test substance with that caused by a standard preparation, all under rigidly controlled conditions (see Fig. 21-17).

Although the unit of measurement for some antibiotics is arbitrary, for others it is established by international agreement or by FDA regulation. For example, the international unit (IU) of penicillin is the amount of

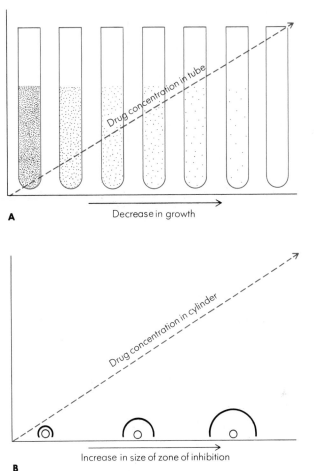

A Decrease in growth

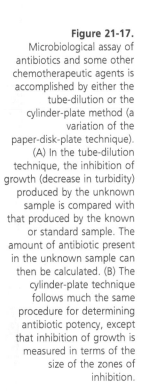

Figure 21-17.
Microbiological assay of antibiotics and some other chemotherapeutic agents is accomplished by either the tube-dilution or the cylinder-plate method (a variation of the paper-disk-plate technique). (A) In the tube-dilution technique, the inhibition of growth (decrease in turbidity) produced by the unknown sample is compared with that produced by the known or standard sample. The amount of antibiotic present in the unknown sample can then be calculated. (B) The cylinder-plate technique follows much the same procedure for determining antibiotic potency, except that inhibition of growth is measured in terms of the size of the zones of inhibition.

B Increase in size of zone of inhibition

activity produced under defined conditions by 0.5988 μg of the International Standard, which is a sample of pure benzyl-penicillin (1 mg = 1,667 units).

The assay of antibiotics in blood serum, urine, tissues, and other similar substances presents some special problems because (1) the amounts present are generally very small, (2) the antibiotic may be bound to proteins in the specimen, and (3) normal inhibitory substances may be present in the blood or other body fluids. There are, however, established procedures whereby one can make these determinations.

Summary and outlook

Chemical compounds which have antimicrobial activity, are nontoxic, and are compatible with the host are of great value for the control of microbial infections. Chemicals which are selectively toxic are called chemotherapeutic agents. Some of these chemotherapeutic agents are synthesized in the chemistry laboratory; however, the chemotherapeutic agents which have had the most dramatic effect and which have revolutionized the treatment of infectious diseases are the antibiotics. These are produced by microorganisms, either in whole or in part. The chemotherapeutic antibiotics that are synthesized in part by microorganisms and then modified by chemical reactions in the laboratory are designated semisynthetic.

Not all antibiotics qualify as chemotherapeutic agents; in vitro antimicrobial activity does not guarantee in vivo action. Furthermore, for an antibiotic to be useful for chemotherapy, it must be selective in its toxicity, that is, inhibitory to microorganisms and nontoxic to the host.

Some chemotherapeutic agents, the so-called broad-spectrum agents, are effective against a wide variety of microorganisms. Others are more specific and limited in their action.

Microbiological techniques are routinely used to assay the potency of commercial antibiotics as well as to measure their concentration in body fluids such as blood, serum, or urine.

One of the major problems associated with the widespread use of chemotherapeutic agents has been the development of resistance to these drugs by microorganisms. Antibiotics once effective for treatment of certain diseases have lost their value for chemotherapy as resistant microbial populations develop. The development of resistance, a basic biological phenomenon, dictates that great care needs to be exercised in the use of chemotherapeutic agents. They should not be used in an indiscriminate manner. Furthermore, it points up the continual need for development of new and different drugs to replace those which become ineffective.

There is a great deal yet to be learned about chemotherapy for virus infections. Developments in this area have not experienced the success that occurred with bacterial diseases. Similarly, the field of chemotherapy for malignant diseases is in its infancy.

Key terms

antibiotic
assay
broad-spectrum antibiotics
chemoprophylaxis
chemotherapy

endocarditis
generic names
MIC
parenteral
penicillins

resistance factor
semisynthetic penicillins
serotype
topical application

Questions

1 Are all antibiotics useful as chemotherapeutic agents? Explain.
2 What is a broad-spectrum antibiotic? Name three.
3 By what mechanism does sulfanilamide inhibit bacterial growth?
4 What is meant by the penicillin "core"? State two ways in which the various penicillins differ. What is a semisynthetic penicillin?
5 Antibiotics are generally more effective against bacterial infections than against virus infections. What are some of the reasons for this?
6 Explain why penicillin is effective only against actively growing bacteria.
7 Name and describe four chemotherapeutic agents that are not antibiotics. What is the antimicrobial spectrum of each?

8 How can the potency (or units) of an unknown sample of penicillin be determined? Describe the procedure.

9 Why is it important to determine the susceptibility of a pathogen to chemotherapeutic agents? How can this be done?

10 In terms of their mode of action, explain why some chemotherapeutic agents are bacteriostatic and others bactericidal.

11 Name three qualities essential in a good chemotherapeutic agent.

12 Which genera of microorganisms produce most of the antibiotics?

13 Compare the status of microbial resistance to penicillin during the 1950s to the situation in the 1970s. Explain.

References for part seven

Block, S. S. (ed.): *Disinfection, Sterilization, and Preservation,* 2d ed., Lea & Febiger, Philadelphia, 1977. *Fundamental principles and practical aspects of the control of microorganisms by chemical and physical methods to achieve disinfection, sterilization, and preservation. The mode of action of antimicrobial agents, bacterial resistance, and methods of testing are discussed. Each major category of chemical substances and physical agents is examined in terms of use and efficacy.*

Borick, P. M.: *Chemical Sterilization,* Dowden, Hutchinson & Ross, Stroudsburg, Pa., 1973. *Contains a selection of previously published articles on the characteristics, application, effectiveness, and mode of action of the major chemical antiseptic, disinfectant, and sterilizing agents.*

Dowling, H. F.: *Fighting Infection, Conquests of the Twentieth Century,* Harvard University Press, Cambridge, Mass., 1977. *A very interesting account of the medical discoveries contributing to improvement of public health, including the sulfonamides and antibiotics and their dramatic impact on control of diseases.*

Hugo, W. B., and A. D. Russell (eds.): *Pharmaceutical Microbiology,* Blackwell Scientific Publications, London, 1977. *Part II of this book covers antimicrobial agents. Antibiotics are discussed under the following headings: types, manufacture, assessment, mechanism of action, and bacterial resistance.*

Lennette, E. H., E. H. Spaulding, and J. P. Truant (eds.): *Manual of Clinical Microbiology,* 2d ed., American Society for Microbiology, Washington, D.C., 1974. *Section V of this volume covers the subject of laboratory tests in chemotherapy, e.g., dilution-test and diffusion-test procedures, assay of antimicrobial agents, and other tests.*

Sykes, G.: *Disinfection and Sterilization,* 2d ed., Lippincott, Philadelphia, 1965. *Contains practical information on the methods available for accomplishing sterilization and the major categories of chemical substances useful as disinfectants.*

Youmans, G. P., P. Y. Patterson, and H. M. Sommers: *The Biologic and Clinical Basis of Infectious Diseases,* Saunders, Philadelphia, 1975. *Chapters 44, 45, and 46 provide excellent coverage of antimicrobial agents and their therapeutic applications.*

PART EIGHT

MICROORGANISMS AND DISEASE—
RESISTANCE TO INFECTION

Infection may be defined as a condition in which pathogenic microorganisms have become established in the tissues of a host. It may be viewed as a type of *parasitism* which occurs when one organism lives at the expense of another, its host.

Infection implies an interaction between two living things—the host and the parasite—with a competition for superiority. If the parasite is successful (because the host's resistance is low), disease results. If the host is superior (because its resistance is high), disease does not occur. Thus the host-microbe interaction governs the outcome of the infectious process.

To produce disease, the parasite must multiply and be metabolically active. It may be prevented from doing this by internal host defense mechanisms. The most important of these are related to the *immune responses* (specific serum and tissue factors that defend the body against foreign agents) of the host. The study of the immune responses is called *immunology*.

It is now recognized that many microbes not ordinarily regarded as *pathogens* (parasites capable of producing disease) do have the capacity to cause infection. This capacity depends not on any of the so-called *virulence factors* (e.g., the ability to excrete a potent exotoxin or poison) but rather on decreased resistance of the host. In recent years, there has been a steady increase in the incidence of infections caused by members of the normal flora of the human body. Lowered resistance brought about by use of modern drugs, such as those which minimize organ rejection after transplantation (immunosuppressive drugs) makes the host more susceptible to infection.

In the healthy individual, however, the normal flora has a protective role. These normal inhabitants suppress the growth of harmful microorganisms. This is most evident when antibiotic therapy has been indiscriminately prescribed over a prolonged period—the result is an overgrowth of harmful parasites.

It seems appropriate for us to begin Part VIII with a chapter on the normal flora of the human body (Chap. 22). Chapter 23 discusses host-microbe interactions. The remaining chapters deal with the immune responses of the host.

Scanning electron micrograph of a human lymphocyte, a cell involved in the cell-mediated immune response. *(Courtesty of Robert A. Good, Memorial Sloan Kettering Cancer Center.)*

THE NORMAL MICROBIOTA
OF THE HUMAN BODY

Humans are constantly in contact with myriads of microorganisms. Not only are microbes in the environment, but they also inhabit the human body. The microbes that naturally inhabit the human body are called the normal flora, or *microbiota.*

It is useful to know the normal microbiota of the healthy human body for the following reasons:

1 This knowledge suggests the kinds of infections that might follow tissue injury at specific sites.
2 It gives an indication of the possible source and significance of microorganisms observed in some clinical infections. For example, *Escherichia coli* is harmless in the intestine but if it gets into the urinary bladder it can cause *cystitis*, an inflammation of the mucous membrane of this organ.
3 It gives a greater appreciation of infections due to microorganisms that constitute the normal or indigenous microbiota of the human host. This is especially important since there is a rising incidence of infections from these microorganisms rather than from an external source. This subject is discussed further in Chap. 30, "Nosocomial Infections."

Origin of the human microbiota

If an animal is delivered by cesarean section, with proper care taken to avoid microbial contamination, and then maintained in a germfree envi-

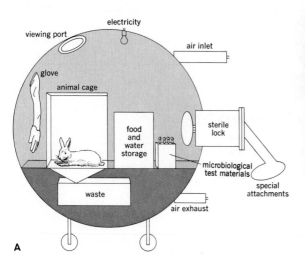

ronment and fed only sterilized food, it does not develop a microbiota (Fig. 22-1). This is evidence that the fetus has no microorganisms up to the time of its birth.

Under natural circumstances, the human fetus first acquires microorganisms while passing down the birth canal. It acquires them by surface contact, swallowing, or inhaling. These microbes are soon joined by other microbes from many sources in the newborn infant's immediate surroundings. Microorganisms which find suitable environments, either on the outer or inner body surfaces, quickly multiply and establish themselves. Thus within hours after birth, the infant is acquiring a microbial flora which will be its indigenous microbiota. Each part of the human body, with its special environmental conditions, has a particular variety of microorganisms. For example, the oral cavity acquires a different natural microbial population than the intestines. In a short time, depending on factors such as the frequency of washing, diet, hygienic practices, and living conditions, the child will have the same kind of normal microbiota as an adult person.

Even though an individual has a "normal" microbiota, it often happens that during his or her life there are fluctuations in this microbiota due to general health conditions, diet, hormonal activity, age, and many other factors.

Indigenous microorganisms and the human host

Most of the indigenous microbes in the human body are *commensals*: they benefit from the association with the host, but the host is not affected. The commensal microbes obtain their nourishment from the secretions and waste products of the human body.

Other indigenous microorganisms have a *mutualistic association* with the host; that is, they benefit the host in some fashion while thriving in the host's body. The advantages to the host in a mutualistic association may be summarized as follows:

1 The microbes are scavengers, using waste material. Many bacteria in the intestines do this.
2 Many intestinal bacteria can synthesize the major B vitamins and vitamins E and K. Vitamins so produced make a significant contribution to the vitamin requirements of the host.
3 The presence of indigenous microbes tends to exclude pathogenic microorganisms and thus serves to protect the host from disease. This exclusion may be due to competition for nutrition or to production of substances inhibitory to the pathogen. For example, lactobacilli in the vagina produce acids which protect the vagina from infection by gonococci, the bacteria that cause gonorrhea. Many strains of *Escherichia coli* in the intestines produce colicins which may protect the intestinal tract from pathogenic intestinal bacteria.

Distribution and occurrence of the human microbiota

Skin

The skin (Fig. 22-2) is constantly in contact with bacteria from the air or from objects, but most of these bacteria do not grow on the skin because it is unsuitable for growth. The skin has a wide variation in structure and function in various sites of the body. These differences serve as selective

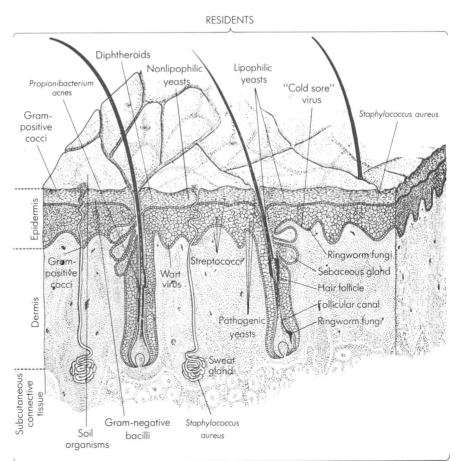

Figure 22-2. The major microbial symbionts found on or in the skin of humans. *(Courtesy of B. C. Block and J. Ducas, Man, Microbes, and Matter, McGraw-Hill, New York, 1975.)*

Table 22-1. Predominant microbial species found in various human anatomic regions

REGION	MICROORGANISM	% INCIDENCE
Skin	*Staphylococcus epidermidis*	85-100
	S. aureus	5-25
	Propionibacterium acnes	45-100
	Aerobic corynebacteria (diphtheroids)	55
Nose and nasopharynx	*Staphylococcus epidermidis*	90
	S. aureus	20-85
	Aerobic corynebacteria (diphtheroids)	5-80
	Branhamella catarrhalis	12
	Haemophilus influenzae	12
Mouth (saliva and tooth surfaces)	*Staphylococcus epidermidis*	75-100
	S. aureus	Common
	Streptococcus mitis and other α-hemolytic streptococci	100
	S. salivarius	100
	Peptostreptococci	Prominent
	Veillonella alcalescens	100
	Lactobacilli	95
	Actinomyces israelii	Common
	Haemophilus influenzae	25-100
	Bacteroides fragilis	Common
	B. melaninogenicus	Common
	B. oralis	Common
	Fusobacterium nucleatum	15-90
	Candida albicans	6-50
	Treponema denticola and *T. vincentii*	Common
Oropharynx	*Staphylococcus epidermidis*	30-70
	S. aureus	35-40
	Diphtheroids	50-90
	Streptococcus pneumoniae	0-50
	α- and nonhemolytic streptococci	25-99
	Branhamella catarrhalis	10-97
	Haemophilus influenzae	5-20
	H. parainfluenzae	20-35
	Neisseria meningitidis	0-15
Jejunum	Gram-positive facultative bacteria (enterococci, lactobacilli, diphtheroids)	Small numbers
	Candida albicans	20-40
Ileum	Distal portion may have small numbers of Enterobacteriaceae and anaerobic gram-negative bacteria	
Large intestine	Gram-negative bacilli: *Bacteroides fragilis, B. melaninogenicus, B. oralis, Fusobacterium nucleatum, F. necrophorum*	100
	Gram-positive bacilli:	
	Lactobacilli	20-60
	Clostridium perfringens	25-35
	Eubacterium limosum	30-70
	Bifidobacterium bifidum	30-70
	Peptostreptococci	Common
	Enterococci (group D streptococci)	100
	Escherichia coli	100
	Klebsiella spp.	40-80
	Enterobacter spp.	40-80

	REGION	MICROORGANISM	% INCIDENCE
Table 22-1. (Continued)	Vagina and uterine cervix	Proteus spp.	5-55
		Candida albicans	15-30
		Lactobacilli	50-75
		Bacteroides spp.	60-80
		Clostridium spp.	15-30
		Peptostreptococci	30-40
		Diphtheroids	45-75
		Staphylococcus epidermidis	35-80
		Group D streptococci	30-80
		Enterobacteriaceae	18-40
		Candida albicans	30-50
		Trichomonas vaginalis	10-25

SOURCE: G. P. Youmans, P. Y. Paterson, and H. M. Sommers, *The Biologic and Clinical Basis of Infectious Diseases*, Saunders, Philadelphia, 1975.

ecological factors, determining the types and numbers of microorganisms that occur on each skin site. Generally, few bacteria that get on the skin are capable of surviving there for long because the skin exudes bactericidal substances. For example, sweat glands excrete lysozyme, an enzyme that destroys bacterial cell walls. Sebaceous glands secrete complex lipids, which may be partially degraded by some bacteria; the resulting fatty acids are highly toxic to other bacteria. (See Fig. 27-2.)

Most skin bacteria are found on the superficial squamous epithelium (outer layer of epidermis), colonizing the surface dead cells. Most of these bacteria are species of *Staphylococcus* (mostly *S. epidermidis* and *S. aureus*) and aerobic corynebacteria, or diphtheroids. In the deep sebaceous glands are found lipophilic anaerobic bacteria, such as *Propionibacterium acnes*, the cause of acne. Their numbers are little affected by washing. The incidence of these organisms is shown in Table 22-1; Fig. 22-3 illustrates the morphology and properties of the predominant microorganisms of the microbiota. The location of these bacteria on or in the skin is shown in Fig. 22-2.

Nose and nasopharynx
The bacteria most frequently and most consistently found in the nose are the diphtheroids. The staphylococci, namely *S. aureus* and *S. epidermidis*, are also common. In the nasopharynx, one can also find bacteria of the species *Branhamella catarrhalis* (a gram-negative coccus) and *Haemophilus influenzae* (a gram-negative rod). (See Table 22-1, Figs. 22-3 and 22-4.)

Mouth
The abundant moisture and the constant presence of dissolved food as well as small food particles make the mouth an ideal environment for bacterial growth. The microbiota of the mouth or oral cavity is very diversified. It is dependent to a large extent on the personal hygiene of the individual.

Acquisition of the oral microbiota. At birth the oral cavity is essentially a sterile, warm, and moist incubator containing a variety of nutritional

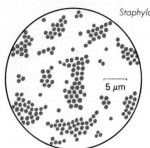

Staphylococcus aureus

Gram-positive cocci
arranged in grapelike
clusters
Nonmotile
Aerobic,
facultatively anaerobic
Coagulase produced
Found in nasal membranes,
skin, hair follicles

5 μm

Propionibacterium acnes

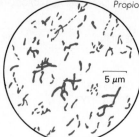

Gram-positive pleomorphic
rods arranged in
short chains or clumps
with V and Y forms
Nonmotile
Anaerobic to
aerotolerant
Propionic acid
produced
Found on skin

5 μm

Branhamella catarrhalis

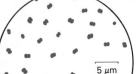

Gram-negative cocci
arranged in pairs
with adjacent sides
flattened
Nonmotile
Aerobic
Oxidase test positive
Found in mucous
membranes

5 μm

Haemophilus influenzae

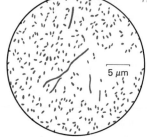

Gram-negative rods
exhibiting filaments
and pleomorphism
Nonmotile
Aerobic
Normally found in
nasopharynx

5 μm

Streptococcus pneumoniae

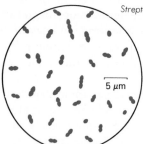

Gram-positive oval or
spherical cells;
typically in pairs or
short chains;
distal ends of cells pointed
Nonmotile
Facultatively anaerobic
Bile-soluble
Found in upper respiratory
tract

5 μm

Neisseria meningitidis

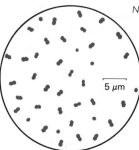

Gram-negative cocci
typically arranged in
pairs with adjacent
sides flattened
Nonmotile
Aerobic
Oxidase test positive
Found in nasopharynx

5 μm

Lactobacillus sp.

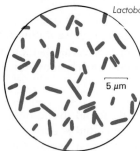

Gram-positive straight or
curved rods
Nonmotile
Anaerobic or
facultatively
aerobic
Complex nutritional
requirements
Found in mouth,
vagina, intestinal tract

5 μm

Candida albicans

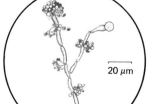

A yeast with budding cells
Shown is pseudomycelium
with clusters of
blastospores and
round chlamydospores
Aerobic
Found in mouth, throat,
large intestine,
vagina, skin

20 μm

Bacteroides fragilis

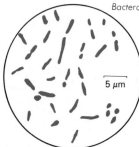

Gram-negative
uniform or pleomorphic
rods
Nonmotile
Anaerobic
Found in lower
intestinal tract and
mouth

5 μm

Fusobacterium nucleatum

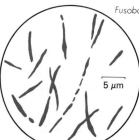

Gram-negative rods with
pointed ends
Nonmotile
Anaerobic
Found in mouth

5 μm

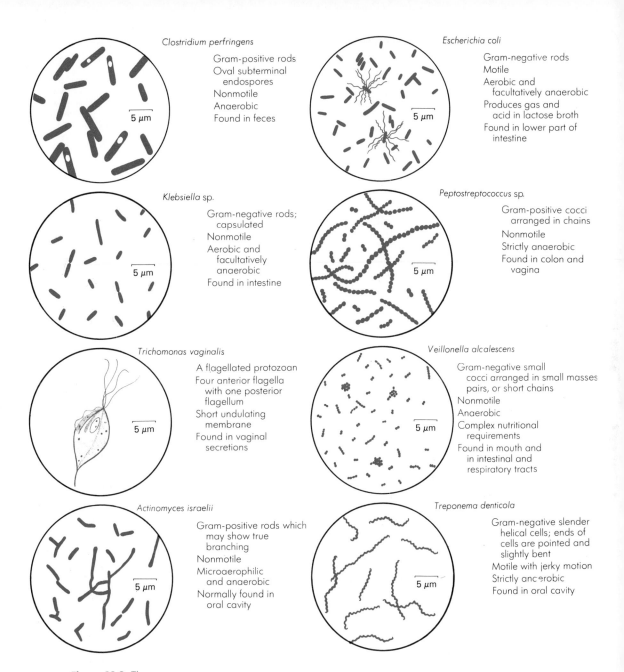

Clostridium perfringens

Gram-positive rods
Oval subterminal
  endospores
Nonmotile
Anaerobic
Found in feces

Escherichia coli

Gram-negative rods
Motile
Aerobic and
  facultatively anaerobic
Produces gas and
  acid in lactose broth
Found in lower part of
  intestine

Klebsiella sp.

Gram-negative rods;
  capsulated
Nonmotile
Aerobic and
  facultatively
  anaerobic
Found in intestine

Peptostreptococcus sp.

Gram-positive cocci
  arranged in chains
Nonmotile
Strictly anaerobic
Found in colon and
  vagina

Trichomonas vaginalis

A flagellated protozoan
Four anterior flagella
  with one posterior
  flagellum
Short undulating
  membrane
Found in vaginal
  secretions

Veillonella alcalescens

Gram-negative small
  cocci arranged in small masses
  pairs, or short chains
Nonmotile
Anaerobic
Complex nutritional
  requirements
Found in mouth and
  in intestinal and
  respiratory tracts

Actinomyces israelii

Gram-positive rods which
  may show true
  branching
Nonmotile
Microaerophilic
  and anaerobic
Normally found in
  oral cavity

Treponema denticola

Gram-negative slender
  helical cells; ends of
  cells are pointed and
  slightly bent
Motile with jerky motion
Strictly anaerobic
Found in oral cavity

Figure 22-3. The morphology and principal characteristics of the predominant microbial species comprising the normal microbiota of the human body. *(Erwin F. Lessel, illustrator.)*

substances. The saliva is composed of water, amino acids, proteins, lipids, carbohydrates, and inorganic compounds. It is thus a rich and complex medium that can be used as a source of nutrients by microbes at various sites in the mouth. (Saliva itself generally contains transient microbes from other sites of the oral cavity, particularly from the upper surface of the tongue.)

A few hours after birth, there is an increase in the number of microorgan-

Figure 22-4. Distribution of normal microbiota of the human body. For a more detailed listing, see Table 22-1.

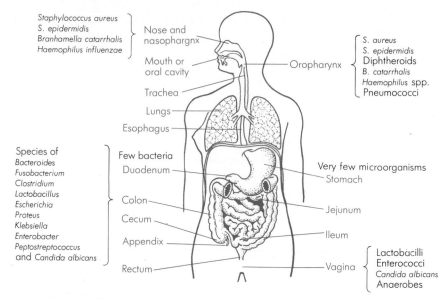

Staphylococcus aureus
S. epidermidis
Branhamella catarrhalis
Haemophilus influenzae
} Nose and nasophargnx

Mouth or oral cavity

Trachea

Lungs

Esophagus

Oropharynx — {
S. aureus
S. epidermidis
Diphtheroids
B. catarrhalis
Haemophilus spp.
Pneumococci
}

Species of
Bacteroides
Fusobacterium
Clostridium
Lactobacillus
Escherichia
Proteus
Klebsiella
Enterobacter
Peptostreptococcus
and Candida albicans
} {
Few bacteria
Duodenum

Colon

Cecum

Appendix

Rectum
}

Very few microorganisms
Stomach

Jejunum

Ileum

Vagina — {
Lactobacilli
Enterococci
Candida albicans
Anaerobes
}

isms so that within a few days the species of bacteria characteristic of the oral cavity have been established. These belong to the genera *Streptococcus, Neisseria, Veillonella, Actinomyces,* and *Lactobacillus.*

The number and kinds of species found are related to the infant's diet and associations, that is, associations with the mother, attendants, and objects such as towels and feeding bottles. The only species consistently recovered from the oral cavity, even as early as the second day of life, is *Streptococcus salivarius.* This bacterium has an affinity for epithelial tissues and therefore appears in large numbers on the surface of the tongue.

Until eruption of the teeth, most microorganisms in the mouth are aerobes or facultative anaerobes. As the first teeth appear, the obligate anaerobes, such as the *Bacteroides* and fusiform bacteria (*Fusobacterium* spp.), become more evident because the tissue surrounding the teeth provides an anaerobic environment.

The teeth themselves become areas for microbial adherence. Two bacterial species are found associated with the tooth surface: *Streptococcus sanguis* and *S. mutans.* The latter is believed to be the primary *etiological* (causative) *agent* of dental caries, or tooth decay. The retention of these two species on the tooth surface is a result of the adhesive properties of both salivary glycoproteins and bacterial polysaccharides. This property of adherence is very important for bacterial colonization in the mouth. Salivary glycoproteins are capable of aggregating certain bacteria and binding them to the tooth surface (Fig. 22-5). Both *S. sanguis* and *S. mutans* produce extracellular polysaccharides called *dextrans* that act like a glue, binding the bacterial cells together as well as adhering them to the tooth surface. Retention of bacteria can also occur by mechanical trapping in the gingival or gum crevices, or in the pits and fissures of the teeth. Such aggregations of bacteria and organic matter on the surface of teeth are called *plaques*. Saliva is continually produced and swallowed and therefore has a cleansing action.

Once teeth are present, the microbiota in infants appears to be generally similar to that in adults. Then, for reasons which are not well understood at present, but probably as a result of hormonal changes, oral spirochetes and *Bacteroides melaninogenicus* colonize the gingival crevices at puberty. (See Table 22-1 for a tabulation of the predominant organisms in the mouth and Fig. 22-3 for the properties of these organisms.)

Oropharynx

The *oropharynx* (back of the mouth) also harbors large numbers of both *Staphylococcus aureus* and *S. epidermidis* as well as diphtheroids. But the most important group of bacteria indigenous to the oropharynx is the α-hemolytic streptococci, also called viridans streptococci. Cultures from the oropharynx will also show the presence of *Branhamella catarrhalis*, species of *Haemophilus*, and avirulent strains of pneumococci (*Streptococcus pneumoniae*). (See Table 22-1, Figs. 22-3 and 22-4.) The deepest portions of the respiratory tract (the finer bronchioles and the alveoli of the lungs) are devoid of microorganisms. This is because the respiratory passages are lined with cilia, hairlike appendages, that sweep microorganisms and other materials from the deeper portion of the tract to the upper portion where they may be expelled. The hairs along with mucus in the nostrils initially help protect the respiratory tract by filtering bacteria from the inspired air.

Stomach

The contents of the healthy stomach are practically sterile owing to the hydrochloric acid in the gastric secretion. Following the ingestion of food, the number of bacteria increases, but it soon falls as gastric juice is secreted and the pH of the stomach's fluid content drops.

Small intestine

The upper portion (or duodenum) of the small intestine has few bacteria. Of those present, the majority are gram-positive cocci and bacilli. In the jejunum (second part of the small intestine, between the duodenum and ileum) there are occasionally found species of enterococci, lactobacilli, and

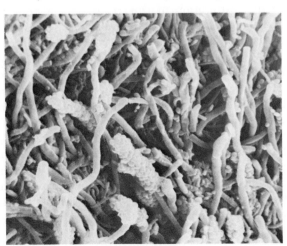

Figure 22-5. Scanning electron micrograph of bacteria adhering to the surface of teeth. Cocci may be seen coating some filamentous bacteria giving "corncob" arrangements. *(Courtesy of Z. Skobe, Forsyth Dental Center.)*

diphtheroids. The yeast *Candida albicans* may also be found in this part of the small intestine. (See Table 22-1 and Fig. 22-4.)

In the distal portion (ileum) of the small intestine, the microbiota begins to resemble that of the large intestine. Anaerobic bacteria and enterobacteria begin to appear in large numbers. (See Table 22-1.)

Large intestine
In the human body, the colon, or large intestine, has the largest microbial population. It has been estimated that the number of microorganisms in stool specimens is about 10^{12} organisms per gram. (Fifty or sixty percent, dry weight, of fecal material may consist of bacteria and other microorganisms.) It has also been calculated that an adult excretes in the feces 3×10^{13} bacteria daily; most of these cells are not viable.

There are about 300 times as many anaerobic bacteria as facultative anaerobic bacteria (such as *Escherichia coli*) in the large intestine. The percentage of incidence of these organisms is shown in Table 22-1.

The anaerobic gram-negative bacilli present include species of *Bacteroides* (*B. fragilis*, *B. melaninogenicus*, *B. oralis*) and *Fusobacterium*. The gram-positive bacilli are represented by species of *Clostridium* (including *Cl. perfringens* which is associated with *gas gangrene*, an infection of tissue with gas bubbles and pus discharge) and species of *Lactobacillus*. (See Fig. 22-3 for the properties and morphology of these bacteria. Figure 22-4 shows their location in the host body.)

It is interesting that the intestinal microbiota of the young breast-fed infant is almost exclusively lactobacilli. With the introduction of bottle-feeding, the lactobacilli number decreases, and finally, with substitution of solid food and an adult-type diet, the gram-negative microbiota predominates.

The facultative anaerobic species found in the intestine belong to the genera *Escherichia*, *Proteus*, *Klebsiella*, and *Enterobacter*. Peptostreptococci (anaerobic streptococci) are common. The yeast *Candida albicans* is also found. (See Figs. 22-3 and 22-4.)

It should be noted that in diarrhea, as a result of rapid movements of the intestinal contents, the intestinal microbiota undergoes considerable change. Alteration of this microbiota also occurs in persons receiving antibiotic treatment; susceptible organisms, unfortunately, may be replaced by resistant ones.

Genitourinary tract
In the healthy person, the kidneys, ureters, and urinary bladder are free of microorganisms, but bacteria are commonly found in the lower urethra of both males and females. Their numbers decrease, however, near the bladder, apparently due to some antibacterial effect exerted by the urethral mucosa and the frequent flushing of its epithelium by urine. The adult female genital tract has a very complex microbiota. The character of this population changes with the variation of the menstrual cycle. The main inhabitants of the adult vagina are the acid-tolerant lactobacilli. The bacteria break down glycogen produced by the vaginal epithelium, producing acid in the process. The accumulation of this glycogen in the vaginal

wall is due to ovarian activity; it is not present before puberty nor after menopause. As a result of the glycogen breakdown, the pH in the vagina is maintained at about 4.4 to 4.6. Microorganisms capable of multiplying at this low pH are found in the vagina and include the enterococci, *Candida albicans*, and large numbers of anaerobic bacteria. (See Table 22-1 and Figs. 22-3 and 22-4.)

Summary and outlook

The human body has a normal microbiota which it begins to acquire as soon as it is born. Each part of the body has its own special environmental conditions and thus a particular population of different microbes. Within a short time, the child has the same kind of microbiota as an adult.

Many of the microbes occurring naturally on the human body are commensal organisms. Others have a mutualistic association with the host. Each ecological site of the body that is capable of supporting microbial life has a microbiota that is unique to the site.

As techniques for the isolation and cultivation of microorganisms improve, the estimates of microbial numbers will become more reliable and the discovery of new microbes more common. For example, with improvements in the techniques for cultivating anaerobes in recent years, we shall no doubt discover new species of anaerobes in the human body.

Key terms

commensal	microbiota	oropharynx
etiological agent	mutualistic association	plaque
lysozyme		

Questions

1 Why do you think it is medically useful to know about the microorganisms in the healthy body?

2 Provide some examples to illustrate that the laws of natural selection also govern the ecology of the normal microbiota.

3 Would it be more advantageous for a symbiotic microorganism in the human host to be a commensal or a parasite? Explain.

4 What benefits does a human host derive from a mutualistic association with its microbiota?

5 Give reasons for the occurrence of the following species of bacteria in their natural habitat:

(a) *Propionibacterium acnes* in skin.

(b) *Lactobacillus* spp. in vagina.

(c) *Bacteroides melaninogenicus* in gingival tissue.

(d) *Streptococcus mutans* on teeth surfaces.

HOST—PARASITE INTERACTIONS

Our entire life is spent in contact with microorganisms. In Chap. 22 we learned that many kinds of microbes grow in or on the healthy human body; they are either commensals or mutualistic organisms. Sometimes, other kinds of microorganisms may invade the human body. They are *parasites* that live at the expense of the host and may harm it by causing disease.

Infectious disease occurs or is prevented depending on the outcome of *interactions* between the parasitic microbe and the host. It is to the parasite's advantage if the disease it causes is not so severe as to kill the host. Killing the host would diminish the parasite's chances of survival.

If the host is to recover, it must eradicate the microbial parasites. Antimicrobial therapy does not always completely destroy the parasites; in most instances it only "buys time" to give the host an opportunity to eliminate the parasite by whatever defense mechanisms the host can bring to bear on the parasite.

Pathogenicity, virulence, and infection

Pathogenicity is the capability of organisms to cause disease. When microorganisms invade a host, that is, when they enter the body tissues and multiply, they establish an infection. The response of the host to infection is an impairment of body function; this is called *disease*. Thus a *pathogen* is any microorganism or macroorganism capable of producing disease. The *Trichinella* worm causes trichinosis, a parasitic disease affecting muscle tissue, and is a pathogen.

The ability of a pathogenic microorganism to cause infection (its *pathogenicity*) is influenced not only by the properties inherent in the microbe, but also by the ability of the host to resist the infection. But the degree of the ability of a microorganism to cause infection is called *virulence*. Thus microbial properties that enhance a microorganism's pathogenicity are termed *virulence factors*. If one microbe is more capable of producing a disease, it is said to be more virulent than another. The virulence factors of some pathogens are known. For example, cells of *Streptococcus pneumoniae* that have capsules are more virulent in causing pneumonia than those without capsules (Fig. 23-1); virulent strains of *Corynebacterium diphtheriae* produce toxins that cause diphtheria. For most pathogens, however, the virulence factors are not so precisely known.

In order to cause infectious disease a pathogen must accomplish the following:

1 It must enter the host.
2 It must metabolize and multiply in the host tissue.
3 It must resist host defenses (see Chap. 25).
4 It must damage the host.

Each process is complex; and all four processes must be fulfilled to produce disease. The means whereby all the above essential processes in disease production are accomplished and those mechanisms that determine host- and tissue-specificity of infection together form the bulk of our knowledge of microbial pathogenicity.

Microbial virulence factors

In most cases, the properties conferring virulence to a pathogenic microorganism are either unclear or unknown. It is known, however, that some bacteria secrete substances, while others have special structures, which contribute to their virulence. Several microbial factors are considered here.

Toxins

Some microorganisms produce poisonous substances known as *toxins*. The capability of a microorganism to produce toxin which has a deleterious effect in a host and the potency of the toxin are important factors in the

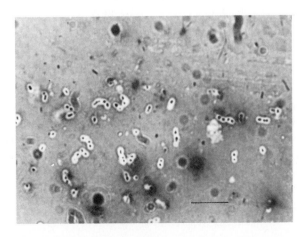

Figure 23-1. Well-defined capsules, made up of complex polysaccharides, are produced by *Streptococcus pneumoniae*, when grown under suitable conditions. The presence of capsules of pneumococci and some other bacteria is related to virulence. Bar equals 10μm. Preparation shown is from the peritoneal exudate of a mouse. *(Courtesty of Liliane Therrien and E. C. S. Chan, McGill University.)*

ability of the organism to cause disease. Many bacteria have not been observed to produce toxin. This may be due to a lack of methods for demonstrating the existence of some toxins or to our failure to understand the problem adequately. The toxins produced by microorganisms may be excreted into the surrounding medium (*exotoxins*) or retained within the cell (*endotoxins*) as part of the cell.

Exotoxins. Exotoxins are diffusible and are excreted from the microbial cells which produce them into the culture medium or into the circulatory system and tissues of the host. The medium might be a can of vegetables contaminated with *Clostridium botulinum*, for example; ingestion of the vegetables containing secreted exotoxin results in a food poisoning called *botulism*. When the diphtheria bacillus *Corynebacterium diphtheriae* grows in the human throat and excretes exotoxin into the bloodstream, the exotoxin causes diphtheria.

Exotoxins are proteins. They lose their toxicity when they are heated or treated chemically. Phenol, formaldehyde, β-propiolactone, and various acids can modify exotoxins chemically so that they lose their toxicity. They are then called *toxoids*. Toxins and toxoids have the ability to stimulate the production of *antitoxins*, substances which neutralize the toxicity of toxins in the body of the host. This ability is important in the protection of susceptible hosts from diseases caused by bacterial toxins. Table 23-1 shows some of the well-known diseases caused by exotoxin-producing bacteria.

Endotoxins. Many organisms, particularly the gram-negative bacteria, do not elaborate a soluble toxin secreted from the living, intact cells but produce an endotoxin that is liberated only when the cells disintegrate. The presence of toxic substances in populations of such bacteria is due to lysis of some of the cells. The endotoxins of gram-negative bacteria are located in the cell wall and are complex substances containing phospholipid and carbohydrate (lipopolysaccharide). Compared with exotoxins, endotoxins (1) are relatively heat-stable, (2) do not form toxoids, and (3) are less toxic. Endotoxins generally play a contributory rather than a primary role in causing disease, being responsible for many of the symptoms, such as fever and shock. The general characteristics of exotoxins and endotoxins are summarized in Table 23-2.

Table 23-1. Some well-known diseases caused by bacteria that produce exotoxins

DISEASE	BACTERIAL SPECIES
Botulism	*Clostridium botulinum*
Cholera	*Vibrio cholerae*
Diphtheria	*Corynebacterium diphtheriae*
Dysentery, bacillary	*Shigella dysenteriae*
Gas gangrene	*Clostridium perfringens*
Scarlet fever	*Streptococcus pyogenes*
Staphylococcal food poisoning	*Staphylococcus aureus*
Tetanus	*Clostridium tetani*
Whooping cough	*Bordetella pertussis*

Table 23-2. Some characteristics of exotoxins and endotoxins

FEATURE	EXOTOXINS	ENDOTOXINS
Bacterial source	Excreted predominantly by gram-positive bacteria	Released from cell walls of lysed gram-negative bacteria
Chemical nature	Protein	Lipopolysaccharide
Heat tolerance	Inactivated easily at 60–100°C for 30 min	Withstand autoclaving
Immunology	Can be converted to toxoids and readily neutralized by antitoxin	Cannot form toxoids; neutralization with antitoxin not possible or possible only with difficulty
Biological effect	Specific for a particular type of cell function	Various effects, but mostly symptoms of generalized shock or hypersensitivity
Lethal dose	Minute amounts	Much larger than that for exotoxins

Extracellular enzymes

The virulence of some microorganisms is partly due to the production of extracellular enzymes (see Table 23-3). Although no single extracellular enzyme has been unequivocably shown to be the sole factor responsible for virulence, there is no doubt that such enzymes play some role in the pathogenic process.

Hyaluronidase. *Hyaluronidase* helps the pathogen to penetrate the tissues of the host by hydrolyzing hyaluronic acid, an essential "tissue cement" which helps hold living cells together. Because of this, the enzyme is referred to as the "spreading factor." It is produced by several of the cocci, some of the clostridia, and some other bacteria.

Lecithinase. *Lecithinase* is an enzyme which destroys various tissue cells and is especially active in lysis of red blood corpuscles. The virulence of *Clostridium perfringens* (the cause of gas gangrene) may be due, at least in part, to the production of lecithinase.

Collagenase. *Collagenase* destroys collagen, which is a tissue fiber found in muscle, bone, and cartilage, and provides the meshwork within which the tissue cells lie. The enzyme may enhance the virulence of *Cl. perfringens* and other organisms that produce it.

Table 23-3. Some extracellular enzymes that contribute to microbial virulence

ENZYME	ACTION	BACTERIA PRODUCING ENZYME (EXAMPLES)
Hyaluronidase	Breaks down hyaluronic acid (a tissue component)	Staphylococci, streptococci, and clostridia
Coagulase	Clots plasma	*Staphylococcus aureus*
Hemolysin	Lyses red blood cells	Staphylococci, streptococci, and clostridia
Lecithinase	Destroys red blood cells and other tissue cells	*Clostridium perfringens*
Collagenase	Breaks down collagen (a tissue fiber)	*Cl. perfringens*
Leukocidin	Kills leukocytes	*Staphylococcus aureus*

Coagulase. Some virulent staphylococci produce an enzyme commonly called *coagulase*; it acts together with an activator in the plasma to transform fibrinogen to fibrin. This causes the deposition of fibrin around the bacterial cells, thus protecting them from the action of host phagocytes. There is some evidence that coagulase is also involved in the walling-off process in boils caused by staphylococci.

Leukocidin. *Leukocidin* is an enzyme produced by some staphylococci and streptococci. It can kill leukocytes (white blood cells).

Hemolysins. *Hemolysins* are substances which lyse red blood cells, releasing their hemoglobin. Generally speaking, hemolytic strains of pathogenic bacteria are more virulent than nonhemolytic strains of the same species. Bacterial hemolysins from various species differ in their chemical nature and mode of action. *Invasiveness*, the ability to penetrate tissue, is enhanced by the production of hemolysins. Bacterial hemolysins are of two types. The first type is extracellular and can be separated from the bacterial cells by filtration. Filterable hemolysins produced by some streptococci are called *streptolysin O* and *streptolysin S*. (Streptolysin O is oxygen-labile and heat-stable; streptolysin S is acid-sensitive and heat-labile.) The second type of hemolysin includes those which produce visible changes on blood-agar plates (Fig. 23-2). On these plates, colonies of hemolytic bacteria are surrounded by a clear, colorless zone where the red blood cells have been lysed and the hemoglobin converted to a colorless compound. This is called β-*hemolysis*. Other types of bacteria can reduce the hemoglobin to methemoglobin, which results in a greenish zone around the colonies. This is called α-*hemolysis*.

Figure 23-2. (Left) α-hemolysis. Enzymes produced by some streptococci, such as *S. salivarius*, only partially hemolyze red blood cells of certain species of animals. Colonies on blood-agar plates are surrounded by a greenish-colored zone which is due to the reduction of hemoglobin in the red blood cells to methemoglobin. (Right) β-hemolysis. Enzymes produced by some streptococci, such as *S. pyogenes*, completely hemolyze red blood cells of certain species of animals. Colonies on blood-agar plates are surrounded by a clear, colorless zone. *(Courtesy of Liliane Therrien and E. C. S. Chan, McGill University.)*

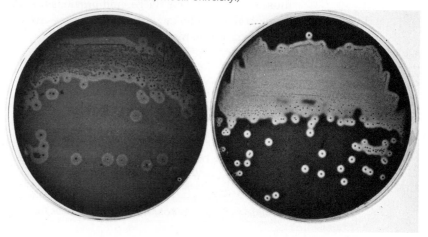

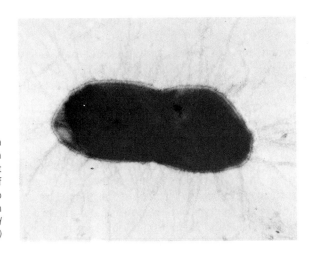

Figure 23-3. A piliated strain of *E. coli* isolated from a patient with urinary-tract infection. The presence of pili may be related to virulence. Final magnification X30,000. *(Courtesy of Jared Fein, McGill University.)*

Capsules

The virulence of bacteria is influenced, in many cases, by the presence or absence of capsules. When pathogens lose their capsules, such as by mutation, they lose their ability to cause disease. This is exemplified by the pneumococci; they are virulent when capsulated but avirulent when not capsulated (see Fig. 23-1). It seems that the increased virulence of capsulated strains is due to the ability of the capsular polysaccharide to prevent *phagocytosis*, or engulfment, by host phagocytes. This ability is probably due to surface properties of the capsule which prevent the phagocyte from forming a sufficiently intimate contact with the bacterium to engulf it.

Pili

Many nonpathogenic bacteria possess pili, just as many nonpathogenic bacteria possess capsules. These organelles may enhance the virulence of some pathogens. It has been reported that the possession of pili helps an organism to adhere better to the surface of host cells and tissues. For example, virulent strains of *Neisseria gonorrhoeae* (causing gonorrhea) and *Escherichia coli* (causing urinary-tract infection) appear to possess pili (Fig. 23-3).

Factors other than virulence influencing infection

Tissue affinity

Some protozoa, such as the malarial parasite, can destroy erythrocytes, yet in certain stages of its life cycle the parasite grows in human blood cells. At other stages, the same parasite has an affinity for the tissues of mosquitoes, in which it develops without causing apparent disease and from which the parasite may be transmitted by bites to humans. Other animals, particularly insects, act as transmitters of many diseases because of such tissue affinities.

Thus it is well established that some microorganisms have a particular affinity for certain cells and tissues, which they may injure and destroy. Their interference with the normal processes of a cell or tissue affects the whole organism, and disease results. For some pathogens, we know the reason for their tissue affinity. For example, brucellae bacteria cause *brucellosis* (resulting in abortion) in cattle, goats, sheep, and pigs. The placentas of these animals contain the sugar erythritol which the brucellae

391

require for enhanced growth. One may ask why the typhoid bacillus chooses the cells of the lymphoid tissue of the intestinal wall (Peyer's patches) in which to grow, or why the poliomyelitis virus has an affinity for nerve cells. These are examples of the many puzzles that must be solved in order to arrive at a better understanding of infection.

Portal of entry

Not only must pathogens enter the body in adequate numbers to produce disease, but many of them must enter by a certain route called the *portal of entry*; this differs for various organisms, depending upon their ability to attack certain organs or parts of the body. The alimentary tract is the portal of entry for the typhoid, dysentery, and cholera organisms, all of which are able to withstand the action of enzymes in saliva and other digestive juices and survive the natural acidity of the stomach. Some microorganisms have a special affinity for the respiratory tract and may set up infections in the bronchi and lungs; the tuberculosis and diphtheria organisms enter by this portal. To produce lobar pneumonia, the pneumococcus, regardless of its virulence, must enter the lungs via the respiratory passages.

The gonococci and some other microorganisms, such as spirochetes, generally enter the body through the urogenital tract, from which they can easily attack the genital organs. Still other organisms enter through abrasions or openings in the skin and set up local infections, such as boils caused by staphylococci. Others, like the anthrax bacillus, enter through breaks in the skin and then spread through the body in the circulatory system. Many pathogens are introduced into the host by the bites of arthropods or larger animals. For example, ticks transmit the rickettsias of Rocky Mountain spotted fever to humans, and the rabies virus is introduced by the bites of rabid dogs or other animals.

The portal of entry of bacterial toxins determines their ability to cause disease in the host. Botulinus toxin taken by mouth causes severe poisoning and possibly death; micrococcus toxin gives rise to serious gastrointestinal reactions when eaten. The toxins of *Clostridium tetani* and the hemolytic streptococci do not cause disease when taken into the digestive system, but do cause a severe reaction when injected into the skin or muscle.

Transmission

Regardless of how virulent an organism may be, it cannot cause disease in a number of persons or establish an *epidemic* (an unusual prevalence of a disease over the normal incidence) unless it can find new susceptible hosts to infect. A very virulent pathogen brings on its own destruction by killing the host that sustains it or by stimulating host immune resistance (see Chap. 25) that destroys it. For this reason, all epidemics are self-limiting, since the low-resistance hosts are eliminated and the highly resistant and immune members of the population survive. But epidemics may cause much suffering and death before they subside. They can, however, be prevented or restricted by appropriate control measures.

Transmission is dependent on two important factors: the escape of the pathogen from its host and its entrance into a susceptible host (Fig. 23-4). If there is a delay between the two events, the organism must be able to

TRANSMISSION OF DISEASE

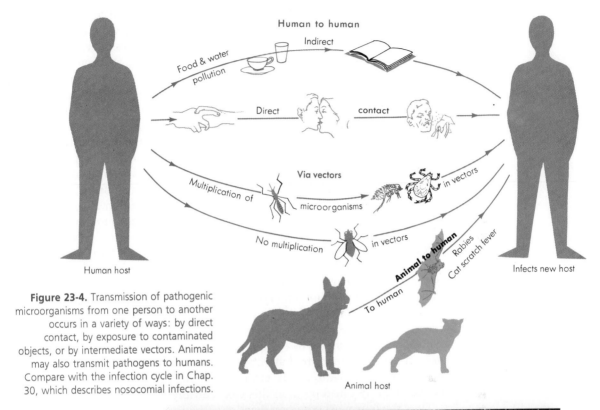

Figure 23-4. Transmission of pathogenic microorganisms from one person to another occurs in a variety of ways: by direct contact, by exposure to contaminated objects, or by intermediate vectors. Animals may also transmit pathogens to humans. Compare with the infection cycle in Chap. 30, which describes nosocomial infections.

Figure 23-5. The microorganisms that cause many respiratory diseases can be spread from one person to another in aerosols generated by sneezing, coughing, or even conversation.. *(Courtesy of Marshall W. Jennison and the American Society for Microbiology.)*

survive in an unfavorable environment. The manner in which pathogens escape depends upon the site of the infection in the host. Agents of diseases of the respiratory tract, such as *Streptococcus pneumoniae*, *Mycobacterium tuberculosis*, *Bordetella pertussis*, and many others, leave the body in exudates from the mouth, nose, and throat. Sneezing and coughing (Fig. 23-5) expedite dissemination of these organisms and enhance their chances of entering another host. The enteric pathogens that cause typhoid fever, salmonellosis, shigellosis, and other intestinal diseases leave the host in fecal excretions and sometimes also in urine. These bacteria may then contaminate food and water and enter a second host which has ingested the contaminated food or water. Some organisms are so delicate that they cannot survive long outside a suitable host, e.g., *Neisseria gonorrhoeae* and *Treponema pallidum*. These organisms are good examples of pathogens transmitted only by direct contact between the infected host and the next victim. *Francisella tularensis* (causing *tularemia*, a systemic disease characterized by an ulcerative lesion at the site of entry) and *Bacillus anthracis* (causing *anthrax*, a rapidly fatal septicemic disease of cattle) are examples of microorganisms that can enter the body through abrasions in the skin of persons who handle infected animals. Some microorganisms must be carried from one host to another by an intermediary of an intermediate host that introduces it directly into the bloodstream of the second host. The agents that cause Rocky Mountain spotted fever and typhus fever are carried by certain ticks and lice, respectively. These agents are not always transmitted directly from person to person. An intermediate host (a rat, for example) is often involved. Malaria is another disease transmitted by arthropods, in this case certain mosquito species.

Thus the four most common types of pathogenic microorganisms, ranked by method of transmission, are (1) airborne, (2) contact, (3) foodborne and waterborne, and (4) arthropod-borne organisms. It must be understood that the success of the pathogen in perpetuating itself depends on departure from its host, and also upon interhost survival and subsequently the gaining of a suitable portal of entry to susceptible individuals. The survival problems of the pathogen do not end here; if all the conditions we have discussed concerning virulence, escape, survival, and entry are favorable, it may still encounter defense mechanisms in the new host that it cannot overcome. These mechanisms will be discussed in Chap. 25.

Summary and outlook

The outcome of interactions between the host and invading microbes determines whether disease occurs. Whether a microorganism produces a disease is governed not only by its properties, but also by the ability of the host to resist infection.

Properties of microbes that enhance their pathogenicity are termed virulence factors. There are many of these factors, including various extracellular substances and specific cellular organelles. Other factors influencing infection include tissue affinity, portal of entry, and mode of transmission.

Even though virulence factors are known for some pathogens (especially those which produce extracellular toxins), most virulence factors of pathogens are as yet unknown. The mechanisms of pathogenicity are generally more subtle and complex than the obvious virulence factors discussed in this chapter. Considerable research is being done to reveal the nature of the pathogenicity, particularly at the biochemical and molecular level, of many pathogens.

Research on pathogenicity at the molecular level may be exemplified by studies bearing on microbial and host tissue surfaces. For instance, other than infections caused by vector bite or trauma, most infections begin on the mucous membranes of the respiratory, alimentary, or urogenital tracts. Surface components of microorganisms could contribute to mucosal-surface (and other surfaces) infection and penetration by promoting adherence, resisting any flushing action, aiding competition with the normal microbiota, and resisting antimicrobial substances and host humoral (serological) and cellular antimicrobial mechanisms.

Key terms

antitoxin	infection	pathogen	toxin
disease	invasiveness	pathogenicity	toxoid
epidemic	parasite	portal of entry	virulence factor

Questions

1 What kind of parasitism would you consider to be the most successful in host-microbe interactions?

2 What must a pathogen accomplish before it can produce a disease? Do these processes depend solely on the inherent properties of the pathogen?

3 What evidence is there to support the thesis that virulence factors possessed by parasites are more the exception than the rule?

4 Explain why exotoxins play a primary role in causing disease while endotoxins only play a contributory role.

5 How do capsules and pili functionally enhance bacterial pathogenicity?

6 Why do you think virulent microorganisms sometimes fail to produce disease and avirulent microorganisms may cause disease?

7 Explain why the method of transmission of pathogenic microorganisms is important in the spread of a disease, and name four methods of transmission.

24

ANTIGENS AND ANTIBODIES

A person who has recovered from a childhood disease, such as measles, mumps, or chickenpox, is usually not susceptible to a second attack of this disease. We say that this person has become highly resistant to the disease caused by a specific pathogen; this state of resistance is called *immunity*. Since this immunity is acquired after initial exposure to a pathogenic organism, we call it an *acquired immunity*.

Acquired immunity may come about through the use of vaccines as well as by natural infection. Thus both vaccines and microorganisms stimulate the host's resistance mechanism (*immune system*).

The basis for acquired immunity lies in the ability of the immune systems of vertebrates to recognize cells or substances of their own bodies ("self") from those of other origins ("nonself"). "Nonself" materials include cells from other animals, viruses, toxins and toxoids, bacteria, and vaccines. Such materials entering the body are recognized as foreign substances, or *antigens*, and will stimulate the production of humoral *antibodies*. These antibodies react with the specific foreign substances in order to eliminate or neutralize them. (Antigens will, however, also stimulate the production of specialized reactive cells.)

This chapter will introduce you to these antigens and antibodies.

Antigens   An antigen is any substance that, when introduced into a vertebrate host, provokes an immune response leading to acquired immunity. The immune response results in the formation of specific antibodies that circulate in the bloodstream (*humoral immunity*) or stimulates the increase in number of specifically reactive cells called *lymphocytes* (*cell-mediated immunity*) or both. These lymphocytes have acquired an enhanced ability to destroy other cells. Both antibodies and the specialized lymphocytes react with the antigen used as the immunizing agent. Immunity acquired in this way enables the body to destroy or neutralize invading microorganisms or their toxins. It is the body's main line of internal defense against pathogenic microbes.

Properties of antigens

In general, the more "foreign" in chemical composition and structure the antigen is to the individual being immunized, the more effective that antigen is in stimulating an immune response.

Only two groups of naturally occurring compounds are clearly *immunogenic*, that is, have the capacity to stimulate an immune response. These are proteins and polysaccharides. Proteins are generally more effective in stimulating antibody production than are polysaccharides. Large complex polysaccharides, however, such as the pneumococcal capsular polysaccharides, are good antigens because they elicit a strong immune reaction. Oligosaccharides, lipids, and nucleic acids do not stimulate antibody production on their own but will do so when combined with protein. They are called *haptens*, substances which in themselves are not antigenic but which when linked strongly to carrier molecules, such as proteins, function as antigenic groups that direct the specificity of an immune response. Haptens will combine with a specific antibody once it is formed.

An antigen may be a soluble substance such as a bacterial toxin or serum (fluid portion of coagulated blood) protein. Or it may be particulate in nature, such as a bacterial cell or virion. Particulate antigens are usually more potent than soluble antigens.

Antigens are, without exception, substances of high molecular weight. A compound having a molecular weight of less than 6,000 daltons can rarely act alone as an antigen. Most antigens possess a molecular weight of 10,000 daltons or greater.

Antigenic determinants

Any substance, whether a virus, bacterium, or body cell, that acts as an antigen, has on its surface, and sometimes in its interior, a number of reactive sites, or *antigenic determinants* (Fig. 24-1). These determinants impart specificity to the immune response and are the sites that react with

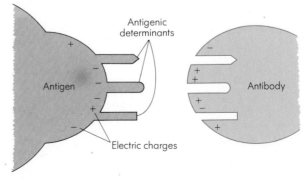

Figure 24-1. Schematic representation of the site of antigen-antibody reaction. The interaction of the antigen and antibody is influenced by antigenic determinants, electric charges, and accessibility of determinants due to the three-dimensional structure of both molecules.

antibody or sensitized lymphocytes. Some antigens have as many as 200 different kinds of antigenic determinants; others have as few as two or three.

Although cells of all species possess distinct antigens specific to a species, some antigens are widely distributed. Groups of related, but not identical, antigenic determinants may be found on the cells of a wide range of unrelated organisms. Such cells, when used as antigens, are called *heterophile antigens*. They stimulate the production of antibodies capable of reacting with tissues of a wide variety of unrelated animals. One such antigen is the much-studied Forssman antigen. Antigens closely related to it occur in numerous species of animals, bacteria, and even plants. Antibodies to the Forssman group of antigens appear in large quantities in the blood of people with certain diseases, such as infectious mononucleosis, and are useful in diagnosis.

Adjuvants

Adjuvants are substances which, when injected together with antigen, increase antibody production. A variety of substances with diverse chemical composition have been found to possess an adjuvant effect. Such substances include alum and other aluminum salts, sodium alginate, bacterial endotoxins, and water-in-oil suspensions with or without killed mycobacteria. For example, Freund's adjuvant has been the most studied and most used in experimental work. The complete Freund's adjuvant consists of a mineral oil, an emulsifier, and killed tubercle bacilli (Fig. 24-2).

Naturally occurring antigens of interest to medicine

Medically important antigens are those found in human tissues and cells and those present in, and produced by, bacteria and other microorganisms. In addition, there are animal and plant antigens which may be injurious to health, such as the ragweed antigens, which cause hay fever.

Human tissue antigens. Antigens present in human red blood cells have been studied extensively. These include the ABO, MN, Ss, P, and Rh

Figure 24-2. Effect of an adjuvant on the antibody response.

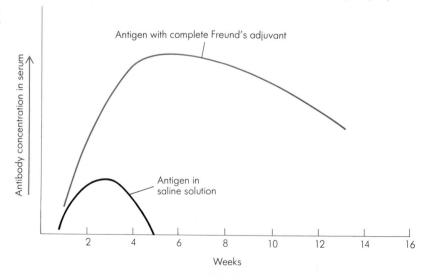

Table 24-1. Antigens and
isoantibodies associated
with the ABO blood
groups

BLOOD TYPE	ANTIGEN ON RED BLOOD CELLS	ISOANTIBODY IN SERUM
A	A	Anti-B
B	B	Anti-A
AB	AB	None
O	None	Anti-A
		Anti-B

antigens, as well as others. The best-known ones are those responsible for the ABO blood groups. These groups are A, B, AB, and O. These antigens present in red blood cells of some persons react with *isoantibodies* (antibodies present in the *same* species) in other persons. For example, individuals of blood group O contain both anti-A and anti-B antibodies in their sera (singular, serum). Individuals of blood group A have only anti-B antibodies, while group B individuals possess anti-A antibodies. It follows that blood group AB individuals have no anti-A and anti-B antibodies in their sera. (See Table 24-1.)

The Rh antigens constitute a complex blood group antigen system. This system is of clinical importance because maternal immunization to an Rh antigen present in the fetus and absent in the mother may result in hemolytic disease of the newborn called *erythroblastosis fetalis*. This disease occurs more frequently in newborns of Rh-negative mothers and Rh-positive fathers and more frequently in babies whose mothers have had multiple pregnancies.

The increase in tissue transplantation and the recognition of *autoimmune diseases* (in which there is production of antibodies against an individual's own tissues because of some malfunction of the immune system such as in rheumatoid arthritis) have in recent years led to an intensive study of tissue antigens. These are called *histocompatibility antigens*, and they stimulate rejection of transplanted organs or tissues which are not compatible.

Bacterial and viral antigens. Bacterial antigens are either excreted as exotoxins and enzymes or are structural components of the cell.

The outermost portion of the bacterial cell may consist of a capsule. The polysaccharide capsule of the pneumococci is antigenic, as is the K or Vi capsular antigen of *Salmonella typhi* and certain other salmonellae (see Fig. 24-3).

In flagellated species of salmonellae, the flagellar antigens are designated

Figure 24-3. External
antigens of gram-negative
bacilli such as the
salmonellae.

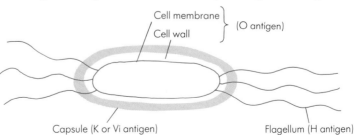

Cell membrane
Cell wall } (O antigen)

Capsule (K or Vi antigen)   Flagellum (H antigen)

as H antigens. The O antigen complex (an endotoxin) of the gram-negative enteric bacilli is situated in the membrane-cell wall complex of these bacteria (even though the O antigen is sometimes called the *somatic,* or body, antigen). These are polysaccharide-phospholipid-protein complexes; their serological specificity lies in the polysaccharide component. Figure 24-3 shows the location of these antigens.

Polysaccharides are also of great importance as antigenic or haptenic components of the cell wall of gram-positive bacteria. For example, they are the basis for the Lancefield grouping of the streptococci by the antigenic differences in their cell-wall polysaccharides. Fifteen different groups of streptococci are recognized on this basis.

Viruses are basically composed of nucleic acid and protein, although some also contain carbohydrate and lipid components (see Chap. 12). The intact structures of the virus exterior make good antigens and are mainly protein or, in some cases, lipoprotein and/or glycoprotein.

Vaccines are suspensions of killed or live *attenuated* (having weakened virulence) cultures of microorganisms or their products which are used as antigens to produce immunity against infection by a specific microorganism. For example, typhoid fever vaccine consists of killed cells of *Salmonella typhi.* The Salk poliomyelitis vaccine is composed of killed poliomyelitis virus; but the oral type, such as the Sabin vaccine, contains attenuated live virus. Toxoids are vaccines made by destroying the toxic portions of toxins without altering their antigenicity. The antibodies (antitoxins) produced will then neutralize the toxins excreted by toxin-producing bacteria which cause diseases, such as tetanus and diphtheria. (See Table 24-2.)

Antibodies

Antibodies have already been defined as specific substances formed by the body in response to antigenic stimulation. All antibody molecules belong to a special class of serum proteins called *globulins,* although not all serum globulins are antibodies. Thus antibodies are also called *immunoglobulins* (abbreviated Ig).

Table 24-2. Types of vaccines

VACCINE	LIVE VACCINES	KILLED VACCINES
Viral	Smallpox	Poliomyelitis (Salk)
	Rubella	Influenza
	Measles	Rabies (for humans)
	Poliomyelitis (Sabin)	
	Rabies (for animals)	
	Yellow Fever	
	Mumps	
Bacterial	BCG	Cholera
	Brucella (veterinary use)	Typhoid
		Whooping cough
		Typhus
Bacterial toxoid vaccines	Diphtheria	
	Tetanus	

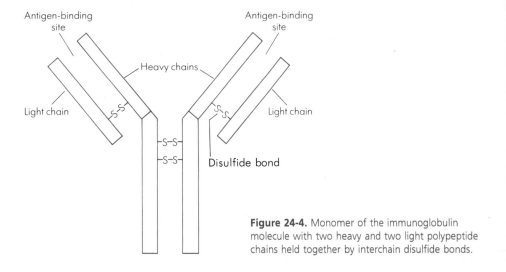

Antigen-binding site

Antigen-binding site

Heavy chains

Light chain

Light chain

—S—S—

—S—S—

Disulfide bond

Figure 24-4. Monomer of the immunoglobulin molecule with two heavy and two light polypeptide chains held together by interchain disulfide bonds.

Classes of immuno-
globulins (antibodies)

There are five classes of immunoglobulins: immunoglobulin G (IgG), immunoglobulin M (IgM), immunoglobulin A (IgA), immunoglobulin D (IgD), and immunoglobulin E (IgE). All consist of the same type of structural, or monomeric, unit made up of two light and two heavy polypeptide chains, as shown in Fig. 24-4. The heavy chains have a molecular weight of approximately 55,000 each; the light chains are about 25,000 each. The chains are joined by disulfide bonds.

The heavy chains are specific for each class of immunoglobulin, containing unique amino acid sequences which specify their type. There are five types of heavy chains; they are designated by the Greek letters γ (gamma), α (alpha), δ (delta), μ (mu), and ϵ (epsilon). The corresponding classes of immunoglobulins are thus named IgG, IgA, IgD, IgM, and IgE. All normal individuals possess varying amounts of immunoglobulins from each of these five classes.

The light chains are of two major types, κ (kappa) and λ (lambda). Both types are found in immunoglobulins of all five classes. In any one immunoglobulin molecule, both light chains are always of the same type. Similarly, each immunoglobulin molecule contains only one type of heavy chain.

The terminal portions of both heavy and light chains (at the antigen-combining site) of each monomeric unit show considerable variation, whereas the remaining portions of the chains are relatively constant in amino acid structure. Thus the amino acid sequence of the constant regions determines the class or biological role of an immunoglobulin, and that of the variable regions determines its specificity.

Each class of immunoglobulin can be characterized not only in regard to structure but also to function. The discussion below is confined to human immunoglobulins.

IgG. Over 70 percent of the immunoglobulins in normal human serum is IgG. It is the most common form of antibody and can pass from mother to fetus before birth (placental transfer). IgG can be subdivided on the basis of

Table 24-3. Some characteristics of the different classes of human immunoglobulins

IMMUNO-GLOBULIN	HEAVY CHAIN DESIGNATION	MOLECULAR WEIGHT	PHYSICAL STATE	J CHAIN	SUBCLASSES
IgG	γ (gamma)	150,000	Monomer	–	4
IgM	μ (mu)	900,000	Pentamer	+	1
IgA	α (alpha)	160,000 (serum) 370,000 (secretions)	Monomers and polymers (serum); dimers (secretions)	+ in dimers and polymers	2
IgD	δ (delta)	180,000	Monomer	–	1
IgE	ε (epsilon)	185,000	Monomer	–	1

antigenic differences into four subclasses designated IgG_1, IgG_2, IgG_3, and IgG_4. Some of the characteristics of IgG and other immunoglobulins are summarized in Table 24-3. All normal human sera contain all four subclasses of IgG. Respectively, IgG_1, IgG_2, IgG_3, and IgG_4 make up approximately 59, 30, 8, and 3 percent of the total human IgG. All occur as monomeric units. The biological properties of IgG and other immunoglobulins are shown in Table 24-4. As may be seen in the table, IgG is the only immunoglobulin that gives protection to the newborn, because not only is it present before birth in the circulation, but it is absorbed from mother's milk (colostrum) into the bloodstream.

IgM. IgM was so named because it is a *macroglobulin,* at least five times larger than IgG. Each IgM molecule is actually composed of five monomeric units (see Fig. 24-5). There is also an additional peptide called the J chain bound to one or perhaps two of the μ heavy chains. It may be responsible for stabilizing the multimeric form of IgM. IgM is usually the first antibody to appear following induction by an antigen; it does not cross the placenta to the unborn child. About 6 percent of the total immunoglobulin is IgM.

Table 24-4. Biological properties of different classes of human immunoglobulins

IMMUNO-GLOBULIN	SITE FOUND	COMPLEMENT-FIXATION*	CROSS PLACENTA	FUNCTIONS
IgG	Internal body fluids, particularly extravascular	+	+	Major line of defense against infection during the first few weeks of a baby's life; neutralizes bacterial toxins; binds to microorganisms to enhance their phagocytosis
IgM	Largely confined to bloodstream	+	–	Efficient agglutinating and cytolytic agent; effective first line of defense in cases of bacteremia (bacteria in blood)
IgA	Serum, external body secretions	–	–	Protects mucosal surfaces from invasion of pathogenic microbes
IgD	Serum, on lymphocyte surface of newborn	–	–	Regulator for the synthesis of other immunoglobulins
IgE	Serum	–	–	Responsible for severe acute and occasionally fatal allergic reactions; combats parasitic infections

*See Chap. 25.

Since it has five monomers, IgM can combine with antigen at more than one site and is very effective in reacting with viruses and bacteria.

IgA. IgA constitutes about 10 percent of the total immunoglobulin in serum. The basic structure is like that of IgG, i.e., two light and two heavy chains. But it also occurs in other polymeric forms bonded together by disulfide bonds. In addition to occurring in the serum as monomers and polymers, it is also found externally in almost all the body secretions, such as tears, saliva, seminal fluid, urine, and colostrum. Therefore its presence in secretions defends external body surfaces; in colostrum it affords protection to the newborn. Secreted IgA is also present in the mucus of the lungs and gastrointestinal tract. The secreted IgA occurs primarily as a dimer, with a small amount found as trimers. Like the IgM molecule, secreted IgA antibodies also possess a J peptide chain. In addition, there is a second polypeptide which has been designated the secretory piece; its function is unknown. (See Fig. 24-5.)

IgA, because of its strategic occurrence in various fluids, may represent one of the first means of defense against bacteria and viruses.

IgA can be divided into two subclasses based on minor antigenic differences. Both are found in all normal individuals.

IgD. IgD is present at a level of only about 1 percent of the total immunoglobulin in normal serums. It was first observed in patients with a

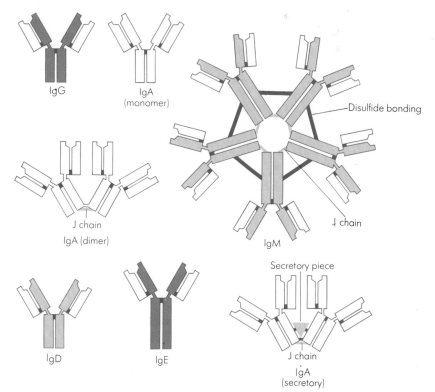

Figure 24-5. Structures of the different classes of immunoglobulins. IgG, IgD, and IgE each consists of a monomer of two light and two heavy polypeptide chains. IgM is a large molecule having five monomers in a star formation joined with a J (joining) polypeptide chain. IgA has three forms. When it appears in the serum, it may consist of one, two, or three monomers (not shown); when it is found in such body fluids as saliva, tears, and nasal secretions, it contains two monomers joined by a special component known as the "secretory piece" (this dimer also has a J chain). The exact location of the J chain in the IgM and IgA molecules is uncertain.

cancer called multiple myeloma; these persons synthesize abnormally high amounts of IgD. If it were not for these patients, IgD might still not have been found as a circulating antibody class.

Little is known about IgD. It has been proposed that the biological role of IgD is that of a regulator for the synthesis of the other classes of immunoglobulins.

IgE. Even though they are present in normal human serum at a concentration of only 0.002 percent of total immunoglobulin, IgE proteins are of great clinical importance. When combined with antigens, they are responsible for allergic (*hypersensitivity*) reactions.

Some hypersensitivity reactions are immediate and dramatic. For example, an allergy to insect stings or to a drug, such as penicillin, can result in a severe generalized reaction which can lead to death if not treated promptly. Less serious allergic reactions occur in persons sensitive to grass pollens, animal danders, house dust, and other agents.

In all such hypersensitivity reactions, contact of the allergen with IgE antibodies releases chemical mediators such as histamine and serotonin which produce the typical symptoms of allergy—sneezing, wheezing, runny nose, and teary eyes.

Even though IgE is noted for its harmful properties, it may have protective functions. The release of histamine may result in the destruction of parasites. This hypothesis arose because it was observed that individuals in tropical areas (where parasites are more prevalent than in northern regions) may have 20 times more IgE antibodies than those who live in colder climates. This, of course, is of little cheer to northerners who suffer the miseries of summer allergies; knowing that IgE may be a beneficial antibody for those who reside in the tropics does not relieve the discomfort.

Functional names of antibodies

Antibodies react against specific microorganisms and their toxic products, and other compounds. They can be used in the treatment of infection caused by the specific microorganisms and, more importantly, they prevent infection and disease caused by these agents. Antibodies are designated by names that describe their reaction in vitro or in vivo when they are allowed to act on certain types of antigens: (1) *antitoxins*, (2) *agglutinins*, (3) *precipitins*, (4) *lysins*, (5) *complement-fixing antibodies*, and (6) *opsonins*. Any antibody may be multifunctional and therefore can be called by more than one of the above terms.

These antibodies are all produced as a result of antigenic stimulus. They are humoral antibodies and can be differentiated as follows:

1 Antitoxins neutralize toxins.
2 Agglutinins cause clumping of the bacterial cells for which they are specific.
3 Precipitins cause precipitation or flocculation of extracts of bacterial cells or other soluble antigens.
4 Lysins cause lysis, or breakdown, of bacterial or other cells that are specifically sensitive to their action.

5 Complement-fixing antibodies participate in the complement-fixation reactions which are described in Chap. 25.

6 Opsonins render microorganisms more susceptible to ingestion by phagocytes.

Summary and outlook

Antigens are substances which when introduced into a vertebrate host provoke an immune response resulting in the production of antibodies and specialized cells. In order to be an antigen, the substance must be foreign to the host, it must be a relatively large molecule, and it must have at least two antigenic determinant groups.

Antibodies are immunoglobulins synthesized in response to an antigenic stimulation. There are five different classes of immunoglobulins designated IgG, IgM, IgA, IgD, and IgE. They may be distinguished on the basis of molecular structure and function.

The success that microbiologists have had in understanding antigens is shown by the many available vaccines that today protect us against various contagious diseases. The future will no doubt witness the availability of others, such as vaccines against hepatitis, gonorrhea, and syphilis. A greater understanding of antibodies may lead to better control of some illnesses such as cancer and hay fever. Other natural phenomena, for example, aging and individual differences in disease susceptibility, may be better understood as fundamental knowledge accumulates on the properties of antigens and antibodies.

Key terms

acquired immunity
adjuvant
agglutinin
antibody
antigen
antigenic determinant
antitoxin
attenuated
autoimmune disease

cell-mediated immunity
complement-fixing antibody
erythroblastosis fetalis
hapten
heterophile antigen
histocompatibility antigen
humoral immunity
hypersensitivity
immune system

immunity
immunogenic
immunoglobulin
isoantibody
lymphocyte
lysin
opsonin
precipitin
somatic antigen

Questions

1 List three criteria that a substance must meet before it can be called an antigen. Which of these criteria is (are) not met by a hapten?

2 Explain why the recognition of "self" and "nonself" by an animal host is the basis for acquired immunity.

3 What are some factors that make an antigen more effective in the stimulation of antibody production?

4 Give some examples to illustrate why human tissue antigens are relevant in the practice of medicine.

5 Provide some examples of the practical uses of microbial antigens.

6 Explain how the heavy chains and light chains of polypeptides are arranged in the structure of the monomeric unit of immunoglobulins.

7 Why do some portions of the monomeric unit of immunoglobulins have constant amino acid sequences while other portions have a variable amino acid sequence?

8 Compare and contrast the structure and biological functions of IgM and IgE.

HOST RESISTANCE AND IMMUNITY

While the parasite uses a variety of means to establish infection, the host possesses a number of defense mechanisms to prevent infection. These defense mechanisms are referred to as *host resistance*.

There are two major types of host resistance: (1) *Specific resistance* directed against particular microorganisms and (2) *nonspecific* or *natural resistance*. The first type is acquired specific immunity and was introduced in Chap. 24. The second type depends on such a large number of factors that only a few of the most important ones are discussed in this chapter. The host's general health and state of nutrition, social and economic conditions, and other nonspecific factors all play a part, but their roles are so interlocked that it is difficult to evaluate their individual importance (Fig. 25-1).

Natural resistance

Species resistance

Resistance or susceptibility to infection by a particular pathogen may vary from one species of animal to another. For example, humans are susceptible to infection by gonococci that cause a *sexually transmitted* (venereal) *disease*. Other animals, such as the rabbit, are resistant to these bacteria. Similarly, humans are relatively resistant to the tubercle bacillus, compared with the guinea pig which is very susceptible to the microbe.

In general, metabolic, physiological, and anatomical differences between species affect the ability of a pathogen to cause infection. But we usually do not know precisely why resistance varies from one species of animal to another.

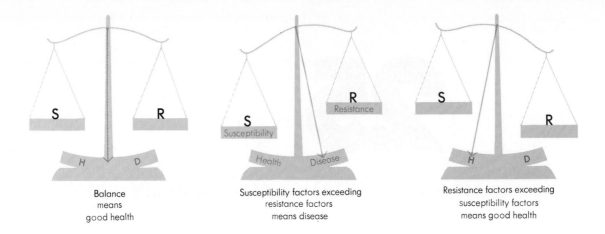

Balance means good health	Susceptibility factors exceeding resistance factors means disease	Resistance factors exceeding susceptibility factors means good health

1 Microbial factors:
Virulence
Invasiveness
Portal of entry
Dosage
Mircoenvironment
Indigenous microbiota

2 Physiological factors:
Injury, constitutional disease
Physiological stress
Psychological stress
Misdirected tissue response,
   inflammation, histamine release

3 Cellular factors:
Phagocytosis by macrophages
   and polymorphonuclear
   leukocytes
Lymphocytes

4 Humoral factors:
Antibodies
Complement
   interferon

5 Constitutional factors:
Nutrition
Age
Tissue tone
Hormonal balance
Vascular condition
Lysozyme
Race

6 Socioeconomic factors:
Housing
Hygiene
Occupational hazards

Figure 25-1. Host-parasite relationship in health and disease: the balance between susceptibility and resistance.

Racial resistance

Just as there are marked differences in susceptibility (lack of resistance) to certain infections among different strains of laboratory animals such as mice, there are also racial differences among humans in their susceptibility to many diseases. American Indians appear to be less resistant to infection by the tubercle bacillus than are people of European background, the ancestors of whom were exposed to tuberculosis many centuries before the Indians.

Also notable is the resistance to malaria (caused by *Plasmodium vivax*) in almost all African blacks and in 70 percent of American blacks. Most whites are susceptible to this parasite. Resistance to the organism is attributed to the lack of a specific receptor on the red blood cells of the resistant host; the parasite must attach to this receptor in order to cause disease.

Other examples include the relatively high resistance of blacks to yellow fever and of Chinese to syphilis. This is explained on the basis of exposure of blacks and Chinese to the diseases before whites came in contact with them. After centuries of association between pathogens and humans, a kind of racial resistance seems to develop in humans, probably as the result of natural selection of the more resistant individuals.

Individual resistance

Some people have one cold after another, while others never "catch" a cold; yet they all have the same chances for exposure. In some families with several children, all but one child may simultaneously have bouts of chicken pox, measles, or other childhood diseases. These are examples of individual resistance; it may be due to a single factor or a combination of several.

Some obvious factors relating to individual resistance are age, state of nutrition, personal hygiene, state of health, sex, and occupation.

Age. Some diseases are called "children's diseases" because they occur with much greater frequency in children than in adults. This is explained by the fact that antibodies against these diseases develop in adolescence in response to infection, so by adulthood a state of acquired immunity has been attained.

Diseases are generally more damaging to the young and to the aged. The underlying reason is, for the most part, the immunological state of the host. From birth to 6 months, infants are protected from many diseases, owing to the presence of maternal antibodies in the blood. But after this time, the maternal antibodies disappear and the young infant must acquire its immunity through natural infection or artificial means (vaccination). In the elderly there is a natural decline in the immune response to infection, so they become more susceptible to some diseases such as pneumonia.

Nutrition. In general, if the state of health of the individual is impaired by inadequate nutrition, infections are more likely to occur.

Occupation. Occupation is a factor in individual resistance mainly because of the opportunity for exposure during work. For example, there are more infections of tularemia and brucellosis in persons exposed to animals and animal hides; these persons include hunters and stockyard workers and are mostly males.

Miscellaneous factors. Many additional factors can also affect an individual's resistance to disease. The compromised patient will be discussed in Chap. 30. Sex may also play a role because of the action of the different sex hormones and endocrines in males and females. Personal hygienic practices are also important in determining individual resistance.

Mechanical and chemical barriers of resistance

Mechanical barriers include the unbroken skin and mucous membranes which generally prevent the entry of microorganisms. However, certain fungi will readily produce skin infections when the skin becomes moist and soft, e.g., fungi that cause athlete's foot. But most bacteria are inhibited by lactic and other fatty acids found in the secretions of the sweat and sebaceous glands and by the low pH which these acids have.

Mucous secretions of the respiratory tract, the digestive tract, the urogenital tract, and other tissues form a protective covering of the mucous membranes, collecting and holding many microorganisms until they can be disposed of or until they lose their infectivity.

In addition to the mechanical action of mucus, saliva, and tears in removing bacteria, some of these secretions contain antimicrobial substances that are important in preventing infections. An example is *lysozyme*, an enzyme that hydrolyzes the cell wall of numerous bacteria; it is found in many body fluids and secretions, especially tears. The acidity or alkalinity of some body fluids has a detrimental effect on many organisms.

Inflammation

When microbes or other foreign materials have penetrated into or through epithelial surfaces, they evoke a complex process called *inflammation* around the site of entry. The affected site now exhibits the four cardinal signs of inflammation; it is *red, warm, swollen,* and often *painful.* These signs are caused initially by the release of some inducers of inflammation by the host tissues and *plasma* components. Plasma is the suspending fluid of blood. The cells suspended in plasma are the red blood cells (erythrocytes), the white blood cells (leukocytes), and the platelets. (See Table 25-1 for a tabulation of the different kinds of cells associated with host resistance.) Plasma without fibrinogen, a protein removed during clotting, is called *serum.* The blood vessels dilate (*vasodilation*) and become more permeable, allowing for a greater concentration of the defensive elements. The affected vessels also become "sticky," causing the adherence of circulating leukocytes called *polymorphonuclear granulocytes* (polymorphs). These polymorphs pass through the walls of the now permeable vessels and engulf the invading foreign cells by phagocytosis, a process which will be discussed in more detail later.

If the inflammation is due to infection with bacteria and the infection continues, then the continued supply of inflammatory products is maintained by the multiplying bacteria because most bacteria form inflammatory materials during their growth in tissues. In this way, vasodilation is maintained, as is the flow of polymorphs to the affected area. Viruses also produce inflammatory products in tissues in the form of *necrotic* (dead) host cells or antigen-antibody complexes. (The fluid product of inflammation containing serum, bacteria, dead cells, and leukocytes is called *pus.*)

Many polymorphs also die since they live only a day or two in tissues. Their death is accompanied by release of inflammatory substances that attract other phagocytes called *macrophages*; these cells phagocytose dead polymorphs, tissue debris, and pathogenic microorganisms.

Inflammation serves to confine infecting microorganisms to the site of entry into the host. If they escape into the blood or lymph vessels, however, the microbes may still be phagocytosed by specialized phagocytic cells.

Phagocytosis

The importance of phagocytosis in protecting the body from infection was first recognized by the Russian zoologist Elie Metchnikoff in 1883 (Fig. 1-12). He believed that certain cells with ameboid motility were the body's main defense against infection, their method of action being the destruction of invading microorganisms. Even though this concept was well entrenched in the minds of people since that time, from the 1920s on, phagocytes were neglected by researchers. Little was added to our under-

Table 25-1. Cell types associated with host resistance to infection

CELL TYPE	LOCATION	DERIVATION	DESCRIPTION AND FUNCTION
Leukocytes	Blood	Bone marrow stem cells	Classified according to structure and affinity for dyes.
Polymorphonuclear granulocytes			Lobular nuclei and abundant cytoplasmic granules.
Neutrophils			No dye preference. Phagocytosis.
Eosinophils			Stain with eosin, an acid dye. Phagocytosis.
Basophils			Stain with basic stains. Binds IgE and produces histamine.
Lymphocytes	Blood, lymphoid tissue, spleen, and thymus	Lymphoid organs, bone marrow stem cells	Smaller than monocytes; large nucleus; scanty cytoplasm. Form T and B cells. (See Fig. 26-9.)
Monocytes	Blood	Bone marrow stem cells	Larger than granulocytes; single horseshoe-shaped or oval nucleus; few cytoplasmic granules. Phagocytosis.
Plasma cells	Lymph nodes and other lymphoid tissues	From B lymphocytes	Produce antibodies.
Macrophages	Tissues throughout body	Transformed from monocytes	Numerous cytoplasmic granules. Phagocytosis.
Wandering			
Alveolar	Lung		
Peritoneal	Abdomen		
Fixed			
Histiocytes	Connective tissue		
Kupffer cells	Liver		

standing of this important process after the initial pioneer studies. In recent years, however, there has been a resurgence of interest in phagocytosis as a central process in infections, and much fundamental information is now being accumulated.

There are two types of specialized cells involved in the phagocytic process: the polymorphonuclear granulocytes (mainly neutrophils) and the macrophages. (See Fig. 25-2.)

The polymorphs constitute the front line of internal defense for the host. They are produced in the bone marrow and are discharged into the blood in vast numbers. The 6×10^{11} polymorphs that are present in normal human blood carry out their functions after leaving the circulation and entering sites of inflammation in tissues. They live for a few days only, and about 10^{11} polymorphs disappear from the blood daily. But they are replaced by new ones from the bone marrow. Polymorphs contain numerous enzymes and antimicrobial substances for the killing and degradation of bacteria. These

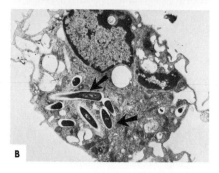

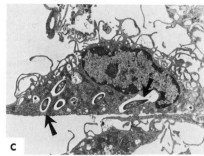

Figure 25-2. (A) Scanning electron micrograph of a peritoneal macrophage from a mouse fixed during locomotion (×2,700). (B), (C). Thin sections of cultured mouse peritoneal macrophage showing phagocytosed cells of *Listeria monocytogenes* (a bacterium) indicated by arrows. (B) Horizontal section (×4,000). (C) Vertical section. (×3,800). *(Courtesy of P. Gill, McGill University.)*

substances are contained in membrane-bound organelles called *lysosomes*.

Macrophages are formed from circulating precursor *monocytes* (blood cells with horseshoe-shaped nuclei; see Fig. 25-11) which also arise from the bone marrow. As soon as the monocytes leave the circulation and begin to carry out their phagocytic duties, they are called macrophages. Unlike the polymorphs, macrophages are long-lived and can persist in tissues for weeks or months. They are widely distributed throughout the body, but they are not as numerous as polymorphs. Under certain conditions, macrophages can synthesize DNA and multiply. When they differentiate in connective tissue, they are called *histiocytes*, in the liver, *Kupffer cells*, and in the lung, *alveolar macrophages*.

Two types of mature macrophages are recognized: (1) those wandering in tissues and body spaces (e.g., alveolar and peritoneal macrophages) and (2) those fixed to vascular endothelium (e.g., Kupffer cells and fixed macrophages of the spleen and lymph nodes).

Macrophages, then, are strategically placed throughout the body to combat invading microorganisms. Macrophages also have lysosomes and bactericidal substances. (See Fig. 25-11 for the morphology of the polymorphs and macrophages.)

Mechanism of phagocytosis. The process of phagocytosis requires a preliminary attachment of the microbe to the phagocytic cell surface. Electrostatic forces are necessary for initial attachment, since divalent cations such as Ca^{2+} and Mg^{2+} are required. The firm attachment is facilitated by serum substances called *opsonins*. Opsonins are antibodies that allow microbes to be more easily ingested by the phagocyte. Phagocytic cells have a special affinity for microorganisms coated with antibody. If complement (a component in blood to be discussed later) is also present on the microbial surface, there are complement receptors on the phagocytic cell which provide additional attachment forces. (See Fig. 25-3.)

In phagocytosis, the phagocyte extends small pseudopods around the microbe after adherence. These pseudopods fuse and form a vacuole by means of invagination of the phagocyte plasma membrane engulfing or surrounding the bacterium. This vacuole is now called a *phagosome*. Subsequent events depend on the activity of the lysosomal granules. These move toward the phagosome, fuse with its wall to form a *phagolysosome*, and discharge their contents (hydrolytic enzymes) into the vacuole, thus initiating the intracellular killing and digestion of the microbe. Within the

Figure 25-3. Attachment of bacterial cell antigen to surface of phagocyte. The phagocyte has receptors on its surface both for the antibody and for the complement. Phagocytosis occurs more readily with this kind of binding.

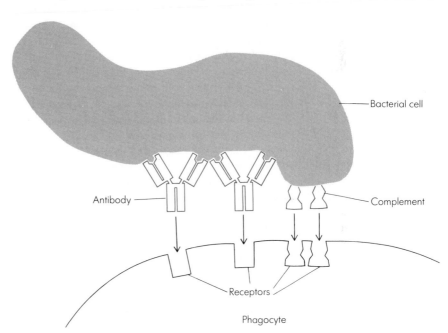

Figure 25.4. Phagocytosis of a microbial cell.

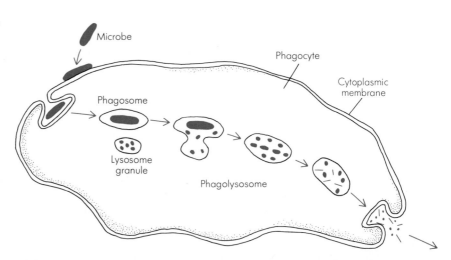

phagolysosome, most microbes are killed within minutes, although the complete degradation may take a few hours. The process of phagocytosis is shown schematically in Fig. 25-4.

Complement system

Complement, which is normally present in serum, consists of a group of related proteins. It is so called because of its complementary effect on certain immune reactions involving antibodies. Such reactions include opsonization, chemotaxis, and cell lysis. Antibody, once it has reacted with its antigen, can do no more. It might precipitate an antigen or cause agglutination of bacterial cells, but other than the neutralization of toxins or of virus infectivity, antibody by itself is a relatively ineffective means of host resistance.

One of the major functions of antibody, however, is to recognize a specific antigen and bind to it. In this way, it provides a site for the initiation of the reactions of the complement system. Activation of the complement system enhances phagocytosis of invading microorganisms (opsonic activity) and leads to the lysis of foreign cells. In the case of a gram-negative bacterium, the integrity of the cell membrane is destroyed, resulting in lysis of the cell (Fig. 25-5). Gram-positive bacteria, however, are not lysed, but fragments of complement components bind to the cells, forming bridges with host leukocytes and enhancing phagocytosis.

Interferon

Interferon is a nonspecific antiviral agent which inhibits intracellular viral replication and is synthesized by cells in response to viral infection. Interferons were discovered in 1957 by investigators working on the mechanism of *viral interference* (resistance of an animal or cell infected with one virus to superinfection with a second unrelated virus). Interferons do not react directly with the virion but rather exert their protective effect by an intracellular mechanism. Thus interferons have no virus specificity.

However, interferons are specific with regard to the species of cells that produced them. For instance, interferons produced by human cells primarily protect human cells and have little capacity for protecting mouse, chick, or other animal cells.

Purified interferons from various sources consist of small proteins unusually stable at low pH and fairly resistant (for example, they can resist a long exposure to pH 2 in the cold).

The mechanism of interferon induction by viruses appears to be related to

Figure 25-5. Mechanism of complement action: the bacterial-killing effects of antibody and complement. *(Courtesy of A. J. Vander, J. H. Sherman, and D. S. Luciano, Human Physiology, The Mechanisms of Body Function, McGraw-Hill, New York, 1970.)*

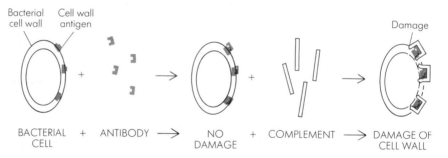

Bacterial cell wall Cell wall antigen

Damage

BACTERIAL CELL + ANTIBODY → NO DAMAGE + COMPLEMENT → DAMAGE OF CELL WALL

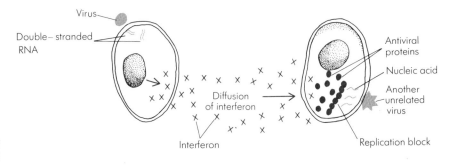

Figure 25-6. Mechanism of interferon induction and action.

the presence of double-stranded RNAs (such as reovirus RNA) and certain synthetic polynucleotides. Double-stranded DNA and DNA:RNA hybrids are relatively poor inducers.

Interferon causes antiviral resistance indirectly by inducing the synthesis of an antiviral protein by cells exposed to another virus. (See Fig. 25-6.) This means that protection by interferon is accomplished through host-cell synthesis of a set of antiviral proteins which block viral replication. Since the inhibition of virus is not a direct one, it is not surprising that interferons are not virus-specific but cell species-specific.

Our interest in interferon, in the context of this chapter, is obviously in its antiviral effect in preventing cell infection. However, this is but one facet of interferon activity. There are other aspects which interest microbiologists; examples are interferon effect on cell-membrane-related events, immune responses, and cell-growth depression and modification.

Unfortunately, interferons are relatively unstable in tissue fluids. For this and other reasons, interferon has not been particularly useful clinically. But interferons play a protective role during naturally acquired viral infections because they are produced more promptly than specific antibodies.

Acquired specific immunity

The study of the immune response, immunology, initially developed in parallel with microbiology because it dealt with the development of host resistance to infectious disease. Today such resistance or immunity is but one facet of immunology. Diverse pathological effects have been shown to arise from immune responses to nontoxic and noninfectious antigens. Examples of these antigens are pollens, certain chemicals, and red blood cells. The immune system is also involved in many diseases, including cancer, rheumatism, and degenerative disorders of aging. Immunology therefore touches on many aspects of modern medicine. Fig. 25-7 shows the main functions of the immune system. As the figure shows, in addition to defense of the body against infectious agents, the immune system also promotes *homeostasis* and conducts *surveillance*. Homeostasis is the maintenance of normal conditions; the immune system contributes to this by removing damaged cellular elements as well as by maintaining immu-

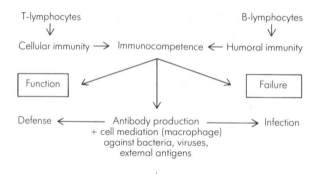

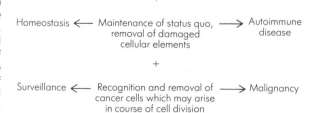

Figure 25-7. The main functions of the immune system are defense, homeostasis, and surveillance. The enforcers of the system are immunocompetent cells of cell-mediated immunity and humoral antibodies.

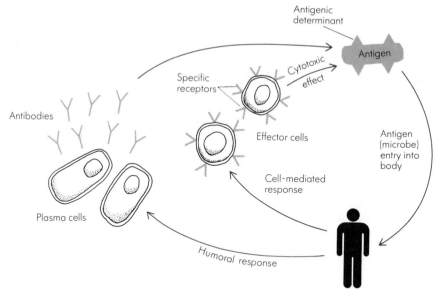

Figure 25-8. An antigen, such as a microbial cell, induces an immune response after entering the host. There are two forms of immune response: humoral response with production of specific antibodies by plasma cells, and cell-mediated response with formation of *effector cells*, sensitized lymphocytes with specific receptors.

nological self-tolerance. When the mechanism of immunological self-tolerance breaks down, a variety of disease states may subsequently occur. Rheumatoid arthritis and hemolytic anemia (deficiency of red blood cells) are examples of autoimmune diseases. Surveillance by the immune system involves the recognition and removal of cancer cells in order to prevent malignancy from occurring.

Types of immune response

Microorganisms that overcome the natural nonspecific resistance of the host (discussed earlier) are then faced with the host's second type of defense, acquired specific immunity. The immune response to an antigen (such as an invading microbe) takes two forms which usually develop in parallel; the humoral response (which involves antibodies) and the cell-mediated response (which involves lymphocytes). This is shown in Fig. 25-8.

Humoral immunity depends on the appearance of antibodies in the blood. *Cell-mediated immunity* involves the production of specifically reactive *effector cells* (also called *lymphoblasts*) which behave as though they have antibody-like molecules on their surfaces. When they react with the inducing antigen, they bring about cytotoxic effects on cells such as those containing viral antigens or on foreign cells from a graft. The precise manner in which this cytotoxic effect is exerted is not well understood.

Specific immunity may be acquired in two main ways:

1 *Actively* acquired by *clinical* (with obvious symptoms) and *subclinical* (no observable symptoms) *infections* or by deliberate artificial immunization. Some infections, such as diphtheria, smallpox, and mumps, usually induce a long-lasting immunity. Others, such as the common cold and influenza, give only a short-term immunity.

2 *Passively* acquired by the transfer of preformed antibodies to a nonimmune person by means of blood, serum components (gamma globulin fraction), or lymphoid cells (see Fig. 25-9). For example, tetanus antitoxin (antibodies against toxin) from an immune person may be used to treat a person who has an accidental wound that may contain the tetanus bacillus.

Development of the immune response

The precursor cells of both cell-mediated immunity and humoral immunity appear to be lymphocytes (white blood cells). There is a pool of recirculating lymphocytes which passes from the blood into the lymph nodes, spleen, and other tissues and back to the blood by the major lymphatic channels such as the thoracic duct. Lymphocytes are found in high concentrations in the lymph nodes and at the sites where they are manufactured and processed (matured): the bone marrow, the thymus, and the spleen.

One class of these lymphocytes is turned on by antigens to make humoral antibody. The interaction of antigens and another class of lymphocytes results in the activation of sensitized cells capable of killing microorganisms or inflicting damage. All the lymphocytes that circulate in the tissues arise from *stem cells*, which are undifferentiated cells in the bone marrow.

Figure 25-9. The various types of acquired specific immunity.

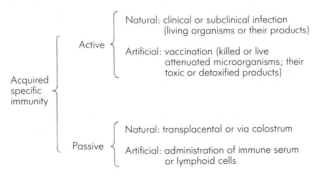

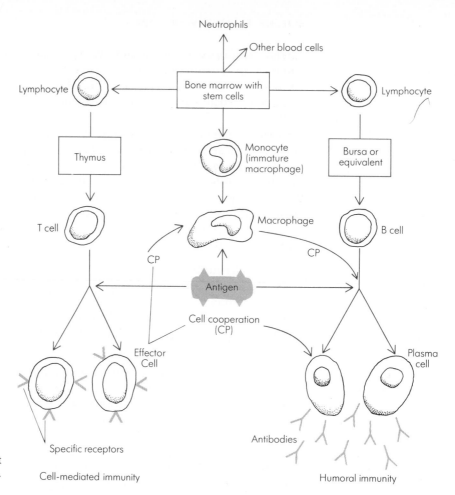

Figure 25-10. Development of the immune response.

Cell-mediated immunity

Humoral immunity

Derivation of the B and T cells. The stem cells have the potential to develop in a number of directions. Depending upon various influences, some develop into red blood cells, and some into different types of white blood cells. (See Figs. 25-10 and 25-11.) The stem cells that become involved in immune responses differentiate into one of two lymphocyte populations. One population is made up of the B lymphocytes, or B cells; the other is the population of T lymphocytes, or T cells.

The B cells are so called because in birds the precursors of the immunocompetent lymphocytes issuing from the bone marrow pass through the bursa of Fabricius (a small lymphoid organ found in birds that is attached to the intestine near the cloaca). It is in this organ that precursor cells develop into B lymphocytes (Fig. 25-10). The organ in mammals corresponding to the bursa of Fabricius is not known with certainty, but there is bursa-equivalent tissue, i.e., gut-associated lymphoid tissue (GALT), such as Peyer's patches in the intestines. Only about 20 percent of circulating lymphocytes are B cells. The remaining B cells are located in lymphoid tissue. B cells usually have a life-span of days to weeks.

The T cells also are developed from precursor cells produced by stem cells in the bone marrow; their development takes place in the thymus (Fig. 25-10), hence the name T cells. They then may leave the thymus and concentrate principally in the spleen and lymph nodes; they are also found in blood. T cells usually have a life-span of months.

Functions of B and T cells. B and T lymphocytes differ markedly in what they do. The B cells are responsible for the humoral response, since after contact with antigen they give rise to antibody-producing *plasma cells*. (Plasma cell does not mean a cell of the blood fluid, plasma, but is the name of a type of cell with a distinctive microscopic morphology as shown in Fig. 25-11.)

Following antigenic activation, the T lymphocytes give rise to the effector cells of the cell-mediated immune response (Fig. 25-10). These participate in eliminating or killing foreign material and invading microorganisms, even cancer cells. T cells are also largely responsible for the rejection of tissue transplants and for such skin allergies as reactions to poison ivy. In addition, T cells can enlist the aid of macrophages in destroying pathogens and stimulate B cells to increase the production of antibodies (a process called *cell cooperation*).

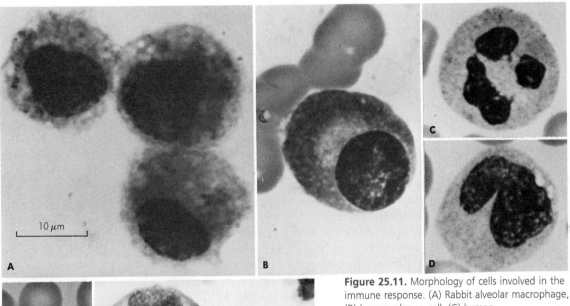

Figure 25.11. Morphology of cells involved in the immune response. (A) Rabbit alveolar macrophage, (B) human plasma cell, (C) human polymorphonuclear leukocyte (neutrophil), (D) human monocyte, (E) human lymphocyte, (F) lymphoblasts from human lymphocyte culture. All cells have same magnification. *(Courtesy of Barbara Bain, McGill University.)*

10 µm

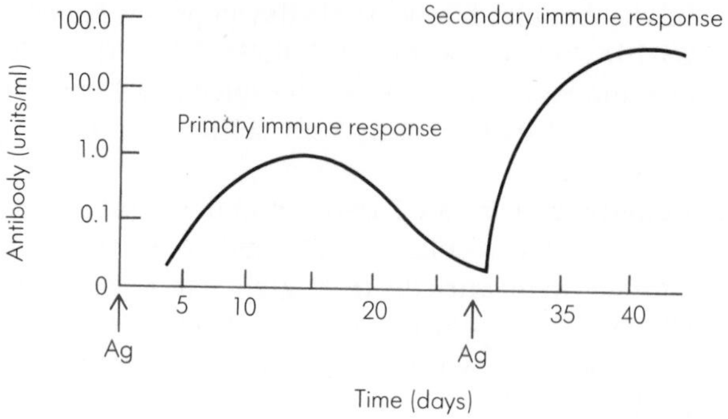

Figure 25-12. Antibody production due to administration of antigen (Ag). Note that the secondary immune response is faster and stronger than the primary immune response.

In addition to forming plasma cells, B lymphocytes can also form *memory cells*, which are long-lived, resting lymphocytes primed by previous contact with an antigen. On renewed contact with the same antigen, they can produce what is termed a *secondary immune response*, a response considerably more rapid and vigorous than the *primary response*. It enables an individual who previously has been exposed to a pathogen to respond more promptly and vigorously in a subsequent encounter. The result is a more rapid and more intense antibody production (Fig. 25-12).

Although harder to measure, cellular immunity is also enhanced by a second exposure to an antigen. This is accounted for by a new or greatly enlarged population of memory T cells specifically able to respond to the second appearance of the antigen. "Immunological memory" or the *anamnestic* (secondary) *response* is of utmost importance in the acquisition of immunity, either by natural infection or by administration of a vaccine.

Antigen-antibody and antigen-lymphocyte interactions

At the molecular level, the two main forces that bind the antigen-antibody complex together are molecular forces of attraction and electrostatic forces. Because the binding is not very firm, spontaneous dissociation may occur.

In vivo the antigen-antibody combination initiated responses by the body. It should be noted again that the mere combination of antigen with antibody, with no subsequent responses of the body, in many cases would not suffice to cope with the offending antigen. The responses of the body are spoken of as the *immunological effector system*. This system releases biologically active substances and thereby activates cells.

The mechanism of interaction between a sensitized T cell and a specific antigen is imperfectly understood. The effector cells can be directly toxic for the antigen. They can also produce *lymphokines*, which include a substance called *migration inhibitory factor* (keeps macrophages from moving away), immune interferon, and other materials. Intracellular parasites (e.g., viruses) are not directly exposed to humoral antibodies and therefore require the antimicrobial action of T cells and their collaborating macrophages to eliminate or destroy them.

Immunodeficiency diseases

We have discussed the protective role of the immune response. However, sometimes the immune response is deficient or malfunctions. This results in a very important class of diseases called *immunodeficiency diseases*. The first immunodeficiency disease was described in 1952. The patient was a young boy afflicted with recurrent severe infections. He lacked the major immunoglobulins and thus had no antibodies. This disease is called *agammaglobulinemia* and is characterized by a deficiency in the B cell component of the immune response, although T cell activity, or cell-mediated immunity, is intact.

In contrast, another disease, *ataxia-telangiectasia*, is characterized by a lack of cell-mediated immunity. Such patients have either no detectable thymus or a small, abnormal one. However, they usually have normal levels of antibody.

In another type of immunodeficiency disease, called *Swiss-type lymph-openic agammaglobulinemia*, the body has little resistance to infecting microbes because neither T-cell nor B-cell immunity is present. That is, such patients do not have cell-mediated immune response or humoral antibodies.

Summary and outlook

Host resistance may be natural resistance or specific resistance. Natural resistance depends on a large number of factors. The inborn resistance factors are species, race, and individual. The external factors include the mechanical and chemical barriers of the body. Among the internal defense factors are inflammation, phagocytosis, complement, and interferon.

Specific resistance is acquired specific immunity and is gained in response to a foreign agent and usually involves specific immune cells or antibodies or both. That is, the immunocompetence of the host depends on the functioning of the immune system involving both cell-mediated immunity and humoral immunity.

The outlook for developments in host resistance and immunity is both promising and exciting. For example, studies on microbial killing and digestion in phagocytes are still in their infancy. These will invariably lead to the revelation of how microorganisms can avoid being ingested, killed, and digested. Consider another example of natural resistance. Even though interferon was discovered in 1957, research on it remains as active as ever. Besides serving as one of the body's more important natural defenses against viral infections, there is strong evidence that interferon may well serve as *prophylaxis* (prevention of disease) against, or therapy for, all viral diseases, including rabies and the common cold. In addition, interferon is being tested as a cancer "drug" in clinical trials. It is also being seriously considered for use to prevent the rejection of transplanted organs or to combat autoimmune disease. This is because the immune interferons (produced by T lymphocytes) function as natural regulators of the immune response by keeping it within limits. Understanding a self-regulatory mechanism of this sort offers exciting possibilities for experimental manipulation of the immune response and the possibility of controlling disease.

Equally exciting are the prospects in the field of immunology. For example, hybrid cells can be produced in the laboratory to become sources of almost limitless supplies of pure antibodies. The hope is that the laboratory-produced, highly purified, and specific antibodies will eventually prove valuable in the discovery, diagnosis, and treatment of disease. To take another example, developments in immunology have made it possible to attempt an explanation of aging on a scientific basis. Our ability to produce new antibodies diminishes greatly in old age because of a decreased supply of stem cells from the bone marrow. Related to this is our inability to eliminate the greater frequency of cancer-cell formation in old age. Future research should produce greater insights into how the vigor of immune function might be extended into old age to improve the health and lengthen the life-span of all people.

In the area of immunodeficiency diseases, until recently little could be done for these young patients who were doomed to die at an early age from overwhelming infections of bacteria, fungi, or viruses. Now, physicians are attempting "immunologic reconstitution." Transplants of fetal thymus have restored health to infants with deficiency in cell-mediated immunity. Also, bone marrow transplants from suitable sibling donors have enabled patients to make antibodies. Immunologic reconstitution is still in its infancy, but as new knowledge accumulates, bringing help to the unfortunate young patients with immunodeficiency diseases should be possible.

Key terms

active immunity
agammaglobulinemia
alveolar macrophage
anamnestic response
ataxia-telangiectasia
cell cooperation
cell-mediated immunity
complement
effector cell
histiocyte
homeostasis
host resistance
humoral immunity
immunodeficiency disease
immunological effector system
inflammation
interferon
Kupffer cell
lymphoblast
lymphokine
lysosomes
macrophage

memory cell
migration inhibitory factor
monocyte
necrotic
opsonin
passive immunity
phagolysosome
phagosome
plasma
plasma cell
polymorphonuclear granulocyte
primary immune response
prophylaxis
pus
secondary immune response
serum
stem cell
surveillance
Swiss-type lymphopenic agammaglobulinemia
vasodilation

Questions

1 Since host resistance factors may be categorized in various ways, set up a table listing host resistance factors under the following headings:

innate factors, external defense factors, and internal defense factors. (Note that the factors may overlap under different headings.)

2 Provide examples of specific microbial infections illustrating species resistance, racial resistance, and individual resistance.

3 Why do you think socioeconomic factors also play a role in host resistance to infections?

4 Explain why inflammation is considered to be a factor of the host's natural resistance.

5 Describe in chronological order the events that befall a microbe once it has gained access into the body of a host.

6 Write a short account of the cells that participate in phagocytosis and describe their mechanism of action in eliminating foreign bodies.

7 In what way is complement considered to be a resistance factor?

8 Explain why you think interferon activity is cell species-specific and not virus-specific.

9 Describe the different ways in which interferon is important in medicine.

10 Provide specific examples to illustrate the meaning of *actively acquired* and *passively acquired* specific immunity.

11 Compare and contrast cell-mediated immunity and humoral immunity with respect to their modes of action.

12 What is the anamnestic response? Why is it important?

DIAGNOSTIC APPLICATIONS OF ANTIGEN—ANTIBODY REACTIONS

Antibodies cannot be seen with the naked eye but are known to exist by the things they do. If an antibody is present in the body, the person is likely to resist infection by the pathogen for which the antibody is specific. For instance, humoral immunity can be observed in laboratory animals by determining the increase in the number of microorganisms required to establish an infection in an immunized animal and comparing it with the number needed to establish infection in a comparable nonimmunized animal.

There are also in vitro tests that show the presence of antibodies in body fluids such as blood serum (the amber-colored fluid that exudes from coagulated blood). These are visible reactions that may cause agglutination or clumping of cells, precipitation or flocculation of cell extracts, or lysis of bacterial and other "indicator" cells, such as red blood cells. For this reason, the antibodies responsible for these reactions are termed *agglutinins*, *precipitins*, and *lysins*, respectively.

The multiplicity of names of antibodies does not mean that there is a different type of antibody for each type of observable antigen-antibody reaction. It merely indicates the type of reaction exhibited, depending on the physical state of the antigen. For example, the same antibody may be involved in both a precipitation and an agglutination reaction. Thus if antitoxin is reacted with a soluble toxin, a precipitation results. If this soluble toxin is adsorbed to latex particles, the same antibodies will agglutinate the particles.

The study of such antigen-antibody reactions in the laboratory is called *serology* because it involves the measurement of antibodies or antigens found in serum. Since the antigen-antibody reaction is specific, if one of the components (either the antigen or the antibody) in a serological test is known, a positive reaction will reveal the identity of the other component.

Serological tests have been used widely for the laboratory diagnosis of infectious disease. When the modern era of immunology was ushered in during the 1960s, the methods of the serological laboratory became very specialized and sophisticated. Immunofluorescence, along with other antibody- or antigen-tagging methods using radioisotopes or enzymes, became routine tools of diagnosis. Also developed were indirect or passive agglutination procedures, including latex fixation and hemagglutination, and a wide variety of techniques designed to quantitate cell-mediated immunity in vitro. The introduction of these newer immunological techniques has extended serology very far indeed from the conventional procedures used in testing for unknown antibodies to known microorganisms, or vice versa.

The detailed methodology of these new techniques is beyond the scope of this chapter, but the student should be aware of their existence and accessibility (such as found in *Manual of Clinical Immunology*, edited by N. R. Rose and H. Friedman, American Society for Microbiology, Washington, D.C., 1980). Our discussion will focus on those basic techniques that are of interest to microbiologists.

Usefulness of applied antigen-antibody reactions

Laboratory tests based on antigen-antibody reactions extend the clinician's diagnostic skill and guide therapeutic efforts. Serological tests represent a large proportion of the laboratory techniques available to aid the clinician. The most important and widely used serological tests include agglutination, precipitation, and complement-fixation reactions. Animal, or in vivo tests, also provide a useful adjunct for diagnostic aid in some diseases. Although not generally called serological tests, skin tests made on the patient help in determining susceptibility or immunity to certain infections. Some serological tests are shown in Fig. 26-1.

Agglutination tests

The antigen in agglutination reactions is a cell (see Fig. 24-3) or a particle, for example, a latex particle onto which soluble antigens have been adsorbed. The addition of homologous antibody will cause agglutination, or

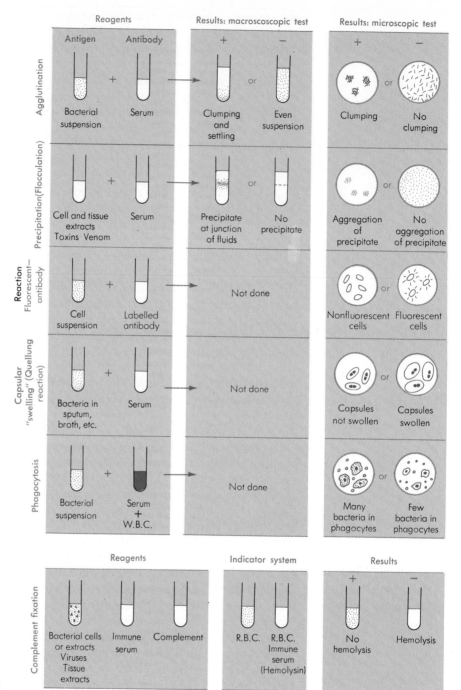

Figure 26-1. Some serological reactions.

clumping, resulting in visible aggregates of the cells or particles. Agglutination occurs because the antibody acts as a bridge to form a lattice network of antibodies and particulate antigens forming clumps. Not only can the diagnosis of certain infectious diseases be confirmed in the laboratory by agglutination of known antigens (for example, bacterial cells) by the

426

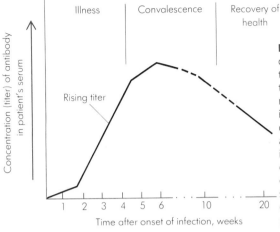

Figure 26-2. Antibody concentration (expressed as a titer) in a patient's serum during the course of an infection. A rising concentration or *rising titer* is good evidence that the specific organism used as antigen in the agglutination test may be involved in the disease. (Note that constant concentration of antibody or a constant or *nonrising titer* would simply indicate a previous exposure which accounted for the presence of the antibody.)

patient's serum (Fig. 26-2), but unknown cultures of bacteria or other microorganisms can be identified by the ability of a *known* serum containing antibodies (*antiserum*) to agglutinate the suspension of *unknown* cells. This is because of the specificity of antigen-antibody reactions. Agglutination reactions are also used in human blood grouping and typing.

Bacterial agglutination tests

Agglutination tests may be carried out in small test tubes or on a slide (Fig. 26-3). Most bacterial agglutination tests are carried out by serial dilutions of antiserum in tubes to which a constant amount of antigen is added. After incubation, clumping is noted by visual inspection, and a *titer* is determined. The titer of antiserum is a relative value and is the reciprocal of the highest dilution that has clumping of cells and antibodies. For example, an antiserum that agglutinates at a dilution of 1:256 (but not at 1:512) has a titer of 256. A higher titer indicates a greater concentration of antibodies.

The *Widal test* was devised specifically to aid in diagnosis of typhoid fever by agglutinating typhoid bacilli with patients' sera, but the term is sometimes loosely applied to other agglutination tests using heat-killed cultures of organisms other that *Salmonella typhi*.

Another agglutination test used in laboratory diagnosis is the *Weil-Felix test* for rickettsial infections. It is unique because heterophile antigens are involved; several of the rickettsias have a common antigen with strains of *Proteus* spp. That is, sera from patients with rickettsial infections agglutinate suspensions of the *Proteus* bacteria. The strains of *Proteus* used most commonly are *Proteus* OX_{19}, OX_2, and OXK. The Weil-Felix reaction is differential, or diagnostic, for certain rickettsial diseases because of the selective agglutination of these strains.

ABO-blood-group typing

In blood transfusion, the major consideration is given to the interaction of the recipient's antibodies and the donor's cells. An incompatible transfu-

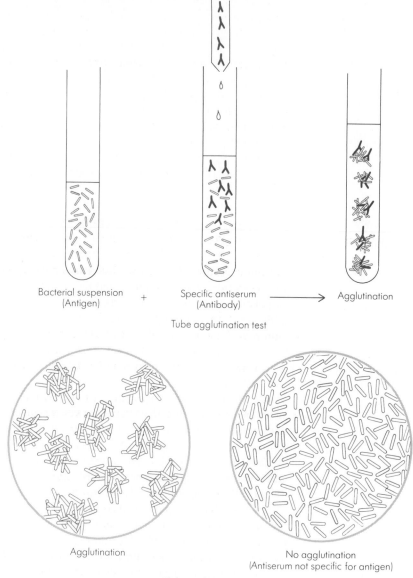

Bacterial suspension
(Antigen)   +   Specific antiserum
                                  (Antibody)   ⟶   Agglutination

Tube agglutination test

Agglutination

No agglutination
(Antiserum not specific for antigen)

Slide agglutination test

Figure 26-3. Agglutination occurs when bacterial-cell antigens are exposed to a homologous antiserum (antibody).

sion results in the clumping and lysis of transfused cells by the patient's isoantibodies and serum complement. This causes fever, prostration, and renal insufficiency (reduced kidney function) in the patient and, as a result, may be fatal.

Blood group O persons (see Chap. 24) are called *universal donors* in blood transfusion because their red cells do not have the A and B isoantigens. AB individuals, lacking anti-A and anti-B isoantibodies, are termed *universal recipients*. In practice, patients are almost always transfused with blood of their own ABO group and Rh compatibility. Commercially available antisera are used in a slide or tube agglutination procedure to determine the blood group of the cells (see Table 24-1).

In addition, further safeguards must be used to detect other incompatibilities between the prospective donor's cells and the patient's serum. For example, after routine ABO and Rh slide and tube tests, *crossmatching* is performed prior to transfusion. In the *major crossmatch*, the donor's cells are tested against the recipient's serum. In the *minor crossmatch*, the donor's serum is tested against the recipient's cells. The major crossmatch is so described because it measures the ability of the recipient's antibodies to attack the donor's cells; this is of prime importance. The reverse reaction (in the minor crossmatch) is rarely serious because the antibody titer in the donor serum is diluted in the body to so low a level that it cannot harm the recipient's cells. Other blood-testing procedures may also be employed to ensure a proper match.

Determination of
Rh blood type

The Rh system is second only to the ABO system in importance. Unlike the latter system, naturally occurring isoantibodies to Rh antigens are not found. However, an Rh-negative person receiving Rh-positive red blood cells (unrelated to the A and B antigens) will respond by synthesizing antibodies to the Rh factor.

In preliminary general testing (commercial reagents are also available), an individual is Rh-positive if his red blood cells have Rh factor D. An individual can also be Rh-positive on the basis of other Rh factors; there are actually 27 known Rh specificities. Fortunately, only antigen D appears to be a sufficiently strong antigen to stimulate the production of destructive hemolytic antibodies.

Hemolytic disease of the newborn, or *erythroblastosis fetalis*, was mentioned in Chap. 24. This disease occurs when maternal anti-Rh antibodies cross the placenta into the fetus and attack fetal erythrocytes. The resulting hemolysis, lysis of red blood cells, may be sufficient to kill the fetus or produce severe anemia; furthermore, irreversible brain damage may occur in these infants. Fortunately, only about 5 percent of Rh-negative mothers actually produce anti-Rh antibodies while bearing an Rh-positive child. It is possible to minimize the isoimmunization of the Rh-negative mother by giving anti-Rh antibodies to the mother within 72 h after the birth of an Rh-positive child. These antibodies will destroy Rh-positive red blood cells, preventing the mother from having a primary antibody response to the Rh factor. This forestalls the development of an Rh-incompatibility problem in a subsequent pregnancy.

Precipitin tests

In the precipitin test a reaction takes place between a soluble antigen and its homologous antibody. The reaction is manifested by the formation of a visible precipitate at the interface of the reactants. Such reactions are usually carried out by using a constant amount of antibody (antiserum) and varying the dilutions of antigen. As may be seen in Fig. 26-4, which shows a typical precipitin curve, the reaction may be inhibited by an excess of either antigen or antibody. Remembering that the concentration of antibody is constant, you can see that only a small amount of precipitate is formed

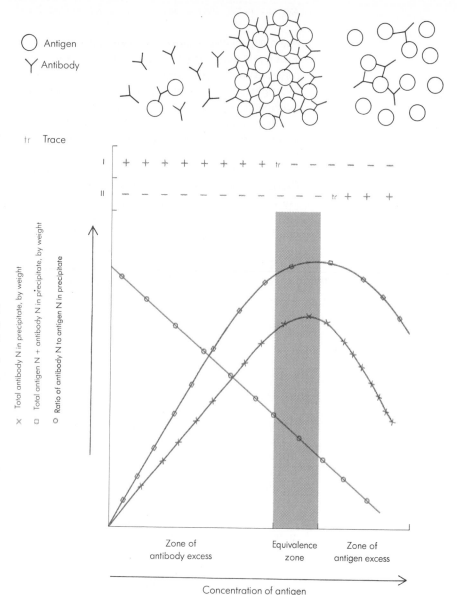

Figure 26.4. Graph and schematic illustration of the events of the precipitin reaction. In this reaction, increasing concentrations of antigen are added to a series of serological tubes containing a constant concentration of antibody (antiserum). The concentration of antigen and antibody present in any tube is determined by assaying for nitrogen (N), a component of both the antigen and antibody. I and II refer to tests for excess antibody and excess antigen, respectively, in the fluid remaining after removal of the precipitate (+ = positive; tr = trace; − = negative).

where there is an excess of antibody. As the antigen concentration is increased, the amount of precipitate increases and reaches a maximum when the proportions of antigen and antibody are optimal. After this zone, as the antigen concentration increases, the amount of precipitate again decreases. Thus there are three zones of antigen-antibody reaction in the precipitin test: *zone of antibody excess*, *equivalence zone*, and *zone of antigen excess*.

In the zone of antibody excess, all antigen has reacted with antibody and has been precipitated (no free antigen in the supernatant). Conversely, in the zone of antigen excess, all the antibody has reacted with the antigen (no antibody in the supernatant), but the complexes remain soluble because the large excess of antigen binds antibody into small complexes which do not crosslink to form large visible aggregates. In the equivalence zone, maximum precipitation of antigen and antibody occurs (no free antigen or antibody in the supernatant) because they are in *optimal* proportions to form a lattice of antigen and antibody which becomes insoluble and visible. For this reason, the most useful precipitin tests provide for diffusion of the reactants until their optimum concentration is reached.

The ring test   The ring test is the simplest of the precipitin tests. In a small-bore tube an antigen solution is layered on a solution of serum containing antibody. The two solutions diffuse until they are in optimum concentration for precipitation, at which point a dense zone or ring of precipitate appears between the two otherwise clear solutions (Fig. 26-5A).

Figure 26-5. Qualitative precipitin tests (Ag = antigen; Ab = antibody).

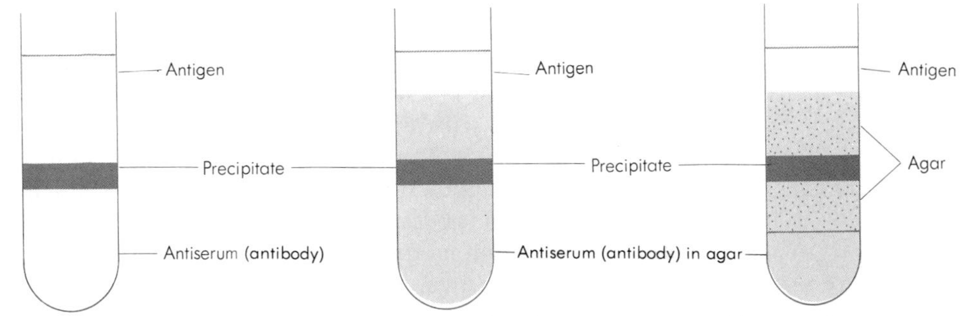

A Ring test
Homologous antigen and antibody react to form a precipitate.

B Single diffusion (Oudin)
Antigen diffuses into agar containing antibody. Positive reaction indicated by band of precipitate.

C Double diffusion in one dimension (Oakley-Fulthorpe)
Antigen and antibody diffuse through a layer of agar until they meet. If they are homologous, they form a precipitate.

D Double diffusion in two dimensions (Ouchterlony)

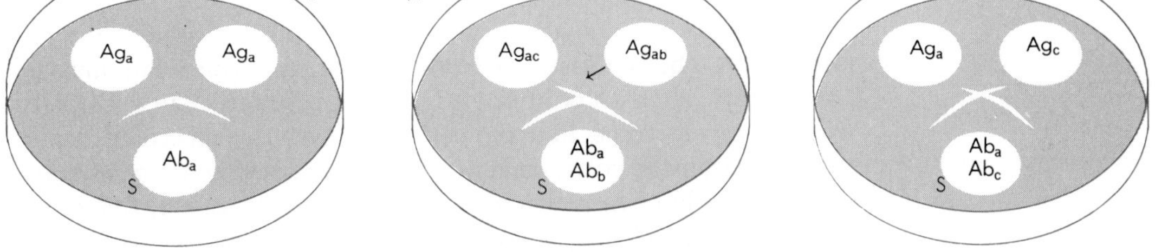

(1) Identity

(2) Partial identity (spur at arrow due to Ab$_b$)

(3) Non-identity

Antigen and antibody are placed in separate wells cut into agar in a petri dish. Homologous reactants produce a line of precipitate where they meet in the agar. Like antigens produce bands which meet exactly as in (1). Unlike antigens produce bands which cross as in (2) and (3).

Agar-diffusion methods

Greater accuracy and separation of components in mixtures of antigens and antibodies can be obtained by allowing the reactants to diffuse together in an agar gel.

The single-diffusion method. In the single-diffusion method devised by Oudin, antigen is overlaid on agar gel containing antiserum in a narrow-bore test tube (Fig. 26-5B). Upon standing for hours or days, the antigen diffuses into the gel, forming bands of precipitate at various levels; depending on the number and kinds of antigen-antibody systems present. Since precipitation takes place as the antigen diffuses through the gel, rings of precipitate first appear near the top of the gel and seem to move slowly downward. This apparent effect may actually be due to increasing amounts of antigen which cause the precipitate to dissolve (since antigen-antibody reactions are reversible). The precipitate reforms at a position farther down the tube where optimal concentrations of antigen and antibody exist. Factors determining the levels at which the reaction takes place are the molecular size and the relative concentration of reactants.

Double-diffusion methods. C. L. Oakley and A. J. Fulthorpe modified the J. Oudin technique by placing antiserum in agar in the bottom of a test tube and overlaying it with agar gel, on top of which is placed the antigen solution (Fig. 26-5C). The two reactants diffuse toward each other in the agar, with precipitation taking place at the point where optimum concentration occurs. This is double diffusion in one dimension.

A double-diffusion method in two dimensions devised by O. Ouchterlony has an advantage over the methods mentioned above in that various antigens and antisera can be compared directly. In this test, reactants diffuse from wells cut in agar in a flat petri dish. Bands of precipitate form in the area between the wells where there is optimum concentration of homologous antiserum and antigen (Fig. 26-5D). If the bands of precipitate formed by the two antigens and the antibody fuse at their juncture, the antigens are the same. If they cross, they are different.

Immunoelectrophoresis. *Electrophoresis* is an electrochemical process in which colloidal (suspended) particles or macromolecules with a net electric charge migrate in a solution or agar gel under the influence of an electric current. A characteristic of living cells in suspension and biological compounds (such as protein antigens) in solution or in a gel is that they travel in an electric field to the positive or negative electrode, depending on the charge on the substances. Positively charged substances travel to the cathode, while negatively charged ones go to the anode; this movement is called *electrophoretic mobility*. When electrophoresis is applied to the study of antigen-antibody reactions, it is called *immunoelectrophoresis*.

When a fluid containing protein antigens is placed in a well in a thin layer of buffered agar and an electric current is applied, antigens will be distributed in separate spots along a line passing through the well and parallel to the direction of current flow (Fig. 26-6A). The spots are

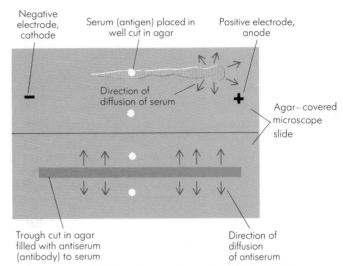

Figure 26-6. *Left*: Agar-gel electrophoresis (A) and immunoelectrophoresis (B) of sera from normal mice and from mice bearing tumors. **1.** Normal mouse serum. **2.** Tumor mouse serum. **3.** Normal mouse serum. **4.** Tumor mouse serum. The precipitin bands were developed with antimouse serum prepared from rabbits inoculated with mouse serum. Those bands corresponding to the tumor protein are indicated with arrows. *(Courtesy of R. A. Murgita, McGill University.) Right:* Explanatory sketches for photographs (A) and (B).

analogous to antigen wells in an Ouchterlony plate. When the current is shut off, diffusion will begin from each of these spots. By filling with antiserum a trench cut in the agar parallel to the electrophoretic distribution of the antigens, the precipitin reaction can be used to demonstrate the nature of the diffusing molecules. In this case, a broad band of antibody diffuses toward the antigens from the linear antibody trench, while the antigens diffuse as expanding disks. This results in a complex pattern of arc-shaped zones of precipitate (Fig. 26-6B). When the antigen molecules are from human serum, there may be 20 or more such zones, each of which represents at least one distinct precipitin system. Precipitin specificity provides a far more critical indicator of the identity of protein molecules than the physical characteristics which control the movement of the molecules in an electrophoretic field.

Radioimmunoassay

Radioimmunoassay is a microtechnique of great sensitivity for determining trace amounts of antigen. The technique is essentially a two-step process. The first step involves a competition between test (unlabeled or nonradioactive) antigen of unknown (variable) concentration and a known (labeled or radioactive) indicator antigen of calculated concentration. The competition is for a known and limiting amount of antibody specific for the radioisotope-labeled antigen. These three reactants are incubated for several hours so that antigen-antibody binding reactions may occur.

The second step involves the addition to the system of antiserum (anti-antibody) specific for, and hence able to bind to, the antibody component of the immune complexes formed during the incubation in the

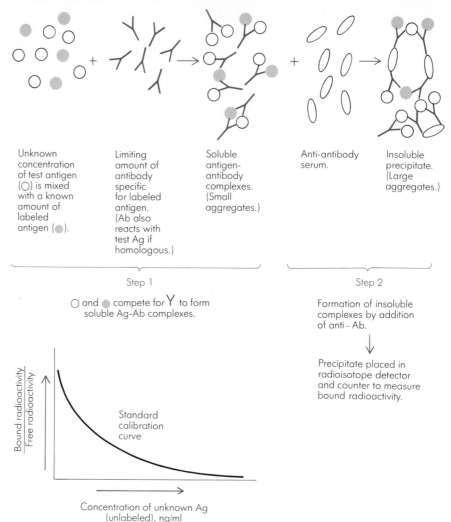

Unknown concentration of test antigen (○) is mixed with a known amount of labeled antigen (●).

Limiting amount of antibody specific for labeled antigen. (Ab also reacts with test Ag if homologous.)

Soluble antigen-antibody complexes. (Small aggregates.)

Anti-antibody serum.

Insoluble precipitate. (Large aggregates.)

Step 1

○ and ● compete for ⅄ to form soluble Ag-Ab complexes.

Step 2

Formation of insoluble complexes by addition of anti–Ab.

↓

Precipitate placed in radioisotope detector and counter to measure bound radioactivity.

$\dfrac{\text{Bound radioactivity}}{\text{Free radioactivity}}$

Standard calibration curve

Concentration of unknown Ag (unlabeled), ng/ml

Figure 26-7. Technique of radioimmunoassay used to detect and measure trace amounts of an antigen. The amount of unlabeled test antigen can be determined from a standard calibration curve. A variation of the technique includes the detection and measurement of antibody using labeled antibody instead of labeled antigen (Ag = antigen; Ab = antibody).

first step. This results in precipitation of the antigen-antibody complexes.

The radioactivity of the precipitate is determined in a radioisotope detector and counter.

If the test antigen in step one has reacted with the antibody, then the radioactive indicator antigen could not react with the antibody. The radioactive count would be low because the indicator antigen is not in the precipitate. However, if the test antigen did not react with the antibody, then the antibody would have been free to bind with the radioactive indicator antigen. The radioactive count would be high because the indicator antigen is bound in the precipitate. Thus the identity and the concentration of the test antigen can be determined by the radioactivity of the precipitate. This quantity of test antigen may be obtained by reference to a standard calibration curve, such as shown in Fig. 26-7.

The radioimmunoassay technique is of importance in detecting the presence of hepatitis B antigen, which may be present in the serum of asymptomatic blood donors [donors who carry the virus (antigen) without exhibiting symptoms]. Other substances measured by radioimmunoassays include insulin, testosterone, estradiol, human IgE, and other materials which are normally present in very small amounts in blood or urine.

Complement-fixation tests

Complement-fixation tests are based on the presence of complement-fixing antibodies in serum. In the presence of complement, these antibodies cause lysis of the specific cells. The purpose of the complement-fixation test is to determine whether specific antibodies are present in serum (see Fig. 26-8). The test involves two systems. One is the test, or *complement-fixing system*, in which serum, bacterial suspension (or other antigen), and complement are mixed. If the antigen and antibody in the serum unite, the complement is said to be *fixed*. The other system in the test is a *hemolytic indicator system*. Hemolytic antibody (hemolysin) is prepared by immunizing rabbits with the red blood cells of sheep. Serum from sheep-cell-immunized rabbits is mixed with red blood cells of sheep. If complement was fixed by being used in the reaction between the test antibody and antigen, no hemolysis will occur. Therefore, a hemolytic reaction indicates

Figure 26-8. The complement-fixation test is based on the fact that if complement, which is found in blood, is "fixed" by reacting with antigen and its homologous antibody, lysis of sheep red blood cells does not occur even though specific hemolysin is present.

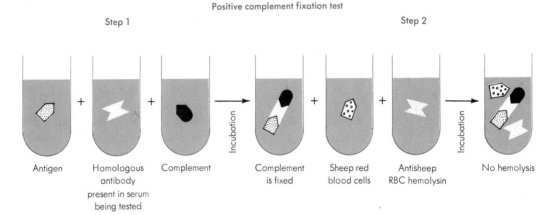

Positive complement fixation test

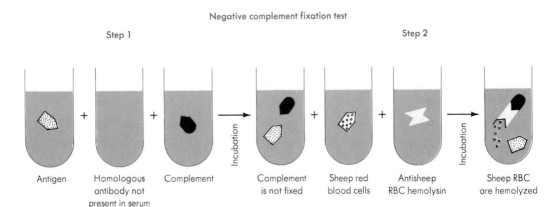

Negative complement fixation test

Figure 26-9.
Fluorescent-antibody technique is used to detect microorganisms. A vaginal smear from a patient infected with *Neisseria gonorrhoeae,* the etiologic agent of gonorrhea, was treated with gonococcus antibody tagged with fluorescein dye. (A) Slide preparation viewed with dark-field microscopy. Note the presence of all kinds of microorganisms. (B) The same field viewed with ultraviolet-light–fluorescence microscopy. Note the presence only of gonococci, which are paired, oval cells. *(Courtesy of W. E. Deacon, Bull WHO,* **24***:349–355, 1961.)*

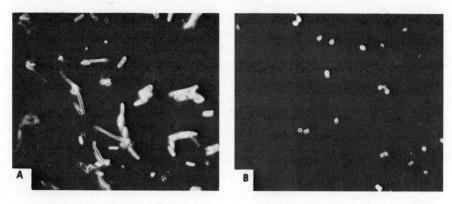

a *negative* test. Obviously, all reactants in the complement-fixation test must be accurately adjusted.

The complement-fixation test is especially useful when the test antigen and antibody combination does not give a visible reaction such as that occurring in agglutination and precipitation. The complement-fixation test is widely used in the laboratory diagnosis of many infectious diseases, including those of bacterial, viral, protozoan, and fungal origin. One of the best known applications of the test is the Wasserman test for syphilis, although this test has been replaced by other tests.

Special serological tests

Fluorescent-antibody technique

The fluorescent-antibody technique is a rapid procedure for the identification of an unknown infectious agent. The technique is based on the behavior of certain dyes which fluoresce (glow) when exposed to ultraviolet light. Examples of such dyes are fluorescein isothiocyanate and rhodamine isothiocyanate. Antibodies can be conjugated, or tagged, with these dyes and are then termed labeled antibodies.

If a mixed culture or specimen is placed on a slide, combined with serum containing fluorescent antibodies, and examined by fluorescence microscopy, only those organisms (antigens) that reacted with the specific labeled antibodies will be visible (Fig. 26-9). Thus only a few organisms need to be

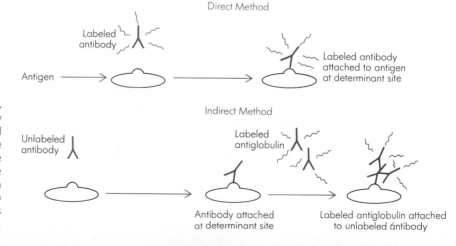

Figure 26-10.
Fluorescent-antibody technique—direct and indirect methods. The indirect method is more sensitive as two or more labeled antiglobulin molecules can be attached to each antibody bound to its antigen.

present to be observed. This is the *direct* method, in which the fluorescent dye is conjugated with the antibody specific for the antigen.

In the *indirect* method the initially applied antibody is not labeled. Instead, a labeled antiserum against the globulin of the animal species used for the preparation of the initial specific antibody is applied. This fixes the fluorescent label to the specific antibody that has already reacted with antigen in the smear (see Fig. 26-10).

Enzyme-linked immunosorbent assays (ELISA)

Enzyme immunoassays, specifically called *enzyme-linked-immunosorbent assays* (ELISA), are used for the immunodiagnosis of viral infections and other microbial antigens. In addition to being as sensitive as radioimmunoassays (antigens and antibodies detectable at levels of about 10^{-10} g, or 1 part in 10 billion), enzyme immunoassays are less expensive, safer, simpler to employ, and just as reliable and accurate as radioimmunoassays.

The principle of ELISA is based on these two observations:

1 Antibodies and some antigens can attach to polystyrene plastic plates (or other solid-phase supports) and still maintain their full immunological capabilities.
2 Antigens and antibodies can be bonded to enzymes and the resulting complexes are still fully functional both immunologically and enzymatically.

It is the enzyme activity which is the measure of the quantity of antigen or antibody present in the test sample. Enzymes used in ELISA include β-galactosidase, glucose oxidase, peroxidase, and alkaline phosphatase.

There are two methods of enzyme immunoassay that have been found to be of most clinical value: the double-antibody-sandwich procedure for the detection and measurement of antigen and the indirect-microplate-ELISA procedure for the detection and measurement of antibody.

Double-antibody-sandwich procedure. In this technique (Fig. 26-11A), the wells or depressions in a polystyrene plate receive antiserum. The antibodies in the antiserum adhere to the surface of each well. The test antigen is added, and if the antigen is homologous, it attaches to the antibody immobilized on the well surface. Enzyme-labeled specific antibody is then added; it will bind to the antigen already fixed by the first antibody. This results in an antibody (with enzyme)-antigen-antibody "sandwich." Finally, the enzyme substrate is introduced for reaction with the enzyme. The rate of enzyme action is directly proportional to the quantity of enzyme-labeled antibody present and that, in turn, is proportional to the amount of test antigen. Enzyme activity may be followed by a color change (brought about by substrate hydrolysis) which can be inspected visually or measured by means of a colorimeter (an instrument used to analyze color changes in a solution). This method has been used to assay hepatitis B antigen with great success.

Indirect-microplate-ELISA procedure. The initial step involves the coating

A Double—Antibody Sandwich Method

Antibody (antiserum) adsorbs to the well surface

Test antigen is added and binds to the antibody (antiserum) if it is homologous

Enzyme–labeled specific antibody is added and binds to the test antigen forming an antibody–antigen–antibody "sandwich"

Enzyme substrate is added. Degradation of the substrate molecules is accompanied by a color change.

B Indirect Microplate ELISA Method

Antigen adsorbs to the well surface

Test antiserum is added and specific antibody will bind to the antigen

Enzyme–labeled antihuman antibody is added and will link to the specific antibody previously bound.

Enzyme substrate is added. Degradation of the substrate molecules is accompanied by a color change.

Figure 26-11. Enzyme-immunoassay techniques. (A) Double-antibody-sandwich method for the detection and the measurement of antigen. (B) Indirect-microplate-ELISA method for the detection and measurement of antibody.

of polystyrene wells with antigen by passive adsorption. (See Fig. 26-11B.) Test antiserum is added and allowed to incubate. If the antibodies in the antiserum are homologous, they will bind to the immobilized antigen. Enzyme-labeled antihuman antibodies are added to the system which will link to the antibody-antigen complexes formed in the previous step. Finally, as in the double-antibody-sandwich method above, the enzyme substrate is added and the rate of its degradation (hydrolysis) is associated with a color change proportional to the concentration of antibody present in the test sample. This color change can be monitored visually or in a colorimeter.

This method has been used for the immunodiagnosis of many infectious pathogens such as viruses, parasites, and fungi. For example, one antiserum sample from a pregnant woman can be screened simultaneously for infections of rubella virus (the agent for German measles, which causes congenital malformations and fetal death), type 2 herpesvirus (causes severe congenital nervous system malformations and small heads), and other infections that can affect the fetus.

Intracutaneous diagnostic tests

Skin tests used in the detection of some infectious diseases fall into one of two categories: (1) those for detecting immunity and (2) those for detecting hypersensitivity.

Tests for immunity. The *Schick test* is a valuable aid for determining susceptibility to diphtheria, a severe infectious disease caused by *Corynebacterium diphtheriae*. The test consists of the intradermal injection of a small standardized amount of diphtheria toxin and of an identical amount of heated toxin as control. In the skin of a person not immune to diphtheria, the active toxin causes a reddening and swelling which reaches maximum about four days after injection and persists for a week or two. The control site shows no reaction. In the skin of a resistant individual who has circulating antitoxin, the toxin is neutralized and produces no reaction. Individuals who have positive Schick tests should be immunized with diphtheria toxoid.

The *Dick test* is used for determining sensitivity to scarlet fever, a disease with a generalized rash caused by *Streptococcus pyogenes*. A standardized erythrogenic toxin of the bacterium is used for intradermal injection. A redness and swelling over 10 mm in diameter occurs in 18 to 24 h in a person susceptible to scarlet fever. This skin reaction is more rapid and transitory than that of the Schick test.

Tests for hypersensitivity. Hypersensitivity, or allergic reactions, can be of two types, the *immediate type* and the *delayed type*. The immediate type, mediated by IgE and accompanied by the release of histamine and serotonin, has been discussed in Chap. 24. The delayed allergy reactions are mediated by T lymphocytes, and not by IgE. These T cells have surface receptors that behave like antibodies. In contact with specific antigen, these cells, along with platelets (protoplasmic disks in blood), release various substances, including cytotoxic materials, that may cause necrosis. Delayed-type allergic reactions are apparent only after about 18 h, persist for days, and are not suppressed by antihistamines, since histamine is not present.

The *tuberculin test* is a typical delayed-type hypersensitivity test. It is performed with a purified protein derivative (called PPD) standardized in terms of tuberculin units (TU). This material is not toxic. The response it elicits is the host's acquired hypersensitivity to the substance.

The important reaction of a positive tuberculin test is a palpable

Table 26-1. Some intra-
dermal tests based on
cell-mediated immunity

TEST	ETIOLOGIC AGENT	DISEASE
Tuberculin	Bacterium: *Mycobacterium tuberculosis*	Tuberculosis—pulmonary disease with necrotic tubercles in lungs; may spread to other parts of body
Lepromin	Bacterium: *Mycobacterium leprae*	Leprosy—chronic disease that affects peripheral nerves, as well as face, hands, and feet to cause disfiguring
Brucellergin	Bacterium: *Brucella* spp.	Brucellosis—undulant (fluctuating) fever in humans
Blastomycin	Fungus: *Blastomyces dermatitidis*	Blastomycosis—lesions develop in lungs, skin, and bones
Histoplasmin	Fungus: *Histoplasma capsulatum*	Histoplasmosis—lesions in lung and lymph nodes, but may spread to other organs, including the liver and spleen
Mumps	Virus: the mumps virus, a paramyxovirus	Mumps—inflammation and swelling of salivary glands, but may have complications like swelling of testes
Leishmanin	Protozoan: *Leishmania* spp.	Leishmaniasis—any disease caused by these parasitic flagellates, e.g., kala azar, a visceral disease with enlarged liver and spleen

induration (a tissue hardening that can be easily felt) about 10 mm or more in diameter occurring in 48 to 72 h following intradermal injection. While a positive Schick test indicates lack of immunity, and a negative reaction suggests immunity, it is the positive tuberculin test that indicates immunity. All persons who have had and overcome tuberculous infections (natural or resulting from vaccination with BCG, or Bacille Calmette-Guérin, a bovine tubercle bacillus strain attenuated by prolonged artificial culture) generally have some degree of immunity to further infection. Since tuberculin hypersensitivity indicates past infection, the assumption is that it also indicates probable immunity.

Similar specific delayed-type hypersensitivities are the bases for a number of skin tests to show past or present infection with bacteria, fungi, or even larger parasites. (See Table 26-1.) There are also skin tests which are based on immediate-type allergy. These include tests for hypersensitivity to horse serum, ragweed and other pollens, and many other antigens; the reactions to these are immediate (within 20 to 30 min) and persist for only an hour or two.

Summary and outlook

Particulate antigens such as bacteria and red blood cells react with their homologous antibodies to form agglutinated aggregates. Such reactions are termed agglutination reactions. Soluble antigens such as bacterial cell extracts and serum proteins react with homologous antibodies to form precipitates. These reactions are known as precipitin reactions. The phenomenon of immune cytolysis forms the basis of the complement-fixation test. The test has two distinct components: the test system and the indicator system. The systems vie for complement, which is an essential component of both systems. The principles for diagnostic applications of these tests were discussed.

The methods of the serological laboratory became very specialized and sophisticated with the advent of modern immunology in the 1960s. It is not practical in this chapter to cover them all in detail. But their underlying principles have been discussed with illustrations from the classical procedures. The future will see greater modifications and refinements of these methods to reveal finer details of diagnostic importance. Radioimmunoassays and enzyme immunoassays are among the newest, most sensitive, and most versatile techniques available for determining trace amounts of antigen and antibody. They are good examples of the kinds of sensitive techniques that will be developed in the future.

Key terms

antiserum	major crossmatch
crossmatching	minor crossmatch
delayed-type hypersensitivity	serology
electrophoresis	test system
electrophoretic mobility	titer
equivalence zone	universal donor
immediate-type hypersensitivity	universal recipient
immunoelectrophoresis	zone of antibody excess
indicator system	zone of antigen excess

Questions

1 Are there different types of antibodies for each type of observable antigen-antibody reaction? Explain.
2 How would you differentiate between the terms "immunology" and "serology"?
3 What is unique about the Weil-Felix test used for diagnosis of rickettsial diseases?
4 Explain why ABO-blood-group typing is important for blood transfusion.
5 Why is there precipitation in the equivalence zone of a quantitative precipitin test? Support your answer with explanatory diagrams.
6 Explain how erythroblastosis fetalis may be prevented from developing in the Rh-negative mother bearing an Rh-positive fetus.
7 By means of diagrams, explain the interpretation of the Ouchterlony method of precipitin testing.
8 Describe the principle of the complement-fixation test.
9 Distinguish between immediate-type allergy and delayed-type allergy.

References for part eight

Bigley, N. J.: *Immunologic Fundamentals*, Year Book Medical Publishers, Chicago, 1975. *A paperback textbook presenting the basic concepts of immunology; directed toward the medical student.*
Eisen, H. N.: *Immunology*, Harper & Row, New York., 1974. *A good reference on the molecular and cellular principles of the immune response.*

Finegold, S. M., W. J. Martin, and E. G. Scott: *Bailey and Scott's Diagnostic Microbiology*, 5th ed, Mosby, St. Louis, Mo., 1978. *Part six of this widely used book in the diagnostic laboratory describes serologic methods in diagnosis.*

Hyde, R. M., and R. A. Patnode: *Immunology*, Reston, Reston, Va., 1978. *A paperback that presents concise coverage of the fundamentals of immunology. It is brief and easily readable. It has a list of instructional objectives at the beginning of each chapter, each of which ends with review questions and proficiency examinations.*

Mims, C. A.: *The Pathogenesis of Infectious Disease*, Academic, New York 1976. *Presents the mechanism of microbial infection and pathogenicity in a unifying theme for all infectious agents. The roles of immune responses and phagocytosis are also well covered.*

Roitt, I.: *Essential Immunology*, 3d ed., Blackwell Scientific Publications, Oxford, 1977. *A concise, yet fairly comprehensive, overview of the principles of immunology.*

Rose, N. R., and H. Friedman (eds.): *Manual of Clinical Immunology*, 2d ed. American Society for Microbiology, Washington, D.C., 1980. *A comprehensive manual containing exact procedures for the immunological detection and analysis of a wide variety of diseases.*

Stewart, F. S., and T. S. L. Beswick: *Bacteriology, Virology and Immunity*, 10th ed., Bailliere Tindall, London, 1977. *A classical British text for medical students on the general principles of microbiology and immunity, and on the aspects of these subjects that are of importance to infectious disease.*

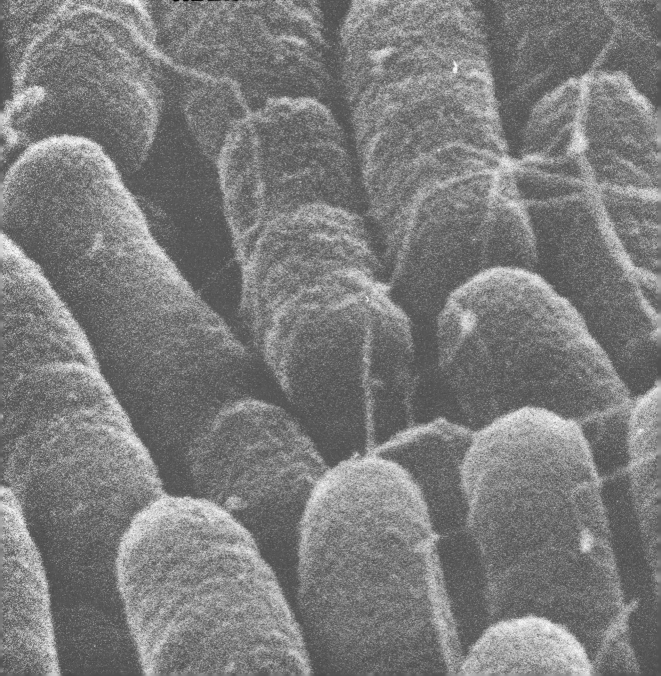

PART NINE

MICROORGANISMS AND DISEASE—
TRANSMISSION OF PATHOGENS

The role of microorganisms in the disease process can be studied in several ways. One is the *organ-systems approach*, in which one learns about the microbes associated with each diseased organ. This approach is appropriate for medical students, who are provided with parallel information on the pathology of the affected organs. Another way is the *diagnostic-microbiology approach*, in which one learns how to cultivate and identify various pathogenic microorganisms from clinical material. This approach is fitting for those who are studying to become clinical microbiology technicians or medical technologists. One other way is the *taxonomic approach*, in which one learns all about the particular pathogens in a specific genus. This taxonomic approach tends to become an exercise in memorization in a course of study. In this text we have adopted another way, the *transmission-of-pathogens approach*, because it provides an epidemiological "knowledge" for an understanding of the *prevention* of disease. At the same time, it affords the opportunity to learn something about the biology of the pathogens and the nature of their pathogenicity.

Except when lowered local or general resistance makes a person susceptible to his or her natural microbiota, pathogenic microorganisms come from outside the patient. They come, directly or indirectly, from soil, animal, or other human beings. These animals and humans may be clinically ill with obvious symptoms, or they may be symptomless *carriers* of the pathogens. Such carriers have no infection or have a subclinical infection (one without symptoms) of the disease. Typhoid carriers are classic examples because their subclinical carrier state may persist for many years; they are very difficult to detect because excretion of the bacterium is intermittent.

There are many routes for the transmission of pathogens. These routes constitute the basis for the organization of the chapters in Part Nine. Chapters 27 to 33 will be concerned, respectively, with airborne infections, foodborne diseases, waterborne infections, nosocomial infections, arthropodborne infections, sexually transmitted infections, and other transmissible diseases.

Scanning electron micrograph of *Vibrio parahaemolyticus*, a pathogen which causes gastroenteritis due to ingestion of contaminated seafood. Magnification ×33,000. *(Courtesy of O. Kowalski and A.G. Clark, University of Toronto.)*

AIRBORNE INFECTIONS

In this chapter some common and important airborne infections or diseases are discussed. They are diseases in which the etiologic agent is transmitted by air and enters the host via the respiratory tract, which includes the nose, pharynx (throat), larynx (voice box), trachea (wind pipe), bronchi, and lungs (see Fig. 22-4). Some of these airborne pathogens cause diseases of the respiratory tract; others infect different areas (such as poliovirus, which affects the brainstem and spinal cord) but share a common portal of entry into the body.

It is characteristic of airborne diseases that they tend to occur in epidemic form, appearing explosively and attacking large numbers of people in a short time. Their *incidence* (frequency of occurrence) usually increases during the fall and winter months when people are more likely to occupy crowded quarters indoors because of the cold temperatures outside.

Typical examples of airborne bacterial infections are diphtheria and the streptococcal diseases. Typical viral diseases spread by the respiratory route are the common cold and influenza. Systemic mycotic diseases are airborne infections too.

Specific mode of transmission

When a person coughs, sneezes, or spits, a spray of small and large droplets is expelled. These droplets may contain organisms that cause airborne diseases. The large droplets may contaminate clothing and other inanimate objects, including dust particles. The fine droplets evaporate and form *droplet nuclei*; these may carry infective agents and can be inhaled directly. (Any inanimate object that is contaminated with pathogens and serves to spread them is called a *fomite*.) Dust-borne spread of infection is enhanced when people move about in poorly ventilated conditions. Any movement that raises dust, such as removal of clothing, bed-making, and dry-dusting, increases the risk of dust-borne infection.

Some organisms that cause respiratory diseases are unable to survive long outside the body. Transmission of these microorganisms therefore depends upon rapid airborne transfer from one person to another or, sometimes, direct transfer (as by kissing). For example, the measles virus is easily inactivated outside the body. But other microorganisms, such as the bacterium of tuberculosis, can survive long periods outside the body.

Control of airborne diseases

It is obvious that control of diseases caused by hardy organisms is more difficult than control of those which die rapidly. Generally, respiratory

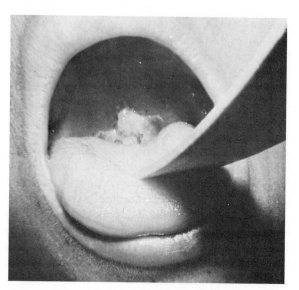

Figure 27-1. Pharyngeal diphtheria. The diphtheritic pseudomembrane forms on one tonsil and soon involves the remaining side. If the infection is untreated, the pseudomembrane will extend into the trachea. *(Courtesy of Center for Disease Control, Atlanta, Ga.)*

diseases are controlled by (1) minimizing contacts between cases, carriers, and susceptible persons, and (2) increasing resistance of individuals to infection. Contacts can be minimized by isolating active cases and disinfecting all articles that have been contaminated by the patient. Resistance to many of the respiratory diseases can be increased by immunization. Good nutrition and personal hygiene reduce the individual incidence of these diseases.

Providing adequate ventilation is a good general control measure in schools, offices, hospitals, and buses, because this reduces the risk of exposure. But in cold climates this is a difficult procedure during some months, and probably for this reason the incidence of respiratory diseases rises with the return of colder weather.

The importance of infectious diseases of the respiratory tract cannot be overemphasized. Acute respiratory illness, including the common cold, accounts for the loss of about 80 million workdays a year in the United States. More than 50 percent of school absences are due to respiratory infections.

Diphtheria

Diphtheria is an infection of the upper respiratory tract caused by bacteria. It is a localized, as well as a generalized, disease caused by diphtherial toxin produced by toxigenic strains of *Corynebacterium diphtheriae*. The organisms localize in the throat, which becomes inflamed as the bacteria grow and excrete a powerful exotoxin. Dead tissue cells, together with leukocytes, red blood cells, and bacteria, form a dull gray exudate called a *pseudomembrane* in the pharynx (Fig. 27-1). Within the pseudomembrane the bacteria continue to multiply and produce toxin. If this pseudomembrane extends into the trachea, the air passages may become blocked and suffocation ensues. Deaths from suffocation occur early in infection, especially in children, but are generally preventable by a tracheostomy. The disease usually becomes apparent 2 to 5 days after infection as a sore throat and fever, accompanied by membrane formation and swelling of lymph nodes.

The severe symptoms and death from diphtheria are due to the action of the toxin, which is carried by the blood to all parts of the body. As a result, damage may occur in the heart, kidney, and nerves. Deaths from toxemia thus occur late in the disease due to failure of vital organ systems.

Biology of *Corynebacterium diphtheriae*

Corynebacterium diphtheriae was isolated in 1883 by Klebs and was shown as the etiologic agent of diphtheria by Loeffler in 1884. Thus it is also known as the Klebs-Loeffler bacillus.

As shown in Fig. 27-2, the bacterium is a *pleomorphic* bacillus, which means that there is variation in size and form between individual cells in a pure culture. Some cells may be straight, while others are curved or club-shaped. Their lengths vary from 1 to 8 μm; their widths are from 0.3 to 0.8 μm. Many cells are formed at right angles to each other. A characteristic feature of the older cells is their granular appearance when stained with

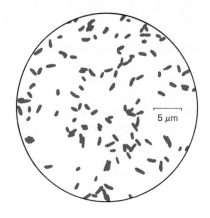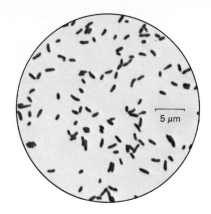

Figure 27-2. Morphology and distinguishing characteristics of *Cornebacterium diphtheria*, the etiologic agent of diphtheria. *(Erwin R. Lessel, illustrator; photomicrograph courtesy of Liliane Therrien and E. C. S. Chan, McGill University.)*

Corynebacterium diphtheriae

Gram-positive pleomorphic rods
Occurs in angular and palisade arrangements
Produces metachromatic granules
Not encapsulated
Nonsporulating
Nonmotile
Aerobic, facultatively anaerobic
Pathogenic, causing diphtheria

some dyes, such as methylene blue or Albert's stain. The granules take on a color different from the stain; these are called *metachromatic granules* and are composed of a polymer of inorganic polyphosphates.

C. diphtheriae grows well on blood-agar medium. But selective media such as Loeffler's coagulated serum-egg-glucose medium and blood agar with potassium tellurite retard the growth of contaminating organisms and favor the growth of the diphtheria bacillus. On tellurite-containing medium, the colonies are dark gray to black in color.

There are three types of *C. diphtheriae* which are distinguished by the morphology of their colonies when grown on tellurite-containing medium: *gravis*, *mitis*, and *intermedius*. The *gravis* colonies are semirough, flat, and gray to black; they are also the largest. *Mitis* colonies are smooth, convex, and black. The *intermedius* colonies are small and smooth with a black center. Other means of differentiating the types are their reactions with carbohydrates and their growth characteristics in broth. At one time the *gravis* type was thought to be the most potent toxin-producer. Although the relationship of virulence to colony form is no longer considered valid, the names *gravis*, *mitis*, and *intermedius* are still used to refer to the three types of the species.

Pathogenicity of
C. diphtheriae

The ability to produce toxin confers virulence on the organism. Toxigenicity depends upon the presence in certain lysogenic bacterial strains of specific, temperate bacteriophages carrying the *tox*⁺ gene. Nonlysogenic strains of bacteria are invariably nontoxigenic and hence incapable of producing diphtheria in natural hosts. But such avirulent strains can be made into virulent toxin-producing strains by infecting them with the *tox*⁺-gene-carrying temperate bacteriophage. This change in the properties of a cell as a result of becoming lysogenic is called *lysogenic conversion.*

The in vitro production of diphtheria toxin (for making toxoid commercially to be used as vaccine) requires the proper concentration of iron in the medium. Toxoid is made by treating the toxin with formaldehyde; it is purified by precipitation with alum. This procedure destroys the toxic portion of the molecule and leaves the antigenic portion intact to induce antibody (antitoxin) formation when it is injected into an animal such as the horse or into a person.

When a horse is immunized, the blood containing the antitoxin is removed aseptically and allowed to clot. The resulting serum contains the antitoxin, which is purified, concentrated, and standardized to proper dosage for use in the treatment of diphtheria patients. Caution must be observed in the use of antitoxin, since some persons become sensitized to horse-serum protein and develop serious allergic reactions when it is injected.

Antitoxin should be administered as soon as possible when diphtheria is suspected because it neutralizes the effect of preformed toxin. Antibiotic therapy is a useful adjunct to, but is not a substitute for, antitoxin in the treatment of the disease. The antibiotics effective against *C. diphtheriae* are penicillin and erythromycin. The latter is preferred for the elimination of the etiologic agent from the throat, especially in the treatment of persistent carriers.

Laboratory diagnosis of diphtheria

Laboratory diagnosis is made by isolating *C. diphtheriae* from the infected area and demonstrating its toxin-producing ability.

Specimens (swabs from nose and throat) are inoculated on Loeffler's coagulated serum slants, blood-agar plates, and tellurite-agar plates. Direct smears (that is, directly from the affected area, such as a smear from a throat swab) are stained with methylene blue and observed for bacilli containing metachromatic granules.

Toxigenicity of the isolates can be tested by the intradermal injection of a suspension of the organisms into the shaved flank of a guinea pig. Four hours later, antitoxin is injected intraperitoneally, and a second inoculation of the test strain is performed in the other shaved flank. If the isolate is toxinogenic, only the initial site of inoculation will become necrotic in 48 to 72 h.

An in vitro test for toxigenicity is also available. It is an agar-gel diffusion test which depends on the formation of a line of precipitation where the optimal proportion of the toxin diffusing from the isolate and antitoxin diffusing from another site meet. The gel-diffusion technique is a rapid method of diagnosis because toxin production can be demonstrated in specimens obtained directly from the infected pharynx.

Epidemiology of diphtheria

Diphtheria occurs worldwide and typically strikes in the form of epidemics. The disease incidence has declined sharply since the introduction of active immunization. In the United States and many other parts of the world, the

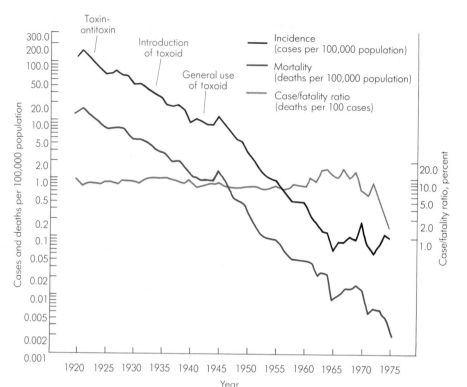

Figure 27-3. Diphtheria annual incidence and mortality rates and case-fatality ratios in the United States, 1920-1975. Note how the incidence of the disease and the number of deaths have declined in response to vaccine use. However, the percentage of persons who died, once they contracted the disease, had not changed much until very recent times. *(Center for Disease Control: Diphtheria Surveillance Report No. 12, July 1978.)*

disease has become rare. The effectiveness of immunization is dramatically shown in Fig. 27-3. From 1920 to 1975, the incidence of diphtheria in the United States has decreased progressively. The mortality rate decreased in a parallel fashion. The case-fatality ratio (deaths over 100 cases) remained relatively constant at about 10 percent for many years, but has declined sharply in recent years.

Children 2 to 5 years old have the highest incidence of disease and death. In adults diphtheria occurs at a low frequency; about 12 percent of all cases in the United States now occur in persons 20 years or older.

Humans are the only natural host of *C. diphtheriae*. It is transmitted from person to person in close contact via droplet nuclei of upper respiratory tract secretions. Sometimes diphtheritic skin lesions (cutaneous diphtheria) may occur, and these may serve as reservoirs for spreading the disease. Clinically ill patients do not generally spread the disease widely; but persons incubating the disease, convalescents and carriers, are much more important disseminators of the bacillus.

Prevention of diphtheria

The important control measures for diphtheria are (1) the *Schick test* for detection of susceptibility to the infection, (2) the use of diphtheria toxoid as vaccine, and (3) the therapeutic use of antitoxin.

The Schick test is made by intracutaneous injection of a very small amount of diphtheria toxin. If the person is immune, the toxin is neutralized by the antitoxin in the body and no reaction occurs. But if the person is susceptible—has no natural antitoxin—a local inflammatory reaction will result that reaches maximal intensity within 4 to 7 days. If the Schick test indicates a susceptibility to diphtheria, even adults should be actively immunized.

All children should be actively immunized with diphtheria toxoid before their first birthday (as young as 2 months). This toxoid is combined with tetanus toxoid and pertussis vaccine (DTP vaccine). The first dose of oral polio vaccine is given at the same time. Three doses are given at 4- to 8-week intervals, with a fourth dose given a year after the third injection (Fig. 27-4). Booster shots should be given when the children enter school. Thereafter, they should get a booster shot of tetanus and diphtheria every 10 years. Toxoid injections should be repeated whenever the Schick test indicates absence of antitoxin in the body. Antitoxin provides a passive immunity of short duration, but it may prevent the disease in susceptible persons who become exposed to diphtheria.

Strict isolation of all active cases of diphtheria, and prompt and thorough disinfection of all discharges, bedding, dishes, and other fomites, help control the spread of the disease.

Streptococcal diseases

Streptococcal diseases can be divided into two types: the *primary* infections (suppurative, or pus-forming, type) and those regarded as *sequelae*, or complications (nonsuppurative type) following primary streptococcal infections.

The most common infective *syndrome* (a group of symptoms which characterize a disease) in primary infections is an acutely inflamed sore throat, with or without exudate. This condition is known as streptococcal

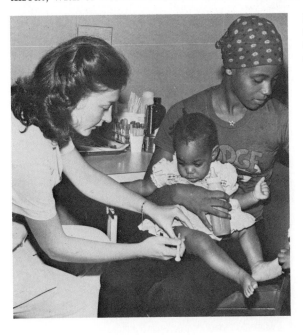

Figure 27-4. Immunization at a well-baby clinic. *(Courtesy of Center for Disease Control, Atlanta, Ga.)*

pharyngitis or *tonsillitis.* In pharyngitis, if the infecting streptococci are capable of producing large amounts of erythrogenic toxin and the host has no antitoxin, *scarlet fever* results. In addition to the sore throat, the patient has a generalized erythematous (reddish) rash (Fig. 27-5). Other suppurative-, or *pyogenic-*, type streptococcal diseases include all kinds of skin infections (pyoderma) such as impetigo (acute inflammation of skin), secondary wound infections, and infected blisters.

Nonsuppurative-type sequelae are *rheumatic fever* (inflammation of the connective tissues of many joints and organs, especially the heart) and *acute glomerulonephritis* (nonsuppurative inflammation of the kidneys, affecting primarily the glomeruli or capillaries). Generally, rheumatic fever is a sequela of pharyngitis, while acute glomerulonephritis is a sequela of pharyngitis and pyoderma. These complications are generally regarded as an immunological manifestation; that is, the organ tissues have antigenic determinants similar to the streptococcal antigens. The latter antigens stimulate host antibodies which can react with the tissues, thus damaging the organs.

Biology of the streptococci

Streptococci are gram-positive bacteria that occur in chains of two or more individual cells. The cells are spherical to ovoid in shape and about 0.5 to

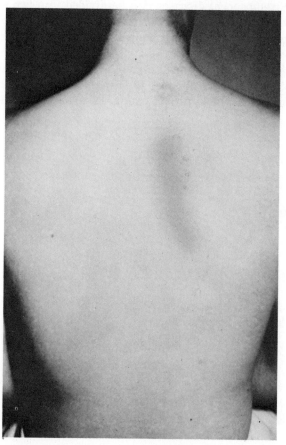

Figure 27-5. Erythematous rash on back of patient due to scarlet fever. *(Courtesy of Center for Disease Control, Atlanta, Ga.)*

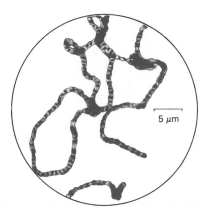 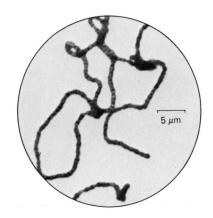

5 µm   5 µm

Figure 27-6. Morphology and distinguishing characteristics of *Streptococcus pyogenes,* the type species of the genus. *(Erwin F. Lessel, illustrator; photomicrograph courtesy of Liliane Therrien and E. C. S. Chan, McGill University.)*

Streptococcus pyogenes
   Gram-positive cocci
   Occurs in pairs and chains
   Not encapsulated
   Nonsporulating
   Nonmotile
   Aerobic, facultatively anaerobic
   β-hemolytic
   Pathogenic

1.0 µm in diameter (Fig. 27-6). The organisms are nonmotile and are facultative anaerobes.

Not all streptococci are pathogenic. Some are members of the normal microbiota, and some are even useful in the manufacture of certain food products, such as Cheddar cheese. Varieties of streptococci are distinguished by means of morphological, cultural, biochemical, and serological tests.

Streptococci may first be divided into those that produce a soluble hemolysin and those that do not. Agar plates containing whole horse or whole sheep blood are inoculated with streptococci in order to get isolated colonies. Those streptococci with greenish zones around the colonies are called α-hemolytic, or *viridans,* streptococci. The greenish zone is due to partially hemolyzed red blood cells. Streptococci with a clear "halo" around the colonies are called ß-hemolytic. Streptococci with no hemolytic zones around the colonies are nonhemolytic.

Reaction on blood-agar medium is the initial step in the identification of a streptococcal isolate from a clinical specimen. The β-hemolytic streptococci are primarily responsible for streptococcal pharyngitis and pyoderma, although some nonpathogenic strains of β-hemolytic streptococci can be found in the pharynx.

Mainly through the initial efforts of Rebecca Lancefield in the early 1930s, it was found that streptococci can also be divided into serological groups on the basis of a group-specific carbohydrate antigen (C substance) that is located in the cell wall (Fig. 27-7). These are called *Lancefield groups* and are designated by the letters A through T. Thus streptococci that belong to group A all share the same antigenic C substance, or

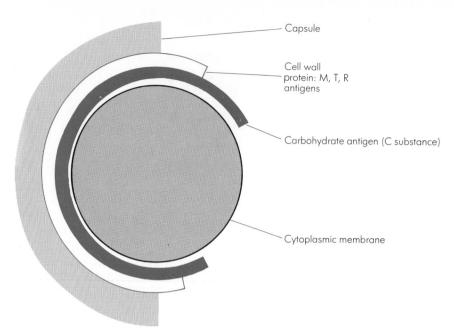

Figure 27-7. Diagram of
cross section of a group A
streptococcal cell.

Labels in figure:
Capsule

Cell wall
protein: M, T, R
antigens

Carbohydrate antigen (C substance)

Cytoplasmic membrane

carbohydrate; all organisms in group B possess another C carbohydrate. Strains belonging to Lancefield group A are responsible for over 90 percent of human streptococcal infections, and their species name is *Streptococcus pyogenes*. From this classification it was found that a single streptococcal species can cause a variety of diseases. For example, *S. pyogenes* can cause pharyngitis, scarlet fever, rheumatic fever, and glomerulonephritis.

In addition to the C carbohydrate for Lancefield grouping, other antigens that are also present are used to subdivide the groups into specific types. The type-specific antigen in some groups is a carbohydrate different from the C carbohydrate, but in group A it is a cell-wall protein called the M protein (Fig. 27-7). These antigens in group A are called Griffith types and are designated by arabic numbers, such as type 1, type 2, and so on. There are more than 50 Griffith types. M protein is also a virulence factor, since organisms possessing it are resistant to phagocytosis.

Other surface antigens of the cell wall are T- and R-protein antigens (Fig. 27-7). T protein is used for typing those strains which do not have the M antigen. R antigen has no clinical or pathogenic importance.

The determination of serologic groups and types is important in diagnostic bacteriology and epidemiology. This determination is done by precipitin tests using specific antisera from rabbits and extracts of cells grown in pure culture.

Pathogenicity of streptococci

The wide variety of diseases caused by group A hemolytic streptococci may well be related to the large number of extracellular products they produce. More than 20 such compounds are elaborated. They are antigenic and most appear to have some role in the production of disease. They are also important in the diagnosis of streptococcal infection. Some of these substances are discussed below.

454

Erythrogenic toxin. The presence of this toxin causes the rash of scarlet fever. It also has other biological activities, including stimulation of fever and damage to heart and liver tissue. Only strains of streptococci elaborating this toxin cause scarlet fever. Three serologically distinct types of the toxin are produced: A, B, and C.

Erythrogenic toxin is antigenic, giving rise to specific antitoxin which neutralizes it. Persons having such antitoxin do not exhibit the rash of scarlet fever, though they are still susceptible to streptococcal infection. Susceptibility to the toxin can be demonstrated by the skin test called the *Dick test*, in which 0.1 ml of standardized erythrogenic toxin is injected intradermally. In susceptible persons (those with an insignificant concentration of antitoxin in the blood), the toxin causes a localized skin reaction: the appearance in 18 to 24 h of erythema (redness) and edema (swelling due to fluid excess) measuring more than 10 mm in diameter.

The *Schultz-Charlton test* is for the diagnosis of scarlet fever. The injection of specific antitoxin into an area of rash will result in blanching and fading of the redness if the rash is due to scarlet fever.

Hemolysins. Two distinct hemolysins, O and S, are produced by group A streptococci, but not all strains produce both; some strains may produce one, or none at all. The presence of antibodies to streptolysin O in the serum is useful in detecting recent infection with streptococci. Streptolysin S is responsible for the zones of hemolysis around colonies on blood-agar plates. (Streptolysin O will also produce β-hemolysis, but under anaerobic conditions only, since it is inactivated reversibly by oxygen.) Streptolysin S is not a good antigen. By causing membrane damage, both hemolysins can kill leukocytes that phagocytose streptococci.

Other products. Other extracellular factors that may contribute to the disease process are the enzymes streptokinase (fibrinolysin), streptodornase (deoxyribonuclease), and hyaluronidase, all of which dissolve tissue components.

Early treatment of patients with streptococcal sore throat or other streptococcal infections is important. The drug of choice is penicillin, since no pencillin-resistant strains of streptococci have been found. For those patients allergic to penicillin, erythromycin is a useful substitute.

Prolongation of treatment for some days after the symptoms have subsided is necessary to eliminate completely the presence of streptococci. This will reduce the risk of complications, especially from the autoimmune diseases, such as acute glomerulonephritis and rheumatic fever.

Laboratory diagnosis of β-hemolytic streptococcal infections

It is important that β-hemolytic streptococcal infections be diagnosed promptly and treated adequately in order to prevent nonsuppurative complications. Laboratory diagnosis consists of isolating the causative organism and identifying it serologically. Laboratory diagnosis is important because it is generally agreed that an accurate diagnosis of streptococcal pharyngitis cannot be made on clinical grounds alone.

Primary identification is made by streaking material from the patient (usually a throat swab) on a blood-agar plate. After about 24-h incubation, the plate is examined for β-hemolytic, tiny, dull colonies. A gram stain of such a colony should show streptococci, giving a tentative positive diagnosis.

Presumptive, or reasonable, evidence that the streptococcus examined belongs to group A can be made by a bacitracin-sensitivity test. Lancefield group A streptococci are more sensitive to this antibiotic than are other groups and are inhibited by the antibiotic concentration used for the test.

Fluorescent-labeled antibodies against the C substance of group A streptococci may also be used to identify the organism as belonging to group A.

In addition to bacteriological confirmation of streptococcal infection, other immunological procedures can be used to detect a patient's prior exposure to streptococcal infection or to diagnose late, nonsuppurative sequelae which usually occur when infecting organisms can no longer be isolated. For example, one can test for the amount of antibody against streptolysin O present in a patient. A high antistreptolysin O titer indicates a recent infection by group A β-hemolytic streptococci.

Epidemiology of streptococcal diseases

Humans are the natural reservoir of *Streptococcus pyogenes*. The organisms are carried in the nasopharynx; this is why they are spread primarily by sneezing and coughing. Unfortunately, pathogenic streptococci may survive in sputum and other body secretions for several weeks, and this helps their dissemination.

Person-to-person spread of group A streptococci is facilitated by the availability of a susceptible population in close contact. This situation exists among school children, military recruits, and institutionalized persons.

Prevention of streptococcal diseases

No active immunization methods are available for the prevention of streptococcal diseases. *Chemoprophylaxis* (use of chemicals to prevent disease) is indicated for at least 5 years in patients with rheumatic fever in order to avert recurrences. Carriers of β-hemolytic group A streptococci should be detected, isolated, and treated to reduce the occurrence of epidemics. It has been shown by epidemiological studies that a school child commonly carries the infections home and spreads them to other members of the family.

Other airborne infections caused by bacteria

Both diphtheria and streptococcal diseases are caused by bacteria and are upper respiratory tract infections. They serve as good examples of airborne diseases caused by bacteria. There are also airborne bacterial diseases that involve areas other than the upper respiratory tract.

Tuberculosis

Tuberculosis is a *chronic*, or long-lasting, infectious disease of humans. In the United States in 1977, there were 30,145 reported cases of tuberculosis and 2,968 associated deaths. In many developing countries, it is still a leading cause of death.

Tuberculosis in humans may affect any tissue of the body, but the lungs are most commonly infected. Since tuberculosis is a chronic bacterial disease, advancing slowly, the infection may go unnoticed until a chance x-ray reveals lung lesions.

Symptoms generally consist of *pleurisy* (lung membrane inflammation) and vague chest pains, often with coughing, afternoon fever, fatigue, and loss of weight. After having been inhaled into the body, the tubercle bacilli initiate small lesions in the lower respiratory tract. These lesions heal to form tiny *tubercles*, or nodules, as a result of a cellular immune response to the organisms and their products. These tubercles continue to harbor the viable tubercle bacilli indefinitely. In some cases, the tubercles may grow in size, producing large abscesses which often drain pus, thus spreading the bacilli; or the tubercles may become walled off and calcified in healed lesions (which still contain viable bacilli).

The causative organism of human tuberculosis is *Mycobacterium tuberculosis* (Fig. 27-8). It is a slender, straight or curved rod, ranging from 0.3 to 0.6 μm by 0.5 to 4.0 μm. It usually occurs singly or in clusters. It is nonmotile and does not form spores or capsules. It is difficult to stain with

Figure 27-8. Morphology and distinguishing characteristics of some airborne bacterial pathogens. *(Erwin F. Lessel, illustrator.)*

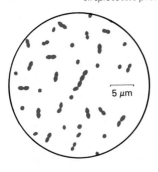

Mycobacterium tuberculosis

Gram-positive
  pleomorphic rods
Acid-fast
Not encapsulated
Nonsporulating
Nonmotile
Aerobic
Pathogenic, causing
  tuberculosis

5 μm

Neisseria meningitidis

Gram-negative cocci
Occurs singly, in pairs,
  or in tetrads
Not encapsulated
Nonsporulating
Nonmotile
Aerobic
Pathogenic, causing
  meningitis

5 μm

Streptococcus pneumoniae

Gram-positive cocci
Occurs in pairs, sometimes
  singly or in short
  chains
Encapsulated
Nonsporulating
Nonmotile
Aerobic, facultatively
  anaerobic
α-hemolytic
Pathogenic, causing
  pneumonia

5 μm

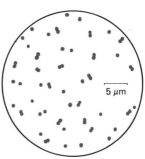

Legionella pneumophila

Gram-negative
  pleomorphic rods
Occurs singly
Not encapsulated
Nonsporulating
Nonmotile
Aerobic
Pathogenic, causing
  Legionnaires' disease

5 μm

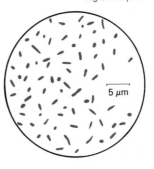

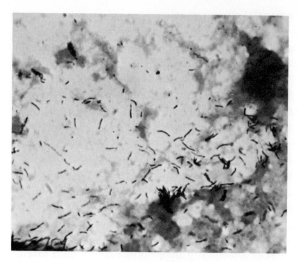

Figure 27-9. *Mycobacterium tuberculosis* in sputum, acid-fast stain (×800). *(From U. S. Army.)*

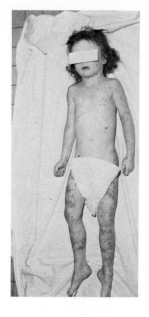

Figure 27-10. Meningococcal septicemia. *(From Armed Forces Institute of Pathology.)*

the usual microbiological dyes, but it stains readily by the Ziehl-Neelson acid-fast technique, in which the initial staining is with carbol-fuchsin heated to steaming and then washed with acid alcohol. Probably because of the high fat content of the organism, it is not decolorized by the acid alcohol (thus it is called an *acid-fast* bacillus) and retains the red color of the initial stain. A counterstain with methylene blue renders all other organisms blue.

A tentative diagnosis of tuberculosis can be made by observing acid-fast rods in smears of sputum (Fig. 27-9), gastric contents, urine, or spinal fluid, for example. But definitive diagnosis depends on the isolation and identification of the bacilli from the patient. This may be done by inoculating the sample material on suitable culture media, or by guinea-pig inoculation.

Treatment of the disease consists of bed rest, an adequate and nourishing diet, surgical intervention when necessary to remove or collapse an infected lung, and chemotherapy. The most effective drug now used is *isoniazid* (isonicotinic acid hydrazide), commonly referred to as INH. Others which may be used are *streptomycin* (SM) and *p-aminosalicylic acid* (PAS). In general, best results are obtained when SM and PAS are given in combination with INH. This delays the incidence of bacterial resistance and reduces the toxicity due to SM.

Tuberculosis may be prevented by improved standards of living and general health. Its control requires the location and treatment of infected individuals. The tuberculin skin test is used for the early detection of the disease; a positive reaction is interpreted as denoting an infected person, regardless of whether the disease is active or quiescent. X-rays are used for the reliable detection of active cases.

A vaccine is available for immunization against tuberculosis. It is made from attenuated live cultures of a bovine strain of *M. tuberculosis* named BCG (bacillus of Calmette and Guérin). It is used widely in some countries such as Canada, but not in the United States. The fear is that the avirulent organisms could revert to a virulent strain, even though no such reversion has occurred in over 50 years of vaccine use.

Pneumococcus pneumonia

Although a lobar *pneumonia* (congestion of the lobes of the lung) may be caused by many different microorganisms, 95 percent of bacterial pneumonias are caused by *Streptococcus pneumoniae*, formerly called *Diplococcus pneumoniae* and commonly called the pneumococcus (Fig. 27-8). Pneumococci are gram-positive oval or spherical cells occurring typically in pairs or short chains. The distal ends of each pair of cells are characteristically lancet-shaped (bluntly pointed). They are nonsporeforming and nonmotile, but virulent strains are encapsulated (Fig. 23-1). They produce α-hemolysis on blood agar and are called bile-soluble, which means that they are lysed in the presence of bile salts and detergents.

The pneumonocci can be divided into over 80 serological types on the basis of antigenic differences in their polysaccharide capsule. The type can be determined by precipitin or agglutination tests, but the quellung reaction—"swelling" of the capsule in the presence of type-specific antiserum—is most commonly used. It is this capsular material that prevents the organism from being phagocytosed. Acute pneumococcal lobar pneumonia is a severe illness with sudden onset and a tendency, in nonfatal cases, toward recovery by "crisis," a sudden change in the course of the disease, usually for the better. Symptoms include chills, fever, and pain in the area surrounding the lungs. The *alveoli* (air cells) of the lungs fill with exudate. Often a bacteremia occurs; pneumococci may also invade other tissues, especially the sinuses, the middle ear, and the meninges.

In laboratory diagnosis, direct smears of sputum can be stained and examined for the organisms. A quellung reaction can be made directly on sputum, or on cultured bacteria, to provide definitive identification of pneumococci. Blood and spinal fluid may also be used for detecting and isolating the organisms.

The most effective treatment of pneumococcal infections is the use of penicillin; pneumococci have not developed a resistance to this antibiotic. Prevention of disease is difficult because the organisms are so widely distributed in the population. Indeed, they are part of the normal flora of the respiratory tract of many healthy individuals. Prophylactic immunization is successful in large-scale trials, but the large number of types make general immunization impractical.

Meningococcus meningitis

Meningococcus meningitis is an acute bacterial infection caused by *Neisseria meningitidis* (Fig. 27-8). It is the major etiologic agent of *meningitis* (inflammation of membranes covering brain and spinal cord) in humans, spread to the meninges by blood from the nasopharynx. It also causes a fulminating septicemia, frequently fatal within 6 to 12 h. Resulting lesions of the skin, bones, and adrenal glands are believed to be due to endotoxin (Fig. 27-10). Each year in the United States, an estimated 18,000 cases of bacterial meningitis occur, with 2,500 associated deaths.

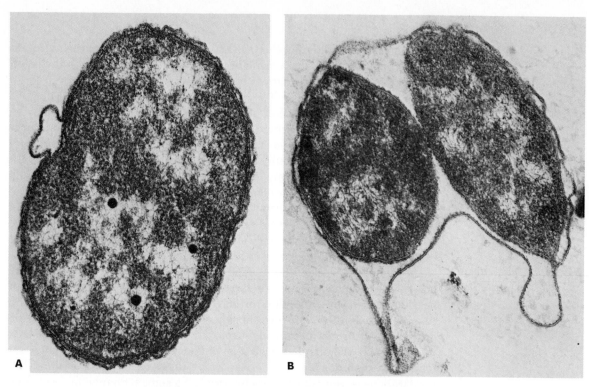

Figure 27-11. The effect of penicillin on *Neisseria meningitidis.* The result of the addition of a minimum inhibitory concentration of penicillin G on logarithmic growth-phase cells of group B meningococci as seen by electron microscopy. (A) after 30 min. (B) after 120 min. Note the early (A) and later (B) damage to the cell wall as seen by its separation from the cell proper (×66,000). *(Courtesy of L. G. Neirinck and I. W. DeVoe, McGill University.)*

The neisserias are nonmotile, gram-negative diplococci whose cells are characteristically kidney-shaped, their concave sides adjacent to each other (Fig. 27-8). They are about 0.6 to 1.0 μm in diameter.

The average incubation period after exposure is 1 week. Early symptoms are excessive nasal secretions, sore throat, headache, fever, pain in the neck and back, loss of mental alertness, and possibly a rash.

Spinal fluid, blood, and nasopharyngeal swabs, as well as smears and cultures from skin lesions, may be stained and examined for the presence of gram-negative diplococci. This examination serves as a quick differential diagnosis for early effective chemotherapy. Final identification depends on sugar fermentation tests and serological tests of isolated cultures.

Penicillin is effective in the treatment of this disease and is now considered the drug of choice (Fig. 27-11). The meningococci may be spread by active cases and by convalescent or healthy carriers. Humans are the only natural hosts known for the meningococcus. Prevention of meningococcal infections may be approached in two ways: by the elimination of carriers and the use of vaccine. Vaccines for serogroups A and C are commercially available. (Meningococci can be classified into serogroups A, B, C, D, X, Y, and Z, based on common antigens.)

Legionnaires' disease

During an American Legion convention held in Philadelphia in the summer of 1976, 221 people associated with the convention hotel developed a severe respiratory disease; 34 of these people died of it. This incident prompted an investigation into the cause of the disease, which the press had named for the conventioneers.

Legionnaires' disease has a syndrome typical of pneumonia. Symptoms begin 2 to 10 days after exposure to the bacterium and include malaise; headache; pain in the muscles, chest, and abdomen; a high fever with chills; dry cough; confusion; and impaired renal function. Mortality for this disease is 15 to 20 percent. Erythromycin is the most effective antibiotic in the treatment of Legionnaires' disease.

The Legionnaires' disease bacterium (Fig. 27-8) is a fastidious microbe which is difficult to isolate by direct culture, although it can be isolated using special enriched media. Isolation by inoculation of eggs or guinea pigs is possible but laborious. Diagnosis is usually made on the basis of indirect fluorescent-antibody tests. In addition, the detection of antibodies to the bacterium can be made with a microagglutination test. Despite the fact that much has been learned about the Legionnaires' disease bacterium through the use of DNA hybridization and other biochemical procedures, microbiologists cannot yet classify the organism. But they have given it the name *Legionella pneumophila*. It is a gram-negative pleomorphic rod ranging from 2 to 20 μm in length and has a diameter of 0.5 to 0.7 μm (Fig. 27-12).

Epidemiological data suggest that the natural habitat of the bacterium may be the soil. From this source, it may contaminate air and water supplies by means of windblown dust. Some other airborne infections caused by bacteria are shown in Table 27-1. Their etiologic agents, effective chemotherapy, and epidemiology are summarized in the table.

The common cold

The symptoms of the common cold are familiar to everyone. The predominant feature of this viral disease is an acute inflammatory infection of the nose, throat, sinuses, trachea, and bronchi, with much fluid exudation.

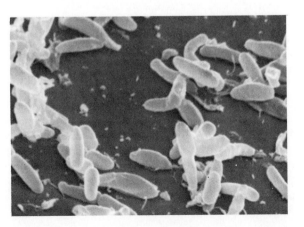

Figure 27-12. Scanning electron micrograph of the Legionnaires' disease bacterium, *Legionella pneumophila* (×16,200). *(Courtesy of D. D. Ourth, D. L. Smalley, and C. G. Hollis, Memphis State University.)*

Table 27-1. Some airborne infections caused by bacteria

DISEASE	ETIOLOGIC AGENT	MORPHOLOGY AND GRAM REACTION	CHEMOTHERAPY	EPIDEMIOLOGY/ PREVENTION
Pneumonia	*Haemophilus influenzae*	Rod, −	Ampicillin, chloramphenicol, streptomycin, and tetracyclines	Active immunization not done; most children have acquired natural immunity at 10 years.
	Klebsiella pneumoniae	Rod, −	Chloramphenicol, streptomycin, and tetracyclines	Vaccination impractical because of large numbers of serotypes. No specific methods of prevention available.
	Staphylococcus aureus	Staphylococcus, +	Penicillin and tetracyclines	Spread in hospitals by fomites, nurses, and doctors. Compromised patients vulnerable.
Whooping cough	*Bordetella pertussis*	Rod, −	Penicillin, chloramphenicol, and oxytetracycline	Mortality high for infants under 1 year. Effective vaccine available; given with tetanus and diphtheria toxoids.
Atypical pneumonia	*Mycoplasma pneumoniae*	Highly pleomorphic, −	Erythromycin and tetracyclines	Occurs frequently where people live in close contact. Rarely fatal.
Psittacosis	*Chlamydia psittaci*	Spherical, −	Tetracyclines	Contracted from psittacine birds, such as parrots. Control includes quarantine of such imported birds.

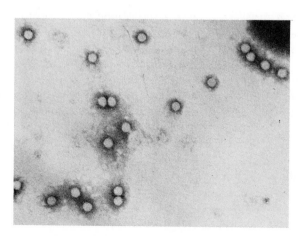

Figure 27-13. Electron micrograph of purified rhinovirus (×125,000). *(Courtesy of Frances Doane, University of Toronto.)*

Fever and other constitutional symptoms usually are absent. The incubation period is from 12 to 72 h; the disease is self-limiting and lasts from 2 to 7 days. Complications are rare, although in children, bronchitis may occur because of a cold infection.

Biology of the rhinoviruses

A number of viruses can cause the common cold syndrome. The most important etiologic agents of the disease in adults are the rhinoviruses, which account for about 30 to 40 percent of the cases.

The rhinoviruses belong to the picornavirus group (see Chap. 12). They contain RNA and are about 30 nm in diameter (Fig. 27-13). There are at present at least 89 serological types of the rhinoviruses.

Nature of pathogenicity in the common cold

The cells that line the nasal passages and pharynx appear to be most heavily infected, since they are the sites of active viral replication. Such replication begins 24 h after infection. During the acute phase of the cold, nasal secretions are produced in large amounts. There is no specific treatment for the common cold. Supportive measures such as adequate rest, warm clothing, and aspirin, which serve to increase the patient's comfort, are recommended. Liberal intake of fluids, such as fruit juices, is helpful.

Laboratory diagnosis of the common cold

Since the disease is easily diagnosed from an evaluation of clinical symptoms, there are no formal procedures for direct examination of clinical specimens or for serological diagnosis. If necessary, however, rhinoviruses can be isolated from clinical specimens by inoculating them into human fetal diploid-cell cultures and observing for cytopathic effect.

Epidemiology of the common cold

The common cold is the most prevalent infectious disease of humans and is worldwide in distribution. It is also costly, since it is a leading cause of absenteeism from work and school. The highest incidence of the disease occurs in the winter months, when there is more crowding in closed quarters, thus affording greater opportunity for spreading the virus. Humans are the natural hosts of the rhinoviruses.

Prevention of the common cold

There is no basis for the popular belief that sharp changes in temperature or wetness induce the common cold. Thus the main strategy for preventing the spread of the disease is to suppress airborne dissemination of the viruses. Persons suffering from this disease should cover their noses when they sneeze, and their contaminated handkerchiefs and other fomites should be disinfected. Hands should be washed frequently, since they are usually contaminated with nasal secretions.

Since the cold syndrome is usually mild and because there are so many serotypes of rhinoviruses, vaccine prophylaxis is not available. Circulating antibodies do, however, appear in response to a rhinovirus infection, and they afford protection against the specific serotype for up to 2 years.

Influenza   Influenza in humans is an acute infectious disease characterized by fever, chills, headache, generalized muscular ache, prostration, and loss of appetite. The virus is generally restricted to the upper respiratory tract. In cases with no complications, the disease is self-limiting, and patients recover completely within 3 to 7 days. Deaths from influenza are usually attributed to severe pneumonia, either due to the virus itself or, more frequently, to secondary bacterial infection of the lower respiratory tract.

Biology of the influenza virus

The etiologic agent of influenza is a virus belonging to the family Orthomyxoviridae. The genus name is *Influenzavirus*. (Some properties of this virus were discussed in Chap. 12.) *Influenzavirus* is a somewhat spherical particle, 80 to 120 nm in diameter, but filamentous forms are produced by some strains (Fig. 12-7). The ribonucleoprotein core, or nucleocapsid, has helical symmetry. The envelope is covered with closely packed projections of two morphological forms each with a different function: one is a spike with a triangular cross section and is a *hemagglutinin*; the other is a mushroom-shaped protrusion and is the enzyme *neuraminidase*. The hemagglutinin permits the virus to bind to various kinds of erythrocytes to form agglutinated masses; neuraminidase breaks down mucoproteins. The viruses can be grown in chick embryos as well as in tissue culture.

There are three known antigenic types of the influenza virus, designated A, B, and C. Influenza A viruses can be classified into subtypes on the basis of two antigens: hemagglutinin (H) and neuraminidase (N). Four types of hemagglutinin (H0–H3) and two types of neuraminidase (N1, N2) are recognized to be among influenza viruses causing widespread disease among humans. (Influenza viruses from animals and birds may have different H and N antigens. The designations are, for example, swine: Hsw1, Hsw2, etc.; duck, i.e., avian: Hav1, etc.) Immunity to H or N antigens reduces the likelihood of infection and reduces the severity of disease in infected persons. There may be sufficient antigenic variation within the same subtype over time, however, that infection or immunization with one strain may not induce immunity to distantly related strains. As a consequence, the antigenic composition of the most current strains is considered in selecting the virus strains to be included in a vaccine.

In order to have uniform and adequate description of influenza viruses, there is a system of notation used which allows comparison of the nature of the virus strains. For example:

A/HongKong/1/68(H3N2)

is a type A from a human source (since there is no designation to indicate otherwise), isolated in Hong Kong; it is the first strain isolated in 1968 with H3 and N2 antigens. An isolation from a nonhuman host (in this case avian) is indicated as in the following example:

A/Turkey/Wisconsin/1/66(Hav5N2).

Nature of pathogenicity in influenza

Virus multiplication is generally restricted to the upper respiratory tract. Spread of the virus is probably facilitated by the ability of the viral neuraminidase to hydrolyze the mucoproteins lining the respiratory tract.

The death and sloughing off of ciliated epithelial cells may be responsible for many of the symptoms, such as congestion and increased secretion.

Influenza virus itself can cause pneumonia without bacterial complications. But bacterial infection complicating influenza is possible; the usual bacteria involved are *Staphylococcus aureus*, *Haemophilus influenzae*, pneumococci, and hemolytic streptococci.

Treatment for uncomplicated influenza involves supportive measures such as bedrest. Bacterial complications may be treated with semisynthetic penicillins; nafcillin is an example.

Laboratory diagnosis of influenza

Although a presumptive diagnosis can be made based on the clinical syndrome, a definitive diagnosis depends on laboratory procedures. These include isolation of the virus from throat washings or sputum and the demonstration of a rising titer of antibody during convalescence.

Isolation of the virus is best carried out by inoculating specimens obtained during the first three days of illness into the amniotic cavity of chick embryos. Primary mammalian tissue culture cells, such as monkey kidney cells, can be used for isolation of the virus. Type-specific antisera can then be employed to identify the viral isolates serologically.

Serum-antibody titer is determined by hemagglutination inhibition, neutralization of virus infectivity, or complement fixation tests. The titer may rise as early as 8 to 9 days after the onset of disease.

Epidemiology of influenza

Epidemics of influenza occur in cycles. Type A strains of the virus commonly follow a 2- to 3-year cycle. B strains have a 4- to 6-year cycle. Type C strains rarely, if ever, give rise to epidemics; they cause subclinical infections or small outbreaks of the disease among children. All *pandemics* of influenza (rapid global spread of the disease) have been caused by type A viruses. During the past hundred years there have been three pandemics, in 1889–1890, in 1918–1919, and in 1957–1958. The 1918–1919 pandemic was the most disastrous; it took more than 20 million lives.

Epidemics of influenza usually occur in the cold months from late fall to early spring. In a community, an epidemic peaks about 2 weeks after onset; it often subsides in about a month. At highest risk are the very young, the very old, the pregnant, and debilitated persons.

The 1957 Asian influenza pandemic apparently started in the central part of mainland China in 1956 or 1957. From there it spread widely in China, then to Hong Kong, and then to other parts of the world. Shortly after September 1957, epidemics swept the United States.

In an influenza outbreak among recruits at Fort Dix, New Jersey, in late January 1976, the virus isolated from four patients proved to have surface antigens resembling those of swine influenza virus. One of the recruits died, and the postmortem findings were compatible with death from influenzal pneumonia. Serological studies showed that other recruits had also been infected. The swine influenza is believed to be derived from, and is possibly identical with, the virus responsible for the most serious influenza pandemic on record, the pandemic of 1918–1919. The virus

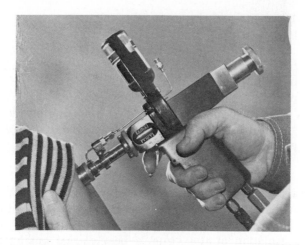

Figure 27-14. A jet-injector used in the influenza immunization project of 1976. With its use an operator can immunize more people in a shorter time. *(Courtesy of Center for Disease Control, Atlanta, Ga.)*

stopped circulating in humans in the early 1930s but has remained endemic in pigs in the United States. It has been identified in a few cases of human influenza in the intervening years, usually in those people who had been in close contact with pigs. The Fort Dix case is the first report of swine influenza case-to-case transmission. If this virus should spread, and if it retains or acquires for humans the virulence of the 1918 virus, then the world could anticipate a similar pandemic. This was the rationale for the planned mass vaccination of the American and Canadian populations in 1976 (Fig. 27-14).

Influenza, like other airborne infections of humans, is transmitted by droplet infection, by direct contact with infected persons (for example, kissing), or by fomites (for example, soiled handkerchiefs).

Prevention of influenza

Active immunization with killed-virus vaccines is the most effective means of prevention. Even though such immunization may not afford absolute protection, it will at least modify or lessen the severity of the disease. Maximum protection would require annual immunization, since the duration of protective immunity is only from 3 to 6 months. High-risk persons should receive annual immunization before November in the northern hemisphere.

Other airborne infections caused by viruses

Both the common cold and influenza are caused by viruses and are infections of the respiratory system. They serve as good examples of airborne infections caused by viruses. There are other airborne viral diseases that involve areas quite apart from the respiratory system. The main features of some of these other viral diseases are summarized in Table 27-2; Fig. 27-15 shows their clinical manifestations.

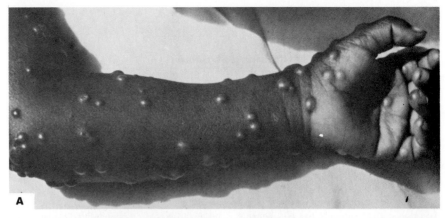

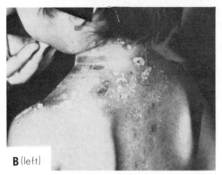

B (left)

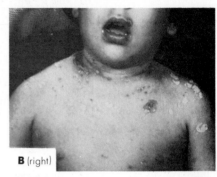

B (right)

Figure 27-15. Clinical manifestations of some airborne infections caused by viruses. (A) Smallpox. Pox on arm and palm of hand. *(Courtesy of Center for Disease Control, Atlanta, Ga.)* (B) Chicken pox. Young patient exhibiting vesicles. *(From Armed Forces Institute of Pathology.)* (C) Measles. Left: Preeruptive measles; third day of illness—inside cheek showing Koplik's spots. Right: Rash on abdomen. *(Courtesy of Center for Disease Control, Atlanta, Ga.)* (D) German measles. Left: Rash on back of patient. *(From Armed Forces Institute of Pathology.)* Right: 3-day-old infant with generalized lesions characteristic of neonatal purpura due to congenital German measles. *(Courtesy of Center for Disease Control, Atlanta, Ga.)*

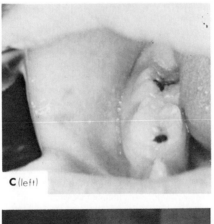

C (left)

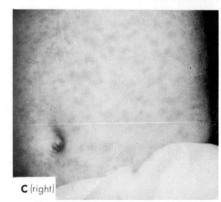

C (right)

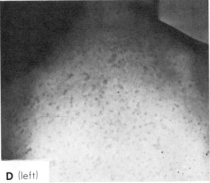

D (left)

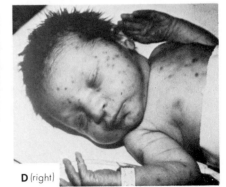

D (right)

Table 27-2. Some airborne infections caused by viruses

DISEASE	ETIOLOGIC AGENT	PATHOGENICITY	LABORATORY DIAGNOSIS	EPIDEMIOLOGY/ PREVENTION
Smallpox (variola)	Variola virus, a poxvirus. Two variants: variola major causes severe illness with mortality rate 25–50%; variola minor (alastrim) causes much milder disease with less than 1% mortality.	Site of entry is upper respiratory tract. Spreads to lymph nodes with febrile illness. Viremia spreads virus to all parts of body. Pox eruption several days later. Incubation 12–16 days. Toxicity generally cited as cause of death.	Diagnosis made on clinical symptoms. Positive diagnosis requires laboratory procedures: inoculation of chick embryo with pox lesion scrapings. Growth of virus shown in 2–3 days by lesions on chorioallantoic membrane. Can also use electron microscopy on material from vesicles, pus, or crusts.	Spread by direct or indirect contact with pox lesions of infected individual. Very contagious. World free of disease as of 1979. Vaccination with vaccinia virus gives long-lasting immunity.
Chicken pox (varicella)	Varicella-zoster virus, a herpesvirus.	Acquired via respiratory route. Virus multiplies in respiratory tract and regional lymph nodes. Released into blood and disseminated throughout body. Fever, malaise and headache precede skin eruption as rashes and eventually as vesicles. Incubation 14–16 days. Very little known about pathogenesis.	Clinical criteria usually sufficient to establish diagnosis, especially with history of exposure. Electron microscopy for virus is reliable and rapid diagnosis procedure.	Occurs worldwide and usually in children; frequently in epidemics in winter and spring. No vaccines available.
Measles (rubeola)	Rubeola virus, a paramyxovirus	Severe, acute, highly contagious disease spread by respiratory secretions. Virus multiplies in upper respiratory tract and conjunctiva (mucous membrane of eye) early in disease. Viremia follows. High fever (may lead to convulsions), eye pain from light, conjunctivitis, cough, and rash over body. Koplik's spots (white patches) on mucous membrane of mouth and throat.	Diagnosis almost always made on clinical syndrome. Infection confirmed by isolation of virus from tissue cultures or an increase in measles-antibody titer. Also presence of giant cells in nasal secretions due to cell fusion of infected cells induced by virus.	Humans are only reservoir of virus. Occurs in epidemics every 2–3 years. Affects children principally. Predisposes to secondary bacterial infections. Transferred by droplet infection early in disease. Live attenuated measles virus vaccine for all children over 1 year.
German measles (rubella)	Rubella virus, a togavirus	Rather mild disease spread by respiratory secretions. Initial viral replication in upper respiratory tract mucosa and cervical lymph nodes. Followed by blood dissemination to whole body with mild fever and variable rash. Incubation:	Clinical diagnosis in sporadic cases difficult. Diagnosis established by viral isolation or serological tests for rubella antibody.	Humans only natural host of virus. Spread through inhalation of droplet nuclei. In United States, highest incidence in 5- to 9-year-old children. Also high in adolescents. Several attenuated live rubella vaccines now available.

Table 27-2 (Continued)

DISEASE	ETIOLOGIC AGENT	PATHOGENICITY	LABORATORY DIAGNOSIS	EPIDEMIOLOGY/ PREVENTION
		ca. 18 days. Disease over in 3–4 days. When acquired in early pregnancy, severe general infection of fetus results with consequent death or variety of congenital defects (mental retardation, cerebral palsy, etc.)		Since vaccine can affect fetus, women should be vaccinated at 11–12 years of age or at least 2 months before pregnancy or immediately after delivery of a baby.
Mumps	Mumps virus, a paramyxovirus	Infection of parotid glands with inflammation, marked swelling behind ears, and difficulty in swallowing. Salivary glands, testes, ovaries, pancreas, and other glands may be involved. Replication of virus in upper respiratory tract epithelium leads to viremia. Most feared complication in males is *orchitis* (infection of testes) accompanied by pain and swelling. Incubation 18–21 days.	Clinical picture is frequent basis of diagnosis. Can be confirmed by inoculation of saliva or spinal fluid into either chick embryos or tissue culture. Rapid diagnosis can be made with electron microscopy of clinical specimens.	Transmitted via respiratory secretions. Live attenuated mumps virus vaccine used with lifelong immunity. Routine immunization of children over 1 year recommended.
Poliomyelitis	Poliovirus, a picornavirus. Three immunological types known: type 1 (Brunhilde), type 2 (Lansing), type 3 (Leon).	Most characteristic acute symptoms result from involvement of both the meninges and motor neurons of spinal cord and brainstem producing permanent paralysis.	Laboratory diagnosis based on isolation of virus from throat swabs, feces, or spinal fluid; or demonstration of increase in neutralizing antibody to one of the three serotypes during convalescence.	Respiratory route most important mechanism of transmission, but also by fecal-oral route. Effective vaccines available: formalin-killed viruses of each of three serotypes as developed by Salk; attenuated live virus vaccine as developed by Sabin and others. Sabin vaccine has largely replaced Salk vaccine.

Systemic mycoses   The systemic, or deep, mycoses are mainly fungus diseases that are often serious or fatal. The organisms invade subcutaneous tissues or the lungs, from which they may spread to other organs of the body, where they become established and produce disease. Many of them are airborne and enter the body through the respiratory tract; they may, however, enter by other portals.

Systemic mycoses appear to be increasing in importance, but their greater apparent incidence may actually be due to improved diagnostic methods and a greater appreciation of their importance. As a result of the great mobility of Americans, more people are exposed to fungi that are only prevalent in and endemic to specific limited areas of the world.

Table 27-3. The systemic mycoses (not always airborne)

DISEASE	CAUSATIVE ORGANISM	CHARACTERISTICS OF THE ORGANISMS	CHARACTERISTICS OF THE INFECTION
Crypto-coccosis	*Filobasidiella (Cryptococcus) neoformans*	These are yeastlike, capsule-producing organisms that reproduce by budding. No hyphae or spores are formed. They grow well on ordinary culture media.	This organism may infect any part of the body but usually starts in the lungs and spreads through the bloodstream. Infections of the brain and meninges usually cause death. Mode of transmission is not known, and spread from known cases in humans or other animals has not been established.
Monilliasis	*Candida albicans* (order Moniliales)	*C. albicans* are yeastlike cells with pseudohyphae. They produce large, thick-walled, spherical chlamydospores. On ordinary culture media, pasty, smooth colonies having a yeasty odor develop.	Monilia may infect any body tissue; it is found on mucous membranes of intestinal tracts of many healthy persons. Infection with *C. albicans* in the mouth is called thrush; may also cause a mycotic endocarditis, pulmonary monilliasis, and vaginitis; may be spread by contact.
North American blasto-mycosis	*Blastomyces dermatitidis*	These are large round cells with a single bud. Intercalary and terminal chlamydospores appear in old cultures. Optimum growth temperature is 37°C. On infusion-blood agar the colonies resemble *M. tuberculosis*. Typical cells are found in body exudates and in cultures.	Infection with *B. dermatitidis* resembles pulmonary tuberculosis with involvement of the lungs and pleura. It is characterized by chronic, granulomatous, suppurative lesions of any body tissue. It occurs only in the United States and Canada, most commonly among rural males aged 30 to 50. Systemic infections are often fatal. It is not transmitted from humans or other animals to humans.
South American blasto-mycosis	*Paracocci-dioides (Blastomyces) brasiliensis*	These are yeastlike cells that are larger than *B. dermatitidis*. (They range from 6 to 30 μm in cultures and to 60 μm in exudates.) Parent cells give rise to many buds. Smooth, waxy, yeastlike colonies appear on blood or meat medium after several days' incubation at 37°C.	South American blastomycosis is clinically similar to the North American blastomycosis. It also resembles coccidioidomycosis. It occurs most frequently in Brazil. Lesions are most commonly found in the mouth and gastrointestinal tract and in the lymph nodes of the neck.

Table 27-3 (Continued)

DISEASE	CAUSATIVE ORGANISM	CHARACTERISTICS OF THE ORGANISMS	CHARACTERISTICS OF THE INFECTION
Histo-plasmosis	*Histoplasma capsulatum*	Small, oval cells found intracellularly in tissues. Colonies on blood agar resemble *Staphylococcus aureus*. Cells have single buds. At room temperature on Sabouraud's glucose agar, delicate, branching, septate hyphae appear with chlamydospores present in old cultures.	This may occur as an acute or chronic, localized or disseminated infection of the reticuloendothelial system. Clinically it may be confused with carcinoma of the nose, tongue, or pharynx; tuberculosis; Hodgkin's disease; and aplastic anemia. Most infections regress spontaneously, but fulminating cases are usually fatal.
Coccidioid-omycosis	*Coccidioides immitis*	In cultures on Sabouraud's glucose agar these organisms develop as typical white- to buff-colored mold colonies that sporulate by arthro-spore formation. In body exudates they are single-celled, thick-walled, spherical organisms filled with endospores. Chlamydospores are present in old cultures.	This disease goes by many other names, such as valley fever, San Joaquin fever, and desert rheumatism. The fungus is highly infectious and widely distributed in the soil of certain areas of the United States. Most cases are mild and transitory, but a few terminate fatally.
Sporo-trichosis	*Sporothrix schenckii*	Organisms are rarely found in tissues, but in experimentally infected rats the organisms are gram-positive, resembling fusiform bacilli, in polymorphonuclear leukocytes. On Sabouraud's glucose agar incubated at room temperature, delicate, branching, septate hyphae with spherical or pyriform microconidia in clusters on lateral branches are found. On brain-heart-infusion agar, colonies are soft and composed of budding, yeast-like, cigar-shaped cells with a few mycelial elements.	This is a chronic infection that usually begins as a subcutaneous nodule at the site of an injury. Initial lesions resemble warts, boils, or chancres. The organisms are disseminated in the body through the lymph channels to various lymph nodes. Pulmonary involvement is infrequent. In rare cases involving spread of the organisms throughout the body, the patient may die.

PARASITIC PHASE in the host and in cultures at 37°C	SAPROPHYTIC PHASE in soil and in cultures at 25°C (room temperature)

Histoplasma capsulatum

Yeast-like
Short oval, very
delicate budding cells
2 to 3 by 3 to 4 μm in size.
Typically intracellular in the host.

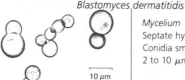

Mycelium
Septate hyphae 2 to 3 μm wide.
Conidia tuberculate or smooth,
short oval to spherical, 2-4 to 8-14
μm in diameter.

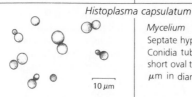

Blastomyces dermatitidis

Yeast-like
Nearly spherical thick-wall
budding cells 8 to 15 μm in
diameter. Buds produced on
broad base.

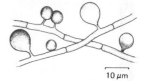

Mycelium
Septate hyphae 2 to 3 μm wide.
Conidia smooth, nearly spherical
2 to 10 μm in diameter.

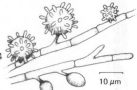

Blastomyces brasiliensis

Yeast-like
Spherical cells 10 to 60 μm in
diameter. Budding single or in
multiples; buds varying in size
from 1-2 to 10 μm in diameter.

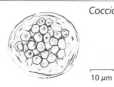

Mycelium
Fine septate hyphae producing
intercalary or terminal
chlamydospores. Conidia rarely
found.

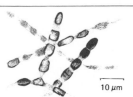

Sporotrichum schenckii (mostly subcutaneous)

Yeast-like
Budding cells, up to 10 μm in
diameter or elongated cigar
bodies (also budding) 1 to 3 by
3 to 10 μm in size.

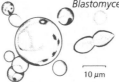

Mycelium
Very fine septate hyphae 1 to 2
μm wide. Conidia 2 to 3 by 3 to 6
μm, occasionally larger, in
bouquetlike arrangement on
sterigmata.

Coccidioides immitis

Spherules
Thick-walled cells 15 to 90 μm in
diameter, packed with small (2 to
5 μm in diameter) endospores.

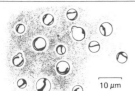

Mycelium
Septate hyphae 2 to 3 μm
wide in nonfertile portions.
Arthrospores alternating with
empty cells. No conidia.

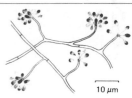

Filobasidiella neoformans

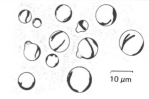

Monophasic growth. A *yeast* at *both* temperatures, in the host *and*
in cultures. Spherical budding cells, typical of true yeasts, 4 to 20
μm in diameter. Buds produced singly on a narrow neck.
Encapsulated. The capsule varies in thickness in different strains.

Figure 27-16. Some
pathogenic fungal species
causing systemic mycoses.
(Erwin F. Lessel, illustrator.)

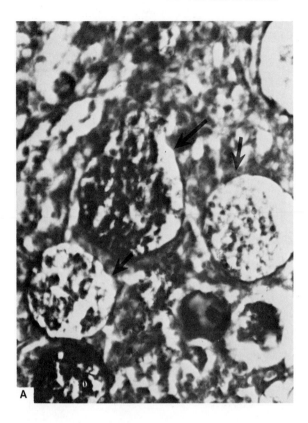

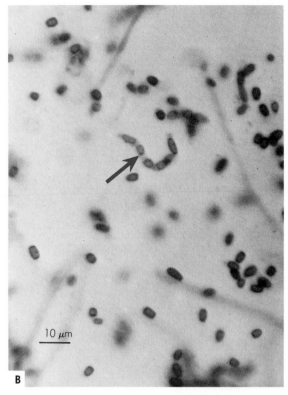

Figure 27-17. *Coccidioides immitis* is the fungus causing the disease coccidioidomycosis. (A) In the tissue phase it forms spherules containing endospores. (B) In culture the saprophytic filamentous phase forms arthrospores alternating with empty cells. *(Courtesy of L. Kapica and E. C. S. Chan, McGill University.)*

Factors that predispose an individual to become more susceptible to systemic fungus infections are:

1 The presence of chronic debilitating diseases such as cancer, diabetes, leukemia, and tuberculosis
2 The use of newer types of drugs, such as antibiotics and hormones, which cause changes in the metabolism of the body or upset the normal relationships among the organisms on or in the body
3 Local lesions caused by vitamin deficiency, irradiation, peptic ulcers, or other factors that allow the fungi to get into the deep tissues

The fungi that cause the deep mycoses are important not only because the diseases they cause can be very serious, but also because the symptoms produced by some of them resemble tuberculosis or other diseases. It is essential that accurate diagnostic procedures be used in order that the most suitable treatment may be employed.

The principal systemic mycoses, their causative agents, the characteristics of these organisms, and the infections they cause are given in Table 27-3. Figure 27-16 shows the morphology of some of these causative organisms. Note that many of them have a biphasic growth habit, for example, a parasitic yeastlike phase and a saprophytic filamentous phase. See Fig. 27-17. Figure 27-18 shows the clinical symptoms of some of these diseases.

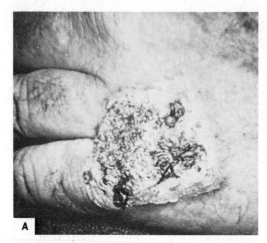

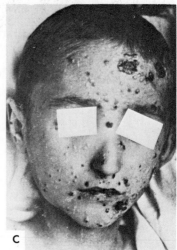

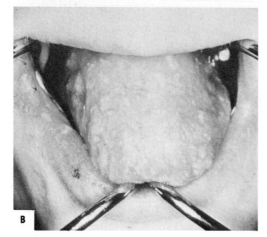

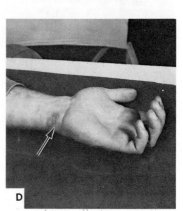

Figure 27-18. Clinical manifestations of some fungal diseases. (A) Blastomycosis of the hand. (B) Candidiasis on tongue and lips of a 19-year-old patient. *(A and B Courtesy of Center for Disease Control Atlanta, Ga.)* (C) Cryptococcosis on the skin of the face. (D) Coccidioidomycosis in a lesion of the arm. *(C and D Courtesy of Armed Forces Institute of Pathology.)*

Summary and outlook

The diseases covered in this chapter are those that are termed airborne infections. Such infections are transmitted through the air, and the vehicles for this transmission include droplet nuclei of saliva and other respiratory secretions, contaminated dust, and fomites. The portal of entry of the etiologic agents is the nasopharynx. Some of these airborne infections affect the upper respiratory tract, but some affect other organ systems of the body, even though they enter the body via the nose and throat.

The outlook for the prevention or cure of the airborne infections discussed in detail in this chapter is as follows:

Diphtheria. Since adequate therapy and effective immunization are available, and since the disease is very well understood, the outlook is promising for elimination of the disease. Such a goal, however, is dependent on social and political action, which may be difficult to achieve.

Streptococcal diseases. The virulence factors of the β-hemolytic group A streptococci are well understood. Even though the pathogenesis of the sequelae to primary streptococcal infections, such as rheumatic fever, is known to be due to autoimmune reactions, much remains to be learned about host-parasite relationships in these complications.

The common cold. This most common of diseases is a nuisance and a personal inçonvenience. It is a benign, or mild, disease. The future does not hold much promise for prevention of the disease by vaccination (even though an effective vaccine can be made) because of the large number of antigenically distinct viruses. For example, a protective vaccine would have to include more than 80 antigenically distinct components!

Influenza. At the present time, immunization with killed vaccines is 65 to 70 percent effective, and the duration is only for 3 to 6 months. The use of a live, attenuated strain of virus as a vaccine and the introduction of the intranasal route of inoculation would be improvements for the future.

In general, the outlook for airborne diseases is for a reduction in their incidence by better education and preventive measures and by better control of the environment, especially in closed quarters. Vaccines will be developed to prevent those respiratory diseases for which effective vaccines are not now available.

Key terms

acute glomerulonephritis
chemoprophylaxis
Dick test
droplet nuclei
fomite
hemagglutinin
Lancefield group
lysogenic conversion
meningitis
metachromatic granule
neuraminidase
pandemic
pharyngitis

pleomorphic
pleurisy
pneumonia
pseudomembrane
pyogenic
rheumatic fever
scarlet fever
Schick test
Schultz-Charlton test
sequelae
syndrome
tonsillitis
tubercles

Questions

1 Discuss some common epidemiological features that characterize airborne diseases.
2 Are all airborne infections respiratory tract infections? Explain.
3 What practices will help prevent the transmission of airborne infections?
4 Why do we say that diphtheria is a well-understood disease?
5 What accounts for the fact that diphtheria has become a rarity in the United States?

6 Explain the conditions necessary for the production of toxin by *Corynebacterium diphtheriae*.

7 How is the toxigenicity of *C. diphtheriae* tested in the laboratory?

8 What kinds of diseases are caused by β-hemolytic streptococci?

9 Explain the relationship of cell structure to serological classification of the streptococci.

10 Describe some specific ways in which extracellular products of streptococci have been used for diagnostic purposes.

11 Explain the problems inherent in the control and prevention of the common cold.

12 Based on your knowledge of the biology of *Influenzavirus*, explain the occurrence of epidemics of influenza.

13 Explain clearly the system of notation of influenza viruses. Why is such a notation necessary?

14 Compare the preventive measures for influenza and the common cold with special reference to immunization procedures.

28
FOODBORNE DISEASES

Foodborne diseases, caused by microorganisms and transmitted by the ingestion of food, result from one of two different mechanisms: (1) the microorganisms in the food may *infect* the host, causing foodborne disease; or (2) the microorganisms may secrete *exotoxins* in the food, causing food *intoxication*, or poisoning, when it is ingested.

The etiologic agents of the common bacterial foodborne diseases are well known. For example, salmonella bacteria cause foodborne infections, while food intoxications are caused by species of clostridia and staphylococci.

Specific mode of transmission   Microorganisms causing acute *gastroenteritis* (inflammation of stomach and intestines) are transmitted by the ingestion of contaminated food. Our

477

food is contaminated with various microorganisms most of the time. But generally we are not infected or poisoned, either because the contaminating microorganisms are harmless, or because the number of microorganisms is low.

Control of foodborne diseases

Much is known about the contributing factors in foodborne diseases, so methods for their control are well established. The contributing factors are:

1 Inadequately cooked food
2 Improper holding temperatures for the food
3 Food obtained from an unsafe source
4 Contaminated equipment
5 Poor personal hygiene
6 Inadequate preservation methods

Inadequately cooked food is usually the cause of *trichinosis* (a parasitic infection resulting from ingesting improperly cooked pork) and *botulism* (a bacterial food poisoning). Inadequate cooking may occur when large birds or roasts are cooked, since their size increases the problem of sufficient heat penetration. Without a high enough temperature for a sufficient time during cooking, either the microorganisms (including parasites) are not killed or heat-labile toxins (such as botulinal toxin) are not destroyed.

Improper holding temperatures for food, such as warm room temperatures, result in microbial growth. To prevent this, adequate chilling and refrigeration facilities are essential. But even at a low refrigeration temperature (4°C, or 37.4°F), some microbes can grow. For example, molds and psychrophilic bacteria, as well as some toxin-producing bacteria, are able to grow. Microbial growth and toxin production are related to time and temperature. Therefore, storage of food for any length of time is best done at freezing temperatures. When holding food for short periods, the microbial behavior at the following temperature ranges should be remembered:

0 to 7°C (32 to 44.6°F): relatively safe range for holding food
10 to 50°C (50 to 122°F): very dangerous range, for it supports the rapid growth of mesophilic bacteria
60 to 100°C (140 to 212°F): another relatively safe range for holding food, since it destroys vegetative cells of bacteria

Food obtained from an unsafe source means food that was poisonous from the start. Examples of this kind of food are poisonous species of mushrooms and shellfish that contain paralytic and neurotoxic poisons.

The contamination of food by microorganisms from equipment can be minimized if proper sanitary practices are employed in the cleaning of equipment. The control of microorganisms by use of chemical agents, such as those used in washing equipment, is covered in Chap. 20.

Good personal hygiene is of great importance in the control of foodborne diseases. It is especially significant in controlling food poisoning by microorganisms which have a fecal-oral route of entry into the host. It is obvious that food handlers should not have open-wound infections such as boils. But not so obvious is that food handlers may be carriers of pathogenic microorganisms. They constitute an important problem in the control of foodborne, waterborne (Chap. 29), and respiratory diseases (Chap. 27). Carriers have usually had the disease caused by the organisms they harbor, but in some cases the attack was so mild that it passed without notice. The organisms are thus no longer pathogenic for the carriers, but they are capable of causing disease in other susceptible people. Carriers are often unaware of their condition until an outbreak of disease is traced to them. The carrier state, fortunately, can be determined by the use of appropriate laboratory tests; carriers may then be treated with drugs to remove the pathogenic organisms.

Epidemiology of foodborne diseases

Local, state, and federal agencies which have responsibility for public health and food protection participate in foodborne disease surveillance. Physicians, hospital personnel, persons involved with food service or processing, and consumers report complaints of illness to health departments or regulatory agencies. State health officials will summarize their

Figure 28-1. Reported foodborne disease outbreaks in the United States in 1977. The four state health departments reporting the largest number of outbreaks were California, Washington, New York, and Pennsylvania. The number of outbreaks in these states probably reflects the interest of the respective state health departments in foodborne disease surveillance. The etiology was confirmed in only 157 outbreaks. Of these, the etiology was bacterial in 101 (64 percent), chemical in 37 (24 percent), parasitic in 15 (10 percent), and viral in 4 (3 percent). While outbreaks of known bacterial etiology accounted for only 64 percent of the total outbreaks, they accounted for 85 percent of the cases. The majority of cases of bacterial etiology were caused by *Salmonella* (42 percent) and *Staphylococcus* (22 percent). *(Center for Disease Control: "Foodborne and Waterborne Disease Outbreaks, Annual Summary 1977." Issued August 1979.)*

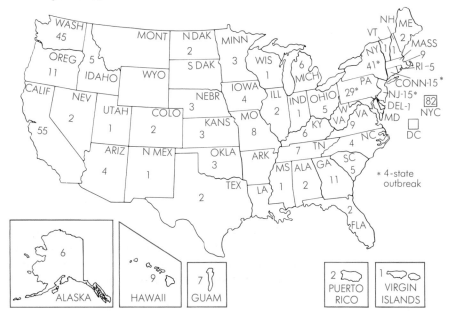

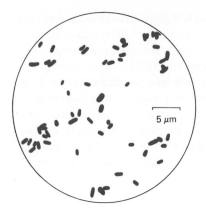 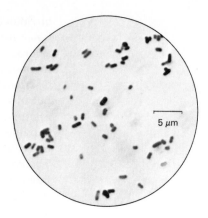

Figure 28-2. Morphology and distinguishing characteristics of *Samonella* sp. *(Erwin F. Lessel, illustrator; photomicrograph courtesy of Liliane Therrien and E. C. S. Chan, McGill University.)*

Salmonella sp.
  Gram-negative rods
  Occurs singly
  Not encapsulated
  Nonsporulating
  Usually motile; peritrichous
  Aerobic, facultatively anaerobic
  Pathogenic

findings and report them to the Center for Disease Control in Atlanta, Georgia.

In 1977, in the United States, there were 436 reported outbreaks of foodborne disease involving 9,896 cases. These outbreaks were reported from 43 states, Puerto Rico, Guam, and the Virgin Islands (shown in Fig. 28-1). No outbreaks were reported from seven states or the Canal Zone. Two outbreaks involved more than one state. There were eight deaths: five were due to eating food containing botulinal toxin; the other three deaths were due to the consumption of herbal tea.

The number of outbreaks reported represents only a small fraction of the total number that occur. The likelihood of an outbreak coming to the attention of health authorities varies considerably from one locale to another, depending largely upon consumer awareness and physician interest.

The epidemiology of some specific foodborne diseases will be discussed below.

Salmonellosis

Infection by bacteria of the genus *Salmonella* (hence the term *salmonellosis*) affects the gastrointestinal tract. The gastrointestinal tract includes the stomach, small intestine, and large intestine, or colon (see Fig. 22-4). Food outbreaks of salmonellosis are explosive in nature, and are associated with weddings, banquets, and other events in which group meals are served. The suddenness of the illness distinguishes it from other gastrointestinal diseases, such as bacillary and amebic dysentery.

Eight to forty-eight hours after eating food contaminated with salmonellas, there is a sudden onset of abdominal pain and loose, watery diarrhea, occasionally with mucus or blood. Nausea and vomiting are frequent; fever of 38 to 39°C (100.4 to 102.2°F) is common. These symptoms are linked in some way to the heat-stable endotoxins of the salmonellas. The symptoms usually subside within 2 to 5 days, and recovery is uneventful.

Biology of salmonellas causing foodborne infection

Several species of *Salmonella* are capable of causing food infection. These include *Salmonella enteritidis* var. *typhimurium* and other varieties, and *S. choleraesuis*. These bacteria are gram-negative, motile, nonsporeforming rods (Fig. 28-2). They ferment glucose, but not lactose or sucrose. Almost all serotypes produce gas when they ferment sugar, the exception being *S. typhi*. (See Table 28-1.)

The salmonellas may be classified by biochemical reactions into three species *S. typhi*, *S. choleraesuis*, and *S. enteritidis* (Table 28-1). The species are further subdivided into serotypes, which are identified by highly specific O (somatic) and H (flagellar) antigens. This method of salmonella categorization is called the Kauffman-White scheme. O antigens are numbered 1 to 65. There are two phases of H antigens: 1 and 2. Phase 1 antigens are lettered a through z, z_1 through z_{59}; phase 2 antigens were originally numbered, but because of cross reactions they now include many phase 1 designations. In addition to these antigens, *S. typhi* also has a Vi (capsular) antigen. One example of an antigenic formula is written as follows:

6,7:r:1,7 representing O antigens 6,7:phase 1 H antigen r:phase 2 H antigens 1,7.

The species *S. typhi* and *S. choleraesuis* have only one serotype each. *Salmonella enteritidis* has over 1,800 serotypes. Each of the large number of serotypes in *S. enteritidis* is designated by a name following the postscript "var." (variety) after the proper name, for example, *Salmonella enteritidis* var. *typhimurium*. Many microbiologists, however, still use the variety name as a specific epithet, for example, *S. typhimurium*.

Nature of pathogenicity in salmonellosis

Large numbers of viable microorganisms must be ingested to produce clinically apparent disease, because many cells may be rapidly eliminated

Table 28-1. Biochemical reactions of *Salmonella* species

TEST OR SUBSTRATE	S. typhi	S. enteritidis	S. choleraesuis
H₂S production	+	+	V
Nitrate reduction	+	+	+
Indole production	−	−	−
Gelatin liquefaction	−	−	−
Lactose	−	−	−
Sucrose	−	−	−
Glucose	A	AG	AG
Maltose	A	AG	AG
Mannitol	A	AG	AG
Dulcitol	−	V	V

V = variable; A = acid; G = gas.

from the gastrointestinal tract. Multiplication of ingested microorganisms in the intestinal tract gives rise to symptoms of gastroenteritis. The intestinal irritation and inflammation are produced by a true infection deep in the mucous membrane.

Generally, salmonellosis is treated by supportive measures without the use of antimicrobials.

Laboratory diagnosis of
salmonellosis

Definitive laboratory diagnosis of the disease depends on isolation of the bacteria from the feces. (Such bacteria should be the same as those isolated from the suspected food.) The use of selective or differential media, such as MacConkey's agar medium, is routine procedure. The identification of the microbes is then carried out by biochemical and serological methods. A rising antibody titer to O antigen within 1 or 2 weeks after infection is also of diagnostic value.

Epidemiology of
salmonella infections

Humans are infected by salmonellas almost solely by the consumption of contaminated food or drink. Foods commonly responsible include cream-containing pastries, ground meats, sausages, poultry, commercially prepared beef roasts, and eggs. Although carriers and sick people may contaminate food or drink, the greatest source of salmonellosis is the reservoir of salmonellas in lower animals (Fig. 28-3). Many species of *Salmonella* occur naturally in chickens, turkeys, ducks, rodents, cats, dogs, turtles, and many other animals. Domestic fowl are often the source responsible for infection of humans (Fig. 28-4).

The incidence of salmonellosis varies with the season. The greatest number of isolates are reported from July through October; the fewest are reported from December through May (Fig. 28-5). This pattern is probably related to the opportunity given the microbes to multiply in food because of the warm temperatures prevailing from July through October.

Prevention of salmo-
nellosis

Since most cases of salmonellosis result from the ingestion of contaminated food, the best preventive measures may be summarized as follows:

1 Proper cooking of foods from animal sources

Figure 28-3. Cycles of salmonella contamination.

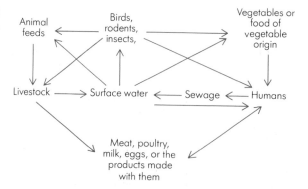

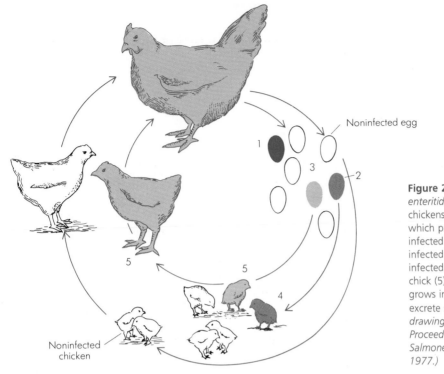

Noninfected egg

Figure 28-4. Cycle of *Salmonella enteritidis* var. *typhimurium* infection in chickens. (1) Badly infected egg (addled) which produces no chick; (2) moderately infected egg which gives rise to an infected chick (4) which dies; (3) lightly infected egg from which an infected chick (5) is hatched, which survives and grows into a bird which may continue to excrete salmonellas. *(Adapted from a drawing by H. E. Marthedal in the Proceedings of the Int. Symp. on Salmonella and Prospects for Control, 1977.)*

Noninfected chicken

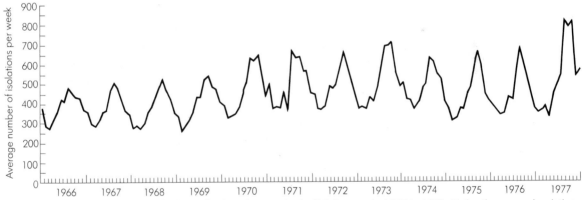

Figure 28-5. Reported isolations of salmonellas from humans in the U.S. by month, 1966 to 1977. Notice the seasonal variation: the greatest number of isolates are reported from July through August, the fewest from December through May. *(Center for Disease Control: "Reported Morbidity and Mortality in the United States," Annual Summary 1977. Issued September 1978.)*

2 Suitable refrigeration temperatures for holding food
3 Protection of food from contamination by rodents, flies, and other animals
4 Periodic inspection of food handlers
5 Proper food production and processing methods
6 Good personal sanitary and hygienic practices

Once a case of salmonella food infection is discovered, it should be reported to public health authorities. Suitable measures can then be implemented to protect the public from an outbreak of food poisoning.

There is no effective immunization against infection by *Salmonella* species (except S. *typhi*, which causes typhoid fever, as discussed in Chap. 29).

Staphylococcal food poisoning

A common food poisoning is caused by the ingestion of a toxin produced by strains of toxinogenic *Staphylococcus aureus* growing in contaminated food. The staphylococci are organisms found normally in various parts of the human body, including the nose, throat, and skin, and hence can enter food easily. They may come from food handlers who are carriers or have pyogenic infections. It is the most common type of food poisoning, and, fortunately, the illness is of short duration (24 to 48 h); complete recovery occurs in almost all cases. In very young children and debilitated persons, shock and death may result, although rarely, from dehydration. It is also usually not recognized as staphylococcal food poisoning unless many people are afflicted at the same time.

Symptoms develop shortly after ingestion of the contaminated food. The amount of enterotoxin consumed determines the time of onset of symptoms, as well as their severity. Generally, symptoms of nausea, headache, vomiting, and diarrhea develop 2 to 6 h after ingestion.

Biology of staphylococci causing food poisoning

Only certain strains of S. *aureus* (Fig. 28-6) produce the enterotoxin. Most of these strains are coagulase-positive, that is, have the ability to coagulate citrated or oxalated blood plasma. There are at least five antigenically distinct types of enterotoxin, designated A, B, C, D, and E. Types A and B are the most common.

Figure 28-6. Morphology and distinguishing characteristics of *Staphylococcus aureus.* (Erwin F. Lessel, illustrator; photomicrograph courtesy of Lilian Therrien and E. C. S. Chan, McGill University.)

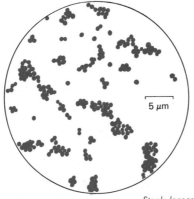 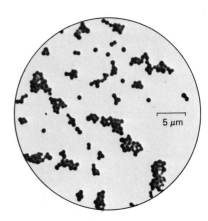

Staphylococcus aureus
Gram-positive cocci
Occurs singly, in pairs, and
  in clusters
Not encapsulated
Nonsporulating
Nonmotile
Aerobic, facultatively anaerobic
Normally coagulase-positive
Pathogenic

Nature of pathogenicity in staphylococcal food poisoning

The enterotoxin is heat-stable; it is unaffected by boiling for 30 min. Eight to ten hours of exposure of contaminated food to room temperature is sufficient to produce adequate amounts of toxin to cause food intoxication. Even subsequent refrigeration of this food for many months will not destroy the toxin. Recooking this food will also not destroy its toxin content.

No antibiotics are indicated for treatment of staphylococcal food poisoning. In cases with severe dehydration, however, administration of intravenous fluids is recommended.

Laboratory diagnosis of staphylococcal food poisoning

Diagnosis can be confirmed in the laboratory by examining a gram-stained smear of the suspected food under the microscope for gram-positive cocci occurring in abundance. The food can also be cultured for staphylococci. Hemolytic coagulase-positive staphylococci can be isolated, and the filtrates from cultures then fed to volunteers or experimental animals. Newer methods of testing for the presence of enterotoxin are based on serological reactions, such as gel-diffusion and fluorescent-antibody techniques.

Epidemiology of staphylococcal food poisoning

Humans are the most important source of staphylococci which produce enterotoxin. In outbreaks of food poisoning by staphylococci, it is usually possible to show that the strain of staphylococcus in the contaminated food and that on the hands of the food handlers are the same.

Foods that have the ability to support good growth of staphylococci are responsible for the disease. Those commonly associated with the disease include custard-filled bakery goods, processed meats such as ham, and potato salads. Unfortunately, foods with sufficient enterotoxin to cause disease are usually normal in appearance, odor, and taste.

Prevention of staphylococcal food poisoning

The best preventive measure is refrigeration (below 6 to 7°C, or 42.8 to 44.6°F) of all perishable food. Food handlers should not have pyogenic skin lesions or be carriers of toxinogenic staphylococci. Reheated foods should not be allowed to stand for several hours at room temperature before serving. Most staphylococcal food poisoning is a consequence of the mishandling of food in service establishments and the home.

Botulism

Botulism is a disease caused by bacterial food poisoning or intoxication. The causative organism is *Clostridium botulinum*, which elaborates a heat-labile neurotoxin. The disease is acquired by the ingestion of preformed botulinal toxin in improperly preserved food, such as found in some home-canned food. But botulism can also be caused by contamination of a wound with *C. botulinum*; the organism produces toxin while growing in dead tissue.

Symptoms of the disease usually begin about 12 to 48 h after ingestion of contaminated food. The symptoms include difficulty in speaking, dilated-fixed pupils, double vision, a dry mouth, nausea, vomiting, and inability to swallow. Paralysis can involve the urinary bladder and all voluntary

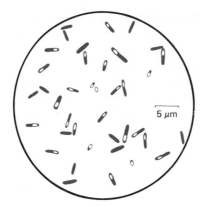

Clostridium botulinum

Gram-positive rods
Occurs singly, in pairs, or in chains
Not encapsulated
Spores are oval and subterminal
Motile; peritrichous
Anaerobic
Produces powerful exotoxin; causes
botulism

5 µm

Figure 28-7. Morphology and distinguishing characteristics of *Clostridium botulinum. (Erwin F. Lessel, illustrator.)*

muscles. Death, caused by respiratory or cardiac failure, is possible a few days after onset of symptoms. Convalescence is a slow process, but there are no permanent aftereffects.

Biology of *Clostridium botulinum*

Clostridium botulinum is an anaerobic gram-positive bacillus which produces heat-resistant spores. It grows well on ordinary culture media. Most abundant growth occurs at 25°C (77°F), but it grows well from 20 to 35°C (68 to 95°F). The organisms, which occur singly or sometimes in pairs or chains, are 0.5 to 0.8 µm by 3 to 8 µm with parallel sides and rounded ends (Fig. 28-7). Spores are oval and subterminal, and only slightly swell and distend the shape of the cells. *Clostridium botulinum* is motile by means of peritrichous flagella; it does not form capsules.

Seven types of *C. botulinum* have been recognized by means of antigenic differences in the toxins they produce. Types A, B, E, and, rarely, F, cause disease in humans. Types C and D cause the disease in birds and

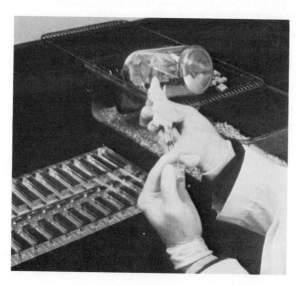

Figure 28-8. Laboratory confirmation of botulism. Mice are injected with an extract from food believed to be contaminated with botulism for detection of botulinal toxin. *(Courtesy of Center for Disease Control, Atlanta, Ga.)*

nonhuman mammals. Type G is not known to cause disease. The toxins are highly type-specific; antitoxins neutralize only their specific toxins.

Nature of pathogenicity in botulism

Botulinal toxin is the most potent poison known. For example, in mice the lethal dose of type A toxin is estimated at 0.000000033 mg; this means that 1 g of the toxin can kill 33 billion mice. The poison affects the nerves, causing a paralysis of the pharynx and diaphragm. When respiratory paralysis develops, a tracheotomy may be performed and artificial respirators employed. The mode of action of the toxin is the inhibition of acetylcholine release by nerve fibers as a nerve impulse passes along a peripheral nerve; this is a consequence of the toxin fixing to the efferent nerve endings.

Since antitoxin cannot neutralize the toxin once it is fixed, treatment by antitoxin should begin as soon as possible once botulism is suspected. Generally, a polyvalent antitoxin consisting of types A, B, and E is employed. Persons suspected of having eaten food containing botulinal toxin should be induced to vomit, or they should have their stomachs irrigated (gastric lavage). An enema also may be given. Patients with neurologic involvement should have the circulating toxin neutralized by the administration of the trivalent (ABE) antitoxin. In cases of wound botulism, there should be thorough debridement, drainage, and irrigation of the wound. Antitoxin should be administered before debridement.

Laboratory diagnosis of botulism

The primary method of confirming a diagnosis of botulism in the laboratory is the demonstration of botulinal toxin in the patient's serum or feces or in the ingested food. Intraperitoneal injection of the serum or water extracts of feces or food into a mouse may result in death of the animal, since mice are very sensitive to the toxin (Fig. 28-8). Also, specimens of the feces and the food should be cultured for isolation of the organism.

Epidemiology of botulism

Clostridium botulinum is widely distributed in terrestrial and marine environments. When the spores contaminate various processed foods or an anaerobic site in a wound, they may germinate into vegetative cells and produce toxin. Recently, true infection of the intestinal tracts of infants have been reported; in this new disease, called "infant botulism," the toxin is formed in the gut of infants, causing weakness, constipation, and paralysis. Such an infection may be due to feeding infants honey containing *C. botulinum* spores.

The Latin word *botulus* means "sausage." The disease derives its name because for years uncooked sausages were associated with this disease. Today, the foods implicated are generally those that have undergone some treatment intended for preservation of the product, such as canning, pickling, or smoking, but which failed to destroy the spores of the bacterium. Some examples are home-canned fruits and vegetables, smoked fish, and spiced meat and fish.

Fortunately, the toxin can be completely inactivated by heating at 100°C (212°F) for 10 min, or 80°C (176°F) for 30 min. This accounts for the relatively low incidence of the disease. In the United States, in 1976, there

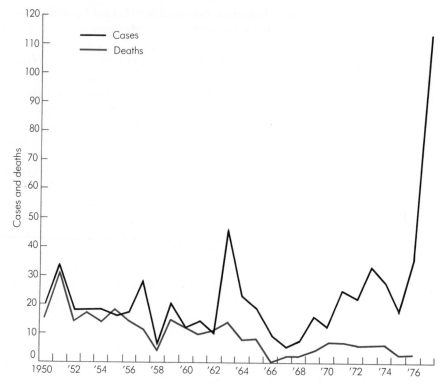

Figure 28-9. Reported botulism cases and deaths in the United States by year, 1950 to 1976. Incidence of this disease is usually low. However in 1976, there were 23 outbreaks of the disease, involving more than 100 cases. *(Center for Disease Control: "Reported Morbidity and Mortality in the United States," Annual Summary 1977. Issued September 1978.)*

were 23 outbreaks of the disease involving many cases, there were five deaths (Fig. 28-9). Botulism remains the leading cause of death by food poisoning.

Prevention of botulism

Careful quality control measures by the food-processing industry have greatly diminished the occurrence of the disease in commercial foods. The greatest danger comes from the home canners who do not employ proper methods for sterilizing containers and food. Unfortunately, food containing toxin does not necessarily appear or smell different from uncontaminated food. In any case, the best prevention is adequate cooking of any preserved food before serving.

Perfringens food poisoning

Clostridium perfringens is commonly found in nature, for example, in raw meats, and the feces of animals and often healthy persons. It is also a major cause of food poisoning. The disease is due to the ingestion of food that was contaminated with this organism and then held at a temperature that encouraged spore germination and vegetative growth. The food must provide an anaerobic environment for the organism; thick gravy is such a food.

Symptoms usually occur 8 to 24 h after ingestion of the contaminated food. Crampy abdominal pain and diarrhea are the major symptoms. The illness is of short duration and is over in less than 24 h.

Biology of *Clostridium perfringens*

This bacterial species is divided into six types, A through F, on the basis of the antigenically distinct toxins produced by each strain. Type A is the strain that causes perfringens food poisoning. The organism is a gram-positive, sporeforming, anaerobic rod (Fig. 28-10).

Nature of pathogenicity in perfringens poisoning

Food poisoning is usually caused by ingestion of type A *C. perfringens*, although other types can also cause disease. The intoxication is produced by the vegetative cells as they sporulate in the lumen of the gut. Treatment is symptomatic, and no specific therapy is indicated.

Laboratory diagnosis of perfringens poisoning

Clinical and epidemiological findings are supported by laboratory diagnosis when large numbers of *C. perfringens* are found in anaerobic cultures of contaminated food. Successful isolation of the same type of organisms from the suspected food and from the feces of patients provides further supporting evidence.

Epidemiology of perfringens poisoning

In the United States in 1976, even though there were only six outbreaks of food poisoning by *C. perfringens*, 509 cases were involved. Many cases probably went unreported because of the mild symptoms of the disease. These outbreaks have been associated with several types of food, especially cooked meat and poultry dishes.

Prevention of perfringens poisoning

The best preventive measure is the avoidance of long periods of holding foods at room temperature that have already been cooked. Spores survive the cooking process; at room temperatures they can germinate and grow as vegetative cells. When these cells are ingested and sporulate in the gut, they produce enterotoxic exotoxin, resulting in disease.

Figure 28-10. Morphology and distinguishing characteristics of *Clostridium perfringens*. (Erwin F. Lessel, illustrator; photomicrograph courtesy of Liliane Therrien and E. C. S. Chan, McGill University.)

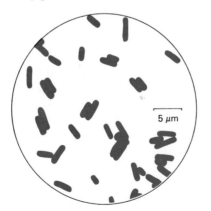 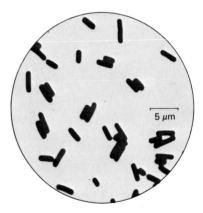

Clostridium perfringens
Gram-positive rods
Occurs singly, in pairs, and
 in chains
Encapsulated
Spores are ovoid, central to eccentric
Nonmotile
Anaerobic
Produces exotoxin, causes gangrene

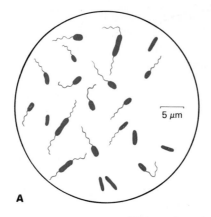

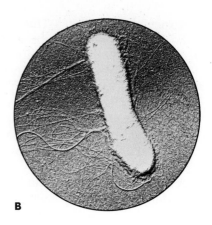

5 μm

A

B

Figure 28-11. (A) Morphology and distinguishing characteristics of *Vibrio parahaemolyticus.* (Erwin F. Lessel, illustrator.) (B) A cell of *V. parahaemolyticus* as seen under the transmission electron microscope. (×10,886) (Courtesy of A. G. Clark, University of Toronto.)

Vibrio parahaemolyticus
  Gram-negative curved or straight rods
  Occurs singly
  Not encapsulated
  Nonsporulating
  Motile; single polar flagellum
  Aerobic, facultatively anaerobic
  Requires salt
  Hemolytic
  Pathogenic, causing gastroenteritis

Foodborne diseases caused by other microorganisms

Vibrio parahaemolyticus food infection

Vibrio parahaemolyticus is a gram-negative facultative anaerobe and a *halophile* (salt-loving). The shape and characteristics of this vibrio bacterium are shown in Fig. 28-11. It is the cause of gastroenteritis after the ingestion of seafood. In Japan, contaminated raw fish is the most important source; in the United States, contaminated cooked seafood is the common source.

The incubation period of this food poisoning is 2 to 48 h. Main symptoms are abdominal pain, diarrhea, nausea, and vomiting. Mild fever and chills are often present. The disease is over in 2 to 5 days. This illness is an infectious one and not caused by toxin.

Laboratory diagnosis is directed toward the isolation of *V. parahaemolyticus* from the feces or vomitus of patients and from the suspected food. Generally, a good preventive measure is proper refrigeration and cooking of seafood.

Other bacteria

Several other bacteria have been implicated as etiologic agents in outbreaks of food poisoning. These include *Bacillus cereus*, certain strains of *Escherichia coli*, and *Proteus* spp. However, when the laboratory isolate is a member of the normal intestinal microbiota, it is extremely difficult to prove that the organism plays a causative role.

Fungi

A fungal toxin called *aflatoxin* is produced by common molds of the genus *Aspergillus* (Fig. 28-12). This toxin can cause acute poisoning in animals,

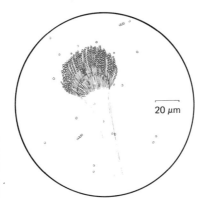

Aspergillus sp.

Filamentous fungus
Septate, branching mycelia
Conidia borne on sterigmata of
conidiophores
Widespread in nature
Some are industrially important
May be pathogenic, causing
aspergillosis

20 µm

Figure 28-12. Morphology and distinguishing characteristics of *Aspergillus* sp. *(Erwin F. Lessel, . illustrator.)*

including humans, when foods contaminated with the molds are ingested. The toxicity of the poison can lead to liver damage as well as the induction of tumors. It is now known that aflatoxin is a group of related compounds. Their precise mode of action on the structure and metabolism of cells remains to be defined. Many other mycotoxins have been discovered. It is important that human foodstuffs, such as peanuts, grains, and the like, be properly dried and stored to prevent growth of the fungi.

Viruses

Viruses also cause infectious food poisoning giving rise to gastroenteritis. Such viruses include echoviruses, coxsackieviruses, poliomyelitis, or hepatitis viruses. Members of the reoviruses, adenoviruses, and others have also

Figure 28-13. Some viruses isolated from cases of gastroenteritis. (A) Rotavirus. (B) Astrovirus. (C) Adenovirus. (D) Calicivirus. (E) "Minireovirus." (F) Picorna/Parvolike virus. All magnifications ×80,000. *(Courtesy of M. Petric and Maria T. Szymanski, The Hospital for Sick Children, Toronto.)*

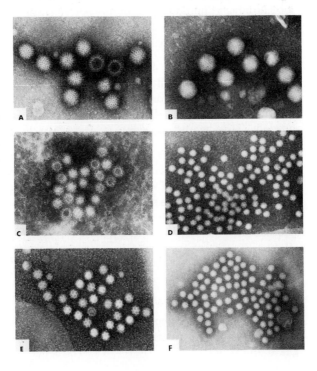

been implicated (see Fig. 28-13). The major symptom is an abrupt onset of diarrhea, which may be associated with nausea, vomiting, malaise, abdominal cramps, or fever. The term "intestinal flu" is often used to describe viral gastroenteritis, even though it has nothing to do with the influenza virus.

Laboratory diagnosis is rarely established, mainly because viral isolation and serological diagnosis are expensive and tedious.

Summary and outlook

Foodborne diseases are transmitted by the ingestion of food contaminated with sufficiently high numbers of certain kinds of microorganisms to constitute an infective dose. There are two mechanisms involved in food poisoning by microorganisms, namely, foodborne infection and food intoxication.

Salmonellosis. The control of the disease in humans requires control of the reservoir of salmonellas in domestic animals. The elimination of species of *Salmonella* in domestic livestock in turn requires the elimination of such organisms from feed. Since salmonellas are ubiquitous, and salmonellosis is well established in some nonhuman hosts, the problem is compounded by the genetic variability of *Salmonella* spp.; thus, an effective vaccine is not available.

Staphylococcal food poisoning. Since enterotoxin is not formed in significant amounts at temperatures below 6.7°C (44.1°F), the best preventive measure against this disease is refrigeration of all perishable foods. Also, food handlers should be free of any active staphylococcal lesions. Thus, the outlook for control of this disease depends on the education of those involved in the preparation and serving of food.

Botulism. This disease is entirely preventable. Home canners especially must be impressed with the importance of maintaining proper techniques in their food preservation process.

A more rapid means of identifying botulinal toxin in serum is very much needed. This would prevent unnecessary administration of antitoxin therapy.

Perfringens food poisoning. This food poisoning is best prevented by avoiding long periods of warming or cooling of foods that have already been cooked. Foods should be stored at temperatures above 60°C (140°F) or below 5°C (41°F). If this practice is adhered to, the incidence of this disease can be decreased.

Other bacteria and some fungi and viruses are known to cause foodborne diseases.

Key terms

aflatoxin
botulism
gastroenteritis

halophile
intoxication

salmonellosis
trichinosis

Questions

1 How is the mechanism of food poisoning caused by staphylococci different from that caused by salmonellas?

2 Explain how different temperature ranges affect the holding of food.

3 Discuss the relationship between the taxonomy of the salmonellas and the Kauffman-White scheme of salmonella categorization.

4 A sample of suspected food is inoculated onto appropriate culture media for the isolation of *Staphylococcus aureus*. No organisms were isolated. Can this food have caused staphylococcal food poisoning? Explain.

5 Compare and contrast the source of food contamination in salmonella and staphylococcal food poisoning.

6 Set up a table comparing the preventive measures for the four food-poisoning diseases discussed in detail in this chapter. Comment on their common and special features.

7 Botulinal toxin is a very potent poison. Why do you think there have not been more cases and deaths from this disease?

8 What simple procedure, that can be carried out in the home, will assure freedom from botulinal toxin?

9 In what way is the physiology of *Vibrio parahaemolyticus* related to the foods it contaminates?

WATERBORNE INFECTIONS

Waterborne infections, like foodborne diseases, are caused by microorganisms that enter and leave the host via the oral-intestinal route. Such infections are also called *enteric infections* because the intestines are infected.

A waterborne disease occurs because contaminated water has been consumed. Actually, the source of the infection is not the water but the fecal material of human (or other animal) origin that has contaminated it. This fecal matter contains the enteric pathogens if it comes from diseased individuals or carriers. (If the water containing pathogens contaminates food, such infections can also be foodborne.)

Transmission of waterborne disease organisms can be more direct than this. For example, transfer of organisms may occur from the excreta of infected persons to the mouths of other persons by hands or by fomites. These fomites could also have been contaminated by insects, such as the common housefly, that had been on excreta.

However, it is by means of water transmission that outbreaks of enteric infections involving many people occur.

494

Control of water-borne infections

The control of waterborne infections depends primarily upon preventing the contamination of water supplies. This can be achieved by means of such sanitary measures as purification of drinking water supplies and proper disposal of human wastes. In other words, the following cycle must be disrupted to effect proper control of waterborne infections:

Human feces→Water→Consumption by humans (who may become ill)→Human feces

It is mainly through improved sanitation and strict control of water supplies that the incidence of these diseases has decreased markedly in Canada and the United States in the past few decades.

The main control measures are summarized below. Chapters 35 and 36 discuss these measures in greater detail.

1 *Public health control of drinking water.*
2 *Sanitary disposal of sewage.*
3 *Pasteurization of milk.*
4 *Exclusion of human carriers from the preparation and handling of food.* The same diseases transmitted by water can be transmitted by food contaminated by carriers. The most notorious of such carriers was a young person named Mary Mallon, better known as Typhoid Mary, a young cook living in the state of New York at the turn of the century. She was responsible for at least 10 outbreaks of typhoid fever involving 51 cases of the disease and three deaths.

Epidemiology of waterborne infections

In 1977, 34 waterborne disease outbreaks involving 3,860 cases were reported to the Center for Disease Control in Atlanta. This was a decrease of 24 percent from 1976. A *waterborne disease outbreak* is defined as an incident in which two or more persons experience similar illness after consumption of water, with the epidemiological evidence implicating the water as the source of illness. As with food disease, the number of outbreaks reported represents only a small fraction of the total number that occur because of similar reporting inadequacies.

Figure 29-1 shows the geographic distribution of outbreaks by state. Eighteen states reported at least one outbreak. Of the 32 outbreaks reported, 16 (50 percent) were of unknown etiology, 2 were due to chemicals (such as copper), 4 to the protozoan, *Giardia lamblia* (giardiasis causes diarrhea, nausea, mild abdominal cramps, and flatulence), 2 to *Salmonella*, 4 to *Shigella* (a bacterial genus), 1 to *Campylobacter fetus* subsp. *jejuni* (a bacterium), and 3 to a parvoviruslike agent. Most outbreaks involved semipublic and municipal water systems.

Typhoid fever

Typhoid fever is an acute infectious disease caused by the bacterium *Salmonella typhi*. The disease is unique to humans. The incubation period is generally from 10 to 14 days. Early symptoms include fever, abdominal

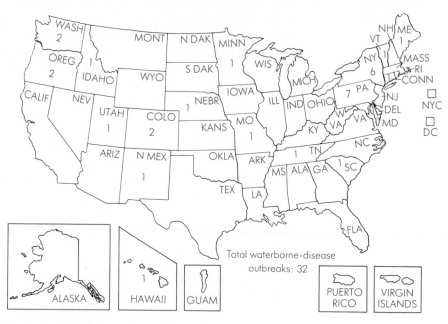

Figure 29-1. Reported waterborne disease outbreaks in the United States in 1978. There were 32 outbreaks of acute waterborne disease involving 11,435 cases reported to the Center for Disease Control. Eighteen states reported at least one outbreak; Pennsylvania reported the largest number, 7 (21 percent). In 16 (50 percent) outbreaks, the etiologic agent was known: *Shigella* (4 outbreaks), *Giardia lamblia* (4), parvovirus-like agents (3), *Salmonella* (2), a chemical (2), and *Campylobacter, fetus* ssp. *jejuni* (1). *(Center for Disease Control, "Morbidity and Mortality Weekly Report," vol. 29, no. 4, Feb. 1, 1980.)*

distention, constipation, headache, apathy, rash, malaise, loss of appetite, nausea, and vomiting. Diarrhea usually occurs during the second week of the infection. Blood may be present in the stool. The bacterium can be found in stools throughout the period of illness, as well as during convalescence.

The course of illness is usually severe. If treatment is not given early, the illness may last several weeks and the patient may die.

Biology of *Salmonella typhi*

Salmonella typhi is a gram-negative, motile bacillus; it was discussed with the genus *Salmonella* in Chap. 28. As indicated there, *S. typhi* has a capsular *Vi antigen* in addition to the somatic (O) and flagellar (H) antigens which are used for serological identification.

Salmonella typhi can also be subdivided into more than 80 definite and stable varieties by means of bacteriophage typing. This property becomes a valuable tool for tracing sources of infection by microbiologists who study epidemics.

Nature of pathogenicity of typhoid fever

After entering the gastrointestinal tract, *S. typhi* organisms rapidly gain access to intracellular sites. For example, they remain alive and multiply within phagocytes, which they may then kill and escape from. Thus this bacterium is an example of a facultative intracellular parasite. The incubation period of typhoid fever may correspond to this phase of invasion from the intestines to the intracellular sites. Clinical symptoms become evident when the bacteria enter the bloodstream from the intracellular sites. Infection of the biliary tract occurs, and multiplication of the bacteria in bile leads to seeding of the intestinal tract with millions of bacteria. This is the reason for the presence of the bacilli in stools.

Salmonella typhi can also gain access to various tissues and organs via the bloodstream. These loci, such as the gall bladder, bone marrow, and spleen, can serve as future sources of reinfection. This accounts for the typhoid fever relapse in many cases.

The Vi antigen apparently interferes with serum bactericidal activity and phagocytosis. Therefore, it appears to be a virulence factor. The pathogenicity of typhoid fever resides in the endotoxins in the organisms.

Chloramphenicol is the antimicrobial agent of choice. But it is not effective in the treatment of chronic carriers; ampicillin is the preferred antibiotic for termination of the carrier state.

Laboratory diagnosis of typhoid fever

Salmonella typhi can be isolated from the blood of about 90 percent of patients during the first week of the disease and from 50 percent of patients by the end of the third week.

The bacterium can be isolated from the feces at any stage of illness, but the period from the third to the fifth week of illness yields the best results for microbiological diagnosis, even with treatment (because many bacilli grow intracellularly). Two or three months after onset of disease, 5 to 10 percent of patients continue to excrete the bacilli; 3 percent continue to excrete S. typhi for periods in excess of 1 year. Those that continue to excrete the bacilli after this time are considered chronic carriers of typhoid fever.

Laboratory diagnosis of typhoid fever can also be performed by specific agglutination of S. typhi with the patient's blood serum. This is known as the Widal reaction.

Epidemiology of typhoid fever

Typhoid fever occurs in all parts of the world, but infrequently where good sanitation—proper disposal of biological wastes and purification of water—is practiced. It is still a disease of major importance in areas of the world that do not have high standards of sanitation. In the United States, the incidence of typhoid fever has decreased progressively since 1900 (except for a transient rise in 1973), as shown in Fig. 29-2. In recent years fewer than 400 cases have been reported annually. Approximately 75 percent of typhoid fever cases occur in persons less than 30 years old, probably because of acquired immunity in the older persons.

The ultimate source of infection with S. typhi is a patient with the disease or a carrier of the organism since typhoid fever is specifically a disease of humans. Water or food contaminated directly or indirectly with human feces is the usual route of infection. This danger is enhanced by the fact that the typhoid bacilli can survive for weeks in water, dust, ice, and even dried sewage.

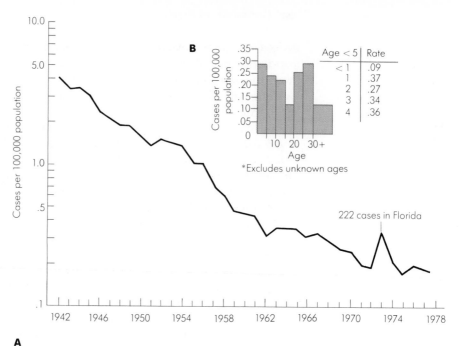

Figure 29-2. (A) Reported cases of typhoid fever per 100,000 population by year in the United States, 1942 to 1977. (B) Reported cases of typhoid fever per 100,000 population by age group in the United States, 1977. *(Center for Disease Control: "Reported Morbidity and Mortality in the United States," Annual Summary 1977. Issued September 1978.)*

The largest single outbreak of typhoid fever in the United States occurred in Florida in 1973 (Fig. 29-2). There were 222 cases among migrant workers who consumed water from a contaminated supply.

Prevention of typhoid fever

At the community level, proper sanitation is the best prevention against typhoid fever. Carriers of the disease must be identified and prevented from processing and handling food.

For individuals, typhoid vaccine is effective in decreasing the incidence of disease. Several different preparations of typhoid vaccine have been shown to protect 70 to 90 percent of recipients, depending in part on the degree of their subsequent exposure.

Routine typhoid immunization is no longer recommended for persons in the United States. Immunization is only indicated when a person has come in contact with a known typhoid fever case in the community, or if a person plans to travel to an area where typhoid fever is endemic.

Salmonellas other than *S. typhi* can cause waterborne disease on occasion, as in a 1966 outbreak in Riverside, California, which had approximately 20,000 cases.

Shigellosis

Shigellosis, or bacillary dysentery, is an acute inflammatory reaction of the intestinal tract caused by bacteria of the genus *Shigella*. It is distinct from amebic and viral dysentery. *Dysentery* is a clinical condition with intestinal inflammation, diarrhea, and watery stools containing blood, mucus, and pus.

The incubation period for the disease ranges from 1 to 7 days; most commonly it is about 4 days. Initial symptoms are fever and cramping abdominal pain. Diarrhea usually occurs after 48 h, with dysentery following about 2 days later. In severe cases, the stools are primarily composed of blood, mucus, and pus. Fluid and *electrolyte* (mineral or salt) loss may be quite significant in very young and very old patients.

Biology of *Shigella* spp.

The genus *Shigella* was named for the Japanese bacteriologist Kiyoshi Shiga, who discovered the dysentery bacillus in 1897. *Shigella* organisms are nonmotile, short, gram-negative rods (Fig. 29-3). Their optimum growth occurs at 37°C (98.6°F) under aerobic conditions. Morphologically they cannot be distinguished from the salmonellas, but they can be differentiated from them on the basis of fermentation reactions and serological tests. Unlike the salmonellas, the shigellae ferment several carbohydrates, with the notable exception of lactose, to produce acid without gas.

Shigellae are essentially restricted to humans as natural hosts. Although shigellae can infect primates, humans are their natural reservoir as well as the main mode of dissemination.

Shigellae have been divided into four major serological groups on the basis of their cell-wall antigens. Group A consists of S. *dysenteriae*, a species seldom encountered in the United States. Group B has S. *flexneri*, a species commonly isolated in the United States. Group C strains are rarely isolated in the United States; the representative species is S. *boydii*. Group D comprises S. *sonnei*; it is the most common cause of shigellosis in the United States (see Fig. 29-4). Except for group D, which has a single serotype, each of the other groups has antigenic subtypes.

The biochemical characteristics for the differentiation of these species are shown in Table 29-1.

Figure 29-3. Morphology and distinguishing characteristics of *Shigella* sp. *(Erwin F. Lessel, illustrator; photomicrograph courtesy of Liliane Therrien and E. C. S. Chan. McGill University.)*

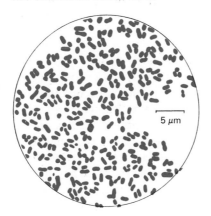 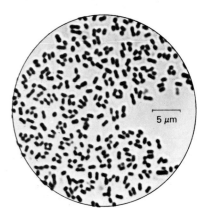

Shigella sp.
  Gram-negative rods
  Occurs singly
  May be encapsulated
  Nonsporulating
  Nonmotile
  Aerobic, facultatively anaerobic
  Pathogenic, causing dysentery

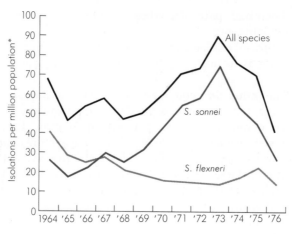

*Includes only persons in states and territories with participating reporting centers.

Figure 29-4. Reported isolations of *Shigella* species by year, United States, 1964 to 1976. *S. sonnei* is the most common cause of shigellosis. *S. flexneri* is also commonly isolated. *S. dysenteriae* and *S. boydii* each accounts for less than 2 percent of the total shigella isolates each year. (*Center for Disease Control: Shigella Surveillance Report No. 39, Annual Summary 1976. Issued October 1977.*)

Table 29-1. Biochemical reactions of *Shigella* species

ORGANISM	PRODUCTION OF HYDROGEN SULFIDE	LIQUEFACTION OF GELATIN	REDUCTION OF NITRATES	PRODUCTION OF INDOLE	FERMENTATION OF CARBOHYDRATES				
					Glucose	Lactose	Sucrose	Mannitol	Dulcitol
Shigella dysenteriae	−	−	+	−	Acid	−	−	−	−
Shigella flexneri	−	−	+	+	Acid	−	−	Acid	−
Shigella boydii	−	−	+	Variable	Acid	−	−	Acid	Variable
Shigella sonnei	−	−	+	−	Acid	−	Acid	Acid	−

Nature of pathogenicity of shigellosis

Shigellae must penetrate the cells of the epithelial lining of the large intestine to induce dysentery. Following intracellular penetration, multiplication of the bacteria occurs. Shigellae are far less invasive than the salmonellas; spread of the shigellae beyond the intestinal tract to cause disease in other organs is extremely rare. The pathogenicity factors in shigellosis are not well understood. They may include an endotoxin with biological activity. However, *S. dysenteriae* causes the most severe disease because it also produces an exotoxin that has neurotoxic as well as enterotoxic properties. (Thus some children with shigellosis may have convulsions.) This exotoxin is a soluble, heat-labile protein.

Specific antibiotic therapy should not be given unless the illness is severe.

Laboratory diagnosis of shigellosis

Blood and mucus in the feces of patients with diarrheal disease of sudden onset strongly indicates shigellosis. For definitive diagnosis, however, the isolation of *Shigella* spp. from the feces is essential.

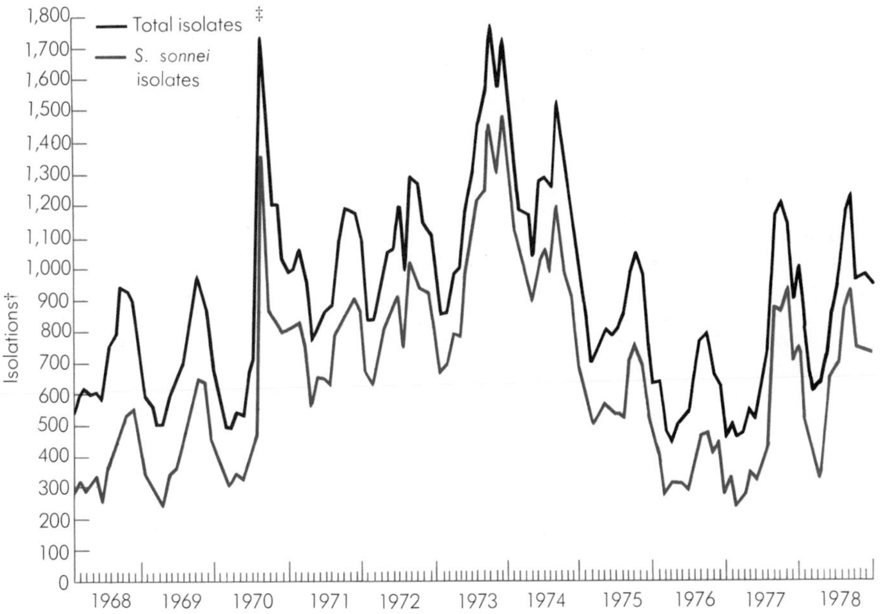

Figure 29-5. Reported isolations of shigellosis in the United States by month, 1968 to 1978. Note how the reported isolations peak in the fall and winter months. *(Center for Disease Control, "Morbidity and Mortality Weekly Report," vol. 28, no. 41, Oct. 19, 1979.)*

*No reports from California or the Virgin Islands after 1969.
†Adjusted to 4-week month.
‡Approximately 400 isolations in August 1970 due to common-source outbreak in Hawaii.

Successful isolation also depends on culturing on selective media (such as deoxycholate agar) as soon as possible after sample collection. This is because the shigellae do not remain viable for long outside the body in the presence of other bacteria. Final identification of the shigellae is based on biochemical and agglutination tests.

Epidemiology of shigellosis

Shigellae have a worldwide distribution. As indicated before, *S. sonnei* is predominant nationwide in the United States (Fig. 29-4). In the Orient and Central America, *S. dysenteriae* is the most common.

Figure 29-5 shows the seasonal distribution of reported isolations of shigellosis, with peaks in fall and winter.

Although all age groups are susceptible to shigella infection, children of 1 to 4 years old are most commonly infected. Fewer than 20 percent of all reported cases in the United States occur in adults (see Fig. 29-6).

Prevention of shigellosis

Over the past decade, as shown in Fig. 29-5, there has been no appreciable change in the number of reported cases. No significant improvement can be expected since there is no effective vaccine against shigellosis.

Since humans are the sole source of the pathogens, the best prevention is the interruption of the anal-oral route of transmission. This requires the sanitary disposal of sewage and the protection of food and water from contamination by carriers, patients, and flies.

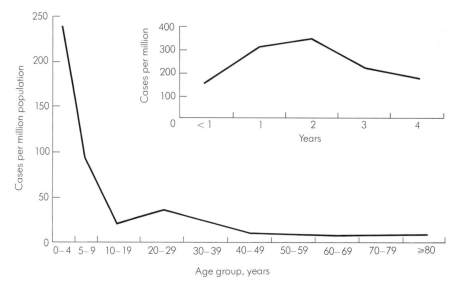

Figure 29-6. Reported cases of shigellosis per 1,000,000 population by age in the United States (excluding California), 1978. Note that children of 1 to 4 years old are most commonly infected. *(Center for Disease Control, "Morbidity and Mortality Weekly Report," vol. 28, no. 41, Oct. 19, 1979.)*

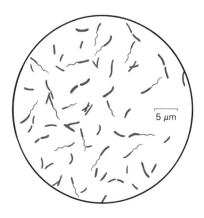

Vibrio cholerae

Gram-negative, straight or curved rods
Occurs singly and in spiral chains
Not encapsulated
Nonsporulating
Motile; single polar flagellum
Aerobic, facultatively anaerobic
Pathogenic, causing cholera

Figure 29-7. Morphology and distinguishing characteristics of *Vibrio cholerae. (Erwin F. Lessel, illustrator.)*

Cholera

Cholera is an acute disease caused by an enterotoxin produced by *Vibrio cholerae* that have colonized the small bowel. The symptoms include vomiting and profuse diarrheal (rice-water) stools resulting in severe dehydration, loss of electrolytes, and increased blood acidity. In severe cases, there is rapid loss of fluid and electrolytes from the gastrointestinal tract. This results in shock, *metabolic acidosis* (accumulation of acid metabolites in blood), and if untreated, death.

Biology of *Vibrio cholerae*

Vibrio cholerae is a short, slightly curved, gram-negative rod (Fig. 29-7). It is motile by means of a polar flagellum. It is readily seen in gram-stained smears of the watery excreta of cholera patients.

Vibrio cholerae occurs in two *biotypes*, or strains: the classical and the El Tor biotypes (the latter was so designated because the organism was isolated at the El Tor quarantine station of the Gulf of Suez in 1905). The

Table 29-2. Biochemical reactions of *Vibrio cholerae* biotypes

TEST	CLASSIC	EL TOR
Voges-Proskauer test for acetylmethylcarbinol	−	+
Indole production	+	+
Gelatin liquefaction	+	+
H₂S production	−	−
Glucose fermentation	+	+
Lactose fermentation	Slow	Slow
Sheep or goat erythrocyte hemolysis	−	+
Chicken erythrocyte hemagglutination	−	+

infections induced by both biotypes are indistinguishable from each other. These two pathogenic biotypes also may each be divided into two serological subtypes called Inaba and Ogawa, on the basis of their O antigen. This distinction is done by slide agglutination with type-specific antisera.

The biochemical characterization of the cholera vibrio biotypes is shown in Table 29-2.

Nature of pathogenicity of cholera

Cholera is a disease of antiquity and has been the cause of untold suffering and death. The incubation period varies from 1 to 3 days. All signs, symptoms, and metabolic derangements in cholera are due to *rapid* loss of fluid and electrolytes from the gut. With prompt fluid and electrolyte replacement, physiological recovery is rapid in spite of continuing diarrhea. (See Fig. 29-8.) Cholera is a self-limiting disease, provided that the patient does not die from dehydration or shock before recovery.

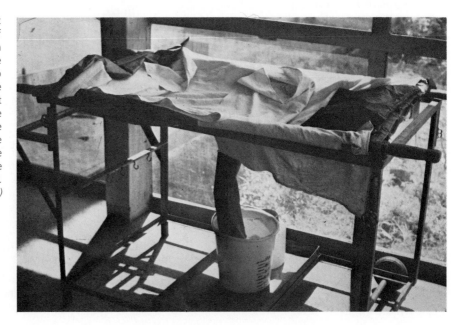

Figure 29-8. A cholera cot as used for the treatment of cholera patients in Bangladesh. It is a simple and yet effective device to collect diarrheal stools, while the patient is kept comfortable and clean. The loss of fluid can also be measured roughly using the collecting bucket so that the same fluid volume can be replaced. *(Courtesy of R. Oseasohn, McGill University.)*

The increased electrolyte secretion is caused by a protein exoenterotoxin (*choleragen*) that is elaborated by *V. cholerae*. The toxin stimulates adenyl cyclase activity in the gut mucosal cells. This causes an increase in intracellular cyclic adenosine monophosphate (cyclic AMP) that leads to the secretion of electrolytes into the lumen of the bowel.

Tetracycline dramatically reduces the duration and volume of diarrhea as well as shortening the recovery time.

Laboratory diagnosis of cholera

Several types of specimens can be used in the microbiological diagnosis of cholera in the laboratory. These include rectal swabs and stool material. These can be cultured directly onto selective and differential media where typical colonies may appear in 18 h. In addition, direct examination under the microscope of stained smears of the stool specimens is employed.

Definitive identification requires agglutination with group- and type-specific antisera as well as characteristic biochemical reactions with the bacterial isolates.

Epidemiology of cholera

Cholera is also known as Asiatic cholera because it is endemic in India, Bangladesh, and other parts of Asia. In recent years, this disease has occurred in Europe, Africa, and North and South America. Worldwide, a total of 74,632 cases of cholera were reported for 1978. There were only four reported cases of cholera in the United States in 1977; one case was laboratory-acquired, two were imported, and only one was indigenous, or locally acquired.

Humans are the only natural host of *V. cholerae*. Water plays a major role in the transmission of the disease in rural areas where cholera is endemic.

Figure 29-9. In some countries, the water supply for a community is a body of water such as a pond. This same water is used for washing clothes, bathing, and drinking. Sewage wastes may also drain into it. Such a water supply is hazardous to health since it transmits waterborne diseases. *(Courtesy of R. Oseasohn, McGill University.)*

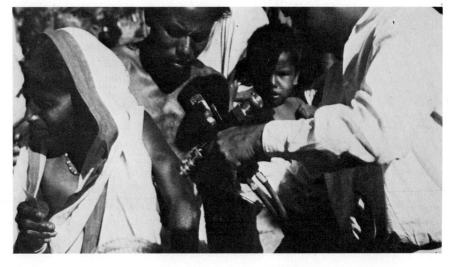

Figure 29-10. In areas where cholera is highly endemic or where there is an epidemic of the disease, mass-vaccination programs are carried out. This picture shows such a program being carried out. Vaccine is being administered using a jet-injector. *(Courtesy of R. Oseasohn, McGill University.)*

Direct contamination of food with infected excreta is also important; houseflies can play a major role in the dissemination of the vibrios. Such modes of transmission may give rise to epidemics.

In areas where the disease is endemic, cholera is predominantly a disease of children. In previously noninvolved areas, however, the incidence rates in adults are as high as those in children.

Prevention of cholera

Careful personal hygiene is the best prevention against cholera. At the community level, water supplies must be purified before drinking and must be protected from contamination by sewage. (See Fig. 29-9.) Further effective measures of prevention include elimination of flies and suitable treatment of patients.

A cholera vaccine composed of heat-killed vibrios can provide some protection. The standard commercial vaccine has 10 billion killed vibrios per milliliter, but it provides only 60 to 80 percent protection for 3 to 6 months. It is used in countries where cholera is highly endemic (Fig. 29-10).

Amebic dysentery (amebiasis)

This is a disease of humans and other animals caused by the ameba *Entamoeba histolytica*. Symptoms range from intermittent diarrhea to severe dysentery, which is sometimes fatal. In mild infections the amebas colonize on the intestinal wall and feed on bacteria and other amebas present in the intestine. In severe cases, the organism attacks the intestinal mucosa producing lesions. Stool specimens reveal ingested red blood cells in the amebas. The pathogen may also cause liver abscesses and may even spread to the lungs, brain, and other organs.

The active amebas or trophozoites reproduce by binary fission. As they are scraped off the intestinal wall and migrate down the colon with the fecal mass they become cysts. These cysts are relatively resistant to the external conditions prevailing after they are excreted from the body. The ingestion of the cyst form spreads and causes the disease. The source of these cysts is usually chronic carriers of the ameba.

In the United States, 1 to 5 percent of the population is infected. The incidence is correlated with the sanitary conditions existing in a given area. In some tropical regions the infection rate is 50 to 80 percent.

Laboratory diagnosis of amebiasis involves the demonstration of trophozoites or cysts in stools and lesions. Effective chemotherapeutic drugs for its treatment include dehydroemetine, chloroquin, and metronidazole.

Summary and outlook

The actual source of waterborne infections is the fecal matter that has contaminated the water. This fecal material contains the pathogenic microorganisms if it comes from infected persons or carriers. Water, unfortunately, is a good vehicle for the transmission and spread of such enteric diseases, which all have a fecal to oral to intestinal route. This route must be interrupted to effect proper control of waterborne enteric infections.

The most common bacterial diseases transmitted by water are typhoid fever, shigellosis, cholera, and amebic dysentery. All these diseases have humans as the natural hosts. The outlook for control of these diseases is discussed below.

Typhoid fever. The present immunization techniques against typhoid fever afford significant protection, but the degree of immunity is not great and can be readily overcome by a large dose of organisms. Oral vaccines with inactivated or even live attenuated bacteria offer better hope for more effective immunoprophylaxis. In countries where there is a high incidence of typhoid fever, raising the socioeconomic standard of living will no doubt lead to a reduction in the incidence of this disease, as well as others with the same epidemiological nature.

Shigellosis. The specific mechanism of immunity against shigellosis needs to be studied and resolved. This may lead to the development of an effective vaccine for this disease.

Cholera. Although treatment of cholera as practiced has been successful (with adequate therapy by prompt replacement of fluid and electrolytes, the mortality rate approaches zero), it is not uniformly available. In areas of the world where the disease is endemic, the economic and logistic difficulties are so great as to render effective therapy unavailable for the masses. There is need for the development of a vaccine that confers long-term immunity in order to surmount these difficulties. In addition, a chemotherapeutic agent should be found that can block the biological activity of cholera enterotoxin and thus reduce the loss of fluid and electrolytes from the gut.

Amebic dysentery. Chronic carriers of ameba are the usual source of feces carrying amebic cysts into a water supply. Treatment of cyst carriers and surveillance and improvement of sanitary conditions and water supplies are measures for the control of amebiasis.

Key terms

biotype
choleragen
electrolyte
enteric infection
enterotoxin

metabolic acidosis
Vi antigen
waterborne disease outbreak
Widal reaction

Questions

1 Are waterborne infections transmitted only by water? Explain.

2 Why would you consider waterborne diseases a public health problem?

3 List species and the major charateristics of waterborne pathogens.

4 Explain how serological identification and bacteriophage typing are used in tracing sources of enteric infections.

5 Provide reasons for the following events in typhoid fever:
   (a) Appearance of clinical symptoms
   (b) Presence of bacilli in stools
   (c) Relapses of the disease

6 Compare and contrast the status of immunoprophylaxis against typhoid fever, shigellosis, and cholera.

7 Which serological group and species of *Shigella* causes the most severe cases of bacillary dysentery? Why?

8 What is responsible for the signs, symptoms, and metabolic derangements of cholera? How is this observation used in the effective treatment of the disease?

9 Explain the biochemical basis for the increased electrolyte excretion in cholera.

10 Why are these practices of importance in the control of enteric infections?
   (a) Elimination of houseflies
   (b) Pasteurization of milk
   (c) Careful personal hygiene

NOSOCOMIAL INFECTIONS

The word *nosocomial*, derived from a Greek root, means "in the hospital"; thus, nosocomial infections are those acquired in the hospital. In testimony before the U.S. Congress in 1976, Dr. David Spencer, Director of the Center for Disease Control, reported that about 5 percent (or approximately 1,500,000) of patients averaged an extra week in the hospital, due to nosocomial infections, at a cost of over $1 billion annually.

Realizing the magnitude of the problem and the importance of preventing and controlling nosocomial infections, both the American Hospital Association and the Center for Disease Control recommended that each hospital develop an infection control plan. The Joint Commission on Accreditation of Hospitals requires all accredited hospitals to have an infection control program.

In this chapter we will discuss the epidemiology of hospital-acquired infections, the surveillance of infections, procedures to prevent their spread, and the components of a program for surveillance and control of nosocomial infections.

Historical background   Nosocomial infections have been problems as long as there have been hospitals. To prevent the spread of disease in the community during the

508

eighteenth century, the sick were isolated in fever hospitals, smallpox hospitals, tuberculosis sanatoriums, or "pest" houses. In these hospitals, which were large dingy halls and overcrowded with several patients to a straw mat, vermin and filth abounded and infection spread rapidly from patient to patient. In 1788 at the Hotel Dieu in Paris, as described by Garrison in 1917, in the *History of Medicine*, "Septic fevers and other contagia were the rule, the average mortality about 20 percent and recovery from surgical operations a rarity." Puerperal fever, the scourge of "lying-in" (maternity) hospitals, was a frequent and often fatal complication of childbirth.

A little more than a century ago developments occurred leading to the evolution of the modern hospital. Oliver Wendell Holmes and Ignaz Philipp Semmelweis instituted the practice of cleanliness, including handwashing between cases. As a result their patients were found to be relatively free of infection.

In 1854, during the Crimean War, Florence Nightingale reorganized the military hospitals and instituted sanitary practices that lowered the death rate of the wounded from 50 percent to 2.2 percent. (See Fig. 30-1.)

Almost 20 years later, as was pointed out in Chap. 1, Pasteur established the germ concept of infection, and Lister applied this concept to organize a system of antiseptic surgery, which formed the basis for the practice of aseptic technique.

The great reduction in deaths from infection, which started with instituting environmental sanitation and practicing antisepsis, was accelerated in

Figure 30-1. Florence Nightingale instituted sanitary practices in military hospitals in 1854. *(Courtesy of National Library of Medicine, Bethesda, Md.)*

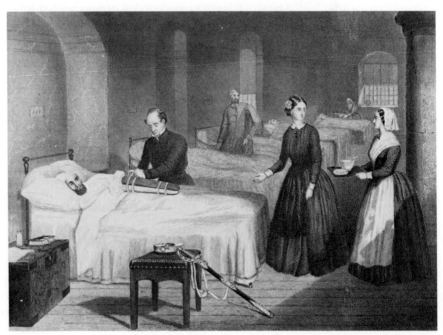

Table 30-1. National nosocomial infection summary for 1976 by hospital category

	COMMUNITY	COMMUNITY-TEACHING	FEDERAL	MUNICIPAL OR COUNTY	UNIVERSITY	ALL HOSPITALS
Number of hospitals	37	23	4	6	13	83
Number of infections	12,892	15,512	1,602	6,098	10,717	46,821
Number of discharges	493,681	434,798	33,224	110,561	243,968	1,316,232
Infection rate/100 discharges	2.6	3.6	4.8	5.5	4.4	3.6
Percent infections cultured	89.3	92.9	94.0	87.7	91.2	90.9
Percent infections causing death	0.8	0.6	0.9	2.4	0.4	0.9
Percent infections contributing to death	2.5	2.2	3.5	5.9	1.5	2.6
Median infection rate	2.2	3.5	5.4	5.2	5.1	3.1
Range of infection rates: Low	0.0	1.6	2.8	2.6	1.6	0.0
High	6.5	7.0	6.4	9.9	8.7	9.9

Source: Courtesy of Center for Disease Control, Atlanta, Ga.

the past 35 years by use of antimicrobial drugs. Unfortunately, this led to the belief, especially among hospital staff, that infection could be controlled by antibiotics and had ceased to be a major problem. Both the widespread use and misuse of antibiotics to prevent surgical infections and the increased reliance on the effectiveness of antibiotics in combatting infection resulted in a relaxation in aseptic technique and isolation precautions. The situation was made more serious with the development of a reservoir of antibiotic-resistant and virulent bacteria in the hospital environment.

In the 1950s frequent outbreaks of staphylococcal disease in hospitals focused attention on infections acquired by patients while in the hospital. A new term for these infections came into use during the 1960s—*nosocomial diseases.*

Although with the use of antibiotics the number of deaths due to nosocomial infections declined rapidly, the number of cases had no corresponding decline.

We should recall the comments of two great hospital reformers of the past century. Florence Nightingale said that "the first requirement of any hospital is that it do the sick no harm." Sir James Y. Simpson stated that "in the treatment of the sick, there is ever danger in their aggregation, and safety only in their segregation." These objectives have not been fully achieved. (See Table 30-1.)

Epidemiology of nosocomial infections

As previously stated, epidemiology is the study of the factors that influence the occurrence and distribution of disease in groups of individuals. The three factors necessary for an infection to occur (including those infections acquired in a hospital) are:

1 A source of microorganisms that can cause infection
2 A route of transmission of the microorganisms
3 A host susceptible to infection with the microorganisms

Sources of infection

The source of microorganisms that can cause nosocomial infections are essentially the same as those in the community—people, objects, substances, air currents, animals, and insects. The most frequent source of microorganisms pathogenic to human beings is people. One of the most serious threats of infection is from our own normal microbiota. For example, the serious infections seen in today's hospitals are most commonly caused by *Escherichia coli, Klebsiella pneumoniae, Candida albicans, Staphylococcus aureus, Serratia marcescens, Proteus mirabilis,* and some *Actinomyces* spp. All these bacteria are found in various parts of the healthy human host. The change in the types of microorganisms causing infections is due to changes in host resistance and modification of the host microbiota. If a patient's resistance is low because of trauma, disease, or an operation, the pathogens may multiply and cause illness. A disease produced by microbes from within an individual is called an *endogenous infection*; an infection caused by germs from an external source is termed an *exogenous infection*. External sources include all the potential reservoirs of the infectious agent. Because of its special purpose, a hospital brings together patients with diagnosed or undiagnosed infections, such as tuberculosis, who may be carriers of other respiratory or enteric pathogens, or who may be highly susceptible to infection. Hospital personnel and visitors may also introduce infectious organisms. Other living reservoirs of pathogens include animals such as rodents and insects; these are of little significance in hospitals in the United States.

Probably the two greatest sources of contamination in hospitals are (1) the hands of personnel and (2) the infected patient. Subsequently, the inanimate environment (air, food, water, and fomites) becomes a source of infectious organisms when contaminated.

The sources of infection to a susceptible host are summarized in Table 30-2.

Susceptibility of host

A number of factors favor the likelihood of infection in hospitalized patients. When admitted, an individual's resistance may be lowered by a disease, wound, or trauma, and after admission, by treatment—surgery, chemotherapy, radiation, intravenous, and other therapy. (See Fig. 30-2.)

A great number of older patients are admitted with established infections and lowered resistance, making them vulnerable to other infections. The frequent use of additional diagnostic, treatment, and anesthetic procedures may impair the normal antimicrobial defense of the host, thus favoring the possibility of hospital-acquired infections caused by *opportunistic* microorganisms. (See Table 30-3.) These microorganisms exist as part of the normal microbiota, but they become pathogenic when transferred from their normal habitat into other areas of the host. For example, *E. coli* normally found in the gut can cause serious infection if introduced into the

Table 30-2. Sources of infection

LIVING RESERVOIRS	
People—patients, personnel, visitors	Other animals and insects
Clinically diseased—ill Subclinically diseased—mild symptoms Carriers—infected with no symptoms	Host of infective organisms Mechanical carriers of organisms

INANIMATE RESERVOIRS	
Air	Soil
Secondary source: Dust from soil Droplets from respiratory secretions	Primary reservoirs: Fungi Clostridia Secondary source: Pathogens from intestinal tract of infected host
Food	Water
Primary reservoirs: Infected animals or their products Secondary source: Contaminated by handling, flies, air	Secondary source: Contaminated or polluted
Fomites (any inanimate object)	
Secondary source: Contaminated by handling, insects, air droplets, dust	

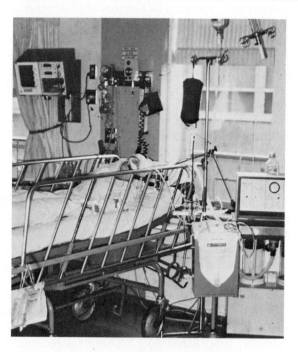

Figure 30-2. A patient undergoing multiple types of intensive care. *(Courtesy of the University of Iowa Hospitals and Clinics.)*

genitourinary tract. Breaking the skin barrier in surgery, biopsy, injection, or venipuncture, and using invasive procedures (catheters, drainage systems, implanted prostheses, and others) increase these patients' risk of infection (Table 30-4). Factors considered in the susceptibility of a patient

Table 30-3. Opportunistic microorganisms commonly causing nosocomial infections and clinical conditions associated with compromised host defense mechanisms

OPPORTUNISTIC MICROORGANISM	CLINICAL CONDITIONS ASSOCIATED WITH COMPROMISED HOST DEFENSE MECHANISMS
Aspergillus sp.	Immunosuppressed (patient receiving drugs to reduce or inhibit antibody formation) kidney transplant recipients
Bacteroides fragilis	General anesthesia; abdominal surgery
Candida albicans	Implanted prosthetic (artificial) heart valve; immunosuppressed graft recipients
Cryptococcus neoformans	Lymphatic tissue malignancies such as Hodgkin's disease
Escherichia coli	Urinary catheterization (a procedure to withdraw urine by inserting a catheter)
Herpes simplex virus	Immunosuppressed allograft (tissue graft from nonidentical donor of the same species) recipient
Klebsiella	Heart transplant recipients on antibiotic or immunosuppressive drugs
Mycobacteria	Immunosuppressed patient; prolonged use of steroids (therapeutic drugs)
Proteus	Urinary catheterization; abdominal surgery
Pseudomonas aeruginosa	Decrease of leukocytes in cancer patients; blood cell–production disorders
Serratia marcescens	Indwelling urinary catheter; continuous intravenous drip
Staphylococcus aureus	Intravenous catheters; implanted prostheses such as heart valve, devices for hand and eye
Streptococcus pneumoniae	Absence of spleen; multiple tumors

Table 30-4. Increased risk of infection in patients with compromised defenses

FACTORS INCREASING RISK OF INFECTION	BASIS FOR INCREASED SUSCEPTIBILITY TO INFECTIONS	MOST COMMON TYPES OF INFECTION
Indwelling catheters or implanted devices	Foreign body	Abscess; bacteremia
Allograft recipients (renal, heart, bone marrow)	Diminished cell-mediated immunity	Pneumonia; urinary tract infection; bacteremia
Extensive skin burns	Diminished cell-mediated immunity; impaired antibody production	Pseudomonas (*Pseudomonas aeruginosa*) bacteremia
Absence or malfunction of the spleen (splenectomy; sickle-cell anemia)	Impaired IgM antibody synthesis	Pneumococcal (*Streptococcus pneumoniae*) bacteremia and meningitis
Bone marrow failure	Agranulocytosis (severe reduction in leukocyte count); neutropenia (low number of large leukocytes)	Bacteremia; pneumonia; urinary tract infection
Malignant disorders (hematoproliferative states; solid tumors)	Diminished or absent cell-mediated immunity; neutropenia; impaired antibody synthesis	Bacteremia; pneumonia; urinary tract infection

Source: Courtesy of G. P. Youmans, P. Y. Peterson, and H. M. Sommers, *The Biologic and Clinical Basis of Infectious Disease*, W. B. Saunders Company, Philadelphia, 1975, with permission.

as host to a nosocomial infection are summarized in Table 30-5. The host's resistance was discussed in more detail in Chap. 25. A summary of nosocomial infections for 1976, by site of infection, is shown in Fig. 30-3.

Table 30-5. Factors affecting a patient's susceptibility to infection

FACTORS	EXAMPLES
Inadequate cellular defenses	Very young patients
	Very old patients
Diseases	Malnutrition
	Diabetes
	Chronic debilitating diseases
	Hematological disorders
	Renal diseases
	Immunological deficiencies
Accidents	Severe burns
	Severe debilitation
	Shock
Surgery	Size of wound and amount of wound drainage
	Duration of operation
	Tissue trauma
	Length of stay before and after operation
	Foreign bodies—sutures, catheters, prostheses
Therapy	Antibiotic
	Radiation
	Steroid
	Immunosuppressive

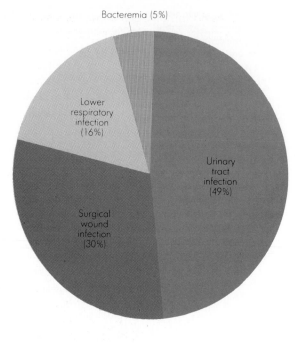

Figure 30-3. Distribution of nosocomial infections by site for 1976. (*Center for Disease Control, Atlanta, Ga.*)

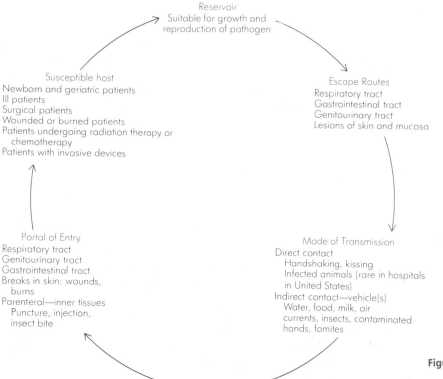

Figure 30-4. The infection cycle.

Transmission of
infectious agent

We have discussed reservoirs of infectious agents in hospitals and the
susceptibility of patients as hosts. In order for transmission to occur, the
cycle of infection (see Fig. 30-4) must be completed. The pathogen must
exit from the reservoir, have a mode or vehicle for transmission, and have a
portal of entry to a susceptible host. If the infectious agent is intercepted at
any stage in the cycle, infection will not occur. To break the cycle is the goal
of infection control.

**Control by breaking
the cycle of infection**

Since host and agent factors are more difficult to control, interruption of the
cycle of infection is aimed primarily at transmission.

Isolation

As practiced in a hospital, *isolation* is the separation of a patient and his or
her care from other people. Hospital isolation policies are designed to
prevent the spread of pathogenic organisms among patients, staff, and
visitors.

The degree of isolation and the procedures to be used, as shown in Table
30-6, are determined by the infection or condition which makes isolation
necessary. The need for a guide for hospital personnel for practical,
effective isolation procedures led hundreds to request the Center for
Disease Control (CDC) for such information. This need resulted in development of a manual, prepared and published by CDC; a group of doctors and
nurses assisted as consultants. The manual, entitled "Isolation Techniques
for Use in Hospitals," describes in detail methods that can be used in large
or small hospitals.

There are seven practices recommended to be used at all times or as
desirable, according to the degree of isolation necessary. The way in which
each is practiced, such as the use of a private room, gloves, masks and
gowns, or the techniques for handling contaminated materials, is described
in the manual. A card system to be used with the manual has been
developed to facilitate rapid use of the appropriate isolation procedure. Five
categories of isolation (see Table 30-7) are designated by a different color of
card. On each card are instructions for people entering the isolated area.
The card for the category of isolation procedure needed should be posted on
the patient's door, chart, or bed. For example, the strict-isolation card
would list the following instructions:

Strict Isolation
Visitors—Report to Nurses' Station Before Entering Room.
1 Private Room—*Necessary*; door must be kept closed.
2 Gowns—Must be worn by all persons entering room.
3 Masks—Must be worn by all persons entering room.
4 Hands—Must be washed on entering and leaving room.
5 Gloves—Must be worn by all persons entering room.
6 Articles—Must be discarded or wrapped before being sent to Central
  Supply for disinfection or sterilization.

Table 30-6. Infectious diseases grouped according to degree of recommended isolation

PRIVATE ROOM	MASK	GOWN	GLOVES	EXCRETA AND EXCRETA-SOILED ARTICLES	BLOOD	SECRETA AND SECRETA-SOILED ARTICLES	
STRICT ISOLATION							
X	X	X	X	X	X	X	Smallpox
X	X	X	X			X	Anthrax, inhalation; pneumonic plague; vaccinia, generalized and progressive, and eczema vaccinatum
X	X	⊗	⊗			X	Burn, skin, or wound infection, major, with *Staphylococcus aureus* or group A streptococcus that is not covered by a dressing or that has copious purulent drainage
X	X	⊗	⊗		X	X	Lassa fever, Marburg virus disease
X	X	⊗	⊗			X	Pneumonia—*Staphylococcus aureus* group A streptococcus
X	X	⊗				X	Diphtheria (pharyngeal or cutaneous)
X	X*	X	X			X	Varicella (chickenpox); herpes zoster, disseminated
X		⊗		X	X	X	Congenital rubella syndrome; disseminated neonatal *Herpesvirus hominis* (herpes simplex)
X			⊗			X	Rabies
RESPIRATORY ISOLATION							
X	X*					X	Tuberculosis, pulmonary (including tuberculosis of the respiratory tract), suspected or sputum-positive (smear)
X	X						Meningococcal meningitis; meningococcemia
X	X*					X	Measles (rubeola); mumps; rubella (German measles); pertussis (whooping cough)
ENTERIC PRECAUTIONS							
D		⊗	⊗	X			Cholera; staphylococcal enterocolitis; gastroenteritis—enteropathogenic or enterotoxic *Escherichia coli, Salmonella* spp., *Shigella* spp., *Yersinia enterocolitica;* typhoid fever

Table 30-6. (continued)

PRIVATE ROOM	MASK	GOWN	GLOVES	EXCRETA AND EXCRETA-SOILED ARTICLES	BLOOD	SECRETA AND SECRETA-SOILED ARTICLES	
D		⊗	⊗	X		X	Diarrhea, acute illness with suspected infectious etiology
D				X	X	X	Hepatitis, viral, types A, B, or unspecified
WOUND AND SKIN PRECAUTIONS							
D			⊗			X	Gas gangrene (due to *Clostridium perfringens*)
D	X*	X	X			X	Herpes zoster, localized
D	⊗	⊗	⊗			X	Burn, skin, or wound infections, limited, including infections with *Staphylococcus aureus* or group A streptococcus, that are covered by and the discharge adequately contained by a dressing; bubonic plague
D		X	X			X	Burn, skin, or wound infections, major (except *S. aureus* and group A streptococcus) that are not covered by a dressing or that have copious purulent drainage; melioidosis, extrapulmonary with draining sinuses

X   Recommended at all times
⊗ With direct contact
X* For susceptibles
D   Desirable, but optional

Source: "Isolation Techniques for Use in Hospitals," 2d ed., Courtesy of Center for Disease Control, Atlanta, Ga., 1975.

Table 30-7. Categories of isolation precautions in hospitals

CATEGORIES OF ISOLATION PROCEDURES	ROUTES OF TRANSMISSION	PURPOSE
Strict isolation	Contact Airborne	Prevent transmission
Respiratory	Direct contact Airborne droplets	Prevent transmission
Protective—strict isolation	Direct Indirect	Prevent contact of pathogen and patient with impaired resistance
Enteric	Direct Ingestion	Prevent disease
Wound and skin	Direct Contaminated fomites	Prevent infection of personnel, patients

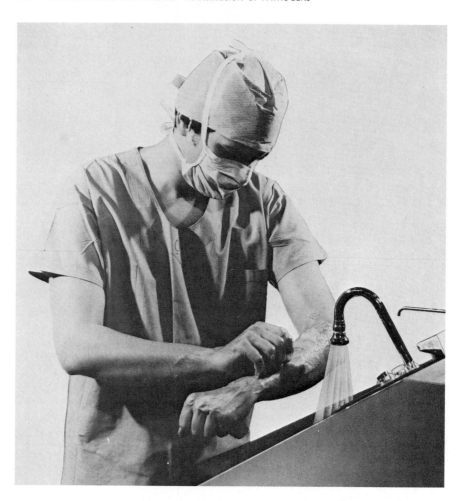

Figure 30-5. A surgical team member performing the "surgical scrub." The areas being scrubbed include the lower third of the forearms. *(Courtesy of American Sterilizer Company.)*

A hospital may adjust such a system to fit its needs. Other cards may be used indicating *discharge*, *excretion*, or *blood precautions*.

Handwashing
Good handwashing technique is the single most important practice to minimize the spread of infection. Vigorous scrubbing with soap or a detergent and water for at least 15 s and thorough rinsing before and after contact with each patient is adequate in most situations. If, however, during patient care the hands come into contact with blood, secretions from wounds, purulent materials, or any other suspect material, they should be washed 2 to 3 min and an antiseptic cleansing agent used.

The "surgical scrub" varies in hospitals with regard to the hand scrubbing procedure. However, all require, the following outlined procedures—scrubbing with a brush for about 10 min, rinsing in running water, and the use of an antimicrobial agent (Fig. 30-5). The comparative effects of various disinfectants on skin are shown in Fig. 20-2. More details about the efficiency of antiseptics and germicides are described in Chap. 20.

Asepsis   As previously defined, asepsis is the avoidance or deterrence of sepsis by exclusion of potentially harmful microorganisms. The goal of asepsis is to prevent or confine infection. Two concepts of asepsis, medical and surgical, are practiced in hospitals. Medical asepsis consists of all practices used to keep personnel, patients, and the environment as free as possible from agents of infection. Handwashing, sanitary practices, and isolation procedures are but a few examples of methods used to achieve medical asepsis. Surgical asepsis includes procedures which prevent microorganisms from gaining access to the wounds and tissues of patients. To achieve surgical asepsis, all instruments, gloves, surgical drapes, sutures, dressings, and other items that come into contact with a patient must be sterile. The environment is sanitized, the microbial flora of the air is controlled by filtering it through high-efficiency filters, known as the laminar-airflow system, and personnel scrub hands and arms and wear sterile caps, gowns, masks, and gloves. (See Fig. 30-6.) By cleaning and applying suitable antiseptics, the number of microorganisms on the surface of the patient's tissues is reduced. Aseptic technique includes the practices designed to prevent contamination by extraneous microorganisms and to reduce the numbers of those already present. The procedures to accomplish asepsis may vary in different hospitals. Surgical asepsis is required in the operating room and in procedures such as the dressing of wounds, catheterization, intravenous therapy, and many other treatments. The maintenance of

Figure 30-6. Hospital room designed with laminar airflow. These high-efficiency particulate air filters in the background wall remove airborne microorganisms. Sterile air passes over the patient whose normal defenses are suppressed (immunosuppressed) because of extensive radiation therapy. Note that attendants wear gloves, masks, etc. (*Courtesy of G. A. Andrews and Oak Ridge Associated Universities, Inc.*)

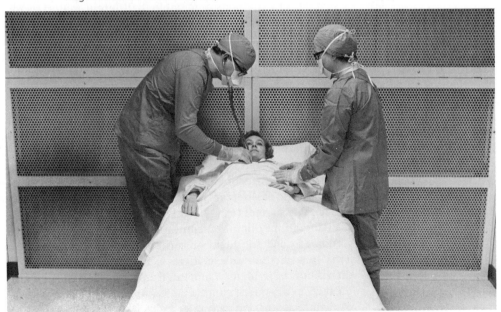

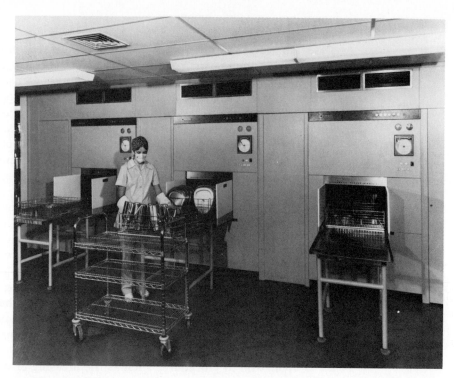

Figure 30-7. Utensils are positioned in baskets, decontaminated, washed, and sterilized in the same autoclave. *(Courtesy of American Sterilizer Company.)*

asepsis is possible only if every person in the sterile area is responsible and conscientious in carrying out aseptic techniques.

Hospital disinfection and sterilization

Many hospitals have central-supply units where most equipment and supplies are cleaned and sterilized (see Fig. 30-7). The efficacy of the process used is monitored by the microbiology laboratory on a regular schedule. The methods applicable to different materials for accomplishing disinfection or sterilization are discussed in Chaps. 19 and 20.

The trend in hospitals toward the use of presterilized, disposable instruments and materials, such as syringes, needles, gloves, and masks, not only decreases the amount of time involved in cleaning, repairing, and sterilizing equipment, but also reduces transmission of pathogens by cross-infection.

Hospital environmental sanitation

The goal of environmental sanitation is to kill or remove microbial contamination from surfaces. To evaluate the procedure and measure the contamination reduction, occasional sampling of microorganisms from surfaces is carried out. Petri dishes showing microbial growth before cleaning and after cleaning are an impressive teaching tool when instructing new personnel (Fig. 30-8).

Reduction of microbial contamination is accomplished best by a combination of good frictional or scrubbing action, and water and detergents. This is sufficient unless there is heavy contamination; then the use of disinfectants is clearly indicated. To be effective, disinfectants must be used in an adequate concentration for a specified time (see Chap. 20). Use of disinfectants, for example, helps to keep mop water uncontaminated.

Mops should be washed and dried thoroughly each day to reduce contamination. A bucket of solution and a wet rag are often used to clean surfaces other than floors. If the same solution is used for an entire daily shift, more microbial contamination may result than before cleaning. Surfaces other than floors can be cleaned best by spraying on a dilute solution of disinfectant detergent and then wiping with disposable wipes or clean cloths.

With a state of cleanliness in the hospital, aseptic conditions can be established more easily.

Surveillance of infections

Surveillance of infections is the systematic observation and recording of the occurrence of transmissible disease; it forms the basis for an active control effort. Identifying and evaluating problems of nosocomial infections and developing and assessing effective control can only be accomplished by organized surveillance of such infections in patients. Not only must data be collected continuously, but it must be tabulated, analyzed, and used. One full-time person is needed for every 300 to 500 beds for adequate surveillance, investigation of clusters of infection, and infection-control activities. This person is usually (but not necessarily) a nurse and is generally called an infection control nurse (ICN) or a nurse epidemiologist.

Surveillance of patients

Patient infection surveillance is initiated on admission by including an infection data card in a patient's medical record. The following information is recorded on the card by the ICN:

1 Type of infection, if any, on admission
2 Hospital-acquired infection, if any, site, and type
3 Organism isolated
4 Surgery or chemotherapy before infection
5 Antibiotic given to control infection
6 Time required to control infection

Figure 30-8. Floor dust yielded these microorganisms. A cotton swab was used to obtain a sample from the floor surface, and the plates were inoculated with the swabs. (A) Sample taken 4½ h after morning mopping (mop water contained a sanitizing agent). (B) Sample taken immediately after midday mopping. (C) Sample taken 3 h after midday mopping. *(Courtesy of Environmental Services Branch, National Institutes of Health, Public Health Service.)*

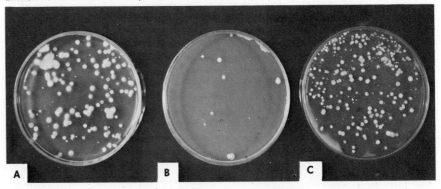

The data collected from the daily printout of cultures from the microbiology laboratory and from laboratory and clinical "rounds" are recorded on each patient's infection data card. Evaluation of this information may uncover new infections, or clusters of infection, and indicate the need for an immediate investigation to discover the source and mode of transmission of the infectious organism. The findings promote the development of better patient-care practices, the most important factor in controlling nosocomial diseases.

Surveillance of hospital personnel

A physical examination should be required for all hospital personnel, and immunization records should be checked. Immunization should be required, if not on record, for poliomyelitis, tetanus, diphtheria, and rubella. Tuberculin-positive personnel should have a chest x-ray to determine the possibility of active tuberculosis. Women of childbearing age, personnel in high-risk categories, and pediatric personnel should be informed of possible susceptibility and offered protection. These are important guidelines for the protection of both staff and patients.

Selective microbial culturing of samples from personnel may be necessary in investigating an outbreak of nosocomial disease.

Surveillance of the hospital environment

When the ICN detects one or more new cases of infection, many cultures of patients, personnel, and the environment may be required to identify the reservoir(s) and eliminate the source(s) of the pathogen. Such microbiologic sampling in the investigation of a specific epidemiologic problem can be helpful.

Components of control program

Effective patient care is the focal point for the prevention and control of nosocomial infections and is the reason for a control program. Figure 30-9 emphasizes the center of any infection control program—*the patient*, who is the concern of an infection control committee, and the starting point for all surveillance and control measures. The interrelation of these elements is shown by the direction of the arrows. Organized and implemented together, they make an effective control program.

Infection control committee

An efficient and effective infection control committee is the most important element of a program for control of nosocomial diseases. The committee, of about 15 members, should include the hospital epidemiologist, microbiolo-

Figure 30-9. Essentials of an infection control program.

Table 30-8. Activities of an active infection control program	ACTIVITY	RESPONSIBLE FOR IMPLEMENTATION
	Surveillance	Infection control nurse (ICN)
	Investigation of epidemics	Microbiology laboratory, ICN
	Formulation of isolation procedures and policies for evaluation of persons exposed to contagious diseases	Infection control committee

	Development of guidelines for patient care practices:
	    Handwashing
	    Urinary catheter care
	    Care of intravenous
	    catheters
	    Handling of respiratory
	    therapy equipment
	    Other practices
		Infection control committee

	Continuing education program	ICN
	Ensuring adequate disinfection and sterilization	Microbiology laboratory
	Environmental sanitation	Central supply, housekeeping
	Employee health service	ICN
	Review of use of antibiotics	Infection control committee

gist, or pathologist as chairman, the infection control nurse, and representatives of the director of nursing services, hospital administration, microbiology laboratory, and principal medical and surgical services.

Some of the major responsibilities of the committee are:

1 Formulating hospital policies regarding nosocomial infections
2 Distributing written guidelines to hospital personnel
3 Assessing the implementation of the policies

In order to perform successfully, an infection control committee requires adequate data to assess the frequency and causes of nosocomial infections.

The major activities of a properly functioning control program are listed in Table 30-8.

Microbiology
laboratory

The microbiology laboratory is a primary resource for an effective surveillance and control program. Its records are a surveillance tool and data source, and they are used for calculating infection rates and compiling nosocomial infection reports. Proficiency in identifying the organisms and reporting culture results is essential for the early detection of nosocomial pathogens.

Environmental microbiological sampling should be carried out when indicated to solve a problem, and the sampling should be coordinated with the objectives of the infection control program. The procedure should give information that can be used to improve policies and methods for controlling environmental contamination. The American Society of Microbiology's *Manual of Clinical Microbiology* recommends that a portion of the environ-

SUPPLIES OR EQUIPMENT	FREQUENCY	METHOD OF MONITORING
Sterilizers:		Bacterial spore strips
Steam	Weekly	(filter paper strips
Hot air	Weekly	impregnated with spores)
Gas	Every load	
Fluids	Monthly or semimonthly— random samples	Biologic test for pyrogens (toxic substances of microbial origin)
Milk formulas	Weekly	Spore colony count (no more than 25 per milliliter)
Instruments	Weekly	Bacterial spore strips
Blood and intravenous fluids	Random sampling of lots or units	Sampling and culturing both aerobically and anaerobically
Inhalation therapy equipment	Intermittently	Sampling and culturing
Water and ice	If contamination suspected	Membrane-filter procedure (sampling and culturing)

mental surveillance program be carried out on a routine schedule. Table 30-9 identifies the sampling that is considered minimal.

Education program

A control program can be effective only if all personnel involved in implementing it know what is expected of them.

General education classes for personnel should be scheduled twice monthly. The presentation should be structured for the group being addressed, such as food service, housekeeping, and laundry staff, two-year nurses, or aides. New staff can be oriented, and policy and procedures reviewed and explained.

In-service teaching by the ICN while making surveillance rounds can be most effective. The most important ideas to convey in teaching are:

1 Personal responsibility
2 Handwashing
3 Containment and confinement of infectious agents and contamination

The continuing education program is one of the primary duties of the ICN. An active educational program for all departments and for all levels of the hospital staff is necessary to ensure the implementation of policies and practices for infection control.

Summary and outlook

Growing concern over nosocomial infections in the past 10 years, because of their morbidity, mortality, and economic consequences, has stimulated hospitals to undertake activities aimed at surveillance and control of such infections. Infection surveillance and control programs have been established in most hospitals. The components of an infection control program include:

1 An infection control committee
2 A clinical microbiology laboratory
3 An infection control person
4 Effective patient care
5 Clearly understood isolation policies
6 Staff education programs

The success of the program requires that all staff members know and practice the recommended policies and procedures. All individuals must act responsibly in carrying out their own duties.

By reevaluating control measures based on surveillance and analysis of data, new measures may be developed to prevent future occurrences of nosocomial infection.

Studies are being done by the Center for Disease Control to measure infection surveillance control program activities related to patient care and practices with regard to changes in infection rate, and to evaluate the efficiency and cost effectiveness of infection surveillance control program activities.

As medical understanding of epidemiologic characteristics of diseases increases and as other infectious agents are identified, methods and practices for controlling nosocomial infections will need to be revised.

Key terms
endogenous infections
epidemiology
isolation in hospital
nosocomial infections
opportunistic microorganisms
surgical asepsis
surveillance of infections
medical asepsis

Questions
1 Name and give the contributions of four people who led the way to the development of the modern hospital.
2 How can you account for a reservoir of antibiotic-resistant and virulent bacteria in hospitals?
3 What three conditions are necessary for nosocomial infections to occur?
4 Describe the conditions under which:
   (a) An endogenous infection might develop
   (b) An exogenous infection might develop
5 Identify several factors which make the hospitalized patient susceptible to infection.
6 Trace two ways in which cross-infection might occur from a patient with a staphylococcus wound infection.

7 What degree of isolation would be appropriate for a patient with leukemia; with a wound infection; with smallpox; with amebic dysentery; with pneumonia?

8 Name several practices or procedures in a hospital which interfere with transmission of infection.

9 Explain the value of an education program and how it should be organized.

10 List the activities and components included in an active infection control program.

ARTHROPOD-BORNE INFECTIONS

Other than humans themselves arthropods are the most important source
of human disease. They are members of the phylum Arthropoda and
constitute the largest assemblage of species in the zoological world. More
than 900,000 species of arthropods are known and thousands more
remain to be classified.

Certain arthropods are of medical importance not only because they are
capable of causing necrotic, traumatic, and allergic injuries, but because
some of them can serve as intermediate hosts for parasites, or as *vectors*
for pathogenic microorganisms. (A *vector* is an organism, such as an
insect, that transports a pathogen. A *biological vector* is one in which
the pathogen undergoes a period of incubation or development.) Their
importance is underscored when the number of people affected and the
number of consequent deaths are considered. For example, each year 150
million people become gravely ill with malaria; in Africa alone, more
than 1 million children die each year of this disease. Arthropod-borne
infections have an extensive distribution over the face of the globe.
Through the centuries, such diseases have produced much suffering,
economic loss, and death in the human population. On innumerable
occasions, these diseases reached pandemic proportions.

527

Arthropods as hosts and as vectors of microbes

The majority of microorganisms using arthropods as vectors are so well adapted to the host that they do not harm it and cause no tissue damage. In these cases transmission of microbes to humans and other animals is accidental and often incidental to the perpetuity of the microbial species.

It should also be remembered that there are many species of microorganisms that are normal parasites of arthropods and are not known to be transmissible to humans or other vertebrates.

As vectors of microorganisms, arthropods can serve merely as mechanical vectors of etiologic agents. The common housefly, *Musca domestica*, is the classic example. The diseases transmitted by it include salmonellosis and other enteric diseases, poliomyelitis, and infectious hepatitis.

However, in most of the arthropod-transmitted diseases, the etiologic agents use the arthropod as a biological vector, requiring a period of incubation or development in this host. Representative diseases of this nature are shown in the accompanying tables. Table 31-1 shows representative diseases of humans caused by protozoa and transmitted by arthropods as biological vectors. Table 31-2 shows representative human diseases of bacterial origin transmitted by arthropods. The majority of these bacterial

Table 31-1. Representative protozoan diseases of humans transmitted by arthropods as biological vectors

DISEASE	ETIOLOGIC AGENT (GEOGRAPHIC DISTRIBUTION)	BIOLOGICAL VECTOR	INTERRELATIONSHIP OF ARTHROPOD-PATHOGEN-HUMAN
Chagas' disease	*Trypanosoma cruzi* (Continental Latin America)	Cone-nosed bugs (*Triatoma* spp., *Panstrongylus* spp.)	Pathogen multiplies in midgut of insect. Inoculated in humans by rubbing onto skin or into conjunctiva.
African trypanosomiasis (sleeping sickness)	*Trypanosoma gambiense* (West and Central Africa) *T. rhodesiense* (East and Central Africa)	Tsetse flies (*Glossina* spp.)	Pathogen multiplies in midgut and salivary glands of fly. Humans inoculated by bite.
Malaria	*Plasmodium vivax, P. malariae, P. falciparum, P. ovale* (Regions with warm climates)	Mosquitoes (*Anopheles* spp.)	Pathogen completes sexual cycle, then multiplies by sporogony in mosquito. Humans inoculated by bite (see Fig. 9-11).
Leishmaniasis	*Leishmania donovani* (China, India, Africa, Mediterranean area, continental Latin America) *L. tropica* (Mediterranean area to western India) *L. braziliensis* (Mexico to northern Argentina)	Sandflies (*Phlebotomus* spp.)	Pathogen multiplies in midgut of fly. Humans inoculated by bite.

Table 31-2. Representative bacterial diseases of humans transmitted by arthropods as biological vectors

DISEASE	ETIOLOGIC AGENT (GEOGRAPHIC DISTRIBUTION)	BIOLOGICAL VECTOR	INTERRELATIONSHIP OF ARTHROPOD-PATHOGEN-HUMAN
Plague	*Yersinia pestis* (Africa, Asia, South America, and Western United States)	Rodent fleas (*Xenopsylla cheopis*), human fleas (*Pulex irritans*)	Pathogen multiplies in gut of flea. Humans inoculated by bite of flea.
Tularemia	*Francisella tularensis* (North America, Asia, and Europe)	Ticks (*Dermacentor* spp., *Amblyomma* spp., etc.) Deerflies (*Chrysops discalis*)	Pathogen multiplies in gut and hemocoele (body cavity through which blood circulates). Congenitally transmitted in some ticks. Humans inoculated through bite or crushing of tick.
Rocky Mountain spotted fever	*Rickettsia rickettsii* (North America, Mexico, Columbia, and Brazil)	Ticks (*Dermacentor* spp., *Amblyomma* spp., *Ornithodoros* spp., etc.)	Pathogen multiplies in wall of tick's midgut; congenitally transferred in tick. Humans inoculated through bite.
Scrub typhus	*Rickettsia tsutsugamushi* (Asia, Australia, and Pacific islands)	Red mites (*Trombicula* spp.)	Pathogen multiplies in gut of mite; congenitally transmitted in mite. Humans infected from bite of larval mite.
Rickettsial pox	*Rickettsia akari* (United States, Russia, Korea, and Africa)	Mouse mite (*Allodermonyssus sanguineus*)	Pathogen multiplies in gut of mite. Humans infected by bite of mite.
Classical typhus fever	*Rickettsia prowazekii* (worldwide)	Body louse (*Pediculus humanus*)	Pathogen multiplies in epithelium of louse's midgut. Humans inoculated by bite, feces, or crushing of louse on skin.
Trench fever	*Rickettsia quintana* (Europe, Africa, and North America)	Body louse (*Pediculus humanus*)	Pathogen multiplies in midgut of louse. Human inoculated by feces or crushing of louse on skin.
Murine typhus fever	*Rickettsia typhi* (worldwide)	Fleas (*Xenopsylla cheopis* and others)	Pathogen multiplies in epithelium of midgut of flea. Humans infected through bite.
Epidemic relapsing fever	*Borrelia recurrentis* and other species (Asia, Africa, and Latin America)	Body louse (*Pediculus humanus*)	Pathogen multiplies in tissues of louse outside gut. Humans inoculated by crushing louse on skin.

Table 31-3. Representative viral diseases of humans transmitted by arthropods as biological vectors

DISEASE	ETIOLOGIC AGENT (GEOGRAPHIC DISTRIBUTION)	BIOLOGICAL VECTOR	INTERRELATIONSHIP OF ARTHROPOD-PATHOGEN-HUMAN
Yellow fever	Yellow fever virus (a togavirus) (Africa and South America)	Mosquitoes (*Aedes aegypti*, *Haemagogus* spp.)	Pathogen multiplies in tissues of mosquitoes. Humans inoculated through bite.
Dengue fever	Dengue fever virus (a togavirus) (Southern and Southeast Asia, Pacific islands, Northern Australia, Greece, Caribbean islands, Nigeria, and Latin America)	Mosquitoes (*Aedes* spp., *Armigeres obturbans*)	Pathogen multiplies in tissues of mosquitoes. Humans inoculated through bite.
Equine encephalitis	Encephalitis viruses (different families of arboviruses) (Eastern United States, Canada, Philippines, Cuba, and South America)	Mosquitoes (*Aedes* spp., *Culex* spp., *Mansonia titillans*)	Pathogen multiplies in tissues of mosquitoes. Humans inoculated through bite.
Colorado tick fever	Colorado tick fever virus (Western United States)	Wood ticks (*Dermacentor andersoni*)	Pathogen multiplies in tissues of tick. Humans inoculated from bite of tick.

diseases, as shown in the table, are caused by rickettsias, a group of bacteria that are obligate intracellular parasites. Table 31-3 shows representative diseases of humans caused by viruses and transmitted by arthropods. Note that there are many kinds of encephalitis diseases with different viral agents and varied geographical distributions. Also notice that the mosquito and the tick are the main vectors involved in the transmission of the arthropod-borne viruses (*arboviruses*).

Arthropods as biological vectors

In some parasitic infections (such as with trypanosomes), the pathogenic microorganism incubates and develops in the intestinal cavity, or midgut, of the arthropod. In rickettsial infections, the pathogen becomes an intracellular inclusion in practically every organ and tissue of the arthropod body. In Rocky Mountain spotted fever and scrub typhus, the rickettsia is transmitted cogenitally, that is, to the offspring of the arthropod. In malaria, the protozoan completes its sexual cycle in the anopheline mosquito, with resultant sporozoite formation (see Fig. 9-12). Arthropod-borne viruses usually multiply in their insect vector and therefore must have the special ability to multiply both in arthropod and vertebrate cells.

With few exceptions, arthropods that transmit disease ingest the pathogenic microorganisms. But the manner in which they actually transmit the disease varies. They may obtain the pathogens in a blood meal from a diseased person and later deposit them in a vomit-drop in the puncture wound (as in plague), or in fecal pellets near the puncture wound (as in typhus) made in the skin of a person. Others discharge the pathogens in minute drops of salivary secretion at the time they procure a blood meal (as in malaria). The specific mode of inoculation of representative diseases is also presented in Tables 31-1 to 31-3: It will not be possible to discuss all these diseases at length; the few discussed in some detail below illustrate the nature of arthropod-transmitted diseases.

Plague

Plague pandemics ravaged Asia and Europe for centuries. The Great Plague, which started in 542, was reputedly responsible for over 100 million deaths in 50 years. The Black Death, the plague pandemic of the fourteenth century, was considered the worst catastrophe to strike Europe, and perhaps even the world. It resulted in the death of an estimated one-third of the world's population. (This pandemic was called the Black Death because of the severe cyanosis, or blue to purple color of the skin, which is characteristic of the disease.) The last pandemic of the nineteenth century began in central Asia in 1871 and spread to other parts of the world. Even today, epidemics continue to occur in many parts of Asia and Africa. Sporadic infections also appear in South and North America.

Plague, caused by bacteria, occurs in humans in three forms. The most common is *bubonic plague*, a highly fatal disease characterized by chills, fever, nausea, vomiting, general weakness, and the enlargement, ulceration, and suppuration of lymph glands. (An enlarged lymph gland is called a *bubo*; hence the name *bubonic plague*.) Bubonic plague is transmitted to

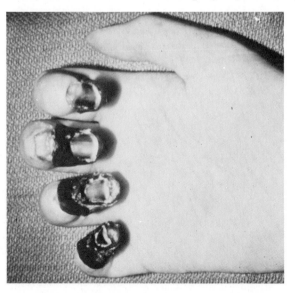

Figure 31-1. Severe finger tissue necrosis appearing during convalescence after acute bubonic plague infection. Affected fingers are shown just prior to amputation of nonviable digits. *(Courtesy of Center for Disease Control, Atlanta, Ga.)*

humans by the bite of infected rat fleas (*Xenopsylla cheopis*). Upon introduction into a human, a progressive infection results with involvement of nearly every organ and tissue of the body; the plague organisms invade the bloodstream, giving rise to septicemia. Patients who survive the disease may have marked necrosis of peripheral tissues, that is, tissues of the toes and fingers (Fig. 31-1).

A second form of plague is called *primary septicemic plague*. It resembles bubonic plague in all respects except that buboes are not formed. Infection resulting in this form of the disease is also transmitted by the bites of infected fleas.

A third form of plague is *primary pneumonic plague*. This form is *fulminant* (sudden and severe), resulting in severe prostration, respiratory distress, and death—often within a few hours after onset. (Without treatment the mortality rate is 100 percent.) The primary vital organ involved is the lung, giving rise to lobar pneumonia. This form of plague is contagious and is spread from an infected person to another person by respiratory droplets.

The incubation period for plague is quite short, being from 2 to 6 days. All forms of plague are caused by the same bacterium, *Yersinia pestis*. Death results from progressive toxemia and even heart failure in severe infections.

Figure 31-2. Morphology and distinguishing characteristics of *Yersinia pestis*. (Erwin F. Lessel, illustrator.)

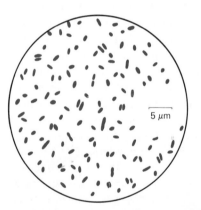

Yersinia pestis

Gram-negative rods; bipolar staining
Involution forms present
Occurs singly
Not encapsulated
Nonsporulating
Nonmotile
Aerobic, facultatively anaerobic
Pathogenic, causing plague

5 μm

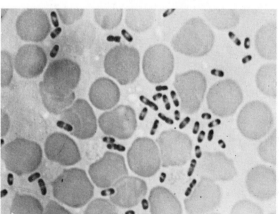

Figure 31-3. *Yersinia pestis* organisms exhibit a bipolar, or "safety-pin," appearance in this smear of mouse blood. Average size of this organism is 1.0 by 2.0 μm. Red blood cells are also seen. *(Courtesy of U. S. Naval Biological Laboratory.)*

Biology of *Yersinia pestis*

Yersinia pestis was first identified as the etiologic agent of plague by A.J.E. Yersin in 1894. It is a plump, gram-negative, non-lactose-fermenting bacillus. It is nonmotile, nonsporulating, and pleomorphic (see Fig. 31-2). A *bipolar* (stained area at both poles of cell) appearance is best demonstrated in smears of animal tissues stained by special methods, such as by Wayson's methods (see Fig. 31-3). The biochemical reactions of the organism are shown in Table 31-4.

Plague bacilli are facultative anaerobes. They are not fastidious and grow easily on most ordinary culture media. Growth is slow, however, even at the optimum growth temperature of 28°C (82.4°F). In broth culture, plague bacilli exhibit a flocculent, or flaky, type of growth with no turbidity.

Nature of pathogenicity of plague

All virulent plague microorganisms produce antiphagocytic antigens and an intracellular toxin that acts on the vascular system, causing irreversible shock and death. But the precise property of the plague bacillus responsible for its high virulence is not known.

Treatment of the disease with streptomycin, chloramphenicol, or the tetracyclines is quite effective.

Laboratory diagnosis of plague

Specimens from patients (such as blood, sputum, and aspirates from buboes) are stained by Gram's stain or Wayson's reagent (a combination of carbolfuchsin and methylene blue). When Y. *pestis*, which is gram-negative, is stained by the latter reagent, it exhibits a characteristic bipolar, or "safety pin," appearance (Fig. 31-3). Fluorescent-antibody technique is also used for rapid diagnosis, especially for cases of primary pneumonic plague.

Confirmed diagnosis is made by isolation of Y. *pestis* from clinical materials or a fourfold rise in specific antibody levels.

Epidemiology of plague

As already indicated, plague is transmitted to humans by bites of rat fleas belonging to the genus *Xenopsylla*. (Recent evidence also implicates the human flea *Pulex irritans*.) These arthropods become infected by biting rats or other rodents that serve as reservoirs of the organisms. Infected fleas generally regurgitate several thousand bacilli into the site of the bite.

The ecology of plague is complex. It exists in nature in two broad but not exclusive ecologic forms: *enzootic* plague and *epizootic* plague. Enzootic plague is maintained in a relatively resistant rodent host population that has no excessive mortality owing to the microorganism. These rodents

Table 31-4. Biochemical characterization of *Yersinia pestis*

TEST OR SUBSTRATE	REACTION
Glucose	A
Sucrose	—
Maltose	A
Lactose	—
Indole production	—
H$_2$S production	—
Oxidase reaction	—

A = acid; — = no reaction.

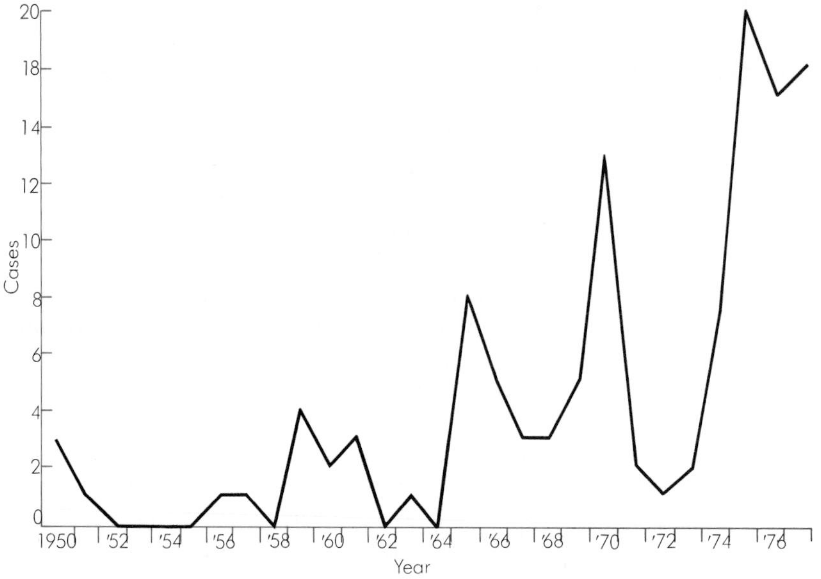

Figure 31-4. Reported cases of plague in humans by year in the United States, 1950 to 1977. A few cases develop each year following exposure to infected wild rodents or their fleas in the western United States. *(Center for Disease Control: "Reported Morbidity and Mortality in the United States," Annual Summary 1977. Issued September 1978.)*

serve as long-term reservoirs of the disease organisms. Epizootic plague occurs when plague bacilli are introduced into rodent and other small animal populations that are moderately to highly susceptible to the lethal effects of the infection.

Plague is endemic throughout the world. In the United States, the organism is established in mammals (such as rats, squirrels, rabbits, and wild carnivores) from the eastern slope of the Rocky Mountains westward. Cases have also been reported from bordering areas of Mexico and Alberta, Canada.

Plague in the United States in the first quarter of this century usually resulted from contact with domestic rats (*Rattus rattus*) and their fleas. The hazards from such infection were reduced by rat-control programs and by higher standards of sanitation in urban areas. Most cases in the past 50 years have resulted from exposure in rural or suburban surroundings. The reported cases of plague in the United States in humans from 1950 to 1977 are shown in Fig. 31-4.

Prevention of plague

At the urban level, prevention of plague depends primarily on control and elimination of fleas and rats. In areas where enzootics and epizootics of plague prevail, an effective flea-control program should precede rodent reduction. Otherwise, the massive release of infected fleas will increase the hazard of human exposure. High standards of sanitation are necessary in order to control plague.

Immunization with plague vaccine reduces the incidence and severity of disease. The plague vaccine licensed for use in the United States is prepared from *Y. pestis* organisms grown in artificial media, inactivated with formaldehyde, and preserved in phenol. Active immunization is recommended only for persons at high risk of exposure to the disease, such as those engaged in activities that are carried out in areas endemic for plague.

Yellow fever

Yellow fever is an acute infectious disease caused by an arbovirus. It was first recognized as a disease entity in the seventeenth century; it was not until 1900 to 1901, however, that Walter Reed and his colleagues discovered the relationship between the yellow fever virus and the mosquito *Aedes aegypti*; this paved the way for control of transmission of the disease.

Yellow fever remains the most serious arbovirus disease of the tropics. For more than 200 years after the first identifiable outbreak in the Yucatan in 1648, it was one of the great scourges of the world. As late as 1905, New Orleans and other southern American ports had an epidemic that involved at least 5,000 cases and many deaths.

As is typical with infections of arboviruses, the severity of the clinical syndrome is very variable, ranging from inapparent infection, through mild fever, to fulminant fatal attacks. Thus in humans, the disease ranges from an almost inapparent febrile reaction to a severe prostration.

The incubation period of yellow fever is usually 3 to 6 days, but it may be longer. The fully developed disease consists of three clinical periods: *infection* (coincident with *viremia*, or presence of virus in blood—headache, backache, muscular pains, fever, nausea, and vomiting), *remission* (infection symptoms subside), and *intoxication* [temperature again rises, intestinal hemorrhage manifested by black vomit, *albuminuria* (high protein level in urine), and jaundice due to liver damage]. Death occurs, or convalescence begins, by the eighth day. The fatality rate is approximately about 5 percent of all cases. Recovery from the disease confers lifelong immunity.

Biology of the yellow fever virus

The yellow fever virus is a small RNA virus belonging antigenically to the flaviviruses (formerly, group B arboviruses). They are members of the family Togaviridae. (It should be noted that arboviruses are taxonomically heterogeneous and belong to different families.)

Togaviruses are single-stranded RNA viruses of icosahedral shape enclosed in a lipid envelope. The virions are 20 to 60 nm in diameter. They multiply in the cytoplasm of cells and mature by budding from cytoplasmic membranes.

Nature of pathogenicity of yellow fever

The arboviruses have a unique ability to multiply within the tissues of vertebrates and some bloodsucking arthropods. These viruses, after their inoculation into the tissues of a susceptible host, multiply rapidly and soon produce a viremia. They may localize in a particular organ, causing tissue damage and organ malfunction, eventually leading to death of the host. In yellow fever, damage to the liver results in the development of jaundice.

There is no specific treatment for the disease except for symptomatic and supportive measures.

Laboratory diagnosis of yellow fever

A blood specimen should be obtained as early as possible; the serum obtained should be used for viral isolation and serological testing. Attempts to isolate the virus by intracerebral inoculation of the serum into suckling mice are likely to be successful only in the first two days of clinical disease.

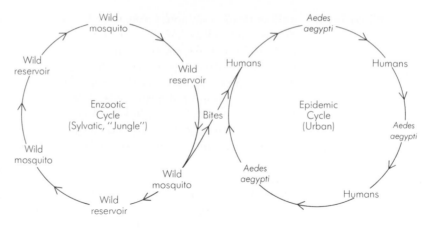

Figure 31-5. Relationship between enzootic and epidemic transmission cycles of yellow fever.

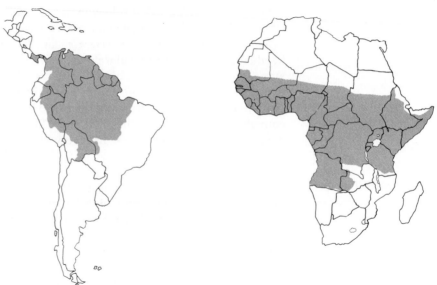

Figure 31-6. Regions where yellow fever is endemic (color screen) in South America and Africa.

The antibody titers of acute and convalescent sera should also be compared for a rise in neutralizing antibodies during convalescence.

Epidemiology of yellow fever

Yellow fever results from two basically different cycles of virus transmission, urban and sylvatic (jungle), as shown in Fig. 31-5. The urban cycle is transmitted from person to person by bites of the *Aedes aegypti* mosquito, i.e., human-mosquito-human transmission. Once infected, the mosquito vector remains infectious for life. Sylvan yellow fever occurs in wild animals; the same yellow fever virus is transmitted among them, and sometimes to humans, by mosquitoes other than *A. aegypti*. In the rain forests of South and Central America, species of treetop *Haemagogus* or *Sabethes* mosquitoes maintain transmission in wild primates. When humans enter the jungle, sporadic cases or local outbreaks may occur because of such mosquito bites. In Africa, the mosquito-primate cycle is maintained by *Aedes africanus*, a species which seldom feeds on humans. However, the mosquito *A. simpsoni* feeds upon the primates encroaching on the village gardens and can then transmit the virus to humans. The threat of yellow fever in urban areas in the tropical and semitropical regions always exists because of the existence of the sylvatic cycles. Once yellow fever is reintroduced into urban areas, the urban cycle can be reinitiated, with the possibility of developing epidemics. The current endemic zones of yellow fever in South America and Africa are shown in Fig. 31-6. Why yellow fever has never invaded Asia despite widespread distribution of human-biting *A. aegypti* mosquitoes is an enigma of medical epidemiology.

Yellow fever no longer exists in the United States. The last outbreak was the 1905 epidemic of New Orleans; the last imported case occurred in 1923.

Prevention of yellow fever

Urban yellow fever can be prevented by eradicating *A. aegypti* mosquitoes or by suppressing their numbers to the extent that they no longer perpetuate infection. But control of the sylvatic form is impractical because of the jungle cycles of virus maintenance. However, jungle yellow fever can be most effectively prevented in humans by immunization. The vaccine is a live, attenuated virus preparation. The licensed vaccine in the United States is prepared from the 17D strain developed by Max Theiler in 1937; the virus is attenuated by growth in tissue culture and is prepared in embryonated eggs. Vaccination is recommended for persons 6 months or older traveling or living in areas where yellow fever infection still occurs.

Malaria

The disease malaria has been known from antiquity and is aptly described as the single greatest killer of the human race. On a global scale, malaria is one of the most common infectious diseases of humans, causing much morbidity and significant mortality. As indicated earlier, each year more than 150 million people become gravely ill with malaria; about 3 million of these victims die of it. The disease has been virtually eliminated from the United States by control of the insect vector, the anopheline mosquito. But in these times of global travel, there is always the threat of contracting the disease in some country where it is prevalent and having it expressed only after returning to the United States.

Of the four species of *Plasmodium* (protozoa) pathogenic for humans, *P. falciparum* and *P. vivax* are the two most common in causing infections. Symptoms usually occur 10 to 16 days after infection by mosquitoes. Prominent symptoms are periodic chills, fever, headache, and sweating. The spleen becomes enlarged and tender (Fig. 31-7); eventually the patient becomes weak and exhausted, and an anemia develops. The pattern of paroxysmal (periodic) illness interspersed with periods of well-being is characteristic of benign malaria, such as is caused by *P. vivax*, *P. ovale*, and *P. malariae*. Paroxysms frequently begin with bed-shaking chills that are followed by high fevers, sweating, headache, and muscular pain. The symptomatic periods usually last less than 6 h.

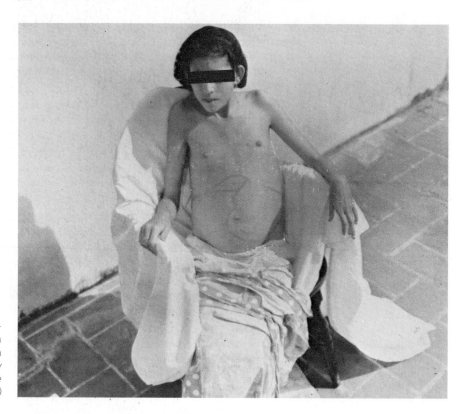

Figure 31-7. Malaria-infected youngster with enlarged spleen (area outlined on belly). *(Courtesy of Center for Disease Control, Atlanta, Ga.)*

In malignant falciparum malaria, the fever and symptoms are usually more persistent. There are also edema of the brain and lungs and blockage of kidney activity.

If not treated, benign malaria usually subsides spontaneously and recurs at a later date. Malignant malaria caused by *P. falciparum* has a high fatality rate, if not treated promptly.

Biology of the malaria
parasites

The malaria parasites are protozoa belonging to the group known as Sporozoa; their genus name is *Plasmodium*. As discussed in Chap. 9, four species of *Plasmodium* cause malaria in humans; the fever cycles vary according to the species causing the infection. The four species are *Plasmodium falciparum*, *P. vivax*, *P. malariae*, and *P. ovale*. *P. falciparum* causes the most serious form of the disease. (Actually malaria is caused by more than 50 different species of *Plasmodium*; only 4 of them attack humans. The rest attack several hundred other animal hosts.)

Plasmodium species have a complex life cycle, as shown in Fig. 9-12. When an *Anopheles* mosquito bites, its saliva, which contains the protozoan at the sporozoite stage of its life cycle, is injected into the bloodstream of the victim. The sporozoites quickly enter the liver, where they divide and develop into multinucleated forms known as schizonts. Within 6 to 12 days, the schizonts rupture and release the form known as merozoites into the bloodstream. These merozoites invade the host red blood cells, where they grow and divide to form more schizonts. These schizonts also rupture, destroying the erythrocytes and releasing more merozoites into the bloodstream to invade more red blood cells. The major symptoms of malaria are associated with rupture of the schizonts.

Some of the asexual merozoites in the patient's bloodstream develop into male and female gametocytes. When a mosquito bites, the gametocytes enter the mosquito's stomach where they become free male and female gametes. After fertilization occurs, the zygote passes to the outside of the stomach lining, where it develops into an oocyst containing many sporozoites. The mature sporozoites migrate to the salivary glands, from which they can be injected into the bloodstream of another victim to begin the cycle all over again.

Nature of pathogenicity of malaria

Disease in malaria is caused specifically by the asexual erythrocytic cycle. The rupture of infected erythrocytes at the completion of schizogony occurs every 48 h with *P. vivax* and *P. ovale*, and every 72 h with *P. malariae*, producing coincident chills and other symptoms. Synchronized, or coincident, schizogony and paroxysms of fever and chills are not, however, common with *P. falciparum* infection. The release of an endogenous pyrogen from injured cells may be the cause of the paroxysms of fever.

The high mortality rate of falciparum malaria is due in part to the high rate of reproduction of the asexual erythrocytic form of the parasite. The small veins and capillaries of the heart are clogged with parasitized erythrocytes; effective coronary blood flow and cardiac function are diminished.

Chemotherapy can be used to suppress malaria symptoms in individuals or to cure the infection completely. Treatment of an acute attack can be made with chloroquine for all types of malaria except drug-resistant falciparum infection. The latter should be treated with a combination of quinine, perimethamine, and one of the sulfonamides.

Laboratory diagnosis of malaria

The typical symptoms of malaria mimic a large variety of other human infections. Therefore the definitive diagnosis of the disease is made in the laboratory by the demonstration of the parasite in blood smears from patients (Fig. 31-8).

The indirect fluorescent-antibody and indirect hemagglutination tests are used in serological diagnosis of malaria. But antibodies are usually not detectable until after the second week of infection.

Epidemiology of malaria

In North America, Europe, and probably Northern Asia, malaria is largely a disease of the past. The cases that occur in North America and Europe have been contracted by persons visiting in Asia, Africa, or Latin America. The influence of military involvement and of control measures on the number of cases reported in the United States is shown in Fig. 31-9. There were 547 cases of malaria in the United States in 1977. In the tropical and subtropical areas of the world, however, malaria is still the single most severe health problem today (Fig. 31-10).

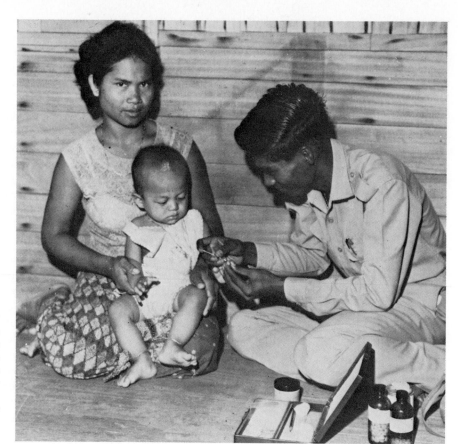

Figure 31-8. Malaria worker taking a blood sample from a child in a Thailand malaria eradication program. The blood sample will be examined microscopically for the presence of parasites. *(Courtesy of Center for Disease Control, Atlanta, Ga.)*

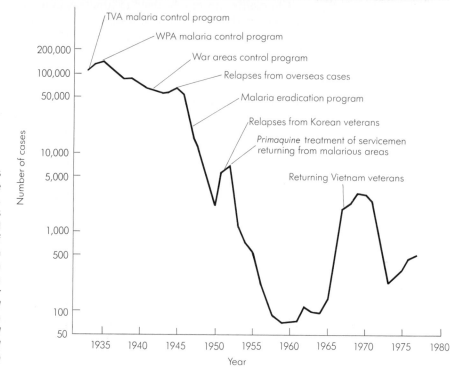

Figure 31-9. Reported cases of malaria by year in the United States, 1933 to 1977. The total number of cases each year has fluctuated with the application of control measures and return of military personnel from areas of world where malaria is common. *(Center for Disease Control: "Reported Morbidity and Mortality in the United States," Annual Summary 1977. Issued September 1978.)*

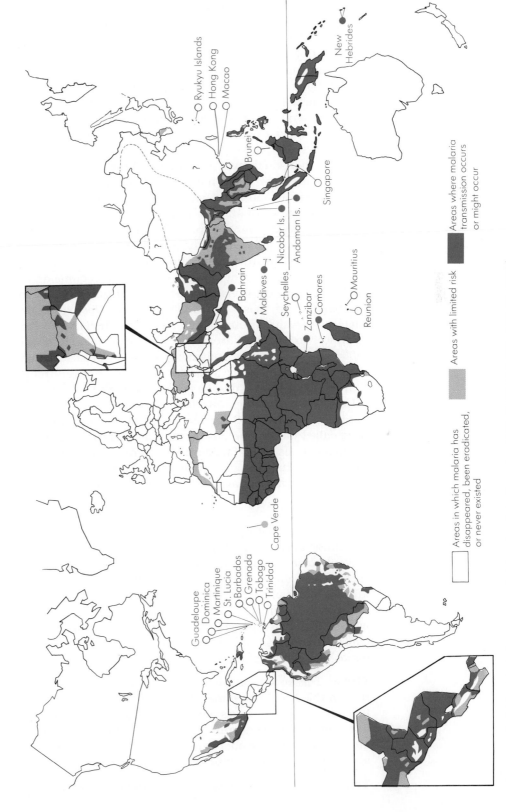

Figure 31-10. Epidemiological assessment of status of malaria, December 1976. (WHO Chronicle, 32:9–17, 1978.)

Areas where malaria transmission occurs or might occur

Areas with limited risk

Areas in which malaria has disappeared, been eradicated, or never existed

Ryukyu Islands
Hong Kong
Macao
Brunei
Singapore
Nicobar Is.
Andaman Is.
New Hebrides
Bahrain
Maldives
Seychelles
Zanzibar
Comores
Mauritius
Reunion
Cape Verde
Guadeloupe
Dominica
Martinique
St. Lucia
Barbados
Grenada
Tobago
Trinidad

Prevention of malaria

The control of malaria depends on the elimination of the insect vector which transmits the disease. Eradication of the mosquito requires destruction of its breeding areas and killing the larval stages and adults. This is not an easy task because some mosquitoes develop resistance to insecticides, the behavioral patterns of others prevent their contact with insecticides, and in certain areas it is physically impossible to eliminate the breeding of mosquitoes.

At the individual level of control, netting can be used around sleeping areas, houses can be screened, and mosquito boots, insecticides, and mosquito repellents can be used. This type of control, of course, can be used for the prevention of all arthropod-borne diseases.

Drug prophylaxis for the prevention of malaria can also be employed. For over 100 years, quinine was the only drug available. In World War II it was replaced by quinacrine, which in turn was supplanted by chloroquine and primaquine, the current drugs of choice. Chloroquine destroys merozoites in the blood, while primaquine destroys schizonts located in the liver. The combination of these two drugs is very effective against susceptible malaria parasite strains found in Africa, India, and Central America. Travelers to endemic areas are advised to take chloroquine phosphate (500 mg) every week starting the week before departure, continuing during their stay in the country, and for 6 weeks after returning to the United States. (The pediatric dosage is 5 mg/kg/week.)

However, beginning in late 1959, chloroquine-resistant strains of malaria parasites appeared in Colombia, Malaysia, Vietnam, and Cambodia. A new drug called mefloquine has been discovered that has been shown to cure chloroquine-resistant falciparum malaria with only one dose. Unfortunately, it is expensive to synthesize, and thus has not seen general use.

There is no vaccine available against malaria at the present time.

Summary and outlook

Arthropods are not only mechanical transmitters of disease (such as the transmission of typhoid fever by the housefly), but they are also biological vectors, since the pathogenic microbes they transmit incubate or develop in them. There are a great number of arthropod-transmitted diseases; they affect millions of people and are widespread over the face of the globe.

Three representative arthropod-transmitted diseases have been discussed. Plague is caused by the bacterium *Yersinia pestis*; yellow fever is caused by a virus; and the etiologic agent of malaria is a protozoan. The outlook for control of these diseases is discussed below.

Plague. The toxic principle of Y. *pestis* has not been identified. Investigation on this should be pursued. The modifying effect of antiserum on the toxemia of severe cases should be determined more precisely. Many questions remain concerning the nature of virulent properties of Y. *pestis*. Much data need to be obtained so that the ecosystems that support enzootic and epizootic plague can be defined more clearly.

Yellow fever. The outlook for eradication of jungle yellow fever is not bright; there is no prospect for achieving it. Transmission of yellow fever from primates to humans will continue to occur. In endemic areas, the only means for preventing outbreaks is by continued maintenance of an active vaccination program.

The southern United States is said to be a receptive or vulnerable area for yellow fever because of the existence of the mosquito *A. aegypti*. But, for unexplained reasons, no yellow fever has spread into this region.

Malaria. Methods other than simple mosquito eradication will be necessary to solve the malaria problem. There is a need for a vaccine against the *Plasmodium* species that cause malaria. The development of such a vaccine has been hindered mainly by the lack of a suitable source of the protozoa (from which a vaccine could be prepared) and the low immuno-genicity of the parasites of malaria (inducing immunity against them is difficult). However, because of recent active research developments the prospects for a vaccine have improved. Cultures of the asexual form of *Plasmodium* (the form that grows in erythrocytes and produces merozoites) have been grown in vitro; such cultures will be good sources of protozoa for vaccine preparation. Rhesus monkeys have been shown to be protected against challenge with the blood forms of *P. knowelsi* by vaccination with merozoites. Furthermore, human volunteers were shown to be protected against mosquito-induced infections with *P. falciparum* and *P. vivax* by vaccination with sporozoites attenuated by irradiation.

Key terms	albuminuria	biological vector	enzootic	fulminant	viremia
	arboviruses	bubo	epizootic	vector	

Questions

1 Describe how arthropod vectors obtain their pathogenic microbes and how these pathogens are transmitted to humans.
2 How are the three forms of plague that occur in humans differentiated from one another?
3 Explain why most cases of plague in the United States in the past 50 years have occurred in rural or suburban surroundings.
4 What is meant by "urban" and "sylvatic" yellow fever?
5 Can yellow fever be completely eradicated from urban areas? Provide epidemiological reasons for your answer.
6 Why is malaria still considered an important disease?
7 Describe the life cycle of the malaria parasite and indicate where sexual fertilization occurs in this life cycle.
8 Distinguish the different forms of malaria in humans caused by the four species of *Plasmodium*.
9 Compare and contrast the nature of the etiologic agents and the modes of transmission of plague, yellow fever, and malaria.
10 Discuss the status of immunoprophylaxis of plague, yellow fever, and malaria.

SEXUALLY TRANSMITTED DISEASES

Sexually transmitted diseases are venereal diseases. The word *venereal* is derived from "Venus," the Roman goddess of love. Thus sexually transmitted diseases (STDs) are infections transmitted through lovemaking, or sexual contact (usually sexual intercourse).

Occurrence of sexually transmitted diseases (STDs)

The incidence of these diseases is very high. The Center for Disease Control estimates that every year at least 8 to 10 million Americans contact some form of STD. STDs occur most frequently in the most sexually active age group—15 to 30 years of age. But anyone who has sexual contact with someone who already has a STD is a potential victim. In general, the more partners one has, the more likely one is to come in contact with venereal disease.

STD is found in every socioeconomic group. Generally, reported male cases of STD exceed reported female cases. This is probably because women are often asymptomatic and are sometimes more difficult to diagnose than men.

544

Transmission of STDs   All STDs are transmitted by intimate body contact. Sexual intercourse with an infected individual carries a risk of infection.

Unfortunately, some infections, including syphilis, can be transmitted to the fetus in the uterus. Others are acquired by the newborn during childbirth, such as in gonococcal conjunctivitis and herpes infection. Infants may also acquire gonococcal conjunctivitis and similar infections from manual contamination and fomites. This is probably because infants are extremely susceptible and because the transfer of pathogens is almost immediate. Infection with STDs in adults by fomites (such as from contaminated toilet seats) is highly unlikely because the pathogenic microbes are very fragile and cannot survive long outside the human body.

Venereal disease is transmitted by any manner of sexual contact. The microbes can be passed from an infected person to another.

The microorganisms of STDs   There are venereal diseases other than gonorrhea and syphilis, both of which are of bacterial etiology. Gonorrhea is the most common, and syphilis the oldest, venereal disease, but there are additional STDs caused

Table 32-1. Microorganisms which may be sexually transmitted and some diseases they may cause

ORGANISM	DISEASE	ORGANISM	DISEASE
Bacteria:		Viruses:	
Neisseria gonorrhoeae	Gonorrhea, conjunctivitis, pelvic inflammatory disease, disseminated gonococcal infection	*Herpesvirus hominis*, type 2	Genital herpes, meningitis, neonatal herpes
		Cytomegalovirus	Infectious mononucleosis-like illness, congenital birth defects
Chlamydia trachomatis	Nongonococcal urethritis, infection of the cervix, conjunctivitis in newborns and adults, trachoma, lymphogranuloma venereum	Genital wart virus	Condyloma acuminatum (genital warts)
		Molluscum contagiosum virus	Genital molluscum contagiosum
Ureaplasma urealyticum (T-mycoplasma)	Possibly nongonococcal urethritis	Others:	
		Trichomonas vaginalis	Vaginal infection, urethritis
Treponema pallidum	Syphilis	*Phthirus pubis*	Pubic lice infestation (crabs)
Haemophilus ducreyi *Calymmatobacterium granulomatis*	Chancroid Granuloma inguinale	*Candida albicans*	External genital and vaginal infection in female, lesions on penis in male
Haemophilus vaginalis	Possibly vaginal infection	*Sarcoptes scabiei*	Scabies
Group B-hemolytic streptococcus	Blood infection in the newborn	Intestinal protozoa	Amebiasis and giardiasis in male homosexuals
Shigella spp.	Shigellosis in male homosexuals		

SOURCE: Dr. King K. Holmes, "Sexually Transmitted Diseases," Public Health Service, National Institutes of Health, DHEW Publication No. (NIH) 76-909, 1976.

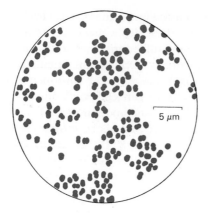 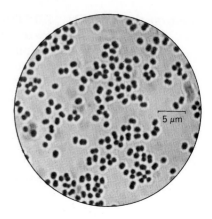

Figure 32-1. Morphology and distinguishing characteristics of *Neisseria gonorrhoeae*. (*Erwin F. Lessel, illustrator; photomicrograph courtesy of Liliane Therrien and E. C. S. Chan, McGill University.*)

Neisseria gonorrhoeae
Gram-negative cocci
Occurs singly, often in pairs with adjacent sides flattened
Not encapsulated
Nonsporulating
Aerobic, facultatively anaerobic
Pathogenic, causing gonorrhea and other infections in humans

by other bacteria, viruses, and protozoa. Some are less common in Canada and the United States than elsewhere. Some of these microorganisms and the diseases they cause are listed in Table 32-1.

Control of STDs

The principal reason for the high incidence of STDs in most countries of the world at the present time (some even call it an epidemic) is contemporary human sexual behavior. But the main control problem lies not in the treatment of venereal disease, but in getting individuals to seek treatment. (Penicillin and other antibiotics can effectively cure most STD.)

The first step toward an effective cure of STD is to recognize that one has it. Unfortunately, in many women and some men the symptoms are not apparent. However, if one partner exhibits symptoms, the other should seek early medical examination. In other words, medical attention after potential exposure to STD and *before* further sexual activity minimizes personal consequences and the risk of transmitting acquired disease. It is important to emphasize that *both* partners should seek medical examination to prevent further dissemination through the community and to minimize "ping-pong" infection, that is, passing the infection back and forth between the two partners.

Gonorrhea

Gonorrhea is a highly infectious, pyogenic disease. It is often referred to as *specific urethritis*. The symptoms of the disease depend on the site of infection, the sex and age of the victim, the duration of the infection, and the occurrence of spread of the bacterial cells.

In men, gonorrhea produces an acute *urethritis* (infection of the urethra, the canal carrying urine outside the body from the bladder). The first sign may be a sudden burning on urination and a pus-containing discharge 2 to 8 days after exposure. In females, infection of the urethra and the *cervix* (the opening of the womb) is usual. This may result in painful urination and a discharge from the vagina, although many women (and a significant number of men) have no noticeable symptoms of early infection. Such symptom-free infections may be one reason for the spread of the disease.

The disease primarily involves the genitourinary tract. However, contamination of the infant during birth can give rise to gonococcal conjunctivitis, affecting the eyes. Various complications of gonorrhea can also develop, among which are endocarditis and meningitis.

Biology of *Neisseria gonorrhoeae*

Gonorrhea is caused by *Neisseria gonorrhoeae*, a gram-negative, nonmotile diplococcus with adjacent sides flattened (Fig. 32-1). Smears made from exudates often show the bacterium inside polymorphonuclear leukocytes (Fig. 32-2). Growth is best at 36°C (96.8°F) on culture media fortified with serum or heated blood, and in an atmosphere containing 5 to 10% CO_2. The round and whitish-gray colonies on chocolate agar (medium containing heated blood) can be identified as being most probably *Neisseria* by testing for *indophenol oxidase* production, or the oxidase reaction. The neisserias give a positive reaction. Gonococci (cells of *N. gonorrhoeae*) differ from other neisserias by their inability to use maltose, sucrose, or fructose, while metabolizing glucose (see Table 32-2). Only glucose, pyruvate, and lactate can be used as energy sources.

The cell also has pili that are about 0.07μm in diameter and up to 2μm long. These appendages appear to be important for the initial attachment of

Figure 32-2. *Neisseria gonorrhoeae* (×630) in urethral exudate. These bacteria range from 0.6 to 1.0 μm and are seen *within* the leukocytes. *(Courtesy of C. Phillip Miller.)*

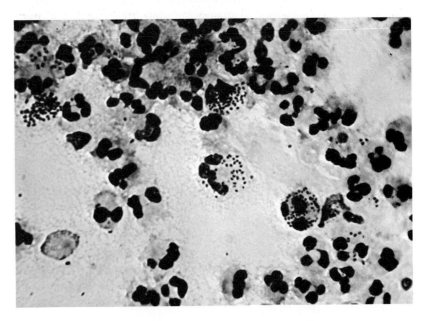

Table 32-2. Some sugar fermentation characteristics differentiating the species of genus *Neisseria*

	N. gonorrhoeae	N. meningitidis	N. sicca	N. subflava	N. flavescens	N. mucosa
Acid from:						
Glucose	+	+	+	+	−	+
Maltose	−	+	+	+	−	+
Fructose	−	−	+	v	−	+
Sucrose	−	−	+	v	−	+
Starch	−	−	v	v	−	+

v = variable.

the bacterium to epithelial cells. Virulence of the organism appears to be related to the presence of pili.

Gonococci are fragile microorganisms. Drying is lethal to them within 1 to 2 h. This is a factor to be considered for successful isolation from clinical specimens. Silver nitrate solution can kill the organism in 2 min. (This is the reason why routine prophylaxis with 1 percent silver nitrate eyedrops has greatly reduced the frequency of occurrence of gonococcal conjunctivitis, also called ophthalmia neonatorum, or acute inflammation of the eyes in newborns.)

Nature of pathogenicity of gonorrhea

Not everyone exposed to gonorrhea is infected with the disease. The reason for this is not clear. The natural microbiota of the genitalia may contribute some resistance to infection by the gonococcus. It is also not known whether naturally acquired gonorrhea confers any immunity to reinfection with the same or other strains of N. *gonorrhoeae*.

The adherence of gonococci by means of pili has been cited as a possible virulence factor. Following mucosal attachment, the gonococci reach the subepithelial connective tissue by penetrating the intercellular epithelial spaces. Gonococci contain endotoxin, and also may excrete a diffusible toxin that induces mucosal damage. The gonococcal cells are present both on and within polymorphonuclear leukocytes. Although most ingested gonococci are killed, many microbiologists believe that some gonococci can survive and multiply within phagocytes. It also has been observed that nonpiliated gonococcal cells are readily ingested and killed, while piliated cells remain extracellular.

In its passage through an infected mother's birth canal, a newborn infant may acquire a potentially blinding infection of the eyes (ophthalmia neonatorum). As mentioned before, silver nitrate solution is placed in the eyes of infants immediately after birth to prevent this kind of infection.

In both sexes, infection may spread along the genital tract. In men, sterility may result if the infection extends to the prostate, *epididymis* (portion of the seminal duct lying posterior to the testes), and testes. In women, spread is more common and is associated with more serious sequelae. In approximately 15 percent of infected females, the infection spreads to the *fallopian tubes* (which carry eggs from the ovaries to the uterus), producing inflammation (*salpingitis*). At the first menstrual period, there is lower abdominal pain. Salpingitis produces partial or complete obstruction of the fallopian tubes, which may result in tubal pregnancy or lead to sterility. Implantation of the fertilized ovum in the fallopian tube is potentially fatal and usually requires early surgery.

Gonorrhea is not always confined to the genital and urinary tracts. In a few patients, the bacteria enter the blood and spread throughout the body to induce fever, chills, and loss of appetite. They then localize in various parts of the body; they can produce small red pustules on the skin or arthritis in the joints (knees, wrists, and joints of fingers and hands). Complications of endocarditis and meningitis have been mentioned previously. Gonococcal infection in the rectum and throat also are possible and may be due to changing sexual preferences and practices.

Gonorrhea may be treated with aqueous procaine penicillin accompanied by oral probenecid or with oral ampicillin together with probenecid; the probenecid retards the urinary excretion of the antibiotics. Unfortunately, in 1976, penicillinase-producing gonococci were isolated from widely separated areas in England and the United States. These strains have the ability to destroy penicillin and are thus resistant to the antibiotic.

Laboratory diagnosis of gonorrhea

Diagnosis is made by microscopic examination of the bacterium in discharge specimens or by growing and observing the gonococcus in material obtained from inside the urethra in males, from the cervix and urethra in females, and from any other suspected site in both sexes. The gram stain of urethral and endocervical exudates is considered diagnostic of gonorrhea when typical gram-negative diplococci are observed *within* leukocytes.

Cultivation of gonococci is done on chocolate-agar medium or Thayer and Martin medium. Cultures should be incubated under CO_2 (5 to 10%) at 36°C (8°F) for 48 h. Typical gonococcal colonies should be confirmed by the oxidase reaction, gram stain, and sugar fermentation tests. (The oxidase reaction is performed by covering colonies with 1 percent tetramethyl-p-phenylenediamine; *Neisseria* colonies turn from a translucent off-white to a purple color.)

Epidemiology of gonorrhea

The rising incidence of gonorrhea and other STDs in the Western world coincided with the introduction of oral contraceptives and contraceptive intrauterine devices in the 1960s. These contributed to increased sexual freedom among females and, naturally, the decreased use of spermicidal preparations and condoms, both of which also afford some protection against gonorrhea. This epidemiological picture is reflected in the rising incidence of gonorrhea in the United States, as shown in Fig. 32-3. As seen in the figure, between 1965 and 1975, reported cases of gonorrhea increased annually by at least 10 percent. However, in 1976, reported cases increased by less than one-half of 1 percent. This was probably due to increased control efforts which were intensified in 1972 when the United States federal government committed itself to a national gonorrhea control program. (The incidence of gonorrhea has also leveled off or declined in the United Kingdom and the Scandinavian countries.) Nonetheless, the true

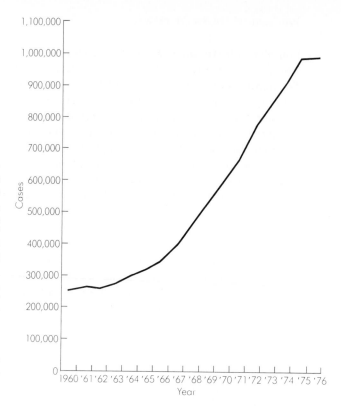

Figure 32-3. Reported cases of gonorrhea by year in the United States, 1960 to 1976. From 1965 to 1975, the number of reported cases of gonorrhea in the United States tripled—from 324, 925 to 999, 937. Since 1975, however, reported cases have leveled off. For 1979, the projected number of cases was 1,010,000, an increase of 1 percent since 1975. *(Center for Disease Control: "Morbidity and Mortality Weekly Report," vol 26, 1977.)*

incidence is probably 10 times higher in the United States because many cases go unreported to the Center for Disease Control.

The only natural host for *N. gonorrhoeae* is the human. Thus the disease is contracted by direct contact with an infected sexual partner who is a carrier.

Prevention of gonorrhea

At the present time, there is no vaccine against gonorrhea. The condom and intravaginal spermicides remain the best means for reducing the risk of infection. Additional preventive measures against gonorrhea are similar to those for other STDs and were discussed previously.

Syphilis

Syphilis is caused by bacteria called spirochetes. It is not as widespread as gonorrhea, but is more dreaded because it is potentially more devastating. Like gonorrhea, it is spread by direct contact with the lesions of someone in the infectious stage of the disease. The spirochete, like the gonococcus, is a fragile microbe outside the human body, so it is very unlikely that the disease can be contracted from inanimate objects.

Treponema pallidum enters the body during sexual contact through minute abrasions of the epithelium, by penetrating intact mucous membranes, or possibly through unbroken skin via hair follicles. The incubation period of syphilis ranges from 10 to 90 days (average 21 days) following infection. Syphilis, if left untreated, runs through several stages.

Primary syphilis. The first symptom is a small firm sore or ulcer called a *chancre* at the site of infection. It is usually at the tip of the penis in men and in the cervix or vagina in women. While it is obvious in the man, it is frequently hidden in the woman. It does not itch or hurt. Thus primary syphilis can pass undetected. Treponemas can usually be found in such chancres by darkfield microscopic examination.

Also during this stage, the spirochetes invade lymph nodes, causing them to become enlarged and firm. After 3 to 5 weeks, the chancre spontaneously heals, and the disease outwardly appears quiescent. But meantime, the organisms are distributed by the bloodstream throughout the body.

Secondary syphilis. This stage of the disease is ushered in by a rash (cutaneous eruption) which appears anytime from 2 to 12 weeks after the chancre disappears (Fig. 32-4). There is now generalized spread of the disease and widespread *lymphadenopathy* (diseased lymph nodes). Syphilis has been called the "Great Imitator" because the symptoms of this stage of the disease are similar to those of other diseases such as the flu or infectious mononucleosis. Besides rashes, the symptoms include sore throat, tender lymph nodes, fever, malaise, and headache. Sometimes the hair may fall out in patches. Lesions occur on the mucous membranes, eyes, bones, and central nervous system; these lesions swarm with treponemas (Fig. 32-5). Victims may suffer from only one or two of these symptoms or all of them. This stage lasts several weeks, and the symptoms, including lesions, disappear without treatment. But, meantime, the treponemas have probably begun to invade other organs in the body.

It is only in the primary and secondary stages of syphilis, which last for up to 2 years, that an individual is infectious to others.

Figure 32-4. Secondary syphilis is ushered in by a widespread rash. *(From Armed Forces Institute of Pathology.)*

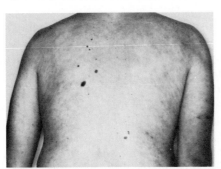

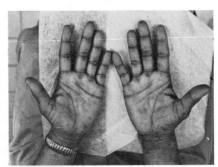

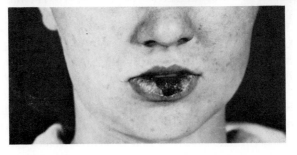

Figure 32-5. In secondary syphilis, lesions swarming with spirochetes may occur on the mucous membranes. Note the lesion on the lip. *(From Armed Forces Institute of Pathology.)*

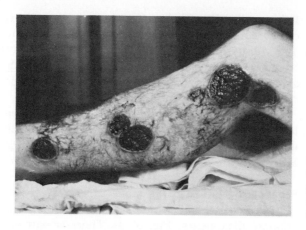

Figure 32-6. In tertiary, or late, syphilis, lesions called gummata rupture and result in ulcers. *(From Armed Forces Institute of Pathology.)*

Latent syphilis. If untreated, secondary syphilis is followed by latent syphilis. During this stage the patient shows absolutely no obvious symptoms. It may last months, years, or even a lifetime. This latent stage can only be detected by blood tests (serology).

Tertiary, or late, syphilis. This stage appears in about 30 percent of untreated individuals and may occur 5 to 40 years after the initial infection. The results of the silent but deadly workings of the spirochetes during the latent stage become evident. Tertiary lesions occur in the central nervous system, the cardiovascular system, the skin, and other vital organs, such as the brain, eyes, bones, kidney, and liver. These lesions, called *gummata*, rupture and result in ulcers, as shown in Fig. 32-6. Thus the patient may be stricken with mental illness, blindness, or heart disease; death may result.

Biology of *Treponema pallidum*

The etiologic agent of syphilis is the spirochete *Treponema pallidum*. It is a delicate, tightly coiled microorganism with pointed ends and 6 to 14 spirals; it measures 0.25 to 0.3 μm by 6 to 15 μm (Fig. 32-7). It can be recognized best under dark-field microscopy of clinical specimens from primary- and secondary-stage lyphilitic lesions; the spiral nature of the organism and its corkscrewlike rotation are easily observed.

Treponema pallidum has an outer membrane, or sheath, called a *periplast* which surrounds the remaining inner components of the cell (collectively termed the *protoplasmic cylinder*). An *axial filament*, composed of three to six fibrils, is located between the periplast and the protoplasmic cylinder.

Virulent *T. pallidum* has not been cultured in vitro. Nonvirulent (nonpathogenic) strains of *T. pallidum*, such as the Reiter and Noguchi strains, have been successfully cultivated in vitro and serve as sources of antigens for diagnostic laboratory tests.

Nature of pathogenicity of syphilis

Syphilis occurs naturally only in humans and is transmitted primarily by direct sexual contact or by an infected mother to the fetus (congenital, or prenatal, syphilis) via the placenta. In untreated cases, 25 percent of fetuses die before birth; another 25 to 30 percent die shortly after birth; the others develop late symptomatic complications (for example, deafness).

Great numbers of treponemas in the blood and tissues are destroyed during secondary syphilis. The mechanism for this is not known exactly. It has been suggested that the spirochetes are destroyed by a lytic process involving specific antibody and complement.

Penicillin is the antibiotic of choice in the treatment of syphilis. Other antibiotics are sometimes used. The effectiveness of treatment must be monitored over a long period of time.

Laboratory diagnosis of syphilis

Diagnosis of syphilis can usually be made from the combined information of symptoms, history of exposure, and a positive blood test, or by dark-field microscopic examination.

A positive dark-field examination (for characteristic morphology and motility of spirochetes) of a lesion is the only means of making an absolute diagnosis of primary syphilis. For secondary syphilis too, an absolute diagnosis depends on dark-field examination of exudate from a moist, cutaneous, nonoral lesion. (The oral cavity may contain many nonsyphilitic spirochetes.) Fortunately, serologic tests for syphilis are reactive or dependable by the secondary stage of the disease. They are also of great importance in laboratory diagnosis. Since the initial use of the complement-fixation reaction by Wassermann and others in 1906, over 200 tests for syphilis have been reported. However, only a few are in common use today (Table 32-3).

The type of antigen used serves as a criterion for initial differentiation of serological tests for syphilis. That is, we have *nontreponemal* and *treponemal* tests using nontreponemal and treponemal antigens, respectively. It

Figure 32-7. Morphology and distinguishing characteristics of *Treponema pallidum*. (Erwin F. Lessel, illustrator; photomicrograph courtesy of Liliane Therrien and E. C. S. Chan, McGill University.)

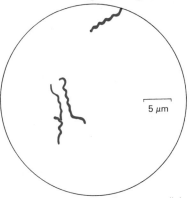 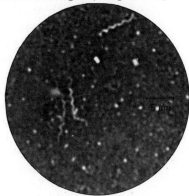

Treponema pallidum
  Gram-negative, slender, helical rods with pointed ends; three axial fibrils are inserted into each end of the cell
  Occurs singly
  Not encapsulated
  Nonsporulating
  Motile
  Not flagellated
  Anaerobic
  Pathogenic, causing syphilis

NATURE OF ANTIGEN	SEROLOGIC REACTION	TEST
Nontreponemal	Flocculation	Venereal Disease Research Laboratory (VDRL) slide
		Rapid reagin (various)
Treponemal	Immobilization	*Treponema pallidum* immobilization (TPI)
	Immunofluorescence	Fluorescent treponemal antibody–absorption (FTA-ABS)
	Hemagglutination	*Treponema pallidum* hemagglutination assay (TPHA)

follows that two kinds of antibodies are measured. One of the two types of antibodies produced by the host in response to *T. pallidum* infection is the reagin, or Wassermann, antibody. Its formation results from the interaction of the treponema with the host's tissues. This antibody reacts with a variety of nontreponemal substances (antigens). One such substance is cardiolipin, a highly purified extract from beef heart. The second type of antibody is produced in response to specific treponemal antigens, which are either live *T. pallidum* grown in the testes of rabbits or the Reiter strain of *Treponema* grown on artificial media.

Many nontreponemal tests have been developed over the years; these include the Wassermann, Eagle, Hinton, Kahn, Kline, and Kolmer tests. They are now of historical importance only. The nontreponemal tests currently used in the United States are flocculation procedures, including the Venereal Disease Research Laboratory (VDRL) slide test and rapid reagin tests. The most widely employed nontreponemal serologic test for syphilis is the VDRL slide test. It is easily controlled and performed, and it can be accurately quantitated. This test mixes a buffered saline suspension of cardiolipin, plus lecithin and cholesterol, on a slide with the patient's serum. The slide is shaken on a mechanical apparatus for several minutes. A positive test is indicated by a clumping, or flocculation, of the cardiolipin. Results are reported as "reactive," "weakly reactive," or "nonreactive."

Treponemal tests are technically more difficult and cost more to perform than nontreponemal tests. They have greater specificity, however, and therefore are primarily employed as verification procedures. Treponemal tests are classified according to the method used to detect the antigen-antibody complex (see Table 32-3).

The *Treponema pallidum* immobilization (TPI) test was the first acceptable treponemal verification test. But it is not as sensitive as the fluorescent-antibody technique. The fluorescent treponemal antibody–absorption (FTA-ABS) test, which is performed manually, has evolved as the standard treponemal test in use today. The absorption step in this test is necessary to eliminate nonspecific reactions which occur as a result of antibodies to common antigens shared by both pathogenic and saprophytic treponemas. This is done by mixing the patient's serum with a standardized extract from nonpathogenic Reiter treponemas. The absorbed serum is then employed to cover a smear of pathogenic, rabbit-testes-grown treponemas. After 30 minutes at 37°C (98.6°F), the slide is rinsed well to remove unreacted serum proteins, and the smear is then covered with fluorescently labeled anti-human gamma globulin. This again is permitted to react for 30 min before rinsing. The slide is then examined under a fluorescent microscope for the presence of fluorescent treponemas. Since fluorescently labeled anti-human gamma globulin could react only with spirochetes coated with human antibody, the presence of fluorescent treponemas indicates the presence of specific antibodies to *T. pallidum*. The FTA-ABS test is reported as "nonreactive," "borderline," or "reactive." This test is the most sensitive serologic test in all stages of syphilis. It is more reactive earlier in primary syphilis and stays positive longer in late syphilis than the VDRL or TPI tests.

Epidemiology of syphilis

Since 1962, reported cases of syphilis have increased annually by at least 4.7 percent (Fig. 32-8). In 1976, however, reported cases of primary and secondary syphilis declined by 7.4 percent compared with cases reported during 1975. Again, like gonorrhea, this decline was probably due to increased control efforts which were intensified in 1972 by the United States federal government.

As with gonorrhea, the number of *reported* cases of early syphilis (primary, secondary, and early latent cases) is not an indication of the *true* incidence because most cases go unreported. The true incidence is probably more than six times the number of reported cases. In 1977, there were 64,473 reported cases of syphilis of all stages in the United States.

Prevention of syphilis

There is no vaccine against syphilis. At the individual level, use of the condom is highly effective. At the community level, the major means of preventing syphilis is through control programs which include serological

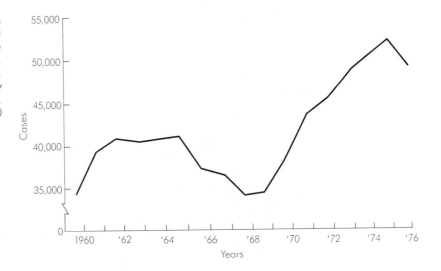

Figure 32-8. Reported cases of primary, secondary, and early latent syphilis in the United States, 1960 to 1976. (*Center for Disease Control, "Morbidity and Mortality Weekly Report," vol. 26, 1977.*)

screening and treatment of contacts. Congenital syphilis can be prevented by adequate prenatal care.

Other STDs

Genital herpes

Genital herpes is a disease of the genitals caused by the herpes simplex virus, *Herpesvirus hominis*. There are two closely related strains of this virus, designated type 1 and type 2. Type 1 is best known as the cause of cold sores or fever blisters on the lips. Type 2 infects the genitals.

First, or primary, herpes infection of the genitals is transmitted by sexual contact. After an incubation period of 10 to 20 days, the affected site tingles or burns and blisters develop. These blisters are on the penis, labia, vagina, and other contact parts. They soon rupture to leave painful shallow ulcers. The clinical syndrome includes fever and other general body infection. Blisters usually heal spontaneously in 10 to 14 days.

However, the symptoms of primary genital herpes may recur without further sexual contact. Recurrent herpes infection varies in frequency over an indefinite period of time; it tends to be less disabling than the primary infection.

The most severe complication of genital herpes is infection of the newborn during birth (Fig. 32-9). This is severe in many infants and may even be fatal. Survivors may have residual damage to the central nervous system or the eyes. If a mother has genital herpes at the time of delivery, cesarean section may reduce the likelihood of neonatal infection.

A link has been demonstrated between genital herpes infection and cervical cancer, although it has not been proven that the herpesvirus actually causes the cancer.

It has been estimated that there are 300,000 new cases of genital herpes each year in the United States.

Nonspecific urethritis

This disease affects about 2.5 million men in the United States each year, and about half the cases are caused by a small bacterium named *Chlamydia trachomatis*. The etiology of the remaining cases is unknown.

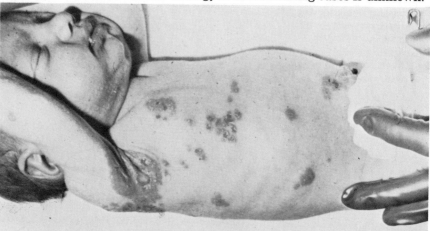

Figure 32-9. Herpes simplex *(Herpesvirus hominis)* skin lesions in a newborn (age 12 h). *(Courtesy of Center for Disease Control, Atlanta, Ga.)*

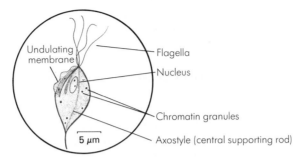

Figure 32-10. *Trichomonas vaginalis,* the causative protozoan of trichomoniasis. It is readily recognized by its characteristic motility in wet smears prepared from vaginal discharge. *(Erwin F. Lessel, illustrator.)*

In men the symptoms of nonspecific urethritis mimic those of gonorrhea. They include frequent and painful urination, a discharge from the urethra, and inflammation of the eyes, joints, mouth, or testes. In women the disease is less well-defined. Many women show no symptoms at all; some may have a discharge, bleeding from the cervix, or infection of the cervix.

Lymphogranuloma venereum (LGV)

LVG is also caused by certain strains of *Chlamydia trachomatis.* It is an important disease in the tropics and semitropics. Many cases have been brought back to the United States from Vietnam.

Symptoms appear in 1 to 3 weeks after exposure. A small lesion appears on the genitals or rectum. As the infection localizes in the lymph nodes, a *bubo,* or suppurating lesion, may appear, and multiple ulcers, rectal stricture, and *elephantiasis* (chronic enlargement) of the genitals sometimes develop. The patient usually experiences fever, chills, headache, and abdominal and joint pains.

Chancroid, or soft chancre

Until recently, this venereal disease was common only in the tropics. However, it has been making increasing appearances in the United States. It is caused by the bacterium *Haemophilus ducreyi,* a very small, gram-negative, nonmotile rod. One to five days after exposure, one or more painful ulcers occur at the site of contact.

Granuloma inguinale

This is a bacterial infection caused by *Calymmatobacterium granulomatis,* a gram-negative pleomorphic rod. The disease involves the development of small, spreading, pus-filled ulcers on the genitals. It takes at least 3 months for the symptoms to appear. In the interim, a great many people can be infected before treatment is sought.

Trichomoniasis

This is a parasitic infection caused by a protozoan, *Trichomonas vaginalis* (Fig. 32-10). An estimated 2.5 to 3 million Americans have this disease each year. In humans this organism infects only the genitourinary tract.

In women the common symptom is *vaginitis* (inflammation of the vagina). Although vaginitis may be asymptomatic, it usually causes severe general itching and an offensive-smelling discharge. In men trichomoniasis is usually without symptoms, and hence it is more difficult to diagnose.

Within 4 to 20 days after exposure, symptoms of trichomoniasis may appear. Diagnosis is performed by microscopic examination of the discharge or by growing the protozoan in laboratory culture. Except for personal annoyance, the infection does not lead to serious complications.

Summary and outlook

Venereal diseases are transmitted by direct sexual contact; thus, they are also called sexually transmitted diseases (STDs). There are more of these diseases than gonorrhea and syphilis, but these two infections are the most familiar. Some of the others are less common in the United States than elsewhere. Altogether there are at least a dozen STDs, as shown in Table 32-1.

The toll of all these infections is very high not only in the United States but all over the world. In many countries, STDs are said to be in epidemic numbers. This situation exists despite the fact that specific etiologic agents have been identified, diagnosis is relatively easy and accurate, and treatment methods are well established. The reasons for this situation must be attributed to ignorance and false confidence, combined with reticence and changing sexual mores. Control of STDs must depend on early detection of cases, tracing of contacts, local prophylactic measures, prompt and adequate treatment, and education.

Gonorrhea and syphilis have been discussed at some length in this chapter. The outlook for their control is described below.

Gonorrhea. In spite of the fact that antigonococcal agents (such as penicillin) have been available for treatment of gonorrhea, they are not sufficient to eradicate the disease. Personal responsibility in illness and sexual behavior have been so poor that gonorrhea remains a serious problem.

The search for gonococcal antigens should continue with the goal of obtaining an effective vaccine; at the least, immunogens might be found that can be useful in serological diagnosis. It has been shown experimentally that antibodies against the pili of the gonococcus seem to have the potential of conferring effective immunity.

Penicillinase- or ß-lactamase-producing *Neisseria gonorrhoeae* were reported in England and the United States in early 1976. Patients infected with these gonococci remained symptomatic even after undergoing routine penicillin therapy. (Antibiotic-resistance plasmids, called R plasmids, had been introduced into *N. gonorrhoeae* from some other bacterial species to make it resistant.) The emergence of such penicillinase-producing strains will pose a major problem for therapy. Other effective antibiotics or a combination of them will have to be used.

Since 1976, penicillinase-producing *N. gonorrhoeae* strains have been detected in many additional countries (more than 16 in 1978) including widely separated ones such as Canada, Australia, Hong Kong, Singapore, the Philippines, Sweden, and Switzerland. (The resistant organism accounts for about 30 percent of all recent gonococcal isolates in the Philippines, and for 16 percent in the Republic of Singapore in 1979.) In the United States, there were 220 reported cases of penicillinase-producing *N.*

gonorrhoeae during 1978 and 554 total cases during the 3-year period ending February 1979.

Syphilis. The outlook for this disease has many remaining problems. One of these is the need for developing an effective vaccine, which would demand the in vitro cultivation of *Treponema pallidum* of appropriate antigenic characteristics.

Key terms

axial filament	fallopian tubes	urethritis
bubo	gummata	vaginitis
cervix	lymphadenopathy	venereal
chancre	periplast	vulva
elephantiasis	protoplasmic cylinder	
epididymis	salpingitis	

Questions

1 What reasons can you give for the high incidence of sexually transmitted diseases in many countries?
2 Can STDs be transmitted by means other than direct sexual contact among adults? Explain.
3 Discuss control measures that can be taken to decrease the high incidence of STDs.
4 What are the characteristic properties of *Neisseria gonorrhoeae*? Discuss those that confer virulence on the organism.
5 Describe the sequelae that may result from untreated gonorrhea.
6 Compare the morphology and nutrition of *Neisseria gonorrhoeae* and *Treponema pallidum*.
7 Explain why it is important that syphilis be effectively treated.
8 Describe briefly the different types of serological reactions involved (methods used to detect the antigen-antibody complex) in the serologic tests for the diagnosis of syphilis.
9 Describe the differences between *primary* herpes (type 2) infection and *recurrent* herpes infection.

OTHER TRANSMISSIBLE DISEASES

In the preceding chapters we have discussed many diseases grouped according to their principal modes of transmission. These are airborne, foodborne, waterborne, arthropod-borne, and sexually transmitted diseases. As is usual with attempts to group (or classify) anything, there are always exceptions. And so there are certain interesting diseases which cannot fit in "nicely" with our scheme of grouping, but which we consider important enough to warrant inclusion in a separate chapter. Such diseases constitute the contents of this chapter.

Gastroenteritis caused by *Escherichia coli*

Even though *Escherichia coli* (Fig. 33-1) is part of the normal microbiota of the intestinal tract, it has now been established that certain strains are capable of causing a moderate to severe gastroenteritis in humans and animals. Although *E. coli* is an indicator organism used in water analysis to test for recent fecal contamination, it is not typically associated with water as a mode of transmission. Rather, it is transmitted by hand-to-mouth activity or by *passive transfer* (no necessity for growth) in food or drink.

Escherichia is now regarded as a genus with only one species in which there are several hundred antigenic types. The types are characterized by

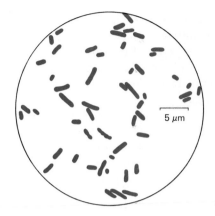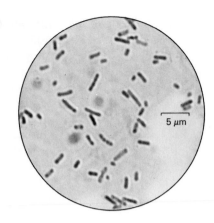

Figure 33-1 Morphology and distinguishing characteristics of *Escherichia coli.* (Erwin F. Lessel, illustrator; photomicrograph courtesy of Liliane Therrien and E. C. S. Chan, McGill University.)

Escherichia coli
  Gram-negative rods
  Occurs singly, in pairs, and in short chains
  Usually not encapsulated
  Nonsporulating
  Motile or nonmotile; peritrichous
  Aerobic, facultatively anaerobic
  Normal inhabitant of intestines; frequently causes infections

different combinations of the antigens O (somatic lipopolysaccharide antigens in the cell wall), K (capsular polysaccharide antigens), and H (flagellar protein antigens). In addition, the K antigens are divided into L, A, or B antigens on the basis of differing physical characteristics.

Escherichia coli involved in acute diarrhea may be grouped into three categories: *enteropathogenic, enteroinvasive,* and *enterotoxigenic.*

Enteropathogenic
E. coli

Enteropathogenic *E. coli* causes acute gastroenteritis in the newborn and in infants up to 2 years of age. The use of detailed serotyping in association with epidemiological investigation suggests that about 17 O serotypes have been implicated in infantile enteritis in many countries. The most common serogroups are O26, O55, O86, O111, O114, O119, O125, O126, O127, O128, and O142.

The mechanism by which this group of *E. coli* causes diarrhea is unknown, but it is known that colonization of the jejunum and upper ileum of the small intestine with the enteropathogenic strain is a prerequisite. Serotyping remains the only way to recognize potentially pathogenic strains among the O serogroups.

Enteroinvasive *E. coli*

Certain *E. coli* serotypes other than the enteropathogenic ones were found to cause diarrhea in older children and adults. These were serotypes O124, O136, and O144. These serotypes of *E. coli* invade the epithelial cells of the large bowel and produce a clinical syndrome similar to that caused by *Shigella* (Chap. 29). These bacterial strains are referred to as enteroinvasive; fortunately, this property is restricted to a small number of serotypes.

Enterotoxigenic *E. coli*

Enterotoxigenic, or enterotoxin-producing strains of *E. coli,* are most frequently of the following serotypes: O6, O8, O25, O78, O148, and O159. They produce one or both of two different toxins. Some strains produce a heat-stable toxin (ST), while others synthesize in addition a heat-labile toxin (LT). A few strains produce LT only. Both toxins cause diarrhea in adults and infants.

Little is known concerning the chemistry or the mode of action of ST, although it appears to stimulate guanylate cyclase in the gut of susceptible animals. It is a small protein which retains its toxic activity even when heated at 100°C (212°F) for 30 min.

LT is destroyed by heating at 65°C (149°F) for 30 min. Its mode of action is identical to that of cholera toxin (Chap. 29). It stimulates adenyl cyclase activity in the intestinal epithelial cells and produces an accumulation of cyclic adenosine monophosphate (cAMP). In turn, cAMP induces a net secretion of fluid into the bowel lumen. This results in the loss of copious amounts of fluid from the intestine.

The production of LT and ST is controlled by transferable plasmids. Thus the possibility exists for any serotype of *E. coli* to become an enterotoxigenic strain.

Hepatitis

Viral hepatitis (inflammation of the liver) is a common infectious disease. It is caused by at least two distinctly different viruses which can be clinically differentiated from one another by the epidemiology of the infection and the length of their incubation periods. One virus is designated hepatitis type A (causing what was formerly called infectious hepatitis); the other is hepatitis type B (causing serum hepatitis). Both viruses cause the same pathology; neither has been grown in cell cultures. In 1977 in the United States, there were 31,153 cases of hepatitis A, 16,831 cases of hepatitis B, and 8,639 cases of unspecified hepatitis. Table 33-1 summarizes the differentiation between viral hepatitis type A and viral hepatitis type B.

Hepatitis A

The major mode of transmission of hepatitis A is by the fecal-oral route

Table 33-1. Differentiation of hepatitis A and hepatitis B

DISTINGUISHING FEATURE	TYPE A	TYPE B
Principal sources	Food, water	Blood products
Major route of infection	Fecal-oral	Parenteral injection
Seasonal incidence	Higher in fall and winter	Year-round
Age incidence	Children and young adults	Any age
Incubation period	15–60 days	50–160 days
Clinical onset	Abrupt	Gradual
Fever over 100.4°C	Common	Uncommon
Gamma globulin useful in prophylaxis	Effective	Ineffective
Viral particles	27 nm	42 nm

through person-to-person transfer or by means of food or water. The ingestion of raw oysters or clams from fecally contaminated water has also caused a large number of heptatitis A infections.

Hepatitis A is endemic in nursery schools, mental institutions, and all establishments and societies where there is a high risk of fecal contamination. The disease is generally limited to humans.

The hepatitis A virus appears to be one of the most stable viruses infecting humans. It is very resistant to many disinfectants. The virion measures approximately 27 nm in diameter and appears to have an icosahedral symmetry. It may be a parvovirus; but since the etiologic agent has not been cultivated, this suggestion is equivocal.

Following ingestion, the acid resistance of virus A allows it to pass through the stomach to the small intestine. The virus infects the mucosal epithelial cells, replicates, and spreads to adjacent cells and then to the liver via the portal circulation. About midway through the incubation period, the feces become infective, and the virus persists in the stools (and blood) for several weeks or months. Before the jaundiced stage of the disease, there is loss of appetite, fatigue, malaise, abdominal discomfort, and fever. With the appearance of jaundice, the patient feels better, but the liver remains tender to the touch. Jaundice persists for 1 to 3 weeks. Complete recovery may take 8 to 12 weeks. Fatalities are rare. During the convalescent period, the patient frequently remains weak and occasionally depressed.

Hepatitis B

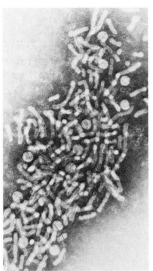

Hepatitis B virus is associated with great numbers of viruslike particles which are found in the blood of both infected individuals and asymptomatic carriers. These viruslike particles (Fig. 33-2) were originally discovered in the serum of an Australian aborigine by Baruch S. Blumberg and his colleagues and until recently have been called *Australia antigen*. (Blumberg received the Nobel prize in physiology and medicine in 1976 for his work on Australia antigen and the biology of hepatitis B.) Three morphological forms of this antigen exist. The predominant form is a spherical particle with an average diameter of about 22 nm; also found are filaments or tubular particles 22 nm in diameter, with varying length; the third form is a 42-nm spherical particle with a complex morphology. The smaller 22-nm particles do not contain nucleic acid and are assumed to be excess viral-coat protein. They have a surface antigen designated HBsAg. The 42-nm particle (originally called the Dane particle) represents the intact virus. It has an inner, double-stranded DNA and at least two antigens—HBsAg and an antigen designated HBcAg which is found within the core of the virus.

As noted in Table 33-1, the incubation period of hepatitis B is longer than that of hepatitis A. The symptoms are the same as those for hepatitis A, but their onset is more gradual. In most cases, viruses disappear from the blood

Figure 33-2. Viruslike particles in the blood of an individual with hepatitis B (×64,000) *(Courtesy of Margaret Gomersall, McGill University.)*

when liver functions return to normal. But in 10 to 30 percent of cases, the viruslike particles may be found in the blood for months or years after recovery. These cases become carriers of the disease, and the estimate is that there are over 100 million carriers of hepatitis B in the world. This is the reason that all donor blood is now routinely examined for HBsAg. But blood transfusion is not the only way that the virus is spread. Most infections of hepatitis B occur through the use of contaminated needles and syringes, blood and blood products, and renal and hemodialysis units. Drug addicts, dialysis patients, and those who receive blood products are most prone to infection. Other high-risk persons include health care professionals (nurses, physicians, laboratory technicians, and dentists) who come in close contact with hepatitis-infected persons. The fecal-oral route, the respiratory route, and insect vectors are also possible modes of transmission of hepatitis B.

Dental caries

Dental caries is one of the most common human infectious diseases and is ubiquitous in civilized populations. In the United States it affects more than 90 percent of the population. The cost of its treatment ranks it among the most expensive of bacterial infections. Besides, it also results in considerable loss of time and productivity.

Dental caries is the localized destruction of teeth by bacterial action. The initiation and extent of this destruction are determined by other factors, such as composition and quantity of saliva, nutritional balance, oral hygiene, fluoride level in the drinking water, and diet. Before infection begins, a superficial *plaque* is formed over the enamel of the tooth. Dental plaque can be defined as aggregations of bacteria and organic material on the tooth surface. The microorganisms are embedded in the matrix of the

Figure 33-3. Anatomy of a tooth, showing sites of microbial attack. *(Redrawn from a diagram in World Health Magazine, World Health Organization, Geneva, 1973.)*

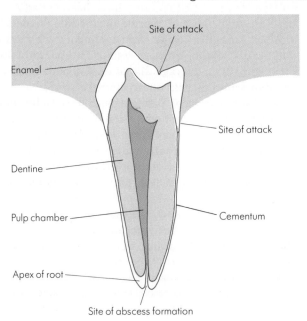

Site of attack

Enamel

Site of attack

Dentine

Pulp chamber

Cementum

Apex of root

Site of abscess formation

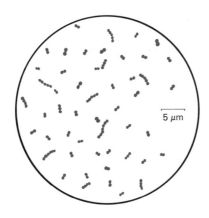

Streptococcus mutans
  Gram-positive cocci
  Occur in pairs and chains
  Not encapsulated
  Nonsporulating
  Nonmotile
  Aerobic, facultatively anaerobic
  Found in dental plaque

5 μm

Figure 33-4. Morphology and distinguishing characteristics of *Streptococcus mutans. (Erwin F. Lessel, illustrator.)*

organic matter; this matter is largely protein and polysaccharide in nature and is derived partly from saliva and partly from bacterial metabolic products. This matrix binds the microorganisms to each other and to the tooth surface.

The infection starts on the outer surface of the tooth, which is composed of enamel. Although enamel is the hardest substance in the human body, it is, paradoxically, one of the most vulnerable because once it is damaged, it cannot repair itself. Early lesions, called *white spots*, form on the enamel surface covering the crown of the tooth. They eventually progress into the dentine and reach the pulp. The anatomy of the tooth, the site of attack by bacterial action, and the site of abscess formation when the bacteria reach the pulp tissue are shown in Fig. 33-3. Bacteria ferment the dietary carbohydrate, predominantly sucrose, producing acid which dissolves the calcium salts (hydroxyapatite crystals) of the enamel in a process known as *demineralization*. This dissolution seems to precede the loss of the organic matrices of both enamel and dentine.

Etiology of dental caries

Bacteria have been proven to be the causative agents of dental caries. In addition, the disease can be transmitted between animals, thus emphasizing its infectious nature. Epidemiological studies in humans have shown that frequent consumption of sucrose-containing foods increases caries development.

As mentioned previously, carious lesions develop under dense masses of bacteria and organic matter, called dental plaques, which are adherent to the tooth surface. Many different bacterial species colonize teeth, and their distribution varies from person to person, from tooth to tooth, and even from one area to another on the same tooth. Therefore, the term *dental plaque* only vaguely describes the nature of these bacterial masses.

Of all the microorganisms investigated, none appears to fill the cariogenic role more adequately than *Streptococcus mutans* (Fig. 33-4). *Streptococcus mutans* colonizes teeth in a highly localized manner. It has been isolated from almost all human-enamel carious lesions examined. The data indicate that *S. mutans* must be considered an important organism in the initiation of carious lesions on enamel surfaces.

However, dental microbiologists are hesitant to rule out other bacterial species. For example, it has been shown that *Lactobacillus* spp., *Actinomyces* spp., and other bacterial species have been isolated from human carious lesions and can induce caries when injected into caries-free animals.

Prevention of dental caries

Since the development of dental caries is due to the interactions of specific types of bacteria, a susceptible tooth surface, and a multitude of other factors, the following combination of methods for caries prevention is recommended:

1 Restriction of the frequency and quantity of dietary carbohydrates ingested, particularly sucrose.
2 Diligent observance of good oral hygienic practices which remove bacterial accumulations from the teeth (for example, brushing and flossing).
3 Fluoridation of water. Topical application of fluorides in the dental office and home use of fluoride dentifrices is especially important for the young during the time of tooth development. (Fluorides decrease the solubility of the inorganic components of teeth and thus prevent caries.)

Periodontal disease

Periodontal disease is actually a group of pathological states or diseases affecting the supporting tissues of teeth (*periodontium*). It is the principal cause of loss of teeth in the North American population and is the most common infection. There is general agreement that microorganisms, specifically bacteria, are the etiologic agents of the various forms of periodontal disease. Support for this agreement comes from the following evidence:

1 Removal of dental plaque prevented or reversed clinical *gingivitis* (inflammation of the *gingiva*—that portion of the oral mucous membrane surrounding the tooth) in humans.
2 Progress of destructive periodontal disease could be halted and partially reversed by surgical procedures accompanied by thorough cleaning.

The conclusion drawn from these observations is that suppression of bacterial action is effective in controlling periodontal disease in humans.

Association of bacteria with periodontal disease

Recent research has shown that different types of bacteria are associated with clinically distinguishable states of periodontal disease. These states, which also may be considered as different diseases, are as follows: (1) gingivitis, (2) periodontitis, (3) acute necrotizing ulcerative gingivitis (ANUG), and (4) periodontosis.

There is a correlation between the state of oral hygiene, the amount of superficial and *subgingival* (below the gums) plaque, and the state of the periodontium. It follows that the worse the oral hygiene, the more plaque that accumulates, and thus the greater the deterioration of the tissues around the teeth.

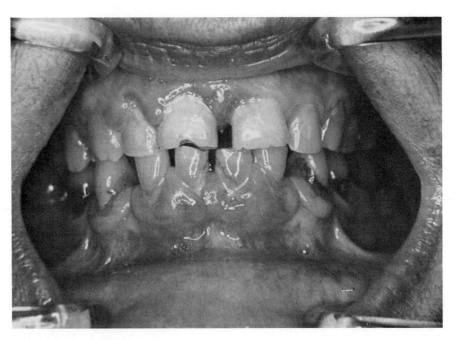

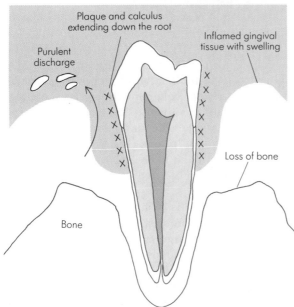

Figure 33-5. Periodontitis. The gums are inflamed and there is plaque and *calculus* (mineralized plaque) on the teeth, especially at their bases. *(Courtesy of R. F. Harvey, McGill University.)*

Plaque and calculus extending down the root

Purulent discharge

Inflamed gingival tissue with swelling

Loss of bone

Bone

Figure 33-6. Diagram illustrating periodontal disease.

Periodontal disease commonly starts as marginal gingivitis. The infection progresses from the gum margin toward the root. All the tissues around the teeth may be affected, or the involvement may be confined to one area or a few isolated areas. Gingivitis, if untreated, progresses to periodontitis (Fig. 33-5) which can become moderate to severe. During

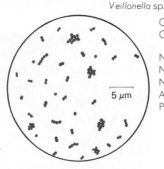

Actinomyces viscosus

Gram-positive rods
Not acid-fast
Occur as branching filaments or as diphtheroid cells
Not encapsulated
Nonmotile
Aerobic, facultatively anaerobic
Found in the mouth; pathogenic, causing periodontal disease in hamsters; pathogenicity for humans not established

5 μm

Veillonella sp.

Gram-negative cocci
Occur in pairs, masses, and short chains
Not encapsulated
Nonsporulating
Nonmotile
Anaerobic
Parasitic in the mouth and intestinal and respiratory tracts of humans and other animals

5 μm

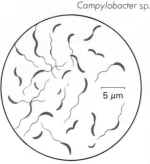

Campylobacter sp.

Gram-negative spirally curved rods
Occurs singly or in short chains forming S and gull-wing shapes
Not encapsulated
Nonsporulating
Motile; single polar flagellum at one or both ends
Microaerophilic to anaerobic
Found in the oral cavity
Some species pathogenic for humans and other animals

5 μm

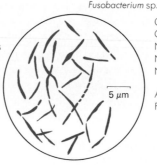

Fusobacterium sp.

Gram-negative rods
Occurs singly
Not encapsulated
Nonsporulating
Nonmotile or motile; peritrichous
Anaerobic
Found in cavities of humans and other animals; some species are pathogenic

5 μm

Figure 33-7. Morphology and distinguishing characteristics of some bacterial species associated with gingivitis. *(Erwin F. Lessel, illustrator.)*

periodontitis, there is progressive infiltration of lymphocytes, accumulation of immunoglobulins, and loss of *collagen* (the major protein component of periodontal tissue). This results in the formation of a pocket between the root and the overlying soft tissue, usually with marked inflammatory changes and exudation of pus (see Fig. 33-6). The disease keeps on advancing with destruction of bone and supporting tissues until loss of teeth occurs. This is the major cause of tooth loss in people over 35 years of age.

Acute necrotizing ulcerative gingivitis is characterized by necrosis of the interdental gingival tissues. In periodontosis, also called juvenile periodontitis because it occurs in the young, there is absence of obvious inflammation, but there is destruction of bone and loss of teeth.

Microorganisms in gingivitis. Members of the genus *Actinomyces*, especially *A. viscosus* (Fig. 33-7), predominate in the supragingival plaque in early gingivitis. They frequently comprise more than 50 percent of the isolates. In general, the predominant flora in early gingivitis are mainly gram-positive organisms. They appear to represent the overgrowth of those organisms found in plaques associated with healthy sites. However, in long-standing gingivitis, there is a shift in the microbiota population; about 25 percent may be gram-negative species in the genera *Veillonella*, *Campylobacter*, and *Fusobacterium* (Fig. 33-7).

568

Microorganisms in periodontitis. The subgingival plaque in rapidly destructive periodontitis has a predominance of gram-negative rods. Two microbial patterns of subgingival colonization have been observed. One is dominated by *Bacteroides asaccharolyticus* and spirochetes (Fig. 33-8). The other pattern is composed of monotrichous bacteria, corroding vibrios (i.e., agar-digesting vibrios), *Eikenella corrodens*, and other bacterioides (Fig. 33-8). Found in both patterns are *Fusobacterium nucleatum*, "gelatin-loving" *Bacterioides*, and other unidentified gram-negative anaerobes. Chronic periodontitis has a predominance of *Actinomyces viscosus, A. naeslundii,* and *B. asaccharolyticus.*

Microorganisms in ANUG. Spirochetes and fusobacteria are found in smears from the ulcerative tissue.

Microorganisms in periodontosis. The lesions in periodontosis reveal a sparse microbial flora predominated by gram-negative, *capnophilic* (CO_2-loving), anaerobic rods. Present are fusobacteria, *Bacteroides* spp., spirochetes, *Actinobacillus actinomycetemcomitans,* and other unidentified gram-negative rods.

Figure 33-8. Morphology and distinguishing characteristics of some bacteria associated with periodontitis. *(Erwin F. Lessel, illustrator.)*

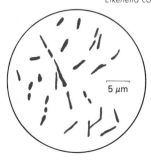

Bacteroides asaccharolyticus

Gram-negative rods
Occurs singly
Not encapsulated
Nonsporulating
Nonmotile
Anaerobic
Oxidase-negative
Produces black pigment
Found in oral cavity and in mixed infections

5 μm

Spirochetes (Treponema denticola)

Gram-negative, slender, flexuous, helically coiled cells
Occurs singly and in chains
Not encapsulated
Nonsporulating
Motile, probably by means of axial filaments
Aerobic, facultatively anaerobic, and anaerobic
May be free-living, commensal, or parasitic

5 μm

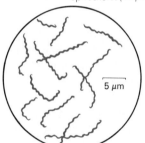

Eikenella corrodens

Gram-negative rods
Occurs singly
Not encapsulated
Nonsporulating
Nonmotile
Aerobic, facultatively anaerobic
Oxidase-positive
Normal inhabitant of the mouth; may be an opportunistic pathogen

5 μm

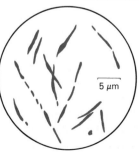

Fusobacterium nucleatum

Gram-negative rods
Occurs singly
Not encapsulated
Nonsporulating
Nonmotile
Anaerobic
Found in the mouth and in infections of the upper respiratory tract and pleural cavity

5 μm

While it is clear that the diseased sites are dominated by microorganisms not dominant in supragingival plaque, and that different types of bacteria are associated with clinically distinguishable states of periodontal disease, the significance of any single species in periodontal disease is still not clear. It is possible that mixed cultures, together with host-predisposing factors, are necessary to elicit the disease state.

Dermatomycoses

Fungus diseases that occur on the nails, skin, hair, and mucous membranes are referred to as *superficial mycoses* (see Figs. 8-11 and 33-9). Many of these fungi cause various forms of *ringworm*, or *tinea*, and the organisms that cause them are commonly called the *dermatophytes* or ringworm fungi (Figs. 33-10 and 33-11). These fungi spread radially in the dead keratinized layer of the skin by means of branching hyphae and occasional arthrospores. Inflammation of the living tissue below is very mild and only a little dry scaling is seen. Usually there is irritation, erythema, edema, and some vesiculation at the spreading edge; this pinkish circle gave rise to the name ringworm (Figs. 33-12 and 33-13). These diseases are widespread and difficult to control, but fortunately they

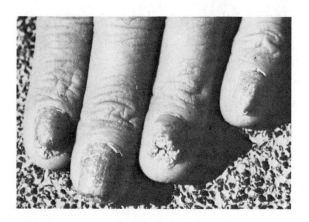

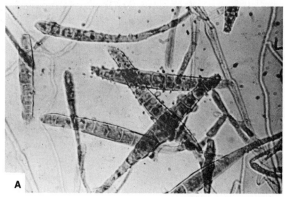

Figure 33-9. Onychomycosis, a disease of the nails, caused by fungi. *(Courtesy of Center for Disease Control, Atlanta, Ga.)*

Figure 33-10. Some ringworm of the scalp is caused by various species of *Microsporum,* such as (A) *M. canis* and (B) *M. gypseum.* The large spindle-shaped conidia (spores) attached to the hyphae are characteristic of species of this fungus. Such conidia are formed in artificial culture and are used to differentiate between the genera of dermatophytes. *(Courtesy of Everett S. Beneke, Michigan State University.)*

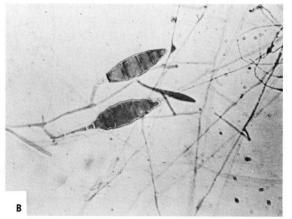

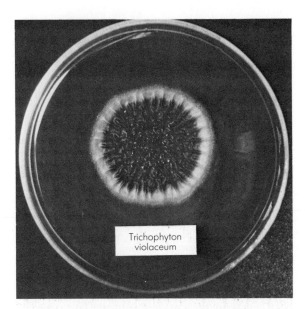

Figure 33-11. A colony of *Trichophyton violaceum,* a dermatophyte fungus. *(Courtesy of Center for Disease Control, Atlanta, Ga.)*

Trichophyton violaceum

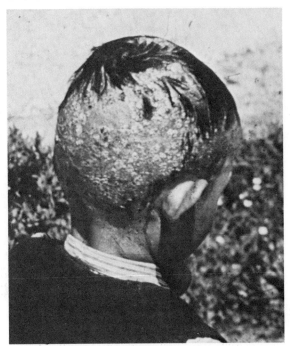

Figure 33-13. Ringworm of the scalp caused by a species of *Trichophyton. (Courtesy of Center for Disease Control, Atlanta, Ga.)*

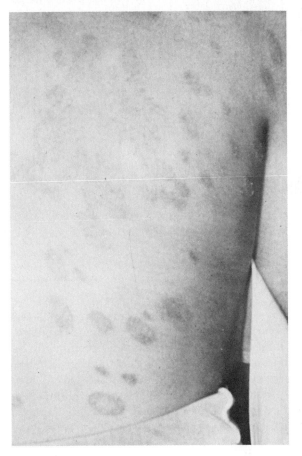

Figure 33-12. Ringworm lesions on the back of a patient caused by *Trichophyton verrucosum. (Courtesy of Center for Disease Control, Atlanta, Ga.)*

Table 33-2. The dermato-phytes

GROUP	ORGANISMS	OCCURRENCE AND DISEASE
Epidermophyton	*E. floccosum*	Causes infections of the skin and nails on fingers and toes
Microsporum	*M. audouinii*	Causes epidemic ringworm of the scalp in children
	M. canis	Common cause of infection of skin and hair on cats, dogs, and other animals; causes tinea capitis of children
	M. gypseum	Occurs as a saprophyte in the soil and as a parasite on lower animals; occasionally found in ringworm of the scalp in children
Trichophyton	Gypseum subgroup:	
	T. mentagrophytes	Primarily a parasite of the hair
	T. rubrum	Causes ringworm on many parts of the body; infects hair and scalp
	T. tonsurans	Infects hair and scalp
	Faviform subgroup:	
	T. schoenleinii	These fungi cause ringworm of the skin, scalp,
	T. violaceum	and glabrous skin in humans;
	T. ferrugineum	*T. verrucosum* causes ringworm in
	T. concentricum	cattle also
	T. verrucosum	
	Rosaceum subgroup:	
	T. megnini	Causes ringworm of the human scalp
	T. gallinae	Causes an infection in chickens
Miscellaneous	*Piedraia hortae*	Causes an infection of the hair and scalp characterized by hard, black concretions; black piedra
	Trichosporon beigelii	Causes an infection similar to above, except that the concretions are white; white piedra
	Nocardia minutissima	The cause of erythrasma, a chronic infection of axillae and genitocrural areas
	Malassezia furfur	Causes tinea versicolor, a generalized fungus infection of the skin covering trunk and sometimes other areas of the body

are often more annoying than serious. The causative microorganisms are sometimes present in the epidermal tissues without producing symptoms. They rarely, if ever, cause fatal infections. Transmission is commonly by direct contact with infected people or animals and by fomites. Dry skin is a fairly effective barrier against such diseases, but a "waterlogged" skin is vulnerable. This is why the sweat-laden, moist feet of athletes get infected with tinea, giving rise to the term "athlete's foot." The most common dermatophytes are listed in Table 33-2.

Anaerobic infections

In recent years there has been a renewed interest in anaerobes and their role in infectious disease. The availability of new methodology and apparatus for the cultivation of anaerobes has contributed to this interest. Until recently, research activity on anaerobic bacteria has lagged far behind that of aerobic microorganisms, which are more easily grown in the laboratory.

It is now generally recognized that anaerobes play an important role in a wide variety of infections in humans. Bacteremias; brain abscesses; ear, nose, and throat infections; and infections of the respiratory tract, intestinal tract, soft tissue, and other organs may be caused by anaerobes. This means that anaerobes may produce infection in any organ or tissue of the body. It is a fact that anaerobic infections of humans are much more frequent than had generally been supposed.

Other than the diseases caused by anaerobic and microaerophilic clostridia, most of the infections caused by anaerobes in humans come from members of the body's own normal microbiota. They are therefore endogenous in origin, and not from external sources. This is indeed a departure from the traditional concept of disease control because transmission of the pathogens from person to person is not a factor. Such classical means of transmission as the vector and fomite need not be considered. The main consideration is the host.

In general, the site of anaerobic infection in the body is compromised tissue, which, because of a traumatic injury, prior infection, reduced blood supply, and other less well-defined factors, has an anaerobic environment favoring the proliferation of anaerobes. These anaerobes abound in the human body and are numerically greater than aerobic microorganisms. There are more than 300 obligate anaerobic species residing in or on our body surfaces; they include 31 genera, 245 species, and many currently unclassifiable organisms.

Before further discussion of other aspects of these endogenous anaerobic bacteria is undertaken, it would be appropriate to discuss first those anaerobes of external origin, namely, *Clostridium* spp., causing tetanus and gas gangrene. These anaerobic infections, unlike those caused by endogenous anaerobic bacteria, have been known for many years.

Clostridial infections

The clostridia are anaerobic or microaerophilic bacterial rods that produce endospores. They are generally gram-positive. Most species decompose proteins and ferment carbohydrates (see Table 33-3), and many produce

Table 33-3. Selected differential characteristics of some clostridia

SPECIES	MOTILE	SPORES	REDUCE NITRATE	GLUCOSE	FERMENT LACTOSE	FERMENT SUCROSE
Cl. botulinum	Yes	Subterminal	No	Yes	No	No
Cl. tetani	Yes	Terminal	Yes	No	No	No
Cl. perfringens	No	Eccentric	Yes	Yes	Yes	Yes
Cl. histolyticum	Yes	Terminal	No	No	No	No
Cl. septicum	Yes	Subterminal	Yes	Yes	Yes	No
Cl. fallax	Yes	Oval	No	Yes	Yes	Yes

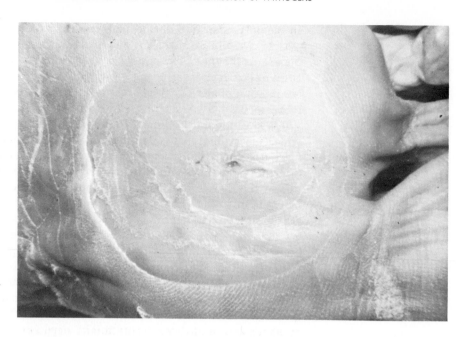

Figure 33-14. Tetanus lesion on the bottom of the foot. *(Courtesy of Center for Disease Control, Atlanta, Ga.)*

exotoxins. Several species are pathogenic, and many occur as saprophytes in the soil and in the intestinal tracts of humans and other animals.

Tetanus. Tetanus is an acute toxemia caused by *Clostridium tetani* bacteria which have gained entrance into wounds that provide conditions favorable for their growth. Such wounds are usually deep and ragged, with devitalized tissue in which aerobic organisms are also growing (Fig. 33-14). This provides the anaerobic state and dead tissue needed by *C. tetani* to produce a potent neurotropic, water-soluble toxin which circulates through the body after an incubation period of 1 to 3 weeks. The toxin, called *tetanospasmin*, causes tetanus when it reaches the central nervous system. The symptoms, caused by the toxin and characteristic of the disease, are painful contractions of the muscles, usually of the neck and jaw (restricting opening of the mouth, hence the term *lockjaw*), followed by paralysis of the thoracic muscles, frequently causing death. The causative organism is a gram-positive bacillus, 0.3 to 0.8 μm by 2 to 5 μm. Spores are formed at one end, giving the organisms a drumstick appearance (Fig. 33-15). Motility in young cultures is by means of peritrichous flagella. It is anaerobic, growing best at 37°C (98.6°F) and at pH 7.0 to 7.5, and does not ferment carbohydrates. The spores are resistant to heat and chemicals. *Clostridium tetani* are found as saprophytes in the intestinal tracts of herbivorous animals and are widely distributed in soil, especially soil fertilized with animal manure.

Infection may occur whenever a laceration or a puncture wound is contaminated with soil or other material harboring *C. tetani*. Contamination frequently occurs in traumatic injuries (deep wounds), for example, automobile accidents and gunshot wounds, or when addicts administer drugs to themselves. Battle injuries were the site of many infections until prophylactic immunization with formalinized toxoid became mandatory for all members of the armed forces. The morbidity and mortality due to tetanus has been reduced significantly, due largely to prophylactic immunization of civilians as well as military personnel. Infant immunization against tetanus concurrent with diphtheria immunization is now routine pediatric practice. Reimmunization every 3 to 5 years and a "booster" injection at the time of injury provide excellent protection against tetanus. The use of antitoxin in unimmunized accident victims is imperative and provides passive protection of short duration. The disease is not transmitted from person to person. Mortality in untreated cases may reach 50 percent.

Gas gangrene. Gas gangrene is similar to tetanus in its mode of transmission. It is characterized by a toxemia, preceding and concurrent with the development of gas in the tissues surrounding the wound. An anemia is caused by the hemolytic action of the toxin produced by the organisms.

Several clostridia, such as *Clostridium perfringens*, *C. novyi*, and *C. septicum*, may cause gas gangrene in humans; closely related species are pathogenic for lower animals. The most virulent of these is *C. perfringens*, a short, gram-positive, sporulating, capsulated, nonmotile, anaerobic bacillus readily distinguished from *C. tetani* by its lack of motility and its fermentation reactions. Human and animal types of *C. perfringens* are differentiated by the exotoxins produced. The organisms are normal inhabitants of the intestinal tract of humans and animals and are found in the soil in large numbers. *C. perfringens* is transmitted only by direct contact with contaminated objects or soil, never from person to person.

Figure 33-15. Morphology and distinguishing characteristics of *Clostridium tetani. (Erwin F. Lessel, illustrator; photomicrograph courtesy of Liliane Therrien and E. C. S. Chan, McGill University.)*

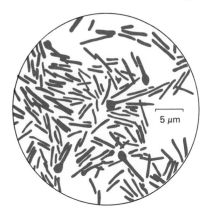 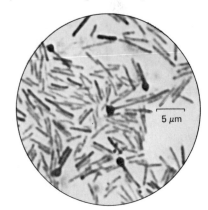

Clostridium tetani
Gram-positive rods
Occurs singly, in pairs, and in long chains and filaments
Not encapsulated
Spores are spherical and terminal
Motile; peritrichous
Anaerobic
Produces potent exotoxin; pathogenic, causing tetanus

Active and passive prophylaxis by immunization has not proved effective, but the prompt use of a polyvalent gas-gangrene antitoxin is recommended. Treatment and prophylaxis with antibiotics and chemotherapeutic agents have not been uniformly satisfactory. Experience with war casualties has shown that prompt surgical removal of dead tissue prevents the appearance of symptoms.

Infections of endogenous origin

Anaerobes and the natural anaerobic environment. There are several ecological reasons why the naturally occurring microbiota in the human body have so many anaerobic members. First, the lumen of the gut and the oral cavity are good examples of environments that have a reduced oxygen tension. Second, in the natural environment, symbiotic aerobes and facultative anaerobes use up the available oxygen, reducing the environment in the process so that anaerobes can survive and grow. It has been shown that most of the anaerobes of medical significance are able to survive brief periods in the presence of oxygen, although they are unable to grow until anaerobiosis is achieved. Third, tissue and body fluids are known to contain reducing, or at least protective, factors conducive to the growth of anaerobic organisms. (Consider, for example, that anaerobes are routinely cultivated in chopped meat or minced tissue broth in some laboratories.) Fourth, in living tissue, impaired blood supply, due to either disease or trauma, provides the anaerobic environment that allows anaerobic growth.

Enzymes such as catalase, the peroxidases, and the superoxide dismutases contribute to the ability of the cell producing them to survive under adverse atmospheric conditions. Both catalase and peroxidase break down hydrogen peroxide produced by the bacterial cell; unless broken down, it is toxic to the cell. Superoxide dismutase is important to anaerobes for the following reason. In the presence of air, the organism reduces oxygen by one electron, resulting in the production of $superoxide$ (O_2^-). The superoxide cannot diffuse out of the cell; it can destroy the cell unless it is converted to the less reactive hydrogen peroxide by the enzyme superoxide dismutase. A positive correlation has been reported between the degree of aerotolerance and the content of superoxide dismutase in anaerobes.

Virulence factors of anaerobic bacteria. Two well-known virulence factors of pathogens are toxin production and invasiveness. Invasiveness leads to damage in the local area. Soluble toxins can be transported to distant sites and thus may cause much more damage than the original lesion site. There are anaerobes which are virulent through one of the two mechanisms and some that are endowed with both mechanisms.

Toxigenicity can be either endotoxin or exotoxin production. The potent endotoxins of gram-negative bacteria that are aerobic or facultatively anaerobic have been discussed previously. However, the gram-negative anaerobic pathogens, *Bacteroides melaninogenicus* and *B. fragilis,* do *not* contain biologically active endotoxin. Important exotoxins are elaborated by anaerobic bacteria, especially those of the genus *Clostridium.* Toxemias of gas gangrene and tetanus have been described. Other toxemias due to exocellular toxins are caused by *Clostridium botulinum* (botulism) and *C. difficile* (*colitis,* or inflammation of the colon).

Invasiveness, the other major mechanism of virulence, has to do with the capacity to invade tissues. There are many factors governing invasiveness. One of these is the role of surface components of the cell, in particular capsules and lipopolysaccharides, and how they interact with the host. Another example is the ability to synthesize enzymes that attack tissue components; such as the enzymes collagenase and hyaluronidase.

The microorganisms of anaerobic infections. Even though more than 300 anaerobic species reside in the human body, only a few species are consistently isolated. Five species (with their subspecies) or groups of bacteria account for two-thirds of all clinically significant isolates of anaerobic bacteria: *Bacteroides fragilis, B. melaninogenicus, Fusobacterium nucleatum, Clostridium perfringens*, and gram-positive cocci. Among the invasive bacteria of endogenous origin, the species of *Bacteroides* may be the most important. *B. fragilis* has high rates of isolation from all kinds of infections. For example, it has been isolated from bacteremias, bone infections, brain abscesses, and abdominal abscesses. *Bacteroides melaninogenicus* has high rates of isolation from respiratory-tract infections. It may also be associated in some way with periodontal disease, the most common infection in the United States. It has been observed that most anaerobic infections are mixed infections; that is, many kinds of microbes are involved. This finding has suggested that bacterial interactions, particularly synergism, may play a determining role in anaerobic infections.

Summary and outlook

This chapter is concerned with those diseases that are not *typically* transmitted by the major modes of transmission such as by air, food, water, arthropods, and sexual contact. Acute diarrhea is caused by *Escherichia coli* transmitted by hand-to-mouth activity or by passive transfer in food or drink. The transmission of hepatitis A is by the fecal-oral route, by passive transfer, or by the ingestion of contaminated raw shellfish. The major route of transmission of hepatitis B is by means of parenteral injection (such as in blood transfusion). Dental caries is not a disease that is transmissible in the usual sense (although this can be demonstrated in laboratory animals). Several interactive factors determine the development of dental caries. Periodontal disease is like dental caries in its mode of transmission. For the superficial mycoses, direct contact and fomite transmission is the rule. Other than the well-known clostridial infections, most anaerobic infections are due to members of the indigenous microbiota.

There are many remaining problems concerning the epidemiology of gastroenteritis caused by *E. coli*. Routine microbiological cultures of feces do not distinguish among the different categories of pathogenic *E. coli*. Thus the prevalence of *E. coli* as a cause of diarrhea has not yet been determined. The apparently limited geographical distribution of disease caused by enteroinvasive *E. coli* (e.g., common in adults in Southeast Asia) needs to be explained. Vaccines against the different types of *E. coli* infections are needed, especially in areas where the disease is endemic.

The inability to grow the hepatitis viruses outside the animal body clouds the outlook for this disease. Without a capability of cultivation in tissue culture, the development of vaccines against the disease is not likely. The state of hepatitis serology is quite well advanced. Unfortunately, serological tests suggest to us that there are other viruses of hepatitis, not related to A and B, that we know very little about at the present time.

We seem to have the key organism, *Streptococcus mutans*, in the etiology of dental caries. But the future may reveal others just as important. However the main preventive measures against caries are not microbiological, but rather have to do with public health procedures which are more difficult to implement. For example, it is difficult to alter the eating habits of the population, to motivate people to clean their teeth properly, and to fluoridate all drinking water supplies.

The weight of evidence implicates bacteria as the cause of periodontal disease. The search for its etiology is still in its infancy. No one particular species has been shown to be the etiologic agent. The descriptive stage of microbiological research in periodontal disease must go on for some time before we will see a real breakthrough. But awareness of this disease is keen at the present time; it is attracting many good workers, and this augurs well for the future.

Anaerobes are now recognized more widely as causative agents of infection. There are available specific procedures and equipment for routine cultivation of anaerobes. Many advances in our knowledge of anaerobic infections are likely to occur in the next several years.

Key terms

Australia antigen	enterotoxigenic	plaque
capnophilic	gingiva	ringworm
collagen	gingivitis	subgingival
demineralization	K antigen	superficial mycoses
dermatophyte	lockjaw	superoxide
enteroinvasive	passive transfer	tetanospasmin
enteropathogenic	periodontium	tinea

Questions

1 Explain why *Escherichia coli*, the favorite organism of biochemists, should not be handled carelessly.
2 How can one study the epidemiology of gastroenteritis caused by one of the groups of pathogenic *E. coli*?
3 Other than by serologic methods, how can hepatitis A infection be differentiated from hepatitis B infection?
4 Are hepatitis B carriers dangerous to the community? Explain and give some suggestions for their control.

5 Describe the course of events leading to the formation of dental caries.

6 Why is *Streptococcus mutans* a prime suspect in the etiology of dental caries?

7 What factors do you think might hinder effective implementation of the methods for caries prevention?

8 Mechanical debridement (surgical cutting of gums) and antibiotic therapy are two procedures used by dental surgeons to treat periodontal disease. Explain why these procedures may or may not be useful.

9 Describe the different kinds of bacteria associated with the healthy oral cavity, early gingivitis, and long-standing gingivitis.

10 Describe the types of microorganisms found associated with periodontitis.

11 What role does plaque play in the initiation of dental caries and periodontal disease?

12 Explain how one can contract "athlete's foot."

13 How do the etiologic agents of tetanus and gas gangrene cause their effect in infection?

14 What is the source of the microorganisms in mixed anaerobic infections? What factors predispose the host to such infections?

15 What conditions favor the anaerobic microbiota in the human body?

16 Give some reasons for the invasiveness of anaerobic bacteria.

17 Name some anaerobic bacteria that are isolated frequently and consistently from anaerobic infections of endogenous origin. Describe their morphology and other characteristics.

References for part nine

Some of the references of Part Eight are also useful for this part.

Burnett, G. W., H. W. Scherp, G. S. Schuster: *Oral Microbiology and Infectious Disease*, 4th ed., Williams & Wilkins, Baltimore, 1976. *This is a classic text that presents the basic principles of the origins, mechanisms, and management of infectious diseases for students and practioners of dentistry.*

Collee, J. G.: *Applied Medical Microbiology*, Blackwell Scientific Publications, Oxford, 1976. *This small paperback serves as an elementary introduction to clinically related areas of microbiology. It has a limited number of examples to present basic concepts and terminology.*

Duguid, J. P., B. P. Marmion, and R. H. A. Swain (eds.): *Medical Microbiology*, 13th ed., vol. I: *Microbial Infections*, Churchill Livingstone, Edinburgh, 1978. *This text is intended for advanced science and medical students as well as practicing physicians. Its strength lies in the contributions by many authors. Besides being a good introduction to microbial biology, it gives detailed and organized discussion of microbial infections according to groups of organisms.*

Hoeprich, P. D., (ed.): *Infectious Diseases*, 2d ed., Harper & Row, New York, 1977. *This is an outstanding reference book on infectious diseases with 97 contributors who are specialists in their fields. Discussion of infections is by organ systems and is extremely well organized and detailed.*

Lennette, E. H., E. H. Spaulding, and J. P. Truant: *Manual of Clinical Microbiology*, 3d ed., American Society for Microbiology, Washington, D.C., 1980. *A guide to the isolation and identification of infectious organisms from clinical specimens by more than 125 specialists. The student should refer to this for details of any clinical microbiology procedure.*

Nolte, W. A. (ed.): *Oral Microbiology*, 3d ed., Mosby, St. Louis, 1977. *Eight contributors to this well-known text for dental students present the role that microorganisms play in oral disease.*

Thorn, G. W., R. D. Adams, E. Braunwald, K. J. Isselbacher, and R. G. Petersdorf (eds.): *Harrison's Principles of Internal Medicine*, 8th ed., McGraw-Hill, New York, 1977. *This superb and classic reference text covers the field of internal medicine. Part Six discusses "disorders caused by biologic agents." This is a book with a clinical approach for both medical students and graduate physicians. It has a long list of expert contributors.*

Turk, D. C., and I. A. Porter: *A Short Textbook of Medical Microbiology*, 3d ed., English Universities, London, 1974. *This short paperback for medical students gives a concise presentation of microbial parasites together with applied aspects of their laboratory diagnosis, prevention, and treatment.*

Volk, W. A.: *Essentials of Medical Microbiology*, Lippincott, Philadelphia, 1978. *This paperback is a comprehensive medical microbiology text. It contains the essentials of microbiology for the medical student, as well as good scientific explanations for observed phenomena. It is well illustrated.*

Youmans, G. P., P. Y. Paterson, and H. M. Sommers: *The Biologic and Clinical Basis of Infectious Disease*, 2d ed., Saunders, Philadelphia, 1980. *A good text on host-microbe interactions, with discussion of infectious diseases organized along organ-system lines.*

PART TEN

ENVIRONMENTAL AND
APPLIED MICROBIOLOGY

Microorganisms are capable of promoting chemical changes in many materials. For this reason, they can be thought of as miniature chemical factories. In the environment, their abundance and diverse chemical abilities make them responsible for many essential processes. For example, some microorganisms decompose or degrade dead plant and animal tissues. The chemicals recycled in this way are used as nutrients for other organisms. Other microorganisms synthesize nitrogen compounds from atmospheric nitrogen or carbohydrates from carbon dioxide, thus producing essential nutrients for other organisms from simple compounds present in the environment.

In our environment, microorganisms can also deteriorate materials used by humans—wood, textiles, metals, foods. Or by their ability to promote chemical changes, they can transform pollutants in our environment to harmless substances. Control or prevention of deterioration of materials and control of pollutants are both important aspects of applied microbiology.

The chemical abilities of microorganisms are also important in industry. Microorganisms are essential for production of foods such as wine, cheese, and vinegar. Drugs such as penicillin are products of microbial synthesis. Microorganisms will become even more important in production of products as more economically important products of microorganisms are identified and as microorganisms are genetically engineered to produce valuable products, to produce higher yields of these products, or to decompose pollutants.

In Part Ten we will look at these roles of microorganisms in the environment and in industry.

A large scale industrial bacterial mutant hunt for super producers of antibiotics or metabolic products. Shaker-flask fermentation tests are carried out in big warm rooms where temperature is precisely controlled by computer. *(Courtesy of CETUS CORPORATION, pioneer research company in biotechnology, Berkeley, California.)*

ENVIRONMENTAL MICROBIOLOGY

Microorganisms occur everywhere in our environment; they inhabit the soil, the water, and the atmosphere of our planet. Their existence has been searched for on other planets, but thus far deep-space probes have not revealed the existence of extraterrestrial microorganisms.

The study of microorganisms in their natural environments is also referred to as *microbial ecology*. Ecology is the division of biology that is concerned with the study of the relation of organisms or groups of organisms to their environment. The inhabitants of a given environment are viewed as part of an ecological system, or *ecosystem*. The largest ecosystem is planet Earth; as such, it is termed the *biosphere*. An ecosystem has two major conponents: (1) the community of organisms within it and (2) the nonliving components (chemicals and the physical conditions). An ecosystem is a *dynamic* system—a fact that becomes apparent when one recognizes the tremendous populations and the great diversity of organisms in it. Of all the organisms in any given ecosystem, microorganisms are usually present in the largest numbers and possess the greatest capacity to effect changes.

The strong public concern for the quality of the environment has contributed to a growing interest in microbial ecology. For example, microorganisms play the key role in decomposing many human and industrial wastes that are dumped into bodies of water or onto soil; they are capable of recycling many materials. The quality and the productivity of natural waters are related, in large part, to their microbial population. Clean, dust-free air has relatively few microorganisms. Thus we see that the assessment of the quality of our environment is intricately related to the microbial flora present.

It is our purpose in this chapter to describe some microbial ecosystems.

Some characteristics of microbial ecosystems

Microbial ecosystems function in many different environments. For example, one ecosystem might be that in a pond or lake. Another could be the microbial community of soil in the region of the root system of a plant. There are, however, several characteristics which ecosystems share.

Diversity of microbial species

Microorganisms, in their natural environment, rarely occur as pure cultures. Various specimens of soil or water are likely to harbor many different species of fungi, protozoa, algae, bacteria, and viruses. Hence the pure-culture concept of microbiology which was stressed earlier must be reassessed when studying microbial ecosystems. Pure-culture techniques are needed in order to identify the various species in a given habitat. However, the chemical transformations accomplished by this menagerie of microorganisms cannot be determined by simply compiling the biochemical properties of each species as determined in pure culture. In this case, the sum of the parts is not necessarily equal to the whole. This is because several types of interactions may occur among the species, resulting in an outcome different from that produced by the individual species in pure culture. In the context of a natural microbial ecosystem, pure cultures represent an artifical situation.

Population dynamics

Every species of microorganism thrives in its environment only as long as the conditions are favorable for its growth and survival. As soon as some physical or chemical change, such as exhaustion of nutrients or radical change in temperature or pH, makes conditions for growth of another species more favorable, the organism that was well adapted to the formerly prevailing environmental conditions yields its place to an organism better adapted to the new conditions. Thus environmental factors have a selective effect on—that is, they *select for*—microbial populations. Figure 34-1 shows the rise and fall of microorganisms (phytoplankton) in an ecosystem at different seasons of the year.

Adaptation and mutation

Survival and continued growth of a species within the biological community require an ability to adjust to changing environmental conditions. *Phenotypic* adaptation is the microorganism's response to temporary changes of limited degree. For example, many species of microorganisms can grow over a wide range of temperatures. Their metabolic activity, however, is not necessarily the same at the extremes of the temperature

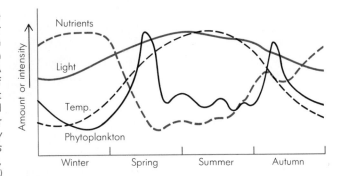

Figure 34-1. The occurrence of "blooms," or phytoplankton "pulses," in northern temperate lakes in the spring and autumn. A combination of conditions—nutrient concentration, light, and temperature—accounts for this phenomenon. *(Courtesy of E. P. Odum, Fundamentals of Ecology, 3d ed., Saunders, Philadelphia, 1971.)*

range. This adaptive ability is within the limits of the genotype of the microorganism, as explained in Chap. 17. There are, in addition, opportunities for a change in the genotype. As you will recall from the discussion in Chap. 17, a genotypic change results in a mutation. The mutant is a permanently changed organism. If capable of thriving in the environment, the mutant will perpetuate itself; otherwise it will perish.

Microbial associations in ecosystems

The microorganisms that inhabit an ecosystem exhibit many different types of associations and interactions among the species. Some of the associations are indifferent or neutral (that is, the species are unaffected); some are beneficial or positive for one or more members; others are detrimental or negative to one or more members. As each different type of association or interaction was elucidated it was given a specific descriptive label. As you might assume, many of these associations do not fall neatly into discrete categories. The general term *symbiosis* is used to designate the relationship that exists when two or more organisms live together in close proximity. We will refer to types of symbiosis as:

> *Neutralism:* Members of an association are unaffected by growing in the same environment.
> *Mutualism:* Both members of the association are benefited.
> *Commensalism:* One member of an association is benefited; that is, it may grow faster, grow to a larger total population, and in general grow "better." The other member is unaffected.
> *Antagonism, competition, or parasitism:* One member of an association is inhibited or destroyed; the other member is benefited.

Examples of various associations are shown in Fig. 34-2; others are described later in this chapter and in other chapters of this part.

Microbiology of soil

Directly or indirectly, the wastes of humans and other animals, their bodies, and the tissues of plants are dumped onto or buried in the soil. In due time, they are transformed into the organic and some of the inorganic constituents of soil. Microbes mediate these changes—the conversion of organic matter into substances that provide the nutrient material for the

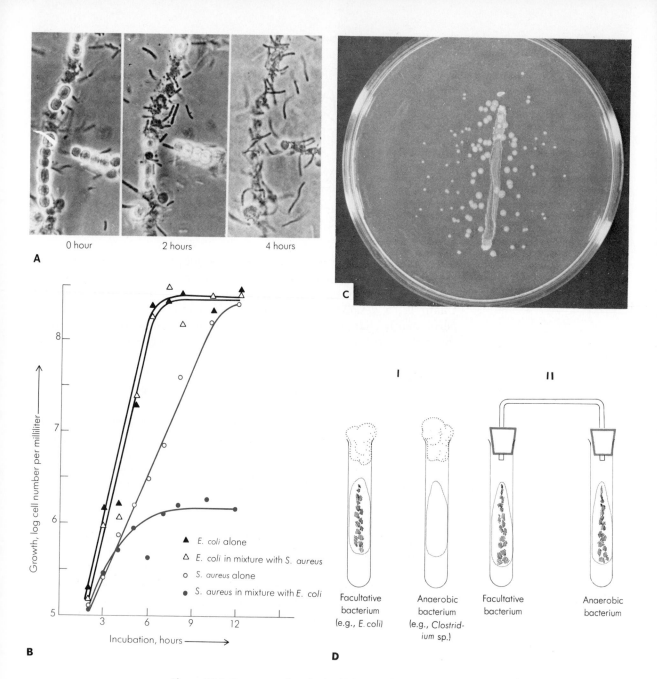

A

0 hour   2 hours   4 hours

C

B

Growth, log cell number per milliliter →

▲ *E. coli* alone
△ *E. coli* in mixture with *S. aureus*
○ *S. aureus* alone
● *S. aureus* in mixture with *E. coli*

Incubation, hours →

I   II

Facultative bacterium (e.g., *E. coli*)

Anaerobic bacterium (e.g., *Clostridium* sp.)

Facultative bacterium

Anaerobic bacterium

D

Figure 34-2. Some examples of microbial associations. (A) Antibiosis. Lysis of blue-green algae by a myxobacter. Shown in this series is a sequence of lysis of the alga *Nostoc* filament by the myxobacter. The myxobacter culture used in this experiment was isolated from fishponds and is capable of lysing many species of unicellular and filamentous blue-green algae and certain bacteria. *(Courtesy of Mirian Shilo, J Bacteriol,* **104:***453,1970.)* (B) Competition between *Escherichia coli* and *Staphylococcus aureus*. Results of growth are shown when each species is grown in pure culture and in mixed culture. Note the decrease in growth of *S. aureus* in the mixed culture. *E. coli* has a shorter generation time than *S. aureus*; it grows faster and depletes nutrients, thus limiting the growth of *S. aureus*. *(Courtesy of T. R. Oberhofer and W. C. Frazier, J Milk Food Technol,* **24:***172,1961.)* (C) Commensalism. Growth of *Arthrobacter citreus* in the

Table 34-1. A comparison of the numbers of various groups of microorganisms in the rhizosphere (region of the root system) of spring wheat and in control soil (minus the plant root system)

MICRO-ORGANISMS	RHIZOSPHERE SOIL	CONTROL SOIL
Bacteria	$1,200 \times 10^6$	53×10^6
Actinomycetes	46×10^6	7×10^6
Fungi	12×10^5	1×10^5
Protozoa	24×10^2	10×10^2
Algae	5×10^3	27×10^3
Bacterial groups:		
Ammonifiers	500×10^6	4×10^6
Gas-producing anaerobes	39×10^4	3×10^4
Anaerobes	12×10^6	6×10^6
Denitrifiers	126×10^6	1×10^5
Aerobic cellulose decomposers	7×10^5	1×10^5
Anaerobic cellulose decomposers	9×10^3	3×10^3
Sporeformers	930×10^3	575×10^3
"Radiobacter" types	17×10^6	1×10^4
Azotobacters	<1,000	<1,000

SOURCE: From T. R. G. Gray, and S. T. Williams, *Soil Microorganisms*, Hafner Publishing Company, New York, 1971.

plant world. Without this microbial activity, all life on earth would gradually be choked off.

The soil environment

For our purposes, soil may be considered as the land surface of the earth which provides the substrate for plant and animal life. The characteristics of the soil environment vary with locale and climate. Soils differ in depth, physical properties, chemical composition, and origin. There are five major categories of soil constituents: mineral particles, organic material, water, gases, and living organisms.

The microbial flora of soil

Few environments on earth provide as great a variety of microorganisms as fertile soil. Bacteria, fungi, algae, protozoa, and viruses make up a microbial menagerie which may reach a total of billions of organisms per gram of soil (Table 34-1). The great diversity of the microbial flora presents a real problem in any attempt to enumerate the total viable population of microorganisms in a soil sample. Culture methods will reveal only those

presence of *Saccharomyces cerevisiae*. The *Arthrobacter citreus* is inoculated throughout the entire medium, but the cells grow only in the region near the yeast growth (*S. cerevisiae*), which is confined to the vertical streak. *(Courtesy of E. C. S. Chan and M. B. Johnson, Can J Microbiol,* **12:***581, 1966)* (D) Commensalism. (I) When grown in pure culture, facultative anaerobic or aerobic bacteria develop in the presence of atmospheric oxygen, but strict anaerobes do not. (II) When culture tubes are stoppered tightly and a tube is inoculated with anaerobic bacteria and connected to a tube containing a culture of facultative organisms, both types can grow, because the facultative bacteria utilize the oxygen trapped in the closed system, thus establishing conditions favorable for growth of the anaerobes.

physiological and nutritional types that can grow in the environment provided in the laboratory. For example, if a soil sample is inoculated on nutrient agar and incubated at 35°C (95°F), some of the types of bacteria that will not grow are the obligate thermophiles, in addition to psychrophiles, anaerobes, and autotrophs. The protozoa will not be recovered, and few algae and fungi will grow. This means that when a sample of soil is cultured in the laboratory, any single culture procedure will recover only a small sector of the total microbial population. Direct microscopic counts theoretically should permit enumeration of all microorganisms except the viruses, but this technique also has limitations, especially in distinguishing living from dead microorganisms.

One way to develop a better understanding of the great diversity of microbial life in the soil is to assess the roles played by various microbial groups in bringing about chemical changes in the soil.

Biochemical activity of microorganisms in soil

The most important role of soil microorganisms is their function in bringing about chemical changes in substances in the soil, mainly the conversion of organic compounds of carbon, nitrogen, sulfur, and phosphorus to inorganic compounds. This process is known as *mineralization*. A large number of chemical changes is involved, and many different species of microorganisms play a role in accomplishing them. The events can be displayed as a cyclic process which starts, for example, with an element such as nitrogen that undergoes a series of changes through inorganic compounds to organic compounds. Later the nitrogen is released from the protein, and the process starts over again. One of the best understood of these cyclic processes is that which describes the transformations of nitrogen and its compounds.

The nitrogen cycle. Because of its important role in crop nutrition, the nitrogen cycle has attracted considerable attention from soil microbiologists and others concerned with agriculture. Nitrates are the form of nitrogen most available to plants. Through a sequence of microbial reactions, organic nitrogen compounds, as well as nitrogen gas from the atmosphere, are transformed into nitrates. The sequence of changes—from free atmospheric nitrogen to simple and complex organic compounds in the tissues of plants, animals, and microorganisms, and the eventual release of this nitrogen back to atmospheric nitrogen—is summarized in Fig. 34-3.

Nitrogen fixation. This is one of the many biochemical processes in the soil in which microorganisms play one of the most crucial roles—transforming atmospheric nitrogen (N_2, or free nitrogen) to nitrogen in a compound (*fixed nitrogen*).

Two groups of microorganisms are involved in this process: (1) *nonsymbiotic* microorganisms, those living freely and independently in the soil; and (2) *symbiotic* microorganisms, those living in roots of leguminous plants (plants that bear seeds in pods, such as peas, soybeans, and clover). The magnitude and importance of biological nitrogen fixation can be appreciated from a recent estimate which states that living organisms fix more nitrogen

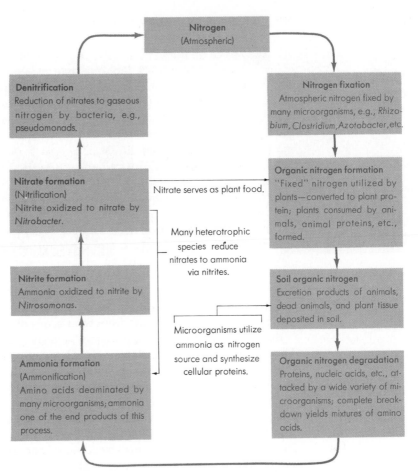

Figure 34-3. The nitrogen cycle. Various species of microorganisms, particularly bacteria, are instrumental in converting nitrogen from atmospheric or free nitrogen to fixed nitrogen and back again to atmospheric nitrogen. These conversions are intimately associated with soil fertility and plant growth.

The boxes in the figure read:

Nitrogen (Atmospheric)

Denitrification Reduction of nitrates to gaseous nitrogen by bacteria, e.g., pseudomonads.

Nitrogen fixation Atmospheric nitrogen fixed by many microorganisms, e.g., *Rhizobium, Clostridium, Azotobacter,* etc.

Nitrate formation (Nitrification) Nitrite oxidized to nitrate by *Nitrobacter.*

Organic nitrogen formation "Fixed" nitrogen utilized by plants—converted to plant protein; plants consumed by animals, animal proteins, etc., formed.

Nitrate serves as plant food.

Many heterotrophic species reduce nitrates to ammonia via nitrites.

Nitrite formation Ammonia oxidized to nitrite by *Nitrosomonas.*

Soil organic nitrogen Excretion products of animals, dead animals, and plant tissue deposited in soil.

Microorganisms utilize ammonia as nitrogen source and synthesize cellular proteins.

Ammonia formation (Ammonification) Amino acids deaminated by many microorganisms; ammonia one of the end products of this process.

Organic nitrogen degradation Proteins, nucleic acids, etc., attacked by a wide variety of microorganisms; complete breakdown yields mixtures of amino acids.

than all the world's factories—175 million tons of nitrogen fixed by living organisms versus 40 million tons by factories in 1974.

Nonsymbiotic nitrogen fixation has been studied extensively in *Clostridium pasteurianum* and species of *Azotobacter.* For many years, these bacteria were the only ones known to be capable of this activity. In recent years, however, it has been discovered that many other bacteria are also capable of fixing nitrogen nonsymbiotically.

It has been estimated that the amount of nitrogen fixed by the nonsymbiotic process may reach as high as 50 lb/acre annually. This estimate is, of course, subject to much variation, depending upon climatic and soil conditions.

Symbiotic nitrogen fixation is accomplished by the association of bacteria of the genus *Rhizobium* with legumes. Before these bacteria can fix nitrogen, they must establish themselves in the cells of root tissue of the host plant. Infection of the root system is closely associated with the formation by the bacteria of an "infection thread" that develops in certain root hairs. The nitrogen-fixing bacteria invade the host plant cells via this infection thread. Some of the cells of the plant are thus infected, causing cell enlargement and an increased rate of cell division; this leads to the formation of nodules on the root system (see Fig. 34-4). The legume, the

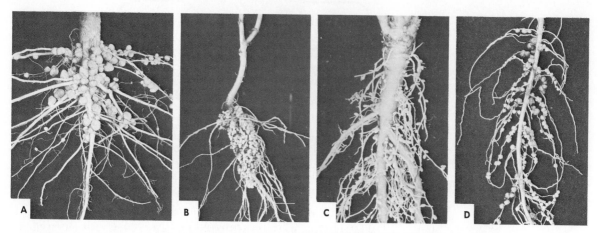

Figure 34-4. Root nodules are produced by effective strains of nitrogen-fixing bacteria on several legumes: (A) soybean, (B) sweet clover, (C) pea, (D) birds-foot trefoil. Effective strains are those which result in nodule formation and fixation of nitrogen. *(Courtesy of The Nitragin Co., Inc.)*

bacteria, and the nodule constitute the system for this type of nitrogen fixation. It is a process where both the bacteria and the plant benefit by the association. The bacteria make nitrogen available to the plant, and in turn, the bacteria derive nutrients from the tissues of the plant.

Preparations of symbiotic nitrogen bacteria are available commercially. Farmers use these to inoculate seeds of legumes prior to planting. However, there is a degree of specificity between the bacteria and legumes; not all species of *Rhizobium* produce nodulation and nitrogen fixation with any legume. For purposes of inoculation with commercial preparations of these bacteria, legumes are divided into seven major categories, as follows: alfalfa, clover, peas and vetch, cowpeas, beans, lupines, and soybeans. *Rhizobium* species or strains effective for one group are less effective or ineffective for other groups. Even within a species, certain strains are more effective than others with a given host plant (Fig. 34-5).

Inoculation of seeds before planting is an economical and efficient way to

Figure 34-5. Different strains of rhizobium have different effects on the growth of clover. Tests are carried out on crimson clover in the following manner: Seeds are planted in sterile sand contained in a jar. The sand is then inoculated with the bacteria. Each jar contains a solution of nutrients—except nitrogen—which diffuse through the sand. Thus the extent of growth is indicative of the amount of nitrogen being supplied by the bacteria. (A) Uninoculated sand; (B), (C), and (D) sand inoculated with different strains of rhizobium. Note the difference in growth response. *(Courtesy of L. W. Erdman, USDA.)*

increase soil fertility and productivity. Most of the commercial preparations consist of selected strains of bacteria dispensed in moist humus. This material is mixed with a specified quantity of water and sprinkled over the seeds prior to planting.

Microbiological degradation of pesticides and herbicides

Pesticides, as the term denotes, refer to chemical substances that destroy pests. *Herbicides* are chemical substances that kill weeds.

The wide-scale application of these chemical agents dramatically increases crop yields. But it raises questions about the short- and long-range effects of these chemicals on the soil. Are these substances degraded by soil microorganisms, and if so, how rapidly? Do they have a temporary, or a permanent, effect upon the soil microflora? Do they constitute a form of pollution of streams and rivers and as such affect aquatic life? These are some of the questions that concern the soil microbiologist as well as biologists, agriculturalists, and environmentalists. Major research efforts are focused on these questions, as well as others. Ideal pesticide compounds would be ones that destroy the pests quickly and are then degraded quickly into nontoxic substances by the soil microorganisms. Figure 34-6 shows the rate of disappearance (degradation) of the herbicide 2,4-D deposited in the soil.

Microbiology of aquatic environments

Aquatic microbiology is the study of microorganisms and their activities in fresh, estuarine, and marine waters, including springs, lakes, rivers, and the seas. It is the study of the viruses, bacteria, algae, protozoa, and microscopic fungi which inhabit these natural waters. Some of these microorganisms are indigenous to natural bodies of water; others are transient, entering the water intermittently from air or soil, or from industrial or domestic processes. These microorganisms and their activities are of great importance in many ways. They may affect the health of humans and other animal life; they occupy a key position in the food chain by providing nourishment for the next higher level of aquatic life. They are instrumental in the chain of biochemical reactions which accomplishes

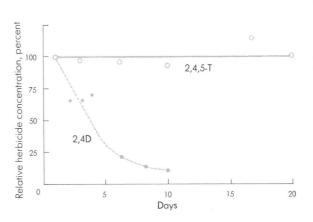

Figure 34-6. Degradation of herbicides by microorganisms in the soil is important in preventing the accumulation of these harmful chemicals in the soil. Some herbicides (2,4-D) are more readily susceptible to bacterial degradation than others (2,4,5-T); the latter compound persists in soil for several months to a year. *(Courtesy of M. Alexander, Introduction to Soil Microbiology, Wiley, New York, 1961.)*

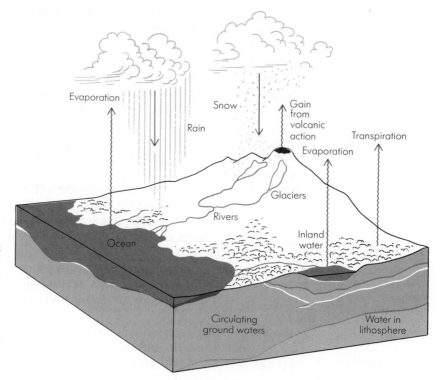

Figure 34-7. The hydrologic cycle, illustrating the system of circulation of water from the earth's surface into the atmosphere and back to earth. *(Courtesy of G. K. Reid, and R. D. Wood, Ecology of Inland Waters and Estuaries, Van Nostrand, New York, 1976.)*

recycling of elements, very much as was described for soil. Urbanization and consequently the growing demand of communities for water, the importance of natural waters as a major food reservoir, the off-shore exploration for oil and minerals, the establishment of the Environmental Protection Agency—these and other developments have given new significance to the subject of aquatic microbiology.

Natural waters

The earth's moisture is in continuous circulation, a process known as the *water cycle*, or *hydrologic cycle*. This term refers to the circulation of water from the oceans and other surface waters into the atmosphere by evaporation and transpiration, followed by precipitation back to earth as rain, snow, or hail. (See Fig. 34-7.) Natural waters are classified as:

Atmospheric water: Water contained in clouds and precipitated as rain, snow, or hail
Surface water: Bodies of water such as lakes, streams, rivers, and oceans
Subsurface water: Water which is subterranean in the region where all pores of the soil as well as all spaces in and among rocks are saturated

It has been estimated that about 80,000 cubic miles of water from oceans and 15,000 cubic miles from lakes and land surfaces evaporate annually. The total evaporation is equaled by the total precipitation, of which about 24,000 cubic miles fall on land surfaces.

Microorganisms in
natural waters

Because the various environments of natural water are so different, it is not surprising that their microbial flora differs greatly. The microbial flora of atmospheric water is contributed by the air. In effect, the air is "washed" by rain; particles of dust to which microorganisms are attached accumulate in the atmospheric water.

The microbial flora of subsurface waters is affected by the process of filtration. Microorganisms are retained by the particulate matter in soil, which functions as a filter. Thus deep subsurface waters are likely to be free of microorganisms. Springs consist of ground water that reaches the surface through a rock fissure or exposed porous soil; these can be of "good" microbiological quality.

Surface waters, such as lakes, streams, rivers, estuaries, and oceans, represent very complex microbiological ecosystems. To a greater or lesser degree, these waters are susceptible to periodic contamination with microorganisms from atmospheric water, the surface runoff from soil, and any wastes dumped into them from either domestic or industrial sources. Surface waters vary considerably in terms of available microbial nutrients, physical conditions, and biological characteristics. Thus it is readily apparent that tremendous differences occur among them in terms of the microbial populations.

Microorganisms constitute an extremely important component of the ecosystem in surface waters. This field of study is known as aquatic microbiology.

Distribution of microor-
ganisms in the aquatic
environment

Microorganisms in an aquatic environment may occur at all depths, ranging from the surface to the very bottom of ocean trenches. The top "layers" and the bottom sediments harbor the largest populations of microorganisms, particularly in deep waters.

Plankton (phytoplankton and zooplankton). The aggregation of floating and drifting life—most of it microbial—in the surface region of the aquatic ecosystem is called *plankton*. The plankton population is made up of algae (*phytoplankton*), protozoa, minute animal life (*zooplankton*), and other microorganisms. Phototrophic microorganisms are regarded as the most important plankton because they are the primary producers of organic matter; that is, they are photosynthesizers. Most planktonic organisms are motile, or contain oil droplets, or possess some structural feature which give them buoyancy; all these features aid the organisms in maintaining their location in the photosynthetic zone (upper layer) of water.

Benthic microorganisms. Microbial inhabitants of the bottom region of a body of water (the soft mud or ooze) are referred to as the *benthic* organisms. The richest region of an estuarine-marine system, in terms of numbers and kinds of organisms, is the benthic region, which extends from the high-tide level to the depth below where attached plants do not grow abundantly. The bottom region of the sea contains millions of bacteria per gram.

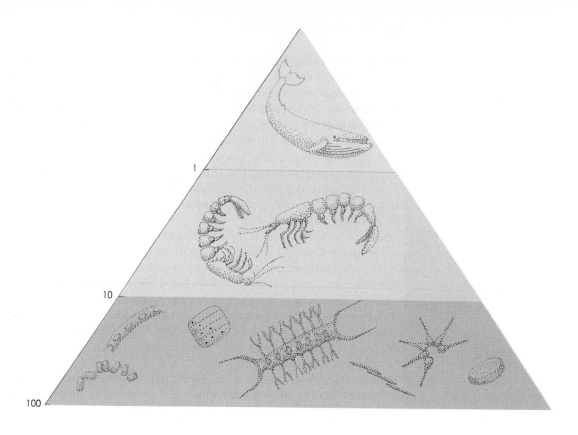

Figure 34-8. Food chain of the Antarctic Ocean, showing the krill as the link between phytoplankton (diatoms) and whales. The pyramidal design illustrates that the overall process involves a "diminishing return." On a unit basis, the yield of organic matter of each successive category of organisms in this example is decreased by a factor of 10. *(Courtesy of R. C. Murphy, Sci Am, 207:186, 1962. Reprinted with permission. Copyright © 1962 by Scientific American, Inc. All rights reserved.)*

The physical conditions and the chemical ingredients which characterize the region of water between the planktonic zone and the benthic zone are so variable that generalizations are not meaningful. Consider for a moment the vast differences between a pond and an ocean! Ponds and lakes also have a characteristic zonation and stratification, and there is considerable information available about their microbiological populations.

The role of microorganisms in the aquatic environment

Aquatic life exhibits a vast complex of interactions among microorganisms and between microorganisms and macroorganisms—both plant and animal. Microorganisms, particularly algae, occupy a key role in the food chain of the aquatic environment.

The *primary producers* in the aquatic environment are the algae, which predominate in the phytoplankton. Through photosynthesis they are capable of transforming radiant energy into chemical energy (organic compounds).

Protozoa (species of *Foraminifera* and *Radiolaria*, as well as many flagellated and ciliated species) are present in large numbers in the region inhabited by the phytoplankton. These zooplankton feed ("graze") upon phytoplankton organisms, bacteria, and other organic or inorganic substances. Observations indicate that many zooplankton avoid light, exhibiting diurnal migrations. At night the zooplankton graze on phytoplankton at the surface; during the day they sink below the photic zone.

Plankton, particularly the phytoplankton, has been referred to as the "pasture of the sea." Fish, whales, and squids feed directly on plankton or on larger plankton-feeding animals (see Fig. 34-8). The term *fertility of oceans* is used to express the ability to produce organic matter by the organisms present in these waters. The terrestrial environment produces 1 to 10 g of dry organic matter per square meter per day compared to 0.5 g for the deep ocean areas. Nevertheless, the oceanic area is so much larger than the productive land area that this difference is inconsequential, and the total productivity of the oceans vastly exceeds that of the land. This fertility depends primarily on the production of phytoplankton. Growth of the phytoplanktonic organisms is dependent upon radiant energy, carbon dioxide, water, inorganic nitrogen and phosphorus compounds, and several elements in trace amounts. The nitrogen, phosphorus, and trace elements are made available through the biochemical activities of microorganisms, particularly the bacteria. This involves the dissimilation of organic substrates (plant and animal tissues) with production of inorganic compounds, the process known as mineralization. The sequence of chemical transformations which yields nutrients to the various species of aquatic life follows a course similar to that outlined for the nitrogen cycle in soil.

Planktonic algae, under certain environmental conditions, may grow into enormous populations that discolor the water. The characteristic color of the Red Sea is associated with heavy blooms of a blue-green alga (cyanobacterium), *Oscillatoria erythraea*, which contains the pigments phycoerythrin and phycocyanin. "Red tides" are likewise due to the explosive growth of certain planktonic species. Brown, amber, or greenish-yellow

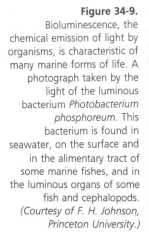
Figure 34-9.
Bioluminescence, the chemical emission of light by organisms, is characteristic of many marine forms of life. A photograph taken by the light of the luminous bacterium *Photobacterium phosphoreum.* This bacterium is found in seawater, on the surface and in the alimentary tract of some marine fishes, and in the luminous organs of some fish and cephalopods. *(Courtesy of F. H. Johnson, Princeton University.)*

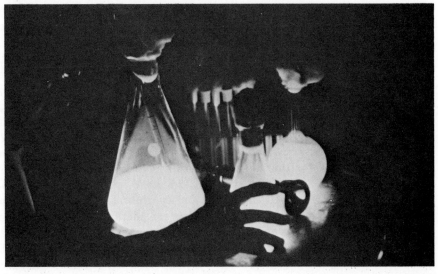

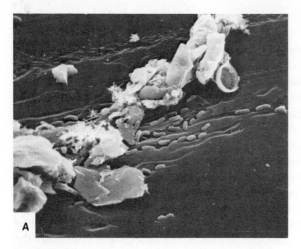

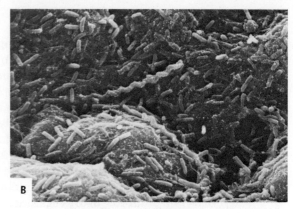

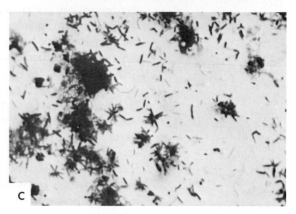

Figure 34-10. (A) Marine bacteria attacking cellulose. The bacteria shown here by scanning electron microscopy are degrading a cellulose (dialysis) membrane (×1,700). Decomposition of cellulose is an important activity in the marine environment. It results in recycling of carbon. *(Courtesy of J. M. Sieburth, Microbial Seascapes, University Park Press, Baltimore, 1975).* (B) Marine bacteria (bacilli and spirochetes) growing on a salmon tail and degrading the tissue. *(Courtesy of J. M. Sieburth, Microbial Seascapes, University Park Press, Baltimore, 1975.)* (C) Marine bacteria attached to cellulose fragments. *(Courtesy of W. A. Corpe, L. Matsuuchi, and B. Armbruster, Proceedings of Third International Biodegradation Symposium, 1976.)*

discoloration of extensive areas of water occurs as a result of blooms by other microorganisms.

Many physiological types of bacteria are present in the different regions of the aquatic environment. Among the psychrophilic forms are certain luminous bacteria, which can produce light in the presence of oxygen (see Fig. 34-9). Some bacteria (*Flavobacterium, Micrococcus, Chromobacterium*) in the surface region of the marine environment are often pigmented, a characteristic which may afford protection from the lethal portion of solar radiation. In areas domestically polluted and rich in organic nutrients, the predominant bacteria include coliforms, fecal streptococci, and species of *Bacillus, Proteus, Clostridium, Sphaerotilus, Beggiatoa, Thiothrix, Thiobacillus*, and many others. Viruses of the enteric group are also likely to be found. In regions of an estuary that are nutritionally poor, one is likely to find the budding and/or the appendaged bacteria, e.g., *Hyphomicrobium, Caulobacter*, and *Gallionella*, in addition to pseudomonads. Some typical marine microorganisms degrading cellulosic material and salmon tissue are shown in Fig. 34-10.

In shallow estuaries, the role of microbial photosynthetic organisms as primary producers is considerably reduced. Plant growth from the shoreline contributes leaves, stems, and other loose organic substances.

Phytoplankton and benthic algae make a small contribution to the food supply in a shallow estuary. The organic vegetation is degraded by bacteria and fungi and converted to microbial protein which may serve as nutrients for protozoa. However, the estuary contains many detritus consumers (herbivorous and omnivorous crustaceans, mollusks, insect larvae, nematode and polychaete worms, and a few fishes). They derive their energy from vascular plant material along the shores.

Microbiology of air

The microbial flora of air is transient and variable. Air is not a medium in which microorganisms can grow, but is a carrier of particulate matter, dust, and droplets, all of which may be laden with microbes. The numbers and types of microorganisms contaminating the air are determined by the sources of contamination in the environment; for example, organisms from the human respiratory tract are sprayed by coughing and sneezing, and dust particles are circulated by air currents from the earth's surface. Airborne microorganisms may be carried on dust particles, in large droplets that remain suspended only briefly, and in *droplet nuclei*, which result when small liquid droplets evaporate. Organisms introduced into the air may be transported a few feet or many miles; some die in a matter of seconds, whereas others survive for weeks, months, or longer. The ultimate fate of airborne microorganisms is governed by a complex set of circumstances including the atmospheric conditions, humidity, sunlight, and temperature; the size of the particles bearing the microorganisms; and the characteristics of the microorganisms, particularly their susceptibility to the physical conditions in the atmosphere.

The importance of airborne pathogens in transmitting diseases was discussed in Chap. 27. Here we are interested in the occurrence of microorganisms in the atmosphere.

Microbiological sampling of air

Air is sampled to determine its microbial content by various procedures. Several instruments, referred to as either *solid impingement* devices or

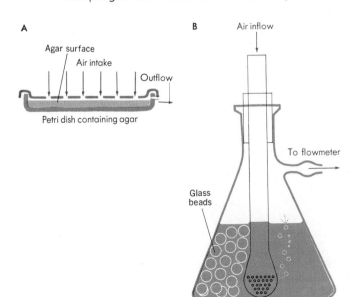

Figure 34-11. Air-sampling devices. (A) A solid impingement device (sieve sampler). Air is drawn through the small holes; particles striking the agar medium adhere to this surface. (B) A liquid impingement device. The air is bubbled through a layer of glass beads in a suitable medium or liquid, which entraps the particles. Samples of the liquid are than cultured to determine the microbial content.

liquid impingement devices (see Fig. 34-11), have been designed for this purpose. In solid impingement devices, the microorganisms are collected, or "impinged," directly on the solid surface of an agar medium or filter disk. Subsequent incubation of the sample results in the development of colonies where organisms impinged. In liquid impingement devices, the air sample is bubbled through a layer of glass beads that are covered by a broth medium or other liquid. Samples of the liquid are then plated or cultured to determine their microbial content.

The microbial content of air

Although no microorganisms are indigenous to air, the air in our immediate environment, as well as that several miles above the earth's surface, contains various species of microorganisms in variable numbers.

Indoor air. The degree of microbial contamination of indoor air is influenced by such factors as ventilation rates, crowding, and the nature and degree of activity of the individuals occupying the quarters. The microorganisms are expelled in droplets from the nose and mouth during sneezing, coughing, or even talking. Droplets expelled from the respiratory tract vary in their dimensions from micrometers to millimeters. Those in the low-micrometer range can remain airborne for long periods of time, but the large droplets settle rapidly onto the floor or other surfaces. Dust from these surfaces becomes airborne intermittently during periods of activity in the room. Results of microbiological sampling of floor dust are shown in Fig. 34-12.

Outdoor air—the atmosphere. The earth's surface—land and sea—is the source of most microorganisms in the atmosphere. Winds create dust from soil; the dust particles carry microorganisms that inhabit the soil. Large quantities of water in the form of droplets are injected into the atmosphere

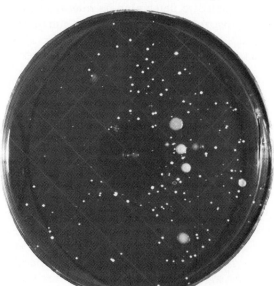

Figure 34-12. Flora of room air, with and without occupants. The left side of the plate represents the sample taken when the room was unoccupied; the right side of the plate represents the sample taken while two people were working in a laboratory. The sampling was performed with a slit-sampling device, an apparatus which draws air through a very fine slit which revolves over the surface of an agar plate. *(Courtesy of Environmental Services Branch, National Institutes of Health, Public Health Service.)*

Table 34-2. Types of bacteria and fungi isolated from upper air

HEIGHT, FT	BACTERIA (GENERA)	FUNGI (GENERA)
1,500–4,500	Alcaligenes Bacillus	Aspergillus Macrosporium Penicillium
4,500–7,500	Bacillus	Aspergillus Cladosporium
7,500–10,500	Sarcina Bacillus	Aspergillus Hormodendrum
10,500–13,500	Bacillus Kurthia	Aspergillus Hormodendrum
13,500–16,500	Micrococcus Bacillus	Penicillium

SOURCE: Courtesy of B. E. Proctor and B. W. Parker, *J. Bacteriol.* **36:**180, 1938.

from the surface of oceans, bays, and other natural bodies of water. In addition to this global origin of microorganisms found in the atmosphere, there are many local or regional industrial, agricultural, and municipal processing facilities which have the potential of producing aerosols of microorganisms. Some examples are:

1 Sprinkler irrigation (with sewage effluent) of agricultural crops or forest land
2 Large threshing operations
3 Trickling-bed filters of sewage disposal plants
4 Slaughtering and rendering plants

Algae, protozoa, yeasts, molds, and bacteria have been isolated from the air near the earth's surface. An example of microorganisms found in an urban atmosphere is shown in Table 34-2. These samples were taken from an industrial area over a period of several months. Mold spores constituted the largest portion of the airborne microflora. The predominant mold spores were of the genus *Aspergillus*. Among the bacterial types were sporeforming and nonsporeforming, gram-positive bacilli, gram-positive cocci, and gram-negative bacilli.

Summary and outlook

Microorganisms are ubiquitous in our environment; they occur on us, within us, and around us. They are an important component of many ecosystems. In their natural habitats, they live in a community of many different kinds of microorganisms, together with other biological species. Within this community, microbes of one species may affect another species in several different ways—some beneficial, some detrimental.

The microbial flora present in natural environments is responsible for numerous biochemical processes which, ultimately, provide for the continuity of life as we know it on earth. Microorganisms, for example, are responsible in soil and in water for mineralization, a process whereby elements are released from complex organic molecular compounds and made available for new plant life which in turn supports new animal life.

This kind of interaction is exemplified in cyclic processes and food chains.

A more thorough knowledge of the processes accomplished by microorganisms in their natural environments can lead to improvements in our way of life. For example, the attractiveness of a nitrogen-fixing process similar to the *Rhizobium*-legume association that could be adaptable to cereal crops (wheat and corn) has been recognized for a long time. Many countries cannot afford the cost of nitrates required for expansion of their agriculture. Through research in microbial genetics, we may see the development of strains of symbiotic nitrogen-fixing microorganisms which would work effectively on wheat and corn crops. It is estimated that if this became a reality, the world's cost for fertilizer could be reduced by one-half and the food supply doubled.

The microbial flora of many natural environments is incompletely understood in terms of the identity of all the species that constitute the population. Many species of marine microorganisms cannot be cultivated in the laboratory by procedures presently available. The same is true of the microbial population in the soil. Hence, as our technology improves, we will witness an expansion of our knowledge of microbiology with the discovery of new kinds of microorganisms.

Key terms

atmospheric water	herbicides	pesticides
benthic organisms	hydrologic cycle	phytoplankton
biosphere	microbial ecology	plankton
commensalism	mineralization	subsurface water
droplet nuclei	mutualism	surface water
ecosystem	neutralism	symbiosis
fixed nitrogen	nitrogen fixation	zooplankton
free nitrogen	nonsymbiotic	

Questions

1 Describe several different mechanisms by which one species of microorganism exerts antagonism toward another species of microorganism.

2 Discuss the microbial flora of soil in terms of the kinds of microorganisms present, with estimates of the magnitude of the microbial flora.

3 If a soil sample is plated on nutrient agar, will the resulting colony count be indicative of the total microbial population? Explain.

4 Assume that some protein material is buried in the soil. Trace the changes it may undergo as a result of microbial attack. Identify bacteria capable of bringing about each of the changes.

5 Tremendous amounts of plant material, largely cellulose, are deposited annually on the earth's surface. Insofar as microbiological events are concerned, what happens to this cellulose?

6 What is the significance of microorganisms in terms of the agricultural use of herbicides and pesticides?

7 Describe the distribution of microorganisms in the aquatic environment.

8 Discuss some of the physical and chemical properties of the aquatic environment in terms of their effect upon microbial growth and survival.

9 What occurs during the process of mineralization?

10 What contributes to the fertility of the ocean? What does the term *pasture of the sea* refer to?

11 What are the sources of microorganisms in the atmosphere?

MICROBIOLOGY OF DOMESTIC WATER AND WASTEWATER (SEWAGE)

The drinking water of most communities and municipalities is obtained
from surface sources—rivers, streams, and lakes. Such natural water
supplies, particularly streams and rivers, are likely to be polluted with
domestic, agricultural, and industrial wastes. Many city dwellers are not
aware that their water has been used at some earlier stage. The reuse of
water is a natural process, as is shown in the hydrologic cycle (Chap.
34). There is, however, a new dimension to water reuse in the present
era. The growth in population, the demand for vast quantities of water by
industry, and the expanded requirements for water for agricultural irrigation have combined to place new demands upon available water
sources. Accordingly, interest has grown in developing acceptable methods
for making "used" water safe and suitable for reuse.

**Water contamination
(pollution)**   Contaminants which pollute water are classified into three categories:
chemical, physical, and biological. Specific entities from each of these

categories can significantly affect the quality of the water. In this chapter we are interested in the biological category.

As a potential carrier of pathogenic microorganisms, water can endanger health and life. The incidence of waterborne diseases in the United States was discussed in Chap. 29.

The pathogens most frequently transmitted through water are those which cause infections of the intestinal tract, namely, typhoid and paratyphoid fevers, dysentery (bacillary and amebic), cholera, and the enteric viruses. The causative organisms of these diseases are present in the feces or urine of an infected person and, when discharged, may enter a body of water that ultimately serves as a source of drinking water.

Accordingly, there must be (1) procedures whereby water can be examined to determine its microbiological quality, (2) water-purification methods that provide safe drinking water, and (3) treatment facilities for wastewater prior to its disposal or reuse. These topics constitute the subject material in this chapter.

Water purification

Water can be perfectly clear, odorless, and tasteless, and yet be unsafe to drink. Water that is free of disease-producing microorganisms and chemical substances deleterious to health is called *potable* water—that is, safe to drink. Contamination with microorganisms or chemicals means the water is polluted and *nonpotable*.

Home water supplies

Underground sources—wells and springs—provide most of the water for individual homes in rural areas. Surface water should not be used for drinking purposes unless it is first treated (or boiled) to remove any contaminants, since it is in constant danger of pollution. Water from wells and springs has undergone filtration as it penetrated through the layers of soil; this process removed suspended particles, including microorganisms. It is of prime importance that the supply of ground water selected be properly located so as to prevent contamination from pit privies, cesspools, septic tanks, and barnyards (Fig. 35-1). Water from home water supplies should be submitted for laboratory examination periodically for assurance

Figure 35-1. A farmstead showing a well properly located to avoid contamination. The arrow indicates the direction in which the ground water moves. *(Redrawn from "Safe Water for the Farm," Farmers' Bulletin no. 1978, USDA, 1948.)*

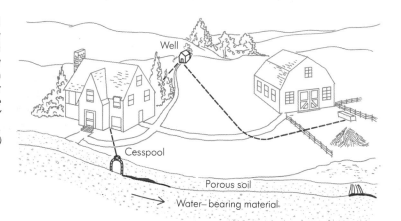

Well

Cesspool

Porous soil

Water–bearing material

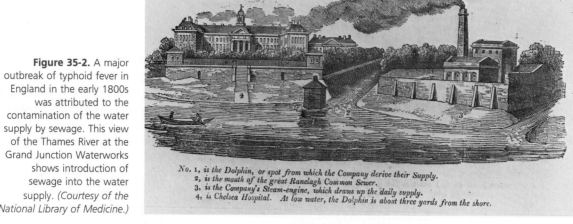

Figure 35-2. A major outbreak of typhoid fever in England in the early 1800s was attributed to the contamination of the water supply by sewage. This view of the Thames River at the Grand Junction Waterworks shows introduction of sewage into the water supply. *(Courtesy of the National Library of Medicine.)*

No. 1, *is the Dolphin, or spot from which the Company derive their Supply.*
2, *is the mouth of the great Ranelagh Common Sewer.*
3, *is the Company's Steam-engine, which draws up the daily supply.*
4, *is Chelsea Hospital. At low water, the Dolphin is about three yards from the shore.*

of potability. The American public water systems are involved in the well-supervised production and distribution of safe water. This service is given some of the credit for all but wiping out the causes of typhoid fever—a scourge 100 years ago (Fig. 35-2), and for virtually eliminating epidemics of waterborne dysentery.

Municipal water purification

The principal operations employed in a municipal water-purification plant to produce water of a quality safe for human consumption are sedimentation, filtration, and chlorination (see Fig. 35-3). Sedimentation occurs in large reservoirs, where the water remains for a holding period; large particulate matter settles to the bottom. Sedimentation is enhanced by the addition of alum (aluminum sulfate), which produces a sticky flocculent (cloudy) precipitate. Many microorganisms and finely suspended matter are removed as this precipitate descends through the water in settling basins. The water is next passed through sand filter beds, a process which removes 99 percent of the microorganisms. Subsequently, the water is chlorinated to kill any remaining microorganisms and to ensure its potability. The chlorine dosage must be sufficient to leave a residue of 0.2 to 1.0 mg free chlorine per liter of water.

The purification process may include additional procedures, such as removing minerals that cause the water to be hard, adjusting the pH if the water is too acid or alkaline, removing undesirable colors or tastes, and adding fluoride for the control of dental caries.

Sanitary surveys

Inspection of a water-producing system by a qualified sanitarian or engineer is called a *sanitary survey*. It includes inspection of (1) the source of the raw water and the conditions that may influence its quality (the frequency of microbiological sampling and the results of analysis are examined), (2) the operation of the water-purification plant, and (3) the mechanism for distributing the water to the consumers. Conditions in a community or municipality that may influence the quality of water are not

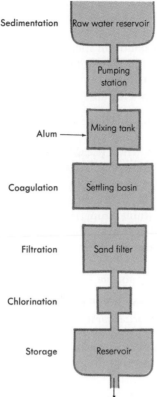

Sedimentation — Raw water reservoir

Pumping station

Alum → Mixing tank

Coagulation — Settling basin

Filtration — Sand filter

Chlorination

Storage — Reservoir

To consumer

Figure 35-3. Flow diagram of usual procedures in municipal water-purification plant.

static. There may be changes in population, types of industry, and the quantity of sewage and the manner in which it is disposed. Consequently, periodic and comprehensive sanitary surveys by personnel from county, state, or federal agencies are necessary.

Microorganisms as indicators of water quality

In the routine microbiological examination of water to determine its potability, it would not be satisfactory to base the test upon the presence of (or isolation of) pathogenic microorganisms for the following reasons:

1 Pathogens are likely to gain entrance into water sporadically, but since they do not survive for long periods of time, they could be missed in a sample submitted to the laboratory.
2 If they are present in very small numbers, pathogens are likely to escape detection by laboratory procedures.
3 It takes 24 h or longer to obtain results from a laboratory examination. If pathogens were found to be present, many people would have consumed the water and would be exposed to infection before action could be taken to correct the situation.

Indicator micro-organisms

The term "indicator microorganisms" as used in water analysis refers to a kind of microorganism whose presence in water is evidence that the water is polluted with fecal material from humans or other warm-blooded

605

animals. This kind of pollution means that the opportunity exists for the various pathogenic microorganisms, which periodically occur in the intestinal tract, to enter the water.

Some of the important characteristics of an indicator organism are:

1 It is present in polluted water and absent from unpolluted (potable) water.
2 It is present in water when pathogens are present.
3 The quantity of indicator organism correlates with the amount of pollution.
4 It has greater survival ability than pathogens.
5 It has uniform and stable properties.
6 It is harmless to humans and other animals.
7 It is present in greater numbers than pathogens (making detection relatively easy).
8 It is easily detected by simple laboratory techniques.

Several species, or groups, of bacteria have been evaluated for their suitability as indicator organisms. Among the organisms studied, *Escherichia coli* and other coliform group bacteria most nearly fulfill the requirements of an ideal indicator organism and are regarded as the most reliable indicators of fecal pollution.

Escherichia coli and other coliform bacteria

Escherichia coli is a normal inhabitant of the intestinal tract of humans and other warm-blooded animals. Normally, it is not pathogenic. Another member of the coliform group is *Klebsiella pneumoniae*, which is widely distributed in nature. It is found in soil, water, and grain, and also in the intestinal tract of humans and other animals. *Enterobacter aerogenes,* a coliform bacterium found in the intestinal tract of humans and other animals, occurs also in soil, water, and dairy products. The coliforms as a group are characterized as gram-negative, nonsporeforming, aerobic and facultatively anaerobic, rod-shaped bacteria that ferment lactose with the production of acid and gas within 48 h at 35°C.

The coliforms have several characteristics in common with members of the genera *Salmonella* and *Shigella*, two genera which have enteric pathogenic species. However, a major distinctive biochemical difference is that the coliforms ferment lactose with production of acid and gas; *Salmonella* and *Shigella* do not ferment lactose. As you will see, the fermentation of lactose is the key reaction in the laboratory procedure performed to determine potability of water.

Bacteriological examination of water for potability

Methods of examining water bacteriologically are presented in the book *Standard Methods for the Examination of Water and Wastewater*, prepared and published jointly by the American Public Health Association, the American Water Works Association, and the Federation of Sewage and Industrial Wastes Associations. The methods are "standard"; the procedural details must be adhered to if the results are to have official significance. It is essential that strict attention be given to the following details when water samples are submitted for bacteriological analysis:

1 The sample must be collected in a sterile bottle.
2 The sample must be representative of the supply from which it is taken.
3 Contamination of the sample must be avoided during and after sampling.
4 The sample should be tested as promptly as possible after collection.
5 If there is a delay in examination of the sample, it should be stored at a temperature between 0 and 10°C.

The routine bacteriological procedures consist of (1) a plate count to determine the number of bacteria present and (2) tests to reveal the presence of coliform bacteria.

Standard plate count. Colony counts are performed after plating samples of the water. Plate-count standards have not been suggested for water because water with a few pathogenic bacteria is obviously more dangerous than water containing many saprophytic bacteria. Nevertheless, water of good quality is expected to give a low total count, less than 100 per milliliter. Plate counts are useful in determining the efficiency of the operations removing or destroying organisms—sedimentation, filtration, and chlorination. A count can be made before and after the specific treatment. The results indicate the extent to which the microbial population has been reduced.

Figure 35-4. General scheme of laboratory testing for detection of coliform bacteria in water.

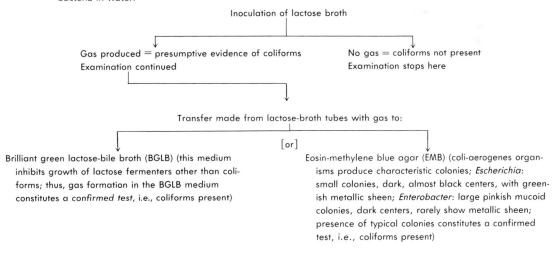

Tests for the detection of coliform bacteria

Several selective and differential media greatly expedite the examination of water for coliform organisms. The examination involves three successive steps: (1) the *presumptive* test, (2) the *confirmed* test, and (3) *the completed* test (see Fig. 35-4).

The laboratory procedure can be performed following the multiple-tube-inoculation procedure or by the membrane-filter technique.

Multiple-tube-inoculation procedure. This test is carried out by inoculating tubes of lactose broth with the water sample, as outlined in Fig. 35-4. In water of good microbiological quality no acid or gas is produced in the lactose broth.

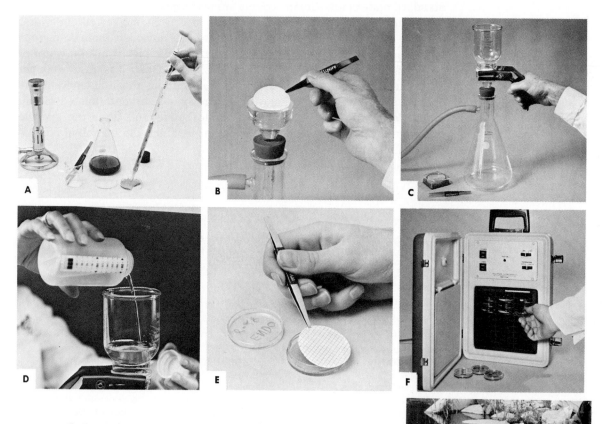

Figure 35-5. Analysis of water with the Millipore filter membrane. (A) Approximately 2 ml of Endo medium is added to the pad contained in the dish. The dish is then covered until the water sample has been filtered through the membrane. (B, C, and D) The filter is placed on a filter holder and clamped in position below the funnel, and the water sample (100 ml for testing potability) is poured into the funnel and passed through the Millipore filter with the aid of a vacuum pump. (E) The funnel is removed, and the filter disk, handled with sterile forceps, is placed on the pad previously impregnated with medium (A). (F) The plates are incubated at 35°C for 20 h, at which time the number of coliform colonies can be determined. (G) A Millipore filter apparatus in use in the field. *(Courtesy of Millipore Corporation.)*

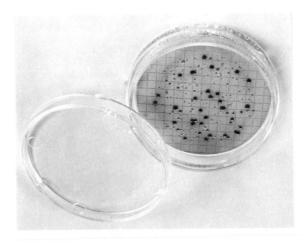

Figure 35-6. Coliform colonies from a water sample grown on a membrane filter with MF-Endo medium acquire an easily distinguishable green metallic "sheen" and are pink to rose-red in color. *(Courtesy of Millipore Corporation.)*

Membrane-filter technique. The membrane-filter technique for the bacteriological examination of water is illustrated in Fig. 35-5; it consists of the following steps:

1 A sterile filter disk is placed in a filtration unit.
2 A volume of water is drawn through this filter disk, the bacteria being retained on the surface of the membrane.
3 The filter disk is removed and placed upon an absorbent pad that has previously been saturated with the appropriate medium. Special petri dishes that will accommodate the absorbent pad and filtration disk are employed for incubation.
4 Upon incubation, colonies will develop on the filter disk wherever bacteria were entrapped during the filtration process (see Fig. 35-6).

This technique has several desirable features, some of which are:

1 A large volume of water sample can be examined; theoretically almost any volume of water could be filtered through the disk, the organisms being deposited on the disk.
2 Results can be obtained more rapidly than by the older technique of inoculating a series of tubes of lactose broth.
3 Quantitative estimations of certain bacterial types, e.g., coliforms, can be made.

Microorganisms other than coliform bacteria

A variety of bacteria are regarded as *nuisance bacteria* in water systems because they create problems of odor, color, and taste, as well as forming precipitates of insoluble compounds within pipes that reduce or obstruct water flow. Algae may also be responsible for the development of odors, discoloration, and other objectionable characteristics. The detrimental actions of some microorganisms are as follows:

Slime-forming bacteria: Produce gummy or slimy conditions.
Iron bacteria: Transform soluble iron compounds to insoluble compounds. Deposition of insoluble iron compounds reduces the flow of water through pipes (see Fig. 35-7).

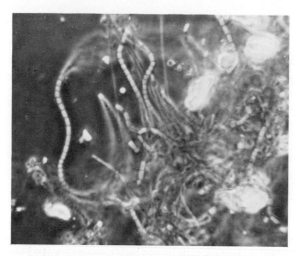

Figure 35-7. Long trichomes of *Sphaerotilus* appear in this photomicrograph of a slime mass. This organism causes many difficulties in industrial water systems because it forms large slime masses and iron deposits. Dark-field phase contrast microscopy (×640). *(Courtesy of J. M. Sharpley, Applied Petroleum Microbiology, Buckman Laboratories, Inc., Memphis, Tenn., 1961.)*

Sulfur bacteria: Produce sulfuric acid and hydrogen sulfide, which can make the water very acid and contribute an obnoxious odor.

Algae: Produce turbidity, discoloration, and unpleasant odor and taste.

Viruses

Viruses, particularly the enteric viruses, can, like the coliform bacteria, be carried by human wastes into water. Waterborne virus diseases were discussed in Part 9.

The analysis of a water sample for presence of viruses requires much more elaborate procedures than that used for isolation of bacteria. Although considerable research is underway in developing and assessing various procedures to accomplish this, no "standard method" has been developed. Better procedures are needed for the isolation of viruses on a routine basis.

Swimming pools

Water in swimming places, particularly in public swimming pools, may be a health hazard. Swimming pools and surrounding areas may transmit infections of the eyes, nose, throat, and intestinal tract; they may spread athlete's foot, impetigo, and other dermatoses. Thus it is imperative that constant attention be given to the sanitary quality of the water. There must be constant surveillance of the disinfection process to ensure the proper level of the chlorine disinfectant.

Wastewater (sewage)

Sewage is the used water supply of a community—the wastewater—and consists of:

1 Domestic waterborne wastes including human excrement and wash waters—everything that goes down the drains of a home and city and into its sewage system
2 Industrial waterborne wastes—acids, oils, greases, and animal and vegetable matter discharged by factories

3 Ground, surface, and atmospheric waters that enter the sewage system

The sewage of a city is collected through a system which carries the used water to its point of treatment and disposal. There are three kinds of sewerage systems: (1) *sanitary sewers*, which carry domestic and industrial sewage; (2) *storm sewers*, which are designed to carry off surface and storm water; and (3) *combined sewers*, which carry all the sewage through a single system of sewers.

Characteristics of sewage

Physical and chemical characteristics of sewage. Sewage consists of approximately 99.9 percent water. The amount of suspended solids in sewage is so small that it is expressed in parts per million (ppm); the solid content ranges from a few to 100 ppm. This amount of solids appears small; however, the tremendous volume of sewage processed daily by a major municipal plant is several hundred million gallons and contains tons of solids. The chemical constituents, also present in low concentrations, nevertheless are extremely important and are subject to variations, between communities as well as within a community, even from hour to hour. Inorganic chemicals initially present in the water supply will likewise be present in the sewage; organic compounds are contributed through human excrement and other domestic wastes, and both organic and inorganic compounds are added by industrial wastes. For example, slaughterhouses, sugar factories, paper mills, and creameries add organic substances; mines and metal industries contribute acids and salts of metals and other inorganic wastes.

Modern technology has produced changes in sewage characteristics. The use of household garbage-disposal units has increased the total organic load. Synthetic detergents, displacing soaps, exert antimicrobial activity.

Biochemical oxygen demand. The biochemical oxygen demand (BOD) is the amount of dissolved oxygen required by microorganisms for the aerobic degradation of organic matter present in the wastewater (sewage). One of the primary reasons for treating wastewater prior to its being returned to a stream or lake is to reduce the drain on dissolved oxygen supply in the receiving body of water. The magnitude of the BOD is an indication of the amount of organic material in the sewage; the more oxidizable organic material present, the higher the BOD. The "strength" of sewage is expressed in terms of BOD level.

The life of any body of water depends to a large extent upon its ability to maintain a certain amount of dissolved oxygen. The dissolved oxygen is needed to maintain marine life. Without dissolved oxygen, for example, fish suffocate; the normal biota is destroyed.

Microbiological characteristics. Fungi, protozoa, algae, bacteria, and viruses are present in wastewater. Raw sewage may contain millions of bacteria per milliliter, including the coliforms, streptococci, anaerobic sporeforming bacilli, the *Proteus* group, and other types originating in the

Table 35-1. Generalized scheme of microbial degradation of the organic constituents in sewage

SUBSTRATES +	ENZYMES OF MICROORGANISMS →	REPRESENTATIVE END PRODUCTS	
		Anaerobic Conditions	Aerobic Conditions
Proteins and other organic nitrogen compounds		Amino acids Ammonia Nitrogen Hydrogen sulfide Methane Carbon dioxide Hydrogen Alcohols Organic acids Indole	Amino acids Ammonia → nitrites → nitrates Hydrogen sulfide → sulfuric acid Alcohols Organic acids $\Big\}$ → $CO_2 + H_2O$
Carbohydrates		Carbon dioxide Hydrogen Alcohols Fatty acids Neutral compounds	Alcohols Fatty acids $\Big\}$ → $CO_2 + H_2O$
Fats and related substances		Fatty acids + glycerol Carbon dioxide Hydrogen Alcohols Lower fatty acids	Alcohols Lower fatty acids $\Big\}$ → $CO_2 + H_2O$

intestinal tract of humans. Additional microorganisms are contributed from ground, surface, or atmospheric waters or from industrial wastes.

Moreover, the effectiveness of the treatment process is dependent upon the biochemical changes produced by a large variety of microorganisms (Table 35-1). The predominant physiological types of bacteria may shift during the various stages of sewage treatment, which range from vigorous aerobic to strict anaerobic conditions.

Sewage-treatment processes

Sewage cannot be disposed of untreated without serious objectionable consequences. Disposal of inadequately treated sewage may produce one or more of the following undesirable situations:

1 Greater possibility for dissemination of pathogenic microorganisms
2 Increased danger in using natural bodies of water for drinking supplies
3 Contamination of oysters and other shellfish by the pollution, making them unsafe for human consumption
4 Large losses in the waterfowl population because of pollution of their winter feeding grounds
5 Increased danger of swimming in the water and diminished value of the water for other recreational purposes
6 Depletion of oxygen supply of the water by unstable organic matter in sewage, killing aquatic life
7 Creation of miscellaneous objectionable conditions such as offensive odors and accumulation of debris, thereby decreasing property values

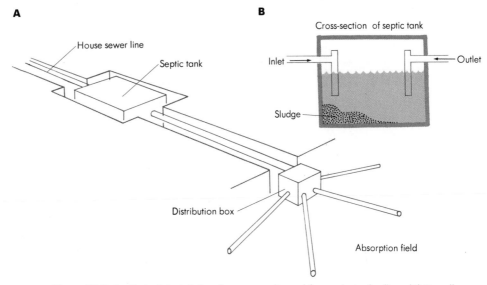

Figure 35-8. Septic-tank installation for sewage disposal from private dwelling. (A) Overall installation including absorption field. *(Redrawn from Public Health Rep., Reprint 2361.)* (B) Cross-sectional view of septic tank. Anaerobic digestion of the sludge converts the solids to liquids and gases.

Sewage-treatment processes are many and varied. We will discuss the treatment processes applicable to two separate situations: (1) a single dwelling or unit structure, and (2) a community or municipality.

Single dwelling or unit structures

Treatment and disposal of wastewater and sewage from individual dwelling or other unit structures, such as motels or shopping centers, can be accomplished by an anaerobic-digestion or aerobic-digestion tank. The *septic tank*, an anaerobic system (Fig. 35-8) is a commonly employed unit for treatment of a limited volume of sewage. A septic tank unit accomplishes two objectives: sedimentation of solid materials and biological degradation of these solids (sludge). As sewage enters the tank, sedimentation occurs in the upper portion, permitting a liquid with fewer suspended solids to be discharged. The sedimented solids are subject to continual degradation by anaerobic bacteria; the end products are very unstable (not oxidized), have a high BOD, and are odorous. The effluent from the septic tank is distributed under the soil surface through a disposal field, as shown in Fig. 35-8. Further microbial degradation, largely aerobic, takes place as the liquid effluent seeps through the drainage field. This kind of treatment cannot guarantee elimination of all pathogens. Consequently, it is imperative that drainage from the system be prevented from seeping into the drinking-water supply.

Aerobic wastewater treatment systems are available commercially for small-unit situations. These tanks are designed with compartments and equipment to reduce incoming solids to small particle size, an aeration chamber, and an effluent settling chamber. Oxygen is pumped into the

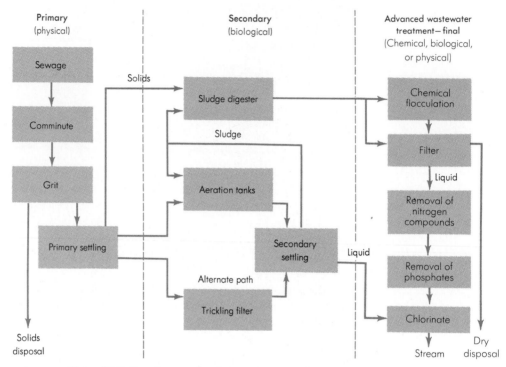

Figure 35-9. Flow diagram of major steps in a municipal sewage-treatment plant.

aeration chamber to allow for continuous oxidation and aerobic decomposition of sewage solids. These units are especially useful in locations where percolation through soil is poor, for example wetlands or regions with lots of stones and rocks.

Municipal facilities

Municipal sewage-treatment plants carry out a series of treatment processes divided into three stages, as shown in Fig. 35-9. Each of these processes are characterized as follows:

1 Primary treatment—*Physical* removal of coarse solids.
  a Screening—Removes largest solids such as boxes, tires, bottles, and cans. Material may be incinerated, ground up, or used for landfill.
  b Grit chambers—Remove smaller solids, such as pebbles.
  c Sedimentation (primary settling)—Removes smaller particulate material, such as fecal matter and paper. This particulate material (sludge) is often treated biologically through anaerobic degradation in a sludge digester.
2 Secondary (*biological*) treatment—Degradation of organic matter to reduce the BOD. One or more of the following methods might be engineered into a facility:
  a Trickling filter—Sewage sprayed (thus aerating it) over rock beds, which contain populations of bacteria that decompose the sewage as it trickles through (see Fig. 35-10).

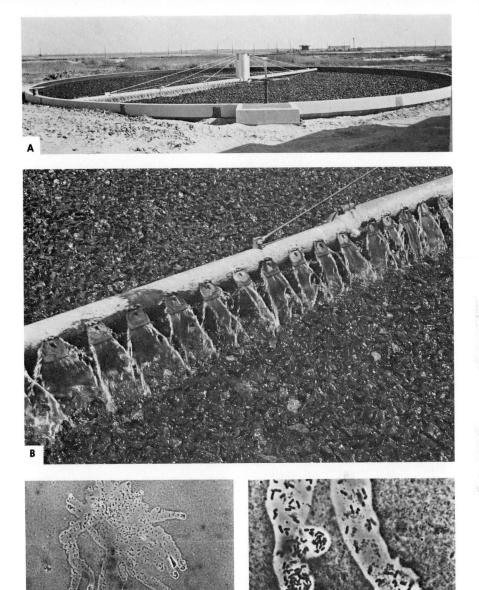

Figure 35-10. A commercial distributor, a trickling filter, used in sewage treatment. (A) The distributor arm applies liquid sewage over a bed of stones. This liquid effluent passes through the stones. (B) Distribution of liquid sewage onto filter stones from a rotary distributor. Filter stones are coated with microbial growth. As the liquid effluent passes through this unit, the microorganisms growing on the stones' surfaces metabolize (oxidize) many of the organic compounds present in the effluent. The liquid effluent is collected at the bottom of the filter bed and piped to the next treatment process. *(Courtesy of R. E. McKinny, Microbiology for Sanitary Engineers, McGraw-Hill, New York, 1962. Used by permission.)* (C) Branched and amorphous zoogloeas (gelatinous masses containing microorganisms) collected from the surface of trickling filters receiving primarily domestic wastewaters. (1) Natural, branched, trickling-filter zoogloea. (2) Portion of specimen shown in (1) illustrating morphological similarity of bacterial cells and their concentration at anterior points of zoogloeal branches. Phase-contrast microscopy. *(Courtesy of R. F. Unz and N. C. Dondero, Water Res., 4:575, 1970.)*

Figure 35-11. An activated-sludge basin with a full load of sewage in the process of aerobic microbiological treatment. *(Courtesy of Washington Suburban Sanitary Commission.)*

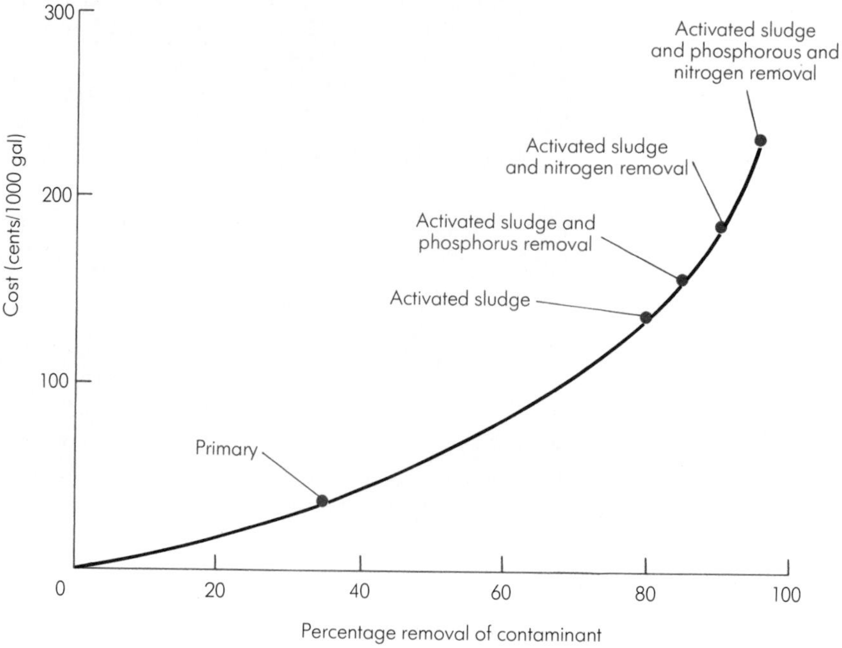

Figure 35-12. Wastewater treatment, total cost versus degree of treatment. Calculations were made from Ronald W. Crites, Michael J. Dean, and Howard L. Sleznick: "Land Treatment vs. AWT—How Do Costs Compare?" *Water and Wastes Engineering*, Sept. 1979. Notes: Costs are based on 1 MGD capacity; costs exclude site specific costs, for general comparison only; costs are for new construction; costs are based on March 1978 dollars. *(Courtesy of Gary C. Beach.)*

b Activated-sludge process—Vigorously aerated sewage coalesces into particles teeming with aerobic microbial degraders. Done in aeration tanks and followed by further sedimentation (see Fig. 35-11).

c Oxidation ponds (lagoons)—In these 2- to 4-ft-deep ponds, algae, such as species of *Chlorella*, use sewage nutrients and provide oxygen for aerobic degradation.

d Sludge digestion (degradation of solids accumulated during primary treatment and sometimes after secondary treatment)—Anaerobes digest sludge in deep tanks, yielding methane (may be used as heating fuel), carbon dioxide, and smaller amounts of nitrogen and hydrogen. Anaerobic sewage degradation is slow, requiring weeks.

3 Advanced wastewater treatment or final treatment processes accomplish further removal of pollutants which may remain after secondary treatment. A wastewater of high quality can be produced that is suitable for many reuses. The advanced treatment may consist of one or more of the following procedures:

a Chemical flocculation—Removes much of the remaining particulate matter.

b Final filtration—The flocculate is filtered, dried, and incinerated. It may serve as landfill or even fertilizer.

c Removal (or reduction in amount) of phosphates and nitrogen compounds.

d Chlorination—The final liquid effluent is chlorinated to kill microorganisms, some of which may have been pathogenic. This low-BOD, nonoffensive, "nonhazardous" liquid may now be drained into natural waterways.

Economics of waste-water treatment

The average cost of treating wastewater in the United States is estimated to be approximately 20 cents per thousand gallons of water. However, when more advanced wastewater treatment processes are employed, the cost becomes markedly higher (see Fig. 35-12). By using the most modern and sophisticated sewage-treatment technology, it is possible to produce an effluent which is potable; the wastewater can be reused. The cost is very high, but some countries in the world have no alternative because of the scarcity of raw water. This is an example of a decision involving a *cost-benefit assessment*; a judgment is made based on the level of quality desired and the availability of alternative sources of water. The increasing demand for water, together with the demand for a high-level quality of the environment, ensures an increase in cost of this commodity once viewed as free and abundant.

Summary and outlook

Tremendous demands for water have been made over the years, and this demand continues to grow. The gravity of the water-pollution problem is reflected in the following quotation taken from an article written by Nandor Porges[1]:

Waste treatment is a means of maintaining or recovering man's most precious and

---

[1]"Newer Aspects of Waste Treatment," in W. W. Umbreit (ed.), *Advances in Applied Microbiology*, vol. 2, Academic, New York, 1960.

most abused natural resource, fresh water. Fresh water supplies were all-important in the establishment and growth of civilizations. Much of man's bitterest fighting has been incited by altercations over water rights, and the course of history may well be written around the theme of primitive and modern man's need for water.

Despite the fact that modern technology has provided effective methods for sewage (wastewater) treatment, many communities and municipalities do not employ adequate treatment procedures; but, even worse, some perform *no* treatments. They simply dump raw sewage into the waterways. Fortunately, significant progress is being made in upgrading sewage-treatment processes to curtail this primitive and unacceptable practice.

The basic function of a sewage-treatment plant is to speed up the natural processes by which water purifies itself. This is accomplished by a host of microorganisms, aerobically in some stages of the procedure and anaerobically in others. The anaerobic degradation of the sewage solids (sludge) solubilizes much of the material and also produces a gas, methane, which can be used as an energy source for the plant. The aerobic processes oxidize organic material and thereby decrease the demand for oxygen (BOD) of the final effluent. Low-oxygen-demand effluents are required in order to maintain normal aquatic life in rivers and streams receiving these effluents.

The worldwide growing demand for high-quality water requires that increased attention be given to the development of more efficient and rapid methods for the improved treatment of wastewater.

Key terms

BOD	pollution	septic tank
chlorination	potable	sludge
effluent	sanitary survey	zooglea

Questions

1 Describe, in order, the operations followed in a municipal water-purification plant. What does each operation accomplish?

2 Why is *Escherichia coli* considered an indicator of pollution? What are coliforms?

3 Is fermentation of lactose with production of acid and gas positive evidence for the presence of *E. coli*? Of the coliforms? Explain.

4 Describe how selective and differential media facilitate the bacteriological analysis of water samples.

5 What are the morphological and cultural characteristics of *Escherichia coli, Enterobacter aerogenes, Proteus vulgaris, Salmonella typhi,* and *Shigella dysenteriae*?

6 What advantages does the membrane- or molecular-filter technique offer for microbiological analysis of water?

7 List several reasons why the routine laboratory procedure to determine potability of water is based upon the presence or absence of coliforms rather than some specific pathogen, such as *Salmonella typhi*.

8 In terms of sewage treatment, what problem has accompanied the wide usage of detergents? The increase in use of home garbage-disposal units?

9 Describe the microbial activity that occurs when sewage passes through a trickling-filter bed.

10 What is activated sludge? Compare the microbial activity in the activated-sludge process with that which occurs in a septic tank.

11 What is the rationale for studying domestic water and sewage concurrently?

MICROBIOLOGY OF FOODS

Early historical records show that nomadic tribes devoted most of their
time to seeking food and shelter. They depended on nature for food; they
gathered wild berries and plants, hunted animals, and caught fish. By
the use of dating techniques, it has been determined that in 9000 to 7000
B.C. in the Near East and in the New World, primitive civilizations had
advanced to the planting and cultivation of seeds—wheat, barley, rice,
rye, and maize—and to the domestication of animals—pigs, fowl, goats,
cattle, and dogs. Since then, there has been a never-ceasing struggle to
preserve enough food to exist from one growing season to the next.
Records of food spoilage date from about 6000 B.C. The practices of
storing food in caves; of smoking, drying, salting, fermenting, freezing;
and of using spices have been traced back to 2500 to 3500 B.C. Ancient
peoples used natural honey as a sweetening agent and preserved juices
from fruit by concentrating them into syrup. Methods of food preservation have developed through the centuries based on practices employed
on the farm. It was not until the nineteenth century, when the relationship between food spoilage and microorganisms was established, that
modern methods of preserving foods began to evolve.

Foodstuffs are composed of proteins, carbohydrates, fats, vitamins, and
minerals. They are a good growth medium for many kinds of microorgan-

620

isms. Microbial action causes protein foods to putrefy, carbohydrates to ferment, and fats and oils to become rancid. Although many microorganisms are harmless to humans, some contaminating microorganisms cause spoilage, and others cause disease or produce toxins which cause food poisoning (Chap. 28). Some microorganisms are beneficial; they produce special food products, such as cheeses and pickles, that are both palatable and resistant to spoilage. Microorganisms are also a food supplement for humans and other animals. In fact, they represent an attractive alternative source of food.

This chapter describes some of the many ways in which microorganisms are involved in food spoilage, food production, and food preservation.

Importance of micro-organisms in food

Microorganisms are important in food for several reasons:

1 Their presence, particularly their numbers and kinds, may indicate the level of food quality.
2 They may be responsible for spoilage of food.
3 Some are used to manufacture specialty food products.
4 Microorganisms are used as food or a food supplement for humans and other animals.
5 Some diseases are foodborne.

Microorganisms as indicators of quality

The microbial content of a food specimen may provide information reflecting the quality of the raw food, the sanitary conditions under which the food was processed, and the effectiveness of the method of preservation.

To help ensure that foods will be pure, healthful, and of the quality claimed, international, federal, state, and private agencies have been established to enforce and control standards, regulations, and inspection of foods. Several agencies of the United Nations have an expressed interest in the world food supply, including the sanitary quality of the food. These agencies include (1) the Food and Agriculture Organization (FAO), (2) the World Health Organization (WHO), and (3) the International Children's Emergency Fund (UNICEF). Although they are not major enforcement or control agencies, they do express a common interest in healthful, safe foods.

The earliest law dealing with food alteration in the United States was enacted by Congress in 1824, and it established inspection of flour in the Capital area. Legislation regulating interstate traffic in foods was first established by the Food and Drug Act of 1906. This was replaced by the 1938 Federal Food, Drug and Cosmetic Act. To this act, amendments and regulations have been added to assure consumers that products are both safe and of good quality. Federal agencies which have authority for enforcement of regulations include (1) the Food and Drug Administration of the Department of Human Health Services (HHS, formerly HEW), (2) the United States Public Health Service, (3) the Meat Inspection Branch,

Agricultural Research Service, USDA, and (4) the Agricultural Marketing Service of the USDA. Federal agencies exercise authority on foods shipped interstate or foods produced in or shipped into territories. States also have food laws to ensure the quality and safety of foods. These laws are enforced by the state departments of health and other state agencies such as departments of agriculture and agencies which have responsibilities for water and wastewater.

Food spoilage by microorganisms

Most foods are good media for growth of many kinds of microorganisms. Under favorable physical conditions, particularly temperatures in the range of 7 to 60°C (45 to 140°F), the organisms will grow and produce changes in the appearance, flavor, odor, and other qualities of foods. These degradation processes may be described as follows:

Protein foods + proteolytic microorganisms →
amino acids + amines + ammonia + hydrogen sulfide

Carbohydrate foods + carbohydrate-fermenting microorganisms →
acids + alcohols + gases

Fatty foods + lipolytic microorganisms → fatty acids + glycerol

The changes that microbes cause in foods, including milk, are not limited to the results of decomposition or degradation; they may also be the result of products of microbial synthesis. Some microorganisms produce pigments that discolor foods. Others which can synthesize polysaccharides produce slimes in or on foods.

Spoilage of food other than canned food. Some examples of noncanned food spoilage and some of the microorganisms involved are given in Table 36-1.

Spoilage of canned foods. Microorganisms that spoil canned foods may be grouped according to the acidity of the product, as shown in Table 36-2. Because of their heat resistance, sporeformers (species of *Clostridium* and *Bacillus*) constitute the most important group of microorganisms in the canning industry. The three most important types of microbiological spoilage of commercially canned foods are:

1 *Flat-sour spoilage.* This spoilage is due to the production of acid. However, the can retains its normal outward appearance; the ends of the can remain flat, hence the term "flat sour." Species of *Bacillus* are the usual causative agent. Spoilage occurs mainly in low-acid foods such as peas or corn. Acid foods such as tomatoes can also be spoiled by growth of *Bacillus coagulans*, which produces additional acid.
2 *TA spoilage.* This type of spoilage is produced by a thermophilic anaerobe, hence the name "TA." The TA bacterium is *Clostridium thermosaccharolyticum*. It ferments sugars, producing acid and gas; the gas may swell the can after a time, producing bulged ends. This kind of spoilage is

Table 36-1. Types of food spoilage (other than canned foods) with some examples of causative organisms

FOOD	TYPE OF SPOILAGE	SOME MICROORGANISMS INVOLVED
Bread	Moldy	*Rhizopus nigricans*
		Penicillium
		Aspergillus niger
	Ropy (stringy)	*Bacillus subtilis*
Maple sap and syrup	Ropy	*Enterobacter aerogenes*
	Yeasty	*Saccharomyces*
		Zygosaccharomyces
	Pink	*Micrococcus roseus*
	Moldy	*Aspergillus*
		Penicillium
Fresh fruits and vegetables	Soft rot	*Rhizopus*
		Erwinia
	Gray mold rot	*Botrytis*
	Black mold rot	*Aspergillus niger*
Pickles, sauerkraut	Film yeasts, pink yeasts	*Rhodotorula*
Fresh meat	Putrefaction	*Alcaligenes*
		Clostridium
		Proteus vulgaris
		Pseudomonas fluorescens
Cured meat	Moldy	*Aspergillus*
		Rhizopus
		Penicillium
	Souring	*Pseudomonas*
		Micrococcus
	Greening, slime	*Lactobacillus*
		Leuconostoc
Fish	Discoloration	*Pseudomonas*
	Putrefaction	*Alcaligenes*
		Flavobacterium
Eggs	Green rot	*Pseudomonas fluorescens*
	Colorless rots	*Pseudomonas*
		Alcaligenes
	Black rots	*Proteus*
Concentrated orange juice	"Off" flavor	*Lactobacillus*
		Leuconostoc
		Acetobacter
Poultry	Slime, odor	*Pseudomonas*
		Alcaligenes

Table 36-2. Bacterial spoilage relationships in canned foods

SPOILAGE TYPE*	pH GROUPS	EXAMPLES
Thermophilic: *Flat-sour*	5.3 and higher	Corn, peas
Thermophilic anaerobes	4.8 and higher	Spinach, corn
Sulfide spoilage	5.3 and higher	Corn, peas
Mesophilic: *Putrefactive anaerobes*	4.8 and higher	Corn, asparagus
Butyric anaerobes	4.0 and higher	Tomatoes, pears
Aciduric flat sours	4.2 and higher	Tomato juice
Lactobacilli	4.5–3.7	Fruits
Yeasts	3.7 and lower	Fruits
Molds	3.7 and lower	Fruits

*Italic indicates bacterial sporeformers.
SOURCE: National Food Processors Association.

Table 36-3. Some
characteristics of
fermented milk products.

FERMENTED PRODUCT	PRINCIPAL MICROORGANISMS RESPONSIBLE FOR FERMENTATION AND CHANGES THEY PRODUCE
Cultured sour cream	Same as used for cultured buttermilk, i.e., streptococci and leuconostocs Acid and flavor
Bulgarian milk	*Lactobacillus bulgaricus* Acid and flavor
Acidophilus milk	*L. acidophilus* Acid
Yogurt	*Streptococcus thermophilus* *L. bulgaricus* Acid and flavor
Kefir	*S. lactis* *L. bulgaricus* Lactose-fermenting yeasts Acid and flavor
Kumiss	Similar to those found in kefir Acid and flavor

most likely to occur in low-acid foods, such as peas, corn, beans, meat, fish, and poultry, and medium-acid foods, such as spinach, asparagus, beets, and pumpkin.

3 *Sulfide spoilage.* This type of spoilage is produced by the bacterium *Desulfotomaculum nigrificans* (formerly named *Clostridium nigrificans*), particularly in low-acid foods. During growth and metabolism this bacterium produces hydrogen sulfide. The odor of this gas is noticeable upon opening a can of food that is spoiled. *Desulfotomaculum nigrificans* is an obligate thermophile; therefore, if heat processed foods are not cooled promptly, this thermophile can grow.

Manufacture of food products using microorganisms

Thus far we have stressed the undesirable characteristics of microorganisms in food. However, there are many useful applications of microorganisms in the food industry. A variety of important products in our diet are produced with the aid of microbial activity.

Fermented dairy products. In the dairy industry, fermented milks are produced by inoculating pasteurized milk with a known culture of microorganisms, sometimes referred to as a *starter culture*, which can be relied on to produce the desired fermentation, thus assuring a uniformly good product. Some fermented milk products and the role of microorganisms in producing them are summarized in Table 36-3.

Several hundred varieties of cheese are manufactured, and with few exceptions, most of them can be made from the same batch of milk. Microorganisms—bacteria or molds—convert the curd of the milk into the desired cheese. For the manufacture of some cheeses, such as Roquefort, it is neccessary to inoculate the curd with the microorganism which brings about the changes (in this case, *Penicillium roqueforti*). Some of the steps in the process of making Roquefort cheese are shown in Fig. 36-1.

Other fermented foods. Imported food items produced in whole or in part by microbial fermentations include pickles, sauerkraut, olives, and certain

Figure 36-1. Roquefort and blue cheese. (A) Cubes of sterile whole wheat bread are inoculated with *Penicillium roqueforti*. After extensive growth of the mold on the bread cubes, the cubes are removed, dried, and powdered to be used as inoculum for making cheese. *(The Borden Company.)* (B) The addition of a lactic culture and rennet curdles the milk. The curd is cut when it becomes firm. (C) The curd particles are removed and placed in metal hoops. The spores of *P. roqueforti* may be added in either of steps B or C. (D) The hoops are placed on a draining board to facilitate whey drainage and matting of the curd, after which the curd is removed, salted periodically, and (E) eventually placed in an area of high humidity (95 to 98 percent) and low temperature (9 to 12°C), where the ripening process occurs over a period of several months. The hoops of cheese shown ripening here are wrapped in foil. [(B to E) *Courtesy of Roquefort Association, Inc.*]

Table 36-4. Some examples of fermented food products

FERMENTED FOOD	STARTING PRODUCT	MICROORGANISMS INVOLVED

Sauerkraut	Shredded cabbage	Early stage:
		    Enterobacter cloacae
		    Erwinia herbicola
		Intermediate stage:
		    Leuconostoc mesenteroides
		Final stage:
		    Lactobacillus plantarum

Pickles	Cucumbers	Early fermentation:
		    Leuconostoc mesenteroides
		    Streptococcus faecalis
		    Pediococcus cerevisiae
		Later fermentation:
		    Lactobacillus brevis
		    L. plantarum

Green olives	Olives	Early stage:
		    Leuconostoc mesenteroides
		Intermediate stage:
		    Lactobacillus plantarum
		    L. brevis
		Final stage:
		    L. plantarum

Sausage	Beef and pork	*Pediococcus cerevisiae* *Micrococcus* spp.

types of sausage (see Table 36-4). Lactic acid bacteria are chiefly responsible for the desirable type of fermentation required for the production of each of these substances. The microorganisms that produce the changes may be the natural flora on the material to be fermented or may be something added as a starter culture. (See Fig. 36-2, which shows *Streptococcus thermophilus* and *Thermobacterium bulgaricum* organisms used as starter cultures in the preparation of yogurt.) Most commercial sour, sweet, mustard, and mixed pickles are made from fermented salt-stock pickles. The other major type of pickled cucumber is the *fermented dill pickle*.

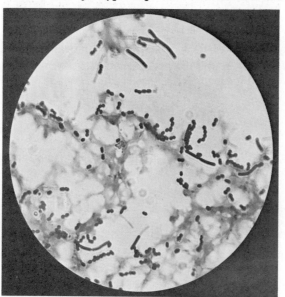

Figure 36-2. Photomicrograph of yogurt, illustrating microbial flora, *Streptococcus thermophilus* and *Thermobacterium bulgaricum* (×800). *(From K. J. Demeter, Bakteriologische Untersuchungsmethoden der Milchwirtschaft, Eugen Ulmer, Stuttgart, 1967.)*

Fungi (molds), in addition to being used to produce several varieties of cheese, are also used in producing some Oriental foods, such as soy sauce. Some are grown as food or feed. Yeasts are important in the baking industry as the agents responsible for making dough rise, thereby imparting a desirable texture as well as flavor to the bread.

Microorganisms as a source of food (single-cell protein)

The world's growing population faces chronic food shortages and, unhappily, periodic famines. Traditional food production and harvesting—that is, conventional agriculture, livestock, and fishing—is not likely to achieve the increase in food necessary if, as estimated by the United Nations Statistical Office, the planet's human population doubles by the year 2011.

Efforts are under way in many countries to increase food production by various unconventional methods. Scientists are striving to develop ways to produce proteins from new sources, including microorganisms. Bacteria, yeasts, and algae, produced in massive quantities, could be important sources of food for animals and humans. These microorganisms can be cultivated on industrial wastes or by-products to yield a large cell crop rich in protein (*single-cell protein*). Bacterial cells grown on hydrocarbon wastes from the petroleum industry are now a source of protein in France, Japan, Taiwan, and India. Yeast-cell crops harvested from the vats used to produce alcoholic beverages have been used as a food supplement for generations. By the next decade, we may witness a greatly expanded use of microorganisms for food.

The microbial flora of food

Natural foods have a normal microbiota. Some of these microorganisms are contributed from the environment; others may be introduced into the food during handling, processing, and storage.

Milk

At the time it is drawn from the udder of a healthy animal, milk contains organisms that have entered the teat canal through the teat opening. They are flushed out during milking. The number present at the time of milking has been reported to range between several hundred and several thousand per milliliter. From the time the milk is drawn from the cow, and until it is dispensed into containers, everything with which it comes into contact is a potential source of more microorganisms.

Microorganisms found in milk may be divided into categories on the basis of three major characteristics: biochemical characteristics, temperature characteristics, and pathogenicity.

Biochemical characteristics. If maintained under conditions that permit bacterial growth, raw milk of a good sanitary quality will develop a distinct sour flavor. This change is brought about mainly by *Streptococcus lactis* (Fig. 36-3) and certain species of lactobacilli (Fig. 36-4). The principal change is lactose fermentation to lactic acid. This type of change is sometimes referred to as the *normal fermentation* of milk. Other organisms may cause changes that produce unpalatable end products.

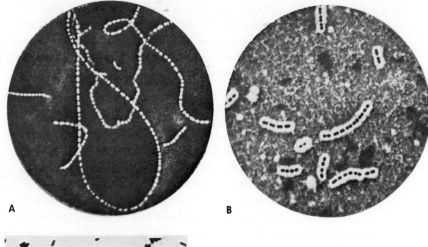

Figure 36-3. *Streptococcus lactis* (A) and *S. cremoris* (B), two important fermentative bacterial species in milk and milk products. These species, along with *Lactobacillus fermenti* (Fig. 36-4), cause the so-called normal fermentation of milk; they are not pathogens. *(Courtesy of S. Orla-Jensen, The Lactic Acid Bacteria, Ejnar Munksgarrd, Copenhagen, 1919.)*

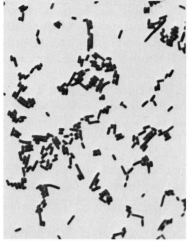

Figure 36-4. *Lactobacillus fermenti*, one of the heterofermentative lactobacilli. It produces a mixture of acids and is involved in the normal fermentation of milk; it is not pathogenic. Its cells vary in length, are grampositive, nonmotile, and nonspore-forming. *(Courtesy of A. P. Harrison.)*

A description of the biochemical changes in milk, together with the microorganisms responsible for these changes, is presented in Table 36-5.

Table 36-5. Biochemical types of microorganisms in milk

BIOCHEMICAL TYPES	REPRESENTATIVE MICROORGANISMS	SOURCE OF MICROORGANISMS	SUBSTRATE ACTED UPON AND END PRODUCTS
Acid producers	Streptococci	Dairy utensils, silage, plants	Lactose fermented to lactic acid (homofermentative) or lactic acid and other products such as acetic acid, ethyl alcohol, and carbon dioxide (heterofermentative)

Table 36-5 (Continued)

BIOCHEMICAL TYPES	REPRESENTATIVE MICROORGANISMS	SOURCE OF MICROORGANISMS	SUBSTRATE ACTED UPON AND END PRODUCTS
Acid producers (continued)	Lactobacilli	Feeds, silage, manure	As above for streptococci
	Microbacteria	Manure, dairy utensils, and dairy products	Lactose fermented to lactic acid and other end products; do not produce as much acid as the streptococci or lactobacilli
	Coliforms	Manure, polluted water, soil, and plants	Lactose fermented to a mixture of end products, e.g., acids, gases, and neutral products
	Micrococci	Ducts of cow's mammary gland, dairy utensils	Small amounts of acid produced from lactose (weakly fermentative); micrococci are also weakly proteolytic
Gas producers	Coliforms *Clostridium butyricum* *Torula cremoris*	Soil, manure, water, feed	Lactose fermented with accumulation of gas; the gas may be a mixture of carbon dioxide and hydrogen, or only carbon dioxide in the case of yeast fermentation
Ropy or stringy fermentation	*Alcaligenes viscolactis* *Enterobacter aerogenes* *Streptococcus cremoris*	Soil, water, plants, feed	Organisms synthesize a viscous polysaccharide material that forms a slime layer or capsule on the cells; milk may become viscous
Proteolytic	*Bacillus* spp. *Pseudomonas* spp. *Proteus* spp. *Streptococcus liquefaciens*	Soil, water, utensils	Proteolytic organisms degrade the casein to peptides which may be further dissimilated to amino acids; proteolysis may be preceded by coagulation of the casein by the enzyme rennin; end products may impart off flavor, color
Lipolytic	*Pseudomonas* spp. *Achromobacter lipolyticum* *Candida lipolytica* *Penicillium* spp.	Soil, water, utensils	Lipolytic microorganisms hydrolyze milk and fat to glycerol and fatty acids; produce rancid condition

Temperature characteristics. Bacteria that occur in milk may be classified according to the temperatures at which they will grow and according to their resistance to heat. This is a very practical consideration, since low temperatures are employed to prevent or retard microbial growth that spoils milk, and high temperatures (pasteurization) are used to reduce the microbial population, destroy pathogens, and in general improve the keeping quality of the milk. On the basis of temperature requirements, the bacteria encountered in milk are of the following types: psychrophilic, mesophilic, thermophilic, and thermoduric.

Since certain psychrophiles grow at temperatures just above freezing and some thermophiles grow at temperatures in excess of 65°C (149°F), it follows that the temperature at which milk is held will determine which species grow and predominate. Pasteurized milk stored in a refrigerator may be satisfactorily preserved for a week or even longer. But eventually, microbial deterioration, manifested by "off" flavor or odor, will become evident because of the accumulation of metabolic products of psychrophilic bacteria. Thermophiles present a problem at the other extreme of the temperature scale. *Bacillus stearothermophilus*, a thermophile, grows at 65°C (149°F), which is above the temperature used to pasteurize milk. Table 36-6 characterizes bacterial growth that predominates in milk held at various temperatures.

In the dairy industry, *thermoduric* bacteria are regarded as those which survive pasteurization but do not grow at pasteurization temperatures. Microorganisms of this category are extremely troublesome. They may contaminate dairy equipment; thus subsequent batches of raw milk processed through the same equipment will become heavily contaminated.

Table 36-6. Effect of holding temperature of raw milk on numbers and types of bacteria

HOLDING TEMPERATURE, °C	CHANGES IN NUMBERS	PREDOMINANT ORGANISMS
1–4	Slow decline first few days followed by gradual increase after 7 to 10 days	True psychrophiles, e.g., species of *Achromobacter, Flavobacterium, Pseudomonas,* and *Alcaligenes*
4–10	Slight change in number during first few days followed by rapid increase in numbers; large populations present after 7 to 10 days or more	As above; changes produced on holding are of the following types: ropiness, sweet curdling, proteolysis, etc.
10–20	Very rapid increase in numbers; excessive populations reached within a few days or less	Mainly acid-producing types such as lactic streptococci
20–30	High populations develop within hours	Lactic streptococci, coliforms, and other mesophilic types; in addition to acid there may be gas, off flavors, etc.
30–37	High populations develop within hours	Coliform group favored

Pathogenicity. Pathogenic microorganisms may gain entrance into milk from several sources and, if not destroyed, be responsible for spreading disease. In recent years, milk has been involved in fewer and fewer outbreaks of illness, to the point that the public and regulatory agencies no longer consider milk a primary source of foodborne illness. Milk and dairy products can now be considered model foods from the standpoint of regulation and surveillance of production, processing, and distribution.

The source of a pathogenic agent occurring in milk may be either a cow or a human, and it may be transmitted by either of the following routes:

1 Pathogen from infected cow→milk→human or cow. The causative agents of tuberculosis, brucellosis, and mastitis are examples.
2 Pathogen from human (infected or carrier)→milk→human. The causative agents of typhoid fever, diphtheria, dysentery, and scarlet fever are examples.

It is also possible for humans to infect cows. For example, mastitis may be caused by a variety of organisms, including *Staphylococcus aureus*. The organism which infects the cow may come from the hands of a person who has a staphylococcus infection.

Vegetables and fruits
Generally, the inner, healthy tissues of plants are free of microorganisms, but their surfaces may be contaminated with a variety of organisms. The magnitude (and type) of the microbial contamination is determined by the environment from which the vegetable or fruit was taken, its condition (freshness), the method of handling, and the time and conditions of storage. Fruits and vegetables are normally susceptible to infection by bacteria, fungi, and viruses. Microbial invasion of the plant tissue can occur during various stages of fruit and vegetable development, and, depending on the extent that the tissues are invaded, the likelihood of spoilage is increased. A second factor contributing to the microbial contamination of fruits and vegetables is their postharvest handling. Mechanical handling is likely to produce breaks in the tissue which facilitates invasion by microorganisms. The pH of fruits is relatively acid, ranging from 2.3 for lemons to 5.0 for bananas. This restricts bacterial growth, but does not retard fungal growth. The pH range of vegetables is slightly higher, 5.0 to 7.0, and hence they are more susceptible to attack by bacteria.

Poultry
Freshly dressed poultry have a bacterial flora on their surface (skin) that originates from the bacteria normally present on the live birds and from contamination during the killing, defeathering, and evisceration. Under good sanitary conditions, the bacterial count has been reported to be from 100 to 1,000 bacteria per square centimeter of skin surface, whereas under less sanitary conditions the count may increase 100-fold or more. Pseudomonad types constitute the major contaminants on the skin of freshly dressed poultry.

Eggs
The interior of a freshly laid egg is usually free of microorganisms; its

subsequent microbial content is determined by the sanitary conditions under which it is held, as well as the conditions of storage, such as temperature and humidity. Microorganisms, particularly bacteria and molds, may enter the egg through cracks in the shells or penetrate the shells when the "bloom" (a thin protein film) covering the shell has deteriorated. The types of microorganisms involved are those present in the environment.

Seafood

The microbial flora of freshly caught oysters, clams, fish, and other aquatic specimens are very largely a reflection of the microbial quality of the waters from which the animals are harvested. Of particular significance is the sanitary quality of the water. Is it polluted? The bacterium *Vibrio parahae-molyticus* has been responsible for a number of seafood-borne gastroenteritis epidemics in the United States. This organism occurs widely in the Atlantic, Pacific, and Gulf Coast waters and has been isolated from seafood samples including fish, shellfish, and crustaceans. *Vibrio cholerae*, the causative agent of cholera, has been found in United States coastal marine waters. Improperly processed shellfish food products from these waters have been contaminated with this organism and, when eaten, have caused infection. Shellfish that grow in contaminated water can concentrate viruses and may be sources of hepatitis infection. Raw oysters and clams harvested from polluted waters have caused epidemics in various parts of the world.

Meats

The carcass of an animal slaughtered for meat and held in a refrigerated room is likely to have surface contamination by many kinds of microorganisms from various sources such as the air, personnel, and equipment. The inner tissues of a healthy animal are sterile. Fresh meat cut from the chilled carcass has each new surface contaminated with microorganisms characteristic of the environment and the implements (saws or knives) used to cut the meat. The ultimate in providing new surfaces for the contamination of meat occurs in the process of making hamburger.

To improve the microbiological quality of meat and meat products, particularly ground beef, cold cuts, and frankfurters, some cities and states have adopted standards, or at least have initiated action to require microbiologic standards for these products at the time of their purchase.

Among the more common species of bacteria occurring on fresh meats are pseudomonads, staphylococci, micrococci, enterococci, and coliforms. The low temperature at which fresh meats are held allows the growth of only psychrophilic microorganisms.

Control of micro-organisms in foods

Most foods are soon decomposed or spoiled by microorganisms, unless they are preserved. Modern methods of food preservation employ elaborate refinements of the primitive processes, such as salting, drying, and smoking, as well as newer techniques. Food preservation methods may be summarized as follows:

1 Aseptic handling
2 Removal of microorganisms
3 High temperatures
  (a) Boiling
  (b) Steam under pressure
  (c) Pasteurization
4 Low temperatures
  (a) Refrigeration
  (b) Freezing
5 Dehydration
6 Increased osmotic pressure
  (a) In concentrated sugar
  (b) In brine
7 Chemicals
  (a) Organic acids
  (b) Substances developed during processing (smoking)
  (c) Substances contributed by microbial fermentations (acids)
8 Radiation
  (a) Ultraviolet
  (b) Gamma

All methods of food preservation are based upon one or more of the following principles: (1) prevention or removal of contamination, (2) inhibition of microbial growth and metabolism (microbistatic action), and (3) killing of microorganisms (microbicidal action).

Methods for preservation must be assessed with an understanding of the extraordinary resistance of bacterial spores to such agents as heat, radiation, chemicals, and dehydration.

Aseptic handling   Keeping spoilage microorganisms out of food decreases food spoilage, makes preservation of food easier, and lessens the possibility of the presence of pathogens. Protective coverings, such as egg shells, fruit and vegetable skins, corn husks, and the skin and fat on meats, are natural barriers to spoilage organisms. Packaging of foods, canning processed foods in sealed cans, and practicing sanitary methods in handling food are all examples of aseptic handling.

Removal of micro-organisms   Liquids forced by positive or negative pressure through a sterilized, "bacteria-type" filter may be used to clear fluids and remove microorganisms. This is a method used with beer, soft drinks, fruit juices, wine, and water.

High temperatures   High temperature is one of the safest and most reliable methods of food preservation. Heat is widely used to destroy organisms in food products in cans, jars, or other types of containers that restrict the entrance of microorganisms after processing.

Figure 36-5. (A) Thermal-death-time (TDT) can designed by the American Can Company Laboratories. Left, can and cover before use; right, sealed can ready for test. For determinations, suspensions of bacteria or their spores are added to foods or media contained in the TDT cans. The cans are then vacuum-sealed and heated to various temperatures and for various time periods in small sterilizers (B). After this treatment, the TDT cans are incubated and subsequently examined for evidence of surviving organisms. *(Courtesy of American Can Company.)*

Canning. This has been the basic method of food sterilization for approximately 170 years. In 1810, Nicholas Appert, a Frenchman, published *L'Art de Conserver*, which described his successful research in food preservation; in the same year, Peter Durand was granted an English patent describing the use of tin containers for food preservation.

Steam under pressure, such as in a pressure cooker, results in temperatures above 100°C in a saturated atmosphere of steam; it is the most effective method of food preservation, since it kills all vegetative cells and spores. Food preservation by heat requires knowledge of many factors, particularly the heat resistance of microorganisms, and bacterial spores. One must also consider the rate at which heat penetrates through foods of different consistencies and the size of the containers in which they are packed. Killing microorganisms by heat involves a time-temperature relationship, as discussed in Chap. 19. Considerable experimentation has been performed to determine the thermal death times of bacteria likely to cause spoilage. From such information, it is possible to establish satisfactory heat-processing conditions. The research that has been done on this subject accounts for the highly successful results achieved in food preservation by canning. Spoilage of commercially canned foods occurs very infrequently. Special laboratory equipment designed to determine with precision the heat resistance of various bacterial species, particularly the sporeformers, is shown in Fig. 36-5.

The most important organism to be eliminated in canned foods is the anaerobe *Clostridium botulinum*, which is capable of producing the very potent, lethal toxin which was discussed in Chap. 28.

Pasteurization. As described in the Milk Ordinance and Code of the U.S. Public Health Service:

The terms *pasteurization, pasteurized*, and similar terms shall mean the process of heating every particle of milk or milk product to at least 145°F (62.8°C) and holding it continuously at or above this temperature for at least 30 minutes, or to at least 161°F (71.7°C) and holding it continuously at or above this temperature for at least 15 seconds, in equipment which is properly operated and approved by the health authority.

Two methods of pasteurization are used commercially, the low-temperature holding (LTH) method and the high-temperature–short-time (HTST) method. The holding method, or vat pasteurization, exposes milk to 62.8°C (145°F) for 30 min in appropriately designed equipment. The HTST process employs equipment capable of exposing milk to a temperature 71.7°C (161°F) for 15 s. In either method of pasteurization, it is essential that the equipment be designed and operated so that every particle of milk is heated to the required temperature and held for the specified time. Precautions must be taken to prevent recontamination after pasteurization. The finished product should be stored at a low temperature to retard growth of microorganisms which survived pasteurization.

Fruit juices, vinegar, and beer are examples of other foods which may be pasteurized.

Sterilization of milk. A sterile milk product has several attractive features—it does not require refrigeration, and it has an indefinite shelf life. The usual methods of food sterilization have not proved satisfactory for the sterilization of milk. However, commercial milk-sterilization techniques have been developed which expose the milk to ultrahigh temperatures for very short periods of time, for example, 148.9°C (300°F) for 1 to 2 s. In addition, the processing includes steps that eliminate any traces of a cooked flavor. The final product is comparable to pasteurized milk in taste and nutritional quality and, if packaged in a sterile container, can be stored safely at room temperature for an extended period of time.

In Belgium most milk sold is sterile. Other countries are starting to follow this practice.

Low temperatures

Temperatures of 0°C (32°F) and lower retard the growth and metabolic activities of microorganisms for long periods of time. Refrigerated trucks and railway cars, ships' storage vaults, and the refrigerators and freezers in

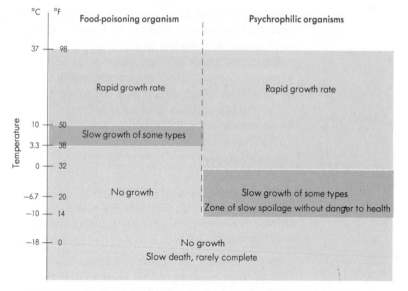

Figure 36-6. Food-poisoning organisms grow in a somewhat higher temperature range than psychrophilic microorganisms. *(Courtesy of R. P. Elliot and H. D. Michener, Review of the Microbiology of Frozen Foods, in "Conference on Frozen Food Quality," ARS-74-21, USDA, 1960.)*

homes and stores have improved the quality of the human diet and increased the variety of foods available. Frozen-food production in the United States is expected to more than double from 20 billion lb in 1975 to 48 billion lb by 1985. Much of this increase will be in precooked ready-to-serve foods such as TV dinners. The quantity of these foods has tripled over the last 10 years and is expected to approach 50 percent of all frozen foods by 1985. The growth and importance of this segment of the food industry, together with the increased use of automatic vending machines for dispensing perishable foods, have made it necessary to increase the study of microorganisms at low temperatures to learn more about their survival, growth, and metabolic activity.

Before freezing, most fresh produce is steamed (blanched) to inactivate enzymes that would alter the product even at low temperatures. Some products which do not need to be blanched are green peppers, onions, and fruits. Quick-freeze methods, using temperatures of −32°C (−25.6°F) or lower, are considered most satisfactory; smaller crystals of ice are formed, and cell structures in the food are not disrupted. It must be emphasized that freezing foods, no matter how low the temperature, cannot be relied upon to kill all microorganisms. The number and types of viable and nonviable microorganisms present in frozen foods reflect the degree of contamination of the raw product, the sanitation of the processing plant, and the speed and care with which the product was processed. The microbial count of most frozen foods decreases during storage; but many organisms, including pathogens, e.g., species of *Salmonella*, survive for long periods of time at −9 to −17°C (14 to 2°F). The temperature ranges at which food-poisoning bacteria and psychrophilic microorganisms are capable of growing are shown in Fig. 36-6.

The growth of food-poisoning bacteria (*Clostridium botulinum* types A and B, *Staphylococcus aureus*, and salmonellas) is prevented by temperatures at or below 5.5°C (42°F). *Clostridium botulinum*, type E, has been reported to grow at a temperature as low as 3.3°C (38°F). Figure 36-7 plots the growth of salmonellas and staphylococci in foods at various temperatures and times of incubation.

Dehydration
Dehydration is the removal of water. It may be accomplished by various methods, such as the sun's rays, applied heat, or high concentrations of sugar or salt.

Preservation by dehydration. The preservative effect of dehydration is due mainly to inhibition of growth; the microorganisms are not necessarily killed. Growth of all microorganisms can be prevented by reducing the moisture content of their environment below a critical level. The critical level is determined by the characteristics of the particular organism and by the capacity of the food item to bind water so that it is not available as free

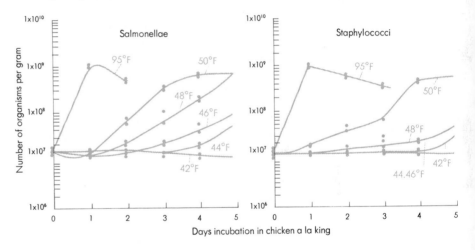

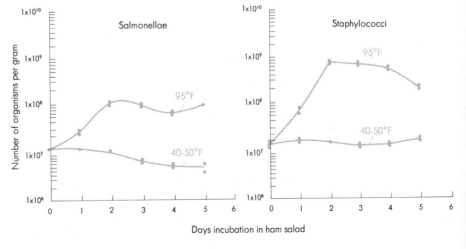

Figure 36-7. Salmonellas and staphylococci multiply rapidly in chicken a la king and ham salad incubated at room temperature. Curves also show growth at other temperatures. *(Courtesy of R. Angelotti, M. J. Foter, and K. H. Lewis, "Time-Temperature Effects on Salmonellae and Staphylococci in Foods, "Am J Public Health, 51:76-88, 1961.)*

moisture that can be removed by dehydration. Recall from Chap. 19 that lyophilized cultures of microorganisms survive for years.

Preservation by increased osmotic pressure. Water is withdrawn from microorganisms when they are placed in solutions containing large amounts of dissolved substances such as sugar or salt. In other words, the cells are dehydrated, and metabolism is arrested, thereby slowing or preventing the growth of the microorganisms. Although yeasts and molds are relatively resistant to osmotic changes, processes of food preservation based on this principle are, nevertheless, very useful. Jellies and jams are rarely affected by bacterial action because their sugar content is high. However, it is not uncommon to find mold growth on the surface of jelly that has been exposed to air. Similar results are obtained by curing meats and other foods in brines. High osmotic pressure may inhibit microbial growth, but it cannot be relied upon to kill organisms.

Chemicals   The addition of chemical preservatives to foods is regulated by the United States Food, Drug and Cosmetic Act as revised in 1972. According to this act, a food is adulterated (which is illegal) if any poisonous or deleterious substance has been added which may render it injurious to health. Only a few chemicals are legally acceptable for food preservation. Among the most effective are benzoic, sorbic, acetic, lactic, and propionic acids, all of which are organic acids. Sorbic and propionic acids are used to inhibit mold growth in bread. Nitrates and nitrites, used in curing meats (primarily for the preservation of color), are inhibitory to some anaerobic bacteria, especially *Clostridium botulinum.* The possibility that nitrites may be carcinogenic for humans raises questions about their continued use.

Foods prepared by fermentation processes, for example, sauerkraut, pickles, and silage for animals, are preserved mainly by acetic, lactic, and propionic acids produced during the microbial fermentation.

Radiation   Radiation sterilization provides an entirely new approach to food preservation; it could bring about a radical change in industrial methods of food processing. Ultraviolet rays have been used to reduce and inactivate microorganisms, especially molds, in the air of storage areas and packaging rooms of bakeries and meat plants. Irradiation of meat, hung for aging and tenderizing, reduces growth of microorganisms on the surface of the meat. This decreases the time of the aging process from several weeks at 2.2 to 3.3°C (36 to 38°F) to 2 to 3 days at 18°C (64.4°F).

High-intensity ionizing radiations, such as *gamma rays* emitted from radioactive cobalt (cobalt 60), are capable of sterilizing various materials, including packaged foods. The application of this technique (radiation sterilization) has been under investigation since the 1950s. There is no question about the effectiveness of high-level radiation (such as the gamma rays of cobalt 60 which reach a level of 100 million electronvolts) to sterilize a product. The effect of high-intensity ionizing radiations on the flavor, odor, color, texture, and nutritional quality of food needs to be more

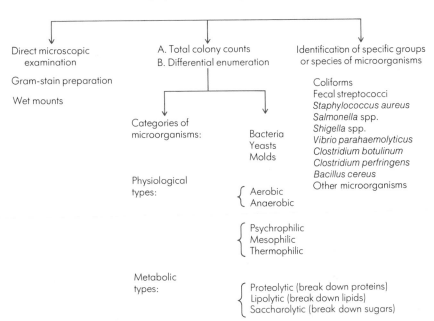

Food sample
(preparation of homogenate)

Direct microscopic
examination

Gram-stain preparation

Wet mounts

A. Total colony counts
B. Differential enumeration

Categories of
microorganisms:

Bacteria
Yeasts
Molds

Physiological
types:

{ Aerobic
  Anaerobic

{ Psychrophilic
  Mesophilic
  Thermophilic

Metabolic
types:

{ Proteolytic (break down proteins)
  Lipolytic (break down lipids)
  Saccharolytic (break down sugars)

Identification of specific groups
or species of microorganisms

Coliforms
Fecal streptococci
Staphylococcus aureus
Salmonella spp.
Shigella spp.
Vibrio parahaemolyticus
Clostridium botulinum
Clostridium perfringens
Bacillus cereus
Other microorganisms

Figure 36-8. Generalized scheme for microbiological examination of food.

completely understood. Likewise, the chemical changes produced in the irradiated food product need to be more adequately evaluated in terms of their effect on humans and other animals.

Microbiological examination of foods

Microbiological examination of foods, including milk and milk products, may provide information concerning the quality of the raw food, the sanitary conditions under which the food was processed, and the effectiveness of the method of preservation. In the case of spoiled foods, it is possible to identify the agent responsible for the spoilage; having discovered the agent, it may be possible to trace the source of contamination and the conditions which permitted spoilage to occur. Corrective measures can then be instituted to prevent further spoilage.

Procedures used in food examination

Microbiological procedures for food examination take advantage of special microscopic techniques and cultural methods. Extensive use is made of selective and differential media to facilitate the enumeration and isolation of certain types of microorganisms. The type of examination performed is determined by the type of food product to be examined and by the specific purpose of the examination. For example, a food sample being investigated for possible contamination by *Clostridium botulinum* would be subject to different laboratory tests than one being examined for coliform organisms. The increasing significance of salmonellas in foodborne disease has made it mandatory to develop more rapid, reliable, and reproducible methods for the detection of salmonellas in foods. An adaptation of the fluorescent-antibody technique is an attractive possibility.

The various procedures and techniques that are used for the microbiological examination of a food specimen are shown schematically in Fig. 36-8.

Summary and outlook

Microorganisms are important in foods for several reasons. In general, the number and kinds of microorganisms present in a food are a reflection of the sanitary quality of the food. Microorganisms are capable of spoiling foods—many foods support good growth of microorganisms; given a chance to grow, they will metabolize the food and produce objectionable metabolic products. Microorganisms are used to manufacture many food products. Microorganisms themselves may serve as food material. Massive quantities of cells are produced which can supplement or substitute for other foods.

The public acceptance of and the increased demand for processed foods have resulted in the growth of centralized food-processing plants. While this may encourage the development of efficient and specialized procedures, it creates a situation whereby any faulty processing of a food item, because of its wide distribution, exposes a large population to spoiled food. To guard against such an occurrence, it is necessary to maintain constant microbiological surveillance of the process and the product.

No doubt in the future, additional pickling or fermentation processes using microbes will be developed. However, the main importance of microbes might be as a bulk food themselves. Microbes grown in vast quantities are already used as supplements for animal feeds. Utilization of microbial cells which have been produced from cheap and abundant substrates may become a major source of human food. It is probably just a matter of time before microbes become accepted widely as an item in the diets of people. Approximately one acre of land is required to feed one person by efficient agriculture, yet a tank on one square yard of land can produce all the caloric, protein, and vitamin requirements for one person.

The cultivation of microbes as a food item (single-cell protein) may be one way of solving three major problems confronting humanity. These solutions are (1) the utilization of waste products as food for producing single-cell protein, thus (2) solving food shortage with the possibility of (3) releasing areas of farmland for other uses.

Key terms

dehydration	single-cell protein	TA spoilage
flat-sour spoilage	starter culture	thermoduric bacteria
pasteurization	sulfide spoilage	

Questions

1 Explain the relationship that exists between the microbial content of food and the quality of the food.
2 What type of spoilage does microbial action cause in foods that are composed primarily of protein, carbohydrate, or fat?
3 List the major sources of the microbial contamination of food such as (a) raw hamburger meat, (b) milk.
4 Describe the various types of biochemical changes brought about in milk by microorganisms. Identify the predominant types of bacteria responsible for each of these changes.

5 Of what particular significance are psychrophilic, thermoduric, and thermophilic bacteria in milk and milk products?

6 What are the likely sources of the pathogens that gain entrance into milk?

7 List several types of microbial food spoilage, and name the organisms responsible in each instance.

8 What physiological types of bacteria are most likely to be present when canned food spoils? What is the significance of sporeformers in food preservation by canning?

9 Compare the types of microorganisms that might be involved in the spoilage of refrigerated foods with those causing spoilage of canned foods.

10 Explain how microorganisms may be used as a source of food. What are the possible advantages of this source of food?

11 List the methods of food preservation and cite an advantage and a disadvantage associated with each method.

12 Compare the antimicrobial action of the following methods of food preservation: canning, refrigeration, dehydration, and increased osmotic pressure.

13 Is pasteurized milk sterile milk? Explain.

14 Outline a procedure suitable for the enumeration, isolation, and identification of a specific group of microorganisms from a sample of food.

INDUSTRIAL MICROBIOLOGY

Industrial microbiology refers to the large-scale growth of microorganisms, under controlled conditions, for the purpose of producing an economically valuable and a useful product. Practically all the developments in this field have occurred during this century. One of the first large-scale uses of microorganisms, under controlled conditions, was for the production of acetone and butanol during World War I. The British exploited this process to obtain chemicals urgently required for the manufacture of explosives. The acceleration in the development of industrial microbiology occurred during World War II when an all-out effort was initiated between the British and the Americans to produce the "miracle drug," penicillin. This adventure and subsequent accomplishments revolutionized the fermentation industry; tremendous advances have followed. Today, many companies are engaged in the large-scale production of chemicals and other products which are in great demand and which have great economic value.

There are other important aspects of industrial microbiology. These include the control of microorganisms that destroy food, wood, fabrics, and other materials, and the extensive use of microorganisms to assay products such as antibiotics, vitamins, and amino acids.

Microorganisms and industry

From the standpoint of industry, the microorganism is a "chemical factory" capable of bringing about a desirable change. The microorganism acts upon a raw material (some component of the medium in which it is growing and which may be regarded as substrate material) and converts this raw material into a new product. The general reaction, then, may be viewed as:

Substrate   + microorganisms→new product(s)
Raw material

Prerequisites of practical microbiological industrial processes

If a microorganism converts an inexpensive raw material into a useful and more valuable substance, it may be feasible to perform this reaction on a large industrial scale. Some of the prerequisites of an economically practicable industrial-microbiological process are:

1 *The organism.* The organism to be employed must be able to produce appreciable amounts of the product. It should have stable characteristics and the ability to grow rapidly and vigorously, and it should be nonpathogenic. Such an organism can even be genetically engineered for a particular purpose by using current technology. (See Chap. 17.)
2 *The medium.* The medium, including the substrate from which the organism produces the new product, must be cheap (relative to the product that is to be produced) and readily available in large quantities. For example, nutrient-containing wastes from the dairy industry (whey) and the paper industry (waste liquors resulting from the cooking of wood) are employed in the production of valuable materials.
3 *The product.* Industrial fermentations are performed in large tanks; capacities of 50,000 gal are not unusual. The product formed by the metabolism of the microorganism is present in a heterogeneous mixture that includes a tremendous number of microbial cells and unused constituents of the medium, as well as products of metabolism in addition to those being sought. Thus a feasible, large-scale method of recovery and purification of the desired end product must be developed.

Microorganisms employed industrially are molds, yeasts, and bacteria. Production of bacteria and viruses on a commercial scale for use as vaccines may also be regarded as an industrial process.

Industrial microbiological products

Microbiological industries fall into several categories, the most important of which are:

1 Alcoholic beverages
2 Food supplements
3 Pharmaceutical chemicals
4 Biologics (vaccines and antisera)
5 Industrial chemicals

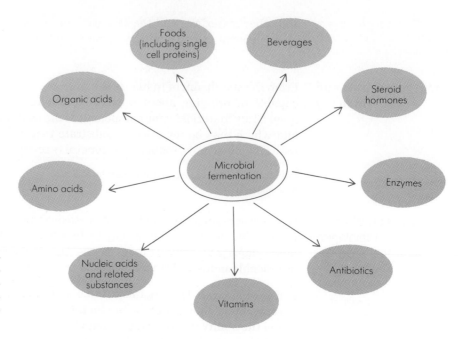

Figure 37-1. Products of microbial fermentation. Commercial microbiological processes are in use for the production of a great many substances.

The variety of products of microbial origin are shown in Fig. 37-1.

Industrial products from bacteria

Bacteria are used on an industrial scale for the production of various chemicals, enzymes, amino acids, vitamins, and other substances (see Table 37-1). One of these processes, the manufacture of vinegar, will be described in detail.

Vinegar production

The word *vinegar* is derived from the French term *vinaigre*, meaning "sour wine." Vinegar is prepared by allowing a wine to "go sour" under controlled conditions.

The production of vinegar involves two types of biochemical changes: (1) an alcoholic fermentation of a carbohydrate and (2) oxidation of the alcohol to acetic acid. There are several kinds of vinegars, and the differences among them are primarily associated with the kind of material used in the alcoholic fermentation, such as fruit juices, sugar-containing syrups, and hydrolyzed starchy materials. The definition and standards for one type of vinegar as given by the U.S. Food and Drug Administration are as follows:

Vinegar, cider vinegar, apple vinegar. The product made by the alcoholic and subsequent acetous fermentations of the juice of apples. It contains, in 100 cubic centimeters (20°C), not less than 4 grams of acetic acid.

A yeast fermentation is used for production of the alcohol. The alcohol concentration is adjusted to between 10 and 13 percent and then exposed to the action of acetic acid bacteria. Many types of equipment have been designed for industrial production of vinegar. All depend upon providing a

PRODUCT	BACTERIUM	USES
Acetone-butanol	*Clostridium acetobutylicum* and others	Solvents; chemical manufacturing
2,3-Butanediol	*Bacillus polymyxa; Enterobacter aerogenes*	Solvent; humectant; chemical intermediate
Dihydroxyacetone 2-Ketogluconic acid	*Gluconobacter suboxydans* *Pseudomonas* spp.	Fine chemical Intermediate for D-araboascorbic acid
5-Ketogluconic acid Lactic acid	*G. suboxydans* *Lactobacillus delbrueckii*	Intermediate for tartaric acid Food products; textile and laundry; chemical manufacturing; deliming hides
Bacterial amylase	*Bacillus subtilis*	Modified starches; sizing paper; desizing textiles
Bacterial protease	*B. subtilis*	Bating hides; desizing fibers; spot remover; tenderizing meat
Dextran	*Leuconostoc mesenteroides*	Stabilizer in food products; blood-plasma substitute
Sorbose	*G. suboxydans*	Manufacture of ascorbic acid
Cobalamin	*Streptomyces olivaceus; Propionibacterium freudenreichii*	Treatment of pernicious anemia; food and feed supplementation
Glutamic acid Lysine	*Brevibacterium* spp. *Micrococcus glutamicus*	Food additive Animal-feed additive
Stretokinase-Streptodornase	*Streptococcus hemolyticus*	Medical use (dissolving blood clots)

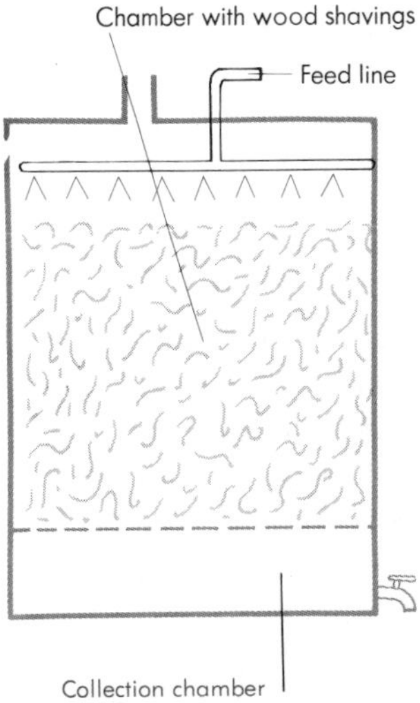

Chamber with wood shavings

Feed line

Collection chamber

Figure 37-2 Frings vinegar generator. A dilute solution of alcohol percolates through wood shavings that are covered with a growth of *Acetobacter*. The bacteria oxidize the alcohol to acetic acid.

suitable environment for the bacterial oxidation of alcohol to acetic acid. The essential features of one of the industrial processes for vinegar production, the Frings method, is shown in Fig. 37-2 and may be summarized as follows. A mix is prepared which consists of an adjusted solution of alcohol acidified with acetic acid and nutrients for the growth of acetic acid bacteria. Acetic acid bacteria, species of the genus *Acetobacter*, are inoculated onto beechwood shavings in the chamber (Fig. 37-2). The mix is applied in a trough at the top of the chamber and allowed to trickle down over the shavings. Air is available in abundance and the temperature is maintained between 15 and 34°C. As the alcoholic solution passes over the shavings, the *Acetobacter* oxidizes some of the alcohol to acetic acid. The mix is collected at the bottom of the unit and may be recirculated over the shavings, resulting in more oxidation of alcohol until vinegar of the desired strength is produced.

Since this is an aerobic process, oxygen is required as shown in the following reaction accounting for the formation of acetic acid:

$$2CH_3CH_2OH + 2O_2 \rightarrow 2CH_3COOH + 2H_2O$$

Ethyl alcohol   Acetic acid

Industrial products from yeasts

The best known and one of the most important uses of yeasts is the production of ethyl alcohol from carbohydrate materials. This fermentation process is used by brewers of malt beverages, distillers, bakers, wine makers, chemical manufacturers, homemakers, and many others.

Alcoholic fermentations

Next to water, alcohol is the most common solvent and raw material used in the laboratory and in the chemical industry. The microbiological aspects of the process of ethyl alcohol production can be summarized as follows.

The substrate. Ethyl alcohol can be produced from any fermentable carbohydrate by means of yeasts. When starches, such as corn, and other complex carbohydrates are used as the raw material, it is first necessary to hydrolyze them to simple fermentable sugars. The hydrolysis can be accomplished with enzymes from barley malt or molds or by heat treatment of acidified material. Corn, molasses, sugar beets, potatoes, rice, and grapes are some of the common raw materials used as substrates throughout the world.

The organism. Selected strains of *Saccharomyces cerevisiae* are commonly employed for the fermentation. It is imperative that the culture be one that grows vigorously and has a high tolerance for alcohol as well as the capacity for producing a large yield of alcohol. Much attention has been directed toward the selection and development of strains of yeasts which excel in these particular characteristics.

The reaction. The biochemical change accomplished by the yeast is as follows:

$$C_6H_{12}O_6 \quad + \quad yeast \quad \rightarrow 2C_2H_5OH + 2CO_2$$

Glucose (fermentable carbohydrate)   (enzymes)   Ethyl alcohol   Carbon dioxide

Alcoholic beverages. Beer, rum, whiskey, wine, and other alcoholic beverages are all products of yeast fermentations. The products are different because of the differences in the material fermented and in the strains of yeast that do the fermenting. For example, rum is a product of blackstrap molasses fermentation; whiskey is a product of corn and/or rye fermentation; wine is a product of grape juice fermentation. In each of these examples, strains of yeasts in the genus *Saccharomyces* are used to ferment the substrate. Companies take great care to develop special strains of yeasts that will produce a high-quality product.

Bakers' yeast

The use of yeast as a leavening, or rising, agent in baking dates back to the very early histories of the Jews, Egyptians, Greeks, and Romans. In those days leavened bread was made by mixing some leftover dough from the previous batch of bread (which contained yeasts) with fresh dough. Another practice, since the Middle Ages, has been to use excess yeasts from

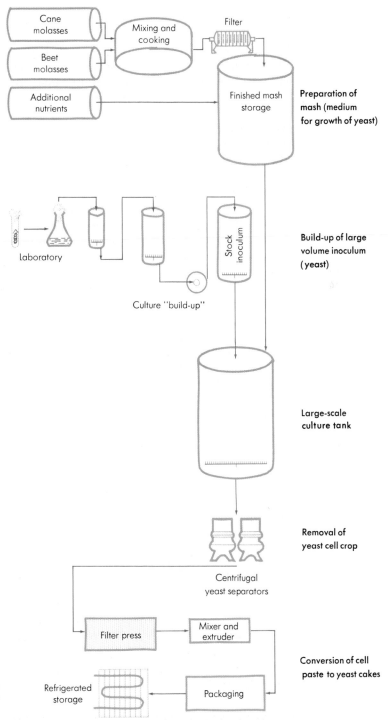

Cane
molasses

Beet
molasses

Additional
nutrients

Mixing and
cooking

Filter

Finished mash
storage

Preparation of
mash (medium
for growth of yeast)

Laboratory

Culture "build-up"

Stock
inoculum

Build-up of large
volume inoculum
(yeast)

Large-scale
culture tank

Removal of
yeast cell crop

Centrifugal
yeast separators

Filter press

Mixer and
extruder

Conversion of cell
paste to yeast cakes

Figure 37-3. Steps in the
commercial production of
bakers' yeast.

Refrigerated
storage

Packaging

brewing and wine-making operations. But the variable quality of such products makes this practice unsatisfactory for large-scale commercial production. In modern baking, pure cultures of selected strains of *Saccharomyces cerevisiae* are mixed with the bread dough to bring about desired changes in texture and flavor. *Saccharomyces cerevisiae* strains selected for commercial production of bakers' yeast have the ability to ferment the sugar in the dough vigorously and to grow rapidly. The carbon dioxide produced during the fermentation is responsible for the leavening of the dough. The quality of the product depends on the proper selection of yeasts, the incubation conditions, and the choice of raw materials. The procedure used for manufacturing bakers' yeast is shown in Fig. 37-3.

Gasohol. The energy shortage has increased interest in the use of alcohol as a supplement to conventional fuels. *Gasohol*, usually a mixture of 90 percent unleaded gasoline and 10 percent alcohol, is in use in several regions of this country and is being produced on a much larger scale in Brazil. Widespread usage of this fuel mixture will most likely be determined by economic factors, that is, the cost of producing the alcohol. Abundant quantities of suitable low-cost raw materials must be available for fermentation, and more efficient methods for the recovery of alcohol must be developed to make gasohol an economically feasible fuel.

Industrial products from molds

Molds are used in the production of antibiotics and various chemicals, enzymes, and food products (see Table 37-2). One of the best-known uses of molds is in the fermentation process utilized for production of penicillin.

Penicillin production

The development of commercial production of penicillin and other antibiotics is one of the most dramatic breakthroughs in the history of industrial microbiology. The antibiotic industry did not exist in 1941, but 10 years

Table 37-2. Some industrial products (other than antibiotics) derived from molds

PRODUCT	MOLD	USES
Citric acid	*Aspergillus niger* or *A. wentii*	Food products; medicinal citrates; in blood for transfusion
Fumaric acid	*Rhizopus nigricans*	Manufacture of alkyd resins; wetting agent
Gluconic acid	*A. niger*	Pharmaceutical products; textiles; leather; photography
Itaconic acid	*A. terreus*	Manufacture of alkyd resins; wetting agents
Pectinases	*A. wentii* or *A. aureus*	Clarifying agents in fruit juice industries
11-γ-Hydroxyprogesterone	*R. arrhizus, R. nigricans,* others	Intermediate for 17-γ-hydroxycorticosterone (steroid)
Gibberellic acid	*Fusarium moniliforme*	Setting of fruit; seed production
Lactic acid	*R. oryzae*	Foods and pharmaceuticals

later net sales of antibiotics had reached $344 million per year. Data reported by the U.S. Tariff Commission revealed that over 25 million pounds of bulk antibiotics were manufactured in 1978.

Penicillin was the first antibiotic to be produced industrially. Much of what was learned in the transformation of Sir Alexander Fleming's laboratory observations into an economically feasible large-scale operation paved the way for successful production of other chemotherapeutic antibiotics as they were discovered.

The mold isolated by Fleming (*Penicillium notatum*), as grown in his laboratory, yielded only a few units of penicillin per milliliter, an exceedingly small amount when one considers that a patient may require treatment with millions or billions of units. The remarkable chemotherapeutic effectiveness of penicillin was demonstrated by Sir Howard Florey and Ernest B. Chain during 1939 and 1941. Because of the pressures of war, British scientists brought the mold to the United States in hope of developing production of the antibiotic on a large scale. An extensive research program having one of the highest wartime priorities was initiated. In a relatively short time the yield of penicillin was increased about a thousand times. The developments contributing to this enormous increase in yield were as follows:

1 Improvements in composition of the medium.
2 Isolation of a better penicillin-producing mold species, *Penicillium chrysogenum*.
3 Development of the submerged-culture technique: cultivation of the mold in large volumes of liquid medium through which sterile air is forced.
4 The production of mutant strains of *P. chrysogenum* which were capable of producing large amounts of penicillin. A series of mutants, produced by x-ray and ultraviolet radiation, resulted in strains with a remarkable capacity for synthesis of penicillin.
5 The addition of chemicals to the medium which served as precursors for synthesis of penicillin.
6 Refinements in methods of recovering penicillin from the fermentation mixture.

The major steps in the commercial production of penicillin are shown schematically in Fig. 37-4; a commercial production facility is shown in Fig. 37-5. The changes which occur during the fermentation process (growth, production of penicillin, changes in medium) are shown in Fig. 37-6.

Most other antibiotics are produced in a similar manner. The major differences are related to the kind of microorganism used, composition of the medium, and method of extraction. It is feasible to use the same fermentation equipment for production of more than one antibiotic.

Enzyme production   Several species of molds synthesize large quantities of enzymes. The amounts they produce and excrete into the medium make it commercially

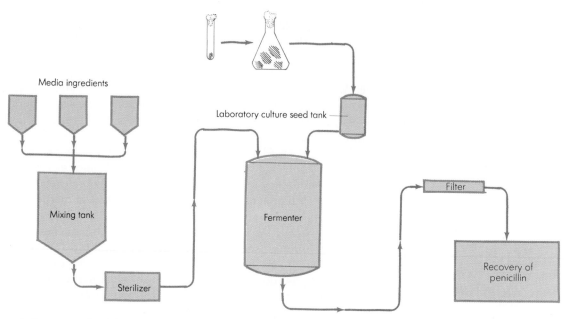

B Preparation of inoculum

Media ingredients

Laboratory culture seed tank

Mixing tank

Fermenter

Filter

Recovery of
penicillin

Sterilizer

A Preparation of medium

C The fermentation

D Recovery of penicillin

Figure 37-4. Manufacture of penicillin shown schematically. (A) A medium of corn-steep liquor, lactose, salts, and other ingredients is mixed, sterilized, cooled, and pumped into the fermenter. (B) The mold *Penicillium chrysogenum* is transferred from slant cultures to bran, and spore suspensions from bran are transferred to a sterile vessel with medium, which in turn is used to inoculate the seed tank. (C) The fermenter is inoculated from the seed tank; sterile air is forced through the fermenter during incubation. (D) After the maxium yield of penicillin is produced, the mold mycelium is removed by filtration and the penicillin is recovered in pure form by a series of manipulations which include precipitation, redissolving, and filtration.

Figure 37-5. Tops of large fermentation tanks of the type used to produce antibiotics. *(Courtesy of Merck and Co., Inc.)*

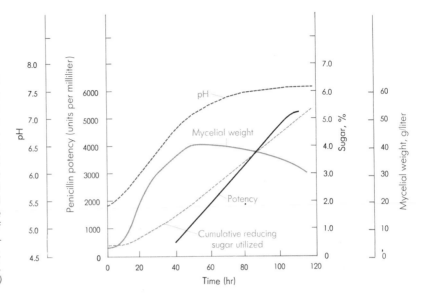

Figure 37-6. Biochemical changes that occur in the fermenter during production of penicillin by *Penicillium chrysogenum*. The data presented in this figure show that the increase in amount of antibiotic is related to the increase in growth (mycelial weight), the increase in pH (acid production), and the decrease in the amount of reducing sugar in the medium. *(Courtesy of R. Donovick, Appl Microbiol, 8:117, 1960.)*

feasible to recover these enzymes and to concentrate them for industrial applications. Pectinases, invertases, amylases, and proteases are some of these enzymes. *Amylases* hydrolyze starch to dextrin and sugars and are used in preparing sizes and adhesives, desizing textiles, clarifying fruit juices, manufacturing pharmaceuticals, and for other purposes. *Invertase* hydrolyzes sucrose to a mixture of glucose and levulose and is widely used in the production of candy and noncrystallizable syrups from sucrose. *Proteases* are used for bating (treatment of hides to provide a finer texture and grain) and other leather-processing steps, manufacturing liquid glue, degumming of silks, clarification of beer protein haze (particulate protein matter), and cleaning in laundries as an adjunct to soap. For centuries—long before the role of enzymes in the bating of hides was understood—this treatment was accomplished by soaking the hides in suspensions of dog or fowl manure. Today, standard enzyme solutions have replaced the concoctions of dung. *Pectinase* is used to clarify fruit juices and also to hydrolyze pectins in the stems of flax hemp or jute plants, thus releasing cellulose fibers for the manufacture of linen and burlap, respectively.

Immobilized enzyme technology. The potential industrial uses of enzymes have been significantly enhanced by developments in *immobilized* enzyme technology. In this technique, the enzyme is "fixed" (immobilized) on some insoluble matrix; the substrate is allowed to pass through the immobilized enzyme layer, during which time the substrate is acted upon by the enzyme. This technique offers attractive features such as:

The enzyme is reusable and recoverable; its usable span of life is greatly extended.

The enzyme does not contaminate the product.

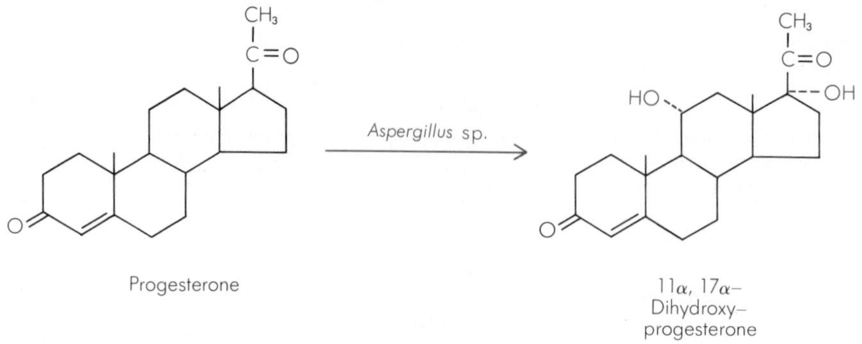

Figure 37-7. One example of the many steroids that can be produced by the action of enzymes of various fungi and bacteria on progesterone. This illustrates the precision and specificity with which a complex organic molecule can be altered by microbial enzymes. Many other important steroids are produced by exposure of progesterone to selected microorganisms. Steroids are used for treatment of disease conditions such as arthritis, inflammatory conditions, and shock.

Steroid conversions

In 1949 it was demonstrated that the *steroid* cortisone produces dramatic effects in the treatment of rheumatoid arthritis. This discovery opened the way for extensive investigations of various steroids as possible therapeutic agents. Today a large number of these steroid hormones have been identified as valuable therapeutic agents in the treatment of arthritis, rheumatism, leukemia, hemolytic anemias, and many other diseases.

Steroids are complex chemical substances. The chemical synthesis of compounds of this type is exceedingly difficult and costly. In the early 1950s it was discovered that certain fungi can cause chemical changes in steroidal substances obtained from plants or animals, converting these substances to therapeutically active steroids. This discovery set the stage for a new approach to making these compounds. Through the combined skills of the organic chemist and the microbiologist, many steroids of considerable therapeutic value have been produced. An example of the type of change produced by a microorganism on progesterone (a steroid from animals, including humans) is shown in Fig. 37-7.

Biologics for immunization

Control of infectious diseases through immunization requires the manufacture, on a commercial scale, of a variety of microbiological antigens. Development of effective immunizing antigens, together with the stringent test requirements to ensure their safe use, constitute major programs in the pharmaceutical industry. The total doses of some biologic products distributed in the United States during 1978 are shown in Table 37-3.

Petroleum and mining microbiology

The role of microorganisms in the petroleum industry and the field of mining has gained considerable attention during the last decade. This can be attributed to our growing concern about our natural resources and the condition of our environment.

Table 37-3. Net distribution of biologics in the United States (1978)*

PRODUCT DESCRIPTION	NET DOSES DISTRIBUTED JANUARY–SEPTEMBER
Influenza virus vaccine, bilvalent§	7,931,320
Monovalent	603,090
Diphtheria toxoid with tetanus toxoid (pediatric)	682,780
Diphtheria and tetanus toxoids with pertussis vaccine	13,084,020
Tetanus toxoid with diphtheria toxoid (adult)	7,390,324
Diphtheria toxoid	880
Tetanus toxoid	8,272,207
Pertussis vaccine	45,120
Poliomyelitis vaccine live, oral, trivalent	19,741,985
Measles virus vaccine, live, attenuated	6,818,692
Rubella virus vaccine, live†	5,899,643
Mumps virus vaccine, live†	3,487,622
Smallpox vaccine	3,901,286
Rabies vaccine	‡
Immune serum globulin, human (reported in cm³)	2,956,885
Tetanus immune globulin, human (reported in cm³)	1,030,002

*Center for Disease Control, Atlanta, Ga. "Biologic Surveillance," Report No. 75, July, August, September, 1978.
†All products containing this antigen.
‡Not shown since fewer than three distributors reported.
§Includes polyvalent and bivalent vaccines.

Petroleum microbiology

Microorganisms are involved in such diverse areas as petroleum formation, exploration for petroleum, clean-up of oil-spills, manufacture of food (single-cell protein—see Chap. 36) from hydrocarbons, and deterioration of products and equipment. Petroleum microbiology is emerging into a field of its own; it is a field that requires a team of investigators—microbiologists, chemists, engineers, geologists—who will conduct research and solve problems.

Mining microbiology

Microorganisms play a role in the recovery of minerals from ores. Their importance as agents in the process of extracting metals from ores is likely to increase for the following reasons:

1 The richer mineral deposits are being depleted. Lower-quality ores are being processed, and they require development of techniques which yield more nearly complete extraction of metals.
2 The traditional method of processing ores, namely smelting, is a major cause of air pollution and is under attack from environmental groups.

Microorganisms are capable of improving both these situations. For example, some autotrophic, aerobic bacteria (*Thiobacillus thiooxidans* and *Thiobacillus ferrooxidans*) when grown in the presence of copper ores produce acid and affect oxidation of the ore with subsequent precipitation (removal) of the metal. This process is known as *leaching* (see Fig. 37-8). This technique improves the recovery of metal from an ore and is nonpolluting to the atmosphere.

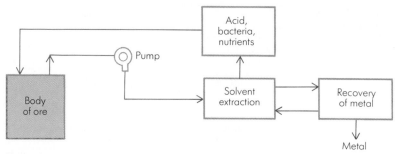

Figure 37-8. Leaching of low-grade ores using bacteria. *Thiobacillus ferrooxidans* plays an important role in the extraction (leaching) of metal from low-grade ores. This scheme shows an arrangement whereby the bacteria-nutrients-acid are pumped into the ore bed. Continued growth of *Thiobacillus ferrooxidans* produces more acid, which solubilizes the metal content, promoting its extraction. The metal is then recovered from this acid solution.

A

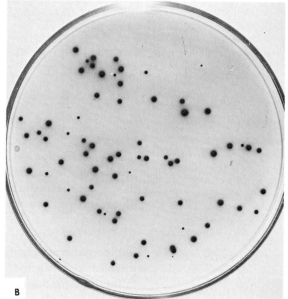

B

Figure 37-9. (A) Corroded cast-iron pipes from a tidal marsh. Corrosion is due primarily to activities of sulfate-reducing bacteria which convert iron to soluble compounds of iron. (B) *Desulfovibrio* spp., a bacterial species associated with the process of corrosion, growing on an iron salts-agar medium. The colonies appear black because of iron-sulfide formation. *Desulfovibrio* spp. occur widely in fresh, polluted, marine, and brackish waters. *(Courtesy of W. P. Iverson and the National Bureau of Standards, U.S. Department of Commerce.)*

Deterioration of materials

The term *materials*, in the sense in which it is used here, refers to all products—paper, petroleum, textiles, wood, rubber, and metals—other than foodstuffs. It has been estimated that deterioration, from all causes, of such materials represents a loss running into several billions of dollars annually. Microbial deterioration accounts for much of this loss.

Several types of microbial deterioration are described in Table 37-4; illustrations of microbiological deterioration are shown in Fig. 37-9.

Analytical microbiology

Many techniques have been developed whereby a specific microorganism is used to assay quantitatively substances such as vitamins, amino acids, and

Table 37-4. Some examples of deterioration of materials by microorganisms

MATERIAL	ACTION OF MICROORGANISMS	RESPONSIBLE MICROORGANISMS
Paper	Slime, spots, discoloration; weakens and destroys fibers	Capsulated bacteria, fungi, algae, and protozoa
Painted surfaces	Mildew, discoloration, and deterioration of paint	Species of fungi: *Pullularia, Aspergillus, Penicillium, Cladosporium,* and *Alternaria*
Textiles and cordage	Cellulose degraded; loss of strength of fiber	*Myrothecium venrucaria* and many other fungus species
Iron pipes	Corrosion	*Desulfovibrio* spp.
Wood	White and brown rots; soft rot; decay	White and brown rot fungi
Jet fuel	Growth of microorganisms in fuel-water interface may cause engine malfunctions	Filamentous fungi and many species of bacteria
Domestic fuel oil	Clogs filters	Filamentous fungi and many species of bacteria

Table 37-5. Selected examples of microorganisms used for the assay of amino acids and vitamins.

MICROORGANISM	SUBSTANCE ASSAYED
Streptococcus faecalis (bacterium)	Several amino acids
Tetrahymena geleii (protozoan)	Folic acid
Saccharomyces carlsbergensis (yeast)	Pantothenic acid
Neurospora crassa (mold)	Biotin
Ochromonas malhamensis (alga)	Vitamin B$_{12}$

antibiotics. Microbiological methods are routinely employed to determine the potency of all antibiotic preparations at various stages of development, from their crude forms to the finished product. One type of assay involves measurement of inhibition of growth of a test organism by the antibiotic. Within limits of antibiotic concentration there is proportionality between the degree of inhibition and the amount of drug, as is shown in Chap. 21.

Another type of microbiological assay is based on measurement of increase in growth or metabolic activity. The principle of this technique is that a *single* nutrient, such as a vitamin or amino acid, may be the limiting factor for growth or metabolic activity of a specific organism in a medium which lacks this essential nutrient. Within limits, the magnitude of the growth or metabolic response is proportional to the amount of the essential nutrient present in the assay medium (see Fig. 37-10).

Microbiological techniques are employed extensively for the assay of vitamins and amino acids in pharmaceutical preparations and in foods (see Table 37-5).

Microbiological assays are highly specific and unusually sensitive. For example, as little as 0.1 nanogram (0.000 000 000 l g) per milliliter of the vitamin biotin can be detected by using *Lactobacillus casei.*

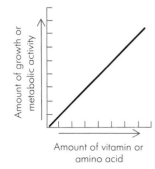

Figure 37-10. Principle of microbiological assays as used for the measurement of vitamins and amino acids. Within limits of concentration of the substance being assayed, the amount of growth of the organism is proportional to the amount of vitamin or amino acid present.

Summary and outlook

Microorganisms can be thought of as chemical factories with an extremely diversified capability of bringing about chemical changes. It is generally accepted that every naturally occurring substance is capable of being degraded by some species of microorganism. When a microbial process is discovered that results in a product that is both needed and has economic value, this process can be developed for industrial production. In this instance we seek to use microorganisms for economic and social gains.

By virtue of their diversity of biochemical ability, microorganisms are capable of degrading numerous materials. Industry, constantly aware of this, seeks to develop products which will be resistant to deterioration, to prevent economic and other losses. Undoubtedly, new ways will be found to harness microorganisms in industrial processes. Likely areas of application include the recovery of metals from ores, conversion of waters into new energy sources for domestic use, and massive production of single-cell protein as food for humans and other animals.

The ability to introduce new genetic material into a microbial cell, a technique referred to as genetic engineering, has opened the door to developing new strains of bacteria capable of producing desired pharmaceutical substances. Thus far, processes for microbiological production of insulin and interferon have been reported. It is very likely that many valuable and important substances will be produced industrially by genetically engineered microorganisms. The future is bright for development of industrial microbiological processes to meet national and global needs.

Key terms

bakers' yeast	gasohol	leaching
fermentation	immobilized enzymes	steroids

Questions

1 Why are some microorganisms uniquely suitable for industrial production of certain products?
2 Identify the microorganisms and describe the general biochemical processes involved in the production of vinegar.
3 Outline the procedure for industrial production of penicillin.
4 What developments contributed to the increase in yield of penicillin over that originally obtainable?
5 The technique of immobilized enzymes may increase the use of enzymes in industry for product modification. Why is this likely?
6 What are the desirable features of the use of enzyme preparations for refinement of a product?
7 Explain the role of microorganisms in the production of steroid hormones.
8 Describe several applications that microorganisms have in the petroleum industry.
9 The deterioration of various materials is caused by microorganisms. Identify some of these materials and describe how they are affected by microorganisms.

10 Name two important characteristics of microbiological assay procedures.

References for part ten

Alexander, M.: *Introduction to Soil Microbiology*, 2d ed., Wiley, New York, 1977. *The microbial flora of the soil is described in terms of the diversity and magnitude of this population as well as the role that these microorganisms perform in the soil.*

Appleton, J. M., V. F. McGowan, and V. B. D. Skerman: *Microorganisms and Man*, World Data Center, University of Queensland, Australia, 1979. *A digest of the role of microorganisms in the production of various chemicals (amino acids, fatty acids, fats, proteins, antibiotics), foods, and alcoholic beverages, and as agents for waste disposal.*

Edmunds, Paul: *Microbiology—an Environmental Perspective*, Macmillan, New York, 1978. *Describes the diversity of microbial groups in nature, their natural habitats, and their biochemical activities.*

Frazier, W. C., and D. C. Westhoff: *Food Microbiology*, 3d ed., McGraw-Hill, New York, 1978. *Microorganisms important in food microbiology, their effects, and methods for their control and detection are discussed.*

Marth, E. H. (ed.): *Standard Methods for the Examination of Dairy Products*, 14th ed., American Public Health Association, Washington, D.C., 1978. *This publication describes laboratory test procedures in meticulous detail and is regarded as the basic reference in the field of dairy products.*

Miller, B. M., and W. Litsky: *Industrial Microbiology*, McGraw-Hill, New York, 1976. *A comprehensive coverage of the fundamentals of industrial microbiology and a detailed treatment of the more important modern processes.*

Ramalho, R. S.: *Introduction to Wastewater Treatment Processes*, Academic, New York, 1977. *Wastewater treatment processes are presented based on the concepts of a series of integrated unit operations.*

Rossmore, H. W.: *The Microbes, Our Unseen Friends*, Wayne State University Press, Detroit, Mich., 1976. *A popular book to inform the reader of the importance of microorganisms to our health, to industry, to a balanced ecosystem, and to our diet.*

Safe Drinking Water Committee, National Research Council: *Drinking Water and Health*, 1977. National Academy of Sciences, Washington, D.C. *Chapter three provides a review of bacteria, viruses, and protozoa and their significance in drinking water.*

Speck, M. L. (ed.): *Compendium of Methods for the Microbiological Examination of Foods*, American Public Health Association, Washington, D.C., 1976. *A comprehensive selection of the methods that have been tested and proved satisfactory for use in the food industry.*

Standard Methods for the Examination of Water and Wastewater, 15th ed., American Public Health Association, New York, 1980. *An extensive compendium of physical, chemical, and biological "standard methods" for the examination of water.*

Underkofler, L., and Margaret Wulf (eds.): *Developments in Industrial Microbiology*, American Institute of Biological Science, Washington, D.C., 1980, vol. 21. *A series of papers representative of the many and diverse facets of the microbiology of industry. Published annually.*

APPENDIX A

CHARACTERISTICS OF SELECTED GENERA OF MICROORGANISMS

Characteristics of selected genera of bacteria[1]

Acetobacter. Cells ellipsoidal to rod-shaped; straight or slightly curved; 0.6 to 0.8 by 1.0 to 3.0 μm; occurring singly, in pairs, or in chains. Involution forms occur. Motile by peritrichous flagella or nonmotile. No endospores. Young cells gram-negative; some old cells gram-variable. Chemoorganotrophs. Oxidize ethanol to acetic acid. Acetate and lactate oxidized to $CO_2 + H_2$. Grow on simple and complex media. Strict aerobes. Optimum temperature 30°C (86°F); range 5 to 42°C (41 to 107.6°F). Found widely in nature on fruits and vegetables and in juices, vinegar, and alcoholic beverages. The G + C content[2] of DNA ranges from 55 to 64 moles %. Type species: *A. aceti*.

Actinomyces. Filaments, varying in length and degree of branching. Diphtheroid forms and branched rods are common; V, Y, T forms occur. Filaments are 1 μm or less in diameter. Chemoorganotrophs. Carbohydrates fermented with acid but no gas. Organic nitrogen required for growth. Facultative anaerobes. Certain species pathogenic for humans and/or other animals. Type species: *A. bovis*.

Alcaligenes. Cells rods, coccal rods, or cocci 0.5 to 1.2 μm by 0.5 to 2.6 μm usually occurring singly. Motile with one to four peritrichous flagella. Gramnegative. No endospores. Chemoorganotrophs. Strict aerobes. Nutritional requirements simple for most strains. Optimum temperature between 20 and 37°C (68 and 98.6°F). Most species are common saprophytic inhabitants of the intestinal tract of vertebrates. Occur in dairy products and freshwater, marine, and terrestrial environments. The G + C content of DNA ranges from 57.9 to 70 moles %. Type species: *A. faecalis*.

Arthrobacter. Cells which in complex media undergo a marked change in form during their growth cycle. Older cultures consist of coccoid cells; other stages are rod-shaped, club-shaped, and other forms. The cells are nonmotile or motile by one subpolar or a few lateral flagella. No endospores. Gram-positive. Chemoorganotrophs. Strict aerobes. Little or no acid from glucose. Temperature optimum 20 to 30°C (68 to 86°F). Typically soil bacteria. Laboratory strains grow on nutrient agar. The G + C content of DNA ranges from 60 to 72 moles %. Type species: *A. globiformis*.

Azotobacter. Large ovoid cells 2 μm in diameter, of varying length down to coccoid morphology. Occur singly, in pairs, or in irregular clumps. Marked pleomorphism. No endospores but do form thick-walled cysts. May produce copious amounts of capsular slime. Motile by peritrichous flagella or nonmotile. Gram-negative with marked variability. Fix atmospheric nitrogen nonsymbiotically. Grow well aerobically. Optimum temperature 20 to 30°C (68 to 86°F). Normal habitat is soil and water. The G + C content of DNA ranges from 63 to 66 moles %. Type species: *A. chroococcum*.

Bacillus. Cells rod-shaped, 0.3 to 2.2 μm by 1.27 to 7.0 μm. Majority motile; flagella typically lateral. Endospores formed; not more than one in a sporangial cell. Gram-positive. Chemoorganotrophs. Metabolism strictly respiratory, strictly fermentative, or both respiratory and fermentative. Strict aerobes or facultative anaerobes. Commonly found in soil. The G + C content of the DNA ranges from 32 to 62 moles %. Type species: *B. subtilis*.

Bacteroides. Rods. Gram-negative. Nonsporeforming. Nonmotile or motile with peritrichous flagella. Chemoorganotrophs. Metabolize carbohydrates or peptones. Products of carbohydrate fermentation include combinations of succinic, lactic, acetic, and propionic acids. Obligate anaerobes. The G + C content of the DNA ranges from 40 to 55 moles %. Found in cavities of humans, other animals, infections of soft tissue, and sewage. Some species are pathogenic. Type species: *B. fragilis*.

Bordetella. Minute coccobacilli, 0.2 to 0.3 μm by 0.5 to

---

[1]These descriptions are selected from material in *Bergey's Manual of Determinative Bacteriology*, 8th ed., Williams & Wilkins, Baltimore, 1974. By permission of the publishers.
[2]Data on G + C content usually for a limited number of species within each genus.

1.0 μm, arranged singly or in pairs. Nonmotile or motile by lateral polytrichous flagella. Gram-negative. Bipolar. Chemoorganotrophs. Metabolism respiratory. Strict aerobes. Temperature optimum 35 to 37°C (95 to 98.6°F). Mammalian parasites and pathogens of the respiratory tract. Cause whooping cough or whooping cough-like disease in humans. Type species: *B. pertussis.*

Borrelia. Cells 0.2 to 0.5 μm by 3 to 20 μm, helical with 3 to 10 or more coarse, uneven irregular coils, some of which may form obtuse angles. Gram-negative. Nutritional characteristics of most species not known. Strict anaerobes. Growth at 20 to 37°C (68 to 98.6°F); optimum 28 to 30°C (82.4 to 86°F). Some are pathogenic for humans, mammals, or birds. Type species: *B. anserina.*

Branhamella. Cocci commonly arranged in pairs with adjacent sides flattened. Gram-negative. Nonspore-forming. Nonmotile. Chemoorganotrophic. Acid not produced from carbohydrates. Catalase and cytochrome oxidase produced. Aerobic. Temperature optimum about 37°C (98.6°F). Parasites of the mucus membranes of mammals. Type species: *B. catarrhalis.*

Brucella. Coccobacilli or short rods, 0.5 to 0.7 μm by 0.6 to 1.5 μm, arranged singly, more rarely in short chains. No capsules. Nonmotile. No endospores. Gram-negative. Chemoorganotrophs, metabolism respiratory. Strict aerobes. Mammalian parasites and pathogens. The G + C content of DNA range from 56 to 58 moles %. Type species: *B. melitensis.*

Calymmatobacterium. Encapsulated pleomorphic rods. Gram-negative. Nonmotile. Nonsporeforming. Genus contains only one species, *C. granulomatis,* which is the type species. It causes granuloma inguinale and other granulomatous lesions of humans. Not pathogenic for laboratory animals.

Campylobacter. Spirally curved rods that do not ferment or oxidize carbohydrates. Gram-negative. Nonsporeforming. Motile with a corkscrewlike motion. Single polar flagellum at one or both ends of the cell. Microaerophilic to anaerobic. Chemoorganotrophs. Found in the oral cavity, intestinal tract, and reproductive organs of humans and other animals. Some species pathogenic for cattle and sheep, causing abortion. Some cause human infections. Type species: *C. fetus.*

Chlamydia. Coccoid microorganisms 0.2 to 1.5 μm in diameter. Gram-negative. Multiply only within the cytoplasm of host cells by a developmental cycle. Obligate intracellular parasites greatly dependent on the host cells for metabolic activity. Only two species recognized. *Chlamydia trachomatis* causes trachoma, inclusion conjunctivitis, lymphogranuloma venereum, urethritis, and proctitis (inflammation of anus or rectum). *Chlamydia psittaci* causes psittacosis, ornithosis, and other infections of domestic animals. Type species: *C. trachomatis.*

Clostridium. Rods, usually motile by means of peritrichous flagella; occasionally nonmotile. Form ovoid to spherical endospores that usually distend the cell. Generally gram-positive. Chemoorgano-

trophs. Some species are saccharolytic, some are proteolytic, some are both or neither. Most strains are strictly anaerobic. Commonly found in soil, marine, and freshwater sediments, and in the intestinal tract of humans and animals. The G + C content of DNA ranges from 23 to 43 moles %. Type species: *Cl. butyricum.*

Corynebacterium. Straight to slightly curved rods with irregularly stained segments and sometimes granules. Frequently show club-shaped swellings. Snapping division produces angular and palisade (picket fence) arrangement of cells. Generally nonmotile. Gram-positive. Chemoorganotrophs. Metabolism is mixed fermentative and respiratory. Aerobic and facultatively anaerobic. The G + C content of DNA is probably 57 to 60 moles %. Widely distributed in nature. Type species: *C. diphtheriae.*

Desulfovibrio. Curved rods; morphology influenced by age and environment. Motile by means of polar flagella. Gram-negative. No endospores. Chemoorganotrophs. Derive energy by anaerobic respiration, reducing sulfates or other reducible sulfur compounds to H_2S. Strict anaerobes. Optimum temperatures for growth 25 to 30°C (77 to 86°F). Type species: *D. desulfuricans.*

Enterobacter. Gram-negative rods, motile by means of peritrichous flagella. Some strains are encapsulated. Citrate and acetate can be used as sole sources of carbon. Glucose is fermented at 37°C (98.6°F) with the production of acid and gas (CO_2: H_2:: 2:1). Found in feces of humans, other animals, sewage, soil, and some natural waters. The G + C content of DNA is 52 to 59 moles %. Type species: *E. cloacae.*

Escherichia. Straight rods, 1.1 to 1.5 μm by 2.0 to 6.0 μm. Motile by peritrichous flagella or nonmotile. Gram-negative. Grow readily on simple nutrient media. Lactose is fermented by most strains, with production of acid and gas. G + C content of DNA is 50 to 51 moles %. Type species: *E. coli.*

Flavobacterium. Cells vary from coccobacilli to slender rods. Motile with peritrichous flagella or nonmotile. Endospores not formed. Gram-negative. Growth on solid media is pigmented yellow, orange, red, or brown, and hue may vary with media and temperature. Incubation temperatures below 30°C (86°F) preferable, and growth may be inhibited by higher temperatures. A few species grow at 37°C (98.6°F). Chemoorganotrophs. Widely distributed in soil, freshwater, and marine waters. Commonly found on vegetables during commercial processing and in dairy products. Two distinct ranges of DNA nucleotide base ratios (G + C moles %) have been reported: 30 to 42 and 63 to 70. Type species: *F. aquatile.*

Francisella. Nonsporeforming rods and cocci that are highly pleomorphic. Gram-negative. Nonmotile. Strictly aerobic. Optimum temperature 37°C (98.6°F). Frequently found in natural waters. Can be parasitic on humans, other mammals, birds, and arthropods. *Francisella tularensis,* the type species, causes tularemia in humans and many other warm-blooded animals and is transmitted by bloodsucking arthropods, inhalation, ingestion, and contact.

Fusobacterium. Spindle-shaped rods. Gram-negative. Nonsporeforming. Nonmotile or motile with peritrichous flagella. Obligately anaerobic. Metabolize carbohydrates to organic acids. Found in cavities of humans and other animals. Some are pathogenic and appear in various human infections. Type species: *F. nucleatum.*

Haemophilus. Minute to medium-sized, coccobacillary to rod-shaped cells which sometimes form threads and filaments, and may show marked pleomorphism. Nonmotile. Gram-negative. Strict parasites, requiring provision of growth factors present in blood, especially X and/or V factors. Aerobic, facultatively anaerobic. May or may not be pathogenic. Optimum temperature usually 37°C (98.6°F). Occur in various lesions and secretions as well as on normal mucous membranes of vertebrates. The G + C content of the DNA (in the species examined) is from 38 to 42 moles %. Type species: *H. influenzae.*

Klebsiella. Nonmotile, capsulated rods, 0.3 to 1.5 μm by 0.6 to 6.0 μm, arranged singly, in pairs, or in short chains. Grow on meat-extract media producing more-or-less dome-shaped, glistening colonies of varying degrees of stickiness. No special growth requirements, and most strains can use citrate and glucose as sole carbon sources, and ammonia as nitrogen source. Glucose is fermented with the production of acid and gas (more CO_2 than H_2), but anaerogenic strains occur. Optimal temperature for growth 35 to 37°C (95 to 98.6°F). The G + C content of DNA is from 52 to 56 moles %. Type species: *K. pneumoniae.*

Lactobacillus. Rods, varying from long and slender to short coccobacilli. Chain formation common, particularly in later logarithmic phase of growth. Motility unusual; when present, by peritrichous flagella. Nonsporing. Gram-positive becoming gram-negative with increasing age and acidity. Some strains exhibit bipolar bodies, internal granulations or a barred appearance with the gram reaction or methylene blue stain. Metabolism fermentative; some are strict anaerobes on isolation. Complex nutritional requirements. Temperature range 5 to 53°C (41 to 127.4°F); optimum generally 30 to 40°C (86 to 104°F). Found in dairy products and effluents, grain and meat products, water, sewage, beer, wine, fruits and fruit juices, pickled vegetables, sourdough and mash. Also parasitic in the mouth, intestinal tract, and vagina of many homothermic animals, including humans. Pathogenicity is highly unusual The G + C content of the DNA ranges from 34.7 ± 1.4 to 53.4 ± 0.5 moles %. Type species: *L. delbrueckii.*

Legionella. Gram-negative rods approximately 0.5 to 0.7 μm wide and varies from 2 to 20 μm or greater in length. Cells contain fat droplets. Aerobic; enhanced growth with 2.5% carbon dioxide. Optimum growth at 30°C (86°F). Cultured on special complex organic media containing cysteine. No acid production from carbohydrates demonstrated. Cause of so-called Legionnaires' disease, referring to illnesses resulting from infection during an American Legion convention in Philadelphia in 1976. The disease is commonly recognized as a form of pneumonia. The one species described is called *L. pneumophila.*

Micrococcus. Cells spherical, 0.5 to 3.5 μm in diameter, occurring singly and in pairs, and characteristically dividing in more than one plane to form irregular clusters, tetrads, or cubical packets. Usually nonmotile. No resting stages known. Gram-positive. Chemoorganotrophs. Metabolism strictly respiratory. Nutritional requirements are variable. Aerobes. Growth optimum 25 to 30°C (77 to 86°F). Common inhabitant of soils and freshwater. Frequently found on the skin of humans and other animals. The G + C content of the DNA ranges from 66 to 75 moles %. Type species: *M. luteus.*

Mycobacterium. Slightly curved or straight rods, 0.2 to 0.6 μm by 1.0 to 10 μm, sometimes branching; filamentous or myceliumlike growth may occur but on slight disturbance usually becomes fragmented into rods or coccoid elements. Acid-alcohol-fast at some stage of growth. Not readily stainable by Gram's method but usually considered gram-positive. Nonmotile. No endospores, conidia, or capsules, no grossly visible aerial hyphae. Diseases produced include tuberculosis, leprosy, and other usually chronic, more or less necrotizing, limited or extensive granulomas. Found in soil, water, and warm-blooded and cold-blooded animals. The G + C content of the DNA ranges from 62 to 70 moles %. Type species: *M. tuberculosis.*

Neisseria. Cocci, 0.6 to 1.0 μm in diameter, occurring singly but often in pairs with adjacent sides flattened. Division in two planes at right angles to each other, sometimes resulting in the formation of tetrads. No formation of endospores. Nonmotile. Capsules and fimbriae may be present. Gram-negative. Chemoorganotrophic. Few carbohydrates utilized. Aerobic or facultatively anaerobic. Temperature optimum about 37°C (98.6°F). Parasites of mucous membranes of mammals. The G + C content of the DNA ranges from 47 to 52 moles %. Type species: *N. gonorrhoeae.*

Nitrobacter. Cells short rods, often wedge- or pear-shaped. Reproduce by budding. Cells possess a polar cap of cytomembranes. Usually nonmotile. No resting stages known. Gram-negative. Cells rich in cytochromes imparting a yellowish color to cell suspensions; void of other pigments. Some strains are obligate chemolithotrophs which oxidize nitrite to nitrate and fix CO_2 to fulfill energy and carbon needs. Strictly aerobic, using oxygen as terminal electron acceptor. Temperature range for growth, 5 to 40°C (41 to 104°F). The G + C content of the DNA ranges from 60.7 to 61.7 moles %. Habitat: soils, freshwater, and seawater. Type species: *N. winogradskyi.*

Pasteurella. Cells ovoid or rod-shaped, 1.4 ± 0.4 by 0.4 ± 0.1 μm, singly or less frequently in pairs or short chains. Nonmotile. Do not form endospores. Gram-negative; bipolar staining common, especially in preparations made from infected animal tissues stained with Giemsa or methylene blue. Chemoorganotrophic, growing best on media containing blood. Metabolism fermentative. Aerobic, facultatively anaerobic. Temperature range 22 to 42°C (71.6 to 107.6°F), optimum 37°C (98.6°F). Parasitic on mammals (including humans) and birds. The G + C content of DNA ranges from 36.5 to 43.0 moles %. Type species: *P. multocida.*

Propionibacterium. Rods. Gram-positive. Nonspore-forming. Nonmotile. Usually pleomorphic. Diphtheroid or club-shaped with one end rounded and the other end tapered or pointed and stained less intensely. Cells of some cultures may be coccoid, elongate, bifid, or even branched. Cells usually arranged in singles, pairs, or V and Y configurations, short chains, or clumps in "Chinese character" arrangement. Chemoorganotrophs. Anaerobic to aerotolerant. Growth most rapid at 30 to 37°C (86 to 98.6°F). The G + C content of the DNA ranges from 59 to 60 moles %. Found in dairy products or from the human skin, and from intestinal tract of humans and animals. Some species may be pathogenic. Type species: *P. freudenreichii.*

Proteus. Usually straight rods, 0.4 to 0.6 µm by 1.0 to 3.0 µm; coccoid and irregular involution forms, and filaments and spheroplasts are frequent under certain conditions. May occur in pairs or chains. Not encapsulated. Motile by peritrichous flagella; swimming motility most pronounced at 20°C (68°F) and often absent at 37°C (98.6°F). Gram-negative. Nonpigmented. Temperature range for growth about 10 to 43°C (50 to 109.4°F). The G + C content of the DNA ranges from 38 to 42 moles % for four species and is about 50 for *P. morganii.* Type species: *P. vulgaris.*

Pseudomonas. Cells single, straight or curved rods, but not helical. Dimensions generally 0.5 to 1 µm by 1.5 to 4 µm. Motile by polar flagella; monotrichous or multitrichous. Do not produce sheaths or prosthecae. No resting stages known. Gram-negative. Chemoorganotrophs. Metabolism respiratory, never fermentative. Some are facultative chemolithotrophs, able to use H_2 or CO as energy source. Molecular oxygen is the universal electron acceptor; some can denitrify, using nitrate as an alternate acceptor. Strict aerobes, except for those species which can use denitrification as a means of anaerobic respiration. Catalase positive. The G + C content of the DNA of those species examined ranges from 58 to 70 moles %. Type species: *P. aeruginosa.*

Rhizobium. Rods 0.5 to 0.9 µm by 1.2 to 3.0 µm. Commonly pleomorphic under adverse growth conditions. Often contain granules of poly-β-hydroxybutyrate, which are refractile under phase contrast, stainable with Sudan black B, and soluble in chloroform. Motile by two to six peritrichous flagella or by a polar or subpolar flagellum. Nonsporing. Gram-negative. Growth on carbohydrate media usually accompanied by copious extracellular, polysaccharide slime. Chemoorganotrophs. Metabolism respiratory. Aerobic. Temperature optima 25 to 30°C (77 to 86°F). Members of this genus characteristically able to invade root hairs of leguminous plants and incite production of root nodules, wherein the bacteria occur as intracellular symbi-

onts. Nodule bacteroids characteristically involved in fixing molecular nitrogen into combined forms utilizable by the host plant. The G + C content of the DNA ranges from 59.1 to 65.5 moles %. Type species: *R. leguminosarum.*

Rickettsia. Short rods, 0.3 to 0.6 µm by 0.8 to 2.0 µm, some species up to 4 µm long prior to cell division. No flagella or capsules, but an outer layer of amorphous material is occasionally seen in electron micrographs of cells subjected to a minimum of laboratory manipulation. Gram-negative. Have not been cultivated in the absence of host cells. Growth generally occurs in the cytoplasm. Human pathogens. Human beings are the reservoirs of the type species, incidental hosts of the other species. Small rodents and other vertebrates serve as reservoirs or disseminate the organisms. Arthropods play primary roles in the life cycles of all species, often as main reservoirs, and almost exclusively mediate natural transmission among vertebrates. The G + C content of the DNA ranges from 30 to 32. 5 moles %. Type species: *R. prowazekii.*

Salmonella. Rods, usually motile by peritrichous flagella; nonmotile mutants may occur. Most strains will grow on defined media without special growth factors, and they can use citrate as C source. Most strains are aerogenic, but *S. typhi,* an important exception, never produces gas. G + C content of DNA ranges from 50 to 53 moles %. Type species: *S. choleraesuis.*

Serratia. Motile peritrichously flagellated rods. Some strains are capsulated. Gram-negative. Citrate and acetate can be used as sole carbon source. Many strains produce pink, red, or magenta pigment. Glucose is fermented with or without the production of a small volume of gas; cellobiose, inositol, and glycerol are fermented without gas production. G + C content of DNA ranges from 53 to 59 moles %. Type species: *S. marcescens.*

Shigella. Nonmotile rods. Gram-negative. Not encapsulated. Grow well on nutrient media and do not need special growth factors. Cannot use citrate or malonate as sole carbon source. Growth inhibited by KCN. H_2S is not produced. Glucose and other carbohydrates are fermented with the production of acid but not gas. Type species: *S. dysenteriae.*

Spirillum. Rigid, helical cells, 0.25 to 1.7 µm in diameter, with less than one turn to many turns. Intracellular granules of polyhydroxybutyrate present in most species. Motile by means of polar polytrichous flagella. Gram-negative. Chemoorganotrophs. Strictly respiratory metabolism with oxygen as the terminal electron acceptor. A few species can grow anaerobically with nitrate. Found in fresh and salt waters containing organic matter. The G + C content of the DNA ranges from 38 to 65 moles %. Type species: *S. volutans.*

Spirochaeta. Helical cells, 0.20 to 0.75 µm by 5 to 500 µm. The cells have axial fibrils. The protoplasmic cylinder is wound around or together with the axial fibrils. The cells do not possess either terminal hooks or cross striations. Motile, probably by means of the axial fibrils. Free-living in H_2S-containing mud, in sewage, and in polluted water. Nonparasitic. Of the five species described, only the type species has not been grown in pure culture. The four species which have been cultured are chemoorganotrophs, and obligate and facultative anaerobes, and ferment carbohydrates, and have a G + C content in their DNA ranging from 50 to 66 moles %. Type species: *S. plicatilis*.

Staphylococcus. Cells spherical, 0.5 to 1.5 μm in diameter, occurring singly and in pairs, and characteristically divide in more than one plane to form irregular clusters. Nonmotile. No resting stages known. Gram-positive. Cell wall contains two main components: a peptidoglycan and its associated teichoic acids. Chemoorganotrophs. Metabolism respiratory and fermentative. Facultative anaerobes, growth more rapid and abundant under aerobic conditions. Temperature optimum 35 to 40°C (95 to 104°F). Mainly associated with skin, skin glands, and mucous membranes of warm-blooded animals. Their host range is wide, and many strains are potential pathogens. The G + C content of DNA ranges from 30 to 40 moles %. Type species: *S. aureus*.

Streptococcus. Cells spherical to ovoid, less than 2 μm in diameter, occurring in pairs or chains when grown in liquid media. Occasional motile strains in serological group D. Gram-positive. Chemoorganotrophs. Metabolism fermentative. Facultative anaerobes. Minimal nutritional requirements are generally complex (but variable). Temperature optimum about 37°C (98.6°F). The G + C content of the DNA ranges from 33 to 42 moles %. Type species: *S. pyogenes*.

Streptomyces. Slender, coenocytic hyphae, 0.5 to 2.0 μm diameter. The aerial mycelium at maturity forms chains of from three to many spores 0.5 to 2.0 μm in diameter. Gram-positive. On isolation, colonies are small (1- to 10-mm diameter), discrete and lichenoid, leathery or butyrous; initially relatively smooth-surfaced but later develop a weft of aerial mycelium that may appear granular, powdery, velvety, or floccose. Produce a wide variety of pigments responsible for colors of vegetative mycelium, aerial mycelium, and substrate. Many strains produce one or more antibiotics. Heterotrophs. Highly oxidative. Aerobes. Temperature optimum 25 to 35°C (77 to 95°F). Type species: *S. albus*.

Thiobacillus. Small rod-shaped cells. Motile by means of a single polar flagellum; two species nonmotile. No resting stages known. Gram-negative. Energy derived from the oxidation of one or more reduced or partially reduced sulfur compounds. The genus includes strictly autotrophic species which derive

their carbon from carbon dioxide, facultative autotrophs, and at least one species which requires both a partially reduced sulfur compound and organic matter for optimal growth. Obligate aerobes, except *T. denitrificans* which grows anaerobically with nitrate as electron acceptor. Optimum temperature about 28 to 30°C (82.4 to 86°F). Found in sea water, marine mud, soil, freshwater, acid mine waters, sewage, sulfur springs, and in or near sulfur deposits. The G + C content of the DNA, for the species tested, ranges from 50 to 68 moles %. Type species: *T. thioparus*.

Treponema. Unicellular, helical rods 5 to 20 μm long and 0.09 to 0.5 μm wide, with tight regular or irregular spirals. Cells have one or more axial fibrils inserted at each end of the protoplasmic cylinder. Motile. Gram-negative. Best observed under darkfield or phase-contrast microscopy. Found in the oral cavity, intestinal tract, and genital regions of humans and animals. Some species are pathogenic. The G + C content of the DNA ranges from 32 to 50 moles %. Type species: *T. pallidum*.

Veillonella. Cocci. Gram-negative. Nonmotile. Nonsporeforming. Anaerobic. Complex nutritional requirements. Carbon dioxide required. A common inhabitant of the intestinal tract of humans, ruminants, rodents, and pigs, and a predominant species in the oral cavity. Type species: *V. parvula*.

Vibrio. Short asporogenous rods, axis curved or straight, 0.5 μm by 1.5 to 3.0 μm, single or occasionally united into S shapes or spirals. Motile by a single polar flagellum, or, in some species, two or more flagella in one polar tuft; very occasionally nonmotile. Spheroplasts frequently present, usually formed in adverse environmental conditions. Gram-negative. Not acid-fast. No capsules. Grow well and rapidly on standard nutrient media. Chemoorganotrophs. Metabolism is both respiratory (oxygen is utilized) and fermentative. Facultatively anaerobic. Temperature optima range from 18 to 37°C (64.4 to 98.6°F). The G + C content of the DNA ranges from 40 to 50 moles %. Found in freshwater and salt water and in the alimentary canal of humans and animals; some species are pathogenic for humans and other vertebrates (fish). Type species: *V. cholerae*.

Yersinia. Ovoid cells or rods. Not encapsulated. Gram-negative. Nonmotile at 37°C (98.6°F); at temperatures below 37°C, two species motile with peritrichous flagella. Facultatively anaerobic. *Yersinia pestis* is the cause of plague in humans and rodents and is the type species. *Yersinia pseudotuberculosis* causes a form of pseudotuberculosis in animals.

Characteristics of selected genera of fungi (yeasts and molds)

Absidia. A genus of fungi of Class Phycomycetes. Pear-shaped sporangia produced in partial whorls (loops) at intervals along stolonlike branches. These

branches produce rhizoids at intervals but not opposite sporangiophores. Zygospores surrounded by curved unbranched suspensor appendages which may arise from one or both suspensors. Most species are heterothallic. Can cause phycomycoses resulting in infections of orbital tissues which rapidly invade the central nervous system. Example of pathogenic species is *A. corymbifera*.

Aspergillus. An unbranched conidiophore arises from a foot cell and terminates in a vesicle which gives rise to bottle-shaped sterigmata. Chains of conidia develop on secondary sterigmata (branches of the primary sterigmata). In some species the spore head is spherical; in others the arrangement of sterigmata gives a fanlike or cylindrical appearance. Ascospores are produced in some species. When formed, there are eight round to oval ascospores in each ascus. The asci are irregularly arranged throughout the perithecium. Spores are variously colored and give these molds their characteristic colors. These fungi are important economically in both a positive and a negative sense. A typical species is *A. niger*.

Blastomyces. This genus is a dimorphic fungus placed in Class Deuteromycetes, although some authorities consider it an Ascomycete. The well-known pathogenic species is *B. dermatitidis*. Within the tissues, the fungus occurs in the form of yeastlike spherical cells 8 to 15 μm in diameter. In this unicellular form, the fungus reproduces by budding. The cells are multinucleate. The yeastlike form can be cultured on blood agar at 37°C (98.6°F). On Sabouraud's agar and other suitable media at room temperature or 30°C (86°F), the organism grows as a white mold with septate mycelia. The mycelial form produces spherical conidia terminally or laterally on short, straight conidiophores. The ascomycetous state is named *Ajellomyces dermatitidis*.

Candida. Cells are of various shapes. Pseudomycelia are produced in abundance. True mycelia and chlamydospores may be formed. Blastospores may be present in a position typical of the individual species. Vegetative reproduction is by multilateral budding. Dissimilation may be oxidative, but in many species also strongly fermentative. Sediment, often a ring, and a pellicle are formed in liquid media. An example is *C. albicans*.

Coccidioides. A genus of fungus placed in Class Deuteromycetes, although some authorities consider it a Phycomycete. The sole species is *C. immitis*. It occurs as a saprophyte in certain soils and is the etiologic agent of coccidioidomycosis in humans and other animals. It is a dimorphic fungus. In culture, the organism forms a grey mycelium which subsequently fragments to form barrel-shaped arthrospores. In humans, the fungus occurs within the tissues primarily as a multinucleate, spherical, thick-walled cell (spherule) which, when mature, has a diameter of about 50 μm. Upon maturation, the spherule wall ruptures, releasing the endospores (sporangiospores) into the tissues to repeat this reproductive cycle. The disease is transmitted by the respiratory route.

Dictyostelium. A genus of the cellular slime molds. The species occur in soil and in decomposing vegetation, particularly in woodlands. A well-known species is *D. discoideum*.

Filobasidiella. Cells are round, oval, long oval, ameboid, or sometimes polymorphic; surrounded by a capsule. Pseudomycelia absent. Reproduce vegetatively by multilateral budding. Dissimilation is strictly oxidative. On solid media, colonies are often mucoid. In liquid media, a sediment, ring, and sometimes a pellicle are formed. An example is *F. neoformans*. Inositol is usually assimilated.

Histoplasma. A genus of a dimorphic fungus; it is considered a deuteromycete by some authorities and an ascomycete by others. Well-known species is *H. capsulatum*, which causes histoplasmosis. Its natural habitat is soil or composted plant material enriched by feces or other nitrogenous substances. It occurs intracellularly as yeastlike ovoid cells, each approximately 2 to 3 μm by 3 to 4 μm, which reproduce by budding. In culture, at room temperature, the organism grows as a mycelium; the mycelial stage reproduces by means of microconidia and macroconidia. The ascomycetous stage is named *Emmonsiella capsulata;* it produces eight-spored asci.

Microsporum. The genus *Microsporum* produces very small conidia (microconidia) on its hyphae, but in addition it produces large, multiseptate, spindle-shaped macroconidia which are considerably larger than the microconidia. Five species are recognized. *Microsporum* generally infects the scalp of children, more rarely of adults, causing a condition called *tinea capitis. Microsporum audouini* usually occurs in epidemics.

Mucor. A genus within Class Phycomycetes. Widespread in soil and on dung and other organic substrata, although some species can behave as pathogens in humans and other animals. Sporangia are globose and borne on branched and unbranched sporangiophores. Sexual reproduction involves the fusion of gametangia, with the subsequent formation of zygospores. Most species are heterothallic. Common species are *M. hiemalis, M. racemosus,* and *M. mucedo*.

Neurospora. Red mold of bread. Characterized by the elaborate receptive neck (trichogyne) on the ascogonidium or female gametangium. Mycelium is freely branched with multinucleated cells in hyphae. Asexual reproduction is by means of macroconidia borne on branches on the end of a conidiophore. These macroconidia usually contain several nuclei. An erect structure resembling a conidiophore on a mycelium may also produce spermatia which are uninucleate sporelike bodies. These elements, which bud laterally from the cells, can function as conidia to form mycelia upon germination. Sexual reproduction is effected by spermatization. The genetics of *Neurospora* has been thoroughly investigated. An example is *N. sitophila*.

Penicillium. Blue-green molds. Mycelium grows on or penetrates the substrate. Hyphae are freely branching and thin-walled, and have two or more nuclei. In some species, mycelia develop into sclerotia. Penicillia reproduce asexually by forming brushlike tufts of conidia on tips of multicellular conidiophores. Twenty or more species produce ascocarps, which may be composed of interwoven hyphae in some species or result in the production of a pseudoparenchymatous peridium in others. Penicillia are important economically in fermentations, antibiotic production, cheese manufacture, food spoilage, and other useful and harmful ways. An important type is *P. notatum,* which produces penicillin.

Piedraia. A genus of ascomycetous fungi. Well-known pathogenic species is *P. hortae,* which causes an infection of hair called black piedra. On hair the fungus forms black, hard, gritty nodules composed of an organized, firmly cemented mass of fungal cells. On Sabouraud's agar, it grows slowly to form a small, conical, dark brown or black colony with short aerial hyphae. Hyphae are of varying diameter with frequent septation and many intercalary chlamydosporelike cells. Asci found within thicker portions of colony and may have two to eight ascospores within a single specimen.

Rhizopus. Bread mold or black mold. Young mycelia are multinucleate and have no crosswalls, and all the hyphae are alike. Hyphae later develop into three types: branching rhizoids that penetrate the substratum, stolons that grow laterally on the surface of the substrate uniting the units where tufts of rhizoids are formed, and sporangiophores that grow upward from stolons. Sporangiophores are unbranched, with sporangia at their tips. Asexual reproduction is by means of aplanospores or, in some species, chlamydospores. Sexually, some species of *Rhizopus* are homothallic, and others are heterothallic. In the latter, hyphae of both sexes join to form progametangia, which develop through stages to form a zygote characterized by a thick, black, warty wall. Zygotes must remain dormant for several months before germinating. An example is *R. stolonifer.*

Saccharomyces. Cells are round, oval, elongated, or threadlike and produce pseudomycelia. Reproduce vegetatively by multilateral budding. Isogamous or heterogamous conjugation may or may not precede ascus formation. Protuberances may be formed. Asci contain one to four spores of various shapes per ascus. Spores may conjugate. Dissimilation proceeds from preferably oxidative to predominantly fermentative. Bottom growth usually occurs in liquid cultures. Ring and pellicle may form after a longer time. Common sugars are usually fermented vigorously; nitrates are not assimilated. Most industrial yeasts belong to this genus. An example is *S. cerevisiae.*

Schizosaccharomyces. Cells are cylindrical, rectangular, oval, round, and semicircular; true mycelia and arthrospores are formed. Vegetative reproduction is by fission only. Asci are formed right after isogamous conjugation; gametangia absent. Asci contain four to eight round, oval spores. Dissimilation is strongly fermentative; also oxidative. Nitrate is not assimilated. An example is *S. octosporus.*

Trichophyton. A genus of fungi within Class Deuteromycetes, although the ascomycetous states of some species have been found. It is included in the category of the dermatophytes. Species occur as soil saprophytes and as parasites and pathogens in the skin, hair, and nails of humans and other animals. The vegetative form is a septate mycelium. Characterized by club-shaped macroconidia 4 to 8 μm by 8 to 50 μm with smooth walls usually not exceeding 2 μm in thickness and with none to four septa. Microconidia are spherical (2.5 to 4 μm in diameter) or clavate (2 to 3 μm by 3 to 4 μm). Known ascomycetous stages have the genus name *Arthroderma.* Common species of *Trichophyton* include *T. mentagrophytes, T. rubrum,* and *T. schoenleini.*

Characteristics of selected genera of protozoa

Ameba. Members of this genus have no fixed body shape. Their cytoplasm may be divided into a granular endoplasm filled with inclusions and a hyaline ectoplasm. The nucleus is vesicular, usually with a central nucleolus; it is located in the endoplasm; many amebas are multinucleated. The pseudopodia are often lobular and sometimes pointed. Most species are free-living, but some are parasitic. Most freshwater species can encyst under unfavorable environmental conditions. They are also found in brackish and salt water. Nutritionally, they are holozoics.

Balantidium. These ciliates are oval, ellipsoidal to subcylindrical; the peristome begins at or near the anterior end. Longitudinal ciliation is uniform. The macronucleus is elongated; a micronucleus is also present. A contractile vacuole is situated terminally. Species occur in the gut of vertebrates and invertebrates. *Balantidium coli* is the sole ciliate human parasite.

Entameba. These are the parasitic amebas. The nucleus is vesicular, with a comparatively small endosome, and is situated in or near the center and contains varying numbers of peripheral nonchromatinic granules attached to the nuclear membrane. Motility is due to broad pseudopodia. Numerous species occur in vertebrates and invertebrates. *Entameba histolytica* is the causative agent of amebic dysentery.

Giardia. Members of this genus are flagellates. Their shape is pyriform to ellipsoid. The anterior end is broadly rounded; the posterior end is drawn out. There is bilateral symmetry. The dorsal side is convex; the ventral side is flat or concave with a sucking disk in the anterior half. Two nuclei are present. There are eight flagella in four pairs. The cysts are oval to ellipsoid. Species occur in the intestines of various vertebrates.

Leishmania. These organisms are all colorless flagellates. In humans or dogs, the organism is an ovoid body with a nucleus and a blepharoplast; the body is 2 to 5 μm in diameter. The organism is an intracellular parasite in the cells of the reticuloendothelial system. Multiplication is by binary fission. In the intestine of bloodsucking insects or in blood-agar cultures, the organism develops into leptomonad form which multiplies by longitudinal fission.

Paramecium. This genus is composed principally of common, free-living, freshwater ciliates. They are described as being cigar- or foot-shaped; i.e., they are blunt at the anterior end, but pointed at the posterior end. The elastic cell membrane or pellicle is covered with lengthwise rows of cilia. Immediately beneath the pellicle is a layer of dense ectoplasm, and inside this is a mass of granular endoplasm. An oral groove extends diagonally back from the anterior end, terminating in a cytostome, or cell mouth. Motility is by beating of the cilia. Reproduction is by binary fission or conjugation. Species of *Paramecium*, especially *P. caudatum*, have been useful in basic biological studies.

Plasmodium. The best-known species of this genus is *P. vivax*, the malarial parasite of humans. There are numerous other species. All members of the order are minute intracorpuscular parasites of vertebrates. All have complex life cycles, including asexual reproductions (schizogony) in the blood of vertebrates and sexual reproduction which results in formation of sporozoites while in the alimentary canal of some bloodsucking invertebrate host. Sporozoites of *P. vivax* formed in the intestinal tract of a female anopheline mosquito are fusiform in shape and are 6 to 15 μm in length. Schizogony takes about 48 h in the human host and is repeated in 3-day cycles. This cycle is described in Chap. 9.

The morphology of the sporozoites varies widely within species. There are numerous species characterized most commonly on host affinity.

Toxoplasma. A genus of protozoa belonging to the Sporozoa. The sole species, *T. gondii*, is an intracellular parasite in a range of hosts and is the etiologic agent of toxoplasmosis. The uninucleate, crescent-shaped cells are 5 to 7 μm in length. *T. gondii* can be grown in tissue cultures.

Trichomonas. Parasitic, flagellated protozoa widely distributed, especially in tropical and subtropical regions. Commonly found in digestive and reproductive tracts of humans. Typically, there are four free anterior flagella and a fifth along the outer margin of the undulating membrane. A well-developed axostyle protrudes beyond the posterior end of the cell. Pyriform in shape, they range from 5 to 20 μm in length. They reproduce asexually by binary or multiple fission. Sexual reproduction has been observed in some species. Cysts are often formed. An example is *T. vaginalis*.

Trypanosoma. These protozoa are parasitic in the circulatory system of vertebrates. Polymorphism is common, but the organisms are characteristically flattened and pointed at the flagellate end. The nucleus is centrally located. A blepharoplast is found near the aflagellate end; the flagellum arises from this organ and forms the outer boundary of the undulating membrane. The flagellum extends well beyond the body of the cell. Reproduction by longitudinal fission is most common, but reproduction may also occur by multiple fission, or even budding, in some species of the order. Reproduction by sexual processes has not been confirmed.

Characteristics of selected genera of algae

Chlamydomonas. A freshwater genus composed of some 325 species. Cells are ellipsoidal, pyriform, subcylindrical, or spherical. They have two flagella located close together and anteriorly. Cell wall is composed of cellulose, but some species have a gelatinous pectic sheath covering the cellulose layer. Most species have a single cup-shaped chloroplast, but some species have laminate or stellate chloroplasts. Usually two contractile vacuoles are located near the base of the flagella. All species are uninucleate. Some species are homothallic, and some are heterothallic. Gametic union may be isogamous, anisogamous, or oogamous.

Chlorella. Members of this genus are cultured for their high protein production, sometimes commercially. Cells are round or oval and are commonly found in moist soil. Species are both marine and freshwater. There is one thin parietal chloroplast either with or without pyrenoids. Autospores are the only known reproductive elements. Growth is rapid and has been investigated for oxygen production in space flights.

Chrysameba. The body is naked. The flagellate stage is ovoid with two chromatophores, sometimes with slender pseudopodia at the same time. The flagellum may be lost, and the organism becomes ameboid, resembling standing freshwater.

Eudorina. This is a genus of green algae. The cells are arranged in alternating rings to form 16- to 32-celled coenobia (colonies in which the number of cells is fixed at the organism's origin and not augmented later). Asexual reproduction is by autocolony (miniature of the parent colony) formation in colonies that are nonmotile, at least in the later stages. In sexual reproduction, clonal populations may be genetically unisexual or bisexual, and the gametes are anisogamus. In both cases the individual colonies are either male or female. All the cells of a female colony may function as gametes, while all the cells of the male colonies develop into packets of small, spermlike male gametes. The sperm packets are released and swim as units to the female colonies into which individual sperms penetrate. After a period of dormancy, the zygotes germinate, usually producing one biflagellate cell or occasionally two or three. A common species is *E. elegans*.

Euglena. Often found in abundance in mud along shores of rivers, ponds, and marshes. Members of this genus are single-celled, normally uninucleate flagellated cells. Other inclusions are chloroplasts and carbohydrate food-reserve granules. A cytostome is located at the anterior end of the cell. Beneath the cytostome is a flask-shaped gullet which connects to a reservoir. There may be one or more contractile vacuoles. Multiplication is by cell division.

Gonyaulax. Cells are spherical, polyhedral, fusiform, elongated with stout apical and antapical prolongations, or dorsoventrally flattened. The apex is never sharply attenuated. Chromatophores are yellow to dark brown, often dense. No stigma is present. It is a dinoflagellate. It occurs in freshwater or brackish or salt water. There are numerous species. Some species are luminescent. *Gonyaulax catanella* contains an endotoxin. Although the toxin is harmless to the fish and shellfish themselves (fish and shellfish feed on this alga), human consumption of the fish and shellfish results in accumulation of the toxin in human tissues, especially the liver, resulting in severe illness and death. During summer months, *Gonyaulax* may be present in amounts of millions per liter of seawater. The toxin is estimated to be 10 times more potent than strychnine, and its effects are not unlike those of botulinum poisoning.

Gymnodinium. Cells have no walls. They are broadly oval with the epicone somewhat narrowed anteriorly. There is a transverse furrow and a short longitudinal sulcus. (Epicone = top part or anterior part of an undeveloped envelope; sulcus = a type of furrow.) The flagella are simple and pleuronematic (with numerous lateral fibrils). Chloroplasts are numerous and usually golden-brown, oval discs. A few species are colorless. Most are marine but many are freshwater organisms. Several species are known to produce a lethal toxin and are the causative organism in the death of fish (red tide) which occurs periodically along continental shelves.

Nostoc. The genus is a blue-green alga with globular, beadlike cells arranged in much-tangled trichomes, many of these enclosed in a copious gelatinous matrix which is bounded externally by a pelliclelike membrane. Colonies may be microscopic, pea-size, walnut-size, or as large as 8 cm in diameter. Olive-green or blue-green colonies may be smooth or warty. A main characteristic of *Nostoc* is the formation of akinetes in long series; sometimes an entire trichome is converted.

Ochromonas. Organisms are solitary or colonial. The body surface is delicate. The posterior end is often drawn out for attachment. One or two chromatophores; usually has a stigma. Organism can encyst. It is a freshwater organism. Some species have been used for vitamin B_{12} assay.

Prototheca. A genus of Division Chlorophycophyta. It is usually considered a chloroplast-free, achlorophyllous strain of *Chlorella*. The spherical or ellipsoidal cells reproduce by autospore (a nonmotile spore that is a miniature of the cell from which it is derived) formation. Some species found to be pathogenic to humans; these are *P. zopfii* and *P. wickerhamii*.

Vaucheria. This genus contains about 40 species. All but a half-dozen, which are marine species, are freshwater or terrestrial inhabitants. The thallus is a sparingly branched tubular cell which may attain a length of several centimeters. Cells attach to soil by means of rhizoidlike branches, which contain few chromatophores. Directly beneath a thin cell wall is a layer of cytoplasm surrounding a single central vacuole which contains many chromatophores and numerous minute nuclei. Asexual reproduction commonly takes place by means of multiflagellate zoospores which are formed singly within club-shaped sporangia. Other types of asexual reproduction may occur in *Vaucheria*. Sexual reproduction occurs frequently among thalli growing on damp soil or in still waters. All freshwater species are homothallic, but two or three of the marine species are heterothallic.

APPENDIX B

GLOSSARY

Pronunciation guide for glossary

Many of the terms in the glossary are followed by informal phonetic spellings in parentheses to facilitate their pronunciation. A precise rendering of the pronunciation of each term "from scratch" is impossible without the use of a full-scale phonetic system, such as can be found in an unabridged dictionary. All that we intended is that a speaker (and speller) of English be able to form a workable idea of a term's pronunciation. Only words or word elements whose proper pronunciation may not be obvious are provided with these informal phonetic spellings.

In these informal phonetic spellings, stressed syllables are set in boldface type. The intended pronunciations of some of the letters and letter combinations we have used are:

Letter(s)	Pronunciation
ah	Broad **a** sound, as in **father**, **rotten**
ay	Long **a** sound, as in **ray**, **fade**
g	Hard, as in **good**, **rug**
i	Short **i** sound, as in **sit**, **pacific**, except when used with final **e** (e. g., **tide**) or in **igh**
igh	as in **sight**
j	as in **jump** (used for soft **g**, as in **rage**)
o alone or preceded by consonant	long **o** sound as in **no**, **boat**
o followed by consonant	short **o** sound, as in **fox** (also rendered by **ah**)
oe (e. g., **doe**)	long **o** sound, as in **foe**
oy	as in **toy**, **soy**
uh	unaccented syllable, as in **biology** (bye•**ahl**•uh•jee), **telephone** (**tel**•uh•fone), **arrest** (uh•**rest**)
ye	long **i** sound, as in **dye**, **rye**, **side**
zh	as in **revision**, **pleasure**

abiogenesis (ay•bye•o•**jen**•uh•sis, ab•ee•o•**jen**•uh•sis). See **spontaneous generation**. (*Abio-* means nonliving; *genesis* means origin.)

abiotic (ab•bye•**ot**•ik, ab•ee•ot•ik). Pertaining to or characterized by the absence of living organisms.

abscess (**ab**•sess). A localized collection of pus.

acid curd. Coagulation of milk protein by acid.

acid-fast. Retaining the initial stain and difficult to decolorize with acid alcohol. A property of certain bacteria.

acquired immunity. Resistance to disease after initial exposure to a pathogenic organism either through vaccination or by natural infection.

actinomycete (ak•tin•o•**mye**•seet, ak•tin•o•mye•**seet**). A member of the bacterial order Actinomycetales.

activated-sludge process. The use of biologically active sewage sludge to hasten the breakdown of organic matter in raw sewage during secondary treatment.

active immunity. Specific resistance to disease acquired by individuals as a result of their own reactions to pathogenic microorganisms or the products of such organisms.

adaptive enzyme. An enzyme produced by an organism in response to the presence of the enzyme's substrate or a related substance. Also called *induced enzyme*.

adenine (**ad**•uh•neen). A purine component of nucleosides, nucleotides, and nucleic acids.

adenosine (uh•**den**•o•seen). A mononucleoside consisting of adenine and D-ribose, produced by the hydrolysis of adenosine monophosphate.

adenosine triphosphate (trye•**fos**•fate). A compound of one molecule each of adenine and D-ribose and three molecules of phosphoric acid, which plays an important role in energy transformations in metabolism. Abbreviation: ATP.

adjuvant (**aj**•oo•vunt). A substance that when injected in addition to antigen increases antibody production.

aerobe (**air**•obe). Any oxygen-requiring organism. Compare **anaerobe**.

aflatoxin (aff•luh•**tahk**•sin). The toxin produced by some strains of the fungus *Aspergillus flavus;* a carcinogen.

agar-agar (ay•gar-•**ay**•gar, ah•gar-ah•gar). A dried polysaccharide extract of red algae (Rhodophyceae) used as a solidifying agent in microbiological media. Commonly referred to as *agar*.

agglutination (uh•gloo•tin•**ay**•shun). Clumping of cells.

agglutinin (uh•**gloo**•tin•in). An antibody capable of causing the clumping or agglutination of bacteria or other cells.

alga, pl. **algae** (**al**•guh, **al**•jee). Any member of a heterogenous group of eucaryotic, photosynthetic, and unicellular and multicellular organisms.

alleles (uh•**leelz**). Two genes that are alternative occupants of the same chromosomal locus on a pair of homologous chromosomes.

allergy. A type of antigen-antibody reaction marked by an exaggerated physiological response to a substance in sensitive individuals.

allosteric enzymes (al•o•**stehr**•ik, al•o•**steer**•ik). Regulatory enzymes with a binding or catalytic site for the substrate and a different site (the *allosteric site*) where a modulator acts.

allosteric site. See **allosteric enzyme**.

amebiasis (am•i•**bye**•uh•sis). Amebic dysentery, an infectious disease of humans and other animals.

amino acid (uh•**meen**•o). An organic compound containing both amino (-NH$_2$) and carboxyl (-COOH) groups.

ammonification (uh•mon•i•fi•**kay**•shun, uh•mo•nif•i•**kay**•shun). Decomposition of organic nitrogen compounds, e.g., proteins, by microorganisms with the release of ammonia.

amphitrichous (am•**fit**•rik•us). Having a single flagellum at each end of a cell.

amylase (**am**•i•lase, **am**•i•laze). An enzyme that hydrolyzes starch.

anabolism (uh•**nab**•o•lizm). The process of synthesis of cell constituents from simpler molecules, usually requiring energy. Compare **catabolism**.

anaerobe (**an**•uh•robe). An organism that grows in the absence of molecular oxygen. Compare **aerobe**.

anamnestic response (an•am•**ness**•tik). The heightened immunologic reaction to a second exposure to an antigen.

anaphylaxis (an•uh•fi•**lak**•sis). Hypersensitivity in an animal following the parenteral injection of an antigen.

anaplasia (an•uh•**play**•zhuh). Structural abnormality in a cell or cells.

antagonism. The killing, injury, or inhibition of growth of one species of microorganism by another when one organism adversely affects the environment of the other.

antibiosis (an•tee•bye•o•sis). Antagonistic association between two organisms in which one is adversely affected.

antibiotic (an•tee•bye•**ot**•ik). A substance of microbial origin that has antimicrobial activity in very small amounts.

antibody (**an**•ti•bod•ee). Any of a class of substances (proteins) produced by an animal in response to the introduction of an antigen.

anticodon (an•tee•**ko**•don). A sequence of three nucleotides (in a *t*RNA) complementary to a codon triplet in *m*RNA.

antigen (**an**•ti•jen). A substance that when introduced into an animal body stimulates the production of specific substances (antibodies) that react or unite with the substance introduced (antigen).

antigenic determinant (an•ti•**jen**•ik). The part of an antigen molecule that, as the structural complement of certain chemical groupings on certain antibody molecules, determines the specificity of the antigen-antibody reaction.

antimicrobial agent (an•tee•migh•**kro**•bee•ul). Any chemical or biologic agent that either destroys or inhibits growth of microorganisms.

antiseptic. Acting against or opposing sepsis, putrefaction, or decay by either preventing or arresting the growth of microorganisms.

antiserum (an•tee•seer•um). Blood serum that contains antibodies.

antitoxin (an•tee•**tahk**•sin). An antibody capable of uniting with and neutralizing a specific toxin.

aplanospore (ay•**plan**•o•spore). A nonmotile spore; an abortive zoospore.

apoenzyme (ap•o•**en**•zime). The protein moiety (portion) of an enzyme.

arbovirus (ahr•bo•**vye**•rus). Arthropod-borne virus.

arthropod (**ahr**•thro•pod). An invertebrate with jointed legs, such as an insect or a crustacean.

arthrospore (**ahr**•thro•spore). An asexual spore formed by the fragmentation of the mycelium.

ascitic fluid (uh•**sit**•ik). Serous fluid that accumulates abnormally in the peritoneal cavity.

ascomycetes (ass•ko•migh•**see**•teez). A class of fungi distinguished by the ascus.

ascospore (**ass**•ko•spore). A sexual spore characteristic of the Ascomycetes, produced in a saclike structure (an ascus) after the union of the two nuclei.

ascus (**ass**•kus). A saclike structure, characteristic of the Ascomycetes, in which ascospores are produced.

asepsis (ay•**sep**•sis). A condition in which harmful microorganisms are absent. Adjective: **aseptic** (ay•**sep**•tik).

aseptic technique. Precautionary measures taken to prevent contamination.

assay (**ass**•ay). Qualitative or quantitative determination of the components of a material, such as a drug.

assimilation (uh•sim•i•**lay**•shun). Conversion of nutritive material into protoplasm.

asymptomatic (ay•sim•tuh•**mat**•ik). Exhibiting no symptoms.

ATP. See **adenosine triphosphate.**

attenuation (uh•ten•yoo•**ay**•shun). Weakening; reduction in virulence.

autoclave (**aw**•toe•klave). An apparatus using steam under pressure for sterilization.

autogenous vaccine (aw•**toj**•uh•nus). A vaccine prepared from bacteria isolated from the patient to be treated.

autoimmune disease (aw•toe•im•**yoon**). A condition in which the body develops an immunological reaction against its own tissues.

autolysis (aw•**tol**•i•sis). Disintegration of cells by the action of their own enzymes.

autotroph (**aw**•toe•trofe). A microorganism that uses inorganic materials as a source of nutrients; carbon dioxide is the sole source of carbon. Compare **heterotroph.**

auxotrophic mutant (awk•so•**troe**•fik). An organism having a growth requirement of specific nutrients not necessary in the parental strain.

axenic culture (ay•**zen**•ik, ay•**zee**•nik). An organism of a single species, e.g., a bacterium, fungus, alga, or protozoan, growing in a medium free of other living organisms.

bacillus (buh•**sil**•us). Any rod-shaped bacterium.

bacteremia (bak•tuh•**ree**•mee•uh). A condition in which bacteria are present in the bloodstream.

bacterial filter. A special type of filter through which bacterial cells cannot pass.

bactericide (bak•**teer**•i•side). An agent that destroys bacteria.

bacterin (**bak**•tuh•rin). A suspension of killed or attenuated bacteria used for artificial immunization.

bacteriochlorophyll (bak•**teer**•ee•o•**klor**•uh•fil). A chlorophyll-like pigment possessed by photosynthetic bacteria.

bacteriocinogenic factor (bak•**teer**•ee•o•sin•o•**jen**•ik). A plasmid in some bacteria that determines the formation of bacteriocins, which are proteins that kill the same or closely related species of bacteria.

bacteriocin (bak•**teer**•ee•o•sin). See **bacteriocinogenic factor.**

bacteriolysin (bak•**teer**•ee•o•**lye**•sin). An agent or substance that causes disintegration of bacteria.

bacteriophage (bak•**teer**•ee•o•fayj). A virus that infects bacteria and causes lysis of bacterial cells.

bacteriostasis (bak•**teer**•ee•o•**stay**•sis). Inhibition of growth and reproduction of bacteria without killing them.

bacterium, pl. **bacteria** (bak•**teer**•ee•um, bak•**teer**•ee•uh) Any of a group of diverse and ubiquitous procaryotic single-celled microorganisms.

basidiomycetes (buh•**sid**•ee•o•mye•**see**•teez). A class of fungi that form basidiospores.

basidiospore (buh•sid•ee•o•spore). A sexual spore produced following the union of two nuclei on a specialized clublike structure known as a *basidium.*

basidium (buh•**sid**•ee•um). A club-shaped specialized structure of the Basidiomycetes on which are borne the exogenous basidiospores.

benthos (**ben**•thahss). A collective term for the organisms living along the bottom of oceans and lakes.

beta hemolysis. A colorless, clear, sharply defined zone of hemolysis surrounding certain bacterial colonies growing on blood agar.

binomial nomenclature (bye•no•mee•ul). The scientific method of naming plants, animals, and microorganisms, so-called because in it, species names are binomial, i.e., consist of two terms.

biochemical oxygen demand. A measure of the amount of oxygen consumed in biological processes that break down organic matter in water; a measure of the organic pollutant load. Abbreviation: BOD.

biodegradable (bye•o•dee•**grade**•uh•bul). Capable of being broken down by microorganisms.

biogenesis (bye•o•**jen**•uh•sis). The production of living organisms only from other living organisms. Compare **spontaneous generation.**

biogeochemical agents (bye•o•jee•o•**kem**•i•kul). Microorganisms that function to accomplish the mineralization of organic carbon, nitrogen, sulfur, phosphorus, and other compounds.

bioluminescence (bye•o•loo•min•

ess•unce). Emission of light by living organisms.

biomass (bye•o•mass). The mass of living matter present in a specified area.

biosphere (bye•o•sfeer) The zone of the earth that includes the lower atmosphere and upper layers of soil and water.

biota (bye•o•tuh). The animal, plant, and microbial life characterizing a given region.

biotype (bye•o•tipe). A group of organisms having the same genotype.

blastospore (blass•toe•spore). A spore produced by a budding process along the hypha or by a single cell.

blood plasma. The fluid portion of blood. Also called *plasma*.

blood serum. The fluid expressed from clotted blood or clotted blood plasma.

bloom. A colored area on the surface of a body of water caused by heavy growth of plankton.

BOD. See **biochemical oxygen demand.**

botulism (bot•choo•lizm). Food poisoning due to the toxin of *Clostridium botulinum.*

Brownian motion (brown•ee•un). A peculiar dancing motion exhibited by finely divided particles and bacteria in suspension, due to bombardment by the molecules of the fluid.

budding. A form of asexual reproduction typical of yeast, in which a new cell is formed as an outgrowth from the parent cell.

buffer. Any substance in a fluid that tends to resist the change in pH when acid or alkali is added.

calorie. A unit of heat; the amount of heat required to raise the temperature of 1 g of water by 1°C.

capsid. The protein coat of a virus.

capsomere (kap•so•meer). A morphologic subunit of a capsid as seen by electron microscopy.

capsule. An envelope or slime layer surrounding the cell wall of certain microorganisms.

carrier. A person in apparently good health who harbors a pathogenic microorganism.

catabolism (kuh•tab•o•lizm). The dissimilation, or breakdown, of complex organic molecules releasing energy. A part of the total process of metabolism. Compare **anabolism.**

catalase (kat•uh•lase, kat•uh•laze). An enzyme that converts hydrogen peroxide to water and oxygen.

catalyst (kat•uh•list). Any substance that accelerates a chemical reaction but remains unaltered thereby in form and amount.

cavitation (kav•i•tay•shun). The use of high-frequency sound waves in liquid to produce small bubbles that collapse violently, disintegrating microbial cells.

cell. The microscopic, functionally and structurally basic unit of all living organisms.

cellulase (sel•yoo•lase, sell•yoo•laze). An extracellular enzyme that yields cellobiose on hydrolysis of cellulose.

cell wall. A rigid external covering of the cytoplasmic membrane.

chancre (shang•kur). The primary ulcerative lesion in syphilis.

chemoautotroph (kee•mo•aw•toe•trofe). An organism that obtains energy by oxidizing inorganic chemical compounds. Carbon dioxide is the sole source of carbon. Also called *chemolithotroph.*

chemolithotroph (kee•mo•lith•o•trofe). See **chemoautotroph.**

chemostat (kee•mo•stat, kem•o•stat). A device for maintaining a bacterial culture in the exponential, or log, phase of growth.

chemotaxis (kee•mo•tak•sis). Movement of an organism in response to a chemical stimulus.

chemotherapy (kee•mo•thehr•uh•pee). Treatment of disease by the use of chemicals.

chemotroph (kee•mo•trofe). An organism that obtains its energy from the oxidation of chemical compounds.

chlamydospore (klam•id•o•spore). A thick-walled, resistant spore formed by the direct differentiation of the cells of the mycelium.

chlorophyll (klor•uh•fil). A light-trapping green pigment essential as an electron donor in photosynthesis.

chloroplast (klor•o•plast). A cell plastid (specialized organelle) in plants and algae that contains chlorophyll pigments and functions in photosynthesis.

chromosome (kro•muh•sohm). A gene-containing filamentous structure in a cell nucleus; the number of chromosomes per cell nucleus is constant for each species.

cilium, pl. **cilia** (sil•ee•um, sil•ee•uh). A hairlike appendage on certain cells.

cistron (siss•trahn). The genetic unit that carries information for the synthesis of a single enzyme or protein molecule; determined by the cis-trans complementation test.

citric acid cycle. See **Krebs cycle.**

classification. The systematic arrangement of units (e.g., organisms) into groups, and often further arrangement of those groups into large groups.

clone (klohn). A population of cells descended from a single cell.

coagulase (ko•ag•yoo•lase, ko•ag•yoo•laze). An enzyme, produced by pathogenic staphylococci, that causes coagulation of blood plasma.

coccus (kock•us). A spherical bacterium.

codon (ko•dahn). A sequence of three nucleotide bases (in *m*RNA) that codes for an amino acid or the initiation or termination of a polypeptide chain.

coenocytic (see•no•sit•ik). A term applied to a cell or an aseptate hypha containing numerous nuclei.

coenzyme (ko•en•zime). The nonprotein portion of an enzyme.

colicin (ko•li•sin). A highly specific protein, released by some enteric bacilli, that kills other enteric bacilli.

coliphage (ko•li•fayj). A virus that infects *Escherichia coli.*

colony. A macroscopically visible growth of microorganisms on a solid culture medium.

commensalism (kuh•men•suh•lizm). A relationship between members of different species living in proximity (the same cultural environment) in which one organism benefits from the association but the other is not affected.

competitive inhibition. Inhibition of the action of an enzyme by a nonsubstrate molecule's occupying the site on the enzyme that would otherwise by occupied by the substrate.

complement (**kom**•pluh•ment). A normal thermolabile protein constituent of blood serum that participates in antigen-antibody reactions.

complement fixation. The binding of complement to an antigen-antibody complex so that the complement is unavailable for subsequent reaction.

compromised host. A person already weakened with debilitating disease.

conidiophore (ko•**nid**•ee•o•fore). A branch of mycelium that bears conidia.

conidiospore (ko•**nid**•ee•o•spore). See **conidium.**

conidium (ko•**nid**•ee•um). An asexual spore that may be one-celled or many-celled and of many sizes and shapes. Also called *conidiospore.*

conjugation. A mating process characterized by the temporary fusion of the mating partners and transfer of genes. Conjugation occurs particularly in unicellular organisms.

constitutive enzyme. An enzyme whose formation is not dependent upon the presence of a specific substrate.

contamination. Entry of undesirable organisms into some material or object.

crossmatching. The determination of blood compatibility for transfusion by mixing donor cells with recipient serum, and recipient cells with donor serum, and examining for an agglutination reaction.

culture. A population of microorganisms cultivated in a medium.

cytochrome (sigh•toe•krome). One of a group of reversible oxidation-reduction carriers in respiration.

cytoplasm (sigh•toe•plazm). The living matter of a cell between cell membrane and nucleus.

cytoplasmic membrane (sigh•toe•plaz•mik). A membrane surrounding the cytoplasm and its contents.

DAP. Diaminopimelic acid, a component of cell-wall mucopeptide in some bacteria.

dark-field microscopy. A type of microscopic examination in which the microscopic field is dark and any objects, such as organisms, are brightly illuminated.

deamination (dee•am•i•**nay**•shun). Removal of an amino group, especially from an amino acid.

decarboxylation (dee•kar•bock•si•**lay**•shun). Removal of a carboxyl group,—COOH.

decimal reduction time. The amount of time at a particular temperature sufficient to reduce a viable microbial population by 90 percent.

dehydration. The removal of water.

dehydrogenation (dee•high•drah•juh•**nay**•shun, dee•high•druh•juh•**nay**•shun). A reaction involving an enzyme that causes oxidation of a substrate by removing hydrogen from it.

demineralization (dee•min•ur•ul•i•**zay**•shun). The process by which acid produced by bacterial action dissolves the calcium salts of tooth enamel.

denature (dee•**nay**•chur). To modify, by physical or chemical action, the structure of an organic substance, especially a protein, in order to alter some properties of the substance, such as solubility.

denitrification (dee•nigh•tri•fi•**kay**•shun). Reduction of nitrates to free nitrogen.

dental plaque (plak). An aggregation of bacteria and organic material on the tooth surface.

deoxyribonucleic acid (dee•**ahk**•see•rye•bo•new•**klee**•ik). The carrier of genetic information; a type of nucleic acid occurring in cells, containing phosphoric acid, D-2-deoxyribose, adenine, guanine, cytosine, and thymine. Abbreviation: DNA.

deoxyribose (dee•ahk•see•**rye**•bose, dee•ahk•see•**rye**•boze). A five-carbon sugar having one oxygen atom less than the parent sugar, ribose; a component of DNA.

dermatotropic (dur•muh•toe•**trope**•ik). Having a selective affinity for the skin.

dextran (**deks**•tran). A polysaccharide (glucose polymer) produced by a wide range of microorganisms, sometimes in large amounts.

dialysis (dye•**al**•i•sis). The separation of soluble substances from colloids by diffusion through a semipermeable membrane.

diauxic growth (dye•**awk**•sik). Growth in two separate phases due to preferential use of one carbon source over another between which a temporary lag occurs.

differential stain. A procedure using a series of dye solutions or staining reagents to bring out differences in microbial cells.

dimorphism (dye•**more**•fizm). Occurring in two forms.

diplobacilli (dip•lo•buh•**sil**•eye). Bacilli occurring in pairs.

diplococci (dip•lo•**kahk**•sigh). Cocci occurring in pairs.

diploid (**dip**•loyd). Having chromosomes in pairs the members of which are homologous; having twice the haploid number.

disaccharide (dye•**sak**•uh•ride). A sugar composed of two monosaccharides.

disease. A state of impaired body function occurring as a response to infection, stress, or other conditions.

disinfectant. An agent that frees from infection by killing the vegetative cells of microorganisms.

dissimilation (dis•sim•i•lay•shun). Chemical reactions that release energy by the breakdown of nutrients.

DNA. See deoxyribonucleic acid.

DNA ligase (**lye**•gase, **lye**•gaze). A specific enzyme that joins fragments of replicated strands together.

DNA polymerase (**pahl**•i•mur•ase, **pahl**•i•mur•aze). An enzyme that adds nucleotides and synthesizes DNA in one direction.

dysentery (**diss**•un•tehr•ee). Disease due to infection of the lower intestine.

ECHO virus. Acronym for enteric cytopathogenic human orphan virus, the causative agent of several diseases in humans.

ecology. The study of the interrelationships that exist between organisms and their environment.

ectoplasm (ek•toe•plazm). The outer layer of cytoplasm.

edema (eh•dee•muh). Excessive accumulation of fluid in body tissue.

effluent (eff•loo•unt). The liquid waste of sewage and industrial processing.

endemic (en•dem•ik). Peculiar to or occurring constantly in a community.

endergonic (en•dur•gahn•ik). Describing or pertaining to a biochemical reaction in which the products possess more free energy than the starting materials.

endoenzyme (en•doe•en•zime). An enzyme formed within the cell and not excreted into the medium. Also called *intracellular enzyme.*

endogenous (en•dahj•uh•nus). Produced or originating from within.

endonuclease (en•doe•new•klee•ase, en•doe•new•klee•aze). An enzyme that excises a damaged segment of DNA.

endophytic (en•doe•fit•ik). Describing or pertaining to algae that are not free-living but live in other organisms.

endoplasm (en•doe•plazm). The inner layer of cytoplasm.

endoplasmic reticulum (en•doe•plaz•mik ree•tik•yoo•lum). An extensive array of internal membranes in a eucaryotic cell.

endospore (en•doe•spore). A thick-walled spore formed in the bacterial cell.

endothermic (en•doe•thur•mik). Describing or pertaining to a chemical reaction in which energy is consumed overall.

endotoxin (en•doe•tahk•sin). A toxin produced in an organism and liberated only when the oranism disintegrates.

energy-rich compound. A compound that contains potential energy that is released when the compound donates a portion of itself to water or another molecule.

enteric (en•tehr•ik). Pertaining to the intestines.

enteropathogen (en•tur•o•path•uh•jen). An organism that causes intestinal disease.

enterotoxin (en•tur•o•tahk•sin). A toxin specific for cells of the intestine. It gives rise to symptoms of food poisoning.

enzootic (en•zo•aht•ik). Describing or pertaining to a disease afflicting animals in a limited geographic area.

enzyme (en•zime). An organic catalyst produced by an organism. See also **adaptive enzyme, constitutive enzyme, endoenzyme, exoenzyme.**

epidemic (ep•i•dem•ik). A sudden increase in the incidence of a disease, affecting large numbers of people over a wide area.

epidemiology (ep•i•dee•mee•ahl•uh•jee, ep•i•dem•ee•ahl•uh•jee). The study of the factors that influence the occurrence and distribution of disease in groups of individuals.

episome (ep•i•sohm). A plasmid which can integrate reversibly with the chromosome of its bacterial host; in the integrated state it behaves as part of the chromosome, but it is also able to multiply independently of the chromosome.

epizootic (ep•i•zo•aht•ik). Describing or pertaining to a widely diffused and rapidly spread disease affecting many animals of one kind.

erythroblastosis fetalis (eh•ree•thro•blass•toe•sis fee•tay•lis). Hemolytic disease of the newborn resulting from maternal immunization to an Rh antigen present in the fetus and absent in the mother.

etiology (ee•tee•ahl•uh•jee). The study of the cause of a disease.

eucaryote (yoo•care•ee•ote). A cell that possesses a definitive or true nucleus. Compare **procaryote.**

exergonic (ek•sur•gahn•ik). Energy-yielding, as in a chemical reaction.

exobiology (ek•so•bye•ahl•uh•jee). Extraterrestrial biology.

exoenzyme (ek•so•en•zime). An enzyme excreted by a microorganism into the environment. Also called *extracellular enzyme.*

exogenous (ek•sahj•uh•nus) Produced or originating from without.

exonuclease (ek•so•new•klee•ase, ek•so•new•klee•aze). An enzyme that excises a damaged segment of DNA.

exospore (ek•so•spore). A spore external to the vegetative cell.

exothermic (ek•so•thur•mik). Describing a chemical reaction that gives off energy.

exotoxin (ek•so•tahk•sin). A toxin excreted by a microorganism into the surrounding medium.

exponential phase (ek•spuh•nen•chul). The period of culture growth when cells divide steadily at a constant rate. Also called *logarithmic phase* (commonly, *log phase*).

extracellular enzyme. See **exoenzyme.**

extrachromosomal genetic element (ek•struh•kro•muh•so•mul). A genetic element, called a *plasmid,* that is capable of autonomous replication in the cytoplasm of the bacterial cell.

exudate (eks•yoo•date). The more or less fluid material found in a lesion or inflamed tissue.

facultative anaerobe (fak•ul•tay•tiv). A bacterium that grows under either aerobic or anaerobic conditions.

fastidious organism (fass•tid•ee•us). An organism that is difficult to isolate or cultivate on ordinary culture media because of its need for special nutritional factors.

feedback inhibition. A cellular control mechanism by which the end product of a series of metabolic reactions inhibits the further activity of an earlier enzyme of the sequence.

fermentation (fur•men•tay•shun). Anaerobic oxidation of compounds by enzyme action of microorganisms; gaseous oxygen is not involved in this energy-yielding process. An organic compound is the electron acceptor.

F factor. The fertility or sex factor in the cytoplasm of male bacterial cells.

fibrinolysin (figh•bri•nahl•i•sin). A substance, produced by hemolytic streptococci, that can liquefy clotted blood plasma or fibrin clots. Also called *streptokinase.*

filamentous (fil•uh•men•tus). Characterized by threadlike structures.

fimbriae, sing. **fimbria** (fim•bree•ee, fim•bree•uh). Surface appendages of certain gram-negative bacteria composed of protein subunits. They are shorter and thinner than flagella. Also called *pili*.

fission (fish•un). An asexual process by which some microorganisms reproduce; transverse cell division in bacteria.

fixed nitrogen. Nitrogen in a compound.

flagellates (flaj•uh•luts, flaj•uh•laits). One of the subphyla of the Phylum Protozoa.

flagellum, pl. **flagella** (fluh•jel•um, fluh•jel•uh). A flexible, whiplike appendage on cells, used as an organ of locomotion.

flora (flore•uh). In microbiology, the microorganisms present in a given situation, e.g., intestinal flora, normal flora of soil. See also **biota**.

fluorescence (floo•uh•ress•unce). Emission of a longer wavelength of light by a substance that has absorbed radiation from a source of shorter wavelength light.

fluorescence microscopy (migh•krahss•kuh•pee). Microscopy in which microorganisms are stained with a fluorescent dye and observed by illumination with ultraviolet light.

fomites (fo•mights). Inanimate objects that carry viable pathogenic organisms.

food poisoning. A general term applied to all stomach or intestinal disturbances due to food contaminated with certain microorganisms or their toxins.

formalin (for•muh•lin). A 37 to 40% aqueous solution of formaldehyde.

Forssman antigens (force•mun). Heterophile antigens widely distributed in nature.

fractional sterilization. Sterilization of material by heating it to 100°C (212°F) on three successive days with incubation periods in between.

free nitrogen. Atmospheric nitrogen.

Frei test (fry). A skin test to determine sensitivity to the agent that causes lymphogranuloma venereum.

fruiting body. A specialized, spore-producing organ.

fulminating infection. A sudden severe and rapidly progressing infectious disease.

fungicide (fun•ji•side). An agent that kills or destroys fungi.

fungus, pl. **fungi** (fung•gus, fun•jye). A microorganism that lacks chlorophyll and is usually filamentous in structure; a mold or yeast.

fusiform (fyoo•zi•form). Spindle-shaped, tapered at the ends.

gamete (gam•eet, guh•meet). A reproductive cell that fuses with another reproductive cell to form a zygote, which then develops into a new individual; a sex cell.

gamma globulin. A fraction of serum globulin that is rich in antibodies.

gasohol (gass•uh•hawl). A mixture of unleaded gasoline and alcohol used for fuel.

gastroenteritis (gass•tro•en•tur•eye•tis). Inflammation of the mucosa of the stomach and intestine.

gelatinase (juh•lat•i•nase, juh•lat•i•naze). An exoenzyme that degrades gelatin.

gene (jeen). A segment of a chromosome, definable in operational terms as the repository of a unit of genetic information.

generation time. The time interval necessary for a cell to divide.

genetics. The science concerned with the study of biological inheritance.

genome (jee•nohm). A complete set of genetic material; i.e., a complete set of genes.

genotype (jee•nuh•tipe, jen•uh•tipe). The particular set of genes present in an organism and its cells; an organism's genetic constitution. Compare **phenotype**.

genus, pl. **genera**. (jee•nus, jen•ur•uh). A group of very closely related species.

germ. A microorganism; a microbe, usually a pathogenic one.

germicide (jurm•i•side). An agent capable of killing germs, usually pathogenic microorganisms.

gingiva (jin•ji•vuh). The mucous membrane and soft tissue surrounding a tooth. Adjective: **gingival** (jin•ji•vul).

globulin (glahb•yoo•lin). A protein soluble in dilute solutions of neutral salts but insoluble in water. Antibodies are globulins.

glucose (gloo•kose). A carbohydrate classified as a monosaccharide and hexose, used as an energy source by many microorganisms. Also called *dextrose* or *grape sugar*.

glycogen (glye•kuh•jen). A carbohydrate of the polysaccharide group stored by animals. It yields glucose on hydrolysis.

glycolysis (glye•kahl•i•sis). Anaerobic dissimilation of glucose to pyruvic acid by a sequence of enzyme-catalyzed reactions. Also called the *Embden-Meyerhoff pathway*.

gnotobiotic (no•toe•bye•aht•ik). Pertaining to organisms living in the absence of all demonstrable, viable organisms other than those known to be present.

Golgi apparatus (gol•jee). A membraneous organelle in the endoplasmic reticulum of the cell.

gram-negative bacteria. Bacteria that appear red after being subjected to the Gram stain.

gram-positive bacteria. Bacteria that appear blue or violet after being subjected to the Gram stain.

Gram stain. A differential stain by which bacteria are classed as gram-positive or gram-negative depending upon whether they retain or lose the primary stain (crystal violet) when subjected to treatment with a decolorizing agent.

growth. In reference to microorganisms, an increase in the total mass or number of cells (e.g., in a culture) rather than in the size or complexity of any individual organism.

growth curve. Graphic representation of the growth (population changes) of bacteria in phases in a culture medium.

guanine (gwah•neen) A purine base, occurring naturally as a fundamental component of nucleic acids.

habitat. The natural environment of an organism.

halophile (hal•o•file). A microorganism whose growth is accelerated by or dependent on high salt concentrations.

hanging-drop technique. A technique in which microorganisms are observed suspended in a drop of fluid.

H antigen. A type of antigen found in the flagella of certain bacteria.

haploid (**hap**·loyd). Having a single set of unpaired chromosomes in each nucleus; having the chromosome number characteristic of a mature gamete of the species. Compare **diploid.**

hapten. A simple substance that reacts like an antigen in vitro by combining with antibody but cannot induce formation of antibodies by itself.

HeLa cells (**hee**·luh). A pure cell line of human cancer cells used for the cultivation of viruses.

helix (**hee**·liks). A coiled spiral form.

hemagglutination (hee·muh·gloo·ti·**nay**·shun). Agglutination (clumping) of red blood cells.

hemoglobin (**hee**·mo·glo·bin). The constituent of red blood cells that gives them their color and carries oxygen.

hemolysin (hee·**mahl**·i·sin). A substance that lyses (dissolves) red blood cells, liberating hemoglobin.

hemolysis (hee·**mahl**·i·sis). The process of dissolving red blood cells.

hemorrhage (**hem**·uh·rij). Loss of blood from the circulatory system, either through capillary walls or from damaged vessels; bleeding.

heterogamy (het·ur·**og**·uh·mee). Conjugation of unlike gametes.

heterokaryon (het·ur·o·**care**·ee·on). A cell having two nuclei that differ genetically.

heterologous (het·ur·**ahl**·uh·gus). Different with respect to type or species.

heterophile antibody (**het**·ur·o·file). An antibody that reacts with totally unrelated species of microorganisms or cells from unrelated species of animals. Agglutination of *Proteus* spp. cells by serum from typhus fever patients is an example.

heterophile antigen. An antigen that reacts with antibodies stimulated by unrelated species.

heterotroph (**het**·ur·o·trofe). A

microorganism that is unable to use carbon dioxide as its sole source of carbon and requires one or more organic compounds. Compare **autotroph.**

histocompatibility antigens (hiss·toe·kum·pat·i·**bil**·i·tee). Tissue antigens that stimulate rejection of transplanted organs or tissues that are not compatible.

holdfast. A suckerlike base that attaches the thallus of certain microorganisms to a surface.

holoenzyme (ho·lo·**en**·zime, halh·o·**en**·zime). A fully active enzyme, containing an apoenzyme and a coenzyme.

holozoic (ho·lo·**zo**·ik, hahl·o·**zo**·ik). Pertaining to or describing protozoa that ingest food as solid particles through a mouth opening.

homeostasis (ho·mee·o·**stay**·sis). Maintenance of normal conditions.

host. An organism harboring another as a parasite (or as an infectious agent).

humoral immunity (**hyoo**·mur·ul). Immunity arising from the formation of specific antibodies that circulate in the bloodstream in response to the introduction of an antigen.

hyaluronidase (high·ul·yoo·**ron**·i·dase, high·ul·yoo·**ron**·i·daze). An enzyme that catalyzes the breakdown of hyaluronic acid. Also called *spreading factor.*

hydrogen bond. A type of molecular linkage that is mediated by a hydrogen atom; hydrogen bonds join the opposite bases in the double-chained structure of DNA.

hydrologic cycle (high·dro·**lahj**·ik). The complete cycle through which water passes, from oceans, through the atmosphere, to the land, and back to the oceans.

hydrolysis (high·**drahl**·i·sis). The process by which a substrate is split to form products through the intervention of a molecule of water.

hypersensitivity. Extreme sensitivity to foreign proteins, e.g., allergens.

hypha (**high**·fuh). One filament or thread of a mycelium.

icosahedron (eye·kah·suh·**hee**·drun, eye·ko·suh·**hee**·drun). A solid formed of 20 triangular

faces and 12 corners; the geometrical shape of many virions.

ID. Infective dose; the number of microorganisms required to infect a host. Compare **LD**.

ID$_{50}$. The dose (number of microorganisms) that will infect 50 percent of the experimental animals in a test series.

immune serum. The liquid portion of blood containing one or more specific antibodies.

immunity. A natural or acquired resistance to a specific disease.

immunization. Any process that develops resistance (immunity) in a host to a specific disease.

immunodeficiency disease (im·yoo·no·dee·**fish**·un·see). A disease occurring because of a deficiency or malfunction in the immune response.

immunogenicity (im·yoo·no·jeh·**niss**·i·tee). The capacity to stimulate the formation of specific antibodies.

immunoglobulin (im·yoo·no·**glob**·yoo·lin). Any of the serum proteins, such as gamma globulin, that possess antibody activity.

immunology. The study of the specific serum and tissue factors that defend the body against foreign agents (immune responses).

imperfect fungi. Fungi that do not have a sexual cycle.

IMViC test. Acronym for a group of tests used to differentiate *Escherichia coli* from *Enterobacter aerogenes.*

inactivate. To destroy the activity of a substance; e.g., to heat blood serum to 56°C for 30 min to destroy complement.

inclusion bodies. Discrete assemblies of virions and/or viral components that develop within virus-infected cells.

incubation. In microbiology, the subjecting of cultures of microorganisms to conditions (especially temperatures) favorable to their growth.

induced enzyme. See **adaptive enzyme.**

induced mutation. Mutation produced by use of a mutagen.

induction. The production of an increase in the rate of synthesis of an enzyme, generally by the enzyme's substrate or a closely related compound.

infection. A pathological condition

due to the growth of microorganisms in a host.

infectious. Capable of producing disease in a susceptible host.

inflammation. A tissue reaction resulting from irritation by a foreign material and causing migration of leukocytes and increased flow of blood to the area, producing swelling, reddening, heat, pain, and tenderness.

inheritance. Transfer of genetic information from parents to their progeny.

inhibition. In microbiology, prevention of growth or multiplication of microorganisms.

inoculation (in•ahk•yoo•**lay**•shun). The artificial introduction of microorganisms or substances into the body or into a culture medium.

inoculum (in•**ahk**•yoo•lum). The substance, containing microorganisms or other material, that is introduced in inoculation.

in situ (in **sigh**•too, in **sigh**•tyoo). In the original or natural location.

interferon (in•tur•**feer**•ahn). An antiviral substance produced by animal tissue.

intoxication. Poisoning.

intracellular (in•truh•**sel**•yoo•lur). Within a cell.

intracellular enzyme. See **endoenzyme**.

invertase (in•**vur**•tase, **in**•vur•tase). An enzyme that hydrolyzes sucrose to glucose and fructose.

in vitro (in **vee**•troe). Literally, "in glass." Pertaining to biologic experiments performed in test tubes or other laboratory vessels. Compare **in vivo**.

in vivo (in **vee**•voe). Within the living organism; pertaining to laboratory testing of agents within living organisms. Compare **in vitro**.

isoantibody (eye•so•**an**•ti•bod•ee). An antibody, found only in some members of a species, that acts upon cells or cell components of other members of the same species.

isoenzyme (eye•so•**en**•zime). Any one of a group of enzymes of different structural forms that possess identical (or nearly identical) catalytic properties. Also called *isozyme*.

isogamous (eye•**sog**•uh•mus). Sexual reproduction in algae with morphologically similar gametes.

isograft (**eye**•so•graft). A graft of tissue from a donor of the same species as the recipient. Also called *homograft*.

isotope (**eye**•suh•tope) Any of several possible forms of a chemical element, differing from other forms in atomic weight but not in chemical properties.

isozyme (eye•so•zime). See **isoenzyme**.

Kahn test. A flocculation test for the diagnosis of syphilis.

Kline test. A microscopic flocculation test for the diagnosis of syphilis.

Koch's postulates. Guidelines to prove that a disease is caused by a specific microorganism.

Krebs cycle. An enzyme system that converts pyruvic acid to carbon dioxide in the presence of oxygen, with concomitant release of energy that is captured in the form of ATP molecules. Also called *citric acid cycle*, *tricarboxylic acid cycle*.

krill. A name applied to planktonic crustaceans.

Kupffer cells (**Koop**•fur). Macrophages that phagocytose in the liver.

lac. See **lactose**.

lactose. A carbohydrate (disaccharide) that is split into glucose and galactose on hydrolysis. Also called *milk sugar*. Abbreviation: lac.

lag phase. The period of slow, orderly growth when a medium is first inoculated with a culture.

laminar airflow (**lam**•i•nur). Flow of air currents in which streams do not intermingle; the air moves along parallel flow lines.

LD. Lethal dose; the number of pathogenic microorganisms required to cause death in a given species of animal or plant. Compare **ID**.

LD$_{50}$. The dose (number of microorganisms) that will kill 50% of the animals in a test series.

leukemia (loo•**kee**•mee•uh). An excessive number of white blood corpuscles (leukocytes) due to a disease of the leukocyte-producing tissues of the body.

leukocidin (loo•ko•**sigh**•din). A substance that destroys leukocytes.

leukocyte (**loo**•ko•sight). A type of white blood cell which is characterized by a beaded, elongated nucleus, Also called *white corpuscle*.

leukocytosis (loo•ko•sigh•**toe**•sis). An increase in the number of leukocytes that is caused by the body's response to an injury or infection.

leukopenia (loo•ko•**pee**•nee•uh). A decrease in the number of leukocytes.

lichen (**lye**•kun). A symbiotic, mutalistic association of an alga and a fungus.

ligand (**lig**•und, **lye**•gand). A molecule that binds to a protein; e.g., one that binds to an enzyme and, through the role it plays in other processes, directly controls enzyme activity.

linkage map. A method of showing the locations of genes on a bacterial chromosome.

lipase (**lye**•pase, **lye**•paze). A fat-splitting enzyme.

lipid. A fat or fatlike substance.

lipolytic enzyme (lip•o•**lit**•ik). An enzyme that hydrolyzes lipids.

liquefaction (lik•wi•**fak**•shun). Transformation of a gas or solid (e.g., a gel) to a liquid.

liter (**lee**•tur). A metric unit of volume containing 1,000 milliliters (ml), or 1,000 cubic centimeters.

lithotroph (**lith**•o•trofe). See **autotroph**.

litmus. A plant extract used as an indicator for pH and oxidation or reduction.

locus. In genetics, the site on a chromosome occupied by a gene, operon, mutation, etc.; in some cases, identifiable by reference to a marker.

logarithmic phase (log•uh•**rith**•mik). Commonly called *log phase*. See **exponential phase**.

lophotrichous (lo•**faht**•ri•kus). Having a polar tuft of flagella.

lymphadenopathy (lim•fad•uh•**nop**•uth•ee). 1. Enlargement of lymph nodes. 2. Any disease of the lymph nodes.

lymphatic (lim•**fat**•ik). Involving the lymphatic system.

lymphocyte (**lim**•fo•site). A cell in the blood tissues, and lymph that has the ability to destroy other cells. Lymphocytes play important roles in immunity.

lyophilization (lye•off•il•i•**zay**•

shun). Preservation of biological specimens by rapid freezing and rapid dehydration in a high vacuum.

lysin (lye•sin). An enzyme, antibody, or other substance capable of disrupting or disintegrating cells (lysis).

lysis (lye•sis). The disruption or disintegration of such cells as bacteria or erythrocytes, e.g., by the action of specific antibodies plus complement.

lysogenic conversion (lye•so•jen•ik). A phenomenon in which a prophage (a temperate bacteriophage integrated into the genome of a bacterium) is able to make changes in the properties of the host bacterium.

lysogeny (lye•sah•juh•nee). The state of a bacterium that is carrying a bacteriophage (often as a prophage—see preceding definition) to which it is not itself susceptible.

lysosome (lye•so•sohm). A closed, intracellular, membranous sac that contains hydrolytic enzymes.

lysozyme (lye•so•zime). An enzyme capable of digesting the cell wall of certain bacteria.

lytic phage (lit•ik). A virulent bacterial virus.

macroscopic (mak•ro•skahp•ik). Visible without the aid of a microscope.

maltase (mawl•tase, mawl•taze). An enzyme that hydrolyzes maltose, yielding glucose.

maltose (mawl•tose). A carbohydrate (disaccharide) produced by the enzymatic hydrolysis of starch by diastase.

marine. Of or relating to oceanic and estuarine environments.

marker. In genetics, chromosomal locus that is associated with a particular phenotypic characteristic.

medium. A substance used to provide nutrients for the growth and multiplication of microorganisms.

meiosis (mye•o•sis). A process occurring during cell division at different points in the life cycles of different organisms, in which the chromosome number is reduced by half, thus compensating for the chromosome-

doubling effect of fertilization. Compare **mitosis**.

membrane filter. A filter made from such polymeric materials as cellulose, polyethylene, or tetrafluoroethylene.

mesophile (mez•o•file). A bacterium growing best at the moderate temperature range 25 to 40°C.

mesosome (mez•o•sohm). A membranous involution of the cytoplasmic membrane.

messenger RNA. The intermediary substance that passes information from the DNA in the nuclear region to the ribosomes in the cytoplasm. Abbreviation: *m*RNA.

metabolic pathway (met•uh•bol•ik). A series of steps in the chemical transformation of organic molecules.

metabolism (meh•tab•o•lizm). The system of chemical changes by which the nutritional and functional activities of an organism are maintained.

metabolite (meh•tab•uh•light). Any chemical participating in metabolism; a nutrient.

metachromatic granule (met•uh•kro•mat•ik). An intracellular body found in some bacteria and yeasts that becomes stained with a color which is different from the color of the dye used to stain it.

metastasis (meh•tass•tuh•sis) The process of a malignant cell's detaching itself from a tumor and establishing a new tumor at another site within the host.

metazoa (met•uh•zoe•uh). Animals whose bodies consist of many cells.

methanogenic bacteria (meth•uh•no•jen•ik). Bacteria that produce methane gas under anaerobic conditions.

microaerophile (migh•kro•air•o•file). Any microorganism that grows best in the presence of small amounts of atmospheric oxygen.

microbe (migh•krobe). Any microscopic organism; a microorganism. Adjective: **microbial** (migh•kro•bee•ul).

microbial cytology (sigh•tahl•uh•jee). The study of the structures and functions of microbial cells.

microbial ecology. The study of microorganisms in their natural environments.

microbiology (migh•kro•bye•ahl•uh•jee). The study of organisms of microscopic size (microorganisms), including their culture, economic importance, pathogenicity, etc.

microbiota (migh•kro•bye•o•tuh). The microscopic flora and fauna.

microcysts (migh•kro•sists). Vegetative cells transformed to resting cells during the fruiting process.

micromanipulator (migh•kro•muh•nip•yoo•lay•tur). A device for manipulation of microscopic specimens under a microscope.

micrometer (migh•kro•mee•tur). A unit of measurement: one-thousandth of a millimeter. Abbreviation: μm.

microorganism (migh•kro•or•guh•nizm). Any organism of microscopic dimensions.

microtome (migh•kro•tohm). An instrument for making thin sections of tissue or cells.

microtubules (migh•kro•tyoo•byoolz). Very thin rods that occur within all types of eucaryotic microbial cells.

mitochondrion (migh•toe•kahn•dree•un). A cytoplasmic organelle in eucaryotic cells, the site of cell respiration.

mitosis (mye•toe•sis). A form of nuclear division characterized by complex chromosome movement and exact chromosome duplication. Compare **meiosis**.

modulator. The regulatory metabolite that binds to the allosteric site of an enzyme and alters the maximum velocity. Also called *effector, modifier.*

mold. A fungus characterized by a filamentous structure.

molecular biology. The study of organisms at the subcellular level.

monera (mo•neer•uh). The procaryotic protists, including bacteria and cyanobacteria.

mononucleotide (mah•no•new•klee•o•tide). The basic building block of nucleic acids (DNA and RNA).

monosaccharide (mah•no•sak•uh•ride). A simple sugar, such as a five-carbon or six-carbon sugar.

monotrichous (mo•**not**•ri•kus). Having a single flagellum.

mordant (**more**•dunt). A substance that fixes dyes.

morphogenesis (more•fo•**jen**•uh•sis). The process by which cells are organized into tissue structures.

morphology (more•**fahl**•uh•jee). The branch of biological science that deals with the study of the structure and form of living organisms.

mRNA. See **messenger RNA**.

mutagen (**myoo**•tuh•jen). A substance that causes the occurrence of mutation.

mutant (**myoo**•tunt). An organism with a changed or new gene.

mutation (myoo•**tay**•shun). A stable change of a gene, such that the changed condition is inherited by offspring cells.

mycelium (mye•**see**•lee•um). A mass of threadlike filaments, branched or composing a network, that constitutes the vegetative structure of a fungus.

mycology (mye•**kahl**•uh•jee). The study of fungi.

mycophage (**mye**•ko•fayj). A fungal virus.

mycoplasma (mye•ko•**plaz**•muh). A group of bacteria composed of highly pleomorphic cells.

mycosis (mye•**ko**•sis). A disease caused by fungi.

mycotoxin (mye•ko•**tok**•sin). Any toxic substance produced by fungi.

myxameba (mik•suh•**mee**•buh). A nonflagellated ameboid cell that occurs in the life cycle of acellular slime molds.

myxospore (**mik**•so•spore). A resting cell of a myxobacterium within a hard slime capsule.

NAD. Nicotinamide adenine dinucleotide, an organic coenzyme that functions in enzymatic systems concerned with oxidation-reduction reactions.

naked virion. A nonenveloped virus.

nanometer (**nan**•o•mee•tur, **nay**•no•mee•tur). A unit of length equal to one-billionth of a meter or 10^{-9} meter; one millimicrometer. Abbreviation: nm.

Negri bodies (**neg**•ree). Minute pathological structures (inclusion bodies) found in certain brain cells of animals infected with rabies virus.

neoplasm (**nee**•o•plazm). An aberrant new growth of abnormal cells or tissue; a tumor.

neurotoxin (nyoo•ro•**tahk**•sin). Any nerve poison, such as that produced by certain marine algae.

nitrate reduction. The reduction of nitrates to nitrites or ammonia.

nitrification (nigh•trif•i•**kay**•shun). The transformation of ammonia nitrogen to nitrates.

nitrogen fixation. The formation of nitrogen compounds from free atmospheric nitrogen.

nitrogenous (nigh•**trahj**•i•nus). Relating to or containing nitrogen.

nomenclature (**no**•men•klay•chur). Any system of scientific names, such as those employed in biological classification.

nonseptate (non•**sep**•tait). Having no dividing walls in a filament.

nosocomial disease (no•so•**ko**•mee•ul). Describing or pertaining to disease acquired in the hospital.

nuclear material. Strands or threadlike pieces of nuclear substance, which is DNA.

nucleic acid (new•**klee**•ik). One of a class of molecules composed of joined nucleotide complexes; the types are deoxyribonucleic acid (DNA) and ribonucleic acid (RNA).

nucleolus, pl. **nucleoli** (new•**klee**•o•lus, new•**klee**•o•lye). A small body in a cell nucleus.

nucleoprotein (new•klee•o•**pro**•teen). A molecular complex composed of nucleic acid and protein.

nucleotide (**new**•klee•o•tide). A compound formed from one molecule each of a sugar (pentose), phosphoric acid, and a purine or pyrimidine base.

nucleus. The structure in a cell that contains the chromosomes.

numerical taxonomy. A method of using computer techniques to determine and numerically express the degree of similarity of every culture to every other culture in a particular group.

objective. The system of lenses in a compound microscope nearest the object being observed.

ocular micrometer (migh•**krom**•uh•tur). A glass disk etched with equidistant lines that fits into the eyepiece of a microscope.

oidium, pl. **oidia** (oh•**id**•ee•um, oh•**id**•ee•uh). A single-celled spore formed by disjointing of hyphal cells.

Okazaki fragments (oh•kuh•**zah**•kee). DNA strands replicated in small pieces.

oligodynamic action (ahl•i•go•dye•**nam**•ik). The lethal effect exerted on bacteria by small amounts of certain metals.

oncogenic virus (on•ko•**jen**•ik). A tumor-inducing virus.

oncology (on•**kahl**•uh•jee). The study of the causes, development, characteristics, and treatment of tumors.

oncornavirus (on•kor•nuh•**vye**•rus). A tumor-inducing RNA virus.

operator. A specific region of DNA at the initial end of the gene, where synthesis of mRNA is initiated.

operon (**ahp**•ur•on). A cluster of genes whose expression is controlled by a single operator.

opportunistic microorganism. A microorganism that exists as part of the normal microbiota but becomes pathogenic when transferred from the normal habitat into other areas of the host or when host resistance is lowered.

opsonin (**op**•suh•nin). A substance in the blood serum that renders microorganisms susceptible to ingestion by phagocytes.

order. In systematic biologic classification, a group of families.

organelle (or•guh•**nel**). A structure or body in a cell.

organotroph (or•**gan**•uh•trofe). An organism that obtains nourishment from the ingestion and breakdown of organic matter.

osmosis (oz•**mo**•sis). The passage of a fluid through a semipermeable membrane due to osmotic pressure.

osmotic pressure (oz•**mot**•ik). The force or tension built up when water diffuses through a membrane.

osmotic shock. Any disturbance in a cell when it is transferred to a hypertonic or hypotonic medium.

oxidase (ok·si·dase, ok·si·daze). An enzyme that brings about oxidation.

oxidation. 1. The process of combining with oxygen. 2. The loss of electrons or hydrogen.

oxidation-reduction (O/R) potential. A measurement of the state of oxidation of a system.

oxidative phosphorylation (ok·si·day·tiv). A sequence of oxidation reactions in the respiratory chain that releases energy at three points sufficient for ATP synthesis.

pandemic (pan·dem·ik). A worldwide epidemic.

paramecium (pehr·uh·mee·see·um). A protozoan ciliate having cilia over the entire cell.

parasite. An organism that derives its nourishment from a living plant or animal host. A parasite does not necessarily cause disease.

parasitism (perhr·uh·sit·izm). The relationship of a parasite to its host.

parenteral (pur·ren·tur·ul) By some route other than via the intestinal tract.

passive immunity. Immunity produced by injecting blood or serum containing antibodies.

pasteurization. The process of heating liquid food or beverage at a controlled temperature to enhance the keeping quality and destroy harmful microorganisms.

pathogen (path·uh·jen). An organism capable of producing disease.

pebrine (pay·breen). A silkworm disease caused by a protozoan.

penicillin. The generic name for a large group of antibiotic substances derived from several species of the mold *Penicillium*.

pentose. A sugar with five carbon atoms; e.g., ribose.

peptide. A compound consisting of two or several amino acids.

peptide bonds. Bonds linking amino acids to form a chain.

peptidoglycan (pep·ti·doe·glye·kan). A large polymer that pro-

vides the rigid structure of the procaryotic cell wall, composed of three kinds of building blocks: (1) acetylglucosamine, (2) acetylmuramic acid, and (3) a peptide consisting of four or five amino acids.

peptone. A partially hydrolyzed protein.

peptonization (pep·tun·i·zay·shun). Conversion of proteins into peptones; the solubilization of casein in milk curd by proteolytic enzymes.

perfect fungi. Fungi with both an asexual and a sexual life cycle.

peridontium (pehr·ee·o·don·chee·um). The supportive tissues around the teeth.

periphytes (pehr·i·fites). Microorganisms that become attached to surfaces, grow and form microcolonies, and produce a film to which other organisms become attached and grow.

periplast (pehr·i·plast). A surface cell membrane or pellicle of certain algae and bacteria.

perithecium (pehr·i·theece·ee·um). A spherical, cylindrical, or oval ascocarp that usually opens by a slit or pore at the top.

peritrichous (puh·rit·ri·kus). Having flagella around the entire surface of the cell.

permeability. The extent to which molecules of various kinds can pass through cellular membranes.

permease (pur·me·ase). Any of a group of enzymes that mediate the phenomenon of membrane transport.

per os (pur ose, pur oss). Through the mouth.

pH. A symbol for the degree of acidity or alkalinity of a solution; $pH = \log (1/[H^+])$ where $[H^+]$ represents the hydrogen ion concentration.

phage (fayj). See **bacteriophage**.

phage-typing. Identifying a pathogenic bacterium by the pattern of lysis caused by different phage-types.

phagocyte (fag·o·site). A cell capable of ingesting microorganisms or other foreign particles.

phenol (fee·nol). A compound that is microbicidal or microbistatic, depending on concentration and temperature.

phenol coefficient. The ratio between the greatest dilution of a test germicide capable of killing a test organism in 10 min but not in 5 min and the greatest dilution of phenol giving the same result.

phenotype (fee·no·tipe). The observable characteristics of an organism. Compare **genotype**.

phosphatase (fahss·fuh·tase, fahss·fuh·taze). An enzyme that splits phosphate from its organic compound.

phosphorylation (fahss·fo·ri·lay·shun). The addition of a phosphate group, such as $-H_2PO_3$, to a compound.

photoautotroph (fo·toe·aw·toe·trofe). An organism that derives energy from light and uses CO_2 as its sole carbon source.

photophosphorylation (fo·toe·fahss·fo·ri·lay·shun). Phosphorylation induced by light in photosynthesis.

photoreactivation. The restoration to full viability by immediate exposure to visible light of cells damaged by exposure to lethal doses of ultraviolet light.

photosynthesis (fo·toe·sin·thuh·sis). The process in which chlorophyll and the energy of light are used by plants and some microorganisms to synthesize carbohydrates from carbon dioxide and water.

photosynthetic autotroph (fo·toe·sin·thet·ik). An organism that requires CO_2 as a carbon source and must have a source of light for energy.

phototroph (fo·toe·trofe). A bacterium capable of utilizing light energy for metabolism.

phycology (fye·kahl·uh·jee). The study of algae.

phylogeny (fye·lahj·uh·nee). The evolutionary or ancestral history of organisms.

phylum, pl. **phyla** (fye·lum, fye·luh). A taxon consisting of a group of related classes.

physiology. The study of the life processes of living things.

phytoflagellate (fye·toe·flaj·uh·lut, fye·toe·flaj·uh·lait). A plantlike form of flagellate. Compare **zooflagellate**.

phytoplankton (fye·toe·plank·tun). A collective term for plants and plantlike organisms present in plankton. Compare **zooplankton**.

pili, sing. **pilus** (**pye**• lye, **pye**•lus). See **fimbriae**.

plankton (**plank**•tun). A collective term for the passively floating or drifting flora and fauna of a body of water, consisting largely of microscopic organisms.

plaque (plak). A clear zone in bacterial agar-plate cultures or in tissue cultures caused by the lysing of bacteria or cells by viruses.

plasma. See **blood plasma**.

plasma cell. An antibody-producing cell.

plasma membrane. A selectively permeable thin layer under the cell wall consisting mainly of lipids and protein.

plasmid (**plaz**•mid). See **extrachromosomal genetic element**.

plasmodium (plaz•**mo**•dee•um). A naked, multinucleate mass of protoplasm in the vegetative phase of slime molds.

plasmolysis (plaz•**mahl**•i•sis). The shrinkage of cell contents as a result of withdrawal of water by osmosis.

plasmoptysis (plaz•**mop**•ti•sis). The swelling of a cell as a result of intake of water by osmosis.

pleomorphism (plee•o•**more**•fizm). The existence of different forms in the same species or strain of microorganism. Also called *polymorphism*.

point mutation. A change in a single base as a result of the substitution of one nucleotide for another in DNA (RNA in some viruses).

polar. Located at one end.

polymorphism (pahl•ee•**more**•fizm). See **pleomorphism**.

polypeptide (pahl•ee•**pep**•tide). A molecule consisting of many joined amino acids.

polypeptide chain. A chain formed of a large number of amino acids joined together by peptide bonds.

polyribosome (pahl•ee•**rye**•bo•sohm). See **polysome**.

polysome (**pahl**•ee•sohm). A complex of ribosomes bound together by a single *m*RNA molecule. Also called *polyribosome*.

potable (**po**•tuh•bul). Suitable for drinking.

pour-plate method. An agar-plate technique used to culture colonies of bacteria.

PPLO. An abbreviation for *pleuropneumonialike organisms*. These organisms belong to the order *Mycoplasmatales*.

precipitin (pree•**sip**•i•tin). An antibody causing precipitation of a homologous antigen.

primary producer. One of the algae that form the beginning or base of most aquatic food chains.

primer. A short sequence of RNA complementary to the DNA.

procaryote (pro•**care**•ee•ote). A type of cell in which the nuclear substance is not enclosed within a membrane; e.g., a bacterium or cyanobacterium. Compare **eucaryote**.

prophylaxis (pro•fi•**lak**•sis). Preventive treatment for protection against disease.

prostheca (pros•**theek**•uh). An appendage that is part of the cell wall of certain bacteria, such as those in the genus *Caulobacter*.

protein. One of a class of complex organic nitrogenous compounds composed of an extremely large number of amino acids joined by peptide bonds.

proteinase (**pro**•teen•ase, **pro**•teen•aze). An enzyme that hydrolyzes proteins to polypeptides.

protein biosynthesis (bye•o•**sin**•thuh•sis). The synthesis of new protein by organisms.

proteolytic (pro•tee•o•**lit**•ik). Capable of splitting or digesting proteins into simpler compounds.

protist (**pro**•tist). A microorganism in the kingdom Protista.

protoplasm (**pro**•toe•plazm). The living matter, living material, or living substance of a cell. The term usually refers to the substance within the cytoplasmic membrane.

protoplast (**pro**•toe•plast). An active metabolizing cell without its cell-wall structure.

prototroph (**pro**•toe•trofe). An organism that is nutritionally independent, able to synthesize all required growth factors from simple substances.

protozoa (pro•tuh•**zo**•uh). Eucaryotic microorganisms with animal affinities.

protozoology (pro•tuh•zo•**ahl**•uh•jee). The study of protozoa.

pseudopodium (syoo•doe•**po**•dee•um) A temporary projection of the protoplast of an ameboid cell in which cytoplasm flows during extension and withdrawal.

psychrophile (**sigh**•kro•file). A "cold-loving" microorganism, capable of growing at 0°C.

ptomaine (toe•main). Substances produced during putrefaction of animal or plant protein that cause food poisoning.

puerperal fever (pyoo•**ur**•pur•ul). A serious infection of the mother after childbirth. Also called *childbed fever*.

pure culture. A culture containing only one species of organism.

purine (**pyoor**•een). A nitrogen-containing ring compound called a *nitrogenous base*.

pus. The fluid product of inflammation, containing serum, bacteria, dead cells, and leukocytes.

putrefaction (pyoo•truh•**fak**•shun). Decomposition of proteins by microorganisms, producing disagreeable odors.

pyemia (pye•**ee**•mee•uh). A form of septicemia in which pyogenic organisms in the bloodstream set up secondary foci in organs and tissues.

pyogenic (pye•o•**jen**•ik). Forming pus.

quaternary (kwah•tur•nehr•ee, kwah•**tur**•nuh•ree). An antimicrobial cationic detergent.

rabies (**ray**•beez). An acute disease of humans and other animals, caused by a virus. Transmitted by direct contact from infected animals to humans through bites.

radioisotope (ray•dee•o•**eye**•so•tope). An isotope that exhibits radioactivity.

recombinant (ree•**kahm**•bi•nunt). A cell or clone of cells resulting from recombination.

recombination. Formation in daughter cells of gene combinations not present in either parent.

reduction. A chemical process involving the removal of oxygen, the addition of hydrogen, or the gain of electrons.

regulator gene. A gene that directs the control of the rate of enzyme synthesis in an operon.

replica plating. Replication of a pattern of colonies from one plate to another; a disk of sterile material (often velveteen) is pressed on the surface of the first plate, and the adhering bacteria are printed on the second.

replication. 1. In molecular biology, the production of a strand of DNA from the original. 2. In virology, multiplication of virus in a cell.

resistance-transfer factor. A factor that confers on microorganisms resistance to a number of antibiotics. Abbreviation: R factor.

resolving power. The quantitatively measureable ability of an optical instrument to produce separate images of different but closely situated points on an object.

respiration. The oxidation processes in the living cell by which oxygen or an inorganic compound acts as the acceptor for electrons removed from the substrate; a process that provides energy for the cell.

respiratory chain. A sequence of oxidation-reduction agents along which electrons are transferred for the generation of ATP.

reticuloendothelial system (ree·**tik**·yoo·lo·en·doe·**theel**·ee·ul). A system of cells in various organs and tissues, such as the spleen, liver, and bone marrow, that are important in resistance and immunity.

reverse transcriptase (tran·**skrip**·tase). An enzyme for the synthesis of a DNA molecule using RNA as a template.

rhizoid (rye·zoyd). A single-celled or multicellular hairlike structure having the appearance of a root.

rhizosphere (rye·zo·sfeer). The soil region subject to the influence of plant roots and characterized by a zone of increased microbiological activity.

ribonucleic acid (rye·bo·new·klee·ik). A nucleic acid occurring in cytoplasm and the nucleolus, containing phosphoric acid, D-ribose, adenine, guanine, cytosine, and uracil. Abbreviation: RNA.

ribosomal RNA (rye·bo·so·mul). RNA of ribosomes in the cytoplasm making up about 90 percent of total cellular RNA. Abbreviation: rRNA.

ribosome (rye·bo·sohm). A cytoplasmic structural unit, made up of RNA and protein, that is the site of protein synthesis.

rickettsia (ri·ket·see·uh). Obligate intracellular bacteria of arthropods, many types of which are pathogenic for humans and other mammals.

RNA. See **ribonucleic acid.**

RNA polymerase (pahl·im·ur·ase, pahl·**im**·ur·ase). An enzyme that synthesizes mRNA on a DNA template.

root-nodule bacteria. Bacteria belonging to the genus *Rhizobium,* family Rhizobiaceae, that live symbiotically in the nodules of roots of leguminous plants and fix atmospheric nitrogen.

rRNA. See **ribosomal RNA.**

rubella (roo·bel·uh). An acute, systemic, infectious viral disease. Also called *German measles.*

rubeola (roo·bee·o·luh, roo·bee·o·luh). An acute, highly infectious viral disease. Also called *measles.*

saccharolytic (sak·uh·ro·lit·ik). Capable of splitting or degrading sugar compounds.

salmonellosis (sal·muh·nel·o·sis). An infection by *Salmonella* spp. that affects the gastrointestinal tract.

sanitizer. An agent that reduces the microbial flora in materials or on such articles as eating utensils to levels judged safe by public health authorities

saprophyte (sap·ro·fight). An organism living on dead organic matter.

Schick test. A skin test used to determine a person's susceptibility to diphtheria.

schizogony (skiz·og·uh·nee). Asexual reproduction by multiple fission of a trophozoite (a vegetative protozoan).

schizont (skiz·ont). A stage in the asexual life cycle of the malaria parasites.

Schultz-Charlton reaction. A skin test used in the diagnosis of scarlet fever.

semiconservative replication. Replication of a complete DNA molecule in such a way that both of the resultant double-stranded molecules contain one original and one new strand.

sepsis (sep·sis). Poisoning by products of putrefaction; a severe toxic state resulting from infection with pyogenic microorganisms.

septate. Possessing crosswalls, such as septate mycelia.

septicemia (sep·ti·see·mee·uh). A systemic disease caused by the invasion and multiplication of pathogenic microorganisms in the bloodstream.

septic tank. A unit using an anaerobic system for treatment of a limited volume of sewage.

septum. A crosswall in a hyphal filament.

sequela (si·kwel·uh). A complication following a disease.

serial dilution. Dilution of a specimen in successive stages. Thus, a 1:100 dilution is achieved by combining one part of a 1:10 dilution (one part of specimen plus nine parts diluent, such as sterile water) with nine parts diluent.

serology (suh·rahl·uh·jee, seer·ahl·uh·jee). The branch of science that treats of serum.

serum. See **blood serum.**

sewage. Liquid or solid refuse (domestic and industrial wastes) carried off in sewers.

sexual reproduction. Reproduction in which two cells (gametes) fuse into one fertilized cell.

sheath. In biology, tubular structure surrounding cells.

simple stain. The coloration of bacteria or other organisms by applying a single solution of a stain to a fixed film or smear.

single-cell proteins. Microorganisms cultivated on industrial wastes or by-products as nutrients to yield a large cell crop rich in protein.

slime layer. A gelatinous covering of the cell wall. The term is sometimes used as a synonym of *capsule.*

slow virus. A virus that causes a disease having a slowly progressive course, usually with a fatal outcome.

sludge. The semisolid part of sewage that has been sedimented or acted upon by bacteria.

smear. A thin layer of material, e.g., bacterial culture spread on a glass slide for microscopic examination. Also called a *film*.

solutes (**sahl**•yoots). Substances in solution.

species. A single kind of microorganism; a subdivision of a genus.

spheroplast (**sfeer**•o•plast). A gram-negative bacterial cell with peptidoglycan and other cell-wall components removed, leaving it devoid of rigidity.

spirillum (spye•**ril**•um). A spiral or corkscrew-shaped bacterium.

spirochete (**spye**•ro•keet). A spiral form of bacterium; most are parasitic.

spontaneous generation. Origination of life from nonliving material. Also called *abiogenesis*. Compare **biogenesis**.

spontaneous induction. A process in which viral DNA is removed from the host's chromosome and the lytic cycle occurs.

sporangiophore (spo•**ran**•jee•o•fore). A specialized mycelial branch bearing a sporangium.

sporangium (spo•**ran**•jee•um). A closed structure within which asexual spores are produced.

spore. A resistant body formed by certain microorganisms; a resistant resting cell; a primitive unicellular dormant body.

sporicide (**spore**•i•side). An agent that kills spores.

sporogenesis (spore•o•**jen**•uh•sis). 1. Reproduction by means of spores. 2. Formation of spores.

sporophore (**spore**•o•fore). A specialized mycelial branch upon which spores are produced.

sporozoite (spore•uh•**zo**•ite). A motile infective stage of certain sporozoans resulting from sexual reproduction that gives rise to an asexual cycle in a new host.

sporulation (spore•yoo•**lay**•shun). The process of spore formation.

stage micrometer (migh•**krom**•uh•tur). An instrument that functions as a ruler for measurement of microorganisms under the microscope.

staphylococci (staff•i•lo•**kahk**•sigh). Spherical bacteria (cocci) occurring in irregular, grapelike clusters.

starter culture. A known culture of microorganisms used to inoculate milk, pickles, and other food to produce the desired fermentation.

stem cells. Formative cells in the bone marrow from which specialized cells, such as lymphocytes, arise.

sterile. Free of living organisms.

sterilization. The process of making sterile; the killing of all forms of life.

steroid (**steer**•oyd). A complex chemical substance containing the tetracyclic carbon ring system of the sterols; steroids are often used as therapeutic agents.

stock cultures. Known species of microorganisms maintained in the laboratory for various tests and studies.

strain. A pure culture of microorganisms composed of the descendants of a single isolation.

streaked-plate method. A procedure for separating cells on a sterile agar surface so that individual cells will grow into distinct, separate colonies.

streptobacilli (strep•toe•buh•**sil**•eye). Bacilli in chains.

streptococci (strep•toe•**kahk**•sigh). Cocci that divide in such a way that chains of cells are formed.

streptokinase (strep•toe•**kigh**•nase). See **fibrinolysin**.

structural gene. A gene coding for the structure of a protein.

subcutaneous (sub•kyoo•**tay**•nee•us). Beneath the skin.

substrate. The substance acted upon by an enzyme.

superficial mycoses (mye•**ko**•seez). Fungus diseases of the nails, skin, hair and mucous membranes.

susceptibility (suh•sep•ti•**bil**•i•tee). The state of being open to disease; specifically, capability of being infected; lack of immunity.

swarm cells. Cells that emerge from a break in a sheath wall.

symbiosis (sim•bee•**o**•sis). The living together of two or more organisms; microbial association.

synchronous growth (**sing**•kruh•nus). Growth in a population of cells in which all cells divide at the same time.

syndrome. A group of signs and symptoms that characterizes a disease.

synthesis. Any process or reaction in which a complex compound is built up by the union of simpler compounds or elements.

syphilis. A venereal disease caused by *Treponema pallidum*.

systematics. The science of animal, plant, and microbial classification.

systemic (sis•**tem**•ik). Relating to the entire organism instead of a part.

taxis (**tack**•sis). Movement away from or toward a chemical substance or physical condition.

taxon, pl. **taxa** (**tack**•sahn, **tack**•suh). A taxonomic group, such as a species, genus, or family.

taxonomy (tack•**sahn**•uh•mee). The science of classification of organisms, based as far as possible on natural relationships.

teichoic acid (tei•**ko**•ik). A cell-wall constituent unique to procaryotes.

temperate bacteriophage. A bacteriophage capable of replicating in step with its host bacterium, being thus transmitted through cell divisions without necessarily causing host lysis.

tetanospasmin (tet•uh•no•**spaz**•min). A toxin that causes tetanus when it reaches the central nervous system.

tetanus (**tet**•uh•nus, **tet**•nus). Lockjaw; a disease caused by *Clostridium tetani*.

thallophyte (**thal**•o•fite). A plant having no true stem, roots, or leaves; the group includes the algae and fungi.

thallospore (**thal**•o•spore). A spore that develops by budding of hyphal or vegetative cells.

thallus (**thal**•us). A plant or microbial body lacking special tissue systems or organs; thalli may vary from a single cell to a complex, branching, multicellular structure.

therapeutic. Pertaining to the treating or curing of a disease.

thermal death point. The lowest temperature at which microorganisms are killed in a given time.

thermoduric (thur•mo•**dyoo**•rik). Capable of surviving exposure to high temperature.

thermolabile (thur•mo•**lay**•bile,

thur•mo•**lay**•bil). Destroyed by heat at temperatures below 100°C (212°F).

thermophile (**thur**•mo•file). An organism that grows best at temperatures of 50°C (122°F) or higher.

thermostable (thur•mo•**stay**•bul). Relatively resistant to heat; resistant to temperatures of 100°C (212°F).

thymine (**thigh**•meen). One kind of pyrimidine.

tinea (**tin**•ee•uh). Ringworm caused by fungi.

tissue culture. A growth of tissue cells in a laboratory medium.

topical application. Application to a localized area.

toxemia (tahk•see•**mee**•uh). The presence of toxins in the blood.

toxin (**tahk**•sin). A poisonous substance elaborated by an organism, such as a bacterial toxin.

toxin-antitoxin. A mixture of toxin and antitoxin containing slightly more toxin than antitoxin. Formerly used to produce an active immunity.

toxoid (**tahk**•soyd). A toxin that has been treated to destroy its toxic property without affecting its antigenic properties.

transcription. The process in which a complementary single-stranded *m*RNA is synthesized from one of the DNA strands.

transduction. Transfer of genetic material from one bacterium to another through the agency of a virus.

transfer RNA. Specific RNA for each amino acid that becomes esterified to the terminal adenosine. Each of the 60 or so *t*RNAs has a specific trinucleotide sequence that interacts with a complementary sequence in *m*RNA. Abbreviation: *t*RNA. Also called *soluble RNA (sRNA)*.

transformation. The phenomenon by which certain bacteria incorporate DNA from related strains into their genetic makeup.

translation. The process in which genetic information in *m*RNA directs the order of assembly of the specific amino acids during protein synthesis.

transverse binary fission (bye•nuh•ree). An asexual reproduc-

tive process in which a single cell divides into two cells.

tricarboxylic acid cycle (trye•kar•bock•**sil**•ik). See **Krebs cycle**.

trichinosis (**trik**•i•no•sis). Parasitic infection from eating infected pork that is insufficiently cooked.

trickling filter. A secondary treatment process in which sewage is trickled over a bed of rocks so that bacteria can break down organic wastes.

*t*RNA. See **transfer RNA**.

trophic stage (tro•fik). The vegetative stage of free-living protozoa.

trophozoite (tro•fo•**zo**•ite). The vegetative form of a protozoan.

trypsin (**trip**•sin). A proteolytic enzyme in pancreatic juice.

tubercle (**tyoo**•bur•kul). A nodule, the specific lesion of tuberculosis.

tuberculin (tyoo•**bur**•kyoo•lin). An extract of the tuberculosis bacilli capable of eliciting an inflammatory reaction in an animal body that has been sensitized by the presence of living or dead tubercle bacilli. Used in a skin test for tuberculosis.

type culture. A particular species considered representative of the characteristics of the species and used as a reference culture.

ultrasonic waves. Sound waves of high intensity (beyond the audible range), used for the destruction of microbes or the cleaning of materials.

ultraviolet rays. Radiations in the part of the spectrum occupied by wavelengths from about 3900 to about 2000 Å.

universal donor. A person in blood group O who gives blood for transfusion. The donor red cells do not have the A and B isoantigens.

universal recipient. A person in blood group AB whose serum lacks anti-A and anti-B isoantibodies.

vaccination. Inoculation with a biologic preparation (a vaccine) to produce immunity.

vaccine. A suspension of disease-producing microorganisms modified by killing or attenuation so that it will not cause disease and can stimulate the formation of antibodies upon inoculation.

vacuole (**vak**•yoo•ole). A clear space in the cytoplasm of a cell.

variability. Differences among members of a species resulting from genetic or environmental causes.

varicella (vehr•i•**sel**•uh). A mild, highly infectious viral disease; chicken pox.

variola (vehr•ee•**o**•luh, vuh•**rye**•o•luh). An acute, infectious viral disease; smallpox.

vector. An agent, such as an insect, capable of mechanically or biologically transferring a pathogen from one organism to another.

vegetative stage. The stage of active growth, as opposed to the resting or spore stages.

viable (**vye**•uh•bul). Capable of living, growing, and developing; alive.

vibrio (**vib**•ree•o). A slightly curved bacterium resembling a comma.

Vincent's angina. An ulcerative condition of the tonsils and gums caused by a spirillum and a fusiform bacillus.

viremia (vye•**ree**•mee•uh) The presence of virus in the blood.

viricide (**vye**•ri•side). An agent that kills viruses.

virion (**vye**•ree•on). The complete mature virus particle.

virology (vir•**ahl**•uh•jee, vye•**rahl**•uh•jee). The study of viruses.

virulence (**vir**•yoo•lunce). The capacity of a microorganism to produce disease; pathogenicity.

virus (**vye**•rus). An obligate intracellular parasitic microorganism that is smaller than bacteria. Most viruses can pass through filters that will retain bacteria.

Wasserman test. A complement-fixation test for syphilis.

Weil-Felix test. An agglutination test for typhus using *Proteus* spp. as antigens.

white corpuscle. Leukocytes of the blood.

Widal test. A slide agglutination test for typhoid or paratyphoid fever.

wild-type cell. A nonmutated cell.

yeast. A kind of fungus that is unicellular and not characterized by typical mycelia.

zonation (zo•**nay**•shun). Distribution of organisms in zones; specifically, a stratification of certain kinds of algae at certain depths and locations in the ocean.

zooflagellate (zo•o•**flaj**•uh•lut, zo•o•**flaj**•uh•lait). An animal-like form of flagellate. Compare **phytoflagellate**.

zoogleal masses (zo•o•**glee**•ul).

Clumps of microorganisms in the form of a film or in suspension in liquid material.

zoonosis (zo•**ahn**•uh•sis, zo•o•**no**•sis). Animal disease transmissible to human beings.

zooplankton (zo•o•**plank**•tun). A collective term for the nonphotosynthetic organisms present in plankton; compare **phytoplankton**.

zoospore (**zo**•o•spore). A motile, flagellate spore.

zygospore (**zye**•go•spore). A kind of spore resulting from the fusion of two similar gametes in some fungi.

zygote (**zye**•gote). An organism produced by the union of two gametes.

Prefixes and combining forms used as prefixes

anti-, ant-. Against, inhibiting.
auto-. Self, independent.
bio-. Life.
cardi-, cardio-. Heart.
cary-, caryo-. Nucleus, nuclear.
centi-. Hundred.
cyt-, cyto-. Cell.
de-. From, removal.
dermato-. Skin.
di-. Two, double.
ecto-. Outside.
endo-. Within.
exo-, ex-. Outside, without.
hem-, hema-, hemo-. Blood.
hetero-. Different.
hist-, histo-. Tissue.
holo-. Complete, homogenous.
hom-, homo-. Like, similar.
hyp-, hypo-. Deficiency, below.

hyper-. Excessive, above normal.
inter-. Between, among.
intra-. Within, into.
iso-. Equality, similarity.
kary-, karyo-. Nucleus, nuclear.
leuk-, leuko-. White.
macro-. Large.
meso-. Middle.
micro-. Small.
mono-. One.
multi-. Many.
myco-. Fungus.
necro-. Dead.
neo-. New.
noso-. Disease.
nucle-, nucleo-. Nucleus, nuclear.
olig-, oligo-. Few, deficiency.
oxy-. Oxygen in a compound.
pan-. All, many.

path-, patho-. Disease, pathologic.
peri-. About, around.
pneumo-. Pulmonary, respiration.
poly-. Many, diverse.
post-. After, behind.
pre-. Before.
pseudo-. False.
pyo-. Pus.
sacchar-, sacchari-, saccharo-. Sugar.
syn-, sym-. With, together.
tax-, taxi-, taxo-. Arrangement.
therm-, thermo-. Heat, temperature.
thi-, thio-. Sulfur present.
tox-, toxi-, toxo-. Poisonous, toxin, poison.
trans-. Through, across.
trich-, tricho-. Hair, filament.

Suffixes and combining forms used as suffixes

-algia. Pain, suffering.
-ase. Enzyme.
-cide. Killer, killing.
-cyte. Cell.
-emia. Condition of blood.
-ia. Condition; abnormal or pathologic condition.
-iasis. Diseased condition.
-ism. Condition or disease.
-itis. Inflammation of a part.
-logy. Field of study.

-lysis. Dissolution or disintegration.
-oma. Tumor, neoplasm.
-osis. Process, disease, cause of disease.
-otic. Related to or causing process or condition.
-otomy. A cutting into.
-ous. Having or pertaining to.
-pathia. Disease.
-penia. Deficiency.

-phage, -phag. Ingest, break down.
-rrhage, -rrhagia. Abnormal or excessive discharge.
-rrhea. Discharge.
-scope. Instrument for seeing or examining.
-taxia, -taxis. Arrangement or order.
-trophy. Nutrition, growth.
-tropic. Turning toward, having an affinity for.

ORGANISM INDEX

SUBJECT INDEX

Page numbers in *italic* indicate illustrations.

Biological classification, 23, 26, 40
Biological oxygen demand (BOD), 611, 613, 614
Biological vectors, 527, 530–531
Biologics for immunization, 652, 653
Bioluminescence, *595*
Biosphere, 583
Biotin, 655
Bisexual algae, 167
Black Death, 531
Black piedra, *140,* 141
Blastomycin test, 440
Blastomycosis, 145, 440, *474*
   North American, *145,* 470
   South American, 470
Blastospores, 134, *143*
Blepharoplast, *157*
Bloody diarrhea, 160
"Blooms," 165, 174, *585*
Blue cheese, *625*
Blue-green algae, 109
BOD (biological oxygen demand), 611, 613, 614
Boiling water, use of, 322, 331
Boils, 390, 392
Bollinger bodies, *206*
Botulinus toxin, 392
Botulism, 12, 318, 388, 478, 485–488, 492
   epidemiology of, 487–*488*
   "infant," 487
   laboratory diagnosis of, *486,* 487
   nature of pathogenicity in, 487
   prevention of, 488
Bovine abortion, 12
Bright-field microscopy, 48–50, *52,* 53, 56
Broad-spectrum antibiotic, 353, 358
5-Bromouracil, *286*
Bromphenol blue, 96
Bromthymol blue, 96
Brown, Rachel, 359, *360*
Brown algae, 164, 169
Brucellae bacteria, 391
Brucellergin test, 440
Brucellosis, 391, 409, 440
Bubo, 531, 557
Bubonic plague, *531*–532
Buccal cavity, protozoan, *159*
Budding, 134, 153
Budding bacteria, 113, *114,* 115
Buffers, 98
Bulgarian milk, 624
Burkitt's lymphoma cells, 209
Bursa of Fabricius, 418
Bursitis, 164
2,3-Butanediol, 645
Butyrate, 255
Butyryl-CoA, 255

C carbohydrate, 453–454
Calcium hypochlorite, 337–338
Calcium ions in phagocytosis, 412
Calicivirus, *491*
Calories, 233
Calvin cycle, *261,* 262
cAMP (cyclic AMP; adenosine-3',5'-monophos-
   phate), *228,* 230
Cancer, 208–210
Candidiasis, 143–144, *474*
Canning, 313, 634–635
CAP (catabolite gene activator protein), 230
Capsids, 183–*185,* 193, *195,* 198, 199
Capsomeres, 183–*185,* 193, *195, 202*
Capsular Vi antigen, 399, 496, 497
Capsule stain, 58
Capsuled bacteria, *78*
Capsules, bacterial, *25,* 29, *74,* 78–79, 83, *387,* 391
Carbohydrates:
   in animal and plant viruses, 198
   breakdown of, 238
   C, 453–454
   fermentations of, 238–239
   metabolism of, *245*
   as substrate for fermentation, 646
Carbol-fuchsin, 58, 458, 533
Carbolic acid, *17,* 335
Carbon dioxide, 92
Carbon dioxide fixation, *261*

Carbon requirement, 92
Carcinomas, 208
Caries, dental, 564–566
   etiology of, 565–566
   prevention of, 566
   sites of attack, *564*
Carnivores, 152
Carotenes, 168
Carotenoids, 168, 247
Carrageenan, 164
Carriers, 444, 479
Catabolism, 232
   of lipids, 243–244
   of proteins, 244–245
Catabolite gene activator protein (CAP), 230
Catalase, 576
Catalysis, direct control of, 225
Catalysts, 216
   biological, 216
Cathode rays, 328, 331
Cationic detergents, 340
Cavitation, 330
Ceepryn, 340
Cell cooperation, 419
Cell death from virus infection, 207
Cell division, 100–102, *101*
Cell-free DNA, 296
Cell lines, 206
Cell-mediated immunity, 397, 417, 419
   intradermal tests based on, 440
Cell theory, 24
Cell walls:
   bacterial (*see* Bacterial cell walls)
   of eucaryotes, *30*–32
   of procaryotes, 28, *29*
Cells, 23, 24
   bacterial (*see* Bacterial cells)
   as basic structural unit of life, 24–26
   in cork, *24*
   eucaryotic, 26
   procaryotic, 26, 28–*29*
   "typical," *25*
Cellular reactions, coupling of, 233–235
Cellular slime molds, 146, 148
   life cycle of, *147*
Cephalexin, 360
Cephaloglycin, 360
Cephaloridine, 360
Cephalosporins, 357, 360
Cephalothin, *357,* 360
Cetylpyridinium chloride, 340
Chagas' disease, *157,* 528
Chain arrangement, *74,* 75
Chancre, 551
   soft, 557
Chancroid, 557
Chemical agents:
   control of microorganisms by, 308, 333–347
   major groups of, 335–345
   selection of, 334–335
Chemical flocculation, 617
Chemicals in food preservation, 638
Chemiosmotic theory, 248, 255
Chemoautotrophs, 92
Chemoprophylaxis, 349, 456
Chemotherapeutic agents, 333, 349
   antibiotic, 353–361
   synthetic, 361–365
   tests for effectiveness of, 366, 367
Chemotherapy, 13, 17–18, 349
   history of, 350–352
Chemotrophs, 92, 93
α-Chemotrypsin, 220
Chick-embryo technique, *202*–204
Chicken cholera, *13*–14
Chicken pox (varicella), *467,* 468
Childbed fever, 10
Cholera, 12, 117, 388, 502–506
   chicken, *13*–14
   epidemiology of, 504–505
   laboratory diagnosis of, 504
   nature of pathogenicity of, 503–504
   prevention of, 505
   vaccination against, *505*
Cholera cot, *503*
Choleragen, 504

Chlamydias, 124
Chlamydospores, *134, 143*
Chloramine-T, 338
Chloramines, 338
Chloramphenicol, 359, 361, 497
Chlorinated lime, 337–338
Chlorination, 604, 617
Chlorine, 337, 344–345
Chlorine compounds, 337, 344–345
Chloromycetin, 359, 361
Chlorophylls, 31, 109, 167, 168, 245, 246
   (*See also* Bacteriochlorophylls)
Chloroplasts, *25, 30,* 31, *157, 166*–168, *170, 172*
Chloroquine, 506, 539, 542
Chlortetracycline, 358, 360
Chromatin body, 84
Chromatinic area, 84
Chromosomes, *25,* 31, 283
   bacterial, 84, *272*
   homologous, 290
Chronic disease, 456
Cilia, *25,* 31, 151, *159*
Ciliates, 151, 156, 159–160
Cinchona tree, 17, 350
Cirri, *159*
Citric acid, 648
Class, 41
Classification, biological, 23, 26, 40
Clostridial infections, 573–576
Clubroot, 146
Coagulase, 389, 390
Cobalamin, 645
Cocarboxylase, 217
Cocci, 5, 72, 75
   arrangement of, *73*–74
Coccidioidomycosis, 145, 471, *473, 474*
Cockroach, intestinal flora of, *55*
Codes of nomenclature, 42
Codons, 274–275
   nonsense, 275
Coenocytic hyphae, *133*
Coenozygote, *140*
Coenzyme A, 242, 244, 245
Coenzymes, 217
Cofactors, 217
"Cold sores," *195,* 200
Coli-aerogenes-typhoid bacteria, 239
Colicins, 296
Coliform bacteria:
   as indicator microorganisms, 606
   tests for detection of, *607*–609
Coliphages, 184–186
   T2, *182*
Colistin, 361
Colitis, 577
Collagen, 389
Collagenase, 389–390
Colonies, *11*–12, 59, *165, 166*
Colorado tick fever, 530
Colorado tick fever virus, 530
Colostrum, 402, 403
Comma bacteria, 73
Commensalism, 585, *586,* 587
Commensals, 376
Commission on Enzymes of the International
   Union of Biochemistry, 218
Common cold, 461, 463, 475
Common names, 43
Common reactant, 233
Competent bacteria, 297
Competition, 585, *586*
Competitive inhibition, 222, *223,* 363
Complement, 413, 414, 435
   action of, *414*
Complement-fixation tests, 435–436
Complement-fixing antibodies, 405
Complement system, 414
Complementary base pairs, 268, 270
Compromised hosts, 141
Computer taxonomy, 43
Condensers:
   for light microscopy, 48, 50, 55
   magnetic, 55
Conidia, *86, 124, 134, 135, 137,* 144
Conidial heads, *144,* 145
Conidiophores, *134, 135, 137, 144*

MICROORGANISMS THAT CAUSE DISEASE

BACTERIA

1. *Staphylococcus aureus*, causative agent of many staphylococcal infections. Gram stain.
2. *Streptococcus pneumoniae*, causative agent of pneumonia. Gram stain.

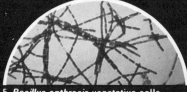

3. *Neisseria meningitidis*, causative agent of meningitis. Gram stain.
4. *Neisseria gonorrhoeae*, causative agent of gonorrhea, present in white blood cells. Gram stain.

5. *Bacillus anthracis* vegetative cells, causative agent of anthrax. Gram stain.
6. *Brucella melitensis*, causative agent of brucellosis. Gram stain.

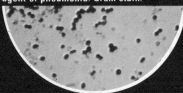

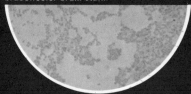

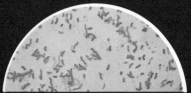

7. *Vibrio comma*, causative agent of cholera. Gram stain.
8. *Shigella sonnei*, causative agent of dysentery. Gram stain.

9. *Clostridium* sp. vegetative cells (pink) and spores (green). Spore stain.
10. *Clostridium perfringens*, one of causative agents of food poisoning. Gram stain.

11. *Clostridium tetani*, causative agent of tetanus. Note terminal spores. Gram stain.
12. *Clostridium botulinum*, causative agent of botulism. Gram stain.

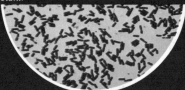

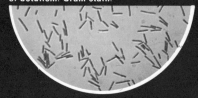

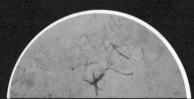

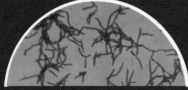

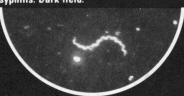

13. *Mycobacterium tuberculosis* (red), causative agent of tuberculosis. Acid-fast stain.
14. *Mycobacterium tuberculosis* (yellow). Acid-fast stain.

15. *Actinomyces israelli*, causative agent of actinomycosis. Gram stain.
16. *Cornyebacterium diphtheriae*, causative agent of diphtheria. Gram stain.

17. *Borrelia recurrentis*, causative agent of relapsing fever, in blood film.
18. *Treponema pallidum*, causative agent of syphilis. Dark field.

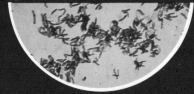